The American Railroad Passenger Car

THE JOHNS HOPKINS UNIVERSITY PRESS   BALTIMORE AND LONDON

The American Railroad Passenger Car

JOHN H. WHITE, JR.

The Johns Hopkins University Press, Baltimore, Maryland 21218
The Johns Hopkins Press Ltd., London
Library of Congress Catalog Card Number 77-1141   ISBN 0-8018-1965-2
Library of Congress Cataloging in Publication data will be found
on the last printed page of this book.

For Christine

CONTENTS

PREFACE

AMERICANS once spent a considerable portion of their lives in railroad cars. From approximately 1860 to 1930 railroads were the most common means of inland transportation, and at their peak just before the First World War, 98 percent of all intercity travel was by rail. No other means of transportation, not even the automobile, has achieved such a monopoly. Each day millions of passengers mounted the end platform steps of steam cars, some to travel a few miles to a suburban home, others to live within the swaying body of a cinder-laden Pullman for a week as it crossed the continent.

This vehicle that was once intimately intertwined with the lives of all Americans has rapidly receded from view. What was commonplace has grown obscure, even unknown. An abundance of surviving artifacts and documents makes it possible to reconstruct everyday life in the American home of that period in great detail, but information about the railroad car, particularly its early years, is difficult to recover. Early cars were junked when they were condemned as obsolete. Engineering offices systematically discarded old drawings, photographs, and specifications (the historian's enemy is always the fastidious housekeeper). The data that have survived are scattered, incomplete, and occasionally contradictory.

It is the purpose of this book to provide a picture of American railroad passenger cars as they were and as they are now. It will try to answer such basic questions as what they looked like at various periods, how they were made, what materials went into their construction, who made them, how the interiors were arranged, how many of each type were in service, and what they cost. The subject is a complicated one, for the passenger car is the product of many crafts and technologies. Even the day coach evolved into a complex conveyance with a maze of auxiliary machinery and systems. Long before steam railroads were envisioned, cabinetmakers, wheelwrights, blacksmiths, upholsterers, foundry workers, drapers, painters, and even silversmiths combined their skills to produce carriages. Some were elegant and light, others plain and rugged. Many were custom-made to suit the buyer's taste or purse. All reflected man's ingenuity in adapting and reshaping natural elements into useful products for human convenience.

Thus the railroad car was born of the established coach maker's tradition. Carpenters and bridge builders soon introduced structural designs that permitted larger cars. In more recent times steelmakers and electrical and aeronautical engineers have contributed their technologies. The evolution of the passenger car has always been dependent on known methods and materials; it has never inspired a major technical breakthrough. Although its progress might be criticized as overly conservative, it has moved forward in a slow but constant series of changes during its 150-year development.

In any technical history it is necessary to deal with the origins of important inventions. Establishing who did what becomes a somewhat arbitrary exercise, but there seems no way to avoid it. Obviously there is little genuine novelty in mechanics, and every inventor has predecessors whose work was close to his. Alexander von Humboldt claimed that an invention goes through three stages: first its existence is doubted, next its importance is denied, and finally the credit is given to someone else.

A more important question about every major invention is when it was accepted into actual practice. The naïve assumption of many popular histories is that once an invention is born, it is immediately put into everyday use. Far more often, a sound idea is initially rejected and then reintroduced at a later date. Determining when an innovation became standard practice is extremely difficult because the event was rarely newsworthy; naturally the technical press tends to emphasize new or unusual ideas. It is the plan of this book, however, to trace the popular acceptance of inventions and to concentrate on ordinary standard designs. Novelties and experiments are included, but they are not emphasized.

My main purpose here is to describe the passenger car as a physical object. I have included some information about the economics and history of travel in order to explain the appearance and disappearance of certain types of car—the railway mail car, for example. Also, the general evolution of the passenger car is tied to the growth of the railroad system, since longer trips in turn called for more comfortable cars. And growth brought increasing competition between parallel lines, which courted passengers by introducing such costly luxuries as dining cars and air conditioning. But the reader should not expect a complete discussion of rail travel history, or a description of each class in every railroad's passenger rolling stock.

Some readers may ask why another book on railroad cars should be added to an extensive bibliography which already includes the work of Lucius M. Beebe, Arthur D. Dubin, Robert J. Wayner, and other authorities. Yet the existing books concentrate on name trains and luxury cars, detailing the joys and splendors of first-class travel. It is easy for a reader to forget that the majority of passengers occupied coach seats, ate box lunches, and sat up all night. The more nostalgic literature deals in velvet-smooth rides with sunset vistas of mountain lakes through picture windows. There is no mention of the incessant rattle and clatter, or of the dust and cinders in pre-air-conditioned days. In fact travel by any mode, even by rail, is often a miserable experience, and I have attempted to cover this side of the story as well.

The existing books do not attempt to describe ordinary cars. In addition, they say little about construction, and they ignore design except for the topic of decor. As early as 1894 *The Railroad Car Journal* pointed out: "Perhaps no branch of the mechanical arts is so poorly supplied with literature as the art of car construction—particularly such as relates to its past history and development." Sixty years would pass before August Mencken's book *The Railroad Passenger Car* (1957) ventured into these uncharted waters. At that time, railroad historical literature was still confined to the evolution of locomotives and the corporate growth of the railroad companies.

This book emphasizes nineteenth-century developments rather than more recent ones, because the nineteenth century was the formative period when the basic shape, arrangement, and types of car came into being. And whereas little meaningful work is being done on the early period, the trade literature and the increasing number of publications for railroad enthusiasts are abundantly documenting twentieth-century developments.

When I began work on the subject in 1968, the railroad passenger car seemed to be fading into the ranks of antique modes of transportation. Few new cars had been ordered since 1955, and the complete abandonment of long-distance trains seemed inevitable. I felt that I might be writing the preface to the end of passenger car history. But Amtrak was created, and a generous public purse has kept the trains running and has even funded a new generation of cars. The new ones do not materially differ from the established lightweight designs, however, and for this reason I saw no need to go beyond the cutoff date of 1970 which I had originally set.

Most travelers fail to appreciate the passenger train for the miracle it is. In essence it forms a small city on wheels, with lighting, heating, air conditioning, food services, toilets, washrooms, a water supply, and sleeping, seating, and lounging facilities for perhaps a thousand people. All these systems must be fitted into the cramped spaces available between the side panels, under the floor, and in utility closets. Space is at a premium, making miniaturization a necessity. Dependability, weight, and cost are also crucial factors. Perhaps even more remarkable, this 1,400-ton city on wheels crosses the countryside at 80 miles an hour through all extremes of weather. It is the perfection of this movable city, a triumph of American technology, that we will survey in the following pages.

Acknowledgments

Many individuals and institutions assisted me in the preparation of this volume, and it is a pleasure to have an opportunity to express my gratitude. For eight years the staff of the Smithsonian Library showed great energy and patience in borrowing the technical journals and books needed for my research. Jack Goodwin, Charles Berger, Frank Pietropaoli, and Lucien R. Rossignol of the library staff were particularly helpful. The manuscript was typed and meticulously proofread by Mary E. Braunagel—to quote George Hilton, her typing efficiency "never ceased to impress me." Dorothy Young, Jean Stensland, Luwan Brown, and Elizabeth James were also helpful in assembling lists, photographs, and tables.

The advice, criticism, and encouragement of Arthur D. Dubin and William D. Edson were invaluable. Their knowledgeable comments did much to improve the text. I am also indebted to Dr. Louis Marre for his critical reading of the final chapter, on rail cars. Many errors and omissions were brought to my attention by these experts.

The following persons provided information or illustrations needed to complete this study: A. Andrew Merrilees, Douglas Wornom, Robert Wayner, Everett DeGolyer, Jr., G. M. Best, Lee Rogers, the late L. W. Sagle, R. C. Reed, C. M. Clegg, Cornelius Hauck, Nora Wilson, John Keller, Lester S. Levy, Ralph Greenhill, George Hart, George Krambles, Lawrence S. Williams, E. P. Alexander, Thomas T. Taber, Herbert H. Harwood, Jr., and Peter B. Bell. Institutions that provided information or illustrations are the National Archives, the Library of Congress, the Historical Society of Pennsylvania, the Lancaster County Historical Society, the Hagley Foundation Library, the Harvard School of Business' Baker Library, the Newberry Library, the Deutches Museum, the Peale Museum, the Baltimore and Ohio Transportation Museum, Pullman, Inc., the Budd Company, the Southern Railway, General Motors Corporation, The Railroadians, and the Hall of Records-Delaware Archives. The special attention over many years of John McLoed and Helen Roland, librarians at the Association of American Railroads, must also be mentioned.

The Simmons-Boardman Company, through its long-time publisher, Mr. Robert G. Lewis, kindly permitted us to reproduce many line drawings from past issues of the *Railway Age* and the *Car Builders' Cyclopedia*.

Direct financial support for the production of this book came from the Allegheny Foundation—Richard M. Scaife, Chairman; the National Museum of History and Technology—Brooke Hindle, Director, Silvio A. Bedini, Deputy Director, and Robert G. Tillotson, Assistant Director; Mr. W. Graham Claytor, Jr.; Morgan Guaranty Trust Company of New York; Mr. John F. Ruffle; and the Smithsonian Institution.

The staff of The Johns Hopkins University Press showed more than the expected professional cooperation in producing the book, and I wish to thank the following persons: Henry Y. K. Tom, Nancy Essig, Patrick Turner, Susan Bishop, Mary Lou Kenney, and finally Mary Barnett, whose skillful copy editing did so much to tighten and smooth the final text.

SEATS 20
Wt. 1.75
1830

SEATS 52
Wt. 6
1835

SEATS 56
Wt. 10
1850

SEATS 50
Wt. 18
1860

SEATS 60
Wt. 25
1880

SEATS 66
Wt. 37
1895

SEATS 88
Wt. 60
1910

SEATS 80
Wt. 70
1925

SEATS 60
Wt. 55
1940

The American Railroad Passenger Car

CHAPTER ONE

The Day Coach in the Wooden Era

1830–1910

THROUGHOUT THE HISTORY of railroad passenger travel, the common day coach has been the largest and most important single class of car. It was not only the mainstay of the passenger fleet, but the archetype of all passenger cars. Yet it has not received the attention that has been given to the splendid cars built for first-class travel. The trade press featured parlor and sleeping cars as attractions far more newsworthy than the humdrum day coach. Railroad managers concentrated on promoting first-class travel as the keystone of prestige and profit. But as one manager, Daniel Willard, president of the Baltimore and Ohio Railroad, said in the 1920s: "We are spending too much time on the problems of the Pullman passenger. Eighty per cent of our patrons ride in day coaches, yet we barely give them twenty per cent of our attention."[1] Because of its central position in the history of passenger cars, the day coach will receive first attention in this book.

Four-Wheel Pioneers

The first American passenger cars were simple four-wheel carriages that can be divided into three basic types: the traditional curve-sided stagecoach, the compartment style with three tandem bodies built as a unit, and the minor subclass of so-called Gothic cars.

A favorite story about these early cars (and one with no basis in fact) is that old stagecoach bodies were hastily mounted on flanged wheels for railway service. At one time or another this anecdote has been pinned on every pioneer car in the country. It is true that most of the earliest cars were built by established carriage makers, who naturally constructed bodies identical with those used for road coaches.

An American pattern of swell-sided road coach was developed in Troy, New York, and Concord, New Hampshire, in the mid-1820s. As the predecessor of the more familiar stagecoach of the mid-nineteenth century, it had a more rounded body profile. By the time railroads came to America, this style of coach was widely used in the United States. Thus the melon-shaped stagecoach body, familiar to many readers as a product of the Wild West, was developed in the Northeastern section of the country long before the West was settled. The swelled sides conformed to the human body, giving maximum space at the middle where seated passengers need it. Cutting away the lower portion of the body not only reduced the weight but also provided better clearance for the high wheels that were necessary for travel on unimproved roads.

The lengthwise curve of the body was designed to accommodate the leather-strap suspension. Steel springs were so expensive and unreliable that carriage builders devised the leather strap or thoroughbrace suspension to avoid them. Heavy strips of leather were sewn and buckled together to form great elastic bands that would give and stretch with the motion of the coach. In addition, the round body could rock to and fro on the limber leather springs, dissipating the jolts from the roadway.

When he was approached on short notice for car bodies, the typical carriage builder would produce what he already knew. Richard Imlay, who began carriage building in Baltimore around 1828, was asked to build the first cars for the Baltimore and Ohio Railroad. The cheapest and simplest approach was, of course, to turn out bodies duplicating the standard mail coach of the period. The design was proved, the patterns were available, and the men in the shop were familiar with the work. Imlay built only the bodies; the railroad provided the running gears, complete with wheels. The first car, one of the very first railroad passenger coaches built in this country, was finished in May 1830 and appropriately named *Pioneer*.[2] Five others made from the same pattern were completed by August. A contemporary newspaper account praising one of the cars spoke of its "beauty, comfortable arrangement and excellence of construction."[3] It mentioned rooftop seats and a canvas awning but gave few other details. A more precise picture of the construction of these cars is found in the mechanical drawing of the car *Ohio* from the B & O's 1831 annual report (see Figure 1.58 in the Representative Cars section at the end of this chapter).

Imlay's production was not limited to local Baltimore railroads. He built two cars for the Little Schuylkill Railroad, delivering them in November 1831. He built six for the Newcastle and Frenchtown Railroad, completing one of them, *Red Rover*, in February 1832.[4] As shown by the two Newcastle and Frenchtown cars in a broadside issued in 1833 (Figure 1.1), the similarity between them and the *Ohio* was unmistakable. Imlay also produced cars for the Paterson and New York (later the Paterson and Hudson River) and the B & O railroads.[5] In 1832 he constructed twenty cars similar to the Newcastle and Frenchtown lot for the Philadelphia, Germantown, and Norristown Railroad.[6] The Baltimore daily press gave the names and color schemes of three of them: the *President* was blue and gold, the *Robert Morris* green and gold, and the *Philadelphia* sage and gold. The practice of naming cars and painting each in different colors was limited to the first years of the railroad car in America.

Another early builder of stagecoach railroad cars was James Goold (1790–1879) of Albany, New York. Goold opened a carriage shop in 1813, and he had produced scores of road coaches when the Mohawk and Hudson Railroad came to him in 1831 for six car bodies. Like Imlay, he built on the existing pattern of road coach. The original drawing furnished the contractor by the railroad's chief engineer, John B. Jervis (Figure 1.2), is close to the cars that Imlay was building in Baltimore.[7] It had larger wheels and a longer wheelbase to improve tracking, but since it, like Imlay's, depended on thoroughbrace suspension, it offered little improvement in passenger comfort. Goold continued building railroad cars until about 1870. While something is known of his later production, no specific instances of other thoroughbrace cars can be found. It is probable, of course, that he made similar cars for other local railroads.

Stagecoach cars were being built in New England as well. The Boston and Providence car of about 1834, described in detail in the Representative Cars section (Figure 1.62), is a surviving

Figure 1.1   A broadside dated June 1, 1833, showing some cars produced for the New-castle and Frenchtown Railroad by Richard Imlay. (Bureau of Railway Economics)

Figure 1.2   James Goold of West Albany built the first cars for the Mo-hawk and Hudson Railroad in 1831. This drawing is contemporary. (The American Railway, 1892, p. 139)

example. It is in fact the only original car of this type in existence, and it represents the final development of the swell-sided car for railway service. The body is larger in all dimensions than the Imlay or Goold cars, but it is far from an ideal design. No other specific examples of New England stagecoach-bodied cars can be offered, though it is probable that the first cars of Osgood Bradley, a carriage builder of Worcester, Massachusetts, who turned car builder in 1833–1834, were of the road coach style. He had produced road coaches for several years before he made his first cars for the Boston and Worcester Railroad.

The shortcomings of stagecoach-body cars were evident from the beginning. They were regarded as a temporary expedient until a more suitable body style could be found. Some lines avoided them entirely, and it might be safely estimated that fewer than one hundred saw service on American railways. The swell-sided body had been developed expressly for horse-drawn road coaches, where size and weight were more important than they were on the railroads. Larger, more commodious cars were not only desirable but possible for railway service. The stagecoach body did not lend itself to infinite expansion. A major reason for the curve-sided body was to provide clearance for the high spoke wheels of the road coach. It was superfluous in railway cars, since wheels larger than 36 inches in diameter were rarely used. Furthermore, it was expensive and difficult to build because of the many compound curves. The curving sides would not accommodate glass windows, and the leather curtains that were used instead proved unsatisfactory with steam locomotives because of the speed and the smoke. The seats were fixed so that some passengers were obliged to ride backwards—a position which many people complain about and which for some reason seems particularly annoying in railway travel. The thoroughbrace suspension only reinforced the galloping motion of the four-wheel car, accentuating the already poor ride of the single-truck carriage. The running gear was entirely unsprung and thus free to clatter and pound over the lightly built track.

COMPOSITE OR TANDEM-BODY CARS

A more rational form of car body for railway service was discovered by uniting several compartments as a single structure. Such bodies might be infinitely extended on a single frame, but for four-wheel cars they were invariably limited to three compartments. Tandem bodies were introduced at the same time as the swell-sided cars and were soon considered the more suitable body style. They appear to have entirely superseded the swell-sided cars by the mid-1830s and were undoubtedly the most numerous four-wheel passenger cars in the United States.[8] The tandem body originated with the European city omnibus of the

Figure 1.3 A lithograph of 1831 showing a four-wheel compartment car built for the Camden and Amboy Railroad. The influence of the swell-sided road coach is still evident. (Smithsonian Neg. 25012-E)

period. The design was copied by early British railways, as shown in the famous Ackermann lithographs of the first trains on the Liverpool and Manchester Railway. These prints were widely circulated, and some of the copies that went overseas were undoubtedly scrutinized by American mechanics. It is also probable that city omnibuses on this pattern were operating in the larger American cities. The early inspection visits of American engineers to British railways form another obvious avenue for the importation of the tandem design.

The earliest American examples of a composite-body car were those built for the Camden and Amboy Railroad late in 1831 by M. P. and M. E. Green of Hoboken, New Jersey (Figure 1.3). A Philadelphia newspaper of the day reported that the cars seated 36 passengers and had wooden-spoked wheels, which the paper described as being safer than imported iron wheels.[9] The lithograph issued by the builder gives a more precise idea of the arrangement. It is not an altogether praiseworthy design. The extreme bowed side panels of the individual compartments retained the defects of the swell-sided body. Window sashes could not be used; side curtains were again necessary. The body was placed high over the large wheels. The short wheelbase encouraged a rocking gait, and the chain coupling was unsatisfactory for a train of cars. The car had leaf springs, however—a decided advance over thoroughbraces. The body ran the full length of the

Figure 1.4 This engraving is often labeled "the first streetcar," but in fact it represents a railroad coach produced by Stephenson in 1832. (Smithsonian Neg. 10861-B)

frame, which was an impossibility with thoroughbrace cars because the body had to be inside the jacks. The body was wider than the usual road coach, so that five to six passengers could sit abreast. The Green shop built other cars of this style for the New Orleans and Carrolton Railroad.[10] In its February 27, 1836, issue the American Railroad Journal said that more cars by the same maker were bound for the Petersburg and Roanoke Railroad.

Just a year after the first Green cars were delivered, John Stephenson, a young coach and omnibus builder of New York

Figure 1.5   John Stephenson's patent drawing of 1833. The name President *is faintly inscribed over the center door. A car of this name was made for the New York and Harlem Railroad. (New York University Engineering Library)*

City, produced an improved style of composite-body car for the New York and Harlem Railroad. In later years Stephenson made his mark as a streetcar builder, and when referring to his original car, the *John Mason*, he called it the first street car (Figure 1.4). This claim can be argued both ways; surely the car was no more a streetcar than the earliest horse-drawn cars used by the other pioneer American railroads. The Harlem railroad did begin streetcar-like operations in lower Manhattan a few years later, but the *Mason* and its counterparts were intended for ordinary railroad service.

A contemporary account spoke of the *John Mason* as resembling an omnibus "or rather several omnibuses attached to each other, padded with fine cloth and with handsome glass windows, each [car] capable of containing outside and inside, fully forty passengers."[11] This description failed to take note of the improvements in Stephenson's design compared with its predecessors. By building the body around the wheels, it was possible to lower it by nearly a foot. The floors became drop wells between the wheels, with the seats over the wheels. Stephenson was issued a patent for this idea in April 1833. He treasured the papers, signed by Andrew Jackson, all his life. The original patent is now in the New York University Engineering Library in New York City. The drawing attached to the papers does not agree with the familiar line cut of the *John Mason*; it is marked *President* in a faint handwritten inscription over the center doorway (Figure 1.5). The *President*, the *Mentor*, and the *For-Get-Me-Not* are among the known names of cars delivered to the New York and Harlem by Stephenson. A comparison of the two drawings shows several differences: the roof seats on the *President* are the most

obvious. Vertical posts indicate that there was a canopy over the roof seats, but the upper part of the drawing is missing. The absence of side doors is a serious and puzzling draftsman's error. The second leather-strap "spring" over the left wheel indicates a supplementary or alternate method of suspension. The general plan is the same in both drawings and represents considerable progress over Green's idea for a compartment coach. Stephenson's body was simpler and more squared off; the body panels were nearly flat. The dropped floor wells provided a lower center of gravity and greater headroom. Except for the leather-strap springs, Stephenson's design was one of the best available to American railroads of the period. He built cars of this type for the Paterson and Hudson River Railroad, the Brooklyn and Jamaica Railroad, and lines in Cuba and Florida.

One of the first engineers to recognize that a slavish adherence to established road coach designs was not necessary or even desirable for railway travel was John B. Jervis. In 1832 Jervis prepared the simplest possible design for compartment cars for the Mohawk and Hudson Railroad. The body was a plain box. The only concession to orthodox styling was the curved panel moldings at the bottom of each car. (A more detailed description of these cars is given in Figure 1.61, in the Representative Cars section at the end of the chapter.) It is probable that Jervis's design is a fair approximation of most of the composite-body, four-wheel cars built in the United States. It is surely the logical simplification of a basic design. Evidence is thin, and it is risky to be dogmatic on any point concerning the earliest cars. But from what information is available, it seems safe to state that of the American four-wheel passenger cars, the three-tandem compart-

Figure 1.6   A Gothic car designed by Asa Whitney in 1834 for the Mohawk and Hudson Railroad. Presumably this is a study drawing.

ment with side doors was the most popular form. Some four-wheelers were built with end doors and platforms and center aisles, but only the sketchiest information exists for cars on this plan. By the time the center-aisle plan was adopted, the four-wheel passenger car was made obsolete by double-truck eight-wheel cars.

GOTHIC CARS

At best a minor subclass of the compartment car, the Gothic car is known only through a few examples. The most familiar of these began operation in 1837 on the Erie and Kalamazoo Railroad running between Toledo, Ohio, and Adrian, Michigan. A drawing prepared in 1881 in consultation with one or more old officers of the road has been widely reproduced.[12] It shows a three-compartment car with an elevated central compartment. The floor of this central area is dropped sufficiently for a baggage space beneath the passenger compartment. The roof and body ornamentation suggests the Gothic style. A second drawing, now in the University of Michigan's Transportation Library (Ann Arbor), is also identified as the Erie and Kalamazoo Gothic car but appears to be of more modern origin. It shows a boxy four-wheel car indistinguishable from an ordinary four-wheel caboose of, say, the 1880s. The validity of this drawing can be seriously questioned. Unfortunately, little concrete information exists on the car, its builder, or its final history.

The Gothic car, or at least the term Gothic cars, originated some five years before the Erie and Kalamazoo coach: Gothic cars were reportedly in service on the Mohawk and Hudson Railroad in 1832.[13] A drawing by Asa Whitney, master mechanic of the road in its early years, clearly shows a car that can only be

described as Gothic (Figure 1.6). If the wheels were removed, it would be a perfect plan for an elaborate transept screen in a Gothic cathedral. But it is difficult to believe that so elaborate a vehicle was constructed for use outside the Vatican railway; the probability is that it was a draftsman's exercise. Gothic cars of some form, however, were in use on the Mohawk and Hudson. The 1840 inventory lists twenty-four "Gothic Coaches" as purchased in 1835 and 1836.[14] The neighboring Tonawanda Railroad built six Gothic cars in its own workshops.[15] F. A. Ritter Von Gerstner, the Austrian engineer who studied American railroads in the period 1838–1840, reported that these seated 24, weighed $2\frac{1}{4}$ to $2\frac{1}{2}$ tons each, and cost $800 each. The cars were known locally as camelbacks, yet because of their classical Grecian exterior they were endowed with a certain architectural grace. But the interiors, with their awkward raised center sections, "afforded all the inconvenience that the most disobliging railroad manager could desire."[16]

The defects of the various four-wheel cars used in this country were not only obvious but numerous. Both the traveling public and the railway managers were loud in their criticism. The wheelbase or spread of the axles was necessarily limited so that sharp curves could be negotiated. This restricted the body length to no more than 20 feet. In 1832 the New York and Harlem attempted to build some long four-wheelers. The journals were apparently made with enough side play for the curves, but the support of only four wheels was not enough, and the body sagged at the middle. A small-wheeled four-wheel truck had to be placed at the center of the car for support.[17] After this discouraging episode no more long single-truck cars were built.

The short body cut up into three sections resulted in dwarfish compartments, which became uncomfortably crowded when they were filled. Roof seats were added to increase capacity, but they were formidable to climb into and provided a disagreeable and dangerous ride. The cars swayed and rolled, low bridges were hazardous to rooftop passengers, and smoke and sparks poured from the locomotive. Seated on high among a jumble of baggage, a traveler often regretted his willingness to "Make room for ladies! Come gentlemen, jump up on the top—plenty of room there."[18] In the opinion of many early railroad travelers, these cars were ugly boxes devoid of even the most meager comfort.

Four-wheel railway vehicles are plagued with a characteristic galloping motion. When the forward set of wheels strikes the rail joint, the body noses down; the springs recoil, and so the pattern is repeated at each joint. This motion, noticeable at low speeds, becomes intolerable and even dangerous as velocity increases. It can be somewhat lessened by spreading the wheelbase, but a rigid wheelbase of much over 7 feet was not possible when relatively sharp curves were to be traversed. The long-wheelbase cars passed through curves with more friction, thus requiring more powerful locomotives because of the additional drag.

The worst fault of the four-wheel car was its inherently unsafe structure. A broken axle or wheel meant certain derailment. In a time of unreliable axles, wheels, and rails, failures were common. A sound-appearing piece of iron might be riddled with fissures or blowholes.

During its first year of operation the Camden and Amboy had a terrible wreck because of a broken axle: the four-wheel cars derailed, killing two people and injuring many others. As late as 1850 the Western Railroad in Massachusetts continued to operate four-wheel cars. In that year a broken axle dropped a car body to the tracks, killing three.[19]

After only a few years' experience, railroad managers realized that something better than the common four-wheel car was needed. It was obvious that the four-wheeler did not lend itself to further improvement; the car was only a makeshift hastily assembled from existing highway vehicle technology. New thinking was needed to develop a conveyance specifically suited to railway requirements.

A car with six-wheel running gears might appear to be the logical successor to the single-truck car. Six-wheelers became popular on European railroads at an early date, and a large number continue in service today. The center pair of wheels stabilized the ride and permitted construction of relatively long car bodies. But this presumably natural transition did not come about in America. The only six-wheel passenger cars observed by Von Gerstner in his exhaustive tour of American railroads were four coaches on the obscure New Orleans and Nashville Railroad. In these the center pair of wheels had no flanges to assist in rounding curves. The only other American railroad known to possess a six-wheel passenger car was the Baltimore and Susquehanna. In 1843 the road had one six-wheeler for passenger service as well as a fairly large number of six-wheel freight cars. From the evidence available it seems clear that the six-wheel car was a rare, almost nonexistent species in this country.

Eight-Wheel Cars—The Beginning of the American Coach

The most stable railway vehicle is that mounted on two independent wheel assemblies known as trucks. These carriages are attached to either end of the car body by center pins and are free to turn or swivel. They provide a flexible arrangement that enables the car to traverse rough track and sharp curves with remarkable ease. The double-truck car has been the ideal running gear for railway vehicles since its introduction. To the present day it has maintained a leading position because of its smooth, easy ride. It was, of course, particularly well suited to the uneven and lightly built tracks of the early American railway.

The eight-wheel car offered more than a flexible running gear, for the car body could be made as long as desired. Double-truck cars provided seating equal to three four-wheeled cars. The saving in cost and dead weight was substantial. Three single-truck cars would have six end bulkheads, twelve wheels, and six couplers. The single double-truck car of equal capacity would have only two bulkheads, eight wheels, and two couplers. If the compartment style of car had continued in use, as it did in Europe, the additional weight of the side doors and interior bulkheads would have been added, making the dead-weight factor of the four-wheel car even more disproportionate to the eight-wheel car. In America the car body was constructed as a single, open compartment. It was the simplest, lightest, and cheapest plan for a railroad coach. The great common compartment was proclaimed as more "republican" in character, and thus more agreeable to the democratic ideals of the American traveling public, than the exclusive compartment car. In the common compartment passengers mingled freely without distinction of class or station. (This sentiment was true enough in ordinary coach travel, but it became something of a myth after the early introduction of first-class sleeping and parlor cars.) By using end doors, the side panels below the windows could be made as a continuous truss. With side doors, extremely heavy floor sills were necessary to support the body. It was argued that multiple side doors permitted quicker loading and discharge of passengers. But in this country, where station stops were less frequent than in Europe, end doors were found adequate.

The double-truck idea was projected in the dawn of the railway era long before passenger carrying lines existed. William Chapman (1749–1832), a British civil engineer, obtained a patent in 1812 for single and double "bogies" for railway service.[20] Chapman's patent specified trucks to spread the weight and improve the trackings of locomotives. He later built several engines with this improvement, and he envisioned the application of the scheme to passenger cars as well. In 1825 he forecast a commodious passenger carriage, complete with dining facilities, rolling grandly across the British Isles. On a more practical level in the same year, Thomas Tredgold, the author of many widely read engineering treatises, suggested that eight-wheel cars would provide more even distribution of weight upon the track. The idea apparently languished in England; the first instance of an eight-

Figure 1.7   R. P. Morgan's double-deck eight-wheel car, patented in July 1827, was illustrated in a Boston newspaper, The American Traveler, *April 14, 1829.* (*Railroadians of America*)

wheel car was the ceremonial carriage made for the Duke of Wellington on the occasion of the Liverpool and Manchester Railway's opening in September 1830.[21] An open car with a red canopy, it was 32 feet long. Apparently it was intended only as a special barge for the celebration, since it was rarely used afterward. Certainly it set no pattern, for British railways remained content with four-wheel cars for many years.

In America a plan for an eight-wheel car was put forward just before the railway age opened. In 1829 Richard Morgan of Stockbridge, Massachusetts, proposed the construction of a huge two-story land barge for railway travel (Figure 1.7).[22] The car was to have sleeping "births" on the first floor and a gay fresh-air promenade on the second. The captain (conductor) was provided with a cupola lookout, which in turn was surmounted by a giant flag. Morgan may have projected his scheme as early as July 27, 1827, when he was granted a patent for improvements in railroad cars. Unfortunately, the papers were lost in the 1836 Patent Office fire and were not among those restored in later years. Morgan did promote a friction bearing, and this may have been all that was covered by the patent. However, newspaper notices with a line drawing appeared in the Boston *American Traveler* in April 1829 and were considered important enough to be copied by a New York paper.

In the same year at Quincy, Massachusetts, primitive eight-wheel cars were being used to carry blocks of granite. Mechanics on the Baltimore and Ohio inadvertently stumbled on the eight-wheel arrangement by loading long timbers on two four-wheel cars. The plan worked so well that double-truck cars were built to haul cordwood and other products. Several were in service late in 1830.

It was not long before the same idea was tried out for passengers. In 1828 or 1829 Joseph Smith of Philadelphia visited the B & O offices with a model of a long, rectangular passenger car.[23] Smith argued that this body style was more suitable for railway service than the ordinary stagecoach. The model, which lacked wheels, apparently made no immediate impression. Construction of the railroad was just getting underway, and rolling-stock considerations were necessarily postponed. Richard Imlay claimed that he suggested long-bodied cars on eight wheels in 1829 to the Newcastle and Frenchtown Railroad, but that the disinterest of railroad officials and the timidity of a partner prevented the construction of a sample car.[24]

These proposals were not entirely lost, for George Brown, treasurer of the B & O, is credited with reviving the idea several years later after passenger service was established. The line's experience with eight-wheel freight cars led to the decision to build a similar running gear for a new long-bodied passenger car. The car would be called the *Columbus*, which seems appropriate for so adventurous an experiment.[25] At a time when the stagecoach style of car was in vogue, the *Columbus* represented a radical departure. The work was performed under the direction of the road's master carpenter, Conduce Gatch, who later insisted that he was also the designer. Although Ross Winans claimed the design and has received popular credit for the eight-wheel car in general, Gatch's assertion was supported by half a dozen old shopmen who were actually engaged in building the *Columbus*. These conflicting claims were aired at length in the famous eight-wheel car case (discussed in the following section).

No contemporary drawings or descriptions of the *Columbus* have survived, but around 1853 Winans prepared a reconstruction sketch which is reproduced here as Figure 1.8. Gatch attacked the sketch, insisting that it had many errors; a second sketch has been prepared to show Gatch's version of the car (Figure 1.9). Winans's drawing portrays a squat, octagonal-ended car with doors at each corner. A wooden railing backed by wire mesh surrounds the flat roof. The drawing is badly proportioned: the window openings are too low and the headroom is insufficient for even a very short passenger.

Gatch, on the other hand, declared that the body measured about 7 feet by 24 feet and was painted yellow and drab. The window openings were 2 feet square. He agreed that the ends were octagonal but insisted that there was only a single door at each end and that iron stairways occupied the opposite triangular spaces. Gatch explained the peculiar octagonal-end entrance of the car by saying that it was simply the only plan that occurred to him at the time. Since no end-entrance or platform-style cars had yet been built, there was nothing to copy. The *Columbus* was an experiment, and many features of the design proved unsatisfactory. According to Gatch, an ornamental iron railing circled the roof—which was arched and not flat, as Winans's sketch indicated. The trucks had roller side bearings and were originally fitted with plain journals. Winans's friction wheels were not placed on the trucks until the following year. Four longitudinal benches provided seating for about 40 passengers. In 1832 roof

Figure 1.8   Winans's drawing of the Columbus, *the first eight-wheel passenger car built by the B & O shops in 1831. The drawing, dating from about 1853, is a reconstruction. (Peale Museum)*

seats, an extended iron railing, an awning, and an interior stairway were added to increase the car's capacity. Other minor differences can be seen by comparing the two drawings.

Work began on the *Columbus* in the spring of 1831. In late June the *Baltimore Patriot* said that "a spacious new carriage running upon eight wheels" would soon be in service. Its capacity was given, with journalistic exaggeration, as 150 passengers. According to another Baltimore paper, the *Columbus* entered service on July 4. Its first trip was horse-drawn, but on July 13 it was pulled by steam to the Ellicott City end of the line. Surprisingly, the car did not take well to the sharp curves on the upper end of the line and went off the track. Gatch blamed the mishap on the unsprung trucks. He placed india-rubber pads between the side frames and bolsters of the trucks, and this elementary suspension improved the car's tracking enough for further operation. It was seen in service by one old employee as late as 1837.

The *Columbus* was successful enough to encourage the construction of more eight-wheel cars. Between 1832 and 1834 several experimental coaches were built at the B & O's Mt. Clare shops. Descriptions of three have survived. They were produced in the following order: the *Winchester*, the *Dromedary*, and the *Comet*.[26] But instead of refining the basically sound square-body plan, the B & O wandered into a blind-alley detour of exotic drop-frame compartment cars with side entrances. The *Winchester* represents a giant step backwards (Figure 1.10). Three swell-side stage bodies with side doors were mounted on an eight-wheel running gear. It was again necessary to use side curtains instead of window glass. Like the *Columbus*, the trucks were unsprung. The second car in the series, the *Dromedary*, had a drop frame that lowered the passenger compartment to about 10 inches from the rails. This arrangement permitted easier board-

ing. The main reason for using it, however, was that of safety: in the event of a broken axle, the underslung car was considered less likely to upset.

Winans's version of the *Dromedary* shows a mechanically unsound scheme (Figure 1.11). The truck mounting is particularly weak; the overhead truss support is as ungainly as it is ineffectual. A rack for cordwood is shown at the right end. The trucks are sprung, which represents some advance over the previous cars. Gatch's sketch also portrays the depressed-center plan, but it shows a long, continuous compartment with a single side entrance (Figure 1.12). Both Winans and Gatch agreed that small glass sashes were placed over the window openings. In Gatch's drawing the wooden frames are clad with sheet iron, and this detail is confirmed in the 1834 B & O annual report, which said that the passenger car frames were "plated with iron."

The *Comet*, the last of the ill-conceived trio, had five narrow bodies on a single drop frame and was mounted on live-spring trucks. It is possible that other experimental eight-wheel passenger cars were built at Mt. Clare. The 1834 annual report mentioned that additional new cars were under construction and that four old single-truck cars were being rebuilt on the eight-wheel pattern. In 1840 Von Gerstner saw abundant evidence of past follies in the Mt. Clare yard. He reported: "At no other railroad station in North America does one find . . . so many old and new cars of every size and shape, so many broken wheels and axles and demolished car bodies."[27] It was obvious, he said, that every conceivable plan for rolling stock had been tried and that the B & O had paid high for the experience gained. As early as 1834 the road's chief officers were convinced that the eight-wheel passenger car was the superior plan. The president's message in the annual report contained a strong endorsement:[28]

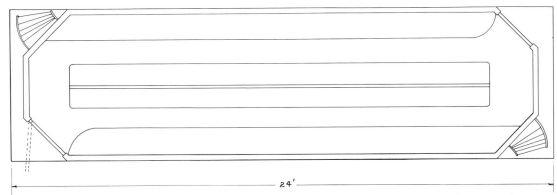

Figure 1.9 *A drawing of the* Columbus *based on the description by Conduce Gatch, who contested Winans's claim to be the car's designer.* (Drawing by John H. White, Jr.)

Figure 1.10 *The* Winchester, *built by the B & O around 1832, was among the earliest eight-wheel cars constructed. It shows the adaptation of swell-sided coach bodies for railroad use.* (Peale Museum)

After the experience of several years, the Board have come to the determination of employing an eight wheel car invented by Ross Winans for the transportation of passengers. This consists of two sets of ordinary running gear with steel springs, each set having what may be called a rose bolt, equidistant from the centre of motion of each wheel. The two sets are placed at the desired distance apart, and connected by a frame, stiffened by a thin iron plate on the sides, on which frame the body of the car rests. The great advantages of this mode of construction consist in the steadiness of the car, when moving on a curved road, or on one whose surface is uneven or slightly out of repair; for the two sets of wheels accommodate themselves to the inequalities of the surface, without affecting the car, resting as it

does on the centre of each set. Another important object, which is attained, is safety: for while, in the ordinary four wheeled cars, the breaking of a wheel or an axle, might be productive of the most fatal consequences, such could not be the case, where, with eight wheels, there would be enough left to support it on the track until the train could be stopped. Again, upon a curved road it is necessary to place the pairs of wheels of a four wheeled car as near as possible, so as to diminish friction, and this makes it necessary to build the body of the car very short, to prevent a disagreeable, and at times, dangerous vibration from side to side while the car is in rapid motion. In the eight wheel cars, on the contrary, the pairs of wheels of each set are placed as near together as can be desirable, under any circumstances,

Figure 1.11   Two other early eight-wheel cars, the Dromedary *and the* Comet, *were built by the B & O in 1834. (Peale Museum)*

Figure 1.12   This drawing of the Dromedary *is based on a sketch made by Conduce Gatch around 1855. (Traced by John H. White, Jr.)*

and the sets themselves may be as far apart as may be necessary to accommodate a car of any length, without increasing the friction on the rails when passing curves of the least radius, or producing the lateral motion complained of. In point of economy too, this mode of construction is much preferable to the one hitherto employed.

The B & O decided to order ten eight-wheel cars for the Washington branch and wisely resolved to return to the rectangular body. Important improvements had accrued since the building of the *Columbus*, however. The octagonal end door gave way to the now familiar end door, with platform entrances. A center aisle was flanked by double cross seats. The classic American day coach was born.

The first Washington branch cars were ready in early July 1835. By October of that year the B & O had twenty-five eight-wheel passenger cars in service (see Figures 1.69 and 1.70 in the Representative Cars section at the end of the chapter).

The B & O deserves full credit for introducing and developing

the eight-wheel passenger car years before most roads thought beyond the simplest style of four-wheelers. In this early period, however, there were also a few other builders who showed a strong interest in experimenting with eight-wheel designs. The first builder outside of the B & O was Richard Imlay, who constructed a number of four-wheel cars at his Baltimore coach factory. By 1833 he had opened a new car works next to the Baldwin and Norris locomotive shops in the Bush Hill section of Philadelphia. Once there, Imlay's interest in eight-wheel cars was renewed. In 1832 he had built a sample car for Jonas P. Fairlamb, who obtained a patent on January 19, 1833, that shows an eight-wheel running gear.[29] The next year Imlay decided to construct an eight-wheel car on his own plan. The body design has been credited to a model built as early as 1829 by either George Fultz or Laban Proctor.[30] It was on a drop-center plan similar to Gatch's reconstruction of the *Dromedary*. Whatever the source of the design, Imlay began work on his first eight-wheel car, the

Figure 1.13   A possum-belly, double-truck car designed by Richard Imlay of Philadelphia. The drawing is based on a lithograph produced around 1836. (Traced by John H. White, Jr.)

Victory, in August 1834. It was described as ready for service in the *Philadelphian* of July 3, 1835:

It is divided into three apartments, two being designed for private parties of ladies and gentlemen, while the third [the central compartment] is a kind of ordinary. The car will be drawn by four matched horses, and run from the corner of Chestnut and Broad to the viaduct or bridge on the Schuylkill, or to any other place on the railroad when chartered by private parties for that purpose.

The model of the *Victory* is illustrated in Figures 1.67 and 1.68 in the Representative Cars section. A lithograph showing a full-sized car of the same design was issued around 1836 by the well-known Philadelphia printmakers Lehman and Duval, but only a poor photocopy of the print has survived. A modern tracing made from it shows the car's construction (Figure 1.13). After a trial on the Philadelphia and Columbia, the 36-foot-long car was sold to the Philadelphia, Germantown, and Norristown Railroad. The line was satisfied with the *Victory* and, according to the directors' minutes, ordered three more eight-wheel cars from Imlay in 1835 at a cost of $1,600 each. Continued approval of eight-wheel cars was indicated in 1837, when the minutes record that two four-wheel cars were rebuilt as the new double-truck *Norristown*.[31]

Imlay's design was defective in many respects. He followed the drop-center or "possum-belly" plan that had already been rejected by the B & O because of the complicated and weak floor frame. The side-door entrance, swell sides, and open windows were obvious defects. The end compartments over the truck could be reached only by mounting two or three steps. The lack of end doors or platforms prevented passage from one car to the next. The leaf-spring truck bolster, patented on September 21, 1837, by Imlay (Patent No. 389), proved unstable and was little used. One notable improvement made by Imlay was the raised or clerestory roof; he elevated the central portion of the roof to admit additional air and light. This idea seemed to have been forgotten by the industry for a while after Imlay was forced to quit the car-building trade, but it was revived with considerable success in the early 1860s.

Despite their shortcomings, Imlay's original cars had many purchasers. W. Milnor Roberts recommended them to the private car operators of the Philadelphia and Columbia after a personal inspection of the *Victory*. A car was promptly placed on the line and named for Roberts.[32] Within a few years the road was largely stocked with Imlay cars, all of them eight-wheelers.[33] The largest carried 60 passengers and cost $2,000. All had water closets, and many were fitted with ladies' compartments—separate rooms set aside for women and their escorts. Years later, an old employee of the Philadelphia and Columbia still remembered the early Imlay cars and praised their easy ride.[34]

Imlay's business flourished during the next years. He not only built cars for the nearby roads but also had customers referred

directly by the prominent locomotive builder M. W. Baldwin. For example, in a typical letter dated February 17, 1837, Baldwin stated:[35]

Mr. Imlay of our city will make passenger car No. 1 on 8 wheels 37 ft. long, projecting roof [clerestory], side seated or shifting backs on cast iron wheels, with axles 3¼ diameter finished in handsome style, and capable of carrying comfortably 60 passengers all seated, for the sum of 1750 Dolls. No. 2 similar to No. 1 but not so well finished, 1600 Dolls. No. 3 1550 Dolls. Passenger on 4 wheels, . . . with shifting backs, cross seats, capable of carrying seated 24 passengers, to be well finished for 800 Dolls. No. 2 also on 4 wheels 725 Dolls.

In addition to day cars, Imlay was among the first to offer sleeping cars (see Chapter 3). These cars, first offered in 1837 or 1838, were not built on the depressed-center plan; they were ordinary square-body cars with end doors and platforms. It is probable that Imlay had recognized the folly of the drop-center cars by this time and had switched over to the square-body plan for day cars as well.

Although Imlay was a pioneer in the field and seemed destined to become a major car builder, his business foundered in the financial depression following the 1837 panic. The shop closed around 1840, and Imlay drifted into other lines of employment.

Through the B & O and Imlay, the gospel of the eight-wheel car quickly spread among the mid-Atlantic railroads. The idea was carried to the Newcastle and Frenchtown in 1834 by a former B & O employee, James B. Dorsey.[36] His new employers were skeptical at first when Dorsey spoke of the superior performance of the eight-wheel cars, but he persisted. To prove his point he built a crude double-truck flatcar from two four-wheel freight cars. The success of this test car led to construction of eight freight cars and finally, in the spring of 1835, to a passenger coach.

Next to fall into line was the Philadelphia, Wilmington, and Baltimore. The southern end of the road was built by Benjamin H. Latrobe between 1835 and 1837.[37] Because of his experiences with the B & O, he faithfully recommended the eight-wheel car to the road's directors. But the directors sided with William Strickland, the engineer for the northern end of the line, in opposing "the large cars." He maintained that they were heavier, more expensive, and more likely to go out of order. Strickland also claimed, incorrectly, that no other Eastern road had adopted them. In the spring of 1835, just before the final pattern of the eight-wheel car was perfected, the Wilmington line decided to continue buying four-wheel cars. With some reluctance the directors agreed to try a double-truck baggage car, and from the beginning its performance was good enough to convince them of their error. Before the year was out their shops were busy splicing short four-wheel bodies together for remounting on eight wheels.

Immediately after the P W & B capitulation came the Camden

Figure 1.14   Camden and Amboy Railroad passenger cars built about 1836, shown at the time of their retirement thirty years later. (Chaney Neg. 14489)

and Amboy Railroad. Construction of eight-wheel cars began in its repair shops in 1836, and by 1840 fifteen double-truck cars were in service. They were patterned very closely on the Washington branch cars, as will be apparent from a comparison of Figures 1.69 and 1.70. The underhung truss remained popular with the C & A for some time; cars of this construction were in service until the early 1870s. Those pictured in Figure 1.14 may be cars of the first lot that were later modernized with extended platform roofs and roof ventilators. Another scene of the 1870s showing a car on a swing bridge illustrates the need for retaining the underhung truss (Figure 1.15). The center side door would have cut an inside-the-panel truss in half.

A second New Jersey line to adopt the eight-wheel car was the Morris and Essex Railroad. The first eight-wheeler was placed on the road in February 1838. Von Gerstner reported that it was a giant vehicle capable of seating 90 passengers and that it had a baggage compartment under the car.

Southern railroads, often dismissed as slow to accept technical innovation, were quick to recognize the merits of the American-style coach. One reason is that many of these roads were founded in the mid- or late 1830s, just as the reputation of the eight-wheel car was being established. In 1837 the Monroe Railroad and Banking Company had two eight-wheel cars under construction in Philadelphia, possibly at Imlay's shop.[38] In the same year the South Carolina Railroad began acquiring double-truck coaches. This pioneering Southern line was already stocked with four-wheelers, but it went ahead with a costly program of substitution after being convinced of the new style's superiority. In the road's 1840 annual report the four-wheelers were contemptuously described as "old and nearly valueless."

Not content with simply copying the invention of the Yankee car shops, the South Carolina proceeded to develop its own style of car body. The hogsheads that were used to transport tobacco throughout the South suggested this form of construction to one

of the road's directors, Colonel James Gadsden. These mammoth barrels, cheap and hardy containers, seemed well suited to the strenuous labor of railway service.[39] On Gadsden's suggestion a freight car was built in September 1837 by the master carpenter, George S. Hacker. The fact that this style of construction was familiar to the local mechanic undoubtedly kindled enthusiasm for the bizarre design. The road produced more barrel-shaped freight cars and then, in 1840, a passenger car in the same style (Figure 1.16). The body was 30 feet long, with 30-inch platforms and twenty 15- by 30-inch windows. Heavy staves 1¼ by 5 inches were held together with iron bands.

The unprepossessing cars were "frequently condemned by strangers at first sight," the 1843 annual report admitted. The directors claimed, however, that once the initial shock was over, the discerning passenger would find that the cars had many advantages. In a masterful series of rationalizations, the merits of the barrel car were summarized in the *American Railroad Journal*.[40] According to this article the cars were lighter, simpler to build, and cheaper than square cars. The round body was nearly fireproof, for it shed live cinders as fast as they fell; lateral wind pressure was dissipated; and in a derailment the car rolled down the embankment without injury. Hacker patented the design on January 21, 1841 (No. 1937), and the South Carolina reports were full of praise for the hogshead car. Happily, the design was confined to the home road; the prospect of barrel cars "flying with the speed of a locomotive" held little appeal elsewhere.

Other Southern roads, while avoiding the South Carolina's misguided digression, showed a strong attraction to the eight-wheel car. Nearly all the Southern lines visited by Von Gerstner in his travels of 1838–1840 were well stocked with American coaches. He reported that the Wilmington-to-Raleigh line in North Carolina had eight very elegant double-truck coaches that were provided with ladies' compartments and water closets.[41] Many of the lines contained half-fare compartments in the baggage car for

Figure 1.15   Camden and Amboy Railroad coach of about 1840, shown in a scene from the 1870s. (E. P. Alexander Collection)

Figure 1.16   Hacker's hogs head car, built in the shape of a gigantic barrel, was produced for service on the South Carolina Railroad in 1840. (Southern Railway)

Negroes, but Negroes could also ride in the coaches at full fare. Von Gerstner pointed out that this was in fact better than the treatment given Northern Negroes, who had to ride in separate cars no matter what the fare.

By the late 1830s the entire car-building industry had begun to approve of the eight-wheel car. In 1837 Stephenson stopped pushing cars with omnibus bodies and developed a distinctive, if not beautiful, design for double-truck passenger coaches.[42] The whole side of the body was made as a lattice truss. This plan, which was among the first to employ bridge-construction methods for car building, created a very strong car. The truss assisted materially in supporting the body and lessened the need for heavy frame sills. Long self-supporting car bodies thus became possible. The cars were popularly known as X- or Diamond frame cars.

One X-frame car built for the New York and Harlem proved too high for the overhead clearances and it was sold to the Auburn and Syracuse Railroad in August 1839. A contemporary newspaper described it as a "traveling palace . . . gorgeously flowered and varnished."[43] The car, renamed *Auburn*, was reserved for through passengers. It was fitted with thirty double reversible seats and had a stovepipe running the length of the car for more even heating. The glass windows were stationary; adjacent wooden panels were opened for air. Cars of the same pattern were built for the Mohawk and Hudson, the New Jersey Railroad and Transportation Company, and the New York and Erie. One of them is pictured in the lower right corner of Figure 1.17, a lithograph showing Stephenson's factory. (Another contemporary, somewhat foreshortened, sketch is shown in Dunbar's *A History of Travel in America*.[44]) Stephenson failed in 1843, however, and that ended the production of Diamond frame cars. He was able to reopen, but he blamed the "bad paper" of the steam roads for his failure and said that henceforth he would build omnibuses and horsecars only when payment was to be in cash.

In Wilmington, Delaware, the car builders Betts, Pusey, and Harlan (later reorganized as the mighty Harlan and Hollingsworth Company) also came to approve of the eight-wheel car. In a printed prospectus of 1839 the firm stated that during the short period of two years since their general adoption in this country, eight-wheel cars had given "universal satisfaction."[45] Thirty-nine passenger and twenty-eight mail, baggage, and freight cars on

*Figure 1.17   Stephenson's lattice frame car, shown in the lower right of this lithograph,
was produced for several railroads, including the New York and Erie. The print dates
from about 1840.*

this plan had already been delivered. The prospectus listed seven
major Middle Atlantic and Southern railroads receiving these cars
and described the various classes of passenger car:

1st Class Car. For 60 passengers, body 32 feet in length and 8 feet
6 inches in width, spring seats crosswise, with shifting backs, trimmed
in best style, with silk or worsted, venetian blinds to the windows, the
platforms protected by an ornamental iron railing, with iron steps to
ascend the same, draw springs and buffers, zinc roof, wheels of cast
iron, (chilled, 3 feet in diameter, effective double acting brakes, the
body secured in the strongest manner by iron bolts and brace rods;
painting in best style, with gilt border and lettering, and the entire
finished in the most workmanlike manner

    for . . . .   493 Liv. St. or 2,400 dollars.
    Wrought iron tires on the wheels
      additional   200   ″
    Private apartment for ladies, completely trimmed and
      carpeted   100   ″

2d Class Car. For 60 passengers, body as above, seats crosswise,
with shifting backs, hair or cloth seating, worsted head lining, festoons,
and sun curtains, iron railing and steps to the platforms, spiral draw

springs and buffers, zinc roof, wheels as above, brakes and brace rods,
neatly painted and varnished

    for . . . .   411 Liv. St. or 2,000 dollars.
    Wrought iron tires on the wheels
      additional   200   ″
    Private apartment for ladies, neatly fitted
      up   60   ″

3d Class Car. For 60 passengers, body as above, seats crosswise,
with shifting backs, cloth or leather seating, plain iron railing, canvass
roof painted and sanded, wheels and brakes as above, trimmed and
painted in a plain neat style,

    for . . . .   339 Liv. St. or 1,650 dollars.
    Wrought iron tires on the wheels
      additional   200   ″

4th Class Car. For 60 passengers, body as above, plain wooden
seats, with shifting backs painted, common iron railing on platforms,
wheels, brakes and canvass roof as above, and plainly painted on the
outside,

    for . . . .   287½ Liv. St. or 1,400 dollars.
    Wrought iron tires on the wheels
      additional   200   ″

Figure 1.18   A predecessor of Harlan and Hollingsworth of Wilmington, Delaware, in-cluded this engraving in an advertising brochure of 1839. (Association of American Railroads)

Printed in English and French, the prospectus was addressed to the railroad companies of Europe. Whether any foreign orders were secured is a question, but it is known that American cars were supplied to European roads by other U.S. builders at an early date. The front page of the prospectus was graced with a fine lithograph of a passenger coach (Figure 1.18). The drawing is well proportioned and mechanical in nature, although the human figures shown at the windows appear to be only about one-half life size.

Acceptance of the eight-wheel car was well advanced in some parts of the country by late 1838, but as the *American Railroad Journal* stated:[46]

Much as this description of Railroad car is in repute in the Southern and a portion of the Middle States—yet in this State [New York] and East of us, there is a singular prejudice in favor of the four-wheel cars, and against the new form.

In general New England was slower to adopt the American car than other settled regions of the United States. At this time the Northeastern states were showing the same conservatism in loco-motives: a major locomotive builder of the area was content with the rigid-frame English style of engine some years after all other domestic manufacturers had turned to the truck locomotive. It should also be recognized that the New England roads had al-ready invested heavily in four-wheel cars, and many simply could not afford to discard them after only a few seasons.

Although resistance to the new form was marked in the North-east, it was not universal there. As early as 1833 the Boston and Worcester was considering buying trucks from Winans with bodies to be furnished by Stephenson. The road decided against

it, however, and did not place eight-wheel cars in service until 1837 or 1838.[47] Even so these cars, built by Davenport and Bridges, were among the very first of their type in New England. A year or so earlier the Hartford and New Haven Railroad had asked Winans the price of eight-wheel cars. He sent the following quotation; it cannot be determined if the road actually placed an order.[48]

The 8 wheels, the axles made of swedes iron faggoted under a trip hammer, the 4 iron bolsters made same as axles, the 8 light & 4 heavy English steel springs, the two small carriage frames or running gear made mostly of wrought iron, the large frame for the body to rest on with iron suspension braces, the four friction rollers & various castings for supporting the springs boxes &c. the iron railing surrounding the platform at either end of the car, the two brakes one at each end of the car so arranged as to brake or apply friction to all of the wheels when necessary for stopping or retarding the motion of the car, and the fitting up of everything complete ready to receive the body cost $1200. the body cost $850. making the sum of $2050.

The Boston and Providence's first eight-wheel car was com-pleted at the company workshops in September 1838 (see Figure 1.75 in the Representative Cars section). The car was so success-ful that the line rebuilt some four-wheel cars on the double-truck pattern. Two or even three bodies were united, and the old run-ning gears were remodeled as trucks and placed beneath them.[49]

In December of the same year the Nashua and Lowell received several American-style coaches from the shop of Jeremiah Meyers in Attleboro, Massachusetts.[50] (Meyers secured the order by underbidding Davenport and Bridges, but it was a profitless contract that put him out of business.) The Boston and Lowell built its first eight-wheel car in April 1840 in the company repair shops, under the direction and to the design of the master mechanic. The drawing made from the original car which was still in service in 1853, was offered in testimony at the eight-wheel car case.[51] It was explained that the left truck shows the first mode of suspension and the right truck that which was in use when the drawing was made.

While these examples demonstrate that some of the New England roads were only a few years behind other regions in adopting the eight-wheel car, there is evidence of continuing resistance to the new form in New England. As late as 1853 the Eastern Railroad defended four-wheel passenger cars, claiming that they required less power to pull them and passed through curves with less friction and hence less wheel wear (arguments that, ironically, were usually reserved for the double-truck car). The Western Railroad, later the Boston and Albany, was purchasing new four-wheel passenger cars some years after they were considered obsolete. One of these, built by the Springfield Car and Engine Company in 1848, was involved in an accident which brought about some pointed criticism of the road for clinging to outmoded equipment.[52] In fairness it should be noted that the Western was also running eight-wheel cars as early as 1842. An engraving made after a daguerreotype taken the same year shows cars of this type (Figure 1.19).

In spite of regional variations, it is fair to say that in only five years the double-truck car displaced the four-wheel carriage. By 1840 it was the unchallenged standard on American railroads. Its rapid spread across the nation must be regarded as one of the most complete and revolutionary innovations in American railroad technology. It is also something of an American phenomenon, for eight-wheel cars were not widely used elsewhere in the world until rather late in the nineteenth century.

THE EIGHT-WHEEL CAR CASE

There is a disturbing tendency in some histories of technology to credit basic inventions to the genius of one man. Among the most fallacious, yet best-established reputations are James Watt's as the inventor of the steam engine, Robert Fulton's as the inventor of the steamboat, and George Stephenson's as the inventor of the locomotive. To this list might be added Ross Winans's as the inventor of the eight-wheel car. Many such claims originate with enthusiastic biographers who puff the work of their subjects. Other men simply become folk heroes and are popularly crowned with greatness. Still others actively sought recognition in their own lifetime for personal gain or the need to satisfy a voracious ego.

Most standard references, including the major encyclopedias, unquestioningly assign the eight-wheel car to Winans. Yet the eight-wheel car, like most such devices, was a conglomerate invention that evolved from the labor of several resourceful mechanics. One man rarely envisions a whole machine at once. A basic idea may occur to any number of artisans when a new need arises, and each studies and refines it in his own way. Winans was

certainly involved in the early development of the eight-wheel car. He deserves partial, but not exclusive, credit for its perfection.[53]

Winans documented his claim in a patent of October 1, 1834, for improvements in railroad cars. The long specification covered many points in the construction of eight-wheel cars. According to his critics, it exhibited as much or more genius than the invention itself. The principal claim specified an extremely short-wheelbase truck with live-spring side frame. In the final paragraph, however, Winans specifically disclaimed inventing the eight-wheel car, referring to "the running of cars or carriages upon eight wheels, this having been previously done."

Winans's patent was not in fact the first in the country that broadly touched upon such cars; Jonas P. Fairlamb (1833) and Ephraim Morris (1829) had previously secured papers on eight-wheelers. The patent law was very liberal at the time that Winans's patent was issued. A properly drafted specification and payment of fees secured the patent. No experts examined the draft for conflicts with previous claims, nor was the "state of the art" considered. It is surprising that the wheel itself was not patented. Winans appended no drawing to the patent until after the first lawsuit in 1838, and then he included an illustration of a boxcar rather than a passenger car. Another major discrepancy was the fact that the drawing showed a wooden-beam truck rather than the live-spring design called for in the specification.

If Winans believed himself the creator of the eight-wheel car, he suffered in silence for more than four years before pressing the claim. If he invented and built the *Columbus*, as he claimed, why didn't he apply for a patent in 1831? And why did the text of his 1834 patent deny that he had originated the eight-wheel car? As a wheel manufacturer Winans was in regular contact with major railroads and car builders across the land, yet nearly all stated that Winans had never hinted at his right to a royalty on the eight-wheel cars for which he supplied the wheels. It was uncharacteristic behavior for a man who was renowned for his outspoken independence.

Winans' critics claimed that the patent could not cover an idea which had been abandoned to the public for nearly four years. They also suggested that Winans did not apply for his patent until his chief rival, Conduce Gatch, retired from railroad service in May 1834. In Winans's defense it should be said that an inventor is entitled to a grace period in which to improve and assess the importance of his discovery. It should also be noted that the president's message in the 1834 B & O annual report acknowledges Winans as the inventor of the eight-wheel car. As an employee of the road, Winans reportedly waived his claim against the B & O early in 1834. The document was not produced, however, during the long years of litigation that followed; either it never existed or it was lost.

The first suit that Winans filed was against the Newcastle and Frenchtown Railroad in July 1838 for violating his patent by operating eight-wheel cars without paying him a fee. It was the beginning of twenty years of legal proceedings. The jury could not agree, but before a new trial was ordered the defendant

EXPRESS TRAIN ON WESTERN RAILROAD.

(From a Daguerreotype, made in 1842, under the direction of Charles Van Benthuysen.)

AFTERNOON TRAIN BETWEEN ALBANY AND SPRINGFIELD.

STILLMAN WITT, Superintendent at Albany.

THOMAS W. ALLEN, Master Mechanic.

D. S. WOOD, Engineer.

JOHN B. ADAMS, Conductor.

HORACE H. BABCOCK, Ticket Agent.

Figure 1.19   Early examples of eight-wheel passenger cars on the Western Railroad, later the Boston and Albany. This engraving was based on a daguerreotype made in 1842.

voluntarily paid Winans a judgment of $500. This half victory was enough to encourage him to start action against others. He approached the Baltimore and Susquehanna next, asking for the modest sum of $500 late in 1839. The railroad countered with an offer of $350, which Winans refused. He waited until the road's president was in Philadelphia and then had that official served with a summons. Faced with an inconvenient suit away from the home office, the Baltimore and Susquehanna quickly agreed to Winans's demands. His claims now established in two states, Winans turned to the Portsmouth and Roanoke, which meekly paid him a license fee. The Philadelphia and Reading also agreed to a friendly, out-of-court settlement for the minimum fee of $500.

The strategy was simple: establish the patent's validity by a series of small settlements. Even if the payments were trivial, the precedents confirmed the soundness of Winans's claim. Curiously, at this juncture Winans did not press his advantage. It was not until seven years later that he unexpectedly renewed legal action against the Troy and Schenectady Railroad. A small, financially weak line, it was in no position to fight off so wealthy and vigorous an opponent as Ross Winans of Baltimore. Apparently the suit was prompted by the pending expiration of the 1834 patent. Winans sued the road for $100 a year for every eight-wheel car in operation, plus a settlement for the past use of such cars. While the suit was underway, a special bill was hurried through Con-

gress in 1848 to extend the potentially valuable patent for another seven years. In June 1850 the U.S. Cicuit Court ruled in Winans's favor, but settlement was delayed by appeals. Eventually, a new trial was denied, and Winans accepted a paltry settlement of $100.

The real prize in the Troy and Schenectady suit was legal confirmation of Winans's claim, for the entire industry was now open to a severe levy. Before initiating the expected barrage of lawsuits, he agreed to share the spoils with his attorneys, C. D. Gould and J. A. Spenser, if they would finance and direct the proceedings. Receipts from his past legal battles had been disappointing, and Winans did not want to proceed alone. Between 1852 and 1853 Gould and Spenser started action against eight New York railroads as well as against Eaton and Gilbert, car builders of Troy. Separate suits were entered against A. & W. Denmead, car and locomotive builders in Baltimore, and the Eastern Railroad.

This massive legal assault was the climax of the Twenty Years' War against the Railroads, as Winans's series of patent suits came to be called. It was now a matter of grave importance to the railroad industry. Ruin lay ahead if Winans's extraordinary claim was not nullified. Winans was characterized as the arch-villain of the railway supply trade, a pariah and an "imposter." He was called the "inventor of everything." These attacks were based on more than the car case, for Winans was also conducting court actions on his alleged patent rights to six-wheel connected loco-motives and to the variable exhaust. Moreover, he was engaged in a vitriolic public debate with the directors of the B & O over the merits of his Camel locomotives. It is not surprising that his railway machinery business declined rapidly after the mid-1850s.

Recognizing the seriousness of the Winans suits, the recently formed New York Central Railroad took on the defense of smaller New York railroads and assisted Eaton and Gilbert. It hired two first-rate lawyers, William Whiting and William E. Hubbell. On its part the Eastern Railroad spent $20,000 in legal preparation and published a 1299-page transcript of its defense. "Neither time, nor labor nor money was spared. The history of railroads was thoroughly investigated from their origin both in the United States and in England. Every old workshop was searched for plans and drawings of early railroad cars—every venerable engineer and machinist was questioned of his experience relative to rail-road carriages. No geologist ever examined more carefully for antediluvian remains."[54]

Chief among the witnesses for the defense in the series of suits was Conduce Gatch, who constructed the first double-truck cars on the B & O. Gatch was steadfast in his claim that he designed and built these cars, including the prototype *Columbus*. Several old workmen supported his contentions. Only one, Oliver Crom-well, sided with Winans, but even he would not credit Winans with more than the body design. Evidence given by Gridley Bryant, builder of eight-wheel freight cars for the Granite Rail-way in 1829, and the abundance of printed and patent literature on eight-wheel cars previous to Winans's 1834 patent, made a shambles of the Winans case. The testimony was long and repeti-tious; one trial ran for 75 days. Winans lost one action after another, but as a railroad pamphlet of the time described it, he kept pressing "with a pertinacity of purpose rarely equalled." At last the New York and Erie case was heard by the U.S. Supreme Court in its December 1858 term. Judgment was awarded the defendants on January 10, 1859. So ended Winans's incredible legal adventure—the Twenty Years' War was over. The weight of the evidence was against him, but the publicity, as misinterpreted by later writers, seems to have ensured Winans's enthronement in history as the father of the eight-wheel car.

Age of the Arched-Roof Car: 1840–1860

After the late 1830s, when the eight-wheel car became firmly established, there were twenty quiet years of contentment with the existing day coach pattern. The same design was repeated again and again with little or no change. Why tamper with a simple, inexpensive plan that served every requirement so well? It seemed unnecessary if not imprudent to experiment. A typical car of 1840 had a rectangular body rarely over 30 feet long by 8½ feet wide. Before 1845 the glass windows were generally station-ary, with sliding panels for ventilation. The ceiling was low, pro-viding headroom of just over 6 feet. The seats were closely spaced and had narrow cushions and low backs. The trucks were on a short wheelbase. The total weight was about 8 tons.

An example of such a car was the *Morris Run*, built in 1840 for the Tioga Railroad by Harlan and Hollingsworth (Figure 1.20). When it was exhibited some forty years later at the Railway Appliance Exposition, it was described as being 36 feet long by 8 feet, 4 inches wide, with a 6-foot, 4-inch headroom clearance.[55] Inside-bearing trucks were seldom used on American cars. The truss rods shown were also rare at this early period, although they became a standard fixture on later wooden cars. The seats had cast-iron frames, and the cushions were covered in leather. Single candle lamps were placed at each end of the car. The original cost was given as $2,000. In 1883 the car had run more than a million miles. It continued in service for another ten years, when it was destroyed by fire.[56] Similar cars are shown in the Repre-sentative Cars section (see Figures 1.70 to 1.81) at the chapter end.

By the mid 1840s a few modest improvements were evident. Some cars had 35-foot bodies. Movable window sashes replaced stationary glazed frames, and more and larger windows replaced the upper panels. Roof canopies extended over the end plat-forms to protect boarding passengers. Roof-mounted ventilators were introduced. But the headroom remained restricted, the seats continued to be narrow and closely spaced, lighting was inade-quate, and many cars did not have water closets or ice-water dispensers. The interior finish of these cars was generally plain. Common woods were typically used and painted over. Head lin-ings (ceiling coverings) were sometimes omitted, and the ex-posed wooden roof structure was simply painted. Seat cushions were of leather or broadcloth, and the cheaper cars had wooden-slat seats.

More luxurious cars were occasionally built, however. In 1845 Davenport and Bridges constructed some large coaches seating 70 persons for the Eastern Railroad.[57] They were described as 9½ feet wide, or a foot wider than the usual day coach. The seats were mounted on pivots, upholstered like fine armchairs, and covered in a rich red-velvet plush. The interior paneling was mahogany, and the floor was laid in a diamond pattern of cherry and black walnut.

The early eight-wheel cars were very lightly built. The floor frame, the basic support of the entire car, was the simplest post-and-lintel form. Two side sills 6 by 7 inches, supplemented only by some cross and short end timbers, composed the frame. Center sills, truss rods, or body-panel truss were not generally used, nor were they needed for the short car bodies of the period. The body bolster was built up from three timbers; two to four short beams parallel to the side sills ran between the body bolster and the end sill. Ten or more cross timbers were placed between the bolsters. An example of such a frame is given in the 1845 Eaton and Gilbert drawing (Figure 1.79). John Kirby (1823–1915), a prom-inent master car builder, remembered these frames on the first cars he helped build in the late 1840s.[58] Another pioneer car builder said that the Boston and Albany was using this simple but obviously weak style of framing in 1870.[59] According to *The National Car Builder*, the old form was still found on some roads five years later.

Figure 1.20   *Harlan and Hollingsworth built the* Morris Run *in 1840 for the Tioga Rail-road. The car was preserved as a relic until its loss by fire in 1893.* (Semi Centennial Memoir of Harlan and Hollingsworth)

The body structure was even more fragile than the main frame. It was composed of thin vertical and horizontal pieces very much like the balloon house frame of the day. Thin wooden panels covered the frame. No truss or cross braces were used. The whole framework was like an egg crate resting on two matchsticks—an exaggeration, of course, but not far off. The pre-1855 passenger cars was a generally feeble structure.

Why did the industry fancy this obviously weak construction? Wood-building technology was an ancient art that offered centuries of experience in elaborate bridge and roof trussing. Good timber was abundant and cheap. The explanation appears to lie in an overriding concern to hold down size and weight. Lack of capital had led to the construction of exceedingly light, cheap railways in this country. Flimsy track and bridges could support only the lightest rolling stock. Locomotive power was accordingly limited; hence the cars that were being pulled had to be under-weight. Low ceilings, closely spaced seats, and skimpy framing produced cars of maximum capacity and minimum weight. The absence of auxiliary apparatus, water closets, ventilating equipment, vestibules, and other devices that characterized later cars held down both weight and cost. Certainly for short daylight trips such cars were satisfactory.

By the late 1840s, however, somewhat larger and better-furnished cars were in demand. The Louisa Railroad of Virginia submitted the following specification to Harlan and Hollingsworth in 1848:[60]

Length of Car Out and Out 43 feet 4 inches, with two drop platforms of best yellow heart pine 2 ft. 8 × 4 feet.

Body of Coach from out to out 38 ft. Long × 8 ft. 8 inches wide.

Bottom Rail of best heart yellow pine 38 ft. long 10×5 inches.

Top Rail 42 feet long 8×3 inches best heart pine or oak, to be confined to the pillars with two rivets to each pillar.

Bolsters to be 12×13 best white oak, 4 knee plates of wrought iron ⅝ thick by 10 inches wide, each knee plate confined to the bolster with 3 bolts running through the bolster ¾ in. diameter and 3 running through the bottom railing of ⅝ diameter.

(This a substitute for the circular bar of iron extension bolts.) An inside plank riveted to each pillar with 3 rivets and bolted to the bottom rail between every other window with a bolt of ⅝ diameter, with a nut at the bottom, the plank to be 2½ inches thick by 13 deep.

The Ladies Apartment 6 ft. 4 in. long with a lamp box each side of the door with glass doors inside and out. Seats in the ladies apartment to be red plush sofas, back of the same.

Gentlemen's Apartment to have a passage 16 inches wide, the floor laid with the best yellow pine dressed to 1¼ inches. Seats put crosswise, made of Mahogany wide backs stuffed with best black hair cloth well stuffed.

Seats 2 feet 7 inches from center to center with a foot piece underneath, water cooler to hold 4 gallons, lamp box on each side of the door in the Ladies Apartment—with a stove fixed with hand railings and brass knobs.

Drop platform of best heart yellow pine for the floor 2 steps on each side 9 in. wide and 1 foot high supports on 3 pieces of iron 2½×½ confined to the platform, upright rod from the bottom to the platform to the top of the coach 1⅛ inches in diameter, balance of rod for hand railing 1 inch diameter 2 ft. high.

Door to be 5 feet 10 inches high 2 inches thick paneled of Mahogany good yellow heart pine or curled maple.

Windows to hoist fastened in from the outside with a wooden moulding instead of putty.

Figure 1.21   The cars shown here in 1883 were built around 1850 for the Cleveland, Columbus, Cincinnati, and Indianapolis Railway. They were remodeled with clerestory roofs in the 1860s. (Thomas T. Taber Collection)

Eves from one end of the car to the other 9 inches wide immediately over the window to throw off rain, etc.

Curtains of strong drab twilled worsted.

The car to be 2 inches higher on the inside than the first one made [for] the Louisa Co.

Painted handsomely on the inside and in a plain substantial manner on the outside.

To have a bar of iron on each side 3½ in. wide by ⅜ in. thick running from the front bolster to the back bolster, bending with a regular bow, so that the Top of the bow of iron just comes to the bottom of the window, the said bar to extend from the bolster to each corner in the manner of a brace, and to extend up the corner posts as high as the bottom of the window and the said bar to be made fast to each pillar with rivets, there is to be an extension bolt of ⅝ diameter between every Pillar with a hook head to hook on the 3½ inch bar and to go through the bottom rail with a nut at the bottom. Every pillar is to be confined to the bottom rail with a pilot point bolt.

Very Respectfully, E. Fontaine, Pr. L. RRCO.

In another letter the exterior color was specified as dark red. The car was built for $1,350.

More substantial passenger cars were also constructed in Cleveland for the Cleveland, Columbus, and Cincinnati Railroad (Figure 1.21). These cars, completed in late 1849 and early 1850, were made on the plan of William F. Smith (1826–1878), who was the line's master car builder for nearly thirty years. The cars were 41 feet long by 6½ feet wide and seated 60 passengers.[61] Ash-frame sills were cut and squared by hand in a nearby forest, as no sawmill was available. No center sills were used, but the body was made sag-resistant by a slight camber or bow in the sills. Continuous siding rather than individual panels, together with "plenty of glue," stiffened the body. The cars were still in service in 1883, and they were reported to be holding up better

than many new cars. Some had derailed several times and a few had rolled down embankments, but all survived these mishaps. The truss rods and clerestory roofs that can be seen in Figure 1.21 were added in later years.

During this period a peculiar style of car was evolved by E. A. Stevens for the Camden and Amboy Railroad. In 1848 Stevens revised the old underhung-truss frame used on the road's first eight-wheel cars by boxing in the truss.[62] This improved the appearance of the cars (though they still could not be called attractive), and the line claimed that it provided two giant sled runners that would support the car in case of a broken axle. The cars, known as Rockersills, bore a strong resemblance to a canal-boat and were said to be among the lightest and strongest for their size in service.[63] They were built in large numbers through the late 1850s; a few were even constructed as late as 1864. A train of Rockersill cars on one of the United Railroads of New Jersey is shown in Figure 1.22.

By the early 1850s increased attention was being given to passenger comfort. Larger, more commodious cars were coming into service, though the simple arched roof style remained in fashion. The growth of the railroad network made longer trips possible. The first east-west trunk lines were opening, and night travel became more common in journeys of 200, 300, and even 500 miles. Car bodies were stretched to 40 feet; headroom rose to 7 feet; car weights increased to 10 or even 12 tons. Longer, heavier cars called for stronger framing. The addition of center sills was the most obvious remedy. Two to four timbers running the full length of the body did much to strengthen the car. These were usually somewhat lighter than the side sills.

A more startling change in construction was the introduction of the side-panel truss. This wooden truss built inside the lower

THE BELVIDERE DELAWARE RAILROAD BRIDGE OVER THE PEQUEST, AT BELVIDERE, N. J.

*Figure 1.22   A scene on the United Railroad of New Jersey about 1870 pictures Rocker-
sill cars constructed between 1848 and 1864. (E. P. Alexander Collection)*

body panel greatly stiffened the side sills, permitting long but relatively light car bodies. Thus the side truss, made up of many light members, took the place of excessively large floor timbers that would otherwise be necessary for long bodies. The idea was, of course, taken directly from wooden bridge technology, but it was an imaginative solution and must be regarded as among the most important American contributions to wooden car construction. It was suitable only for the American style of coach; side-door cars did not lend themselves to the side-truss design.

The origins of the side-panel truss are obscure. They might be traced back to the underhung frames on the B & O Washington branch cars. Stephenson's X-frame design was another predecessor. But who first thought of the neat, simple side-panel scheme is unknown.

Specific examples of American coaches for this important transitional period are unfortunately rare. Several pictures from the early 1850s (Figures 1.23 to 1.26) show nothing of the cars' construction. In 1853 the Michigan Central built a number of 60-foot cars, giants for their day, which would require strong framing, and it can be speculated that side trusses were employed.[64] The first concrete example of this construction so far uncovered is a group of 65-foot cars built for the New York and Erie Railway in 1855 and 1856. Curiously, despite its great track gauge of 6 feet, the Erie had been content with short cars in its early years. Some were as much as 11 feet wide, it is true, but their length was held to 36 feet. The new long cars were built to the design of Calvin A. Smith, master car builder at the company repair shops. The top

cord of the truss was arched at the center. The diagonal braces were iron rods. The Erie used this simple form of trussing for many years; after 1855, however, the top cord and window sill were made as one rather than as two independent members, and the diagonal braces were wooden instead of iron rods. Drawings and more information are given in the Representative Cars section (Figures 1.83 to 1.85).

The trend toward big cars was well underway by the mid-1850s. The weight of some cars was over 15 tons. Most roads accommodated the load by increasing journal bearing sizes and wheelbases, but in general they found the ordinary four-wheel truck adequate. Others contended that more wheels and axles were needed to spread the load and ensure safe movement of heavy cars over the existing roadbeds. Improved riding quality was another justification offered. The adherents of multiple-axle trucks created something of a vogue for twelve- and sixteen-wheel cars between about 1855 and 1865.

Isolated examples of twelve-wheel cars can be found a decade earlier in the mid-1840s, when the Camden and Amboy began fitting cars with six-wheel trucks. The main reason was safety: spreading the load over extra axles lessened the possibility of a fracture or derailment. The South Carolina Railroad also operated twelve-wheel cars, according to a note in the *American Railroad Journal* of April 1846. And the Michigan Central's annual reports show that it began building twelve-wheel cars a year or two later. Like the Camden and Amboy, it became largely stocked with these cars; by 1855 forty were in service, with six-

Figure 1.23 *The* Illustrated London News, *April 10, 1852, depicted the interior of an American railway car of that period.*

Figure 1.24 *The Number 164, dating from about 1855, is shown in this photograph taken some forty years later. Note the small ventilators in the letter board. (Library of Congress, Van Name Collection)*

Figure 1.25   Kimball and Gorton produced this coach for the Pacific Railroad (of Missouri) around 1852. The illustration is a contemporary lithograph. (Historical Society of Pennsylvania)

teen more under construction. The Illinois Central was another Western line that adopted twelve-wheelers (drawings for one of them are in Figures 1.86 to 1.87).

If twelve wheels were good, it seemed to some builders that sixteen would be better. They attached two four-wheel trucks to a secondary pivoting framework at each end of the car. It is reported that Davenport and Bridges built such a car in 1839, though it cannot be confirmed. Osgood Bradley built cars of this type for the Norwich and Worcester in 1857 and for the Eastern Railroad.[65] A German visitor described these or identical cars running on one of the New York–Boston roads as having bodies 52 feet long, truck centers of 37 feet, and truck wheelbases of 9 feet.[66] The arrangement was particularly favored for sleeping cars; both Woodruff and Pullman had sixteen-wheelers. But this style was never widely used, and no sixteen-wheel cars are known to have been built for passenger service after the late 1860s. Among their many defects were the extra cost, the dead weight of the complicated running gear, the added friction in passing through curves, and the complicated brake rig. The Old Colony

purchased some sixteen-wheel cars and regretted it, complaining that they did not take curves well and even occasionally turned over.[67]

While the trend to larger cars was unmistakable, not all major trunk roads could adopt them. Some Eastern lines that double-tracked at an early date placed the tracks too close together, with clearance for narrow cars only. The Philadelphia and Columbia, which laid its second track in 1834, was forced to operate cars nearly 2 feet narrower than those on other standard-gauge roads. The Reading built its second track on too narrow a plan in the mid-1840s.[68] The resulting cars were too narrow for ordinary double seats; like the narrow-gauge cars of the 1870s, they had double seats on one side of the aisle and a single on the opposite side. Between 1856 and 1860 the line was rebuilt at considerable expense to allow another 2 feet of clearance. Even this was not enough, however, and the Reading was noted for its narrow cars until it was rebuilt again in later years. The Philadelphia, Baltimore, and Wilmington was another road cursed with a restricted loading gauge. New cars that it received from Harlan and

Figure 1.26 *The Central Railroad of New Jersey's Number 159, dating from about 1855, is shown in this scene at the time of its retirement around 1895. (Library of Congress, Van Name Collection)*

Figure 1.27 *Jackson and Sharp produced this handsome clerestory roof car around 1867. It illustrates the earliest style of clerestory with monitor ends. (Hall of Records, Delaware Archives)*

Hollingsworth as late as 1865 were only 8 feet wide (Figure 1.92). But most American roads were built late enough or wisely enough to avoid loading-gauge problems, and car widths of 9 to 10 feet were common by the late 1850s.

Other noticeable changes came in the same period. There were improvements in seating, decoration, lighting, and heating (these are treated in Chapter 5). The outside appearance of the cars also changed. Day cars had usually been finished with horizontal rectangular panels, but now batten board (narrow vertical wooden strips covering each plank joint on the side of the car) came into favor. In a few cases plain, narrow vertical sheathing was used, although it was not widely adopted until after the 1870s.

By 1860 the arched-roof car reached its ultimate development. The overall size had materially increased, as had the facilities for passengers' comfort. The oppressively cramped, lightly built eight-wheel cars of the 1840s had given way to a solid, roomy conveyance of reasonable if not extraordinary accommodation. The

discomforts of rail travel by this time were largely the result of defects in the roadbed rather than in the cars.

The Clerestory Roof

To the casual observer an American railroad car of 1840 differs little from a car of 1860. Experts could point out many changes, but in general the layman's view is correct: in outer appearance, cars made during this twenty-year span looked much alike. However, a major change in their external configuration began in the early 1860s, when the roof profile was radically altered. The central portion of the roof was raised about 18 inches, and its sides were provided with openings for light and air. This centuries-old architectural device is properly known as a clerestory, but it was referred to at the time under many names—among them the lantern, monitor, raised, deck, clear-story, elevated, and steamboat roof.

Figure 1.28   More monitor roof passenger cars outside the builder's shop in 1867. (Hall of Records)

Richard Imlay's clerestory roof cars of the 1830s were in the vanguard of raised roofs, of course, but they had no lasting influence. In England the Great Western Railway tried a rudimentary raised roof for a few first-class parlor cars in 1838, but again the scheme failed to catch on.[69] The clerestory plan went into dormancy for over twenty years and was then revived by Webster Wagner, who was trying to improve the ventilation of his sleeping cars. Passengers complained about the bad air in Wagner's first sleepers of 1858. When the central portion of the roof was raised, the stale air, which rose to the top, could be exhausted out of the side vents. Wagner's first clerestory roof was reportedly built in 1859. In the spring of the following year his plan, now fully developed, was praised in the American Railway Review: "The roof is raised in the centre, and in the sides of the elevated part are inserted twenty-eight ventilators, which preserve the air pure and sweet, and keep up a steady and healthful circulation— a desideratum which has long been desired but which until now has never been supplied."[70]

Wagner's cars, which he called Monitor roof cars, were operated on the New York Central and its associated lines. One of these, the Hudson River Railroad, was quick to notice the merits of Wagner's roof for ordinary day cars. Four coaches with the "double deck" roofs were under construction at the shops of Eaton and Gilbert in August 1860.[71] The scheme spread quickly across the country. The Michigan Central built some raised-roof cars the following spring. Several years later the company decided that to "sustain the reputation of the road" all its day cars must be rebuilt with the new, fashionable roof. The Burlington shops were likewise busy between 1862 and 1865 fitting clerestories to their cars. The Cleveland, Columbus, and Cincinnati Railroad updated some of its original cars during the same years. And in 1862 the Pennsylvania Railroad adopted its first standard-class car, the PA, which was designed in the current style with a clerestory roof.

Wagner was almost too late in recognizing the value of his

innovation. Together with an able mechanic and owner of several important patents, Alba F. Smith, he hastily prepared a patent application, which was granted on September 23, 1862 (No. 36536). However, the specifications spoke only of the ventilating aspects of the clerestory roof. No windows were shown in the accompanying drawings. Yet one of the chief values of the raised roof for day cars was the additional natural light provided by the side windows. Wagner and Smith tried to establish their ownership of the design through lawsuits, but with little success. The idea had been in the public domain too long to be claimed by any one inventor.[72]

The appeal of the clerestory went beyond its practical benefits of additional air and light; it agreed with the expansive mood of the times. As one journal said, "Loftiness and width are more and more in demand in railroad cars just as they are in steamboats, churches, hotels and theatres."[73] An opportunity for architectural grandeur never before possible had entered the realm of railroad car design. The cramped, tunnel-like appearance of the arched roof car gave way to the open dignity of a more spacious edifice. No more bending or crouching for low ceilings; at last a properly attired Victorian gentleman could stand erect, top hat and all. The means were at hand for more stylish deck lights, moldings, and ornamental art-glass windows. Elaborate center lamps were now possible, and these added greatly to the decor as well as to the lighting of the interior.

During the early years of the clerestory, no one could have anticipated the variety of design treatments it would receive in the course of time. At first Wagner's Monitor design prevailed: the raised roof running nearly the full length of the body terminated in half circles at each end. Several illustrations of Monitor roof cars have already been shown (among the better views are Figures 1.27 and 1.28). A variation on this design is the Buffalo, Corry, and Pittsburgh car Number 7, which has prow-shaped clerestory ends (Figure 1.30). The deck or Monitor style continued in favor until the late 1860s and was then succeeded by the

Figure 1.29   Interior photograph of a Jackson and Sharp monitor roof car. Note the decorative painting in the clerestory. (Hall of Records)

Figure 1.30   The New Haven Car Company built this car around 1866 with an unusual prow-shaped clerestory roof end. (Railway and Locomotive Historical Society)

Figure 1.31   This heavy coach dating around 1870 exhibited the broken bull-nosed roof end once popular with Pullman. (Railway and Locomotive Historical Society)

so-called duckbill roof (for examples see Figures 1.32 and 1.33). Here was an attempt to blend the clerestory ends with the main roof line. The first experiments, made with straight lines, were not too successful; later, graceful curved lines more smoothly integrated the two roof profiles. To achieve a greater decorative effect Pullman and others adopted the broken duckbill pattern, which employed several compound curves (Figure 1.31). The duckbill roof was popular in the seventies, and some roads, notably the Pennsylvania, continued to use it for almost twenty years. It was succeeded by the crescent or bullnose end, known in street-railway circles simply as the railroad roof. The crescent end was the final form of the clerestory. It prevailed from 1880 until the end of the raised roof some fifty years later (Figures 1.32 and 1.33). It was the most logical and to many, the most pleasing style of clerestory roof.

Although the clerestory roof was wholeheartedly accepted from 1860 forward, its value, beyond that of style, was challenged from the beginning. It had hardly come into use when Henry Ruttan, the promoter of a ventilating system for arched-roof cars, bitterly attacked it as a foolish, ostentatious device that served no purpose.[74] Ruttan's self-interest was apparent, but he correctly pointed out that the car builders' mania for the new design was fed by fashion and not utility. His chief complaints were the added first cost of $600 per car, the increased heating space, and the top-heavy bodies of cars with clerestories. The top-heaviness, however, was a matter of appearance rather than weight, for the roof structure was very light.

More serious criticisms could be leveled at the clerestory. Its many joints and corners offered more places for leaks and dirt-catching, so that it certainly added to maintenance costs. It never really solved the ventilating problem; complaints continued on this subject until the advent of mechanical air conditioning in

modern times. But by far its greatest defect was its structural weakness. Each carline (roof rafter) was broken by four right-angle joints. Builders did not use stiffening angle brackets because they did not want to break the clean sweep of the car's interior. Metal reinforcements cut to the exact outline of carlines provided some support for the structure, but the roof remained the weakest part of these cars. It had no lateral strength and did little to support the sides of the body. In fact, the clerestory roof was said to spread the sides, for the weight of the roof caused the carlines to be pressed down and out.

The structural deficiencies of the clerestory prompted periodic clamorings for a return to the simple and strong arched roof. In the early 1880s the *National Car Builder* said that the clerestory had prevailed long enough—that a return to correct mechanical construction was overdue. The Boston and Albany began building arched-roof cars again. They were made with an extremely high arch, however, so that there was ample headroom and space for center lamps. By 1887 this design was being copied by the Boston and Lowell and the Chicago and Eastern Illinois.[75] With their introduction into the United States in 1883, William D. Mann's European-style sleeping cars were also built with plain high arched roofs. The Philadelphia and Reading adapted the old X-frame brace, which had been recommended since the 1870s as a strengthener for the flimsy clerestory.[76] The X frame ran from the side of the car to the underside of the clerestory without a break, thus avoiding the right-angle joints of the conventional construction. This plan was very strong, but it was also ugly. The Reading modified the idea by eliminating the upper arms of the X brace and forming an inverted V truss. This improved the interior appearance somewhat, but not enough to attract imitators.

For the majority of car builders and railroads, however, the

Figure 1.32 *Conventional bull-nose clerestory roof end shown in a drawing of 1873. The design for these cars was prepared by the Cleveland, Columbus, Cincinnati, and Indianapolis Railway. (Railway and Locomotive Historical Society)*

Figure 1.33 *Coach Number 119, dating from about 1880, is another example of the bull-nose clerestory roof end. (Railway and Locomotive Historical Society)*

esthetic appeal of the clerestory overruled all mechanical considerations. The raised roof was to remain a fixed part of American passenger car design until the 1930s.

The Tranquil Years: Design in the Seventies and Eighties

With the clerestory's acceptance in the early 1860s came a period of relative calm in passenger car manufacture. No radical changes in construction or arrangement took place during the next twenty-five years. It was a time for acceptance of things as they were, a time for regrouping, a time for refining the ideas of previous decades. The overall size, weight, and capacity of cars were re-

markably stable throughout the period. Typical day cars were 50 to 60 feet long, weighed 20 to 25 tons, and seated 50 to 60 passengers. New appliances and auxiliary apparatus were introduced, of course. Existing auxiliaries such as the air brake and the Miller platform came into wider use, steam heating experiments began, and decorative treatment developed in ever changing directions. But no fundamental alterations in structure or radical advances in size occurred.

With the clerestory came a greater awareness of the passenger car's architectural potential. Arched windows and doors in Roman or Tudor styles became popular. Small, square windows were out of fashion, although they were still used for cheaper cars. Arched molding over the windows ended in delicate finials. Oval picture-frame moldings, called "name panels," were placed

Figure 1.34 The handsomely decorated coach Number 31 was produced by Jackson and Sharp around 1875. (Hall of Records)

in the center of the body panel. Side panels were invariably done in batten-board style. Interiors of first-class cars were finished in rare cabinet woods such as black walnut, cherry, and mahogany. Only the very cheap coaches, such as emigrant cars and excursion cars, had painted interiors. Gaudy oilcloth ceiling covers or head linings prevailed. Large, showy brass or nickel-plated center lamps were now possible because of the extra headroom provided by the clerestory. Examples of these features are shown in Figures 1.34 to 1.36.

The tranquillity of the seventies was mildly ruffled by the narrow-gauge movement, which seemed to promise a revolution in railway construction and operations.[77] The idea started in Great Britain and spread to this country through the writings of its zealous proponents. It was seized upon here by equally enthusiastic apostles, most of whom were innocent of any engineering knowledge or else chose to ignore what they did know. The major argument for narrow-gauge lines was their low first cost. Presumably the revolution would extend itself to rolling-stock design, where mystical improvements were promised. The laws of physics would be confounded; cars running on rails set at a 2- or 3-foot gauge would weigh less and carry more than standard-gauge equipment. Cars of 8 tons could be built that would carry a dead weight of only 450 pounds per passenger, compared with a standard-gauge car of 25 tons with a dead weight of 700 pounds per passenger. What the narrow-gauge partisans failed to consider was the great difference in overall size and the cramped seating and legroom. In these respects the narrow-gauge car represented a throwback to the diminutive primitives of an earlier time. The standard-gauge car was 60 feet long, 10 feet wide, and 13 feet high, with 24-inch aisles and seats 42 inches wide. The narrow-gauge car was 35 feet long, 8 feet wide, and 10½ feet high, with 17-inch aisles and seats 35 inches wide. A contemporary account attempting to gloss over the skimpy accommodations of the narrow-gauge coaches said that two persons of ordinary size could sit together comfortably; "fat men, however, are generally allowed . . . to occupy a whole seat in peace."[78] To many people who have ridden in narrow-gauge cars, it is questionable that passengers of normal dimensions can share these seats in comfort for any extended trip.

As the narrow-gauge movement developed, it was restricted to secondary lines; few trunk lines were built. Most of these secondary lines operated short local trains at very slow speeds. Consequently the great overhang of the cars presented no substantial danger, although the cars surely swayed more than their standard-gauge counterparts. Again, because these lines did not offer express service, the danger of high-speed collision was improbable and the cars could be lightly framed. Thus they could eliminate even more dead weight.

The first narrow-gauge passenger car built in this country was completed in July 1871 by Jackson and Sharp of Wilmington, Delaware, for the Denver and Rio Grande Railway (Figures 1.37 and 1.38). Named the *Denver*, it was 35 feet long and 7 feet wide.[79] It was divided into two compartments (one for smokers)

and seated 34. The narrow width would not permit double seating throughout. Each side had double and single seats for half its length. The arrangement was reversed at the center of the car to maintain balance. The exterior was painted chocolate brown, with gold and yellow letters and ornamentation. The interior was black walnut. The seats were covered in scarlet plush, and the hardware and lamps were silver-plated. The total weight was 7½ tons. The road's first idea had been to order four-wheel cars on the British plan, but Jackson and Sharp insisted that a practical eight-wheel car could be built. In the following years the Wilmington builder supplied large numbers of coaches to the D & R G. By 1876 the firm had produced over 400 narrow-guage cars.

Other contract builders were quick to sight this fresh market unsullied by a supply of second-hand cars. Billmeyer and Small proclaimed themselves especially devoted to the narrow gauge and built some 2,000 cars by 1878. An example of their work is the *Eureka*, constructed in 1875 for the Eureka and Palisades Railroad in Nevada (Figure 1.39). It was 41 feet long, weighed 8½ tons, and seated 36. The richly furnished interior was said to give it "the appearance of some fairy boudoir rather than a temporary convenience for the traveling public."[80]

Harlan and Hollingsworth, Jackson and Sharp's rival in Wilmington, produced a first-class passenger car for the narrow gauges (Figure 1.40). It was similar to those turned out by other makers, but the outside width was pushed from 7 feet to 7 feet 11 inches. By cutting down the double-seat size and narrowing the aisle, the builder installed double seats the full length of the car. The capacity rose from 36 to 46.

In the Middle West, Ohio Falls and Barney and Smith sought a share of the new market. These builders offered first-class cars and some very cheap coaches as well. Ohio Falls built light excursion cars for a Kentucky road; they seated 64, weighed just under 7 tons, and could carry 125, including standees (Figure 1.41).[81] The seats were wooden slabs. For short commuter lines Barney and Smith built cheap, plainly finished cars with longitudinal benches and candle lighting for as little as $2,500. Far Western narrow-gauge roads ordered from Carter Brothers of Newark, California, who specialized in narrow-gauge rolling stock. Although the firm was never a major builder, it supplied cars for many Western, Mexican, and Central American lines.[82]

The narrow-gauge idea was pushed to its extreme in this country by the construction of 2-foot-gauge trackage. The cars for these roads, almost all located in Maine, were extremely small.

Figure 1.35   Interior photograph of the Providence and Worcester car shown in the preceding illustration. (Hall of Records)

Length of Body 50 ft. 10¼ in.

Figure 1.36   A standard passenger car designed by George Hackney, car superintendent of the Santa Fe Railroad. The car design dates from 1880. (National Car Builder, January 1881)

Figure 1.37   A narrow-gauge car produced by Jackson and Sharp in 1871. (Hall of Records)

They were built on the conventional eight-wheel pattern and were often as well finished as any standard-gauge cars, but in size they were much on the order of today's zoo railways for children. The first 2-foot-gauge car, the *Sylvan,* was built at Laconia, New Hampshire, for the Billerica and Bedford Railway in Massachusetts.[83] It measured 40 feet long by 6 feet, 2 inches wide, weighed 4½ tons, and had wheels 18 inches in diameter. The cost was $2,000. It offered seating for 30; single seats on either side of the aisle were all that the very narrow body could accommodate. In later years the width of these cars was increased to 6 feet, 7 inches, but double seating was still impossible.

The history of narrow-gauge rolling stock is short. After the initial enthusiasm, the defects of the cars, notably their limited carrying capacity and the impossibility of interchange, ended their construction for all practical purposes by the mid- and late 1880s. Narrow-gauge cars contributed little to general developments in car building. The designs were copied directly from existing technology and in essence were scaled-down versions of standard-gauge practice. It appears that little more than 10,000 miles of narrow-gauge line were ever built in this country. Thus even at its peak it accounted for only a small percentage of total mileage. Abandonment or conversion to standard gauge began swiftly; orders for new equipment evaporated, and the market was glutted with second-hand cars.

Most narrow-gauge lines surviving into this century made do with existing stock. Old cars were upgraded and repair parts were taken from surplus equipment. The D & R G kept its wooden cars in service for eighty years and more by periodic rebuilding. Steel frames, new window sashes, wheels, draft gears, and brakes, together with normal maintenance repairs, did their part in extending the service life of these cars well beyond that of their contemporaries. In the course of upgrading, their weight was raised to 16 tons, or nearly twice that of the original cars. Continued tourist traffic on the Silverton branch did force the construction of several new steel coaches in 1963 and 1964. But these were built in the old style, with scribed steel side plates to simulate wooden cars, and must be thought of as replicas rather than as attempts at modern coach design.

If the narrow-gauge movement can be credited with little else, it did rekindle the debate on the dead weight of passenger cars. Designers were admonished to develop new patterns, and the iron car was perennially put forward as a solution. Car builders, however, faced conflicting demands for coaches that were light,

Figure 1.38   Interior of the narrow-gauge car in the preceding illustration. (Hall of Records)

yet strong. Heavy cars were expensive to haul, required heavier locomotives and bridges, and damaged the roadbed. Railroad operating departments called for lighter construction. The daily press, aggravated by the "horrors of travel," clamored for stronger cars. The newspaper crusade was more than yellow journalism; because of increased traffic and speeds, accidents were rising. It was not possible to serve both demands, and designers were inexorably pushed toward the side of safety. The trend since the beginning of the American railroad car had always been toward increased size and weight. Within practical cost considerations, no way was found to lessen dead weight without sacrificing strength. The paradox was never resolved; the talk of lightweight construction continued, but the practical men of the industry were resigned to ever heavier cars.

The steady increase in weight is shown in Table 1.1, which reveals that weights doubled within a decade. Before about 1870

BILLMEYER AND SMALLS'

NARROW GAUGE

EUREKA & PALISADE RAILROAD.

EUREKA

ROLLING STOCK.

YORK, PENN'A.

Figure 1.39   Narrow-gauge car produced by Billmeyer and Small of York, Pennsylvania, in 1875. The car seated 36 and weighed only 8½ tons. (Smithsonian Neg. 46860-A)

THE HARLAN & HOLLINGSWORTH CO.

AMERICA.

─ Stove ─

Figure 1.40   A narrow-gauge car design by Harlan and Hollingsworth, 1875. (N.C.B., August 1875)

Figure 1.41 This diminutive narrow-gauge car was produced in 1875 by the Ohio Falls Car Company for summertime or excursion service. (N.C.B., November 1875)

the increase can be explained by simple growth in the size of cars, but after that date it was the result of adding new auxiliary equipment. Considerable extra weight came from hot-water or steam heating systems, air brakes and the stronger foundation gears they required, gas lighting with its storage tanks and piping, safety end platforms, and couplers. Seats with cast-iron frames, more glamorous interior furnishings, and larger, more fully equipped water closets also increased weight. The Boston and Albany, noted for light and rather sparsely fitted out day cars, found that half the weight of a complete body was in the stove, seats, and miscellaneous hardware.[84] In a car weighing 45,000 pounds, the body—floor frame and all—accounted for only 13,600 pounds. The hardware and trucks made up the remaining weight. In later years steel-reinforced bodies and platforms, electric lighting (batteries and generators), and six-wheel trucks added new burdens. Considering all this auxiliary apparatus, car builders were remarkably skillful in holding the total weight within reasonable bounds.

Table 1.1 Wooden Passenger Car Weights

YEAR	RAILROAD	BODY LENGTH, FT.	SEAT-ING	WEIGHT, LBS.	WEIGHT/PASS-ENGER, LBS.
		COACHES			
1835	Baltimore & Ohio	30	50	12,000	240
1850	Many U.S. roads	35	56	20,000	357
1860	Many U.S. roads	48	50	36,000	720
1870	Many U.S. roads	50	50	40,000	800
1880	Boston & Albany	57	68	45,000	661
1885	Chicago & North Western	55	64	60,000	936
1890	Southern Pacific	54	62	56,300	908
1890	Fitchburg	57	73	46,000	630
1895	Northern Pacific	52	56	64,900	1,133
1900	Southern	57	60	79,500	1,325
1905	Great Northern	72	86	111,200	1,293
		SUBURBAN AND ELEVATED CARS			
1870	Pittsburgh, Ft. Wayne, & Chicago	(unknown)	30	21,500	716
1872	New York Elevated	30	48	9,400	235
1880	Illinois Central	(unknown)	40	29,000	725
1895	New York Elevated	40	48	29,300	614
		PULLMAN SLEEPERS			
1870	All Pullman clients	60	32	52,000	1,625
1880	All Pullman clients	66	27	75,000	2,777
1890	All Pullman clients	69	27	90,000	3,333
1895	All Pullman clients	70	27	111,900	4,144
1905	All Pullman clients	74	30	120,000	4,000

Concern for safety and comfort led to the construction of heavy cars for main-line service. Here the strongest cars were needed because speeds and traffic density were greatest. Most trunk lines justified their heavy coaches by pointing to the necessity for strong framing. But a few roads managed to produce lighter cars: the 6-foot-gauge Erie, where the heaviest cars might be expected, built a surprisingly light 60-foot car that weighed nearly 7 tons less than an equivalent car on the Pennsylvania. How this was achieved is not obvious from comparing the drawings. Late in the

century the New Haven produced a lightweight car in which about 2½ tons of dead weight was saved by substituting Oregon fir for the usual oak frame, and aluminum for the usual cast-iron seat frames.[85] The car body framing was lightened as well (see Figure 1.42).

For suburban, emigrant, or branch-line service, light, cheap cars were considered acceptable. Speeds and traffic density were low enough to justify more lightly built rolling stock. Undersized framing, furnishings, and plain finish reduced the weight and cost. Closely spaced seats raised the seating capacity. For its branch lines the Pittsburgh, Fort Wayne, and Chicago used extremely light cars that weighed just over 10 tons.[86] They were not run faster than 15 miles an hour; at higher speeds they jumped so badly that there was danger of their leaving the track. The road's coaches for main-line service weighed 23 tons. The Illinois Central's suburban service out of Chicago grew so rapidly after 1880 that the line developed a special car for it patterned after those used on the New York Elevated. Here again, very light framing and reduced overall size held the dead weight under 15 tons (Figure 1.43).

In fact, the New York Elevated cars were the established model for lightweight construction. No writer on the dead-weight issue failed to mention them. The earliest cars weighed only 9,400 pounds, but they were so undersized and so unorthodox in design (representing the old depressed-center idea revisited) that no application to steam railroad practice was possible.[87] They were in essence little more than eight-wheel horsecars. But even the Elevated began to build heavier cars, and by the mid-1890s weights were up to 15 tons.

Special lightweight cars were constructed to meet temporary needs as well. For the U.S. Centennial of 1876, the Reading and Pennsylvania railroads built a number of suburban cars to accommodate visitors traveling from Philadelphia to the exhibition grounds. The Pennsylvania produced 100 cars of the same overall size as its standard PB coach, but the seats were crowded together so that the cars carried twelve more passengers (Figure 1.44). Their exterior and interior finish was very simple. After the Centennial the cars were used as emigrant and excursion trains. Even cheaper and more temporary cars, little more than boxcars fitted with windows and wood-slat seats, were manufactured for the Columbian Exhibition in 1893. Built for easy conversion to freight service, they were forerunners of the boxcar-like troop carriers of World War II.

The trend toward heavier construction was evident even in suburban cars. Thus in the one area where it was generally agreed that light cars were possible, builders gradually drifted toward the heavyweights. By 1904 the average weight of America's 3,919 suburban cars was given as 30½ tons each.[88] While this was considerably below the tonnage of a day coach of the period, it represents a marked increase over earlier suburban cars.

Lip service was paid to lightweight construction, but in practice, strength was the designer's goal. It is wrong to dismiss the

Locomotive Engineering

Figure 1.42   A lightweight car with an extremely sparse body and floor frame. The design was prepared by the New York, New Haven, and Hartford Railroad in 1896. (Railway and Locomotive Engineering, June 1898)

Figure 1.43   The Illinois Central suburban fleet running out of Chicago was equipped with lightweight cars on the same general plan used for New York Elevated coaches. This example dates from about 1880 and was produced by Jackson and Sharp. (Hall of Records)

wooden car as a flimsy matchbox or egg crate—a popular flippancy so often expressed that there seems little hope of refuting it. It would be equally wrong to pretend that the wooden car was faultless; it had many defects, but the case against it has been overstated. In the more spectacular wrecks, it did fail as its worst critics predicted. Wooden cars could not withstand the driving force of a locomotive—but neither could all-steel cars, much to the dismay of those who predicted that metal cars would usher in an age of absolute safety for railway travel.

Derailments or collisions of wooden cars, however, did not mean certain death among shattered timbers and splintered wood. In ordinary mishaps wooden cars did well, though occasions like these were not as newsworthy as the Revere, Angola, or

Spuyten Duyvil disasters. The monthly accident reports in *Railroad Gazette* during the years of the wooden car are not a continuous chronicle of death and destruction. Apparently the cars offered reasonable protection to the traveler. A freak event that showed the surprising strength of wooden cars occurred in Philadelphia in the fall of 1878.[89] The train shed at the Pennsylvania Railroad's West Station was blown down by a 75-mile-an-hour gale. The cars not only supported the giant 70- by 800-foot roof, but withstood the battering effect of the 6-ton cast-iron supporting columns. One column broke in half when it struck the side of a car. None of the cars were seriously damaged.

Two years before that, a passenger train derailed on the Richmond and Danville. Several cars rolled down an embankment;

Figure 1.44   *The Pennsylvania Railroad built a special class of light car for the Philadelphia Centennial Exhibition in 1876. The cars were produced in the railroad's own shops in 1875 and 1876. (Pennsylvania Railroad)*

Figure 1.45   *Interior of a New York Central and Hudson River Railroad coach of 1886. (Hall of Records)*

Figure 1.46   Western and Atlantic coach of about 1885. Note that the interior wooden window blinds are all closed. (Hall of Records)

Figure 1.47   A Norfolk and Western coach of 1888 shows the strong influence of its parent line, the Pennsylvania Railroad. (Hall of Records)

one landed 150 feet from the tracks. Yet no one was killed, only 20 injuries were reported, and the cars remained intact. The trucks were lost, as might be expected, but no major damage occurred. Late in the wooden car era, a coach on the Lake Shore and Michigan Southern was thrown from the rails. It rolled over and straddled an opening in a washed-out embankment. The underside of the car was stripped, including the truss rods, yet repairmen discovered that the body was straight and sound despite rough handling in its removal from the accident site. Four hundred dollars put the car back in service.

To build sturdy wooden cars required the skill and experience of the master carpenter. As already suggested, he borrowed directly from the traditional technologies of the architect, the bridge builder, and the millwright. It is doubtful if any unique methods of fabrication can be credited to car builders. But the solutions that they developed to practical design problems were often subtle and ingenious. The various styles of floor and body framing are perhaps the best examples of this inventiveness.

It is true that a civil engineer might criticize the wooden railroad car as a hopelessly inept design. Certainly as a bridge it was terrible. A wooden bridge should support ten times it own weight; the passenger car was rarely required to carry half its weight. Instead the car had to absorb the draft and buffering

impacts of train service. Vertically the framing needed to support only its own weight and the relatively light load of the passengers. The most severe stresses on the structure were longitudinal, from the push and pull of the train. Accordingly the designer aimed to build strength into the floor framing.

After 1870 the standard floor frame consisted of six sills. The outside members forming the main support for the body were the heaviest cross section and often measured 5 by 8 inches. Round iron truss rods an inch or more in diameter were attached to the underside of the outside sills. These rods greatly stiffened the body. If a sag developed, a turnbuckle at the center of the rod could be adjusted to draw the body straight again. The floor frame was in fact often built with a camber of 1½ inches to avoid the optical illusion of a sag. Generally the truss rod terminated in an anchor block bolted to the side sills near the body bolster. A stronger scheme nearly always used for freight cars but apparently not favored for passengers was to run the truss rod over the bolster and through the end sill. The rod extended the full length of the body and tied the frame together as well as jacking it up. It also served to support the ends of the car body which overhung the trucks.

The intermediate and center sills were slightly smaller in size, with the center sills placed close together to make an easy attachment for the drawbar. Because the frame was subject to few transverse stresses, massive cross bracing was not necessary; it was made only strong enough to space the sills and resist end shocks. End pieces were mortised to the sills. A body bolster or transom, as it was sometimes called, was bolted to the underside of the sills at each end of the car. Center plates for receiving the truck were fitted to the bolster. The bolsters were generally made of two or three timbers. After 1870 iron bolsters became more common, but the wooden bolster seems to have prevailed almost to the end of wooden construction.

Closer to the center of the body were the needle beams, which were fitted to the underside of the sills and formed a place of attachment for the queen posts. Short bridging joists were placed between the sills more as convenient nailing pieces for the floorboards than as structural members. Cross iron rods tied the floor together. Builders concerned with maintaining the squareness of body (the Pennsylvania Railroad, for example) used diagonal braces in an X pattern. Planks 1½ by 4 inches were attached to the top of the frame. An objection to this practice was that the sills had to be notched so that the floor could be laid over them. Critics said that the notching weakened the frame. The underside of the frame was covered with a wooden board ceiling, which served to deaden the sound of the running train, exclude dust, and insulate the floor. For many years the air space between the sills was filled with wood shavings for better insulation. The practice was condemned as a fire hazard in an age of oil lighting. Rock wool was sometimes substituted, but it proved too expensive for many roads.

The ordinary floor frame, simple and strong as it was, suffered from one grievous defect. The platforms of the car were made of short timbers bolted beneath the floor frame, and this weak cantilevered structure had to support the coupler, draft gear, end rails, and steps. It invariably sagged. A special committee on the subject created by the Master Car Builders in 1884 admitted that a first-hand inspection had discovered few cars without drooping platforms. Worse still, the dropped platform was vulnerable to the most serious of all possible hazards: telescoping of cars. In the event of a collision, the dropped platform allowed the car behind to climb up on it and slide directly into the first car's body.

How did such an unmechanical arrangement come about, and why was it tolerated? This form of construction existed from the beginning of the eight-wheel car. At first it was adopted to facilitate entrance to the car: the lowered platform was easier to mount than a platform at the level of the car floor. One possible reason for its perpetuation is that as new, larger cars were introduced, the dropped platform was necessary to keep drawbar heights level with the old cars still in service.[90] Another theory is that the larger wheels which became popular for passenger cars raised the level of the body; and the dropped platform was used to maintain a standard coupler height. None of these arguments is entirely convincing. It seems more likely that the dropped platform continued not because of any mechanical merit but because it had become an entrenched traditional design.

It remained for a designer outside the profession to correct this basic and long-standing defect in railroad car construction. Ezra Miller, a civil engineer only casually associated with the railroad industry, recognized the need to raise the platform to the same level as the main floor frame of the car. Miller is usually remembered as the inventor of a popular car coupler antedating Janney's design (see Chapter 7). But he should also receive great credit for the elevated platform—a relatively simple reform that did much to promote safe railway travel. Examples of level platforms, both in actual construction and in the patent record, can be found before Miller's 1865 patent. However, he became the advocate of the level platform combined with a buffered coupler, and it was adopted largely through his efforts. Miller retained the suspended platform timbers, but he inserted short spacers between the end sills and the buffer beam (the end cross timber of the platform) and raised the level to agree with that of the main frame. The entire structure was tied directly to the body bolster by iron rods. By the mid-1870s Miller's platform was widely used, and in another ten years it was considered standard equipment.

There was general agreement among builders on floor framing, but no such consensus can be found for the body frame. Though everyone agreed that there must be some form of panel side trussing, "there is such a wide difference in the ideas of the officers of different roads with regard to the amount of material . . . and the form in which it should be placed that it seems almost impossible to bring their ideas together."[91] So wrote Thomas Bissell, chairman of the Master Car Builders' committee on passenger car framing, in 1884. Since the mid-1850s car builders had delighted in perfecting new and variant forms of side-panel trussing. The history of this elusive art is difficult to trace, because few drawings exist for the early years. Some idea of the variety of forms can be gained from Figures 1.48 and 1.49.

Before 1870 car builders favored the bastard Pratt and Howe trusses, which were simple and did the job for short bodies. The Pennsylvania remained loyal to this form through the 1880s, mainly because it was adjustable during the service life of the car. But it appears that most other roads were attracted to a modified queen-post truss, generally called a combination or compression truss. It may have been suggested to some observant car builder when passing through a covered bridge, where the structure was exposed on the inside. Or possibly the external hog bracing of a river steamer inspired the anonymous inventor of the side-panel truss. The first known example of a compression-truss car is the 1870 Chicago and Alton coach described in the Representative Cars section (Figure 1.103). It is not claimed that this is actually the earliest example; the compression truss may have been used for some years, but no evidence is available. The car follows Pullman's overall design and is unusually large for a coach of that day. Pullman or Wagner may have originated the compression truss in building their heavy sleepers during the 1860s. It seems to have caught on rapidly after 1870, and a survey of existing drawings indicates that it was the most popular style by the end of wooden car construction.

The compression truss grew out of the distortion of the simple

Figure 1.48   Some idea of the variety of side-frame trussing in wooden cars can be gained from this and the following illustration. (Railroad Car Journal, *September 1894*)

Figure 1.49 More examples of wooden passenger car side trusses. Detail third from bottom shows Chalender's iron plate truss. (Railroad Car Journal, *September 1894*)

Figure 1.50   The intricate woodwork of passenger car framing is exposed in a photograph taken in 1907. (Pullman Neg. 10281)

queen-post truss when it was unevenly loaded. The short auxiliary counterbraces were inserted to stabilize the truss. From frame drawings shown throughout this chapter, it might be supposed that the side truss was made of heavy square timbers. This is not the case; the structure must be seen "in the round" to be understood. The photographs in Figures 1.50 to 1.52, and a more careful examination of the drawings, reveal that the truss members are deep but relatively thin planks. The usual size was 1½ by 6 inches. The vertical posts were notched so that the truss planks could be set flush with the car side.

A great quantity of glue was used in car fabrication—50 to 100 pounds was not unusual. Glue was needed not only for frame assembly but also for applying outside sheathing. The glue stiffened the structure and became one of the hidden sources of strength for the lightweight wooden frame. Once the car sagged, however, it could not be straightened without breaking the glue joints. Wooden screws and nails were used as well. Iron rods also helped tie the body and floor frame together. The ends of the body overhanging the body bolsters were supported by a short rod (shaped like an inverted V) or a long iron hog brace. The hog brace, properly known as an overhang brace rod, extended the length of the body just under the belt rail (window sill). It ran on a 30-degree angle from one corner of the side sill to just under the belt rail. At this point the rod was flattened so that it would fit

more easily inside the body panel. The overhang brace rod does not appear to have been used much before the long-bodied cars of the late 1880s.

The compression truss, combined with the six-sill floor frame and iron truss rods, formed a strong, relatively light wooden structure. The designer's ingenuity was taxed to devise a wooden truss stiff enough to support a 70-foot body within a 30-inch panel, but it was done. All elements of this form of construction were available by 1870, and they continued in use until the advent of the steel passenger car some forty years later.

Wood was the obvious material for car building because of America's great timber resources. Virgin stands of oak, pine, ash, and walnut offered the choicest woods, and the supply was bountiful. Iron, on the other hand, was scarce and expensive. The fledgling American mills could not compete with those of Britain, and much iron and steel were imported. The cost of the imported metals was artificially raised by tariffs meant to protect home suppliers.

Native woods seem to have been used by local builders in the early years. Locust was the principal wood in the *Columbus* (1831), local ash in the first cars made at Cleveland. The Eastern hardwood forests provided a good supply of white oak, which was long favored for car frames. With the voracious demand for hardwood, New England's woodlands soon vanished. The loggers

Figure 1.51   The timbers used in floor framing were massive, as shown in a photograph of about 1890. (Pullman Neg. 3327)

then began on the oak of New York, Ohio, and Tennessee. Wherever the white man moved, the timber disappeared. Standards declined, and timber that would earlier have been cast on the firewood pile was later accepted as staple lumber.

By the early 1870s builders were turning to yellow pine as the choice stands of hardwood disappeared. In 1872 Gilbert, Bush, and Company was reported to be importing yellow pine from Georgia and Florida.[92] Its high resin content was considered an effective insect deterrent. Yellow pine proved an excellent wood for floor sills; it was as strong as oak and was available in long, nearly flawless sections. Norway pine was another favorite for car sills. Pine eventually became the basic wood in car construction. Pullman estimated that half of the wood used in its cars was pine.[93] By the middle 1890s, however, all kinds of prime timber for long sills were becoming scarce. Builders turned to the great Pacific Northwest fir forests, where a plentiful supply of best-quality sills was available at half the cost of Eastern pine. The Northern Pacific opened a car shop in Tacoma; the Southern Pacific imported Oregon fir for its big Sacramento shops. But fir did not take over the car sill market, for it could not match the strength of oak or the better pines.

Securing long, perfect sills had been difficult since the 1880s.[94] Good-quality timber in 40-foot lengths was easy to come by, but longer pieces were generally defective at one end. That end was from the top or small part of the tree and was commonly riddled with sap pockets, shakes, and knots. Even when very tall, solid trees were available, it was not always possible to harvest them in one piece. As the timberlands receded into remote or rough terrain, the logs had to be cut into shorter lengths for haulage to the mill. Thus as longer cars were coming into fashion, the supply of prime framing timber was declining. Splicing was the only alternative. The preferred form was the 24-inch lock or ship splice, in which the ends of the timber were cut at an angle of about 20 degrees. A notch was made near the center to prevent slipping. A

6-foot timber was bolted on the spliced sill overlapping the joint. Iron plates were sometimes used as well, but most car builders felt that they were too rigid. The wooden splice bar would bend and give with the sill and thus not loosen the connecting bolts. The splice was best made near the center of the car, preferably over the needle beams but never over the bolsters. In 1901, to the surprise of many car builders, the traditional lock splice was proved inferior to the plain right-angle splice. In a test conducted for the Master Car Builders Association that year, the lock-splice breaking load was 82,000 pounds, while the plain splice broke at 140,000 pounds. But this revelation was beyond the belief of the majority of the members, who chose to stay with the old form.

The best building timber is naturally seasoned by one to two years of air drying. Pullman maintained a vast 60-acre yard for the purpose, but not all contract builders could afford to tie up so much capital. Sometimes lumber was even sawn from logs that had been in the mill pond when the order was received.[95] Such an outrageous practice may have been limited to a few cases, but the use of green lumber by unscrupulous builders was a frequent complaint. One thousand feet of unseasoned lumber might contain as much as a ton of water. Cars built from it seemed tight enough at first, but as the wood dried out and shrank, mortise joints loosened and the whole structure sagged or fell to pieces. An alternative to the expensive process of natural seasoning was to dry the lumber in a kiln. Artificial drying in heated ovens produced lumber of less strength and with more susceptibility to rot than the naturally dried product, but it was certainly preferable to green wood. As early as 1881 Jackson and Sharp placed two kilns in its Wilmington yard.[96] An unexpected upsurge in new car orders had cleared the market of seasoned lumber. When this happened, cynical builders turned to freshly sawn logs, while the honest shops erected kilns.

The blame for using inferior wood cannot be laid entirely upon the contract shops; the railroads themselves were lax in insisting

Figure 1.52 In this Chicago and West Michigan coach, under construction sometime around 1888, the side paneling has already been applied on the right.

upon the best material. Some master car builders made a first-hand selection, but too many delegated this important duty to the purchasing agent.[97] Engineering departments issued vague specifications, often only a few words, which named only the variety and grade of wood. Competing lumber dealers were tempted to supply inferior stock to win the order. The wood they furnished might be "yellow pine, grade B," or it might be sapwood or some poor variety of the species. The density, strength, and decay-resistance properties varied enormously within the subvarieties of a species. Thus the most detailed and careful specifications were necessary in order to obtain the best timber.

Rough guidelines for grading lumber were established through public law in some states by the 1840s. But it was not until the early part of this century—at the close of the wooden car era—that scientific standards were developed for use by the railroads.[98]

Another criticism leveled at car builders was their narrow-minded insistence on a few species of wood. Corner posts and window sills were made of tough, rot-resistant black walnut. But in general the principal framing timber was limited exclusively to oak or pine. Although beech, elm, cypress, and even sycamore were suggested as cheaper alternatives in some locales, they never won general acceptance.

Outside paneling or sheathing presented no unusual supply problems. It was light, conventional lumber, always available in good supply. The earliest references speak vaguely of white wood, which was presumably clear white pine or poplar. Before about 1855, rather large, rectangular panels were the rule. They tended to split, however, and for this reason, as well as from changes in fashion, were superseded by vertical batten-board or narrow tongue-and-groove siding. Batten-board paneling was favored through the 1880s. The boards were 12 to 15 inches wide; narrow strips covered each joint. But even with a carefully built body frame, it was difficult to lay on a perfectly smooth side panel. Minor defects were bound to exist in a 50- or 60-foot panel, and sighting down it at a close angle revealed many irregularities. Some leveling was possible with furring strips or blocks. The Pennsylvania Railroad shops developed a method all their own, in which thick panel boards were securely fastened to the car sides, and skillful cabinetmakers planed and scraped a quarter inch off the entire side of the car to make it true and level.[99] By the early 1890s narrow tongue-and-groove boards, 1½ to 2 inches wide, came into favor. One edge was beaded with a V-shaped chamfer. The boards were glued and blind-nailed to the car frame. They were always made of yellow poplar, a clear, close-grained wood that offered a smooth surface for painting. (The rare decorative woods used for interiors will be treated in Chapter 5).

Perfecting the Art of Wooden Car Construction

By the end of the century car building entered its Augustan age. As the *Railroad Car Journal* said, "The days of building cars by the rule of thumb have passed and men of high scientific training are rounding out the experience of the untutored carpenter."[100] The wooden car reached its high point of development in the middle 1890s, when a masterful system of spidery wood trussing produced sleek coaches. The 80-foot cars manufactured at this time exhausted the potential of wooden construction; the ultimate practical limit had been reached.

The truss plank frame, at first considered practical only for bodies up to 50 feet, proved capable of appreciable expansion. Demand for longer, more luxurious cars in the late eighties added another 10 feet, so that by the middle of the next decade, bodies of 60 feet were common for first-class cars. The New York Central created a sensation in 1893 by building the first 80-foot day coaches; the bodies were just an inch under 72 feet.[101] These were the largest day coaches constructed to that time and marked in fact the final limit for wooden car sizes. The New York Central cars were the fulfillment of a new trend in day cars that might best be described as palace day coaches. A year or two earlier the larger trunk lines had begun to purchase unusually luxurious coaches for express-train service that were modeled on the most expensive parlor and sleeping-car designs. As competition for passengers grew, these lines bought ever more extravagant cars. In 1898 the *Railroad Gazette* protested that the rivalry had gone beyond reason: "The builders are the only parties who can possibly benefit by costly trains."

The longer, heavier cars created problems that wooden construction could no longer safely handle. Car builders turned to more extensive use of iron reinforcement. Composite construction was the basic development of the 1890s. It was a compromise scheme that prolonged the wooden car age into the early twentieth century. Iron reinforcing had been used for years; iron truss rods, carlines, and corner brackets were established materials. The Illinois Central was installing iron body bolsters in the early 1850s, and their use spread in the 1870s. But builders were now working heavy steel and iron plates into the very fabric of the structure.

The earliest important application of composite construction was the Chalender plate truss, a sheet-iron plate that was substituted for the ordinary side-panel wooden truss.[102] Made of boiler iron about ¼ inch thick, it was fastened inside the body frame below the windowsill and ran the full length of the car. An angle iron riveted to the bottom of the plate was screwed to the side floor sill. Possessing the rigidity of a plate girder bridge, the Chalender truss was said to be the strongest possible style of trussing. Installed, the plates weighed 1,600 pounds and cost $90. This was somewhat more expensive than a wooden truss and, being an unfamiliar style of construction, was not favored by wooden car builders. It was also criticized because the vibration and twisting of the car body eventually loosened the screws, rendering the truss ineffectual. The idea was patented October 7, 1873 (No. 143,498), by George F. Chalender, who was then master mechanic for the Burlington and Missouri River Railroad. As a blacksmith and locomotive builder, Chalender was familiar with working boiler iron. He succeeded in promoting his idea when he

became superintendent of the Chicago, Burlington, and Quincy's repair shops. The Burlington remained loyal to the Chalender truss for years after the inventor's departure, building cars on this plan through the 1890s. Few other roads appeared interested, however.

The Pullman and Wagner superintendents, Henry H. Sessions and Thomas A. Bissell, began work on rival schemes for composite construction at the same time. Sessions appears to have completed his plans first and applied for a patent in 1889 (see Figure 1.53). His major concern was to strengthen the ends of the car, an objective he achieved by bolting a ⁹⁄₁₆-by 20-inch plate under the end sill in order to form a solid connection between the floor and end sill. A 3- by 4-inch angle-iron framework formed in the shape of the car's end profile was attached to the steel plate. His method was reported in use by the trade press in February 1890, but he did not receive his patents until the following year.[103] The first patent was issued June 2, 1891 (No. 453,403); the second, more general patent was granted on July 21, 1891 (No. 456,291). Pullman claimed that cars built on this plan could withstand a collision at 40 miles per hour without serious injury.[104] No test of the claim has been recorded, but it seems safe to conclude that it was little more than bravado on the part of Pullman's press agent. Sessions's scheme became standard at Pullman and was recommended by the Master Car Builders Association. It is known to have been adopted by the Boston and Maine to strengthen its day coaches, and it may well have been used by other roads.

In 1894 Sessions turned his attention to strengthening the side-panel trussing. A cantilever plate girder made of flat ⅝- by 3-inch iron bars was placed at each end of the panel truss to prevent the ends of the body from drooping. Two years later Sessions expanded his ideas for composite construction in a patented scheme which advocated an X-brace angle-iron side truss.[105] The iron web plate described in the 1891 patent was much enlarged and ran nearly the full length of the platform from the end body sill. It is not known if Sessions's side truss was ever used.

Bissell's plan for composite framing was completed late in 1890.[106] Like Sessions, he concentrated on strengthening the end bulkheads, but he also effectively tied the floor frame into the end structure (Figures 1.54 and 1.55). Six ¾- by 3½-inch flat iron bars were bolted to the end frame of the body. They were sandwiched between the wooden upright posts, and the ends were twisted at right angles for easy fastening to the frame's end sill and deck sill. A horseshoe-shaped iron plate ⅜ by 6 inches ran across the top of the bulkhead (the end deck sill). The clerestory and roof were reinforced with iron carlines which had lugs welded on for more secure attachment to the wooden framing of the car. The best features of Bissell's plan were the heavy iron plates that ran from the end sills along the side sills just past the bolsters—usually 18 feet. This plate was sandwiched between the 5- by 8-inch side sill and a 3- by 8-inch supplementary timber. In some cases the reinforcing plate ran the full length of the frame. Bissell's stated purpose in the patent specification was a judicious strengthening of the conventional wooden frame. He warned that

Figure 1.53 Reinforced end framing is shown in this construction view of 1907. (Pull-man Neg: 10280)

care must be taken not to overdo it. Too much iron plating would add needless weight and cost; just the right amount of reinforcing was needed.

Bissell received a patent on August 11, 1891 (No. 457,486), and saw his idea widely used for Wagner sleeping and parlor cars. Moreover, because of the Wagner Company's close association with the Vanderbilt interests, Bissell's plan was adopted by the New York Central and its associated lines. After trying the composite frame on some baggage cars, the Lake Shore and Michigan Southern was so pleased with the results that it adopted the scheme for coaches in 1892. Within four years the road had 150 cars on this plan in service. Bissell's scheme seems to have been popular, and as late as 1909 it was used for some new cars by the Central Railroad of New Jersey.

In 1896 composite construction had become respectable enough to be recommended by the Master Car Builders Association.[107] In fact, it seemed a necessity; the Post Office Department was insisting on safer mail cars to protect its employees. Short of going into all-steel construction (a distasteful conversion seemingly to be put off as long as possible), composite construction was an attractive alternative. The association admitted that it could produce no designs which entirely avoided the Sessions and Bissell patents, but a British patent of 1884 seemed to cover the same general features and would undoubtedly void the American claims if they were taken to court. Old-line car builders like F. D. Adams, however, objected strenuously to composite fabrication because it added useless ironwork to cars already overbuilt. According to the association's report, Adams contended that the Bissell and Sessions schemes increased the weight by 5 tons. He also stated that conventional wooden construction was satisfactory on the Boston and Albany. Light, all-wooden cars built twenty-five years ago were turning in good service, Adams said, and would not be improved by the addition of so much scrap metal.

Figure 1.54   Thomas Bissell's reinforced end framing, devised in 1890, was used in many Wagner palace cars. (Railway Review, *December 6, 1890*)

Figure 1.55   Bissell's patent of 1891 for reinforcing wooden car frames with steel plates.
(U.S. Patent Office)

After the introduction of composite construction, the next notable design reform was the steel platform. The vestibules, automatic couplers, and elaborate draft gears that were coming into use by the early 1890s resulted in a greater load than wooden platforms could handle. Platform timbers were stiffened with iron plates, but the advent of wide vestibules and ever heavier draft stresses called for stronger construction. Pullman resorted to 6-inch I beams in 1895; steel construction was coming to the passenger car in gradual steps.[108] Within two years about fifty Pullmans were fitted with steel platforms. Pullman's Henry H. Sessions developed the design and secured two patents (No. 562,343 of June 16, 1896, and No. 575,994 of January 26, 1897). Just weeks after the second patent was granted, the Standard Coupler Company put Sessions's steel platform on the market. The response was good; by 1900 eighty railroads were using it. The Rock Island thought so well of the steel platform that the road installed it in some 400 cars. Thomas A. Bissell also designed a steel platform for Wagner that used Z rather than I beams. After 1899 it was manufactured by the Gould Coupler Company.[109]

End of Wooden Construction

Although composite construction extended the wooden car age into the early years of the twentieth century, pressures were mounting for all-steel fabrication. Acrimonious editorials against wooden cars had been standard fare in the daily press for years. Wooden cars were anathematized as tinderboxes, rolling death traps, and coffins-on-wheels. A popular notion developed that iron or steel cars would miraculously end the hazards of railway travel. And while it is difficult to present a single clear-cut reason

why wooden cars were abandoned, the argument for safety appears to have been the most potent.

Another strong reason was that the industry had already accepted steel freight cars. By the late 1890s they had proved clearly superior to wooden cars in capacity and strength. Few wooden-frame freight cars were built after 1905, and manufacturing and repair shops accordingly converted to steel fabrication. Throughout the nineteenth century, car builders had thought of themselves strictly as a wood working industry. Proposals for iron passenger cars were dismissed as visionary and impractical. The wooden car was perfected to a high point; it was cheap, easy to build, serviceable, and most important, familiar. But the success of steel freight cars could not be denied. Once the trauma of conversion was over, traditional car shops saw little reason to resist the introduction of steel for passenger cars. The equipment was on hand and their employees were practiced in the new construction. What had formerly been a major obstacle was now an academic consideration.

The advantage of economy, long the main justification for the wooden car, was fading. Prime framing lumber was becoming both scarce and expensive. By pushing car lengths to 80 feet, builders had arrived at heavy construction that increased the weight and cost nearly to that of an equivalent steel car. Meanwhile steel was steadily declining in cost as American industry grew. Steel-working tools were not only well developed but becoming common. In addition, the railroads were ripe for the conversion. As the carriers that dominated land transportation in a wealthy manufacturing nation, they could afford and were expected to adopt the latest improvements. Whenever one trunk line purchased steel cars, competition forced other lines to follow suit.

Figure 1.56   A handsome arch-window coach was produced by Pullman in 1907. So-called art glass, a wavy, opaque green or tan material, was used in the arch sashes. (Pullman Neg. 10020)

Figure 1.57   When the Number 274 left the Pullman shop in 1909, wooden passenger cars were in their final years of production in this country. The last were produced in 1913. (Pullman Neg. 11470)

The Pennsylvania Railroad's commitment to steel cars in 1907 was the beginning of the end of wooden construction. New orders for wooden passenger equipment plummeted. In 1909 just over one-half of new orders were for wooden cars.[110] By the next year, orders were down to 29 percent of the total. In 1912 only 276 all-wooden cars were built, and many of these were for Canadian lines. The following year the last all-wooden passenger cars were produced for domestic service. After this time the only cars with wooden bodies were built for export, and most of them had steel frames.

The abrupt suspension of production did not mean that wooden cars were taken out of service. Steel and steel-frame cars rapidly preempted name-train assignments while wooden cars were downgraded to less prestigious trains, but there was no wholesale scrapping. The conversion was in fact remarkably languid. In 1912 over 90 percent of passenger cars were wooden; three years later the figure was 77 percent—comprising some 40,000 cars that made up the backbone of the fleet. In 1920 60 percent of cars were still wooden. Age took its toll during the next ten years, when more than half of the wooden cars were retired. Cutbacks in service during the early years of the Depression resulted in a dramatic reduction, so that by 1935 less than 6 percent of the fleet was wooden. It should be noted, however, that in the preceding years many wooden cars had been steel-framed, and just over 20 percent of the passenger cars in service in 1935 were of this type. By 1950 a few persistent holdouts, notably the Boston and Maine Railroad, had 276 all-wooden cars. Just three years earlier, the B & M was still assigning wooden cars to through trains. The wooden car has since disappeared—a good fifty years after its predicted demise.

Representative Cars

Figure 1.58   The Baltimore and Ohio Railroad's swell-sided coach, the Ohio, *was completed in 1830 by Richard Imlay. The drawing is from an original lithograph appended to the road's 1831 annual report.*

IMLAY CAR, 1830

The *Ohio* was the last of six four-wheel passenger cars completed by Richard Imlay for the Baltimore and Ohio Railroad between May and August 1830. The first car was named the *Pioneer*; others in the group were the *Frederick*, the *Maryland*, and the *Virginia*. They were the first cars built for regular passenger service on the Baltimore and Ohio and were, of course, among the very first rail passenger cars in the country. The B & O purchased another twenty-five or so four-wheel cars before adopting eight-wheel cars in 1834–1835.

This working drawing of the *Ohio* was included in the 1831 annual report of the Baltimore and Ohio. The influence of the

road coach is obvious in the swell-sided body style, the suspension, and the frame. Two transverse stationary seats were fitted opposite one another inside the body. Collapsible jump seats were placed astride the doors, to be opened up once the other passengers were seated. Leather straps were buckled in place for seat backs. This was an established arrangement in road coaches of the day. While it was not the most comfortable seating possible, it did increase the inside capacity to 12 seats. Six more passengers could be accommodated on the outside end seats. The roof bench held another 12, making a total seating of 30 passengers—an extraordinary load for a 13-foot car. The cars were intended for slow-speed horse operation, but once steam locomotives were in use, the increased speed and the smoke and sparks caused passengers to abandon the roof section. The B & O annual report for 1836 gave the capacity of the four-wheel cars as 17. A brakeman stationed on one of the outside end seats took the eighteenth seating place.

The seats were upholstered, but there were few other concessions to passenger comfort—no lighting, heating, or toilet facilities. In bad weather, heavy leather curtains were dropped to cover the window openings. One of these is shown in place on the left side of the car. Cloth curtains could be raised or lowered to block out the sun. This simple arrangement was tolerable at very slow speeds, but when the car was pulled by a locomotive at even 25 miles per hour, the open windows must have let in a stream of smoke and cinders. Or when the side curtains were closed against the wind and rain, the interior of the little car as it lurched over the line must have been gloomy and claustrophobic.

The car's sole suspension was the leather thoroughbraces. Heavy leather straps (about 2 inches wide by 7 feet long), built up of several thicknesses of leather sewn together, formed each thoroughbrace. The ends were attached to jacks firmly bolted to the undercarriage frame. Crescent-shaped adjustment wheels made it possible to take up the slack as the leather stretched. Since the wheels and axles were not equipped with springs, the leather straps alone absorbed the inequalities of the track. The wheels were chilled cast iron, 30 inches in diameter. Mudguards were placed over the tops of the wheels. Winans's friction-wheel journal bearings were employed. The brakes were operated by a lever at either end of the car (the brake shoes are visible between the wheels). A perch pole drawbar not unlike that found on a road coach of the period can be seen in the drawing.

The overall length of the car was 13 feet, 2 inches; the wheelbase, 3 feet, 10 inches; extreme body width, 4 feet, 10½ inches. The weight was 1¾ tons when the coach was empty. (See Figure 1.58)

GOOLD CAR, 1831

In the spring of 1831 James Goold of Albany, New York, contracted to build six stagecoach bodies for the Mohawk and Hudson Railroad. Goold had been a carriage and coach builder since 1813. It is obvious from the specification that he was asked to furnish ordinary bodies without special modifications for railway service:[111]

To the Commissioners of the Mohawk and Hudson Railroad Company:
Sirs—I propose and agree to furnish for said railroad company six coach tops; that is to furnish jacks, and jack bolts, and braces, with thorough braces, and put them on the frames of the company's railroad carriages to support the coach tops; the coach tops to be finished and hung in the style of workmanship generally adopted in Albany and Troy, for post coaches; the materials and workmanship to be first quality; a baggage-rack and boot to be hung at each end; the length of coach body to be seven feet and four inches; five feet wide in the center, and three feet eight inches between the jacks. The general plan of the coach to conform to the plan and explanation given by the engineer of the company; to have three inside seats, the back of the

end seats to be stuffed with moss, and all the seats to be stuffed with hair; to have a door on each side; to have an outside seat on each end across the top of the coach with suitable foot-board; also a seat at each end for driver or brakeman, to drop below to a suitable heighth to make the rack his foot board. An oil cloth to be rigged to the center rod on coach top, to cover baggage, and one at each end rolled to the back of the seat to protect it from rain. The whole completed, and to be hung on the carriage frames at some point on the line of said railroad, as follows:

Two coaches to be hung by the first day of July next, and the remaining four by the first day of August next; the work to be subject to the inspection of the engineer of the said railroad company. The whole to be completed as aforesaid for the sum of three hundred and ten dollars each.

It is understood that the above coaches are not to be provided with lamps or mud leathers.

JAMES GOOLD.

The written proposition is adopted on the part of the Mohawk and Hudson Railroad Company, by order of the commissioner.

JOHN B. JERVIS,
Engineer Mohawk and Hudson Railroad Company.
Albany, 23d April, 1831.

A contemporary side elevation reproduced in Figure 1.2 was acknowledged by Goold a few years before his death to be a true representation of the original design. A reconstruction drawing, shown here, was prepared in 1892 so that full-sized replicas might be built to accompany the *De Witt Clinton* locomotive at the Columbian Exhibition. The locomotive and cars are presently exhibited at the Ford Museum in Dearborn, Michigan.

These cars are very similar in construction and size to the Imlay car in Figure 1.58. Notice, however, the longer wheelbase, larger wheels, and inside axle bearings. The leather thoroughbraces are attached to stationary jacks. The seating capacity was smaller than the Imlay cars because steam operation was contemplated from the beginning, so that the roof seating was limited. A Mohawk and Hudson inventory of 1840 lists nine 9-passenger and two 6-passenger thoroughbrace cars.[112]

The overall length was 13 feet, 9 inches; the wheelbase, 5 feet; wheel diameter, 36 inches; extreme body width, 5 feet. (See Figures 1.59 and 1.60.)

JERVIS CAR, 1832

While he was chief engineer of the Mohawk and Hudson Railroad, John B. Jervis took a direct hand in designing the rolling stock. His improvements in locomotive design, notably the leading truck, brought him national attention. His approach to car construction appears more conservative than his locomotive work, as shown by this drawing. A three-compartment composite body with side doors was mounted on a simple four-wheel running gear. The body was simplified and more box-like, but a gesture toward the orthodox road coach body was made in the curved moldings on the lower panels. The driver's seat and long brake lever were also vestiges of road coach construction. The drum brake is a distinct departure, however, and possibly the

Figure 1.59   *This Mohawk and Hudson Railroad four-wheel coach was built by James Goold in 1831. (New York Central Railroad)*

most interesting detail of the car. The leaf springs are also worthy of notice as an alternative to the leather thoroughbrace suspension. The wrought-iron journal pedestals are, incidentally, very similar to those used on locomotives of the period.

The original drawing from which this tracing was made is dated 1832 and initialed by Jervis. It was apparently the plan followed for several cars built in Schenectady for the Mohawk and Hudson. The *Buffalo Journal* of April 25, 1832, described the cars as being made in a new and convenient "square form" with separate compartments. They were 15 feet long and seated 18; one of the cars was named *Utica*. The exact number that were built is not known, but Von Gerstner's report states that 48 four-wheel passenger cars were in service at the time of his visit in 1840.[113] He described them as having three compartments with seating for 20 first-class passengers. Second-class ticket holders were carried in freight cars.

The basic dimensions are clearly noted on the drawings. Cars of this general style weighed 2 to 2½ tons. (See Figure 1.61.)

BOSTON AND PROVIDENCE CAR, 1834–1836

The only surviving original stagecoach-style passenger car was built at the Boston and Providence Railroad repair shops in Roxbury, Massachusetts, between 1834 and 1836. It was designed and constructed by John Lightner (1811–1896), a Baltimore carriage builder who was master car builder for the railroad from 1834 to 1888.

The survival of an original swell-sided railroad passenger car of the 1830s is a fortuitous incident made even more improbable by the career of the car itself. It became obsolete just a few years after construction, for it could not operate with larger eight-wheel cars and was downgraded to branch-line service. From there it declined to hauling track laborers on work trains. Most swell-sided, thoroughbrace cars were found unsuitable for any service and were scrapped, possibly excepting the running gear, which could be salvaged for a track-construction car. But by some curious fluke, this car survived until the body was smashed in a wreck. According to one story, the pieces were gathered up and stored for many years at the Roxbury repair shops of the Boston and Providence.[114] Whether from thrift or sentiment, the car was preserved. In 1893 it was reassembled with a new running gear under the direction of its builder, John Lightner, for exhibit at the Columbian Exhibition. Around 1905 the car was given to Purdue University for its collection of historic rolling stock. It was returned to the New Haven Railroad in 1939 for exhibit at the New York World's Fair.

For years the New Haven displayed the car at various fairs and exhibitions as a prize publicity tool, but hard times forced its sale in 1951 to the Danbury Fair in Connecticut. The car is presently on exhibit there—under cover, fortunately.

The car differs from the two previous examples of swell-sided cars in that the body is wider and longer and it is hung high to clear the wheels. In general it represents a more positive departure from existing road coach design. If the modifications seem slight, consider that the railroad age had only begun, and that within three or four years builders such as John Lightner recognized that coaches were subject to fewer size limitations on the railway and should be designed differently. This car illustrates the conservative, straight-line development of an existing design for a new technology.

*Figure 1.60   A reproduction of an 1831 Mohawk and Hudson coach built around 1892
for the Columbian Exposition. (New York Central Railroad)*

The car's construction is evident from the drawings. The continuous draft sill or perch pole is similar to the Imlay car. The brake mechanism shown in section EE could brake only one set of wheels unless brakemen were stationed at both ends of the car. The wheels, it should be noted, are cast iron of an exceedingly fragile pattern.

The overall length is 15 feet, 2 inches; the wheelbase, 5 feet, 9 inches; wheel diameter, 36 inches; extreme width of body, 6 feet. The weight is estimated at 2 to 3 tons. (See Figures 1.62 to 1.64.)

NOVA SCOTIA CAR, ABOUT 1838

The history of this small four-wheel coach, now on display at the Baltimore and Ohio Transportation Museum in Baltimore, is very uncertain. According to one story, it was produced in England for the General Mining Association of Nova Scotia. But the late Robert R. Brown, a well-informed Canadian railroad historian, declared that it was turned out by the carpenters of the General Mining Association in Nova Scotia. The date of construction appears to be 1838. A railway was built to connect the Albion coal mines of the General Mining Association with Pictou

Harbor, and the car was used by officials and distinguished guests of the mining company.

By 1883 it was considered enough of a relic to show at the Chicago Railway Appliance Exhibition. Ten years later it was returned to Chicago for the Columbian Exhibition. Apparently abandoned in Chicago, the car was acquired by the Baltimore and Ohio Railroad when the exhibit closed. During the B & O's Centennial in 1927, the fanciful "bridal car" story was hatched: according to this piece of puffery, the car originally carried the governor of Nova Scotia's bride to her wedding, and any young girl seeking a husband need only seat herself in the car. The story has been so often repeated that it is now a hallowed part of the car's history.

The car, with its Bombay-shaped body, seats 6. Two upholstered seats are placed facing one another across the car. The cast-iron wheels are individually suspended by leaf springs. Since there is so little information, and no illustrations of the car dating before 1883, it can be questioned whether the running gear is original. It is possible that the body was originally mounted on thoroughbraces.

The car's dimensions are given in the drawing. (See Figures 1.65 and 1.66.)

Figure 1.61 *A Mohawk and Hudson coach designed in 1832 by John B. Jervis. The tracing made from an original drawing now in the Jervis Library, Rome, N.Y.*

VICTORY MODEL, 1835

Before Richard Imlay constructed the *Victory*, an eight-wheel car that he completed in July 1835 for the Philadelphia, Norristown, and Germantown Railroad, he built this model in order to check out the design. The model was constructed to a large scale, roughly 1½ inches to the foot, and is over 4 feet long. It was expertly made and finished as a showpiece; Imlay probably displayed it to prospective customers. He kept it for years after his car-building career was over, and on his death in 1867 it was purchased by the Eastern Railroad Association for exhibit at its offices. The model was sent to Chicago for the Columbian Exhibition, and through circumstances not entirely clear, it came into the possession of the Baltimore and Ohio Railroad. It is now exhibited at the B & O Transportation Museum. The drawing and the following description were prepared after first-hand inspection of the model.

The body is divided into three compartments. The center underslung section is the passenger compartment. One end compartment has a water closet, but the reason for that section's division into three spaces is not clear. The other end compartment is fitted out as a pantry. Its door has an opening and a serving shelf; on one side is a desk, and at the rear are several shelves for provisions. The passenger saloon is ringed with a continuous side bench broken only at the door openings. Each side of the car has a single door. The benches are not uphol-

Figure 1.62 *A Boston and Providence coach built in the railroad's shops around 1834.*

Figure 1.63 This drawing of a Boston and Providence coach of about 1834 was made from the original, which is now exhibited in Danbury, Connecticut. (Clark Williams)

stered, but it is possible that cushions were used. The interior of this area is covered in leather. The windows are entirely open, but it is probable that there were at least leather side curtains for protection. The lithograph in Figure 1.13 shows window sashes for the straight portions of the body, with curtains at the curved-end windows. The lack of windows in end compartments is another obvious defect of the model, and one that was corrected in the full-sized car.

The rudimentary clerestory runs the full length of the car and is open except for the light iron bars. Some provision (such as curtains) must have been made against the weather. It should be noted that the carlines run completely across the roof and do not break at the clerestory opening. The railing above the clerestory roof has been explained as a seat back, but the roof does not appear to be constructed for such a load. Lightly fabricated from thin wooden strips covered with canvas, it may have been a handrail for trainmen passing over the top of the car.

The trucks follow Imlay's leaf-spring bolster, which he patented in 1837. The instability of this arrangement is evident, but for very slow speeds it was probably satisfactory. Because the bolster was equipped with springs whereas the axles were not, the entire dead weight of the truck was free to damage the track. At some point in the model's history the spoked wheels were lost from one truck; the plate wheels are a later repair.

As a record of the exterior finish, this model is without parallel. The striping and decorative painting are finely executed, and despite successive coats of dark varnish and the scars of a long life, there is evidence that bright colors were used. The panels of the passenger saloon are bright yellow. Broad gold leaf and black bands outline the body shape. Fine brown and black lines outline the gilt bands and the window posts. The heraldry in the body panels is executed in gilt, red, and green, and finely picked out in black. The panels of the end compartments are a rich brown-red. They are striped in black and gold leaf. The figure of the phoenix is in gilt shaded in a brown wash. The car frame is naturally finished mahogany, striped in gilt. The roof is now black, but there is evidence that it was originally painted dark green. The trucks and wheels are black.

SECTION "FF"
RIGHT HALF SHOWS UPHOLSTERY REMOVED

SECTION "EE"

SYMMETRICAL ABOUT ℄
SECTION "GG"

Figure 1.64   Boston and Providence coach: cutaway detail. (Clark Williams)

The general dimensions of the model are clearly shown in the drawing. It should be noted that in certain proportions the model appears out of scale with the track gauge of 7¼ inches. This is particularly true for the end compartments, where insufficient headroom is provided. (See Figures 1.67 and 1.68.)

WASHINGTON BRANCH CARS, 1835

By 1834 the Baltimore and Ohio Railroad had become convinced of the superiority of eight-wheel passenger cars. The experiments were finished, and they had made it clear that a simple box-like body carried on two four-wheel trucks was the safest and most economical passenger car possible. The road therefore decided to build cars of this pattern for the new Baltimore-Washington branch. Construction began at the company shops late in 1834. Ten cars were completed in time for the Washington branch's opening in August 1835.

The cars incorporated all the basic characteristics of the American style of passenger car and may be regarded as the prototype of that enduring pattern. Two drawings exist of these historic

cars. The first (Figure 1.69) was prepared around 1853 for the series of lawsuits known as the eight-wheel car case. It was apparently taken from original drawings or from an original car then in existence but now vanished. And a contemporary drawing (Figure 1.70) of incredible detail was found at the Deutsches Museum in Munich. The origins of the drawing and the reason why it was in Germany cannot be determined, but it might be traced back to one of the numerous European engineers who came to study American railways early in the nineteenth century. In addition to the splendid engineering information, the drawing portrays the subtle elegance of American passenger cars of the period. If the 1853 drawing had been the only one to survive, it would be possible to dismiss the Washington branch cars as simple, even crude, boxes on wheels. But the contemporary drawing from Munich shows finely cut moldings, elaborate scrolls, panels, and louvers.

The car was supported by a simple box frame placed entirely under the body. The frame was stiffened by an iron truss rod. Cross sills were used. A continuous iron drawbar running the full length of the car took the place of the center sills. The body was

Figure 1.65   A coach of about 1838 operated by the General Mining Association, Nova Scotia. The photograph was taken around 1930 at Bailey's roundhouse, Baltimore.

Figure 1.66   Diagram drawing of a Nova Scotia coach, 1838c.

lightly framed and was not self-supporting. Solid shutters and glass panels alternated at the window openings. Presumably the glass was stationary and the shutters could be raised to admit air. Unaccountably, the seats were made with stationary backs even though reversible seats were in use on the B & O before this time. They were well upholstered, and an early newspaper account correctly described them as high-backed and well inclined. The wooden ends were treated in the bold Empire style of the period.

The trucks were of the "live" or leaf-spring side-frame type, with wrought-iron truck bolsters. A secondary leaf spring attached to the frame of the car bore on the ends of the body bolster (item X). This complex double suspension appears

unique to the B & O. The 31-inch-diameter wheels were cast iron made on Winans's "drumstick" pattern. Although not shown in the drawing, they would have had split hubs banded with wrought-iron rings. The axles were tapered, measuring $3\frac{3}{4}$ inches at the heaviest point. The journals were small, measuring 2 by 4 inches. The truck's wheelbase was 40 inches. The safety frames, shown in detail 25 of Figure 1.70, were a variation on Kite's safety beams. If the side-frame spring should break, the bracket Z' would fall on the wheel treads and support the car until it could be stopped. The bar frame shown in the same detail was meant to support a broken axle and hold it in line.

The brake is the earliest mechanical braking mechanism that

Figure 1.67   Richard Imlay's Victory, *constructed about 1835. The model is now in the B & O Museum, Baltimore. (Smithsonian Neg. 68359)*

Figure 1.68   Drawing of the eight-wheel car model Victory.

Figure 1.69 *Washington branch car of the Baltimore and Ohio Railroad, 1834–1835. The drawing was made around 1853. (Peale Museum)*

has been discovered. A light chain attached to the top of the upright staff (O′) at the right end of the car was stretched across the top of the train from one car to the next. Pulling the chain moved the staff, freeing a heavy counterweight lever which in turn set the brakes. The idea for this brake was credited to the brother of the president of the railroad, Evan Thomas, in the B & O's 1835 annual report.

The overall length was 37 feet, 2½ inches; the truck centers, 24 feet; the body, 32 feet, 2½ inches long by 9 feet wide by about 7 feet high. The car seated 44 and weighed an estimated 6 tons. (See Figures 1.69 and 1.70.)

CAMDEN AND AMBOY RAILROAD, CAR NUMBER 3, ABOUT 1836

This is the oldest existing eight-wheel passenger car. It was built around 1836 at the Camden and Amboy shops by Edwin Lockwood (1810–1881), the road's master car builder for over forty years. In 1840 the Austrian engineer Von Gerstner reported fifteen eight-wheel passenger cars in service on the Camden and Amboy.[115] The dimensions given in his report agree almost exactly with those of this car.

Some thirty years later, when the Pennsylvania Railroad took over operation of the New Jersey railroads, much of the obsolete equipment of the Camden and Amboy was sold. Car Number 3 was purchased for a chicken roost by Samuel Newton of South Amboy, who discarded the trucks and set the body on posts. It

remained on his farm until 1892, when the Pennsylvania Railroad retrieved it for exhibit at the Columbian Exposition. The remains of an old truck were uncovered in a marl pit; the wooden members were gone, but the metal parts were salvaged to construct a partial replica. A second matching car was found hidden away in the New Jersey Meadows repair shops. It had been discovered, derelict and half buried under rubbish in a lumberyard, some years earlier by an official of the railroad, who had farsightedly removed it for safekeeping.

Many parts were missing from both cars. The interiors were stripped; much of the hardware was gone; some wooden parts undoubtedly had rotted and been replaced. Probably there had been some rebuilding over the years as well. Thus the existing car must be viewed critically, at least in its details, and not accepted without qualification as a representative example of antique car construction. After restoration both cars were sent across the country in train with the venerable *John Bull* locomotive—a magnificent publicity stunt to advertise the Pennsylvania Railroad exhibit at the 1893 Chicago fair. The history of the cars after the fair is cloudy. Both were in existence as late as 1907, when it was suggested that they be shown with the *John Bull* at the Smithsonian Institution. By 1927, however, one of the cars was gone; the Pennsylvania exhibited only the Number 3 at the Fair of the Iron Horse in Baltimore. The fate of the other car cannot be discovered. In more recent years the Number 3 was given to the Smithsonian, where it has been on display in Railroad Hall since 1963.

Figure 1.70 B & O Washington branch coach of 1835, from an original drawing in the Deutsches Museum, Munich, Germany.

Figure 1.71   Camden and Amboy Railroad car Number 3, built in the road's own shops around 1836. Now in the Smithsonian Institution. (Pennsylvania Railroad)

The overall proportions of the car are small. A modern passenger, even one accustomed to the barbaric poverty of tourist air travel, would feel uncomfortably cramped in this low, narrow car. The ceiling is a bare 6 feet, 4 inches high. The seats are 1 foot deep; the center aisle is 18 inches wide. But it should be remembered that the cars were meant only for short trips. A way passenger could endure a short ride down the line, and a through passenger going the full 60-mile length would surely find the ride of the eight-wheel car more agreeable than that of a pitching road coach. The seats, if narrow and low-backed, were upholstered, and the backs were reversible.

Ventilation received some attention. The small-paned glass windows were stationary, but the larger wooden panels could be opened and dropped down between the sides of the car body. The row of horizontal shutters at the top of the body (where the letter board would normally be) were hinged at their centers and could be tilted open. Lighting, however, received less attention than ventilation. A candle lamp at each end of the car, built into one of the end windows, provided the only illumination. It served to light the doorway of the car and doubled as a running light, since it was made to shine in both directions. It is possible that there once were individual side candle lamps, but no evidence of them survives in the present car.

The decorative treatment is plain and somber, possibly because of the 1893 restoration rather than the original finish. The exterior is painted olive drab outlined in black. The truck frames are cream yellow, while the wheels and hardware are black. The interior woodwork is painted a dull tan. The headlining is a rich, figured grey-green brocade. An ornamental fringe runs around the ceiling. The seats are covered in a heavy light-grey cloth, as are the armrests. Lighter, livelier colors would actually seem to be a more authentic finish for this car.

The most striking structural feature is the underslung wooden truss, which provides the car's main support. The center sill runs only between the body bolsters (which partly accounts for the sag at either end of the body). Light cross pieces and end sills form the remainder of the floor frame. The platforms are supported by separate timbers. A continuous iron plate ½ inch thick is bolted to the underside of the center sill. A clevis attached to the body bolsters receives the end of the long drawbar. The platform end of the drawbar is split to make the coupler pocket. Thus a continuous metal draft rig was created to relieve the wooden frame from this duty.

The body framing is exceedingly light, composed of thin vertical and horizontal members. The side panels are of clear white wood. The roof is fabricated in the usual manner, with arched carlines and narrow wooden sheeting covered over with painted canvas.

The trucks are wooden beams with stationary bolsters that are supported by iron truss rods, one on either side. These rods pass through the side frames, holding the assembly firmly together. A nest of seven light coil springs is fitted over each journal box,

Figure 1.72 *General arrangement of the Camden and Amboy Number 3, from a drawing prepared by the Pennsylvania Railroad. (Pennsylvania Railroad)*

Figure 1.73 *Interior of the Camden and Amboy Railroad coach. (Pennsylvania Railroad)*

Figure 1.74   Truck detail of the Camden & Amboy Number 3 showing the original iron spoked wheels. (Drawing by John H. White, Jr.)

which is made following John Lightner's 1848 design. The journal bearings measure 2¾ by 4¼ inches. The present wheels are a curious mixture of single- and double-plate styles; no two are duplicates, and all date from the 1870s. What the original wheels were like is impossible to say, because the Camden and Amboy tried every conceivable style during its early years. In 1840 Von Gerstner said that cast-iron chilled wheels were used under the road's eight-wheel passenger cars.[116] These were probably spoked wheels.

Brakes are fitted to only one truck, and wooden brake shoes are hung outside of the wheels. The brake beams are supported by leather straps provided with buckles to adjust the position of the shoes. The brake staff has a hinged handle which folds down out of the way when not in use.

The overall length is 35 feet, 7 inches; body width, 8 feet, 6½ inches; height, 10 feet; seating, 48; weight, 14,250 pounds. (See Figures 1.71 to 1.74.)

BOSTON AND PROVIDENCE RAILROAD, EIGHT-WHEEL CAR, 1838

The Boston and Providence was among the first New England railroads to adopt eight-wheel cars. George S. Griggs, master

mechanic of the line (best known as a locomotive designer), claimed credit for the road's first eight-wheel passenger car. In testimony given in the 1850s at the eight-wheel car trial, Griggs said that he began work on the car in March 1838.[117] He claims that he had not seen any other eight-wheel railroad cars except possibly those used on the Granite Railway to haul stone blocks. He completed the car in September of that year, and it proved so satisfactory that the road rebuilt a number of four-wheel cars as eight-wheelers and constructed all new passenger cars on the double-truck plan.

Together with his testimony in the lawsuit, Griggs offered this drawing of his first car, which was reproduced in the published court transcript. The drawing was made from measurements of the existing car several days before Griggs testified. It shows a narrow, low-ceilinged car in sketchy detail, though there is enough to convey the general arrangement. The posts and window sash were omitted. The extremely narrow body of 7 feet might be explained by insufficient passing clearance on the railroad tracks. (Many early roads placed their tracks close together, never anticipating the need for wider cars.) The end platforms are mounted on extended beams of the truck's side frame. The load of the draft was then placed on the truck's kingpins. The platform would swing, of course, when the car passed around a

Figure 1.75 Boston and Providence eight-wheel coach, 1838. The window sash and window posts have been omitted from the drawing. (Winans vs. Eastern R.R., 1854)

curve. Evidence that at least one other American car used this arrangement is shown in an 1845 English drawing (Figure 1.78). The plan was not satisfactory, for Griggs said that the platforms were later attached to the car body in the conventional manner. He also testified that the original 36-inch-diameter wheels were replaced by 33-inch wheels around 1845, and that the car roof had since been extended over the platforms. This car was still in service in 1853.

The car length over the couplers was 38 feet, 6 inches; the body measured 30 feet, 6 inches by 7 feet. The truck had a rather long wheelbase for the period: 4 feet, 11 inches. The seating was an estimated 45 to 50, the weight an estimated 6 to 7 tons. (See Figure 1.75.)

BALTIMORE AND OHIO RAILROAD, 48-PASSENGER CAR, 1835–1840

The drawings for this car are from a report by Michel Chevalier on internal communications in the United States, published in Paris in 1841. The design comes from the period 1835–1840. From what little information the text offers and from the measurements scaled from the drawing, it appears that this was one of the lightweight eight-wheel cars built for the old main line to Frederick. After a visit to the B & O yards in 1840, Von Gerstner said that fourteen slightly smaller cars were used on the main line, which could not accommodate the heavier Washington

branch cars.[118] He reported that the main-line cars weighed 4.85 tons and seated 48. The body of the Chevalier car measured nearly 5 feet shorter than the Washington branch car in Figure 1.70.

It differed from the Washington branch cars in several minor particulars, although both followed the same general plan. The Chevalier car had a side door in addition to the end doors, and it had narrow running boards the full length of the car. The ceiling height was even more cramped; judging from the drawing, a 6-foot man could not stand erect. Two center sills were used in place of the continuous drawbar. The truck's wheelbase was only 36 inches. The counterweighted train brake was replaced by a simpler lever brake, which was actuated by stepping down on a foot treadle attached to the vertical staff (see detail 6 of Figure 1.70).

The overall length was 33 feet, 6 inches; the length of the body, 27 feet, 6 inches; width, 8 feet, 6 inches; truck centers, 22 feet. (See Figure 1.76.)

BALTIMORE AND OHIO RAILROAD, 72-PASSENGER CAR, 1840

Karl Von Ghega, another European engineer who visited the United States to inspect America's railway system, published his report in 1844.[119] The only car pictured in it was this large passenger car built in 1840. It represents a startling growth in

Figure 1.76 Baltimore and Ohio Railroad coach of about 1840, from a contemporary French text on American railroads. (M. Chevalier, Voies de Communication des Etats Unis, 1841, plate V)

ACHTRÄDIGER PERSONEN-WAGEN ZU 72 SITZEN.   *Taf. XVIII.*

Fig. 1. Seiten Ansicht.   *Fig. 2. Hintere und vordere Ansicht.*

Fig. 3. Innerer Grundriß.   *Fig. 4. Grundriß eines Untergestelles.*

Fig. 5.   *Fig. 6.*   *Fig. 9.*   *Fig. 7.*   *Fig. 8.*

Maaßstab für Fig. 5–9.

Figure 1.77   A Baltimore and Ohio Railroad coach of about 1840, from a contemporary
German text. (Karl Ghega, Die Baltimore-Ohio eisenbahn, Vienna, 1844, plate 18)

capacity, if not in size, over the first eight-wheel cars. That 72 passengers could be crammed into a 36- by 7-foot body seems impossible, but it was achieved with extremely narrow seats and a 14-inch center aisle. The seats measured only 14 by 33 inches and were placed on approximately 2-foot centers. A new style of individual seat with an adjustable back is shown in details 5 and 6, although the mechanism for adjustment is not illustrated. Another indication of the car's meager dimensions is the inside ceiling clearance of 5 feet 11 inches.

Structurally the car shows some improvement over the earliest eight-wheel cars. The under-mounted outside frame truss has given way to ordinary side sills braced by an iron truss rod. The spring-framed truck has been retained, but in a simpler form the secondary bolster and spring were eliminated. Although a metal bolster is used, it is made up of two light bars in place of a heavy single piece. The braking mechanism is still the foot-treadle brake, which is independent for each truck. Wood-block brake shoes can be seen between the wheels.

The builder of the car is not known; possibly it was con-

structed at the railroad's Mt. Clare shops. However, the Gothic arch above the windows and the general style of the body bear a strong resemblance to the Harlan and Hollingsworth drawing of 1839 (Figure 1.18).

The overall length was 42 feet; the width, 7 feet, 6 inches; truck centers, 30 feet, 9 inches; truck wheelbase, 39 inches; wheel diameter, 30 inches. (See Figure 1.77.)

AMERICAN PASSENGER CAR, 1845

The British trade journal *The Practical Mechanic and Engineer's Magazine* published this drawing in its January 1845 issue. Unfortunately, only the most general commentary accompanied the drawing; neither the builder nor the railroad was mentioned. The car was described simply as representative of "the best of these commodious vehicles," meaning American double-truck passenger cars with ladies' compartments. The only clue to its builder is the double leaf-spring draft gear, which follows the specifications of Charles Davenport's 1835 patent. However,

Figure 1.78 This very complete drawing of an American passenger car of 1845 was published in a British journal. Unfortunately, the builder and railroad were not identified.
(Practical Mechanics Journal, *January 1845*)

there is no assurance that the car was made by Davenport and Bridges.

The car is longer and slightly wider than the Davenport and Bridges car in Figure 1.81. The ladies' compartment is more generously arranged, with elegant, curving Empire sofas, a dressing table and mirror, and a larger water closet. The truck under the passenger end of the car has extended side beams to support the end platform. At the other end of the car a more conventional truck is shown. The platform at the ladies' end of the car is carried by the sills of the car frame. A roller side bearing is cushioned by an underslung leaf spring. The axles are individually sprung—no equalizing was used. The wheels appear to be on Winans's drumstick plan.

Assuming that the wheels are 33 inches in diameter, the following dimensions can be scaled from the drawing: the length over end sills, 37 feet, 6 inches; body, 32 feet by 8 feet, 6 inches wide; truck wheelbase, 4 feet; overall height, 10 feet. The passenger compartment seated 28, the ladies' room about 10. (See Figure 1.78.)

EATON AND GILBERT CAR, 1845

This one-tenth-sized model of a second-class passenger car on the Württemberg State Railway is exhibited in the Deutches

Museum, Munich, Germany. The model is expertly made, showing the construction of the original car in full detail. It is thought to be a contemporary model, possibly created by Eaton and Gilbert as an exhibition piece for prospective customers. A portion of one side is open, and the roof is removable to show the interior arrangement.

The initial rolling stock of the Württemberg Railway was purchased from American contractors. The Baldwin and Norris companies both supplied locomotives to the line in 1845, and Eaton and Gilbert was called upon to provide cars. No description of these cars has been found, but here is an earlier account prepared by the builder that agrees with the Württemberg car model:[120]

Cars to seat 52 persons are 33 feet long × 8½ wide. Roof and platform each projecting 2½ feet at each end. These [are] handsomely painted outside and well finished inside with mahogany seats. Backs to turn over. Trimmed with hair cloth, Morocco or braid cloth. The sides either cushioned with the same or finished with mahogany panels. On the floor a sutable oil cloth carpet. The inside of the roof either painted or lined with damask as may be desired. We have on each side 12 windows, the upper half of the sash stationary. The lower half to slide up with 4 steel springs on each, frames of mahogany well glazed. With a curtain to cover the windows inside. The spaces between the windows trimmed and sluffed inside and an imitation blind on the

Figure 1.79   Eaton and Gilbert of Troy, New York, produced this eight-wheel passenger car for the Würtemburg State Railway in 1845. (Deutches Museum)

outside. A door in each end will furnish a passage through the center. We make a variety of trucks but the most approved is Williams Patent Iron Truck which are the only kind we can dispose of at present. There is a very good and permanent fixture for center bearing which is so hung on a stirrup as to give a side motion or rather to take away the effect of the side motion. A car of this kind we can furnish delivered on a boat at Troy for $1700. One of the same capacity for second class may be finished for from $1000 to 1200 depending entirely on the kind of finish. We have made but one of this kind and that a plain one for $1000. Those that will seat 52 are the most used. 4 wheel cars to carry half the number of passengers finished in corresponding styles may be built for $900 the first class and $600 the second class although no 4 wheel cars are called for any more here.

The iron truck and imitation outside shutter panels agree exactly with this description. From the model it can be seen that the side sills alone support the car; no center sills or iron drawbars were used. The draft gear was attached to end sills and appears to have been cushioned with a rubber spring. The body framing was equally simple, composed of light vertical members not unlike the balloon frame of a cheap wooden house. No trussing is evident. Small corner lamps with chimney vents through the roof lighted the interior and served as marker lights. The same general plan was used on the 1836 Camden and Amboy car in Figure 1.73.

The iron-frame trucks are of a particularly advanced design and refute the idea that all such assemblies before the Civil War were hard-riding, wooden-frame primitives. The neat, light, wrought-iron frame was cross-braced to hold it square. The swing bolster is after Davenport and Bridges's 1841 patent. A cast-iron pad with a side-bearing roller extends the bolster ends

to the extreme width of the body in order to further stabilize the car's ride. The wheels are double-plate chilled cast iron and have cast raised letters reading "Bush and Lobdells patent March 17, 1838."

Eaton and Gilbert's design apparently suited its South German customers, for an elaborate drawing of nearly identical cars appeared in *Organ Fortschritte Eissenbahn-Wesens*, 1847, Volume 2. The description noted that the American plan of car was being followed by the Württemberg Railway for its new combination first- and second-class cars.

This drawing, prepared in 1906 at the time the model was placed in the Deutsches Museum, was made in metric scale. In English feet, the general dimensions are overall length, 41 feet; width, 9 feet; height, 10 feet, 9 inches. The truck wheelbase is 4 feet. Seating was provided for 56 passengers. (See Figure 1.79.)

DAVENPORT AND BRIDGES CAR, 1845

Davenport and Bridges of Cambridgeport, Massachusetts, was among the first railroad car builders in New England and became one of the largest car makers in the country. Before closing in the mid-1850s, it had supplied cars to railroads throughout New England, New York, and Pennsylvania. In the 1840s the firm mounted an ambitious advertising campaign with regular half- and full-page notices in the *American Railroad Journal*. One engraving that showed a working drawing of a typical car, with a longitudinal cross section and plan view, is among the best illustrations available for a car of the period (Figure 1.80). This view is complemented by an exterior perspective drawing (Figure 1.81).[121] A description of some Davenport and Bridges cars

Figure 1.80 *Davenport and Bridges was one of the largest builders of passenger cars in the United States in 1845 when this advertising drawing appeared.* (American Railroad Journal, *November 27, 1845*)

DAVENPORT & BRIDGES CONTINUE TO MANUFACTURE TO ORDER, AT THEIR WORKS, IN CAMBRIDGEPORT, MASS.
Passenger and Freight Cars of every description, and of the most improved pattern. They also furnish Snow Ploughs and Chilled Wheels of any pattern and size. Forged Axles, Springs, Boxes and Bolts for Cars at the lowest prices. All orders punctually executed and forwarded to any part of the country
Our Works are within fifteen minutes ride from State street, Boston—coaches pass every fifteen minues.   1y1

Figure 1.81 *A perspective illustration of a Davenport and Bridges car of 1845.* (American Railroad Journal, *August 7, 1845*)

made for the Auburn and Rochester Railroad a few years earlier agrees closely with the illustrations:[122]

There are six cars, designed to form two trains. The cars are each 28 feet long and 8 feet wide. The seats are well stuffed and admirably arranged—with arms for each chair, and changeable backs that will allow the passenger to change "front to rear" by a manœuvre unknown in military tactics. The size of the cars forms a pleasant room, handsomely painted, with floor matting, with windows secured from jarring, and with curtains to shield from the blazing sun. We should have said *rooms*; for in four out of six cars, (the other two being designed only for way passengers,) there is a ladies' apartment, with

luxurious sofas for seats, and in recesses may be found a washstand and other conveniences. The arrangement of the apartment for ladies, we consider the greatest improvement; and it will remedy some serious objections that have hitherto existed against railroad travelling on the part of families, especially where any of the members are in delicate health. The ladies can now have their choice either of a sofa in their own apartment, or a seat in the main saloon of the cars, as their health and inclination may require.

These cars are so hung on springs, and are of such large size, that they are freed from most of the jar, and especially from the swinging motion so disagreeable to most railroads.

The lamp of each car is so placed as to light inside and out; and

*Figure 1.82   An arched-roof, eight-wheel passenger car built for the B & O to the designs of its master mechanic, Henry Tyson, in 1856. (*Douglas Galton, Report on the Railways of the United States, London, 1857–1858*)*

last though not least, the breakers are so arranged as to be applied readily and with great power—thus guarding against the danger of collisions, etc.

We almost forgot to mention that these beautiful cars were made by Davenport and Bridges, of Cambridge, Massachusetts, and cost at low prices about $1700 each—or $10,000 for the six.

Although this account stresses the luxury of the cars, it can be seen from the drawing that the cramped ceilings and low-backed seats of ten years earlier still prevailed. The ladies' compartment, a water closet, roof ventilators, and a more generous supply of windows (with movable sash) were new concessions to passenger comfort. Extending the roof to cover the platforms was another improvement over the earlier eight-wheel cars.

Continuous sills supported the car and platforms. Short timbers attached beneath the main floor sills were used to support the platforms in later years. The drawbars, which were carried back to the body bolsters, were fabricated from wrought-iron plates riveted with a wood filler piece. The trucks were inside-bearing wrought iron made after a patent issued to Charles Davenport on August 10, 1844 (No. 3697). The trucks were of all-iron construction except for the bolster. Outside rocker side bearings stabilized the car body. Davenport did not claim credit for the iron side frames or the side truss frames with diagonal braces (the predecessor of arch-bar trucks); he conceded that these improvements had been introduced before his patent. He did maintain that his truck was lighter by 1,200 pounds and about $100 cheaper to make than the existing styles of car trucks.

By assuming a wheel diameter of 33 inches, the following dimensions can be scaled from the drawings: overall length, 34

feet; height, 10 feet; truck wheel cast, 4 feet. Seating was for 43. (See Figures 1.80 and 1.81.)

BALTIMORE AND OHIO RAILROAD, 50-PASSENGER CAR, 1856

Improvements in the B & O passenger rolling stock since the 1830s and 1840s are apparent from the new style of arched-roof car in these drawings. The car was designed in 1856 by Henry Tyson (1820–1877), master of machinery for the B & O from 1856 to 1859. The car shows no extraordinary increase in size, nor was the general construction plan altered. However, more thought was given to passenger comfort: the ceiling was raised to 7 feet, and seat spacing was more generous. Deep cushions, high padded backs, and footrests were added. A water closet, dressing room, and water cooler are shown in the plan view detail of Figure 1.82. Tiny ventilators were placed high on the car body in panels between the windows (best seen in the section detail of the side elevation), so that air could be admitted near the ceiling without opening the regular windows. Adjustable wooden blinds were used for sunshades. No information on lighting or heating apparatus is available.

The first group of Tyson cars were completed at the Mt. Clare shops late in 1856. The *Wheeling Times* wrote: "They are beautiful, substantial and safe. There is an air of comfort observable in the interior of the cars which is only surpassed by the extreme beauty and masterful skill displayed in the outside execution as to painting, carving, etc. The interiors of the cars are supplied with beautiful landscapes and light paintings in a profuse manner."[123] A few indications of the interior decoration can be seen in the drawing.

The mechanical drawing of Tyson's car is from Douglas Galton's report on American railroads to the British Parliament (1857). Truss rods were not used, and it is probable that there was no body-panel wooden truss. However, the 40-foot body was about as long as floor sills of reasonable size could support without the assistance of one or both truss reinforcement methods.

The exterior of the car was fashioned in batten-board trim, forming a pleasing series of panels. The platforms were supported by timbers fastened beneath the floor sills. This remained the typical method of construction until the end of the wooden car era.

The trucks are set well back from the platforms to shorten the total wheelbase; the truck wheelbase was only 51 inches—rather short for the period. Most other lines used the so-called "square" trucks, in which the wheelbase equaled the track gauge. The B & O's adherence to close-coupled wheels may be due to the sharp curves on the tracks of its mountain division.

The suspension of the car was sufficiently elaborate to bring joy to any spring manufacturer. Each truck had eight leaf springs (Figure 7.2). The journal boxes were individually sprung; the swing bolster was cushioned by two heavy, double elliptic springs; a pair of half elliptic springs acted to center the swing bolster. Yet with all the complexity of the suspension, no equalizing was attempted.

The iron bolster and live-spring side frame were at last abandoned; the B & O had turned to the conventional wooden-beam truck. A rather light pattern of cast-iron pedestal was used, which can be better understood from the identical trucks pictured under the camp or office car in Figure 4.95. The plate wheels shown in the photograph here appear to be of the corrugated pattern developed by Asa Whitney's company.

F. A. Stevens's brake gear was used, so that all eight wheels could be braked together from either brake staff. A single rod, visible just below the frame of the car, connected the levers on the trucks. In this car wooden blocks had given way to wrought-iron brake shoes.

The overall length was 45 feet, 6 inches; the width over body, 8 feet, 9 inches; height, 10 feet, 7 inches; truck centers, 26 feet; truck wheelbase, 4 feet, 3 inches. (See Figure 1.82.)

NEW YORK AND ERIE RAILWAY, BROAD-GAUGE CAR, 1856

This car was not typical of the period, but it is representative of the huge vehicles running on the 6-foot-gauge lines after 1855. The drawings here are the earliest concrete examples of truss framing—which, next to the eight-wheel car itself, was the single most important contribution of American designers to the art of car building.[124]

At the urging of Master Car Builder Calvin A. Smith, construction was begun on six giant passenger cars in mid-1855 at the Piermont, New York, repair shops.[125] Completion of the first lot

of cars, which represented a radical departure from the short, wide cars formerly built, was reported the following year, together with a wonderfully complete description:[126]

Splendid Passenger Cars.

The completion of a portion of the number of new passenger cars built at the Piermont shops for the Erie road, renders a notice of them appropriate and interesting at this time.

These cars are not equalled in size, accommodation, or elegance of finish, by any other public railroad conveyances in the world. Whatever may be the opinion as to the policy of running such heavy cars, universal admiration must be yielded to their comfort and luxurious finish.

The Master Car Builder at the Piermont Shops, urged the construction of some large sized passenger cars, more than a year since, and suggested the general design of their finish and furniture. Mr. McCallum, the superintendent, adopted the suggestion and ordered the construction of the cars. He dictated the plan of the frame, employing a bridge truss between the windows and sills, and making the upper portion of the frame independent of the lower.

The body of the car is 60 feet 3 inches long outside, and 10 feet 9 inches wide. The posts are 7 feet high and the height of the cars at the center is about 11 inches greater than of any others now on the road. There are 20 windows on each side of the car, with a single plate of double thick French plate glass, 17 by 21 inches in each. There are two windows also in each end. The car seats 74 passengers, including two in the saloon.

The outside finish of the car is singular and neat. There are three plain vertical panels under each window, and one long panel reaching from the sills to the cap, between every two windows.

The seats, throughout, are C. P. Bailey's patent reclining, day or night seats. They are very large and easy and splendidly upholstered, the covering being a rich velvet plush, costing five dollars a yard. Over one hundred yards of plush were used. This was furnished from Doremus & Nixon's, of this city.

The cars are ventilated on Foote & Hayes' patent plan, and an improvement has been made in working the pump, used in the ventilator, by a friction wheel on the axle. This will obviate the main difficulty before had with this ventilator, apart from which it is the most comfortable means of purifying, cooling, and equalizing the temperature of the air in passenger cars.

The side "pedestals," as they are called, in which the ventilating water fountains play are elegantly finished off, and a large plate glass in each gives a view of the cooling spray when the car is running. Through these "pedestals" the air is conducted from the air cap, on the roof, to the registers in the floor.

There are four pivot single seats, on each corner of each pedestal, moulded and upholstered in a most luxurious style. They are decidedly a new feature in passenger cars, although a very few are already in use in one or two other new cars on the same road.

The head-lining is an expensive and rich pattern, although not quite in good taste with the rest. In some of the other cars, a new pattern will, perhaps, be adopted. The gilt cornice around the inside is 4 inches deep and very rich.

The trucks and running gear, built by Harvey Rice, Esq., are well got up, and although the same kind are already in use in other cars on the road, they are a novelty off of the road. The axles are spread six feet apart centers, and have both outside and inside journals. The inside journals have grooved collars on each end, and are fitted with

Figure 1.83   A broad-gauge passenger car produced about 1856 for the New York and Erie. (Galton's Report on U.S. Railways)

Harvey Rice's improved box. The outside journals have the Lightner box. The axles are from the Brunswick works in England, and the wheels, 33 inches in diameter, are Bush & Lobdell's pattern. The worst fault with the trucks is in using Rubber Springs, which, if they stand the great weight at all, will become rigid or freeze.

The cars have Hopkins' patent coupling, an improvement beyond value as respects safety, and measurably cheaper and far more convenient in use than the old coupling.

The entire weight of these cars is not far from eighteen tons, empty. It will cost more to draw passengers in these cars than in the light cars. The very sacrifice made in this respect, for the comfort and pleasure of its patrons, should commend the Erie road to the traveling public.

Some of the cars will very soon be running regularly on the night express trains.

The top cord of the truss frame described above was 3- by 4½-inch timber, slightly arched toward the center of the car. The bottom cord was the outside frame sill, composed of a 5¾- by 7½-inch timber. The wooden vertical posts, about 2¼ inches square, were set on 18-inch centers and mortised into the top and bottom cords. The truss was stiffened and drawn together by ⅜-inch iron rods placed diagonally between the posts. It was necessary, of course, to run a light parallel stringer over the arched top cord for a level window setting. Additional support to the floor sills was provided by truss rods. The upper car framing was tied together by vertical rods placed directly against the window posts. In all, it was a remarkably sophisticated framing plan, and one that continued into the 1870s.

The trucks were not equal in design competence to the general plan of the car. The suspension seems particularly defective: individual rubber springs were used for each bearing, and the stationary bolsters were carried on rubber springs nearly a foot in diameter. Inside and outside bearings were used as a safety measure in the event of a broken axle. This was a far more expensive plan than safety beams, and it entailed the added disadvantage of wear and friction should one or more of the journals be misaligned.

Dimensions taken from the drawing: overall length, 65 feet; width, 10 feet, 9 inches; height, 12 feet, 8 inches. The weight, empty, was 18 tons; the seating was for 72 if the saloon seats were included. (See Figures 1.83 to 1.85.)

Figure 1.84 Framing details of a New York and Erie car. This illustration is the first concrete record of a side-frame truss, although the idea may have been used earlier. (Galton's Report on U.S. Railways)

FUNNEL TO ADMIT AIR

SECTION OF AXLE BOXES FOR INSIDE AND OUTSIDE

BEARING

JOURNAL

OIL AND COTTON WASTE

WHEEL

LEATHER COLLAR

JOURNAL

OIL AND COTTON WASTE

ENTRANCE DOOR

SEATS

Passage
down centre of Car

Half Section of Carriage to shew
the Ventilating apparatus used on
the New York & Erie Railway.
The Truck is supported on inside
and outside bearings.

A   Pump
B   Connecting pipes
C   Valves
D   Branch pipe
E   Sprinklers
F   Water Tank
G   Air Channel
H   Air Valves

Gauge 6 Feet

Inches 6 9 6 3 0   1   2   3 Feet

Figure 1.85   Cross section of a New York and Erie broad-gauge car of about 1856. It
shows details of the Foote ventilating system. (Galton's Report on U.S. Railways)

Figure 1.86   An Illinois Central twelve-wheel car of about 1857, built in the railroad shops. (Galton's Report on U.S. Railways)

ILLINOIS CENTRAL RAILROAD, TWELVE-WHEEL CAR, 1857

This heavy, first-class car represents the finest passenger equipment of the period. It was a large car for the time, with generous overall proportions. In contrast to the 6-foot ceilings of American cars a decade earlier, it provided 7 feet, 3 inches of headroom. The big windows and wide body are signs of generous seat spacing. However, the arched roof, paneled sides, and fancy exterior ornamentation clearly mark it as a car of the 1850s. It was probably built at the Illinois Central's Chicago car shops, which were started in 1852 by the American Car Company and sold to the Illinois Central four years later for a car repair and construction depot.

The six-wheel trucks are the most distinctive feature of the car, but as Chapter 1 mentioned, some twelve- and even a few sixteen-wheel passenger cars had been operating on American railroads ten to fifteen years earlier. However, these drawings for the Illinois Central's Number 71 are the earliest construction plans available for such a car. They are in fact the oldest authenticated illustrations of a twelve-wheel car that could be discovered. The drawings are from Douglas Galton's 1857 report on American railroads.

The body was so long (about 50 feet) that truss rods were required. It is probable that a wooden truss was built into the car sides as well. Another feature that looked forward to later cars was the letter board. Sheet-iron roof ventilators and wooden blinds are clearly shown in the drawing.

The trucks were wooden beams with cast-iron pedestals and light wrought-iron tie bars. The axles were on 42-inch centers. Each axle was sprung by half elliptic springs, and the spring hangers were cushioned by india-rubber cylindrical snubbers mounted under side-frame beams. The most significant feature of the truck's suspension was the equalizing levers connecting the springs—early examples of this important arrangement. Cast-iron plate wheels are shown. The brake gear was on Stevens's plan, and wooden-block brake shoes were used.

The overall length was 57 feet; the width, 9 feet, 6 inches; height over the roof, 10 feet, 6 inches. Seating was for about 65. (See Figures 1.86 and 1.87.)

CHARLESTON AND SAVANNAH RAILROAD, ARCHED-ROOF CAR, 1860

This drawing is one of some twenty original passenger car drawings from the Harlan and Hollingsworth Car Company of

Wilmington, Delaware. (Several other drawings from the collection are reproduced later in the book.) All date from the 1860s. They are exquisitely colored in pale wash tones of yellow, grey, blue, and red. It is assumed that they were prepared for perusal by prospective customers. They were exhibited at the Columbian Exhibition in 1893 and afterwards came into the possession of the Baltimore and Ohio Railroad. In 1942 they were transferred to the United States National Museum in Washington, D.C.

The example shown here appears to be a standard-gauge car that was large for its time. The high ceiling would clear about 8 feet, offering generous headroom for an arched-roof car. The interior width was nearly 9 feet, again roomy for the period. The body was short enough so that it did not require a side-frame truss or truss rods, although a number of iron tie rods were used to strengthen the cross framing. Small, square ventilators with louvers hinged at the centers are visible in the letter board. The narrow vertical sheeting in place of the customary side body panels or batten boards was another advanced feature of the car. The trucks were ordinary wooden-beam style, with india-rubber springs over each journal and full elliptic springs for the bolsters.

The overall length was 49 feet; the height, 11 feet, 9 inches; width, 9 feet, 8 inches; truck centers, 31 feet, 6 inches; truck wheelbase, 5 feet; seating capacity, 52. (See Figure 1.88.)

LONG ISLAND RAILROAD, ARCHED-ROOF CAR, 1861

Although it was built eight months after the car in Figure 1.88, this product of Harlan and Hollingsworth was in many respects more antique. The paneled side, low ceiling, and small windows with fretwork details at the tops of the window openings are typical of earlier practice. The end lamps and roof chimneys are reminiscent of the Camden and Amboy car of 1836 (Figure 1.73).

The use of lighter sills apparently called for reinforcing truss rods. The trucks were sprung with india-rubber and elliptic springs as before, but a fully developed equalizer is shown.

The overall length was 49 feet; the height, 10 feet, 10 inches; width, 9 feet, 5 inches; truck centers, 31 feet, 3 inches; truck wheelbase, 4 feet, 8 inches. Seating capacity was 57, counting the saloon sofa. (See Figure 1.89.)

PENNSYLVANIA RAILROAD, CLERESTORY ROOF CAR, 1862

The Pennsylvania Railroad began a program of standard passenger car design in 1862 with its class PA. This style of car was

Figure 1.87   End elevation of the Illinois Central passenger car Number 71. (Galton's Report on U.S. Railways)

Figure 1.88 An arched-roof car, completed in December 1860 by Harlan and Hollingsworth. The car measures 49 feet over the couplers.

Figure 1.89   An arched-roof car completed in 1861 by Harlan and Hollingsworth.

SEATING CAPACITY, 52 PERSONS
WEIGHT, 39,300 LBS.

PASSENGER CAR, CLASS PA.

TRACED BY DOROTHY BRIGGS

Figure 1.90   A Pennsylvania Railroad class PA coach. The design was introduced in 1862.

Figure 1.91   A Pennsylvania Railroad class PA coach. The Monitor roof car design was adopted in 1862.

to be used system-wide and built to an exact common standard, whether it was constructed in the company shops or in a contract shop. The exact number produced has not been determined, but photographic evidence indicates that there were a great many PAs and that they were in service on nearly every portion of the system.

The design incorporated the most advanced styling elements of the period, including the clerestory roof. The short-lived Monitor plan was used, however. The body had a wide letter board, small, square windows, and batten-board sides. The platform canopies or hoods were made as separate structures and were not integrated with the roof—a peculiarity of construction common on the Pennsylvania until the end of wooden car construction. (See Figures 1.90 and 1.91.)

Figure 1.92   A 59-passenger coach constructed by Harlan and Hollingsworth in 1865. (Historical Society of Pennsylvania)

PHILADELPHIA, WILMINGTON, AND BALTIMORE RAILROAD, ARCHED-ROOF CAR, 1865

Here is a late example of an arched-roof passenger car, for the clerestory roof was rapidly displacing the arched roof when this car was built by Harlan and Hollingsworth. Some roads and builders were content with the old-style roof, but it was not used by any self-respecting trunk line very long after this date. The body panels, decorative painting, and short-wheelbase trucks were other indications of the anachronistic design. The total dependence on india-rubber springs (even for the bolster mounting) again showed the conservative bent of the P W & B's management. The very narrow body can be explained by restricted clearances. The Wilmington line double-tracked early, and like some other Eastern roads, built the tracks close together on a narrow roadbed. This initial economy not only necessitated major reconstruction in later years but required the use of unusually narrow cars in the interim.

The overall length was 49 feet; the width, 8 feet; the height, 11 feet (over the roof); the truck centers, 32 feet, 3 inches; the truck wheelbase, 4 feet, 3 inches. Seating capacity was 59, including the saloon. (See Figure 1.92.)

LONG ISLAND RAILROAD, CLERESTORY ROOF CAR, 1867

This car, built by Harlan and Hollingsworth for the Long Island, was an early departure from the Monitor style of clerestory. The straight-line sloping ends of the clerestory lacked the grace of the smooth-flowing duckbill ends that came into use at about this time. The small ventilator side openings show that Harlan and Hollingsworth were not going beyond Webster Wagner's

original conception of the raised roof as simply a ventilating apparatus.

The plain, straight lines, square windows, and general lack of architectural ornament were in the simple plan of the old arched-roof car. The raised roof and narrow side sheeting alone classify it as a car of the 1860s.

The length over the couplers was 51 feet; the body length, 44 feet, 6 inches; width, 9 feet, 6 inches; overall height, 13 feet; truck centers, 33 feet, 9 inches; wheelbase, 66 inches; seating, 58. (See Figure 1.93.)

PENNSYLVANIA RAILROAD, CLASS PB COACH, 1867

In 1867 the Pennsylvania's class PA was succeeded by a new common standard design, the PB, which was built until 1878. It was the exact overall size of the PA, but it had a round-end clerestory roof, large arched windows, an oval number panel, and other more decorative features characteristic of the period. The truck wheelbase was extended to 6 feet. The weight was increased by 3,200 pounds, owing to the larger trucks, heavier framing, and new air brake apparatus.

The common six-sill frame was used, with the side sills measuring 5 by 8 inches and the intermediates 4½ by 7 inches. The simple side-panel truss and other framing features of the car are clearly shown in the drawing. The clerestory roof was stiffened by six wrought-iron carlines. The roof was sheathed in thin tin-plated iron sheets with soldered joints. The end hoods were made of heavy sheet iron bolted to the ends of the car. The interior was richly paneled in cherry, ash, and maple. Sliding louvered wooden sashes set inside the windows were used for sunshades. The two gaslight fixtures had two burners each. Gas tanks were attached under the car. A detailed list of costs printed in James

Figure 1.93   A Long Island Railroad coach with an early form of duckbill roof. The car was built in 1867.

Dredge's *The Pennsylvania Railroad* (London, 1879) gives the total cost as $4,423.75. Of this figure the labor charges were about $1,290.

With minor modifications, the PB class continued to serve as the Pennsylvania's standard coach until 1891. The intervening classes followed the size and weight of the PB. The roof profile, sash, side panels, and interior treatments changed with fashion, but the structure was faithfully built to the established pattern of 1867. No other American railroad exhibited such discipline or such steadfast devotion to a common standard.

The PBs were 53 feet, 6 inches over the couplers; 14 feet, 1⅜ inches high overall; and 9 feet, 4 inches wide. The truck centers were 33 feet; truck wheelbase, 6 feet; weight, 42,500 pounds; seating, 54. (See Figures 1.94 to 1.97.)

UNION PACIFIC RAILROAD COACH, 1868

The prospect of through passenger travel to California led Western roads to acquire additional rolling stock as the iron rails pushed westward. The Union Pacific's earliest cars had been rather primitive affairs put together in the Omaha repair shops, which could produce vehicles adequate for the provisional traffic going to the end of the track. Now, however, the Union Pacific began to buy more finished cars from the professional Eastern car shops. Among the suppliers were Harlan and Hollingsworth, whose work is shown in Figure 1.98.

Figure 1.94   The interior of a class PB coach around 1867. (*Pennsylvania Railroad* Guide Book, *1876*)

*Figure 1.95   A Pennsylvania Railroad class PB coach built between 1867 and 1878.
(Chaney Neg. 25524)*

*Figure 1.96   A Pennsylvania Railroad class PB coach, designed and
introduced in 1867.*

At that time the Monitor roof was being replaced by the duck-bill or bullnose clerestory. This car represents the last of the old-style clerestory roofs built in this country. The stained-glass windows between the ventilator openings showed some appreciation of the dual light and ventilation function of the raised roof. The arched window was also more in sympathy with design trends of the late sixties.

The length over the couplers was 51 feet, 2 inches; the body width, 44 feet, 6 inches; width, 9 feet, 6 inches; height, 13 feet, 5 inches; truck centers, 34 feet, 3 inches; wheelbase, 5 feet, 6 inches; seating, 56. (See Figure 1.98.)

VIRGINIA AND TRUCKEE RAILROAD, COACH NUMBER 3, 1869

If it were not for this Nevada short line whose conservative operating policies preserved much of its original rolling stock, we would be lacking several prime examples of Western American railroad equipment. Fortunately, the ancient cars and locomotives of the Virginia and Truckee survived until their unique historic value was recognized in the late 1930s. At that time the main line to Virginia City was abandoned, and most of the older pieces became surplus property. Nearly every car of value was acquired by the movie industry or private collectors. In 1937 Paramount Studios purchased the line's two original coaches, Numbers 3 and 4, together with the cars in Figures 1.106 and 6.20.

Numbers 3 and 4 were the first coaches purchased by the Virginia and Truckee. Moreover, they were built by the pioneer West Coast car-building firm, the Kimball Manufacturing Company of San Francisco. Car building was at best a minor industry in the Western states; relatively few cars were manufactured in that part of the country at any period. It is extraordinary that these obscure relics have been shown to millions of moviegoers, few of whom realize how rare they are. It is hoped that one day the cars will be placed in a museum.

As can be seen from the photographs and drawings, car Number 3 is short and has a plain finish and design. The diagram drawing was based on measurements and photographs supplied by Gerald M. Best of Beverly Hills, California. (See Figures 1.99 to 1.101.)

CHICAGO AND ALTON RAILROAD, COACH NUMBER 39, ABOUT 1870

These drawings, published in the *Railroad Gazette* for June 10, 1871, illustrate one of ten new day coaches built in the Bloomington shops of the Chicago and Alton Railroad. For the first time they show the compression side-panel truss, one of the most important American contributions to railroad car construction.

The drawings also furnish detailed information on an unusually heavy, well-finished day coach of the period. The Chicago

Figure 1.97   General arrangement of class PB coaches in 1867.

Figure 1.98   A Monitor roof coach built by Harlan and Hollingsworth in 1868.

Figure 1.99   This Virginia and Truckee coach was produced by the Kimball Car Works of San Francisco in 1869.

and Alton had a tradition of providing luxurious cars; George M. Pullman refitted his first cars at the road's Bloomington shops. The road extended this policy to the day traveler, and Number 39 is obviously built much on the lines of Pullman's sleepers. It was very large for the times and was outfitted as luxuriously as any first-class car in service, except possibly for the most elaborate drawing-room cars.

In addition to its spacious interior, it was equipped with a patented ventilating system that pulled air in through screened roof hoods. The air was drawn across an open tank of water to cool it and remove dust. The clerestory, built in the old Monitor style but curiously finished with duckbill ends, served largely as an auxiliary light source. Artificial lighting was provided by five center lamps (the gas storage tanks are not shown). As a special safety feature, the car was fitted with a lock-tight antitelescoping coupler and platform invented by the railroad's president, T. B. Blackstone. Two heavy ironclad beams that projected from under the platforms and rode against corresponding timbers from the following car were designed to steady the car's motion and prevent telescoping. The couplers, fitted with adjustable

Figure 1.100   *Virginia and Truckee coach Number 3 by the Kimball Car Works, 1869.* (Gerald M. Best)

screws, could be drawn tight, and this eliminated the slack usually associated with link-and-pin couplers. The entire arrangement, an obvious attempt to avoid Ezra Miller's patents, appears to have been original with the C & A.

The overall length was 58 feet over the safety timbers; the body length, 50 feet; width, 10 feet, 3 inches; truck wheelbase, 8 feet, 3 inches; seating, about 60. (See Figures 1.102 and 1.103.)

BALTIMORE AND OHIO RAILROAD, CLASS A COACH, 1870

Many railroads divided their acquisition of new cars between their own shops and contract builders. The B & O depended heavily on its Mt. Clare shops in Baltimore for much of its new rolling stock. In 1870 the road began to assign serial letters to its passenger equipment. The coaches were designated class A, but unlike the Pennsylvania, the B & O made no strenuous effort to adhere to a single standard design. Small lots of cars were purchased, so that within a few years there were thirteen subdivisions of class A coaches. Cars obtained from outside builders largely followed the designs of the contractors; only the Mt. Clare cars revealed much attempt at standardization. A description of one of these coaches was given in the *National Car Builder*:[127]

The passenger work is in charge of Mr. Jacob S. Schryack. There are two erecting-shops; in one, the floor-framing of a number of passenger and postal cars are in progress, and consist exclusively of Southern pine. The length of these cars is 51 feet 7 inches by 10 feet wide outside of body. The side-sills are 5½ × 7½, the four intermediates 3¾ × 7½, the end sills 6 × 7½, and the truss-planks 2¾ × 12. Wrought-iron plates in the form of an L are placed in the corners and strongly bolted to the end and side sills. Every thing about the framing is designed to secure the greatest strength compatible with the bulk and weight of material. Our attention was specially called to the iron body-transom—as near as we can name it while waiting for the forthcoming "Dictionry of Terms"—which is attached to all the passenger-coaches built at these shops. Instead of the plain piece of timber ordinarily used, a wrought-iron plate 12 inches wide and ¾ thick is bolted to the sills, with a lip turned over at each end to hold more firmly an angular plate underneath of the same width and thickness, the central part of the lower plate resting on the truck bolster, where it is held by the king-bolt, as shown in the cut. This iron frame-work has great strength, and is not liable to shrink, warp, or split, as is often the case with wood.

In the adjoining shop were a number of cars unfinished and showing the character of the side and roof framing, the sills and plates being connected by 44 ⅝ iron rods, and the sides further strengthened by 18 ash spur-braces held by iron bolts.

Figure 1.101   *End detail of Kimball coach Number 3, 1869.* (Gerald M. Best)

In the paint-shop, which is a model structure of its kind, with room for thirty cars at a time, were a number completely finished and ready for the track. They have 6-wheel trucks with strong check-chains attached. Each car has two plain wood burning stoves inclosed in iron screens. The doors are locked, so the fire can not escape in case of overturn. This method of heating has proved very satisfactory. Cobb's elliptic spring seats are used, upholstered with crimson plush and provided with back-iron locks. The roofs have Creamer's exhaust ventilators, the air being admitted by supply ventilators underneath, on a line with the racks. The windows, doors, and panels have semi-circular tops, and there are sliding sashes in the doors for ventilation. The ornamental finishing is very rich and tasteful, and consists of the usual variety of choice woods. The raised panels are French walnut, and the doors are mahogany. The head-linings are made by Howard, of Hartford, and exhibit in variety, tone, and adaptation a progressive improvement in this important feature of car decoration. In one of the cars we specially noticed the admirable effect of a head-lining, so perfectly harmonizing in its subdued tints, graceful lines, and artistic blending and juxtapositon of color, with the upholstery and cabinet-work of the rest of the interior as to form in the combination a genuine work of art, not merely accidental, we are inclined to think, but the result of a right perception of what is required. Each car is supplied with water-coolers, enameled inside, and manufactured by John A. Goewey, of Albany. About three passenger-cars per month are built in this department of the works.

As can be seen from the two illustrations, the class A cars had duckbill roofs and Roman arch windows. Some were in service until the 1920s. (See Figures 1.104 and 1.105.)

PASSENGER CAR,

WITH BLACKSTONE'S IMPROVED PLATFORM AND COUPLER,

Built in the shops of the Chicago & Alton Railroad Company, Bloomington, Ill. R. RENIFF, Master Car Builder.

Figure 1.102   The Chicago and Alton's coach Number 39, built in the Bloomington shops around 1870. (Railroad Gazette, June 10, 1871)

Figure 1.103   Chicago and Alton coach, longitudinal cross section, about 1870. (Railroad Gazette, June 10, 1871)

Figure 1.104   B & O class A coach, constructed about 1870, shown some fifty years later when it was in work service. (Smithsonian Neg. 50742)

Figure 1.105   B & O class A coach. This design was introduced in 1870.

VIRGINIA AND TRUCKEE RAILROAD COACH, 1874

J. G. Brill of Philadelphia is usually thought of exclusively as a streetcar builder, but the firm was active in steam railroad car construction as well. The coach shown here was one of two that Brill constructed for the Nevada short line. They survived because of the motion picture industry's desire for historic props; both cars were purchased by Paramount Studios in 1938.

The peculiar bow-top clerestory is the most distinctive feature of these cars. It may be a unique roof style; no other examples can be located. The square windows appear small for the middle 1870s, and the large-radius end corners are another peculiarity. A partition divides the car into two compartments at the center. One section may have been reserved as either a ladies' or a smokers' compartment. Total seating was for 56 passengers. The

photographs and measurements needed for preparation of the drawing were supplied by Gerald M. Best. (See Figures 1.106 to 1.108.)

ERIE RAILWAY, BROAD-GAUGE COACH, ABOUT 1874

This car, built in the railroad's own Elizabeth, New Jersey, shops, represents the final development of 6-foot-gauge construction in America. A few years after its completion the Erie suspended broad-gauge operations. Because conversion to standard gauge was underway at the time, the body was deliberately made similar to that of an ordinary day coach. Its 10-foot, 4-inch girth was in accordance with the standard-gauge lines that could handle wide cars.

Figure 1.106   Virginia and Truckee coach Number 11, built by J. G. Brill of Philadelphia in 1874. (Gerald M. Best)

Figure 1.107   Interior of the Virginia and Truckee Brill coach, 1874. (R. B. Jackson)

The car was designed by Calvin A. Smith (who also built the 1856 New York and Erie car in Figures 1.83 to 1.85). Smith did not use a truss plank but continued with the bastard Howe. The framing, arranged with economy, produced a remarkably light car for its size: only 34,600 pounds. According to a contemporary description every ounce of material was carefully calculated, and wrought iron was used in place of heavier castings to further reduce dead weight.[128]

Admirable as Smith's design was in avoiding overly heavy construction, he continued to use the long-since-discredited rubber blocks as snubbers in place of coil springs. The equalizing bar was fitted with ¾-inch rubber blocks at each end. The truck bolster was carried by the usual elliptical springs. Miller platforms and couplers were installed.

The car seated 68 in the winter; with the two stoves removed in the summer, the capacity increased to 72. (See Figure 1.109.)

CENTRAL RAILROAD OF NEW JERSEY COACH, ABOUT 1875

In 1927 the Baltimore and Ohio Railroad acquired two aging wooden coaches from the Central Railroad of New Jersey to display in its centennial pageant, the Fair of the Iron Horse.[129] Some light wooden cars of an antique pattern were needed to play a Civil War sequence presented at the fair. Ever since, these cars have been labeled as cars of the early 1860s and are popularly held to be of authentic Civil War vintage. But the arched windows, richly paneled sides, and duckbill roof in Figure 1.112 are more characteristic of the 1870s.

The Wason car works was clearly producing vehicles of a very advanced design when it began offering coaches of this pattern in the mid- and late 1860s (see Figure 1.113). A line engraving of a similar car built for the Central Pacific Railroad appeared in Poor's *Manual of Railroads* for 1869. A photograph of a sister car is reproduced in Gerald M. Best's *Iron Horses to Promontory*. According to records uncovered by Warren B. Crater, vice president of the Railroadians of America, Wason constructed about 200 cars on this design for the C R R N J between 1865 and 1882. Crater contends that two of these cars, originally numbered 255 and 261 but renumbered 1092 and 1095 in 1905, are the present B & O Museum cars 20 and 21. They were built in 1868.

The cars are strikingly similar to the C R R N J coach drawings

Figure 1.108 Diagram of the Virginia and Truckee Brill coach, 1874.

in the 1879 *Car Builders' Dictionary*, shown here as Figures 1.110 and 1.111. The cars now at the B & O Transportation Museum do differ from the drawings in the following particulars. The clerestory stops short of the body ends, and while the number of side windows (sixteen) agrees with that in the published engravings, the length of the bodies differs by 30 inches. The differences can be explained as design alterations to the same plan.

The drawings show small ventilator openings with screen covers placed between every second window through the sides of the car at the letter board. These permitted a flow of air when it was necessary to close the windows. A leather apron at each end of the car, on a level with the platform, helped to prevent dust from blowing up between the cars. Long rods, fitted with coil springs that reached back under the platform, held the aprons together. The platforms were of the old drop pattern that invited telescoping. The platform timbers were secured to the main frame only by stirrup bolts; no tie rods on the Miller plan are in evidence. The coupler attachment, bolted to the platform timbers with no other fastening to the main body frame, is also noticeably weak.

The overall length of the cars was 51 feet; the width, 9 feet, 10 inches; height, 13 feet, 8 inches; truck wheelbase, 6 feet; seating, 52; estimated weight, 22½ tons. (See Figures 1.110 to 1.113.)

PENNSYLVANIA RAILROAD COACH, CLASS PD, 1878

Not long after the U.S. Centennial, the Pennsylvania Railroad decided that it was time to prepare a new standard coach design. The design had to be suitable for service on the Keystone system's vast property, which extended from the United Railroads of New Jersey to the far Middle Western reaches of the Vandalia. The Pennsylvania operated over six hundred coaches on some 1,700 miles of line. The old class PB had been the company standard since 1867, but it was now considered obsolete, at least in appearance. The overall size and plan of construction was retained. The new class PD, adopted in 1878, had fifteen windows on a side like its predecessor, but it seated two less passengers than the PB. Stylistically the car was modified to reflect the purity of design favored by the great English art critic John Ruskin. The arched windows gave way to a square sash with a subtle rounded upper corner. The boldly curved sheet-metal canopy of the PB was replaced with a more straightforward wooden-frame platform roof.

The purity of design was best exhibited by the interior, which the August 19, 1881, *Railroad Gazette* described as "solid, square and honest," and without the jugglery, wood puzzles, and veneers of the old-fashioned cars. Quartered oak or ash panels

engraved with the simple patterns advocated by the critic Charles Eastlake replaced the baroque mixture of ash, cherry, maple, and pine used in the PBs. The seats had wooden rather than cast-iron frames and were covered in golden brown or cherry red material. The head linings were muslin painted in deep red, grayish-brown, or light-drab tones. The windows were glazed, with polished plate glass imported from France. The hardware was polished bronze, again decorated in the simple, straight-line Eastlake style. Two stoves with hot-air registers running under the seats heated the car, and gas lighting was provided. The exterior was painted Tuscan red and striped in gold and black. This finish required sixteen coats of filler, color, and varnish. The Altoona shops could produce twelve coaches a month of this design; each car took forty days to assemble.

The length over the couplers was 53 feet, 6 inches; the width, 9 feet, 4¼ inches; height, 14 feet, 1⅜ inches; trucks, 33 feet on the center; wheelbase, 7 feet; seating, 52; weight, 21 tons. (See Figures 1.114 and 1.115.)

CINCINNATI SOUTHERN RAILWAY, COACH NUMBER 40, 1879

The diagram drawing shows one of the original coaches of the Cincinnati Southern Railway. It was constructed by Barney and Smith of Dayton, Ohio; unfortunately there are few hard engineering data on this important builder. The drawing is of interest as one of the few surviving examples of a Dayton-built car, even though the sketch exhibits no unusual mechanical features. (See Figure 1.116.)

BALTIMORE AND OHIO RAILROAD, CLASS A6 COACH, 1881

Five series and eleven years removed from the original class A coach, the A6's general dimensions and construction agreed closely with those of its predecessors. But several stylistic changes can be seen: the Roman arch windows gave way to square sashes; the duckbill roof was replaced by a bullnose shape; the window posts were made narrower and the exterior finish more subdued and unified. The first A6 was built in November 1881. Over thirty cars were made on this plan before it became obsolete in 1884. The drawings here are from the *National Car Builder* of June 1883.

The supplementary plank truss just below the belt rail was used by some builders who wished to reinforce the regular side-panel trussing. The car also had iron body bolsters and Janney

Plate XCV.

Fig. 4 Transverse Section at the Bolster.

Fig. 8. Plan.

Fig. 6 Horizontal Section above Seats.

Fig. 1. Side Elevation.

Fig. 7. Plan of floor.

Fig. 5. Bottom View of Floor framing.

Fig. 2. Side View of Frame.

Fig. 3. End View.

G. WEISSENBORN'S ENGINEERING OFFICE,
269. Pearl St. New York.

Scale ¼ inch to the foot.

*Figure 1.109 An Erie Railway coach, built at the Jersey City Shops in about 1874.
(Gustavus Weissenborn, American Locomotive Engineering and Railway Mechanism,
New York, 1871)*

Figure 1.110   A Central Railroad of New Jersey standard passenger car, constructed about 1875. (Car Builders' Dictionary, 1879)

PASSENGER CAR-BODY, CENTRAL RAILROAD OF
NEW JERSEY.

End View.

With INTERMEDIATE FLOOR *but no* DEAFENING-CEILING.

Transverse Section.

Figure 1.111   End and transverse section drawing of a Central Railroad of New Jersey passenger car of about 1875. (Car Builders' Dictionary, 1879)

Figure 1.112   A Central Railroad of New Jersey passenger car, relettered as Baltimore and Ohio Number 21 for the 1927 Centennial Pageant of the B & O.

WASON

MANUFACTURING COMPANY,

ESTABLISHED 1845.

Builders of all descriptions of

RAILWAY CARS,

And Manufacturers of

Car-Wheels and Railway Castings.

Particular attention paid to the shipment of Cars, and Materials for Cars, to all parts of the world.

GEORGE C. FISK, President.
H. S. HYDE, Treasurer.
G. T. WASON, Secretary.

SPRINGFIELD, MASS.

couplers. The sills were yellow pine; the other framing pieces were yellow pine or ash.

The interior was paneled in mahogany. The burl walnut panels and gilt highlights of the early cars were omitted, and the painted oilcloth ceilings were replaced by oak-veneer panels. The only extravagance was in the nickel-plated and gilt luggage racks. The restrained decorative treatment was meant to avoid needless display and suggest "good, honest intelligent work," as the *National Car Builder* said.

The car seated 58 passengers and weighed 51,150 pounds. (See Figures 1.117 and 1.118.)

CHICAGO & ATLANTIC RAILWAY, COACH NUMBER 68, 1883

This car formed part of Jackson and Sharp's exhibit at the 1883 Railway Appliance Exposition in Chicago. Structurally it followed the usual pattern of American day coaches except that its finish was more dazzling, as befits a trade-fair display piece. The exterior was painted yellow, with a lake-blue border and gilt decoration. The hand railings were silver-plated. The mahogany interior was carved in the shallow Eastlake style and depicted leaves, flowers, and other "graceful and artistic specimens of hand carving," according to a description in *The Engineer*. One panel showed a spider web with a fly trapped in its fine-chiseled lines.

The best auxiliaries were installed: Foster gaslights, a Searles hot-water heating system, and Sax and Kear steel-tired wheels.

Figure 1.113   Advertisement from Poor's Manual of Railroads, *1869–1870. Note the similarity of this car with the preceding illustrations.*

Figure 1.114 A Pennsylvania Railroad passenger coach class PD, 1879. (Smithsonian Neg. 71261)

Figure 1.115 Assembly drawing of a Pennsylvania passenger coach class PD, 1880c. (Railroad Gazette, *August 5, 1881*)

The car seated 58 passengers and cost about $8,000. The drawing is from *The Engineer* of August 29, 1884 (p. 164). (See Figures 1.119 and 1.120.)

CHICAGO AND NORTH WESTERN RAILWAY, TWELVE-WHEEL COACH, 1882

The drawings and description of this heavy twelve-wheel coach are from the January 1883 *National Car Builder*. The framing was yellow pine except for the ash door and end posts, car-

lines, and end plates. The interior was mahogany, and the ceiling was bird's-eye maple veneer. The body bolsters were iron. Six oil lamps provided the lighting. Heat from two coal stoves was distributed by hot-air registers running the full length of the car. Miller couplers and platforms were used, and the 33-inch wheels were cast iron. One unusual feature was the absence of end windows.

The body was lemon yellow, the letter board was in ultramarine blue, and the lettering and striping were gold, picked out in red. (See Figures 1.121 and 1.122.)

Open Coach.
Non Vestibuled.

No. 40.

16.

wide

wide

9'-3"

36'-10".

42'-10" wheel base

6'-0"

47'-0"

48'-4"

10'-0"

9'-2"

JOURNALS 4¼" x 8"

Seating Capacity 60 Persons.
Weight of Car and Trucks 47,300 lbs.
" Trucks alone 15,660 "

Figure 1.116   A Cincinnati Southern coach built by Barney and Smith of Dayton, Ohio, in 1879. (Southern Railway)

Figure 1.117 A Baltimore and Ohio coach class A-6. The design dates from 1881. (National Car Builder, *June 1881*)

ERIE RAILWAY COACH, 1884

The Erie had suspended broad-gauge operations by the time this car entered service, but the construction of new passenger equipment continued as before at the road's Jersey City shops. The drawings from the *National Car Builder* of September 1884 show the road's loyalty to a proved design. The side-panel truss system, laid out some thirty years before by Calvin A. Smith, was still followed. Owing to the strength and light weight of Smith's plan, cars of this pattern were reported to stand well, with their sides straight and true, after years of service.

The outside panel was covered by narrow 2-inch sheeting rather than the batten board typical of the period. The square window sash and modest belt rail created an austere outside appearance. The interior was equally plain, according to the *National Car Builder*:

Plainness and durability have not been lost sight of in a desire for showy elaboration, and the result is a pleasant, cheerful and attractive interior. Birch is used in the ceiling and white ash and mahogany in the sides, the ash being obtained along the line of the road of an excellent quality, light color, and free from blemishes. Upon these ash panels, which are absolutely free from streaks and hearts, is placed a belt of mahogany. Over the windows the mahogany is ornamented with a sunken rosette, and between the windows with a rosette and engraved lines, as shown. The birch ceiling is put in with mahogany moldings stained to represent old wood, thus affording a better contrast with the ash than could be obtained from the new wood. The window moldings are ¼ round of about an inch radius, presenting no sharp corners. The seat-arms are of ash, and are strong, shapely and

handsome. The seats slide upon the frames so as to be lower at the backs than at the front edges. The roof curves have been carefully studied with a view to harmony between those of the lower and upper roof. The two cross sectional elevations show the points from which the curves of the carlines are struck, and also the radii. The effect obtained is that of an elliptical head, although the curves are struck from centres. The raised roof, in consequence of the thickness of its sides being only 1½ in., is much lighter than usual, without any sacrifice of strength. The ash panels, only about ¼ in. thick, are glued directly upon the rails and posts.

Figure 1.118   A Baltimore and Ohio Railroad passenger car truck, 1881. (National Car Builder, *June 1881*)

Figure 1.119   A Chicago and Atlantic coach built in 1883 for display at the Chicago Exposition of Railway Appliances. (Smithsonian Neg. 71260)

Figure 1.120 A Chicago and Atlantic passenger car, 1883. (Engineer, August 29, 1884)

Figure 1.121 A Chicago and North Western Railway passenger car of about 1882.
(National Car Builder, *January 1883*)

Pintsch gas was used, with oil side lamps for emergency lighting. The hot-water heating was of an undetermined make. The car measured 58 feet, 4 inches long over the platforms. (See Figures 1.123 to 1.125.)

PHILADELPHIA AND READING RAILROAD, ROUND-END COACH, 1884

It is not surprising that the Reading, long famous for its distinctive style of locomotives, should develop a singular day coach. These cars differed from conventional designs in two particulars: the round-corner ends and A-frame roof, which were used by few other American roads. They were first built around 1882 and continued to be produced until the end of the decade. The design attracted much comment, but no other roads seem to have copied it.

The round corners, with a radius of 2 feet, required a complicated framing built up of several heavy timbers joined by ship splices, glue, screws, and bolts. While the design had no practical justification, the road contended that the round corners would divert the truant car in a wreck. In one accident, another car glanced off the corner and only scraped the panel. It was said that the same blow would have caved in the corner of a regular

car.[130] The rounded sash and glass of the corner windows was expensive, though architecturally interesting.

The A-frame roof was stronger than the conventional design but not esthetically pleasing, as can be seen from Figure 1.127. The additional moldings and the basket-like gas fixtures created a busy, cluttered ceiling. The design was seldom used by other roads.

Extra seating was provided by the corner seats and the underbody mounting of the stove. The body width was held within the undersized clearances of the Reading. The car weighed 47,360 pounds. (See Figures 1.126 to 1.129.)

CLEVELAND, COLUMBUS, CINCINNATI, AND INDIANAPOLIS RAILROAD COACH, 1884

This car had no obvious peculiarities of design such as those in the Reading car of Figure 1.126. Outwardly it was a conventional design of the period, but it did employ an unusual method of body support. In place of the customary lower panel truss was a 3- by 12-inch plank that was fastened inside the car to the posts and rested on top of the side sills. It was analogous to a plate girder although, of course, far less stiff than a metal plate. Generally such planks were used as supplementary supports for the

Section Through C D.

Section and End View.

Figure 1.122 End elevation and framing details, Chicago and North Western Passenger car. (N.C.B., January 1883)

Figure 1.123   Erie Railway passenger coach framing details, 1884. (N.C.B., September 1884)

truss, as in the B & O class A6 coach of Figure 1.117. In the B & O coach, however, the plank was placed at the top of the panel just under the window rail.

The interior of the car was paneled in oak, cherry, and mahogany. The outside was praised as being in good taste "and notably devoid of the excessive fire-works style of decoration which becomes an eye-sore as soon as cinders and smoke have destroyed the original luster of the varnish and gilding."[131] The exterior was painted Tuscan red, with gilt striping and lettering. The car was fitted with Allen paper wheels and Miller couplers and platforms. (See Figures 1.130 to 1.132.)

BALTIMORE AND OHIO RAILROAD, CLASS A9, 1890

The Royal Blue was a plush day train intended to attract first-class travelers between New York and Washington. From 1890 to 1893 it was outfitted with luxurious new Pullman-built cars, among them Numbers 1051 to 1063. These were some of the largest and handsomest day cars operating anywhere in the United States. The majestic double windows, scroll platform railings, and enclosed vestibules reflected the latest thinking in first-class coach design.

The separate smoking room and large toilets at each end were

Figure 1.124   Erie Railway passenger car end and interior details, 1884. (N.C.B., September 1884)

Sections through A B and C D.

Figure 1.125   Interior details of an Erie Railway passenger car of about 1884. (N.C.B., September 1884)

Figure 1.126   A Philadelphia and Reading round-corner coach of about 1884. (N.C.B., April 1884)

Figure 1.127   Interior of the Philadelphia and Reading round-corner coach, about 1884. (N.C.B., April 1884)

an extravagant use of space permissible only in first-class cars. The interiors were paneled in light mahogany; the upholstery was old-gold plush. A dazzling dark-blue exterior finish was highlighted by silver striping, with the Maryland coat of arms in full color. Vestibule ends, steam heat, and gas lighting completed the first-class furnishings of these cars.

A car approximately equal in size and vintage is displayed today in the B & O Transportation Museum, Baltimore. It is painted and lettered in the style of the Royal Blue cars, but here the similarities end. The car does not show the quality of the A9 class; it is finished as an ordinary day coach of the period. It has small rectangular windows in place of the large double openings of the original series. A photograph of the museum car taken in 1927 shows its number as 445. No further identification can be

Cross Section Showing End, of Saloon.

Fig. 1.—Corner Framing.

Side View of Saloon.

Figure 1.128 Philadelphia and Reading round-corner coach details, about 1884.
(N.C.B., April 1884)

made; it does not agree with any of the B & O standard cars and may, like the "Civil War" cars in the same collection (Figures 1.110 to 1.113) have been created from available second-hand equipment to play a specific role in the 1927 Centennial celebration. Its classification as a Royal Blue car was established over the years by repeated exhibitions, until now its true history has been lost. (See Figures 1.133 and 1.134.)

BOSTON AND ALBANY RAILROAD COACH, 1890

In a departure from the usual American car of the period, this coach's severe, plain lines, single windows, thin-beaded sheathing, and arched roof give it a deceptively modern appearance. These features and several other unusual characteristics were

developed by the road's master car builder, Fitch D. Adams. Compared with locomotive builders, most car builders remain shadowy figures. A few received acclaim in their lifetime, but almost none achieved the legendary status accorded to scores of locomotive men. Adams, an influential and outspoken leader in his field, deserves more than the token notice given him here.

Adams assumed direction of the B & A shops, just west of Boston at Allston, Massachusetts, in 1870. He found that the cars being built were both poorly framed and grossly overweight. He immediately adopted the six-sill floor frame and in the following year began using an arched truss plank. Adams was alone in his choice of this form of side-panel trussing, though it clearly demonstrated its superiority over the regular system. According to Adams, the arched truss weighed only one-eighth as much as a

Figure 1.129   Philadelphia and Reading round-corner coach diagram drawing, about 1884. (Drawing by John H. White, Jr.)

truss plank. Old cars maintained their stiffness after twenty years of service, and the arched truss continued in use on the Boston line under Adams and his successor at least until 1899. It may well have survived until the steel car era.[132]

In the competition for lightweight construction, Adams won top honors. The trade press regularly featured his successes in this area.[133] In 1879 he fabricated a 68-passenger coach that weighed only 18 tons. Adams built drawing-room cars fully equal to Wagner's, but saved 10 tons by careful construction. The car pictured here was produced late in Adams's career and in effect summarizes his final ideas for passenger car construction. William Voss thought it important enough to include in his book, *Railway Car Construction*. Its description first appeared in the March 1891 *National Car Builder*.

The floor frame was a conventional six-sill variety made of yellow pine, except for the needle beams. The bolsters were iron. The spaces between the sills were packed with mineral wool for insulation.

The arched truss plank is clearly illustrated in the drawing. Like the posts and many other parts of the body, the trusses were made of ash. The car had four $7/8$- by 5-inch truss planks, of which the outside plank was set into the posts. The double planks were apparently considered necessary for a body measuring 58 feet, 6 inches.

The arched roof is the most notable single feature, though the turtleback roof was not original with Adams. Mann sleeping cars with such roofs began operating on the Boston and Albany in the early 1880s. Adams adopted the plan in 1886, although it found little favor elsewhere in the country. The turtleback shape of the roof was designed to avoid the defects of the old arched roof by providing a high ceiling with ample headroom. The roof rose 2 feet over the sides of the car. The wooden carlines were three pieces of $1/2$- by 1-inch ash glued together and steam-bent to the profile of the roof. Nine iron carlines were used for added strength. The absence of a clerestory made it necessary to install thirty globe ventilators.

The exterior sheathing was white wood cut into narrow strips and faced with a half-round front, creating an effect which very much resembled corrugated iron. This decorative scheme was suggested by H. H. Richardson, the architect for some suburban cars built by Adams in 1881. The window sills were capped with a thin, neatly sculptured iron casting—another example of Adams's willingness to develop new solutions for every detail of car construction.

The car was equipped with Miller couplers, Westinghouse brakes, steam heating, and 42-inch steel-tired wheels. The interior was finished in a plain Eastlake style. The head lining was Mexican mahogany. The seats were covered in golden-brown plush. The overall length was 65 feet, 6 inches; width, 10 feet, 3 inches; height over the ventilators, 13 feet, 10½ inches. (See Figure 1.135.)

CHICAGO, PEORIA, AND ST. LOUIS COACH, 1891

This first-class coach or chair car was built for through service between Chicago and St. Louis on the Chicago, Peoria, and St. Louis (Jacksonville Southeastern Line) and the Santa Fe Railroads. It and its companion cars were constructed by the St. Charles Car Company of Missouri. The plant did not begin passenger car production until 1886, but in later years it became the main passenger car works of the American Car and Foundry Company.

The general arrangement is clearly shown in the drawing.[134] The interior was paneled in mahogany with a quartered-oak headlining. The seats (in a Scarritt-Forney pattern) were covered with plush, while the smoking-room sofas were upholstered in leather. The washstands were of Tennessee marble. Heating was provided by a Baker heater. Oil lamps were used, but wiring was installed in case electric lighting should be desired at some future date. Miller couplers and Boyden air brakes completed the special equipment. The smoking room, water closets, and utility

Half Side Elevation.

Half Plan of Floor and Section Through Side Frame.

Figure 1.130 A Cleveland, Columbus, Cincinnati, and Indianapolis coach of about 1884. (N.C.B., July 1884)

closets restricted seating to only 42 passengers. (See Figure 1.136.)

PENNSYLVANIA RAILROAD, CLASS PH, 1892

The Pennsylvania broke away from its traditional use of the duckbill roof in 1892 to adopt a more modern clerestory profile. Vestibules and advanced appliances were used as well, but here the road's design progress seemed to halt. The body was kept short; batten-board sides and small windows reflected the conservative outlook of the Pennsylvania management.

The interior decoration was somewhat more flamboyant than the usual P R R treatment, perhaps partly because of the approaching Columbian Exposition and the anticipated rivalry for that traffic among Eastern lines. The road's sensible oak paneling was retained, but the ceiling was covered in light-blue silk tapestry. The seats were upholstered in red plush, and the floor was carpeted. The extra-wide clerestory gave the interior an agreeable feeling of space.[135] Frost gas lamps and steam heat provided an extra luxury for the 64 passengers.

The car appears to be overtrucked, yet its weight (81,100 pounds) and the desire for a smooth ride and limited axle load-

Longitudinal Section Showing Frame.   Inside Finish.

Half Plan of Floor and Floor Frame.

*Figure 1.131   A Cleveland, Columbus, Cincinnati, and Indianapolis coach of about
1884, detail.* (N.C.B., *July 1884*)

ings account for the decision to install twelve wheels on so short a car. (See Figures 1.137 and 1.138.)

ILLINOIS CENTRAL SUBURBAN CAR, 1893

Many railroads depend on old cars for suburban service, but some lines with extensive suburban operations, such as the Illinois Central's in the Chicago area, built special lightweight cars for this traffic. The general design was very similar to that of a New York Elevated car. Space and luxury were held to a minimum, because rides were short and fares low. Loading and un-

loading were facilitated by the wide aisles provided by longitudinal seats at either end.

Ironically, on these cars designed for utility more than appearance, featuring rattan seats and offering no toilets, the builders felt it necessary to install a solid-mahogany interior with carved panels over the end doors. A Baker stove furnished heat; Miller couplers and Westinghouse brakes rounded out the equipment. The form of lighting was not given. The car seated 48. The drawings and data are from the May 1893 *Railroad Car Journal.* See Figure 1.43 for an earlier Illinois Central suburban car of similiar design. (See Figure 1.139.)

Figure 1.133   Baltimore and Ohio Railroad passenger car class A-9; a narrow-vestibule car built by Pullman in 1890. (Traced by John H. White, Jr.)

Figure 1.132   End elevation of a Cleveland, Columbus, Cincinnati, and Indianapolis passenger coach of about 1884. (N.C.B., July 1884)

NEW YORK CENTRAL RAILROAD, 80-FOOT COACHES, 1892–1893

In the competition for passenger traffic the New York Central attempted to outdo its rivals by the creation of supercoaches. Surely the largest all-wood coaches ever built, they were inspired more by a desire for a spectacular exhibit at the Columbian Exposition than by any practical operating considerations. The West Albany car shops built two 80-foot cars for the Empire State Express. These cars were shown at the 1893 Chicago Fair

together with the celebrated 999 Locomotive. To viewers familiar with car construction, they were a sensation, as was the 112-mile-per-hour run made several months earlier with the same equipment.

The framing followed the iron-plate reinforced plan developed several years earlier by Thomas A. Bissell for Wagner sleeping cars. The floor framing was particularly heavy, as can be seen in the drawings.[136] Massive as the wood structure seems, it was apparently not sufficient to maintain the body's stiffness. Supplementary truss rods were required at either end of the body (Figure 1.141).

The exterior was painted Tuscan red, with gold and black lettering. The trucks were finished in dark brown. The interior was paneled in mahogany, and carving was held to a minimum. The seats were covered in copper-red plush. The head lining was painted white and gold. Major equipment included Gould vestibules, platforms, and couplers; Pintsch gaslights; and steam heating. Seating was provided for 84; the approximate cost was given as $8,720. The body weighed 62,200 pounds, the trucks 33,200 pounds, and the total weight was 95,400 pounds. (See Figures 1.140 and 1.141.)

NORFOLK AND WESTERN RAILWAY COACH, 1894

This drawing, one of the most complete graphic records ever prepared for an American passenger car, shows a fully equipped coach with all appliances in detail.[137] The interior was finished in oak paneling, the hardware was nickel-plated, and the aisles were covered with Brussels carpeting. The exterior was painted Tuscan red, with gold trimming and letters. The car closely followed the standard Pennsylvania Railroad design, which is not remarkable in view of the Keystone system's financial involvement in the N & W.

Frost gasoline lamps, Spear hot-air heaters, and Janney-Buhoup couplers were among the important auxiliaries used. The complete car weighed 69,350 pounds. (See Figures 1.142 to 1.144.)

NEW HAVEN RAILROAD, COACH NUMBER 1203, 1901

Serving the populous area between New York and Boston, the New Haven has always been one of the nation's largest passenger carriers. Its present service reputation hardly reflects the time when the road took pains to maintain a model fleet of cars. Representative of that more fastidious past was coach 1203, one

Figure 1.134 Baltimore and Ohio Railroad passenger car class A-9, Pullman, 1890.
(Pullman Neg. 1506)

Figure 1.135   Boston and Albany arched-roof car of about 1890. (N.C.B., March 1891)

of twenty built in 1901 by Osgood Bradley. Like other New Haven cars of the period, it was sheathed in copper. This unusual form of exterior finish was favored by the road's master car builder, W. P. Appleyard (see Chapter 5). The car weighed 71,750 pounds.[138] (See Figures 1.145 and 1.146.)

NICKEL PLATE, COACH NUMBER 43, 1907

The New York, Chicago, and St. Louis Railroad, better known as the Nickel Plate, acquired six wooden coaches in 1907 from

the American Car and Foundry Company's Jeffersonville, Indiana shops (the former Ohio Falls Car Company). The next coaches, purchased in 1910, were the last wooden cars acquired by the Nickel Plate. The road turned to steel with its subsequent order.

The Number 43's service career is typical of many late wooden cars. Bought almost at the end of the wooden era, it was a mistaken investment that the road tried to salvage by periodic rebuilding. Electric lighting and steam heating were added, and between 1923 and 1925 steel underframes were applied. In later

Figure 1.136 A Jacksonville Southeastern coach built by the St. Charles Car Company of St. Charles, Missouri, about 1891. (American Railroad Journal, October 1891)

Figure 1.137 The Pennsylvania Railroad coach Number 950, class PH, designed and introduced in 1892.

Figure 1.138 Pennsylvania Railroad passenger coach class PH, 1892.

Figure 1.139   An Illinois Central Railroad suburban car built about 1893. (Railroad Car Journal, *May 1893*)

Figure 1.140   The New York Central and Hudson River Railroad coach Number 999. This 80-foot car was one of the longest wooden passenger cars constructed. (New York Central Neg. 9493)

Half Plan of Frame of Baggage, Buffet and Smoking Car.

Half Plan of Frame of Passenger Car.

Half Side Elevation.

Half Side Elevation of Body Frame.

PASSENGER CAR.

Half End Elevation of Frame.

Half Section through Center of Frame.

End Elevation.

Plan of Truck.

Half Side Elevation of Body Frame (Baggage Compartment.)

BAGGAGE, BUFFET AND SMOKING CAR.

Figure 1.141 A New York Central and Hudson River Railroad passenger car built in 1893. (Engineering News, December 14, 1893)

Figure 1.142 A Norfolk and Western Railway narrow-vestibule car built about 1894.
(Railroad Gazette, August 16, 1894)

Figure 1.143   A Norfolk and Western Railway passenger car truck of about 1894. (Railway Master Mechanics' Magazine, *March 1894*)

Figure 1.144   This Norfolk and Western Railway passenger car interior of 1905 is very similar to the cars illustrated in Figures 1.142 and 1.143. (Pullman Neg. 7780)

Side Elevation.

Half Plan of Framing.   Half Side Elevation of Framing.

Figure 1.145   The New Haven coach Number 1203, built by Osgood Bradley in 1901. (Car Builders' Dictionary, 1906)

Cross-Section of
First-Class Coach.
N. Y., N. H. & H.

years the 43 was downgraded to work service. Many of its companion cars remained in the active passenger fleet until after World War II.

Seating was provided for 72 passengers. The weight was 46.2 tons.[139] (See Figures 1.147 and 1.148.)

NEW YORK CENTRAL RAILROAD, 1700 SERIES COACH, 1909–1911

Time was fast running out for the wooden passenger car when the American Car and Foundry Company produced the last coach in this order sometime in 1911. Steel cars had been commercially manufactured for several years. To maintain its position with rival lines, the New York Central began to take delivery of steel cars during the following year. The arch-windowed 1799 was the last new wooden car acquired by the Central, and was among the last purchased by a major domestic railroad.

The drawing here is from a diagram book published by the New York Central Railroad in 1913. By this date the car had been fitted with a steel underframe and electric lighting. Seating was for 84 passengers, and the total weight was 130,300 pounds. (See Figure 1.149.)

Figure 1.146   Cross-section drawing of the New Haven's Number 1203. (Car Builders' Dictionary, 1906)

Figure 1.147   The New York, Chicago, and St. Louis coach Number 43, built by the Ohio Falls Car Company in 1907. (John Keller)

Figure 1.148   New York, Chicago, and St. Louis coach diagram drawing. (John Keller)

COACHES

Cars 1700–1749—Lot No. 745   **Cars 1750–1799—Lot No. 775**

Figure 1.149   New York Central coach built in 1911 by the American Car and Foundry Company.

CHAPTER TWO

The Day Coach

ERA OF
THE METAL CAR

THE HISTORY OF METALLIC PASSENGER CARS can be divided into four major periods: experimentation, introduction, full production, and streamlining. During the extended period of experimentation (1845–1902), individual inventors worked with little support or interest from the railroads or commercial car builders. In the end these inventors failed to persuade the industry to accept iron or steel cars. In the second phase, a short, intensive period of introduction (1902–1910), the industry itself came to support the movement. With the establishment of an acceptable design, which was largely an imitation of the general form of existing wooden cars, the period of full production began. It was to continue through the so-called standard or heavyweight era of American railroading until the early 1930s, when more attention was given to lightweight construction. This last phase, the streamlined era, took hold in the mid-1930s and continues to the present time. After 1955 the decline of railroad passenger traffic, except for suburban or other specialized services, reduced the market for new equipment and thus retarded further developments.

Few improvements have been longer heralded or slower in coming than the metallic passenger car. Steel cars gained acceptance at a measured pace; half a century separated the first experiments and the initial production models. Every major railway accident evoked a flurry of newspaper and magazine articles advocating collision- and fire-resistant cars. The defects of the wooden car were laboriously repeated: it was weak, likely to splinter, and highly combustible. Iron construction would end the hazards of railway travel—from the mid-1840s on, this was a standard editorial theme.

Naturally the publicity stimulated experimentation. Between 1850 and 1900 over thirty patents were granted for iron or steel passenger car construction. Some twenty sample cars were actually built, and countless others were projected. But nearly all this work was done by amateurs, whose efforts drew indifference or ridicule from professional railroad circles. These inexperienced and imprudent inventors generally produced bizarre designs that did not appeal to the practical mechanics who governed the nation's railways. And the complexity and impracticality of most designs was only one obstacle to the development of the iron car, for there were also the obvious economic disadvantages already mentioned in Chapter 1. Wooden cars were cheap, light, and for the most part, adequate for the service of that day. Even the best iron cars proved far heavier and more costly than ordinary wooden coaches.

Before American railroads had ventured more than a hundred miles from the Atlantic coast, iron passenger cars were being built in Great Britain. They were few in number, however, and most were used for the poorest class of travel. The Great Western Railway, one of the largest investors in iron, built ten cars with metal bodies in 1845 and acquired another ten three years later.[1] In the following decade, iron-frame passenger cars were introduced in both England and Germany and within a few years became relatively common in Europe.[2]

By the mid-1850s iron-bodied coal cars in considerable number were running on the Baltimore and Ohio and the Reading railroads. Many had been in use for ten years. Although durable and workaday, they appear to have attracted little attention beyond their home districts. America was far slower than Europe to accept iron for passenger conveyances.

The earliest proposal in this country for a passenger car of iron was offered in 1846 by H. L. B. Lewis of New York City.[3] Lewis envisioned an iron car body suspended between a pair of 6-foot-diameter wheels, with a smaller pair of wheels to guide the truck. A handbill distributed by the inventor described the projected car as 45 feet long and 8 feet square, with a capacity for 80 passengers. The weight less the truck was grossly underestimated at only 4 tons. The idea did not merit a patent, much less a prototype. In fact, its absurdity set the movement back a good five years.

A more practical design came from the iron-making center of Troy, New York, where in 1851 Thomas E. Warren devised a sheet-iron car following a plate girder truss.[4] The side sheets were stiffened by pressed sheet-iron columns set at regular intervals. Iron rods tied the structure together. It is possible that Warren's design was inspired by architectural ironwork fabricated in the Troy mills. Certainly the classical and Venetian mountings gave the design a strong architectural flavor. A patent (No. 10142) was granted on October 18, 1853, but as far as is known, no test cars were built.

The most enduring champion of the metallic passenger car was Bernard Joachim La Mothe, a physician who settled in New York City sometime before 1853.[5] This persistent Frenchman argued for the adoption of safe passenger cars for nearly forty years. In that time he obtained nine patents and built four passenger cars as well as countless freight cars of iron. At first La Mothe concerned himself with designs for a fireproof building. There was considerable interest at the time in such structures—particularly those of cast iron. La Mothe envisioned a latticework or basket-weave framing of iron bands.

In 1853 La Mothe adapted his scheme to railway cars and displayed a model at the New York Crystal Palace Exhibit.[6] A patent (No. 10721) issued on April 4, 1854, outlined his plan, which was an early attempt at unitized fabrication. Light iron bands were interwoven to form a strong basket-like cage. Each crossing was riveted. The cage was fastened to a foundation ring, also made of band iron, that encircled the car and formed the end and side sills. Sheet-iron panels covered the roof and sides, with windows set in at the appropriate openings. Since the center sills, floor, seats, and interior paneling were of wood, La Mothe could not claim that the car was fireproof.

La Mothe engaged a civil engineer, Alfred Sears, to put the design in a form more acceptable to the industry and to help with the promotion, and they prepared a series of pamphlets and advertisements.[7] The trade press was generous, praising the design and uncritically accepting La Mothe's exaggerated claims. It became known as the Life Preserving Car that would safeguard

LA MOTHE'S PATENT IRON RAILROAD CAR.

WE are now prepared to furnish this Car to railroad companies at short notice and reasonable rates.

Notwithstanding its extraordinary advantages, the prices will be arranged wholly with reference to the cost of construction—without regard to patent rights.

We are now building passenger and freight cars for several companies; and it is desirable that parties ordering give early notice of their wants.

The striking features of this principle are :—simplicity—cheapness—durability—superior safety in cases of accident—facility of repairing when damaged—and less weight compared with the wooden cars of the same capacity ; these cars for 60 passengers are more than two tons lighter than the ordinary cars, while the strength is immeasurably greater.

We guarantee these points in the acceptance of orders.

The advantages may be tested by personal observation in this city. Detailed descriptions of the cars will be forwarded to parties wishing them.

ALFRED SEARS,

Civil Engineer and Architect,

Agent.

OFFICE—9 SPRUCE ST., NEW YORK.

Figure 2.1   La Mothe's patented iron car, as advertised in the American Railroad Journal *in 1855.*

passengers from any calamity. An engraving prepared for the *American Railroad Journal* showed a coach of the most ordinary appearance, indistinguishable from a wooden car of that day except for the tiny rivet heads (see Figure 2.1).

La Mothe's leaflets offered a strong, no-nonsense economic argument. For example, *How to Make Railroads Pay a Good Dividend, and at the Same Time Give Security to Life in Railroad Travel* (published in 1856) declared that iron cars would drastically reduce maintenance, renewal, and damage costs. According to the leaflet, iron cars would never wear out, burn, or permit serious injuries to passengers. Moreover, they could be built so that they were lighter than wooden cars, thereby reducing operating costs and saving wear on roadbeds and bridges. These sensible-sounding claims, together with the sensible-looking design, were buttressed by a number of models. The 1854 patent specification mentions a 4-pound model able to support 1,000 pounds. The leaflet *How to Make Railroads Pay* speaks of a model 32 inches long, 7½ inches wide, and 6½ inches high that could support 1,400 pounds. A third model 42 inches long carried a load 196 times its own weight, and although the side bulged somewhat, it sprang back to its original shape when the weight was removed.

Still no orders came, and La Mothe decided to build a test car. For the sake of economy it was a small streetcar constructed by the New York Locomotive Works of Jersey City in the spring of 1855.[8] It weighed 3,300 pounds and followed the general form of a city omnibus. Experience gained in its production suggested that this weight could be cut by 300 pounds; indeed, La Mothe insisted that he could pare the weight to a single ton with a more careful arrangement of the material.

During the next fall, Passavant and Archer, iron fabricators of

New York City, were busy producing six iron horsecars on La Mothe's plan for the city of Boston.[9] Even more important, work was underway in Paterson, New Jersey, on a steam railroad coach designed by La Mothe. The job was taken not by a regular car builder but by a tinsmith, William Cundell (1804–1879), who constructed a shed on one side of his shop to accommodate the project. The Boston and Worcester Railroad was the main purchaser, but several other Boston lines held an interest for test purposes. The work went slowly, then stopped altogether with the panic of 1857; but Cundell survived the panic and had resumed construction by the spring of 1859. The specimen car was finished in July 1859. A drawing from La Mothe's 1856 pamphlet is thought to be a reasonable likeness (Figure 2.2). Press accounts offered the following information: The overall length was 51 feet, 6 inches; the body was 46 feet by 8 feet, 4 inches. Each side had fifteen windows; seating was provided for 60 passengers. The car contained 6,200 feet of band iron and over 9,200 rivets. The roof was of galvanized iron. The window sashes were brass; dark glass was used in the transom windows and end doors. The interior was handsomely outfitted. The curtains were of a rich English rep, the moldings were black walnut, and gilt mirrors occupied the corners. Scenic and patriotic views by a local artist ornamented the papiêr-mâché panels between the windows. Among them were paintings of John Smith's rescue, Evangeline, Niagara Falls, and George Washington's tomb. The exterior of the car had been coated with fish oil and white oxide of zinc to prevent rusting, then finished with yellow and brown paint.

In all this detail the actual weight of the car was never given. During construction La Mothe had claimed that it would weigh 5 tons less than an equivalent wooden coach, and before construction he had claimed that he could build a 60-passenger car of 9

Figure 2.2   This drawing from a pamphlet published by La Mothe in 1856 is believed to resemble an iron car actually built between 1857 and 1859.

Figure 2.3   La Mothe's fourth iron car, built by William Cundell of Paterson, New Jersey, in 1861. The car ran briefly on the Hackensack and New York Railroad. (Walter Lucas)

tons. The body of that hypothetical car was to weigh 4½ tons, but a rough calculation of the weight of the strap iron and rivets used in the Cundell car nearly equals this figure. When the weights of the sheet-iron roof, side panels, floor, and furniture are added, the total equals or exceeds the weight of a wooden car. After delivery, La Mothe reduced his claim to the statement that the car was 2 to 4 tons lighter than a wooden coach, while a New York newspaper said that it was "at least" 1 ton lighter. In truth La Mothe was probably not eager to divulge the true weight, even if he knew it. Several years later the *American Railway Times* stated that the experimental iron passenger cars that had been built so far weighed more, not less, than wooden cars.[10] The journal claimed that they exceeded their wooden contemporaries by 6 tons.

Following its completion in the summer of 1859, La Mothe's specimen car ran on the New York and Erie for six months. During this time it was closely scrutinized by the engineering world. It was greeted with enthusiasm—one admirer predicted that it was "destined to work a revolution on our railroads."[11] It is true that there were complaints about the dwarfish windows and doors, but critics admitted that perfection should not be expected in a test model. In March 1860 the car was sent away to Boston, and its final disposition is unknown.[12] However, an iron car of a "crude and ill-digested pattern" was running on local trains of the Eastern Railroad out of Boston in 1870.[13] It had come into the Eastern's possession after trials on several other lines, and La Mothe himself told one of Cundell's protégés that it was his car.[14]

Obviously this car was not impressive enough to draw an avalanche of orders. Cundell's car shop remained idle until a local textile manufacturer, Robert Rennie, ordered two small cars for his switching line. These combination freight-passenger cars were delivered early in February 1861. At the same time Rennie used his influence as a director of the Hackensack and New York Railroad to arrange for purchase of a full-sized La Mothe car. This car was more conventional in appearance and more generous in overall proportions, including windows of standard size (Figure 2.3). It seated 56 and had a smoking compartment at one end. The exterior was bright red with gold lettering; the interior was plain and neat. Instead of the lavish paintings of the first car, there was wooden paneling of grained oak, relieved by black walnut and gilt. The roof and sides were insulated with felt.

This new car seemed to ensure a fair trial for La Mothe's invention. But it had been in service for only a month when the car, together with the locomotive *Bergen*, plunged through an open drawbridge into the Hackensack River with a roar heard 2 miles away. The damage was slight to the car and its passengers, and it was sent to Cundell's shops for repairs. On the night of April 14, 1861, however, the final disaster struck when a fire reduced the shop to ashes. The car did not survive. Nor did Cundell's patience; after five years in the iron car business, he had built only four cars.

While La Mothe was seeking a new manufacturing agent, he revised the general plan of construction. He abandoned band iron for tube framing, which is used today for lightweight racing cars. In La Mothe's time, however, gas or electric welding was not available. Looking for an effective fastening, he settled first on a riveted clamp, then in later designs turned to U-bolt fastenings. He rejected the obvious choice, threaded couplings, because of their cost and because cutting the threads weakens the structure. He chose common gas pipe for economy. A patent (No. 33,350), issued September 24, 1861, specified flat iron beams and some wooden cross members in the frame. It was reissued late the next year with notable improvements, including more use of large-diameter pipes and the elimination of wooden members.

A sample freight car was built on the new plan for the U.S. Military Railroad.[15] The specifications were the same as those for a pipe-frame passenger car. Inside, the car was to be 44 feet long, 9 feet wide, and 7 feet high. The projected weight, with trucks, was 15,900 pounds, but again it seems that La Mothe was grossly underestimating. When the sample freight car was completed in January 1862, it weighed only 1,500 pounds less than an armored car built of angle iron and boiler plate. La Mothe hoped that the sample car would convince General D. C. McCallum to use iron

No. 23,333.   Patented Mar. 22, 1859.

Figure 2.5   Joseph Davenport's 1859 patent, from which several iron-bodied steam rail cars were produced in 1860. (U.S. Patent Office)

Figure 2.4   Merrick, Hanna, and Company of New Brighton, Pennsylvania, built several iron passenger cars between 1860 and about 1863 based on this 1859 patent. (U.S. Patent Office)

cars on the military lines. A second prototype freight car was also built, but the Army was not sufficiently impressed to carry the experiments further.

The following year the New York Central Railroad agreed to test La Mothe's idea. The road had recently purchased a large number of iron boxcars, which were performing well enough to stimulate interest in a metallic passenger coach. John M. Davidson, an iron safe manufacturer of Albany, was given the job because he had experience in constructing iron freight cars. The passenger car, built according to the gas-pipe plan, was ready by July 1863.[16] La Mothe declared that it was as stout as a man-of-war, yet 5 tons lighter than a wooden car. In an effort to make good on this claim, however, the car had been too lightly framed. After a special excursion to Niagara Falls the sides were notice-

ably dented; the body had twisted out of shape. Davidson went on to build more iron freight and express cars, but presumably on his own pattern. La Mothe retired, at least for the moment.

While La Mothe's first car was being finished, other mechanics were busy with similar plans. Experimentation in the Pittsburgh area was supported largely by the Pittsburgh, Fort Wayne, and Chicago Railroad. John Miner and Silas Merrick of the New Brighton, Pennsylvania, firm of Merrick, Hanna, and Company obtained a patent (No. 26,282) on November 19, 1859, for improvements in railroad cars. The drawing showed a straightforward design with an angle-iron frame riveted in a cellular form and with embossed sheet-iron side panels (Figure 2.4). The pressed panels afforded a reasonably stiff member made from lightweight sheet iron. They also provided additional embellishment in an age when any plain surface was regarded as neglected. A few months later Miner and Merrick obtained a second patent elaborating on the embossed-panel idea, and in the fall of 1860 they built a car on this plan at their New Brighton shops.[17] The body, which accommodated 56 passengers, was 40 feet, 6 inches long and 9 feet wide. The panels were made of no. 20

Figure 2.6   One of the Merrick, Hanna, and Company iron coaches is shown many years afterwards when it was in service on the South Carolina Railroad.

English stovepipe iron. Assuming that this was some form of planished iron, it is possible that the car was not painted. The seats and interior paneling were of wood. The weight was given as 26,350 pounds, or roughly the same as that of an ordinary wooden car. The builders stated that the weight could be reduced by 1 ton, but they wisely refrained from more exaggerated claims. The cost was given as $3,000, which again was about equal to that of a first-class wooden car.

The Pittsburgh, Fort Wayne, and Chicago Railroad was sufficiently pleased with the Miner and Merrick car to order four more, and all were in service by 1862.[18] Meanwhile the railroad turned its attention to reducing the operating costs of accommodation trains (short-haul trains run for local patrons and employees). Self-propelled steam cars were thought to be the answer. A car on this plan was built in 1860 by Russell and Company, car builders of Massillon, Ohio.[19] Its iron body was made after the plan of a partner in the firm, Joseph Davenport (1815c.–1912), who had received a patent (No. 23,333) on March 22, 1859, for a system of iron stanchions and bars (see Figure 2.5). The center sill was an iron latticework girder. The resulting 77-foot car was considerably longer than any coach then in service. It was divided into three compartments: the engine and boiler at the front, a baggage section next, and at the end a space for 94 passengers.

Within a year three of these cars were in service, and in 1863 the railroad ordered six more coaches from Merrick, Hanna, and Company at a cost of $3,600 each.[20] Some reports say that twenty-four cars were built, but only ten can be accounted for from contemporary sources.[21] The railroad's president used one for a private car, and it continued to serve as Number 100, the

road's official car, through the 1870s. The sides were dented and scarred from countless derailments and unscheduled runs over embankments that would have wrecked a wooden car. But the regular passengers showed less liking for the iron cars than management did. Insufficient insulation made them roaring ovens in the summer and freezing boxes in the winter, and complaints were so frequent that the cars were downgraded to emigrant service. The emigrants would have none of them, however, and they were downgraded again to cabooses for local freights.

During the Civil War the government appropriated the three iron steam cars and at least one of the coaches. Evidence that the cars were in service on the military railroads is found among the wet-plate negatives often reproduced in picture histories.[22] One of the coaches was reportedly used by General George Henry Thomas as an office car. After the war it was sold to the South Carolina Railroad as surplus property (Figure 2.6). In 1893 it was hauling laborers to phosphate beds outside of Charleston. By 1925 it served as a toolhouse near the tracks of the Southern Railway (successor to the South Carolina Railroad) in Richmond, Virginia. Its final disposition is unknown. Another of the Merrick, Hanna cars had a career almost as long, for it was used as late as 1898 on the Sharpsburg and Oakland Railroad, a Pennsylvania short line.

Another experiment with iron car construction took place during the early 1860s on the Louisville and Lexington Railroad.[23] It is probable that there were earlier attempts to construct composite iron-wood cars; tin-clad mail cars have been reported as early as 1847. However, Theodore G. Shaw, master car builder for the Louisville and Lexington, took the idea to its outer limits in three coaches produced at the company's shops. The basic

Figure 2.7   John A. Roebling's ideas for iron car construction are shown in this composite drawing made from several of his sketches of 1859. (Traced by John H. White, Jr.)

structure was wooden, but the exteriors were covered with sheet iron bound with heavy iron moldings ($1\frac{1}{4}$- by $\frac{3}{4}$-inch bars). The base trim was a $\frac{1}{4}$- by 8-inch iron plate that completely circled the body. The Civil War appears to have ended Shaw's experiments.

The only engineering celebrity who was attracted to the problem of iron passenger cars at an early date was John A. Roebling, who had prestige, practical experience as a civil engineer, and the capital and industrial plant to fabricate sample cars. His energies were never fully devoted to the idea, however; it appears to have been a fleeting interest. Among his papers are notes and sketches dated 1859 for a tubular lattice-frame car.[24] One plan calls for a double-lattice tubular framework, in which one frame would be inside the other and the connection would be made by stay bolts, as in a locomotive firebox (Figure 2.7). Additional stiffness was provided by U-shaped channel irons; some were to be 10 inches deep. The straps for the lattice were to be $\frac{1}{8}$ inch by 3 inches. The window frames, like those in Stephenson's Diamond cars of twenty years before, were set at a 45-degree angle. Roebling calculated the weight of the complete car, with trucks, to be just over 11 tons. The drawing developed from the inventor's sketches shows a design of surprisingly modern outlines. The fabric of the structure, however, was not much more than an elaboration of La Mothe's band-iron frame. Roebling did propose a pipe frame, and

thus as far as the written record reveals, he was ahead of any other mechanic.

Roebling obtained a patent (No. 30,426) on October 16, 1860, though it did not cover the lattice-frame scheme. It specified a double-wall construction, with a bowing outer sheet for strength and added air space for insulation. A letter from Roebling's son, Washington Roebling, dated June 17, 1860, indicates that some of his father's patented ideas may have been incorporated in the Miner and Merrick cars previously described. The letter, now in the Roebling Collection at Rutgers University, briefly alludes to a visit made by the Roeblings to the New Brighton plant. The phrase *"our* car Factory" ("our" was underlined in the original) suggests more than a casual interest in the matter, but the Roeblings' main efforts were concentrated on wire rope and bridges. The evidence suggests that if Roebling had applied himself, the iron car problem might well have been solved long before it was.

Charles S. Munn of Chicago was also experimenting with iron car manufacture. It does not appear that Munn followed any patented design, but he is known to have delivered a 60-passenger iron car to the Chicago and St. Louis Railroad in the spring of 1865.[25] The car was all iron except for the doors, window sash, and interior paneling. It was said to be 3 tons lighter by actual test than an equivalent wooden car. A second car was in the

Figure 2.8   During the 1870s La Mothe developed a new design for iron or steel tube framed passenger cars. This drawing dates from 1876. (Scientific American Supplement, September 16, 1876)

process of manufacture in Munn's shop, which had already produced iron freight and express cars for the Chicago, Burlington, and Quincy and the Michigan Central. It appears that Munn's efforts were not favored with notable success, for there is no record of subsequent experiments.

The iron passenger car seemed thoroughly discredited. No other specimen cars were built for many years; the industry appeared content with wooden construction. La Mothe concentrated on developing a plan for iron freight cars, which had a moderate success. In 1870 he did obtain a patent (No. 105,699) for a bent-rod and -tube frame car. The design incorporated the clerestory roof and the inventor's latest ideas for uniting a framework made of round shapes. In 1877 his plan, with steel rods specified for greater security, was presented to the Master Mechanics Association (Figure 2.8), but no cars were built. La Mothe's interest in the subject continued throughout his life. In 1892 he was still active in promoting metal cars, and it is possible that he lived long enough to see the first commercial steel cars.[26]

A few freakish metal cars were produced during this time of general stagnation. In 1877, for example, Billmeyer and Small built a bullet-proof car for use in Cuba.[27] Its sides were armored with $\frac{3}{8}$-inch iron and $\frac{3}{16}$-inch steel; iron curtains could be drawn over the windows of this 31-foot, 12-ton tank. Other scattered examples might be uncovered, but there was little substantive movement in metal car engineering until the eighties.

Coming of the Steel Car

In their 1859 patent Miner and Merrick had suggested that steel be used, but they acknowledged that wrought iron was much cheaper and possessed ample strength for all practical purposes. La Mothe's 1870 specification called for soft steel rods as the preferable material. By 1880, with the declining production costs that followed the introduction of the Bessemer process, steel had become competitive and in some cases was actually cheaper than wrought iron. Now the emphasis was on perfecting a steel car, and a new wave of experiments began.

By the 1880s European lines had turned largely to steel underframes for both freight and passenger cars, and some of these cars were exported to South America and Mexico. The trade press was filled with comment by American railway men who had seen the cars while abroad and thought that the general idea of metallic construction was worth reexamination. At least one American

car builder produced steel underframe cars for export: in 1884 Harlan and Hollingsworth turned out some cars for Argentina with frames built up from I beams, channels, and tubes. One was furnished for the private use of the Argentine president. At this time Harlan and Hollingsworth was converted largely to the manufacture of iron and steel ships. As a combination ship and car manufacturer it seemed to occupy a logical position of leadership for the introduction of steel cars, but it developed no such program. The promotion of metallic cars was to remain the obsession of amateurs, cranks, and poor mechanics.

The first steel passenger car actually constructed originated with Edward Y. Robbins of Cincinnati, Ohio. Robbins was a sheet-metal fabricator who specialized in stove making. It is possible that the fearful reputation of the railroad car stove attracted his attention to the general subject of safety in railway travel. His first thoughts were probably limited to the perfection of a safe car heater, but it was plain to anyone who began studying the issue that the entire concept of railway car design called for reform. Past efforts to develop an indestructible, fireproof car were patterned after a traditional plan: a box-like form whose framework was covered with a thin wrapper. Robbins envisioned a radical departure: the basic structure would be a self-supporting tube, and the frame and basic body fabric would be a single sheet-iron unit (Figure 2.9).

Robbins expanded the idea in his patent of November 3, 1868 (No. 83,731). He favored the truly cylindrical cross section as the strongest possible form, although he conceded that a more elliptical form might be preferable. The tubular body was to be stiffened by additional attachments: U irons riveted along the length of the body just below the centerline, a sheet-iron floor with additional support from V-shaped plates or keelsons riveted under the deck, similar plates set transverse to the body to form the bolsters, and hollow ribs, to serve as structural members and air ducts, riveted inside the body at spaced intervals around its full circumference. Additional security was offered by cushioned platforms that were intended to absorb the force of most collisions. The platforms would be made of iron plates set on end, with layers of springs, cork, or rubber between them. Should the force of a collision be greater than the capacity of the end platforms to sustain it, passengers might still escape injury, for the end walls were to be made of double plates, and the interior would be thickly padded so that travelers could be thrown about like projectiles, yet come out unharmed. The stuffing might be wool, shoddy, sponge, hemp, caoutchouc, oakum, or any other

Figure 2.9   *Edward Y. Robbins of Cincinnati patented a radical tubular design for metal passenger cars in 1868.* (Engineering, *April 22, 1870*)

elastic material. If the stuffing was covered in fine cloth, Robbins declared, the interior would be as handsome as any car running, and certainly safer.

Robbins specified a dual hot-air heating plant suspended under the car, with ducts running under the floor. Fresh air would be fed to the furnace from the roof through the hollow, angular ribs of the body. As with all under-the-car heaters, the fire hazard was reduced because the fire was outside of the passenger compartment.

Although this idea for heating was not original, Robbins's general plan certainly was. The design was strong and reasonably practical, but incredibly ugly. The bulbous, top-heavy body, the receding roof, and the awkward union of the platform and the round end bulkhead produced a clumsy silhouette. The dwarfish side windows—not more than 15 inches square, according to the patent sketch—were among the worst defects.

Robbins's efforts to produce this cylindrical car continued for over twenty years. They exhausted as many promoters and involved as many heartaches as the misadventures of La Mothe. The first talk of producing a prototype appeared in the Cincinnati newspapers in July 1869. A large-scale model was put on exhibition, and it was announced that the full-sized car would be made

Figure 2.10   This model of Robbins's iron or steel car was exhibited to attract support for a full-sized prototype. (Engineering, April 22, 1870)

of ⅛- to ³⁄₁₆-inch sheets with interior padding 4 inches thick. The promotion came to nothing, however; possibly it was a casualty of the gold panic later that year.

Robbins then tried to introduce the car in Great Britain, taking out a patent in the name of a straw party, W. Newton Mac Cartney of Glasgow.[28] A model (possibly the one exhibited in Cincinnati the year before) was shown to the Board of Trade, a national body charged with maintaining safe operations on British railways (Figure 2.10). Although the model represented a 36-foot body, for British lines Robbins proposed to construct 26-foot cars on four wheels. The usual arguments for safety, long life, and comfort were offered and quietly accepted, but the Board questioned Robbins's claim that the cars would be light in weight. It was obvious that they would exceed the estimated 4½ to 5 tons; like La Mothe, Robbins refused to recognize that weight would never be a strong point of the metallic car. One American trade paper said that a full-sized prototype was built in Glasgow, but this report appears to have confused the model with an actual car.[29]

Nothing is known of Robbins's activities during the next fourteen years. In 1884 he reappeared in Boston as founder of the Robbins Cylindrical Steel Car Company, which had somehow raised capital to construct a sample car. The existing car builders were unable or unwilling to produce such a maverick, and Robbins turned to an iron ship builder, the Atlantic Works in east Boston. This yard was well equipped to fabricate the tubular car, for it had rolls, bulldozers, punches, and skilled metalworkers. By July 1884 the car was half finished.[30] Although the 1868 patent had specified either iron or steel, this was now the age of cheap steel, and that metal was used. The 54-foot body was made with ⅜-inch bottom plates. The sides were ³⁄₁₆ inch thick, while the roof and clerestory monitor were made of ⅛-inch and thinner sheets. The floor was ⅛-inch plate. The general fabrication followed the specifications of the patent. The interior padding, to be 6 inches thick, would consist of a layer of felt placed against the

shell of the car, a layer of woven wire, a layer of tufted hair, and a cloth cover. The padding would be made in sections so that they could be removed for repair or cleaning. Robbins claimed that the car would be 4 to 8 tons lighter than a wooden one. He estimated the total weight of the finished car at 22 tons. But the car was not finished; Robbins could not raise the money for completion. Some parts went to scrap, and the unfinished hulk eventually found storage in the old repair shops of the Eastern Railroad.[31]

The assets of Robbins's company were acquired by Byron A. Atkinson, a rich Boston furniture dealer whose success in the furniture business was attributed to free delivery, generous credits, and volume selling.[32] He selected Henry D. Perky to promote the enterprise. Perky, a health crank, an organizer of trade fairs, and a general speculator who was said to profit during the panics that ruined other men, became the driving force of the Steel Car Company.[33] A respectable front man, General Adna Anderson, was appointed president. Anderson had had a long career in railroad engineering. He had become a war hero as an officer with the U.S. Military Railroad. Later he served as general manager of the Wabash, engineer of the St. Louis bridge, and finally, chief engineer of the Northern Pacific. By 1885 he was exhausted, his mind was wandering, and he was retired early. He committed suicide not long after his appointment to the Steel Car Company, and it seems doubtful that he did much to promote the venture.[34]

Work resumed on the car body, now in the Eastern Railroad repair shops, in the spring of 1887.[35] Charles M. Smith was engaged to improve Robbins's designs. His most basic change was the substitution of large, dormer-like windows for the small, flush openings of the original design. A sheet-metal framework with an arched top permitted the use of large plate-glass windows. A patent (No. 366,519) for tubular railway cars that were remarkably similar to Robbins's plan was granted to Smith on July 12, 1887. The work went slowly, however. By midsummer of the next year it was reported that the car would be running in a few days,

Figure 2.11   *The* City of St. Joseph, *built in Chicago in 1889 after a modified version of Robbins's 1868 patent. Like one of La Mothe's cars, it was destroyed by fire shortly after delivery.* (Burlington Northern)

but nothing happened. By 1889 it was obvious that the job would never be finished at the Eastern shops. The car was sent to the Laconia Car Company in New Hampshire, where professional car fitters could complete it.[36]

While construction of the Boston car languished, Perky was talking of establishing a great steel car plant. At first he spoke of Chicago; a few months later in August 1888, he mentioned Lincoln, Nebraska, as the site and said that the Boston car would be called the *City of Lincoln*.[37] This scheme failed, and Perky shifted the proposed site to St. Joseph, Missouri.[38] With local backing he secured a tract of land east of the city and put up a building 80 by 900 feet. It was completed by the beginning of 1889, and to celebrate the new enterprise Perky organized an industrial exposition to be held in the autumn at the plant.

Sometime before the fair opened, a second specimen car was built for exhibit in St. Joseph. It was constructed in Chicago, apparently at a bridge works whose identity has not been determined.[39] The Chicago car, named the *City of St. Joseph* (Figure 2.11), was superior to the Boston car in that the roof and windows were more graceful; the awkwardness of the Boston car's projecting clerestory and window dormers was avoided. This was accomplished by modifying the shape of the body and setting the sash well inside the outer walls of the car. The *City of St. Joseph* was large; its overall length was 73 feet, 8 inches (some 12 feet longer than the Boston car), and it was mounted on six-wheel trucks. The interior was arranged as a private car, with observation, dining, and sleeping rooms.

At St. Joseph's "New Era" Exposition, the steel car was a major attraction. Ten more cars were being built in the adjacent shops to form an exhibition train that would be sent on a national tour. But during the evening of September 15, 1889, fire leveled the plant, and the *City of St. Joseph*, the cars under construction, and the prospects of the Steel Car Company vanished in the flames.

After the St. Joseph debacle the affairs of the cylindrical car company were allowed to rest for a year or two. Perky was not easily disheartened, however; he meant to give the scheme one

last push. The old Boston car, which had since been finished for Atkinson, the furniture dealer, to use as a private car, was borrowed for a transcontinental tour (Figure 2.12). Perky took not only his family but also a reporter from the *New York World* to churn out releases. For some time the car stopped over in Chicago, where it was reviewed by several trade papers. The *Railway Review* called it "not nearly so business like" as the *City of St. Joseph*; certainly it had many shortcomings. Apparently it was next shown at the Columbian Exposition of 1893 in Chicago, where it attracted no orders either. More than $40,000 had been invested in the car, but it was abandoned on the fairgrounds. It was taken over by the wreckers who dismantled the exposition, and they sold it to a showman for a traveling repertory company.[40] Around 1907 the car was purchased by another showman, Leo Blondin of Oklahoma City, who traveled extensively in it over the next twenty-two years. During Blondin's ownership several fanciful stories appeared about the car. He moved the date of construction back to 1879, claimed that the car was first used under the name *Mildred* by President Garfield, and said that it was Garfield's funeral car. The validity of these tales is extremely doubtful. After Blondin's retirement in 1929, the car continued to work the show circuit. According to John Roberts of the National Museum of Transport, the car was in existence until sometime in the 1940s, when it was destroyed by fire in Hannibal, Missouri.

The Robbins car was not the only attempt to introduce steel passenger cars during the last twenty years of the nineteenth century. A dozen or more promoters were active, although few contributed more than patents, and many of their designs were derived from Robbins's cylindrical pattern. The lack of novelty indicates a certain laxity on the part of the Patent Office, which granted one license after another based on the same idea. The universal appeal of the tubular design was explained by one of the inventors, George J. Porter, in his specification of October 13, 1891 (No. 461,173): "It has been found that a car-body of sub-

Figure 2.12   The first Robbins demonstrater was constructed by fits and starts between 1884 and 1890. It is reported to have survived until the 1940s. (Railway and Locomotive Historical Society)

Figure 2.13   John W. Post combined La Mothe's and Robbins's ideas in this 1883 patent. (U.S. Patent Office)

Figure 2.14   Post's 1883 patent attempted to combine the round and rectangular cross-section shape into a workable plan. (U.S. Patent Office)

stantially rectangular form in cross-section, when composed of metal is unduly heavy, and, besides this, cannot readily be made to withstand the torsional strain to which car-bodies in service are continually subjected. These facts have led inventors and artisans in devising and constructing metallic car-bodies to adapt the cylindrical form." In Porter's patent the difficulty of cylindrical fabrication was lessened by making the tube in a polygonal form of eleven straight panels.

Colonel John W. Post of New York City offered another variety of tubular body construction in his patent of September 25, 1883 (No. 285,765) by combining both a rectangular and a cylindrical shape in a single profile (Figures 2.13 and 2.14). This form of construction is best shown in a second patent (No. 366,249) granted Post in 1887. In this and several of his other patents, steel tubes were to be used for the framework. In the summer of 1890 it was announced that the Steel Tubular Car Company had been formed to exploit Post's design.[41] A plant as large as the Pullman factory was to be built near Bradford, Pennsylvania, on a 5,500-acre tract, which would contain a model town for 5,000 employees. The grandiose plans were never executed, however.

Another zealous advocate of the cylindrical car body was Casper Zimmerman of Chicago. His 1895 patent (No. 542,746) calls for nothing less than a perfectly round body with spherical ends. Like Robbins's, the interior was to be heavily padded. Zimmerman planned to add mattress springs to the interior walls and to mount a false floor. The cushioned interior, like Robbins's design nearly thirty years before, was for passenger safety, but the air space provided by the layer of springs would also contribute to the car's insulation.

More conventional-minded inventors rejected the tubular body concept and offered orthodox rectangular designs. Some of their plans were surely conscious efforts to provide a scheme that railroad managements would not immediately discount because of its bizarre appearance. Between 1887 and 1896 Chester W. M.

Smith, a West Coast mechanic, prepared several patents for steel passenger cars with conventional outlines. Smith felt that the requisite strength could be gained by the use of very large one-piece sheets combined with a box girder built into the side panels near the floor. During the same period two separate inventors, Carleton B. Hutchens of Detroit and George M. Bird of Boston, offered patent designs for steel cars that were conventional in appearance (Figure 2.15).[42] None of the designs was given an actual test, however.

Two Middle Westerners, William W. Green and James Murison, reasoned that since most railroad men had shown hostility to the idea of steel cars over many years, someone should offer them a compromise scheme. Concluding that such a design should look exactly like a wooden car, Green and Murison took out a patent (No. 372,615) on November 1, 1887, for a composite wood-metal vehicle. Murison had seen these materials being combined during his youth in a Scottish shipyard at a time when, with prime timber long since exhausted, British ship builders turned to iron-frame ships.[43] A model of Green and Murison's car was exhibited in Chicago. The plan appeared so practical that the Louisville, New Albany, and Chicago Railway agreed to build a prototype mail car at its New Albany, Indiana, shops in the summer of 1888.[44] The road's master car builder, Charles Coller, worked out the details and supervised construction, and the car was finished in January of the following year.

Except for a row of rivets, it could not be distinguished from an ordinary wooden car (Figure 2.16). It was a faithful reproduction in iron and steel of the most common postal car of the day. Actually this cunning reproduction contained a fair amount of wood—so much, in fact, that it cannot properly be called fireproof, as its promoters boldly claimed. The frame was oak clad with steel plates. The window sash, doors, and flooring were wooden. The remainder of the fabric was steel. The body frame was assembled from flat plates, gas pipe, and rods. The vertical

C. B. HUTCHINS.
METALLIC PASSENGER CAR.

No. 442,894.   Patented Dec. 16, 1890.

WITNESSES

INVENTOR
C. B. Hutchins
By
Rowan Bushneler
Attorney.

Figure 2.15   C. B. Hutchens of Detroit was one of several early inventors who hoped to perfect a practical plan for steel passenger cars. His 1890 patent was one of the more sensible schemes offered, but no test car was built. (U.S. Patent Office)

nanced by a Philadelphia syndicate, was to be erected in the coal and iron region of Alabama at Bridgeport. These prospects faded, but Green and Murison's hopes were rekindled by a wealthy lumber dealer, T. W. Harvey, and a plant was built near Chicago, 3 miles south of the Pullman factory.[45] Large orders for mail cars on the pattern of the Louisville, New Albany, and Chicago prototype were expected. The inventors claimed that their car was burglar-proof as well as fire-resistant and could be sent across the country in perfect safety. The Burlington reportedly placed an order for a steel coach. But market prospects for steel passenger cars were an illusion in 1890, and Harvey, a practical businessman, recognized the futility of the promotion and ordered the company to concentrate on developing a steel freight car. Green and Murison were dropped, and the project was entrusted to George L. Harvey.

The second concentrated effort to introduce metallic passenger cars can be said to have ended by the early 1890s. Again the attempt seems to have been frustrated by too much emphasis on novel designs. More conventional execution might have led the industry to take the steel car seriously. And as in earlier periods, the work was not carried out by established railroad car builders and so was classified out-of-hand as the foolish pursuit of mechanical cranks. Metallic passenger cars were not to receive serious consideration for another ten years.

Introduction and Acceptance of Steel Cars

In the first years of the twentieth century the railroad industry came to accept steel as the preferable material for passenger car construction. With this commitment, steel cars were on their way. The amateur inventors whose experiments with steel had long been ignored were again bypassed; the railroads turned to their own mechanical staffs or to engineering consultants. It was thus a purposeful, well-financed effort toward developing a commercially feasible metallic passenger car. The experiments of the past proved futile; the professionals saw nothing of merit in the old patents. The new steel cars were modeled on wooden cars, and orthodoxy prevailed.

For years the industry had been censured by the trade press for its conservative position on metal passenger cars. This criticism grew sharper after the introduction of steel freight cars in the late 1890s, for they made it obvious that steel framing provided a wonderful increase in capacity. Fewer cars could carry the same traffic, and unusually heavy shipments were more readily handled. The benefits were immediate and obvious.

A steel coach, on the other hand, could carry no more passengers. Being heavier and more costly, it offered only the prospect of greater safety. This advantage could be realized only in the event of a serious accident, which few officials cared to admit as a possibility on their roads. Yet it was safety that became the main argument for the adoption of steel cars. In addition, the railroad industry displayed a growing disenchantment with the complexity of wooden car construction. *Railway Age* remarked, "A more

posts were ³⁄₁₆- by 3½-inch steel plates that had been bent edgewise when cold to conform to the profile of the car's cross section. The bottoms of these posts were bent over to form feet. Bolts passing through the feet and the wooden sills tied the body to the frame. Gas pipe fitted with flanges served as spacers between frames. The exterior sides were covered in corrugated sheet metal shaped in a convincing imitation of battens. This did much to relieve the uneven appearance so common to hand-fitted sheet-metal work. The inside was covered with a ¼-inch layer of asbestos. The interior panels, also sheet metal, were lined with a second layer of the same material.

After nearly fifteen months of service over some 135,000 miles, no repairs were necessary except for replacement of a worn set of wheels and revarnishing of the exterior. Contrary to the usual complaints about excessive noise and temperature extremes, a clerk who worked in the car during the entire period said that it rode quietly and easily and was perfectly comfortable both summer and winter.

The American Fire Proof Steel Car Company was organized not long after Green and Murison's sample car entered service. The firm's plan was to manufacture passenger cars. A plant, fi-

Figure 2.16   Murison and Green's composite wood and steel mail car, built in the Monon's shops in 1889, was intended to be the forerunner of a fleet of passenger cars. (Poor's Manual of Railroads, 1891)

Figure 2.17   The first production-model steel cars were built in 1904 and 1905 for New York's Interborough subway line and the Long Island Railroad by ACF.

heterogeneous collection of trussing and bracing would be difficult to find . . . would it not be equally cheap and more satisfactory to do away with this arrangement at once and put in all-metal construction?"[46] In 1905 the *American Railroad Journal* said that passenger car design had been stagnant for twenty years.[47] It blamed the contract builders as self-serving bastions of reaction who opposed all progress, and criticized the railroads for placing so many of their orders with private concerns. To depend on what was essentially a woodworking industry for a new design was folly, the journal said; the railroads should perfect a steel passenger car in their own shops. The journal failed to note that there were some new contract builders, such as Pressed Steel Car Company, who were determined to capture the market by introducing a successful steel car.

As it happened, the action began off the main line. By 1901 construction was underway on New York's first subway, the Interborough Rapid Transit. George Gibbs was hired to prepare designs for the rolling stock.[48] A graduate engineer, Gibbs had acquired considerable railroad experience in the Milwaukee Road's mechanical department. He had done pioneering work in electric traction and railway signaling, had held several important consulting positions, and was an associate of prominent men in

the engineering field. Thus when Gibbs suggested steel for the IRT subway cars, the idea was not lightly dismissed. His recommendation was not initially accepted, however, because time was short and cars were needed for the opening. The IRT ordered 500 steel-frame cars, made fire-resistant by sheathing the wooden exterior in copper and lining the floors with asbestos sheets.

The need for fireproof cars was underscored by a tragic accident in the Paris subway, in which a burning car was inaccessible to firemen and heavy smoke added to the injuries of the people who were trapped underground.[49] The IRT then decided that the composite steel-frame cars were not safe enough, and with the support of George Westinghouse and A. J. Cassatt, president of the Pennsylvania Railroad, Gibbs was asked to prepare designs for an all-steel fireproof car. The plans were finished in October 1902. The Pennsylvania Railroad began tunneling under the Hudson River to Manhattan Island in June 1903. Although electric locomotion was planned, Cassatt felt that the Pennsylvania must have fireproof cars for the tunnel operation. Toward the end of developing such cars he offered to build a sample at cost for the IRT in his Altoona shops. Completed in December 1903, the car, Number 3342, was clumsy-looking and heavier than desired. The design used only standard steel shapes, because it would have delayed delivery to create the special shapes and details necessary for a lighter, more handsome car. As it was, the car was in production for fourteen months. After tests on the Second Avenue Elevated, the IRT's president, August Belmont, was convinced of the merits of an all-steel design.

Gibbs then perfected his plan, shaving off pounds by the use of pressed shapes and aluminum panels and fittings for the interior. Only a small amount of wood was used; the floor was a poured composition material called Monolith cement. The Berwick, Pennsylvania, plant of the American Car and Foundry Company built 300 of these cars, most of them in time for the subway's opening in the fall of 1904 (Figure 2.17).[50] The worst fears about the composite cars were realized the following year, when five were totally burned following an accident.[51] After this incident, only steel cars were permitted in the subway. A number of the original Gibbs cars were in service for sixty years; one of them is preserved in the Seashore Electric Railway Museum of Kennebunkport, Maine.

The success of the IRT cars encouraged the Pennsylvania Railroad to develop a design for steam road operations. Other firms were also working on this project. As Chapter 1 mentioned, by the end of the nineteenth century the practice of reinforcing cars with steel was widespread, and it naturally suggested that cars could also be given steel underframes. In 1903 the North Western Elevated Railway of Chicago took delivery of thirty-five steel-underframe cars from the Pressed Steel Car Company. In June of the same year the Jeffersonville, Indiana, shops of the American Car and Foundry Company fabricated a 75-foot combine car for the Monon line.[52] The floor frame was built up from 8-inch I beams with timbers bolted to either side of the steel members. The body was of the usual wooden construction. Charles Coller, master car builder of the Louisville, New Albany, and Chicago, who built Green and Murison's mail car in 1889, was identified with this project as well. A more widely publicized undertaking was the steel-framed, side-door suburban cars of the Illinois Central.[53] The design was patented by two officials of the line, A. W. Sullivan and William Renshaw, on November 19, 1901 (No. 686,959). The floor frame was built up from 9-inch I beams with truss rods for additional stiffness. The $\frac{1}{2}$-inch steel-plate floor was overlaid with wooden planking. The body framing was made of 3-inch channels. Except for the usual hardware, the remainder of the car was wooden. Its weight was just over 42 tons; the road claimed a saving of 3 to 5 tons compared with an equivalent wooden car. The first car entered service in September 1903, and another sixteen were built at the Illinois Central shops (Figure 2.18). The novel design attracted considerable attention, but its ultimate effect on steel car construction was negligible. A more significant development that year was the introduction of cast-steel body bolsters by the American Steel Foundry. Meant for wooden cars, these bolsters became an important feature of the steel passenger car.

During the next three years the pace quickened. Delivery of the New York subway cars was the biggest news of 1904. Other significant developments included an order for 134 steel suburban cars from the Long Island to A.C.F.'s Berwick plant. Most of these cars, which were similar to the IRT design, were placed in service during 1905. Of greater consequence, steam roads began to experiment with full-sized cars. The Erie acquired an all-steel baggage car (Number 599) from the Standard Steel Car Company in 1904.[54] The 60-foot car was very heavy: the deep fish-belly underframe (probably the first use of this style of passenger car framing) and the $\frac{3}{16}$-inch side plates and $\frac{1}{8}$-inch roof plates brought the total weight to 53 tons. The same builder delivered a steel mail car of similar outline to the Erie in March 1905 (Figure 6.55). The Erie acquired a third all-steel car for express service in November. A large number of steel cars were built for the London subway by A.C.F. The Santa Fe acquired thirty-nine steel-underframe mail cars.

A paper calling for the adoption of all-steel passenger cars was read at a meeting of the Master Car Builders, who had not heard a report on this subject for twenty years.[55] In that interval, dead weight in passenger cars had advanced 100 percent, while seating capacity had risen only 30 percent. The most important part of the paper was the description of a plan for a 70-foot steel passenger car designed by George I. King of the Middletown Car Works. The center sills were 12-inch channels, and the general conformation was that of a standard car. The paper also noted that the Pressed Steel Car Company had been working on a similar plan for some time.

But not all designers were thinking along conventional lines. Around 1905 William R. McKeen, Jr., mechanical superintendent for the Union Pacific, began making steel-bodied gasoline rail cars. These distinctive vehicles, with their circular windows, side doors, and knife-edge fronts, are unlike the creations of any other designer. In 1907 McKeen modified his rail car design for use as an ordinary coach 68 feet long, which retained all the usual features of his earlier cars, including an arched roof.[56] The ends were slightly rounded and fitted with end doors and vestibules. Only one car was manufactured from this extraordinary design.

Several trunk lines had now become convinced that steel passenger coaches were a desirable investment, and in 1905 and 1906 three major railroads decided to buy some. The Southern Railway turned to a major builder of metal freight cars, the Pressed Steel Car Company, which was eager to move into the market for passenger equipment. After several years of planning, the firm erected a special shop for passenger car work. It delivered the first steel coach, the Southern's Number 1364, in time for exhibit at the June 1906 convention of the Master Car Builders (Figures 2.19 to 2.21).[57] Two more cars on the same plan followed. In this design the 66-foot body was supported by two fishbelly center sills fabricated from $\frac{3}{8}$-inch plate. While the floor and body framing and side sheathing were of steel, considerable wood was used in the construction. The entire roof was wooden, as was the interior. Although it was not fireproof, the structure was considerably stronger than a wooden-frame coach. The mixed construction held the overall weight to 55 tons, or just 2 percent more than an equivalent wooden car. These cars were considered satisfactory at first, but in later years they were criticized for insuffi-

Figure 2.18 The Illinois Central built a number of steel-frame, side-door cars for its Chicago suburban service in 1903. (Railway Age, September 4, 1903)

Figure 2.19 The Southern Railway, rarely recognized for its pioneering efforts in this field, acquired a steel passenger car in 1906. (Car Builders' Dictionary, 1919)

Figure 2.20 *The Southern Railway's steel coach Number 1364 was built by the Pressed Steel Car Company of Pittsburgh.* (Car Builders' Dictionary, 1919)

cient cross bracing of the underframe.[58] Apparently the defect was minor, for the first car remained in service until 1953; the two others were retired in 1951.

In the Far West the Southern Pacific pioneered in introducing the steel passenger car.[59] A 60-foot coach was ordered from the road's Sacramento repair shops late in 1905 and was completed the following summer. This car, Number 1806, measured 59 feet 10 inches long and weighed 53½ tons (Figure 2.22). The center sills were 12-inch I beams stiffened by truss rods; the side sills were 3½- by 7-inch angles. As with most experimental steel cars, it was fabricated largely from stock shapes. In this instance the fabrication was carried out in the boiler shop; the car shop was responsible for the assembly and the wooden interior. The arched or elliptical roof was the most distinctive design feature, and one that was to remain the hallmark of Southern Pacific passenger equipment for many years.

Because of the wooden interior, of course, the car was not fireproof. This was a defect that the industry eventually came to accept, as will be shown later. The other principal defects were the floor frame and the overall weight. The framing was at once heavy and weak, and the weight per passenger was the heaviest of any experimental car. This was the result of a complete dependence on the floor frame for support: the side panels were not part of the structure and thus did nothing to take up the static forces. In addition, complicated and massive cast-steel body bolsters added unnecessary weight, although these were later replaced in an effort to lighten the car. Despite its defects, Number 1806 performed well enough to stay in service until 1957.

In 1908 the Sacramento shops completed a second steel coach that showed considerable improvement, with several important modifications that had grown out of a conference of mechanical officers of the Harriman Lines.[60] Over 5 tons of dead weight had been trimmed from the new design. A total of 3,390 pounds was saved by using smaller bolster castings; the center sills were reduced to 10-inch I beams; and more care in the design of other small components made other reductions possible. The wooden interior was replaced by metal, but in overall size and appearance the car was identical to Number 1806. In service the underframing proved inadequate, and in 1911 it was replaced by the fishbelly design. The plan was generally satisfactory, however, and the Harriman Lines began a program of acquiring steel cars on a large scale. These companies became major champions of the steel passenger car, second only to the Pennsylvania Railroad in

Figure 2.21 *A few years after acquiring its first metal coach, the Southern began to buy steel-underframe coaches with cast-steel end frames from the Jackson and Sharp Car Company of Wilmington, Delaware.*

their commitment to the elimination of wooden rolling stock (Figure 2.23). By 1910 the Harriman Lines had some 650 steel passenger cars in service or on order. During the same year they received their last deliveries of wooden cars. Once a sufficient stock of steel cars was on hand for first-class trains, however, the Southern Pacific's enthusiasm appears to have waned. In the early 1920s nearly half of the line's 1,700 passenger cars were still wooden, and it was some years before the last was retired.

The history of the steel passenger car is usually told as if it belongs exclusively to the Pennsylvania Railroad, though as this chapter has shown, many other individuals and firms also participated. Nevertheless, the Pennsylvania's role was a major one. A coach design was prepared in 1904 by the road's engineers, notably William F. Kiesel, Jr., and Charles E. Barba, in collaboration with George Gibbs, designer of the subway car. In June 1906 a sample car, Number 1651, was produced at the Altoona workshops and tested between Philadelphia and Lancaster (Figure

Figure 2.22   The Pacific Coast's steel pioneer was produced by the Southern Pacific's Sacramento shops in 1906. (American Railroad Journal, January 1907)

2.24). The 67-foot car's main support was a large box girder fabricated from two 18-inch channels and two 24-inch-wide steel plates. The interior paneling was made of composition board or treated lumber. The roof was wood with copper sheathing. In all, some 1,540 pounds of wood were used. The floor was concrete, the doors pressed steel. Yet the quantity of combustibles used, even though they were treated to resist flame, qualified the car as slow-burning but not fireproof.

Later in the same year the Long Island Railroad, a subsidiary of the Pennsylvania, took delivery of a steel car from American Car and Foundry Company. Like Number 1651, it seated 72 passengers and was just over 67 feet long. Its framing was based

on a principle more like the traditional side-panel wooden car truss. Here, however, a plate girder (formed by the plates of the side panels) supported the car. The center sill, composed of 10-inch I beams, was intended to handle the draft and buffering stresses. The Long Island car contained far less combustible material than the 1651. The roof sheeting and headlining were made from a noncombustible sheet with the trade name Durite. This car was borrowed for tests on the parent company's main line so that it could be compared with the home-built car under actual road conditions. In the end, neither car was found satisfactory. They were too small to accommodate the luxury fittings required for long-distance travel. They were not fireproof, and the framing was not considered substantial enough for passenger safety in case of a wreck. The company decided that a fireproof, 80-foot passenger car seating 88 and capable of withstanding a minimum end force of 200 tons was necessary for trunk-line service. For suburban service, designs were prepared for a lighter car capable of conversion to electric traction if desired.

By early 1907 the Pennsylvania had completed plans for an 80-foot, 88-passenger car destined to be the prototype of the most famous single group of passenger coaches in this country: the class P-70 (Figure 2.25). A box-girder frame similar to that in the Number 1651 was used to support the car. The 56.75-ton car was to be carried on eight wheels. All wood was eliminated except for the seat armrests and a few other incidental parts. The interior paneling was painted sheet steel with extruded aluminum moldings. The general outline was conventional, sparse, and unasham-

edly mechanical. Though they had obviously made an effort to create a neat and workmanlike structure, the P-70 designers were not overly concerned with aesthetics. Nor did they show Pullman's preoccupation with disguising the method of fabrication; in the P-70 rivets, bolts, and welt plates were in plain view.

More important, the P-70 was not another experimental car but a production model. Car builders were asked to bid on an order for 500, but none quoted a price that the railroad thought fair. Negotiations continued through the spring, and in June the momentous announcement was made: 200 steel passenger cars were to be purchased.[61] This, the first large-scale commitment, truly opened the age of the steel passenger car. The historic order was divided among American Car and Foundry (which received the bulk of the work), Pressed Steel, and the Altoona shops. It was estimated that 1,000 coaches, diners, and baggage cars, together

Figure 2.23   The basic arched-roof, heavyweight coach of 1906 was duplicated in subsequent Harriman Lines orders. The Number 1866 was completed by Pullman in 1911. (Pullman Neg. 13677)

Figure 2.24   The Pennsylvania Railroad's first effort at steel coach building for mainline service was the Number 1651, completed in 1906. It was not duplicated.

Figure 2.25 In 1907 the first of the famous class P-70s entered service. Cars on this general plan continued in production for over twenty years. (Railway Review, June 8, 1907)

Figure 2.26 Steel coaches such as the Number 1721 were considered essential for the safe operation of trains through the long tunnels leading to Pennsylvania Station in Manhattan.

with 500 parlor and sleeping cars, would be needed to reequip the main-line trains expected to operate in and out of the New York terminal. The projected investment was staggering, even for the giant Pennsy. The initial order would cost over 20 million dollars. Five years after the first order, twice this amount had been spent and the conversion was still incomplete.

The first P-70, Number 1652, was finished at the Altoona shops in December 1907.[62] (Unfortunately this historic car was sold for scrap in 1966, while the experimental and relatively unimportant 1651 was preserved.) The contract shops began deliveries during the next year. More orders followed; by February 1910, 324 steel cars were running and that many more were under construction or on the books.[63] Late in 1912 a third of the Pennsylvania cars

were steel; some 2,800 were in service, with another 394 on order (Figures 2.26 to 2.28). Wooden passenger cars were banished from regular service on the Pennsylvania in 1928.[64] The great steel fleet, numbering 5,501, represented an investment of 100 million dollars.

The Pennsylvania's example encouraged other railroads to accept the steel car. Contract builders now felt confident enough to invest in a new plant. With some reluctance, the Pullman Company allowed itself to be pushed into the steel era by the Pennsylvania. All cars, including sleepers, running into the New York terminal had to be of the new form. The age of experimentation was over; the period of full production had begun.

Figure 2.27   A class P-70 coach interior dating from about 1910.

Figure 2.28   A late-model class P-70 produced by the Standard Steel Car Company of Butler, Pennsylvania, around 1928.

Other trunk roads were forced to follow the Pennsylvania's example if they wished to remain competitive. Their name trains, at least, had to be converted at once. As already mentioned, the Southern Pacific was close behind the Pennsylvania. By 1910 another ten major roads had steel cars on order or in service. Among these were some relatively small lines. The Lehigh Valley, for example, ordered forty steel cars late in 1910 and publicly committed itself to the purchase of steel cars only after this date (Figure 2.29).[65] The Rock Island placed a small order but would make no statement about its future plans. The Chicago and North Western was ready to convert to steel, and after receiving its first steel cars in 1910, moved rapidly toward retiring its wooden stock.[66] Within three years nearly a third of its passenger equipment was steel or steel-framed. Of its 2,085 cars, 391 were all-steel and 255 were steel-framed. At this point management's zeal waned. The road had established its image as a modernized operation, since enough steel cars were on hand for the fast name trains, and it continued to use wooden cars on the secondary, commuter, and branch-line trains. Other roads also adopted this strategy. The Southern Pacific, so quick to accept the new form,

Figure 2.29 Even relatively small lines such as the Lehigh Valley began to buy steel cars exclusively after 1910. The Number 974 was built by Pullman in 1914. (Pullman Neg. 17736)

Figure 2.30 For a few years early in the steel era, the Santa Fe and some other lines purchased steel-frame, wooden-body cars. The Number 948 was built by Pullman in 1910. (Pullman Neg. 12534)

was criticized in later years for its reluctance to complete the conversion.[67]

Some lines that wished to enter the steel era even more cautiously were content to use only steel underframes. One was the Santa Fe, which had been a pioneer in developing steel-frame baggage cars.[68] In 1909 and 1910 a large number of composite cars of all types were built for the Santa Fe by Pullman (Figures 2.30 and 2.31). The framing followed an original plan prepared by the railroad's mechanical department. The center sill was an enormous fishbelly steel-plate fabrication 32 inches deep at its heaviest point. The side sills were double 10 ¾-inch channels fabricated from ¼-inch plates and angles riveted to one side. The lower channel was visible under the car's side; the upper channel was buried inside the wooden car body. More attention to cross bracing was given in this design than in plans for other early steel

cars, as can be seen from the drawings reproduced elsewhere in this volume (for example, Figure 2.19). No steel castings or special shapes were used; only a light angle-iron framework tied the upper wooden structure to the frame. By 1913 the road had about 550 steel-frame cars of this construction in service, which constituted one-third of its entire fleet. Among them were the finest cars on the line, including those used on the luxury train, the Santa Fe De Luxe.[69] They were criticized by all-steel advocates, but the Santa Fe pointed out that it had a special need for well-insulated cars to withstand the heat of desert routes. Its cars regularly encountered temperatures of more than 100 degrees, and conventional wooden-bodied cars gave the best protection. The Santa Fe contended, with considerable justification, that no effective method of insulation had been devised for steel cars. In addition, the road had already minimized the fire hazard by adopting

Figure 2.31   This composite car under construction for the Santa Fe in 1914 featured a steel upper frame. (Pullman Neg. 17867)

steam heat and electric lighting. But public criticism persisted, and with obvious misgivings the road began to introduce all-steel cars in 1913.

The Chicago, Milwaukee, and St. Paul was another company that first decided to invest in composite cars. The last of the great transcontinental lines, it needed new cars to inaugurate through service to the Pacific coast. Late in 1909 Barney and Smith delivered twenty-seven steel-framed cars, which were similar to the Santa Fe's except that they had an even more substantial underframe.[70] Like the Santa Fe cars, they combined an extremely deep fishbelly center sill with plate side sills. In this case the side plates were built up inside the car body so that they were entirely hidden. Several heavy cast-steel cross members were used. A more complete angle-iron upper framework, combined with the extraordinary underframe, made them the heaviest composite passenger cars built. At 76 tons they outweighed most all-steel cars and for this reason were considered a failure.

In 1911 the Milwaukee turned to all-steel construction, placing an order for 243 cars which, according to new designs, would each weigh only 67½ tons (Figure 2.32).[71] The composite car was dead; those that appeared in the future were conversions from wooden cars. (See Table 2.1.) Railroads were naturally eager to salvage their investment in wooden equipment, which could not be safely run with steel cars. Some lines worked out

their own design of underframing, but many purchased ready-made conversion units from the Commonwealth Steel Company (Figures 2.33 and 2.34).

For example, Indiana's struggling Monon line, with modest financial resources and minor passenger traffic, still wanted to offer "as good equipment as our neighbors are running."[72] In 1918 the line began to fit Commonwealth underframes on its relatively modern fleet of thirty-five wooden cars. All-steel trucks

Table 2.1   Wood and Steel Passenger Cars in Service

	WOOD	STEEL FRAME	STEEL	SUB-TOTAL	PULLMAN	TOTAL
1910	44,900	1,100	1,100	47,100	5,200	52,300
1915	40,000	5,200	10,800	56,000	7,300°	63,300
1920	34,500	6,500	15,100	56,100	7,800	63,100
1925	26,200	9,400	21,200	56,800	8,800	65,600
1930	15,100	10,400	29,000	54,500	9,900†	63,900
1935	5,000	8,400	29,000	42,400	8,000‡	50,400
1940	2,000	6,400	29,900	38,300	6,900	45,200
1945	1,400	5,400	31,800	38,600	8,600	47,200
1950	300	3,500	33,500	37,300	6,200	43,500

° One-third of Pullman cars were all steel by this date.
† 80 percent of Pullman cars were all steel by this date.
‡ All cars were steel or steel underframe; 98.2 percent were all steel.
Source: I.C.C. reports.

Figure 2.32 *A.C.F.'s Ohio Falls plant in Jefferson, Indiana, built this coach in 1910 for the Milwaukee Road. (Railroadians of America)*

Protects Life
and Property

The Old Car Body is Set Down on This Under
frame Without Alteration

Easily Applied

Can Also be
Arranged for
Fishbelly Sills

It Fits Wooden Cars
for Equal Service
with Steel Cars

(PATENTED)

Figure 2.33 *Thousands of late-model wooden cars were salvaged for extra years of service by the application of steel underframes.*

Figure 2.34 *Jackson and Sharp added this steel underframe to a car being produced for the New York Central in 1909.*

and clasp brakes were installed as well. By the late 1920s the truss rods and wooden bodies seemed too ancient for the times, and a second major modernization was undertaken. A new fishbelly underframe eliminated the truss rods and provided a stronger foundation. But many changes were only cosmetic, such as sheathing the exterior of the body with no. 12 steel sheets attached by large, roundheaded wooden screws. The screw slots were filled with putty and the screws were placed exactly to imitate rivets. A deep letter board, again of sheet iron, hid the old-fashioned upper sash windows. The clerestory windows were covered over. A modern interior completed the masquerade, which few travelers could distinguish from the modern steel car. Many of these coaches remained in service until after World War II, when the Monon again improvised a new passenger fleet by remodeling surplus Army hospital cars.

The Monon's efforts to salvage wooden cars may have been extreme, but to a lesser degree some other roads followed its example. Most resorted to steel underframing only as an expedient for secondary commuter trains, and the majority of these were discarded as soon as possible. The Kansas City Southern, like the Monon, steel-sheathed some cars, but the main purpose was to disguise their antiquity—in this case by covering the arched window transoms. The Delaware and Hudson did major surgery on some of its wooden cars, not only installing steel underframes but also replacing the clerestory roof with one of arched construction. Light steel framing was introduced into the body, and the interior was completely redone. Most of the work was on commuter cars, but the line rebuilt at least one parlor-café car in this fashion.[73]

After 1912 thousands of steel cars were in service, orders for wooden cars had ceased, builders' shops were producing record numbers of new cars, and trunk lines were volubly determined to convert to steel in the interests of safer travel. Yet the pace of conversion did not satisfy certain members of Congress, for some lines displayed a noticeable leisureliness about the changeover. The New York Central and the New Haven had no steel cars until 1912, the Jersey Central waited until 1914, the Northern Pacific until 1915, the Boston and Maine until 1916. Some roads pleaded poverty; some argued that the steel car was unsuitable for their operations; some were simply too conservative. Was a push needed? Congress decided to investigate the matter in 1913. Several bills had already been introduced at the urging of the Interstate Commerce Commission. A decade earlier, the state of New York had considered but rejected a bill requiring fireproof, steel-sided cars.[74] Around 1910 a bill sponsored by the reform-minded Congressman John J. Esch of Wisconsin required rapid conversion to all-steel construction and specified the exact nature of the construction and materials. This somewhat premature action was followed by other bills. In 1912 six similar proposals were pending that allowed periods of three to six years for the conversion.[75] One of the most popular (H.R. 6142) stipulated four years. A year earlier, legislation had been enacted requiring steel-frame mail cars, thus establishing a precedent for a more general law. However, the 1911 law in effect covered government employees (mail clerks). In addition, only 1,500 cars were involved. The railroads could live with this law because it was so limited; H.R. 6142, however, would encompass all railroad passenger cars. It was visionary, impractical, and unnecessary, and it would cost over 650 million dollars. The railroad industry was determined to stop it.

The roads first argued that the adoption of steel cars, even if they completely ensured passenger safety, would have little effect on overall railroad fatalities because passengers accounted for only 3 percent of such deaths. Most of the people killed in accidents were trainmen, tramps, and pedestrians. The enormous capital that Congress would force the roads to invest in steel cars could be more effectively spent to improve rails, signals, and grade crossings. A representative of the railway supply industry put particular emphasis on this point in the Congressional hearings. He declared that a myopic concentration on rolling stock alone would prevent other needed safety improvements and possibly result in lower standards of railroad service and passenger security. Furthermore, industry management argued, the railroads could not possibly meet the four-year deadline even if they could raise the cash, because the shops could not produce cars that quickly. As it was, they were hard-pressed to build 3,200 cars a year. Conceivably 6,000 a year could be constructed, but it was unlikely that many new investors would care to enter the field for a short-run market; the capital costs were too great. According to industry estimates, ten years would be the shortest conversion schedule possible. The railroads admitted, however, that at the present rate of replacement a complete conversion would require thirty-seven years. As it happened, this seemingly outrageous estimate was a correct prediction of the actual course of events.

Several railroad officials presented their arguments against the pending legislation. Nearly all agreed that steel cars were a desirable improvement and that they should be adopted for through express trains. To some, the matter of first cost was important. A good wooden coach cost $5,500 to $7,000, whereas one of steel cost $12,500 to $14,000. The Santa Fe maintained that steel underframing was sufficient and made an issue of the discomforts of all-steel cars in hot climates. The Northern Pacific admitted that it did not have a single steel car and explained that it was overstocked with passenger equipment. The road had bought heavily in preparation for the 1909 Yukon Exposition and, now that traffic had returned to normal, owned more cars than it needed. To restock at that point seemed a foolish waste of capital.

The case of the short lines was presented by C. W. Pidcock, proprietor of numerous small roads in the South. He said that the proposed law would work an incalculable hardship on the small railroads, for they had no money for such luxuries. Anyhow, steel cars were totally unnecessary for slow-moving trains, and short lines should be exempted from any such legislation. Pidcock admitted to a personal prejudice against the newfangled "battleship cars;" they were ice-cream freezers in winter and bake ovens in summer.

Figure 2.35   The fishbelly underframe was the most common form of passenger car framing in the heavyweight era, which roughly spanned the years 1910 to 1930.

Figure 2.36   The entire foundation of the car was its underframe. The body contributed almost nothing to the car's strength; hence the massive dimensions of the underframe. (Pullman Neg. 12484)

The industry won the battle. The law was not passed, although the I.C.C. continued to sponsor similar bills as late as 1918. The arguments against the law were generally reasonable; forced conversion at the proposed rate was unjustified. As one witness acknowledged, however, there were "some men who need urging to get up to date."

Age of the Heavyweights—Refinements in Steel Passenger Cars

So far, the steel car had brought about a revolution in materials and construction but not in form or style. Efforts to break away from existing patterns had for the most part died quietly in the storage cabinets of the Patent Office. After turning out a few experimental models, car builders settled on a general design which was seldom seriously challenged from about 1910 to 1930. Thousands were built during this period, the so-called standard era. Their main features were a clerestory roof, a riveted steel body, a pair of four- or six-wheel trucks, and a fishbelly underframe. Solid and extremely heavy, they proved wonderfully durable. Many saw more than fifty years of main-line service.

A massive steel underframe formed the backbone of the car; it was in effect a bridge on which the car body rested (Figures 2.35 to 2.37). Unlike the wooden car, the floor frame in steel cars composed almost the entire structural foundation. The side panels were generally not expected to absorb much of the static (weight) or dynamic (buffering) forces. Of several designs tried

Figure 2.37   Fabricated from heavy steel plate, castings, and bent shapes, most underframes weighed 15 to 20 tons.

out in the early steel car period, most roads preferred the fishbelly, a huge center sill made of two trapezoidal steel plates. The plates, usually ⁵⁄₁₆ inch thick by 26 inches deep, were set about 18 inches apart. They were stiffened by angles riveted at the upper and lower edges. Sometimes a top and bottom plate were used as well. The outer ends were tapered to about 12 inches to clear the axles and truck bolster. Longer cars, 70 feet and over, almost always had fishbelly underframes. (Short or very light suburban cars were given a variety of underframes—sometimes the fishbelly, but often much simpler I or channel beam frames. Channel beams were particularly noticeable in suburban cars after the mid-1920s.)

The fishbelly design appears to have been first used in 1904 by Standard Steel. It was then adopted by Pullman, the largest passenger car builder in the world. Because of Pullman's prestige and huge production, the fishbelly was soon to be found under the majority of passenger cars operating in North America. For all practical purposes it was the standard underframe. And there was much to be said in its favor. It was enormously strong and proved effective in spanning the great lengths required for long double-truck cars. It was relatively simple and cheap to fabricate.

Its principal defect was its weight of 13 to 15 tons, which was 15 to 20 percent of the car's total weight. Yet no competing design could support an 80-foot car without employing side-panel truss that nearly equaled the fishbelly's weight and far exceeded its complexity of construction. A less important disadvantage was that the fishbelly cut the underside of the car in two and thus precluded any easy arrangement of the brake rigging, piping, wiring, and other underbody gear. The low-hanging frame also prevented complete inspection of the running gear from one side of the car. Maintenance and operating departments somehow came to live with these handicaps.

The chief exception to this general plan of framing was the Pennsylvania Railroad, which stayed with the box girder it had developed for its first series of 80-foot cars in 1907 (Figure 2.38). That too was a center-sill plan, although it did not thicken at the center but remained uniform in size for the full length of the car. To ensure the strength of this member, special 80-foot plates were produced by the Cambria Steel Works. In addition, a 1-inch camber was built in. After several years' service the design was pronounced a failure by the authoritative Gutbrod Report because the cars sagged in the middle and the cross framing was

Figure 2.38 Before 1913 the Pennsylvania refused to adopt the fishbelly underframe and followed the plan shown in this drawing. (Railway Age, June 7, 1907)

Figure 2.39 Lewis B. Stillwell patented and oversaw the construction of about 900 passenger cars—mostly for suburban or subway service—using his pressed-steel designs. Note the side-panel truss.

inadequate.[76] In 1913 the road was reported ready to adopt the fishbelly plan, but apparently the report was untrue, for all subsequent P-70s were built with the box-girder frame.

As might be expected, several alternatives to center-sill framing were proposed. A.C.F.'s combined floor and side-panel framing has already been described. In addition to the experimental Long Island car of 1906, A.C.F. built several hundred cars, a number of them for the Rock Island, with this style of framing.[77] A more radical scheme was offered in 1906 by Lewis B. Stillwell, then

consulting engineer with the Manhattan Tubes.[78] Stillwell envisioned a lightweight car that depended on pressed-steel trusses built into—and as deep as—the car body itself. The light and shallow floor frame was a secondary support. The side trusses were formed from 1/4-inch pressed-steel panels. These units were riveted or butt-welded and could form a car of any desired length. The upper portion of the panels was open for windows. The roof members, also of pressed steel, were active parts of the structure and not merely a light, independent framing (Fig-

Figure 2.40   The Standard Steel Car Company built this Stillwell car for the Erie around 1928.

Figure 2.41   An interior photograph of the Stillwell car shown in Figure 2.40.

ure 2.39). Here, then, was the old wooden car's side-panel truss, now adapted to modern steel fabrication.

Fifty cars on this plan were built for the Manhattan Tubes in 1907 by Pressed Steel and A.C.F.[79] A much larger group of subway cars was completed several years later. The line seemed pleased with the design and continued to add Stillwell cars to its fleet as late as 1928. The inventor also succeeded in selling his design to several suburban operations. The Erie, for instance, began buying it in 1915 and filled out its New York suburban service largely with Stillwell cars, many of which were still in use in 1970. A late example is shown in Figures 2.40 and 2.41. In 1917

the Erie purchased a small group of Stillwells for main-line service.[80] The design was somewhat modified to replace the arched Gothic windows and peculiar ogee roof with more conventional forms (Figure 2.40), but the cars were barely full length: only 78 feet. Their impressively light weight of 59 tons was admired, but it did not lead to wide acceptance of Stillwell's design. The inventor lived to see some 900 cars, mostly short suburban ones, built on his plan; the last order arrived in 1934.

The body of the standard steel passenger car was in effect a light steel box resting on a plate-girder bridge. Its light frame of rolled shapes was covered with $\frac{1}{8}$-inch steel plates. The side panel was usually divided into six plates. The joints were covered by a welt plate. A heavy pressed-steel belt rail was riveted below the windows. In all it was a coarse-looking, rivet-studded structure. Compared with the elegant wooden cars, it was as heavy and brutish as a battleship. Pullman, which in particular considered the rivets unsightly, attempted to produce a smooth-sided car by countersinking them, but this proved more expensive and less secure than roundheaded rivets. Next Pullman tried a hollow, pressed-metal siding made in exact imitation of wooden car siding. This plan effectively hid the rivets, and the company used it in a fair number of cars. By about 1912, however, even Pullman gave up the fight and came to accept organic steel car fabrication. Thus little concession to style or fineness was made in the heavyweights; they were plain and utilitarian.

The heavyweights' weakness of body was a more serious disadvantage. Since all of the basic structure was in the floor frame, the lightly built bodies provided little better security for passengers than wooden cars with reinforced ends. Because of the great weight of the floor frame and trucks, designers were tempted to skimp on other parts of the structure to hold down overall weight. A series of disastrous wrecks between 1911 and 1917 drew attention to this inexcusable oversight in the first generation of steel cars (Figure 2.42). Ten people were killed on the Milwaukee Road in a rear-end collision that forced a diner one-third of the way through a sleeper.[81] The wreck was described as a high-level telescope, for the diner invaded the upper portion of the sleeper's body, crashing through the end bulkhead and rolling back the roof. The frames of both cars were undamaged—a clear indication that this portion of the structure was overconstructed. Six years later even more people were killed when a Pennsylvania

Figure 2.42 *Steel cars were not indestructible, as shown by this unidentified wreck photographed at the Bradley Car Works around 1920.*

Figure 2.43 *The steel car's vulnerability was tragically demonstrated by this telescope disaster on the Pennsylvania Railroad at Mt. Union, Pennsylvania, in 1917.* (Robert C. Reed)

Commonwealth Upright End Frame in One Piece, and Commonwealth Combined Platform and Double
Body Bolster for Vestibuled Cars.

*Figure 2.44   Cast-steel end frames provided one solution to the problem of telescoping
accidents.* (Car Builders' Cyclopedia, *1928*)

*Figure 2.45   In 1913 the Barney and Smith Car Company offered a collapsible car end
meant to absorb the impact of rear-end collisions.* (American Railroad Journal, *February
1913*)

Figure 2.46   Steel I beams hidden by the exterior plates reinforced the ends of most cars. (Pullman Neg. 27989)

freight train running at 40 miles per hour rammed a passenger train at Mount Union, Pennsylvania (Figure 2.43).[82] The sleeping car *Bellwood* was completely telescoped by the sleeper ahead. The floor of the *Bellwood* passed under the forward car, *Bruceville*. The body of the *Bellwood* was slit open like a tin can; twenty people died. The *Bruceville* sustained little damage beyond a few broken windows, and none of its passengers were injured.

Stronger end roof construction was needed to prevent a recurrence of such horrors, although there was no design that could provide absolute safety while staying within some reasonable weight limit. As early as 1905 William F. Kiesel, Jr., of the Pennsylvania Railroad patented an I- and Z-beam reinforced-end design.[83] The good showing of the Pennsylvania steel cars in the Tyrone, Pennsylvania, wreck of July 1913 was probably attributable to Kiesel's design. In that major accident, which involved a number of cars, there were no passenger fatalities.

Around 1910 Commonwealth Steel offered a massive cast-steel

end combined with the end portion of the floor frame (Figure 2.44).[84] Its strength was unquestionable, but weight and cost were against it. Even so, a number of roads used this design, and it was offered throughout the standard steel car era.[85] A less monumental solution to the telescoping danger was offered in 1913 by Barney and Smith (Figure 2.45).[86] A collapsible platform and vestibule were meant to absorb as much of the collision shock as possible. If the entire train were fitted with collapsible platforms, the major impact of a collision could be dissipated through the failure of the platforms. The end bulkhead was braced with heavy steel beams, and the structure was carried up into the roof to prevent a high-level telescope. Whether this plan was widely used is questionable, for Barney and Smith's time as a major builder was nearly at an end.

Stillwell, of course, was advocating a stronger body frame for both light weight and passenger safety during these years. A similar scheme that was presented to the New York Railroad Club in 1921 by F. M. Brinckerhoff suggested adding loops of

Figure 2.47   Heavy blankets of insulation were needed to stabilize car temperatures in all seasons.

Figure 2.48   The mounting of car insulation is shown in this drawing of about 1925. (Car Builders' Cyclopedia, 1928)

wire cable to the ends of cars. The loops, elastic but strong, were to be firmly anchored to the frame so that they would act as giant shock absorbers. It is unknown whether this invention ever received a trial. No complete solution to the telescoping hazard was found. The various partial solutions that emerged were, like so many contributions in the everyday world of engineering, anonymous and largely unnoticed refinements of conventional patterns (Figure 2.46). In general, end construction was reinforced by heavier posts and plates; wide letter boards strengthened the upper body structure.

Good insulation was a basic necessity in transforming the uninviting steel body of the heavyweight car into a comfortable passenger compartment. Because of the high heat conductivity of the steel plates, it was difficult to maintain an agreeable temperature in both hot and cold weather. Here was a new problem for the passenger car builder that did not exist in the wooden car era. Fortunately, designers could draw upon the experience accumulated from the refrigerator car. By then this technology was well advanced, and many effective insulating materials were on the market.

At first builders tried gluing thin sheet insulation materials inside the body panels, leaving a large dead air space. But a completely dead air space was impossible to maintain. In service the body joints loosened, and this together with the movement of the train set up air currents that caused an unwanted heat transfer. Loose insulating materials were equally ineffective, for they tended to pack down and leave large unprotected gaps. The ideal material should have mechanical integrity, should not attract vermin, and should be odorless and nonflammable. To add to the

difficulties, the more fire-resistant materials tended to provide the poorest insulation. One of the best commercial materials was a three-ply, ¾-inch-thick substance developed by the Johns-Manville Manufacturing Company under the trade name of Salamander (Figures 2.47 and 2.48). It was in wide use before 1920.[87] Other hair felts and asbestos board materials were used as well. In extremely cold climates a triple layer of insulation was necessary. The Northern Pacific used an insulated divider panel inside the body walls of its cars.[88] This cut the air space in two, but even so, the temperature differentials experienced along this line (where temperatures sometimes dropped to 50 degrees below zero) caused so much working of the interior and exterior

Figure 2.49   Workmen spread concrete over a pressed-metal subfloor in this 1916 construction scene.

panels that rivets were loosened and tap screws stripped their threads.[89] Wooden interior paneling solved the problem.

Another construction method that improved insulation was introduced in about 1913.[90] Insulating strips were placed between all exterior and interior plates and structural members to retard heat transfer. Double or false flooring was used to provide a large insulating space. But even these techniques did not seem to solve the problem, for there was a continuing demand for better car insulation throughout the 1920s.

Floor construction also underwent a radical change with the coming of the steel car. The wooden floor was largely eliminated, except in baggage cars and Canadian cars. Most American roads felt that wood was a fire and health hazard and adopted various forms of thin concrete floors (Figure 2.49). Such floors had had limited use in toilets before the steel era opened. In constructing them, a corrugated pressed metal or wire mesh was first laid down. The metal was generally galvanized. Over this, composition mortar composed of magnesium or asbestos and cement was poured in a layer ¾ inch thick. An underflooring formed a space

both for insulation and for deadening sound. This secondary role of damping down noise was also performed by the car's body insulation.[91] One of the traditional complaints about metal cars was the racket they made. The side-panel sheets produced a drumming sound, while track and undercarriage noises were transmitted through the floor. The fact that the car's heavy insulation filters out noise as well as controlling the temperature is clear to anyone who steps from the passenger compartment to the end platform of a moving train. The din that assails the ears, though some of it is caused by the clatter of the deck plates, draft rigging, and so on, largely represents the contrast between an insulated and an uninsulated area.

The inadequate roof framing of the earliest steel cars, the consequence of efforts to reduce weight, was never really corrected except in the Stillwell car. Part of the problem was the continued use of the clerestory, an inherently weak structure. Here style prevailed over engineering. Designers could argue persuasively in favor of the economy and strength of the arched roof, which they sometimes tried to make more appealing by calling it the semi-

Figure 2.50   Lightweight, arched-roof cars were used by many lines for suburban service. The Number 2601 was built around 1925 by the Standard Steel Car Company.

Figure 2.51   Interior view of the commuter car shown in Figure 2.50.

elliptical or turtleback roof. But to no avail; railroad managements preferred the clerestory. By 1920 it was an even greater anachronism, the "deck lights" (clerestory windows) had been replaced by blank panels, and exhaust ventilation was handled by sheet-metal ventilators mounted at intervals along the blind sides of the clerestory.

A few roads, notably the Harriman Lines, did break away from the clerestory roof. Hundreds of arched-roof cars built to a standard design were operated on the Rock Island, the Union Pacific, the Southern Pacific, and the Illinois Central. But with the breakup of the Harriman empire, most of the associated lines returned to the safe convention of the clerestory.

A new interest in the arched roof appeared in the 1920s, when many roads adopted the design for suburban cars. Among them were the Boston and Albany, the Chicago and North Western, the Rock Island, and the Missouri Pacific (see Figures 2.50 and 2.51). The Delaware and Hudson rebuilt some old wooden cars with turtleback roofs. But what was acceptable for suburban service was not considered good enough for main-line trains. In 1924 Charles E. Barba, chief mechanical engineer of the Bradley Car Company, presented a reasoned argument in favor of the arched roof for all classes of passenger cars.[92] His chief points were not only its strength and cheapness, but also the broader necessity for achieving mass production through standard designs. Although the arched roof did come into common use during the following decade, car building remained a custom business.

During the early years of the steel car, the mechanics of roof construction posed some special problems. The insistence on fireproof cars forced designers into all-metal roof construction. The elaborate clerestory was easily translated from wood to steel, but making it watertight was another matter. The Pennsylvania first tried soft solder joints, but these failed because of flexing of the roof, and around 1910 the road resorted to rivets and welding.[93] Pullman devised a cap joint that proved watertight, though not beautiful.[94] The ends of the short roof plates ($\frac{1}{16}$ inch thick) were flanged and riveted together, and the joint was filled with red lead paint. A pressed-steel cap strip was then placed over the ribs and riveted in place. This roof style was a distinctive feature of Pullman-built equipment until about 1915. The extra initial cost and the maintenance problems connected with an all-metal roof caused many roads to stay with wood-sheathed, canvas-covered roofs. The treated canvas was waterproof and fire-resistant, and the insulation problems were more easily solved.[95]

Among the more interesting developments during the standard steel era was the evolution in the materials used. Some designers of the first-generation heavyweights believed that a flat-sided car body could be made only if precise cold-rolled sheets were used.[96] These plates were indeed true and straight, but they were distorted in the process of fabrication no matter how much care was taken in lining up the frames or how many shims were used. Even if the body left the erecting shop dead true, it soon developed the minor dents, creases, and distortions characteristic of steel cars after a few months' service. The high cost of the cold-rolled plates soon led car builders to substitute common hot-rolled steel sheets. This material was strong, cheap, and readily available, but it had little resistance to corrosion and deteriorated rapidly. By the late 1920s, cars built fifteen or twenty years before were rusting out at an alarming rate.[97] Floors, roofs, sides, and even structural members had rusted through. Regular painting was not enough; a rust-resistant material was needed. In 1916 the American Society for Testing Materials conducted a series of experiments to find a steel alloy that would fill the need.[98] The researchers discovered that even a trace of copper was an effective rust inhibitor. Special copper-alloy steels containing .20 to .25 percent copper were developed for this market. Since the alloys added only about 1½ percent to the cost of a car, they were considered good economy and were widely adopted by the late 1920s.[99] In the following decade U.S. Steel introduced a chromium-copper-silicon steel alloy called Cor-Ten—a low-carbon steel with a durability twice that of the copper-bearing steels.

During the same period there was some talk but little action on the use of high-tensile steel alloys, which would permit the installation of lighter structural members and thus reduce the dead weight. One advocate claimed that passenger car weight could be cut as much as one-third,[100] but other authorities were skeptical and at least one thought that such alloys could be dangerous in practice. Charles E. Barba of the Bradley Car Company feared that shop crews could not be trusted to make repairs with the right grade of steel.[101] What if in their haste or ignorance they replaced a critical structural piece with an ordinary steel plate? Perhaps the alloys actually were tested during this period despite Barba's misgivings, but no evidence of their use can be found. However, alloy steels were incorporated into car construction during the next decade.

A more radical solution to rust and weight problems was the proposal to introduce aluminum. It had received some use since the beginning of the modern steel car era, although it was largely restricted to nonstructural parts such as seat frames, handles, and other bits of hardware. The Pennsylvania Railroad and Pullman tried aluminum sheets for the interior panels of their 1907 experimental cars, but they did not repeat the scheme in the production cars that followed.[102]

The first large-scale use of aluminum was in 1923, when Pullman delivered twenty-five suburban cars to the Illinois Central Railroad.[103] The roof sheets, doors, conduits, junction boxes, and other small fittings were aluminum, with a resultant weight saving of 7,750 pounds. Over the next five years the I C acquired another 235 cars on the same pattern. In 1926 the Pennsylvania Railroad pushed the cause of lightweight aluminum cars even further by building eight 64-foot suburban cars in its own shops. The entire body, including posts, carlines, doors, and interior and exterior sheeting, was aluminum. Heat-treated aluminum was used in the body framing. Over 6 tons of dead weight was saved. It is not known if the experiment was successful enough to lead to additional cars, but six years later Altoona did build a double-deck suburban car for the Long Island Railroad that made extensive use of aluminum. However, it seems that the industry was

Figure 2.52   *Large quantities of aluminum were used in many suburban cars to reduce weight. This car was built in 1929 by the Standard Steel Car Company.*

Figure 2.53   *This coach typical of the heavyweight era was produced in 1921 by the Standard Steel Car Company.*

Figure 2.54   *A coach built by Pullman for the Central of Georgia in 1925. Steel cars were incredibly durable; many stayed in service for more than forty years. (Pullman Neg. 29243)*

not ready for all-aluminum cars, because the Illinois Central's composite model was the example that others followed. The Chicago and North Western acquired 120 semi-aluminum cars in 1927, realizing a saving of 5,700 pounds in each car (Figure 2.52)[104] The Lackawanna acquired a large fleet of semi-aluminum cars for its electric suburban operations in the New York City area. But aluminum made little progress outside of the commuter car; the industry seemed convinced that all-steel construction alone was suitable for main-line service.

The twenty years following the prototype production of steel cars was a period of adjustment. In some ways it was also a time

Figure 2.55   *The Chesapeake and Ohio's coach Number 756 was completed in 1929 by the Standard Steel Car Company.*

of retreat from the rigid tenets of early steel car construction. The compulsive demand for *all*-steel, fireproof cars was relaxed, and many compromises were accepted. Wooden interiors and roofs reappeared, though not universally. A general softening of the interiors, particularly in first-class cars, was evidenced by a return to carpeting, draperies, and wooden furniture. And the cold, utilitarian, enameled, monochromatic interiors gave way to brighter colors and softer lines toward the end of the 1920s. During the same period the deluxe coach was introduced. Since the first steel coaches had aimed to maximize capacity, seats had been bunched tightly together. Now more thought was given to legroom and comfort. Instead of 80 passengers, a car might seat only 60. Individual chair seats were introduced. Possibly the finest day coaches of the period were the Imperial Salon Cars operated on the Chesapeake and Ohio's *Sportsman*. On one side of the car was a row of single parlor seats; on the other, the usual double seats. The car was thus open and uncongested, with wide aisles and few patrons. David P. Morgan captured the feeling of traveling in one of these luxurious coaches: "It was grand to doze in them over lazy Kentucky miles while the handsome Mountain-type up ahead threw her weight over the ridges. It was travel that savored of good, rich, comfortable tradition."[105] Few coach travelers were treated to such spacious accommodations, of course, but the trend was in this direction, and many older cars were upgraded to meet the new standard.

In general the years between 1910 and 1930 were not notable for substantive changes in passenger car design. It was a period much like the 1870s and 1880s, when orthodox design inspired great contentment. This is reflected in the almost total silence of the railroad trade press on passenger car design. There simply was nothing to report. One order of cars followed another with little or no variation. A slight difference in the arrangement of a new parlor-café car might inspire some notice, but the day coach remained the same. Only in the suburban cars was there some interesting design experimentation in the lightweight aluminum construction, arched roofs, and unusual seating already described.

Figure 2.56   *The day coach interior of the 1920s was plain and straightforward, without pretension or unnecessary ornament. (Pullman Neg. 31384)*

Figure 2.57   Late in the standard era, some lines began to install more comfortable seating and more decorative interiors as automobiles and intercity motor buses drew away rail traffic. But chair seating was not a new idea even in 1929. (New York Central Neg. 1891)

However, little of this thinking found its way into main-line cars until the following decade.

Age of the Lightweight

The effects of the Great Depression bore heavily on the railroads and their supply industries. Traffic was down, particularly first-class travel; passengers who still had money for luxuries were attracted to the highway and airway, both glamorous new avenues of travel. The car shops that had struggled to complete orders a few years before were now idle.

Some hope during these gloomy years was offered by a new design concept called streamlining. It presented a sleek, modern image of speed and innovation. What had been an obscure technical term in aerodynamics was made into a household word through an astute publicity campaign mounted by several railroad traffic departments. It succeeded in creating a general interest in railroading practically unknown since the opening of the first transcontinental line. In the early and mid-1930s streamlined trains appeared one after another and proved to be effective traffic generators. The attention to styling was wisely buttressed by emphasis on passenger comfort and faster schedules. According to *Railway Age*, "For the first time in many years, the words 'sold out' re-entered the ticket clerk's vocabulary."[106] The new trains were expected to add at least 20 percent in new business,

and on certain major runs, such as the Broadway Limited, traffic actually jumped 37 percent. How much of the increase was due to an overall improvement in the economy was never discussed. The reports on "Streamline, Light-Weight, High Speed Passenger Trains" prepared by Coverdale and Colpitts, consulting engineers of New York and for many years high-level advisers to the railroad industry, offer one of the most complete accounts of this subject. The general mechanical, financial, and traffic facts are given in a train-by-train account. By 1950 the firm's eighth report had grown to eighty-nine pages and offered particulars on 260 trains.

From an engineering viewpoint streamlining was little more than "a dramatic gesture" aimed at the public.[107] Its true aerodynamic functions were largely abandoned after the first few automotive trains. These undersized, super-lightweight trains did have scientifically designed front and rear profiles meant to reduce air friction, but they were little more than glamorized rail cars and were not found suitable on heavily traveled runs. For major trains the roads favored full-sized cars capable of interchange service. (The City of Salina, the Zephyr, and other super-lightweights are treated in Chapter 8 on rail cars.)

The more fundamental revolution that occurred in railroad car construction during the 1930s was in fabrication and materials. Many writers have called this period in railroad history the age of the streamliner, but a more accurate title might be the age of the lightweight. Streamlining became a vague word casually applied

Figure 2.58   In 1930 the Chesapeake and Ohio offered even more luxurious and spacious seating with a two-and-one arrangement of reclining chairs. (Pullman Neg. 36050)

to any appliance with rounded corners and airflow lines. In reality, railroad cars remained rectangular boxes with only the most superficial vestiges of streamlining in such incidentals as seats, lamps, letter boards, and exterior sheathing. The arched roof—hardly a product of streamlining—was an old form that this book has discussed repeatedly in connection with earlier cars. Possibly the round-end observation car can be fully credited to streamlining, but it was more spectacular than functional.

Streamlining, however, had a prestigious scientific heritage. Its practical application in aerodynamics had been repeatedly demonstrated since the Wright brothers. Now engineers were calling for its application to land vehicles. In 1930 Professor F. W. Pawlowski of the University of Michigan conducted wind tunnel tests for the Brill Company's projected high-speed interurban car. A year later Dr. Oscar Tietjen of the Westinghouse Laboratories carried out similar experiments and concluded that streamlining could save nearly one-third of a train's power at speeds of 75 miles per hour. These and several other wind tunnel tests confirmed the results of experiments by Professor W. F. M. Goss at Purdue University in 1898.[108] Actually, there had been pioneers in railroad streamlining since the 1830s.[109] Prow-fronted locomotives had been tested in this country and in Great Britain during the first decade of the public railway. In 1865 the Reverend Samuel Calthrop (1829–1917) patented a streamlined train of remarkably modern outlines. Some years later a Denver newspaperman, Fredrick U. Adams, patented an equally futuristic

design that was given an actual road test in May and June of 1900 (Figure 2.59). The B & O Railroad remodeled several old coaches according to Adams's plan, but they were soon stripped of their fluted shrouding and returned to service as ordinary wooden cars. William R. McKeen of the Union Pacific did his part in perpetuating the wind-splitting idea during the early years of this century with his knife-edge gasoline rail cars. These ancient adventures were forgotten until the 1930s.

The railroads had been slow to capitalize on the potential of streamlining; the pioneers were ignored because they were ahead of their time. But by the thirties the times were indeed favorable, for the railroads were seeking a new image. Enthusiasm for streamlining was clothed in the sacred vestments of science. It seemed a natural—the perfect combination of glamour and utility.

But what was convincing in the laboratory became much less so when tested in the field. For short trains, molded front and rear profiles did have a measurable advantage in speed and fuel economy. But the railroads' express trains were long, not short, and their larger surface areas created a drag or "skin friction" that substantially diminished the promised savings of streamlining.[110] The more zealous advocates of streamlining insisted on flush windows and doors, full-width vestibules, smooth underbody sheathing, and even air-flow shrouds for the trucks. These features did reduce air turbulence, but at a considerable price both in initial cost and in maintenance expense. Most of them became

Figure 2.59   Adams's experimental train, shown here on the B & O in 1900, was visually remarkably advanced. It anticipated the basic designs followed thirty-five years later in the streamline era.

the pet hates of the repair crews. Of the relatively few cars ever treated to full streamlining, most were assigned to the articulated motor trains. The only streamlined feature to be adopted as standard practice was the full-width vestibule covering. After the initial enthusiasm, the excesses of streamlining were abandoned. Superficially the style was perpetuated in the new cars, but many of the true elements of streamlining vanished by about 1937.

Lightweight Construction

Streamlining in a corrupt form governed the appearance of the modern passenger car, but its real essence was lightweight construction. Weight, not air friction, was the chief obstacle to economical operation. Lighter construction had obviously been of long-standing concern to the master car builder, yet cars grew heavier with each major advance in comfort or safety. Every improvement seemed to add that many more tons to the train's groaning load. With the coming of the steel car, weights zoomed. Soon coaches were 65 to 80 tons each, sleepers and diners 85 to 90 tons. Train weights in the period 1910–1930 totaled as much as 1,000 to 1,200 tons. There was some talk, to be sure, of the need for lightweight cars, but little was done about it. The only major weight savings were achieved in suburban cars, where arched roofs, lighter framing, a slight reduction in overall size, and aluminum fittings pared away as much as 6 tons per car. These compromises were not thought safe or suitable for main-line cars. But

the light-car advocates persisted, and by the end of the 1920s more serious thought was being given to the problem. Besides the obvious economies possible with lighter cars, some roads were facing serious operational difficulties because of their heavy trains.[111] On densely traveled lines, schedules were congested by the slow starting and stopping of the heavyweights. The alternatives were larger locomotives or more tracks or both, and these were expensive solutions. Lightweight cars began to look very attractive.

The car builder could choose among several remedies to the weight problem. He could:

1. Reduce the overall car size.
2. Reduce the auxiliary equipment.
3. Adopt articulated design.
4. Turn to new structural designs.
5. Use new methods of fabrication.
6. Exploit new materials.
7. Reduce truck weights by using fewer axles.

The first three possibilities were rejected by car builders, while the last four became standard practice. To summarize the consensus of railroad opinion on these points: (1) Undersized cars could effect considerable weight savings—as much as 10 tons in cars of otherwise equal construction.[112] But a major attraction of railway travel was the roomy, full-sized car—the most commodious form of land transportation. (2) The trend toward greater luxury in travel was steadily increasing the amount of auxiliary equipment. Electric lighting and air conditioning in particular added considerable weight. Stripped-down models might be acceptable for short-run suburban cars, but they could never satisfy the long-distance traveler. (3) Articulation eliminated the weight of one truck, which was a substantial saving in a train of fourteen cars. But since it fixed the consists and made it necessary to shorten individual body lengths to keep within swing-out clearances, it was not acceptable in American operating practices. A limited number of "married pairs" (two cars semipermanently coupled together) did come into use in this country. (4) New structural designs were accepted that featured the arched roof and side-panel trusses in place of the traditional clerestory roof and fishbelly underframe. Both had already been used in the wooden car era, but the development of strong alloys made them possible in metal construction. (5) New methods of construction, such as welding that displaced riveting, made light-metal fabrication practical. (6) New materials that were exploited included strong aluminum alloys and high-tensile, rust-resistant steel alloys (Cor-Ten and stainless). It was these, combined with new designs and new methods of fabrication (4 and 5), that produced the modern lightweight car. (7) Improved four-wheel trucks (weighing 18 tons a pair) displaced the heavy six-wheel design (24 tons a pair) typical of the heavyweight car.

All seven remedies were based on techniques and materials available long before their adoption in the 1930s. Why weren't they used sooner? Why did the heavyweight car survive for so long? Tradition, of course, was unquestionably a factor. Once a dependable plan is in use, operating men are reluctant to throw it over. Years of costly development go into every workable design, and practical men rightly distrust those who would discard a proved plan for an unproved one no matter how superior the new plan might seem on first examination. Also, as with the wooden car a few generations earlier, plants and personnel were geared to handle the heavyweights. The commercial builders were equally committed to them. The railroads themselves had an enormous investment in the standard steel car. They immediately asked: Will the lightweights work with our existing stock? Even if they are designed to interchange, will it be safe to run light and heavy cars together?

In addition to these general difficulties, the car builder of the standard era might not be able to obtain the special materials needed for lightweight manufacture. The heavyweight cars were made of common carbon steel, which was available in quantity, in a variety of sizes and shapes, and at low cost. The design was tailored to suit this relatively weak, highly corrosive material. High-tensile steels were specialties, supplied in small quantities, in limited sizes, and at high cost. Like carbon steel, they were highly corrosive; thus their lightweight potential was partially negated by the fear that thin-section structural members would quickly rust out. Availability and first cost, however, remained a large factor in the continued production of heavyweight carbon-steel cars. Only when specialized rust-resistant steels and aluminum alloys became cheaper and more widely available did the lightweight car come into its own.

The first lightweight coaches that appeared in the mid-1930s made the old standard cars look gross. Large weight savings had been effected; body shells were 45 percent lighter. The Milwaukee Road built some coaches of ordinary carbon steel that weighed only 48 tons. Pullman's experimental aluminum observation coach of 1933 probably set an all-time record at 37 tons. After the experimental stage was passed, the weight of the production-model lightweight coach was stabilized at about 50 tons.[113] But this total gradually began to creep up as more extras were added. Ironically, some fully outfitted lightweight cars began to *exceed* the weight of the old standard cars (the dome coaches on G.M.'s Train of Tomorrow, for example, weighed 73½ tons). The reason was the ever increasing size and number of auxiliaries; air conditioning alone added 4 to 5 tons. In general, the modern car's increased power demand greatly increased battery and generator size. There was also the added weight of the massive tightlock couplers, folding steps, high-speed brakes, and roller bearings. By the 1940s some savings were made in running-gear redesign. Hollow axles could save as much as 500 pounds per axle. Mounting brake cylinders on the trucks simplified brake riggings and thus saved 1½ tons per car. Even greater reductions were anticipated because of new truck designs, but outside of the Milwaukee Road's efforts, little was accomplished until composition brake shoes eliminated the need for clasp brake arrangements. Trucks continued to represent one-third of total car weight. The first steel coaches carried only 11,000 pounds of auxiliary systems, while the lightweight car was encumbered by some 48,500 pounds, which equaled 60 percent or more of body weight.[114] Because the heavyweights could seat more people (80), the dead weight per passenger was only about 1,500 pounds. A 60-passenger lightweight car, on the other hand, carried about 2,100 pounds per seat space.

Even after the lightweights were accepted, some railroads continued to buy new heavyweight cars. In the 1930s these purchases might be explained as inertia or the preference of an old-line mechanical officer. Among the last of the heavyweights built completely on traditional lines (with a clerestory roof, fishbelly underframe, and six-wheel trucks) was a group of twenty-five cars completed in 1938 for the Missouri, Kansas, Texas Railroad by American Car and Foundry. Most were scrapped before 1964.

However, it is more difficult to explain the acquisition of such equipment during the following decade. In 1942 the Chesapeake and Ohio took delivery of twenty 70-ton heavyweights from A.C.F.[115] Several concessions to modern design were made, such as air conditioning and an arched roof, but otherwise all the old features were present, including six-wheel trucks and double vestibules. At about the same time the Atlantic Coast Line received some similar cars from Bethlehem Steel (Figure 2.60). Weighing just under 78 tons, they were among the heaviest passenger coaches ever used by an American railroad and hence the heaviest ever used anywhere in the world. Even more improbable, A.C.F. delivered eleven coaches and four parlor cars to the Alton in 1948. They were built to the outline dimensions of lightweight cars, but much of the car body material was ordinary copper-

Figure 2.60 The heavyweight era officially ended in 1934, but a few standard steel cars were still being produced more than a decade later. Bethlehem Steel's Harlan division built several heavyweights from the plan shown here for the Atlantic Coast Line in 1943. (Car Builders' Cyclopedia, 1943)

bearing steel. The rivets and six-wheel trucks were a bold testament to their antique design.

The first streamlined motor trains, the U.P. M-10000 and the Burlington Zephyr, naturally monopolized the public's attention, but more lasting developments were underway at the very time that these spectacular trains were stealing the headlines in their 1934 cross-country tours. In 1933 Pullman completed two experimental, all-aluminum observation cars—full-sized vehicles meant to run with conventional equipment. In May 1934 the Milwaukee Road shops turned out their first lightweight coach, and large-scale production immediately followed. Early in the following year, the New Haven began receiving the first of its lightweight coaches from Pullman's Worcester, Massachusetts, plant. A.C.F. entered the lightweight field at midyear with two full-sized trains for the B & O, both of which soon moved to the Alton. Budd, which was initially committed to motor trains, completed its first full-sized coach early in 1936. By this date all the major car builders were producing lightweight cars. By late 1946, 6,500 lightweights were in service, but this figure represented only about 15 percent of all passenger cars running on American railroads.[116] In all, an estimated 9,000 lightweights were built before the passenger car market temporarily vanished in the late 1950s.

ALUMINUM

Of all the new materials available at the outset of the lightweight era, aluminum was the most attractive. It was, in fact, called the light metal. It already had an established reputation among railroad men because of its wide and successful use in suburban cars since the mid-1920s. The automotive industry had used it in vehicle construction even earlier. It was light, corrosion-resistant, plentiful, and reasonably cheap. It weighed only one-third as much as steel. It seemed the ideal material for car construction.

The aluminum producers were anxious to develop a large market for their product and sought every opportunity to promote it to the railroads.[117] Their continuous selling effort is responsible in large measure for the periodic "revivals" in aluminum car construction.

However, the inherent shortcomings of aluminum far outweighed its advantages. Common aluminum was, of course, much too weak for car construction. The wrought or heat-treated alloys that were developed did equal carbon steel in strength (17 ST, a popular alloy, was equivalent to 58,000 pounds of high-tensile steel), but they were somewhat short of Cor-Ten steel alloys (60,000 to 75,000 pounds) and woefully short of stainless (110,000 to 175,000 pounds).[118] As aluminum alloys became more refined, they grew not only heavier and more expensive but also more difficult to work. In the 1930s they cost 35 cents a pound, or ten times the price of Cor-Ten steel. (Part of their cost could be recovered in salvage, however.) The overall cost of an aluminum car was nearly 25 percent above that of a steel one,

though aluminum cars were slightly cheaper than those made of stainless steel.

Despite the softness of common aluminum, the alloys first offered for car building were surprisingly hard to work with.[119] They tended to crack when worked cold, and the inconvenience of heating each sheet and member was a major drawback. Because riveting was the only practical method of fabrication, there were countless holes to be punched. Welding was suitable only for nonstructural fastenings. Hot riveting was impossible, because aluminum's coefficient of expansion was so great that without extreme care, the assembly might buckle. Alloy aluminum rivets could not readily be worked cold, common aluminum rivets were too weak, and steel rivets reacted and rusted out quickly. In addition, welding had become the modern metalcrafter's principal tool, and a metal that could not be fabricated by this method was thought seriously deficient.

Another less severe objection to aluminum was that of appearance. Aluminum does not weather well, and it was found necessary to paint the exterior surface to maintain the spruce look expected of passenger equipment. Only when it was anodized could car aluminum be kept bright without regular polishing. In 1946 the Louisville and Nashville received some anodized cars from A.C.F. that had polished, fluted sides which seemed to the unknowing eye like stainless-steel-sheathed cars.[120] These polished panels were fitted in order to hide the unsightly body rivets. An alternate method, used as early as 1941 in a group of Union Pacific cars manufactured by Pullman, was to countersink and putty over the rivets before painting.

The most serious failing of aluminum was that corrosion played havoc with the structural parts and plates of aluminum cars, particularly where they were joined to steel members.[121] These connections were often at such crucial junctions as the center sills and the bolster. Deep pitting from electrolytic action resulted in structural failures. It most often developed on the car's underside, well hidden from view. The usual accumulation of roadway grime prevented detection until after the damage was well advanced. The Union Pacific undercoated its aluminum-framed cars to retard this action.[122] The aluminum industry assured the railroads that it had developed a new alloy to correct the problem. The original copper-bearing 17 ST alloy was replaced by 61 ST 6, a magnesium-aluminum alloy that was a more workable, less corrosive metal.[123] In the future better alloys than the ones that exist today will doubtless be developed, but there is skepticism among some mechanical men that the problem will ever be completely solved. Whether their pessimism is justified will not be known until the latest aluminum freight cars have completed a longer term of service.

Whatever the problems of the aluminum car, nearly 1,000 were built of this material, mostly for the U P, L & N, and K C S. The success of aluminum suburban cars led to planning for a mainline car in 1930. During the following year Pullman received an order for an experimental car, but actual construction was delayed for a year. The upcoming Century of Progress Exposition in Chicago rekindled interest in the project, and Pullman decided to

Figure 2.61   Pullman's ultra-light aluminum coach-observation car, completed in 1933, was one of the first streamline cars built in the United States.

Figure 2.62   The art deco-inspired interior of Pullman's 1933 coach-observation car.

produce two all-aluminum cars. The more famous, an observation sleeper named *George M. Pullman*, is described in Chapter 3 on sleeping cars. The other car, reportedly named the *City of Cheyenne*, was an observation coach (Figures 2.61 and 2.62). This 78-foot, 10-inch car seated 50 passengers.[124] Except for wheels, axles, springs, and couplers, it was completely aluminum. The truck side frames were pressed sheet aluminum. Even the body insulation was of crumpled aluminum foil. The center sill was formed from two shallow U-shaped pressings. It was capable of withstanding the Railway Mail Service buffering load, which was then 400,000 pounds. The most remarkable single feature of the car was its total weight of only 73,880 pounds. Of this, the air conditioning accounted for 6,880 pounds. After the close of the Century of Progress Exposition, the car was briefly in service on the Illinois Central and the Union Pacific railroads. It reportedly rode hard and was apparently returned to Pullman, where its ultimate disposition cannot be determined.

Pullman, undiscouraged by the generally cool reception given its aluminum twins, became something of a champion of this style of construction. W. H. Mussey, Pullman's research engineer, voiced his preference for aluminum cars in a talk before the Western Society of Engineers.[125] He noted his employer's long-standing interest in that material; the Illinois Central's semi-aluminum cars of 1923 were Pullman products. In addition, the City of Salina and other Union Pacific motor trains, all made in the same shops, were light-metal fabrications. Pullman was, of course, encouraged in its loyalty to aluminum by the Union Pacific, a steady customer. After the motor train craze had subsided in 1937, the U P began to purchase full-sized aluminum cars. Over the next few years it became the largest operator of such cars in the country. By 1955 it had a fleet of 450 aluminum cars, most of them built by Pullman.

Figure 2.63 *Among the earliest production lightweight cars in the United States were the aluminum cars built in 1935 for the Alton by the American Car and Foundry Company.* (Railway Mechanical Engineer, *May 1935*)

The American Car and Foundry Company also found itself in the aluminum car business, though perhaps with less outspoken enthusiasm. It supplied aluminum cars to the Kansas City Southern, the Louisville and Nashville, the Union Pacific, and the Missouri Pacific. Its first experience was in making the cars for the Alton's Abraham Lincoln.[126] The Alton's owner, the Baltimore and Ohio, ordered a nearly identical train from A.C.F. to be built of Cor-Ten steel for service on the Royal Blue.[127] The two trains provided a rare opportunity to compare the same equipment made from competing materials. The aluminum train entered service in June 1935, the steel train a month later. The cars of both trains were slightly undersized in all dimensions: they were 70 feet long, 12 feet, 7 inches high, and 9 feet, 9 inches wide. This is an important reason for their good showing as lightweights, and comparison with full-sized equipment is unfair. Comparison between the cars of the two trains, however, revealed that the aluminum coaches were 5 tons lighter. The table below gives other details.

	ALUMINUM COACH	COR-TEN COACH
Body weight	62,300 pounds	72,500 pounds
Trucks	24,600 pounds	24,600 pounds
Total	86,900 pounds	97,100 pounds

The complete eight-car steel train weighed 780,800 pounds, the aluminum train 699,540 pounds—a difference of 81,260 pounds in favor of the Abraham Lincoln.

There were steel center sills in the aluminum cars. Other framing members were aluminum, but steel rivets were used in the assembly (Figure 2.63). Although the body and roof sheets were riveted on, the plate joints were butt-welded and ground smooth (Figures 2.64 and 2.65).

Some attempt was made by both A.C.F. and Pullman over the next few years to use even more aluminum in the underframe structure. Extruded side sills of a complex cross section best described as a combination U and L were a feature of the Abraham Lincoln's frame. Z bars were employed for cross framing, and in later years the side posts were given the same shape. These became rather common structural elements in aluminum cars. A.C.F. favored an all-aluminum U-shaped center sill built up of plates and angles. The top plate of this giant 16- by 16-inch channel was ⅜ inch thick. An example of the plan is shown in the cross section of a car from the Missouri Pacific's Eagle (Figure 2.66). However, even in these cars the bolsters, cross bearers, and end sills were of steel.[128] It was also usual to make the body end posts of steel as a precaution against telescoping. After the A.A.R. raised its yield strength requirements for passenger car underframes in 1946, production was suspended on aluminum-frame cars. Steel frames then became a necessity, and some roads specified cast-steel end frames.

If aluminum was not entirely successful for railway car construction, it did continue as an important secondary material. It was widely used in nearly all modern cars for small fittings, accessories, and interior decoration. And the aluminum manufacturers are not yet willing to admit that an end has come to their railroad adventure. Several thousand aluminum-bodied freight and rapid-transit cars have been built in recent years. United Aircraft's sensational Turbo Train of 1967 was aluminum. And the L.R.C. (Light, Rapid, Comfortable), which was projected in 1967 as the very latest development in railroad passenger cars, was to be of the same construction.[129] A prototype based on the Canadian patent was completed in 1971 by the Montreal Locomotive Works, and the design was said to offer a comfortable and safe ride over existing tracks at speeds as high as 120 miles per hour. But although development work continues in Britain on a similar project, the A.P.T. (Advanced Passenger Train), the L.R.C. appears to be dead in North America. Amtrak's most recent orders have been for stainless or lightweight steel equipment. The aluminum passenger car appears to have no future on American railways.

Figure 2.64   A.C.F.'s 1935 riveted-aluminum cars for the Alton were somewhat smaller than conventional equipment of this period.

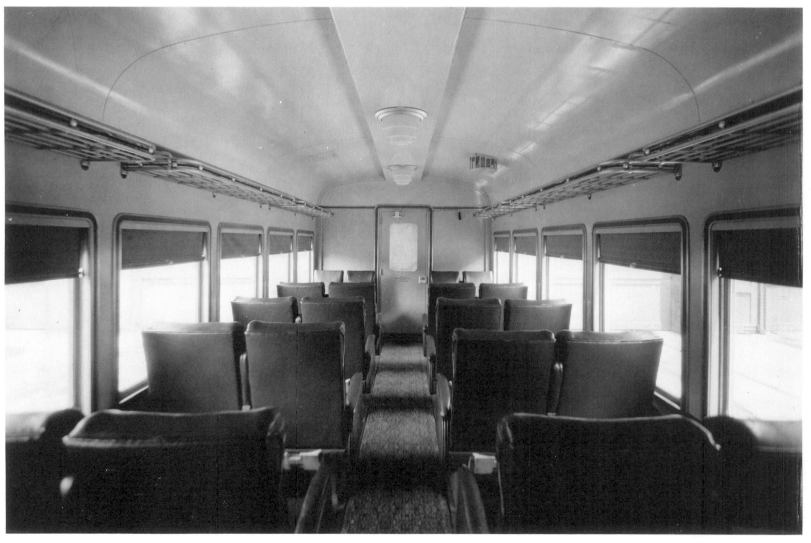

Figure 2.65   The 1935 Alton cars were traditionally furnished, in keeping with the views of the line's conservative management.

Figure 2.66 A.C.F. built this aluminum-bodied car in 1940 for the Missouri Pacific train, the Eagle. *(Railway Age, February 24, 1940)*

STAINLESS STEEL

At the beginning of the lightweight age the revolution in car building was advanced by an entirely new material: stainless steel. As a miracle metal it outclassed aluminum on several counts. It was incredibly strong, yet ductile. It did not rust, and its bright silver-grey appearance was most appealing. It had been introduced in 1912 by Krupp, but manufacturers had thought it suitable for little more than cutlery and decorative novelties. The metal, a combination of 18 percent chrome, 8 percent nickel, and low-carbon steel, seemed destined for only limited use.

Then the noble properties of stainless steel came to the attention of a Philadelphia automotive parts manufacturer, Edward G. Budd.[130] Budd's specialty was auto body stampings, but earlier he had been in the railway supply field and was looking for an opportunity to reenter it. He first heard about stainless steel in 1928 from a German contact. It did not sound particularly promising, but he investigated. With his knowledge of the sheet-metal trade, Budd soon realized that stainless steel had great potential for lightweight structures. A surprisingly strong body could be fabricated from thin sheet metal, but because of the rapid corrosion of common steel, it was not considered suitable for railway car construction. In addition, the higher-strength steel alloys needed for such fabrication required heat-treating, which would result in warpage in so complex a structure. With a strong, rust-proof material, however, extremely light structures would be possible. A beam of the correct proportions made of 30-gauge stainless-steel sheets was equal in strength to a solid I beam nearly three times as heavy. The stainless beam was a closed box, but there was no need for access to it in order to paint; it would never rust out. Budd also discovered that stainless steel was ductile: it could be pressed, drawn, and rolled without difficulty. A deep cup equal in length to its diameter could be formed with no metal failure. It was like working copper, yet this metal was nine times stronger than carbon steel and three times stronger than Cor-Ten. Even subzero temperatures have little effect on its strength or ductility. Cold-rolling resulted in a significant increase of tensile strength (up to 180,000 pounds), but again without sacrificing its easy-forming qualities.

Despite these positive findings, the matter of fabrication presented a serious difficulty. In thin-wall construction, rivets or sheet-metal screws were not considered satisfactory because of the large number of fastenings necessary for an adequate bearing surface. The matter of appearance was a second objection. Conventional gas or electric welding, even spot welding, created too much heat and ruined the metal, reducing it to little more than an inferior mild steel. A special welding technique was needed. The problem was given to Colonel Earl J. W. Ragsdale (1885–1946), Budd's chief engineer. His assignment was to devise a method of quick, extremely high-temperature welding followed by instant cooling, meanwhile preventing the weld from coming to the surface of the pieces being joined. By 1933 Ragsdale and his staff had perfected a method of electric, controlled-energy welding that is known today as the Shotweld process. Precision control is

essential; the current, applied for $\frac{1}{60}$ to $\frac{10}{60}$ second (at 2600 degrees), melts the metal of the two sheets to be joined, allowing them to intermingle and thus bond. The surrounding cold metal chills the weld. In the union of two $\frac{1}{8}$-inch sheets, the diameter of the weld is about $\frac{1}{4}$ inch.

The combination of stainless steel and Shotwelding added a new dimension to railroad car construction. In Budd's words: "The designer is therefore free to place this metal just where it is in the line of strain, and in just a sufficient amount to withstand the stresses." Andrew Carnegie was fond of saying, "Pioneering don't pay." Budd set out to prove that it can.

Even before the Shotweld process was ready, Budd wanted to demonstrate the possibilities of lightweight stainless fabrication.

Figure 2.67   The Union Pacific was the largest operator of aluminum passenger cars in this country. An unfinished body being fabricated for the U P is shown at A.C.F.'s plant in 1949. (American Car and Foundry Company)

Figure 2.68   Interior of a 1949 Union Pacific aluminum car before insulation was installed. (American Car and Foundry Company)

Figure 2.69 *A Union Pacific aluminum car of 1949 after insulation was added. (American Car and Foundry Company)*

Figure 2.70 *An aluminum car delivered to the Union Pacific in 1942 by Pullman. (Pullman Neg. 45837)*

Figure 2.71   Santa Fe coach Number 3070 was the first full-sized stainless-steel car built in the United States. It was manufactured by the Budd Company in 1936. (Robert Wayner)

In 1931 the firm built an airplane named the "Pioneer" that it claimed was the first all-steel aircraft. The plane was retired to the Franklin Institute after its demonstration days ended. In the same year Budd constructed a small gasoline-powered rail car, and over the next two years he produced several more of this type. Unfortunately, most of them had a rubber-tire drive, which placed them in the freak category and did little to advance the cause of stainless-steel rolling stock among practical railroaders. But in 1934 Budd's construction of the Burlington Zephyr quieted these misgivings, for here was a major motor train purchased by a class-one railroad.

Next Budd ventured into the regular passenger car market, a move which the established car builders regarded as an outrageous intrusion. More detached observers must have thought Budd mad for entering a market already overstocked with producers who had seen few orders since the 1929 crash. Nevertheless, his first full-sized car was delivered to the Santa Fe early in 1936. It was a 52-passenger coach, numbered 3070 (Figures 2.71 to 2.73). The 79-foot, 8-inch car weighed just under 42 tons—the lightweight record for an all-steel car.[131] The general design was a joint effort of the Budd and Santa Fe engineering departments.

Except for the side framing, the Number 3070 followed Budd's normal method of construction: All structural members were made of thin-gauge stainless steel, rolled or folded to shape and

Shotwelded. The outer sheathing was formed by fluted, snap-on stainless strips. The center sills were 12- by $1\frac{1}{2}$- by $\frac{1}{16}$-inch channels; the cross bearers measured $8\frac{1}{2}$ by $1\frac{1}{2}$ by $\frac{1}{16}$ inches and were set at 27-inch intervals. In place of the side-panel truss were large, upright channels ($10\frac{1}{2}$ by 3 inches) formed of $\frac{1}{16}$-inch stainless-steel sheet. (This method simplified fabrication, but it was apparently not stiff enough; in the next group of Santa Fe cars Budd returned to the Pratt truss side frame.[132]) As was typical of the lightweight cars, the roof served as part of the structure. Like a bridge truss, it was the upper chord or compression member, and was meant to take one-third of the total bending moment of the body. Besides the carlines, no heavy supports were used other than two stringers running the full length of the car and attached to the collision posts at both ends. The corrugated roof sheeting itself was the main roof stiffener. Being the largest unbroken surface area of the car, this great panel provided some strength for the body structure even though it was made of light-gauge sheet. In the words of Colonel Ragsdale: "It was not merely an umbrella to keep the weather out."[133] As an extra precaution against high-level telescoping, a collision bulkhead was formed by a 4-foot sheet welded to the carlines at each end of the roof.

The floor was corrugated stainless, but in place of concrete, a cork composition was laid in. Linoleum was used for the top

Figure 2.72   This drawing of a stainless-steel coach of 1936 was redrawn from an original tracing too badly soiled for reproduction. (Traced by John H. White, Jr.)

Figure 2.73   The snap-on fastening devised by Budd for fluted car sides. (Railway Mechanical Engineer, March 1936)

surface. The underbody was insulated with 3-inch Dry-Zero airplane blanket. As a gesture to earlier times, the interior was paneled in wood, but the paneling was only veneer mounted on presswood sheets.

The Number 3070 was an experiment, and its extra-light framing was found inadequate. Even so, the car continued on the Santa Fe until 1969, when it was sold to the Erie Lackawanna for

commuter service. Budd did decide, however, to return to the modified Pratt side-frame trussing and a more substantial center sill. The side truss was taken directly from freight car service and could be traced by the historian (but not the modern car builder) to the dark ages of the wooden car era. If the public could have been sold on triangular windows, a simple Pratt truss would have done the job nicely. But the rectangular form was enshrined in

Figure 2.74   A 1938 construction scene in Budd's shop showing the modified Pratt side-framing truss. (The Budd Company)

Figure 2.75   The fabricated Shotweld center sill devised by Budd around 1938. (Railway Mechanical Engineer, April 1939)

public taste, and it was necessary to use the modified Pratt or Vierendeel variety (Figure 2.74). The old-fashioned belt rail was retained to further ensure the squareness of the window openings. Budd had employed the side-frame truss since the time of the firm's earliest rail cars, but it introduced a new style of center sill in 1938 or 1939. The new sill was a complex weldment formed of six 3/16- and 1/4-inch stainless-steel plates drawn to shape specifically for this purpose (Figures 2.75 and 2.76). It was capable of withstanding a 900,000-pound buffering load. These changes materially improved the strength of the Budd cars, but they in-

evitably increased weight. Coaches built for the Seaboard Air Line in 1939 with the new center sill weighed 51 tons—up nearly 9 tons from the Santa Fe car.[134] Part of the additional weight came from the greater length of the S A L cars (84 feet, 8 inches versus 79 feet, 8 inches), but the sturdier structure accounted for much of the extra burden.

By the late 1930's it was clear that Budd had made a success of the stainless-steel passenger car. Silver cars of its manufacture were rolling on the Reading, the Santa Fe, the New York Central, and other major railroads. By 1944 the company had produced nearly 500 cars. The established car builders were understandably annoyed by the success of this upstart. Pullman's pique was fanned into outright hatred in 1937, when Budd began producing sleeping cars for the Santa Fe; this was sacred territory reserved exclusively for Pullman. Dark rumors were circulated deploring the unreliability of Shotwelding and Budd's general lack of experience, but they appear to have carried little weight outside Pullman's own sales office.[135] Pullman then decided to meet the competition head-on; it would produce its own version of the stainless-steel car.

Shotwelding, however, was a patented process held by Budd, and no satisfactory alternate method of fabrication had been devised. Therefore Pullman decided to produce an ersatz stainless-steel car, in which the structure and major fittings would be of Cor-Ten steel and only the side body panels and other bits of decorative trim would be stainless. A side-panel truss helped support the body; the outer corrugated side-panel covering was "loose-sheathed." In Pullman's lightweight steel cars the side-panel sheets were skin-stressed and thus served as part of the structure. The Southern Pacific was an early customer for stain-

Figure 2.76 A stainless-steel underframe being assembled upside down at Budd's factory in 1938. The entire car—not just the body or exterior surface—was of stainless steel, which made it extremely durable but expensive. (The Budd Company)

Figure 2.77 Pullman's answer to the competition from Budd was to produce a cheaper stainless-steel–sheathed car. Shown here is an articulated coach built for the Southern Pacific by Pullman in 1937. (Pullman Neg. 40558)

Figure 2.79   *After the weakness of the stainless-sheathed car was exposed, Pullman began to produce all-stainless cars, such as this model completed in 1954 for the Missouri Pacific. (Pullman Neg. 63160)*

Figure 2.78   *Interior of Pullman's 1937 articulated coach. (Pullman Neg. 40638)*

less-sided cars, which were painted only for color. They were attractive and obviously cheaper than all-stainless cars. The first order came in 1936 for the Daylight trains (Figures 2.77 and 2.78).[136] In later years the Southern Pacific acquired several hundred similar cars from Pullman. Stainless-sided sleepers were built for the Pullman Company's own use on various railroads; by 1944 about 280 cars of this type were in service on the Southern, the Santa Fe, and the Southern Pacific.[137]

After a few years, however, the cars exhibited a disastrous flaw.

To avoid unsightly fastenings, Pullman had copied Budd's snap-on method of applying the stainless-steel side-panel bands. But there was no effective way to seal the cracks, so that water and cleaning compounds gradually seeped inside the cars' lower body cavities. High-pressure car washers only aggravated the situation. The bodies rusted away out of sight, and it was not until major repairs were underway years later that the damage was discovered. The Southern Pacific noticed the condition fairly early but could find no way to seal the sides. The road stopped buying these cars after 1946.[138] Starting in 1958, it was necessary to rebuild the cars with flat sides. Other roads such as the B & O–C & O were too late in discovering the internal-rusting characteristic of stainless-sided cars and were forced to scrap a number of relatively modern ones.[139] Despite this basic defect, however, the cars continued to be built. The New Haven was a late and enthusiastic purchaser, acquiring 200 from Pullman in the late 1940s. The general decline in the passenger car market may have hastened the end of the stainless-sheathed car, but its reputation was badly tarnished by then. Pullman itself gave up on what had at first seemed such an admirable compromise and in the end built some all-stainless cars. These were produced in the 1950s for the Frisco and the M K T. (Figure 2.79).[140] Naturally it was still unthinkable to pay Budd's licensing fee; spot welding and rivets were used instead.

All-stainless cars were not without critics. High cost was the main objection; stainless steel was the most expensive material employed for railway car construction—40 cents a pound even in the 1930s. Budd's argument, however, was that the equipment would last for all time. After thirteen years and 4.5 million miles, the Denver Zephyr was overhauled in 1949 and showed no sign of deterioration in the stainless body or framing. In contrast, significant decay could be expected in an ordinary lightweight car after such extended running. But was this longevity necessarily a desirable characteristic? The useful life of a coach was generally set at twenty-five years. Even if it was not decrepit by this time, it

Figure 2.80   A production-model stainless coach built by Budd in 1938 for the New York Central. (New York Central)

Figure 2.81   Budd produced the Seaboard Airline Railroad's Number 6218 in 1947. The 52-seat coach was used on the Silver Meteor trains. (The Budd Company)

Figure 2.82   Interior of Seaboard coach Number 6218, completed by Budd in 1947.
(The Budd Company)

Figure 2.83   An exterior end view of Seaboard coach 6218. (The Budd Company)

Figure 2.84 *In 1956 Budd produced an ultra-lightweight demonstration model, the* Pioneer III. *The car was intended for quantity production, but no orders were received except for electric commuter cars.* (L. S. Williams, Inc.)

had become obsolete and needed to be replaced by a more advanced design. Overly durable cars only encouraged the railroads to make do with existing equipment.

Other critics challenged Budd's claim that stainless-steel bodies were maintenance-free. It was true that painting was not required, though the Pennsylvania and the Norfolk and Western both insisted on Tuscan red exteriors. However, keeping stainless cars bright was sometimes almost as troublesome and costly as cleaning painted cars.[141] More serious was the problem of major body repairs. Few railroad shops were equipped to deal with stainless-steel fabrication or Shotwelding, and after a smash-up it was often necessary to ship the body back to Philadelphia. Foreign purchasers found this particularly discouraging; indeed, the Portuguese National Railways gave it as their major reason for declining to accept deliveries on additional stainless cars.[142]

Despite these minor complaints, Budd made steady progress in obtaining orders and advancing the cause of the stainless-steel car. After World War II the company took over a war production plant in Red Lion, Pennsylvania, for a new car shop. Five assembly lines and 4,000 employees were soon at work, and by 1949 they had delivered more than 1,000 cars. Budd developed a good export trade and granted several licenses to foreign builders for the manufacture of stainless cars. The firm even successfully revived the rail car in 1949, calling it the RDC (Rail-Diesel Car).

Seven years later Budd developed a cheap, extra-lightweight coach that it hoped to mass-produce.[143]

Budd saw the new coach as a bench-mark development in its company history and, recalling the "Pioneer" airplane and the Pioneer Zephyr train, called the prototype car the *Pioneer III* (Figure 2.84). A full-sized (85 feet, 2 inches long by 11 feet, 6 inches high), full-strength car, it weighed an incredible 26½ tons, or about one-half of a standard lightweight car. This saving produced by a spartan interior, head-end power, and a radical inside-bearing truck design. The weight per passenger was down to 600 pounds, a figure not approached since the days of wooden construction. In some ways the car was a regression, since passenger comfort was sacrificed somewhat for economy and the elimination of dead weight. Seating was provided for 88, with a noticeable reduction of legroom. Small, prefabricated toilets, patterned after the one-piece plastic units made for airliners, replaced the spacious lounges to which passengers had become accustomed. The interior was assembled from precast plastic units. There were no reading lights, footrests, or ashtrays. The *Pioneer III* was a stripped-down economy model which even the manufacturer conceded was not suitable for long-distance travel. Still, the cost was only $90,000, and more luxurious versions were possible if desired. But the *Pioneer III* returned from a two-year, 120,000-mile tour with no customers. The year was 1958, and the railroads

Figure 2.85   The material most often used for lightweight cars was copper-bearing, rust-resistant steel known by the trade name Cor-Ten. The Seaboard's Number 830, built by Pullman in 1936, was an early example of Cor-Ten construction.

were in no mood to acquire more passenger equipment as the big push to abandon all passenger service was beginning.[144] Budd had developed the new design in the twilight of the long-haul passenger train. The company received several orders for electric suburban units but none for conventional Pioneer IIIs, and Budd's car-building division was soon in trouble. Suburban and rapid-transit orders kept the shop open long enough to land the Metroliner contract in the mid-1960s. But this prestigious job was not a profitable one, nor was an earlier order from the Long Island. After a season of steady losses, Budd decided to close its rail division in 1971. Yet only two years later, Amtrak placed an order for more cars and the division was kept open.

COR-TEN STEEL

If aluminum and stainless steel were the glamour metals of the lightweight era, Cor-Ten steel was the workhorse.[145] Developed by United States Steel, it was ready for commercial production in 1934. This combination of carbon steel with traces of copper, chromium, silicon, and phosphrous produced a strong, cheap, rust-resistant steel alloy. With a tensile strength of 70,000 pounds, it also possessed a yield strength double that of common steel and a fatigue resistance 80 percent greater. It could be worked as easily as conventional steel and, most important to the modern car builder, it could be welded. In addition, Cor-Ten was four to six times as corrosion-resistant as carbon steel and two to three times as resistant as copper-bearing steel. U.S. Steel regarded freight cars as the principal market for Cor-Ten, but car builders were quick to apply it to passenger car construction. (In recent years, architects have also found decorative as well as structural applications for Cor-Ten.)

Its low price (3.4 cents a pound) was a major recommendation for a structure as large as a railway passenger car.[146] Cor-Ten represented a cost advantage roughly ten times that of aluminum or stainless steel. Much as Pullman preferred aluminum, it recognized that the price of the new steel alloy was not only an effective argument against Budd's stainless cars but also avenue for the production of low-cost lightweight cars. This conclusion proved correct; more Cor-Ten cars were built than any other type. It was used for more than two-thirds, or about 6,500, lightweights.[147]

Among the earliest passenger cars of Cor-Ten were a set of fifty arched roof cars built in the period 1934–1935 for the New Haven at Pullman's Worcester, Massachusetts, plant.[148] A tubular cross section was partially achieved by the arched roof and convex body sides. The smooth exterior was marred by rivets, but the heads were kept as small as possible. In deference to the high fashion of streamlining, the windows and doors were made flush with the body panels, and deep underbody skirts reached down to within 26 inches of the rails. The large 29- by 60-inch windows were trimmed with aluminum. The seat spacing was not generous; 84 seats were crowded into the 84-foot, 6-inch car. The interior space was even less roomy than that of most light cars, since vestibules were fitted to both ends. This arrangement was a necessity on the New Haven, where stops were frequent and double facilities for loading and unloading were essential. The center sills were formed from two 14-inch channels, which were reinforced and tied together by numerous pressed-steel shapes. The framing members were made of another U.S. Steel alloy called Man-Ten. Somewhat cheaper and less durable than Cor-Ten, it was considered durable enough for the thick cross-section frame pieces. The body was of Cor-Ten. The side posts were box-shaped; the side plates were $3/32$-inch thick, and the roof sheets measured $1/16$ inch.

Weighing $54\frac{1}{2}$ tons, the New Haven cars were comparable to the best lightweights of aluminum or stainless steel. More cars on the same plan were produced during the next three years, and by 1938 the New Haven had 200 coaches in service. The neighboring Boston and Maine and Bangor and Aroostook also placed small orders. Nearly identical cars appeared on the Seaboard Air Line and the Cotton Belt (St. Louis Southwestern). Those built for Seaboard are shown in Figures 2.85 and 2.86.

While the industry was willing to employ Cor-Ten at an early date, the first groups of cars were riveted rather than welded. Like Pullman, American Car and Foundry stayed with the old method of fabrication in its first lot of Cor-Ten cars, which were built in 1935 for the B & O's Royal Blue. Within a year or two after their first orders, however, both Pullman and A.C.F. decided to capitalize on steel's great advantage by producing welded cars. On the other hand, Pressed Steel Car Company stayed with rivets. Once the sole producer of all-steel railroad cars in this country, the firm had lost most of its passenger car orders after about 1915,

Figure 2.86   Interior of Seaboard coach Number 830, completed in 1936 at Pullman's Osgood Bradley car plant.

Figure 2.87   Some New York Central Cor-Ten steel coaches under construction at Pullman's main passenger car shop in Chicago in 1941. (Pullman Neg. 45298)

when the older companies reequipped to produce steel rolling stock. In 1940 Pressed Steel made an effort to recapture a portion of that market with a new design using Cor-Ten.[149] The Pittsburgh builder turned out a pilot car for the New York World's Fair, but the firm had failed to keep up with the most recent methods in car building. The ungainly demonstration model won few admirers with its riveted body and frame. Only the center sill was welded; a molding covered the belt-rail rivets. The Pennsyl-

vania bought this pilot, and twenty-five cars were produced for the New York Central, but no other roads seemed impressed by the design. It was too far outclassed by the all-welded cars then in production.

Pullman not only began to perfect its all-welded plan in the mid-1930s but also developed two styles of framing built of thin-section Cor-Ten steel that were suitable for lightweight cars.[150] First the company tried a self-supporting, welded-truss side fram-

Figure 2.88   A Cor-Ten steel underframe at the Pullman car works in 1946. (Pullman Neg. 48169)

Figure 2.89   An arched-roof forming and welding jig at the Pullman car works, 1946. (Pullman Neg. 48482)

Figure 2.90   This Chicago and North Western coach was built from Cor-Ten steel in 1946 by Pullman. It was one of forty-five lightweight coaches purchased by the North Western to upgrade its postwar car fleet. (Pullman Neg. 48485)

ing, with outer sheathing that was "loosely attached" and had no structural function. This scheme became the permanent design for stainless-steel-sheathed cars. In about 1937 Pullman settled on a girder-sided car, where a simple rectangular framing was stiffened by spot welding a stress skin of .078-inch Cor-Ten sheet. The framing members were thin-walled Cor-Ten channels, boxes, and Z bars (Figures 2.87 and 2.88). The underframe was a similar fabrication. The side, roof, and floor frames were welded in jigs (Figure 2.89). A vertical welding machine spot welded the sheets to side frames. These subassemblies were then riveted together. There was some difficulty in producing smooth side panels, as might be expected because of the very lightweight sheets used. Cupping and waves were inevitable, but this visual defect was somewhat overcome by welding ¾-inch channels at close intervals.[151]

WELDING

Since the beginning of the steel era, car builders had been seeking a quick, strong, and cheap method of fabricating railroad rolling stock. For structural and most major body construction, hot riveting was the only technique available. It required drilling or punching, careful alignment of the holes, and maintenance of a steady supply of hot rivets along the entire assembly line. It was tedious and expensive, and strong though the finished product

was, the rivet heads were an ugly blemish. The rippled bead of a heavy welded joint might not be beautiful, but it was less noticeable and could always be ground level with the surrounding surface if desired. Gas and electric welding, however, was in its infancy during the early years of steel car building, and rivet construction became established with little competition. Light roof plates were welded at an early date and a welded freight car was built in 1911, but it was clearly experimental and attracted little attention. It was not until the lightweight era that welding came into large-scale use in the car-building industry.

In 1933 the Lincoln Electric Company, a major producer of electric welding gear, sponsored a contest for an all-welded passenger car.[152] One suggested design was for a conventional-looking heavyweight with a clerestory roof, a fishbelly underframe, and the other traditional features; yet a weight saving of over 30 percent (17 tons) was claimed for it. During the following year the Frisco built two lounge cars on this general plan in its Springfield repair shops.[153] Both realized a weight saving of 15 tons, thus vindicating Lincoln Electric's claim. Three more such cars were constructed in 1936 and 1937 by the Frisco.

Yet the real potential of welded fabrication was hardly exploited in these cars. A far more imaginative application was being tried at the Milwaukee Road's shops. The practical genius behind these efforts was Karl F. Nystrom, a Swedish mining engineer who had been attracted to car building soon after coming

Figure 2.91   *The Illinois Central's Number 2639 was another Pullman Cor-Ten product of 1947. (Pullman Neg. 48674)*

Figure 2.92   *American Car and Foundry also produced cars from Cor-Ten steel. The Great Northern 60-seat coach shown here was delivered in May 1950. (American Car and Foundry Company)*

to the United States as a young man in 1905.[154] As an employee of the major builders he was schooled in the traditional methods of car design. In 1922 he joined the Milwaukee, where he spent the rest of his engineering career. Nystrom's long service in the railroad industry did not stifle his willingness to try new methods. Past fifty and a seasoned, shop-oriented mechanical expert, he came to champion the then-radical scheme for all-welded freight and passenger cars.

Nystrom's first effort can hardly be called radical, however; he remodeled an old heavyweight sleeper as a smooth-sided coach. It was a slow beginning and to a degree a false start, because the car actually weighed more than many heavyweights of the day. But it proved that the technique of welding was practical, and Nystrom was ready to produce a lightweight prototype. Numbered 4400, it was finished in May 1934.[155] As one of the earliest

streamlined passenger cars in this country, it was considered worthy of display at the Century of Progress Exposition. It is unlikely that many visitors were interested in the 4400's construction, but the designer was satisfied and production began at the Milwaukee shops during the same year. The 80-foot car weighed only 48 tons—a remarkable achievement, considering that no specialty steels such as Cor-Ten were used. The framing was made up of 12-inch center sills and 4-inch Z bars. Large rest rooms and wide seat spacing limited the seating to 54 passengers. The 1934 production cars weighed 56.5 tons because of the addition of air conditioning and double-sash windows.[156]

In a group of cars produced in 1936, a 9-ton weight saving was achieved by using Cor-Ten steel and reducing the depth of the center sills by 2 inches.[157] Otto Kuhler was brought in to help with the appearance of the cars, but Nystrom's basic design was not

Figure 2.93 *Interior of the Great Northern car shown in Figure 2.92. (American Car and Foundry Company)*

Figure 2.94 *This Kansas City Southern coach represents the end of conventional lightweight car construction. The ten cars delivered in 1965 were the last of the breed. (Arthur D. Dubin)*

Figure 2.95   An interior view of the 1965 Kansas City Southern coach. (Arthur D. Dubin)

altered. The rib-sided, flat, arched-roof bodies remained the Milwaukee's distinctive feature in the United States, although there are remarkably similar cars in service on Russian railways.

While Nystrom made good progress in paring down weight, refining the general design, and holding the body cross section slightly under size, he encountered a notable setback at the beginning of World War II. Construction was underway on thirty-one cars in 1942, and permission had been granted by the government to complete the lot. As industry began to gear up for the war, however, tight restrictions were placed on critical materials.[158] The aluminum used for conduits and air ducts was unavailable, only small quantities of Cor-Ten could be found, and standard fixtures were unobtainable. These shortages forced Nystrom to make a series of compromises, each apparently small, but all contributing a total increase of 15 tons in car weight. Here again was proof that lightweight car design was a matter of subtleties.

An interesting feature of the first few groups of Milwaukee lightweights was the enclosed underbody.[159] Streamline advo-

cates favored smooth car bottoms for aerodynamic reasons, but this was not the Milwaukee's motive. Because of the extreme winter conditions on its line, the Milwaukee wanted to protect the underbody equipment from ice and snow and avoid the added weight of accumulating ice. A smooth surface offered fewer spots where snow and ice could lodge.

The car's balance and center of gravity was also improved by placing all apparatus under the center sills inside the sheet-metal "gondola." What servicing problems this arrangement caused are not recorded, but apparently they were offset by the advantages.

Between 1934 and 1948 about 300 new lightweight passenger cars were built at the Milwaukee Road shops. Not only had this line pioneered in the construction of welded equipment, but it also produced the cars. Few other roads attempted lightweight car construction, and certainly no other, with the possible exception of the Pennsylvania, built in such quantities. Like a commercial builder, Nystrom had an assembly line. Subelements of the sides, floor, and roof were assembled in horizontal jigs, and a special automatic welding machine with a range of 50 feet was worked three shifts to speed manufacture.[160] The long service life of these cars is a testament to their careful manufacture. In 1961, for example, a group of the 1934 cars was sold to Mexico, where they are still running.

The Revolution that Never Happened

Despite the success of the lightweight cars, they never displaced the existing heavyweight fleet (or at least, not until the near-abandonment of U.S. railway passenger service in recent years). While the conversion from wood to steel seemed slow, the change from heavy to light cars never really happened.

Once the name trains were reequipped, the railroads seemed to lose interest in purchasing more lightweights. The best trains were periodically upgraded with a set of new cars, the older ones being reassigned to lesser trains. In this way slightly worn lightweight cars filtered down to some secondary trains, and this happened with greater frequency as the number of passenger runs declined. New lightweight cars were rarely purchased at the rate of more than 300 cars per year by the entire industry in the period 1935–1955. After this time, of course, almost no purchases were made. The high cost of new cars—$100,000 and up—discouraged the cash-poor railroads from new purchases almost as much as the declining traffic did. Hence the lightweights never represented more than 15 percent of the total passenger car fleet.

The lightweights appeared during the Great Depression of the 1930s at a time when railroads could least afford a major investment in new equipment. The existing car fleet was adequate for the remaining traffic. Some new trains were put on to maintain the roads' public image, but once this cosmetic need was fulfilled, the acquisitions generally stopped. By the time the economic situation improved enough to encourage more buying, the war was on; production of such "nonessentials" as railroad passenger cars was halted. The big buying splurge that began after the war ended by 1950, again after the top-ranking name trains were reequipped. The older cars that were taken off the top trains satisfied most other needs. Robert Young, a passionate believer in modernizing the passenger car fleet, complained in the New York Central's 1955 annual report that at the present rate of acquisition, it would take 115 years to replace the old cars.

But perhaps the main obstacle to a lightweight takeover was the existing heavyweight fleet. It was simply too good to abandon. Many of these cars were relatively new in the 1930s, and few were much over twenty years old. The fears about rapid corrosion of steel cars were unfounded; in fact they proved to be so durable that with reasonable care they were practically indestructible. In addition, the great steel fleet represented a major investment. It was obvious that a more economical alternative to a large-scale lightweight acquisition was to upgrade the existing cars. The average passenger could not distinguish a lightweight from a modernized heavyweight. The degree of refurbishing varied from one railroad to another, as it did from one car class to another. A complete job included an arched roof, new trucks, air conditioning, reclining seats that were widely spaced, enlarged toilets, double-sash windows, new lighting fixtures and baggage racks, and of course, modern decorating, inside and out. Some roads added body skirts and fanciful paint schemes, so that the exterior was convincingly disguised as a streamlined car. But even if the interior alone was redone, most passengers were satisfied that they were riding in streamlined cars. Probably very few noticed or cared that their train was a World War I–vintage battleship thinly veneered in the modern style.

One reason why the railroads were forced to improve their passenger fleets was that their own publicity had raised the expectations of the traveling public. It was not enough to treat the name-train passenger to luxury travel; everyone (except perhaps the lowly commuter) expected greater comfort. The fast, cheap way to meet this demand was to upgrade the heavyweights.

The New York Central was early and prominent in the drive to remodel its old cars, as might be expected from one of the largest passenger operations in the world. Between 1934 and 1937 the line remodeled 150 coaches, according to a statement issued by its engineering department in January 1938. The most extensive work was done on a group of cars for the Cleveland-Detroit train, the Mercury. Seven lightweight suburban cars were thoroughly rebuilt at the Beech Grove, Indiana, shops in 1936.[161] Over and above the usual improvements, the following features were added: single vestibules, tightlock couplers, air conditioning, fluorescent lighting, and special noise-damping vestibule diaphragms. One car was rebuilt with a round observation end. Its interior was redesigned by Henry Dreyfuss, one of the finest designers in the early streamline period. Three years later the Central remodeled a more prosaic set of cars for its low-fare New York–Chicago Pacemaker. They were not given the opulent treatment of the Mercury cars, but the basic amenities—air conditioning and good seating—were provided.[162]

The B & O shops were upgrading old cars during the same period, but here the intent was to use the stock on the road's best trains. Reduced circumstances required this approach; the company simply could not afford to spend freely. Nor was its traditional-minded President, Daniel Willard, convinced that light eight-wheel cars could ever match the ride of a twelve-wheel heavyweight. When the lightweight Royal Blue cars (built by A C.F. in 1935) were sent to the Alton for the Ann Rutledge, the Mt. Clare shops modernized a set of elderly twelve-wheel heavyweights rather than spend the million or so dollars required for new equipment.[163] Other conversions were carried out at the same time for the National Limited, the Columbian, and the Capitol Limited (Figures 2.96 and 2.97). In 1946 the B & O scooped the C & O by introducing a steamlined daylight train, the Cincinnatian, with modernized cars.[164] While the C & O awaited its new equipment from commercial builders, the B & O worked over some existing thirty-year-old stock and had a new train ready in months. Here was another point in favor of remodeling. The Cincinnatian was more than a paper-and-paste job; the Mt. Clare shops could be proud of their homemade train. Indeed,

Figure 2.96   Many railroads hoped to resuscitate their serviceable but dated heavy-weights by remodeling them. An example is this coach modernized by the B & O in 1940.

their official records claim that the cars were built new (Figure 2.98). The B & O was apparently well satisfied with its remodeled heavyweights, and except for the two-year operation of the A.C.F. cars, the Royal Blue never acquired lightweight equipment.

The Pennsylvania set out to modernize its existing fleet in a more analytical way that was appropriate to a road dominated by engineers. In 1936 it remodeled three P-70s on alternate plans.[165] Scheme 1 involved the fewest modifications consistent with a modernized car. Seating was reduced from 88 to 60. New seats and wider spacing offered greater comfort. Air conditioning was provided by the ice system—the cheapest plan available. No changes were made to the body or roof, but aluminum window sashes were installed. Scheme 2 was identical to scheme 1 except that the windows were larger. Scheme 3, the most radical, involved extensive changes to the existing body to incorporate an arched roof, wide windows, and rest rooms at the center of the car (Figure 2.99). Seating in this car, which was intended for long-distance travel, was reduced to 42. Scheme 3 seems to have been applied to only one car. In the entire program, however, about 900 P-70s were modernized between 1937 and 1952. Some were extensively remodeled, with arched roofs, electromechanical air conditioning, single end vestibules, reclining seats, and wide metal-sash windows. Others received more modest upgrading that consisted of little more than an ice-cooling system and new window sashes. The program is summarized in Table 2.2.

By 1963 the Pennsylvania was again actively rebuilding and modernizing. The average age of its coach fleet was thirty-seven years, far worse than the national average of twenty-nine years.[166] To improve this condition the Altoona shops remodeled 186 cars, while the line returned fifty modern sleepers to Budd for conversion to coaches at a cost of $70,000 each. A few years later it was reported that the Pennsy was buying up second-hand heavyweights for renovation.[167]

Western roads also upgraded their heavyweight stock. In general they tended to buy more new equipment, at least after 1950, than the Eastern lines, but most of them also had active remodeling programs. The Santa Fe, obviously concerned about the "mixed consist" appearance of its trains, developed a cheap method of harmonizing the old with the new. In 1947 the road came out with a shadow-line paint scheme that effectively simulated the fluted sides of a modern stainless-steel car.[168] The old car was painted silver, and the shadow lines were air-brushed on by a skilled painter. It was said to be completely convincing at 50 feet. The Denver and Rio Grande Western and other lines copied this paint scheme (Figure 2.100).

By 1946, 17,000 modernized cars were in service. They represented about 30 percent of the entire passenger car fleet.

Safety

Throughout the history of the railroad passenger car, conflicting requirements for light weight on the one hand and safety on the other have challenged the best talent in this area of engineering design. The solution has always been a compromise. An abso-

Figure 2.97   The number 3556 was remodeled by the B & O in 1940 from the standard heavyweight car Number 5287. It had been built in 1924 by Pullman.

lutely safe railroad car is almost as unattainable as a crash-proof airplane. It cannot be achieved within practical weight or cost limitations; consequently the car designer has always aimed for a crash-resistant plan. The goal has been to perfect a design that, with economical use of materials, will produce a car body capable of withstanding any reasonable impact or buffering. The wooden car builder developed his plans empirically, going by what had been done before and what "looked right" to the chief draftsman. The early steel car builder based his designs on a few general formulas, but again one suspects that much of the design was empirical. With the coming of the lightweight car there was a greater scientific input in car design, because each part was expected to play a critical role in the structure. In addition, structural safety standards were refined and enlarged during the years of the lightweight, as will be discussed later.

Just how to measure the effect of car construction on the safety of railway travel is a difficult problem and one that has never been solved. Too many factors are involved, because at the same time that car design was changing, so were operating practices, signaling, traffic density, track construction, and other matters which would affect railway safety. That the coming of the steel car reduced the hazards of travel is obvious, but by just what percentage cannot be determined. Although steel cars cut down the number of casualties in minor mishaps, they certainly did not provide impenetrable shelters in major collisions. Telescoping remained a grim possibility for the railway traveler.

Naturally there were some misgivings about the safety of the new lightweight cars. In May 1936 the I.C.C. said in a preliminary report: "Questions have been raised as to the capacity of these light weight trains to withstand shocks, colli-

Table 2.2   P R R Class P–70 Coach Modifications

CLASS	PASSENGER CAPACITY	WEIGHT LBS.	ROOF HEIGHT	A-C	SEATS	WINDOWS	VESTIBULE
P-70 Original	88	122,000	?	None[a]	Walkover	Narrow	2
P-70R	76	142,000	?	Ice	Walkover	Narrow	2
P-70R "Deluxe"	formerly 39[b] later 80	142,000	?	Ice	Indiv.	Narrow	2
P-70R (Schemes 1 & 2)	60		14'0"C[c]	Ice	?	Narrow	2
P-70ER (Scheme 5)	76	141,200	13'2"A[d]	EM	Rot.	Narrow	1
P-70ER (Scheme 3)	42[e]	141,000	13'2"A	Ice	Rot. Recl.	Wide	2
P-70ER (Scheme 4)	72	141,000	13'2"A	EM	Rot. Recl.	Narrow	1
P-70FR	84	143,800	13'2"A	EM	?	Narrow	1
P-70FR (Scheme 6)	84	144,900	14'0"C	EM	Walkover	Narrow[f]	1
P-70FAR	80	144,200	13'6"A	EM	Walkover	Wide	2
P-70FBR	72[g]	155,900	14'0"C	?	Walkover	Narrow	2
P-70GR (Scheme 4)	68	143,700	13'2"A	?	Rot. Recl.	Narrow	1
P-70GSR (Scheme 4)[h]	68	145,700	13'6"A	EM	Rot. Recl.	Narrow	1
P-70GSR (Scheme 4)[h]	68	146,500	13'6"A	EM	Rot. Recl.	Wide	1
P-70HR	80	131,300	13'6"A	Ice (9 EM)	Walkover	Narrow	2
P-70HR	80	133,700	13'2"A	Ice	Walkover	Narrow	2
P-70KR[j]	56[i]	146,100	13'6"A	EM	Rot. Recl.	Wide	1
P-85GR	84[g]	161,750	13'6"A	EM	Walkover	Narrow	1

[a] An exception was the No. 3558, which had electromechanical air conditioning.
[b] Two dressing rooms.
[c] C—clerestory.
[d] A—arched roof.
[e] Center washrooms.
[f] Exceptions were the Nos. 3906–3935, which had wide windows.
[g] Cars had 8-foot, 6-inch trucks.
[h] Streamlined.
[i] Lounge rooms.
Source: This table was compiled by W. D. Edson, former car superintendent for Amtrak, from material in his files. It may not be a complete record.

Figure 2.98 In 1947 the B & O manufactured a new deluxe train, the Cincinnatian, using old heavyweight steel cars. At 86 tons each, they were lightweight in appearance only.

Figure 2.99 In 1936 the Pennsylvania considered a radical scheme for remodeling its vast old-fashioned P-70 fleet. The road decided that this sample car was too extreme and costly, but it did rebuild many P-70s along more moderate lines.

sions and other accidents."[169] Apparently this mild comment created a stir in the industry, because it was struck out of the final report. It was also deleted from the printed versions reproduced in *Railway Age* several months later. In the same report the Commission said that it was too soon for a definite appraisal of the new cars, that only their record in regular performance could provide the answers. At the time of the report, accidents that involved lightweight trains were relatively minor ones connected with running-gear failures or road-crossing mishaps that proved nothing about the integrity of the cars' structure.

One of the first major accidents to a lightweight train occurred in August 1939 at the Humbolt River bridge in an isolated part of Nevada.[170] The motor train City of San Francisco, running at high speed, was derailed just before it reached the bridge. Several cars fell down the embankment; others went ahead to sideswipe the bridge structure, causing it to collapse. The remainder of the train piled up on the wreckage. Twenty-four passengers and crew members died. Several of the cars were gravely damaged, but it is impossible to say how well heavyweights would have withstood this disaster. Certainly the lightweights were substantial enough to bring down a steel bridge.

A few months later the Pioneer Zephyr collided head-on with a steam locomotive at 45 to 60 miles per hour.[171] The front 28 feet of the Zephyr were demolished and its engine was driven back into the mail room, killing two clerks. There were 38 injuries, but no passengers died.

Neither of these two wrecks furnished a strong case for or against the lightweight car. A more significant test was an accident in January 1948, when the second section of a Missouri Pacific train rammed into the leading train near Syracuse, Missouri.[172] At the rear of the first train was the lightweight Pullman sleeper *Golden Cloud*. The car was totally demolished, twelve people were killed, and thirty-seven others were injured elsewhere on the trains. The locomotive ran 53 feet into the *Golden Cloud*; the front end of the sleeper was telescoped 12 feet by the preceding car, thus effecting an almost complete telescope. The frame of the car was nearly destroyed by the collision. This wreck rekindled the I.C.C.'s suspicions about lightweight construction, and the commission duly noted the inability of a car to withstand "heavy buffering stresses." As the last car of the first train, the *Golden Cloud* received the full impact of the rear-end collision. One can only wonder how well a standard-weight car would have

Figure 2.100   The Louisville and Nashville attempted a faux paint scheme in a rather desperate effort to duplicate the glistening silver of modern stainless-steel equipment.

held up against it. Pullman maintained that the seven-year-old car was built on a design tested to withstand the end buffering standards then existing.

A more direct comparison of heavyweight and lightweight construction was provided by a head-on collision between two cars of an electric interurban line known as Speed Rail which operated out of Milwaukee. In 1950 two excursion cars plowed into each other at high speed, killing ten persons. The lightweight car was demolished. But what might appear to be damning evidence against lightweight construction must be tempered by the fact that the heavyweight car's main frame was a foot or so higher than the other's. This is not the case in regular steam railroad cars, where the floor levels are all built to one standard. (Exceptions, of course, are the motor trains and the more recent experimental underslung units of the Talgo variety.)

Increasing public concern over passenger safety was reflected in the ever tougher standards adopted by the railroads. Standard setting began outside of the industry in 1912, when the Railway Mail Service issued a set of general strength requirements for Railway Post Office cars. the The Post Office hoped to force the early adoption of steel or steel underframe cars. The main requirement was the ability to absorb a buffering impact of 400,000 pounds. For years this figure was unofficially recognized as a basic minimum for all cars in interchange or main-line service. The Post Office, usually at Congressional prodding, periodically revised and strengthened its requirements. In July 1938 a major revision was put into effect covering lightweight cars. The follow-

ing year the Mechanical Division of the Association of American Railroads decided to adopt the R.M.S. specifications as recommended practice.[173] The required static end load was doubled to 800,000 pounds, and body deflection was not to exceed 1 inch. In 1945 the A.A.R. revised the specifications for new passenger car construction and advanced them to standard practice. All railroads and car builders were now required to follow the specifications. The most important of the new rules were Section 2, item f, and Section 18, item b. The first stipulated that no material could be used in which the yield strength exceeded 80 percent of the tensile strength. This eliminated the choice of aluminum for framing members. The second rule provided for two antitelescoping end posts, each to have a section modulus of not less than 24.375 and an ultimate shear value of not less than 300,000 pounds at a point even with the top of the underframe member to which it was attached. The specifications—composed of twenty-two sections and over 100 rules—covered everything from couplers to insulation.

It was said that the Budd Company actively promoted the tougher strength requirements in these standards. Presumably the firm's motive was to confound its rivals and undercut the position of the aluminum car.[174] It is obvious from the printed record that Budd made a great deal of noise about the superior strength of its stainless-steel cars. Shortly after the 1945 specifications were adopted, Budd dismissed them as a "distinct minimum."[175] Ragsdale, Budd's chief engineer, never failed to mention Budd's superior strength, pointing out that it was in "multiples" of the industry's requirements. Actually, Budd's interest in the matter was substantial. The company installed the first test machine for a full-sized railroad car at the old north Philadelphia plant in 1938 or 1939.[176] Built by the Baldwin Locomotive Works, it could develop 2 million pounds of compression. Its 500 strain gauges could show stresses in every body member of the car. This machine put car design and testing beyond paper calculation or in-service road tests. It was not until 1954 that the Association of American Railroads acquired a testing machine at its Chicago Research Center—ironically, just in time for the virtual end of new passenger car construction in this country.[177]

Standardization

As this book has shown, there was general agreement on car design almost from the beginning of the railroad age in the United States. Details might vary, but materials, methods of construction, overall size, configuration, and interior arrangement were generally in accord from one car to the next during any given period. In the broad sense this can be called standardization, and the word is used here in that sense. Standardization, however, is often misinterpreted. A recent conversation with a prominent author in the field of economic history revealed that he had misunderstood the 1880 Census Report on standardization in the locomotive industry, which stated that parts were made to gauge or pattern so that they might be interchangeable. The historian had taken this to mean that all American locomotives were identical, or that only a few classes were offered, and that parts could be freely interchanged between them. In fact the great diversity of locomotives during this and later periods, and the existence of fifteen or so locomotive manufacturers, each dedicated to its own plan, make the concept of interchangeable parts seem naïve. Some people are misled by looking at illustrations of railroad rolling stock, which do have the same general arrangement or basic design. But there were no Model T Fords in railway equipment, except perhaps for freight cars. A single, identical class of locomotives or passenger cars that numbers even several hundred is notable; any class counted in the thousands is very rare.

A second misconception concerning standardization is the idea that it involves the manufacture of identical duplicates. Such a practice may be possible for certain products, but it does not seem to have been seriously considered for the American railroad passenger car. There was a correct resistance to the cookie-cutter method of production because it would stifle the innovation necessary for the continuous improvement of this complex machine. A more desirable goal, and the one eventually achieved by the Association of American Railroads, was the establishment of certain general standards that all cars should conform to. Under the A.A.R. specifications both the purchaser and the designer had many options. These rules seemed the best compromise between the extremes of custom design and regimented mass production.

There were advocates of standard car design long before the lightweight period. At the beginning of the steel era at least one authority, F. Gutbrod, saw the coming conversion as a grand opportunity to adopt a standard design. A major obstacle was the greater cost of the steel cars, but Gutbrod maintained that a cheap steel car would become possible through the mass production of a few standard models.[178] The lower unit price would then permit a speedy retirement of the existing wooden fleet. However, no one seemed ready to adopt the idea. Each railroad and car builder held to its own plan, claiming special needs or superior requirements. The taste and influence of individual officials and designers was surely a powerful factor contributing to the variety of patterns produced.

The main deterrent to production-line standardization, however, was the rivalry of competing roads for passenger traffic. Speed, comfort, safety, and luxury were the drawing cards. The very last thing competitors wished to admit or practice was the duplication of the service and equipment offered by the opposition. Stylish cars of novel interior arrangement with many special (that is, nonstandard) features were the goal of every major railroad's passenger department. The cars of rivals might have the same basic construction, but there was a conscious effort to make them as individual as possible. One of the most powerful forces in the anti-standardization camp was the realpolitiks of travel promotion.

Government participation in setting passenger car standards began in 1918, when the U.S. Railroad Administration took up the question. The agency had inspired excellent designs by outside designers for locomotives and freight cars, but there seemed less urgency about a similar program for passenger equipment. Relatively few such cars had been built during World War I; the existing fleet was considered adequate, and priorities were given to more necessary products. Even so, the agency prepared a standard design by early 1919.[179] No cars were built on this plan as far as can be determined. Some head-end cars were produced for the U.S.R.A., but within a year the agency's control of U.S. railroad operations was terminated.[180] The railroads were not prodded again by a Federal body until 1935, when the Federal Coordinator of Transportation issued its *Report of the Mechanical Advisory Committee*. It was suggested that cars built to the specifications it outlined (pp. 602–612) might be purchased for as little as $40,000 if they were bought in reasonable quantities. But the Federal Coordinator could only hope to persuade; although the railroads listened politely, they were not bound to act.

Standardization might seem to be in the best interests of railroad car manufacturers, yet they showed little enthusiasm for it. Cars were built to order; they were not off-the-shelf goods, and the builders were in no position to force a ready-made product on their patrons. Like it or not, theirs was a custom trade, and as *Railway Age* said, "so long as the market is not standardized," there was little hope of standardizing the product.[181]

In many ways the custom nature of the business was a nightmare to builders. Between 1945 and 1950, for example, 3,781 cars were produced to 424 different floor plans.[182] In this period A.C.F.'s St. Charles, Missouri, plant received orders from nineteen railroads for a total of 98 lots, or 460 cars. They represented twenty-two different types of car and eight methods of construction, from riveted carbon steel to aluminum. Each lot of cars required 500 to 2,500 drawings and 500 to 1,000 requisition sheets.[183] The effect on cost and delivery time is apparent.

Over the years the car builders had gently pressed the railroads to consider a production model. In the mid-1920s Charles E. Barba, Bradley's chief engineer, suggested a production-line arched-roof car. Twenty-five years later A.C.F.'s mechanical head, Allen W. Clarke, recommended the adoption of standardized floor plans and components. Budd's *Pioneer III* prototype of 1956, described earlier in this chapter, was a singular example of a car builder willing to translate its belief in standardization into an operable prototype.

The larger railroads usually tried to maintain a certain degree of on-line standardization. The Pennsylvania was probably most successful in doing so between 1907 and 1929, when it built or bought passenger stock to a relatively uniform design. From the 1870s forward, in fact, it had followed a rational policy of standard acquisition for all its rolling stock. But statistics have not been found to show exactly how well the policy was carried out in the nineteenth century. The acquisition of existing roads and the semi-independence of the Lines West undoubtedly made it difficult for Altoona to maintain a single common standard. Central authority was stabilized, however, by the time of the all-steel car, and systemwide standardization was imposed. This excellent program was abandoned during the lightweight period, when an increasingly mongrelized fleet of cars was assembled from various makers.

The New Haven adopted a standard body design in 1948 after a five-year study.[184] The plan was capable of accommodating nine different interior arrangements. How far the road was able to follow the scheme before financial difficulties overtook it is not certain; new car purchases stopped before the plan was fully developed.

A large-scale program of standardization was also undertaken by the Harriman Lines in the early years of the twentieth century. Plans for a standard steel car were agreed upon by 1908, and soon arched-roof cars of an unmistakable pattern were running on the Southern Pacific, Union Pacific, Illinois Central, and other roads controlled by E. H. Harriman. But after the system broke up several years later, only the Southern Pacific remained loyal to the common standard design.

The Pullman Company, the largest single passenger car operator in the United States, also struggled to standardize its fleet. Yet at all periods of its history it had a bewildering array of cars with countless floor plans and window groupings. Its acquisition of competing operators and of new and rebuilt equipment added to the variety of stock in service. Basic frame and structural designs were kept uniform as far as possible, and a serious effort was made to standardize mechanical equipment, trucks, and hardware. Even so, Pullman's 1945 *Catalog of Materials and Supplies*,

a massive printed document of 600 to 700 pages, furnishes clear evidence of the variety of parts necessary to maintain its 8,600 cars.

While standardization may have been an impossible dream, it is unfair to portray railroad mechanical officers as irrational on the subject. They devoted considerable effort to standardizing heating systems, seats, wheels, journals, and other systems as far as possible for all cars on the line. While floor plans or manufacturers might differ, basic components were all of one style on the majority of cars in service. Consequently a degree of standardization was achieved.

One of the strongest points in favor of standardization was cost. Yet how effective would standardization actually have been in lowering the cost of passenger cars? Not very, according to Paul Kiefer, chief mechanical engineer for the New York Central.[185] Because relatively few new passenger cars were built each year, the savings on standard cars might be insignificant. Kiefer argued that the biggest price break was given to large single orders, and that even bigger savings were available to roads which carefully drafted their specifications and resisted the luxury of requesting changes once the work was underway.

A strong case for standardization can be made for freight cars, and in this area the Master Car Builders Association and its successor, the A.A.R.'s Mechanical Division, had an excellent record. Major components were standardized at an early date, and standard overall designs were generally adopted by the 1930s. But the chief concern of both organizations was freight cars. Passenger cars (except Pullman cars) were rarely involved in unrestricted interchange service and hence were within the jurisdiction of the owning road.

The official reports of either association rarely touch upon the passenger car problem. A few standards were offered and one or two reports were issued on construction, but in general the subject was ignored before 1900. Finally in 1910 a letter classification was adopted to designate various styles of passenger cars: PA was the symbol for short-haul passenger cars, PB for standard coaches, PS for sleepers, and so on. During the next years, standards for handrails and steps were also adopted.

Over time the coupler, axles, and wheel specifications were upgraded and modified, but general design considerations were left to the judgment of the individual railroad. Finally in 1940 it was decided to invade this private territory with overall standards. They were rather mild and general, however, and involved only exterior contour and roof-clearance lines. The purpose was little more than the creation of a more uniform exterior appearance. The following dimensions were established in consultation with Pullman, A.C.F., and Budd:

Overall height from railhead	13 feet,	6 inches
Roof arch radius from center of coupler	10 feet,	9 inches
Width over side posts	10 feet,	0 inches
Height to lower edge of window glass	6 feet,	11½ inches

Other than these and a few minor measurements, car builders were free to construct the car as they wished. Even the contour profile could not be enforced, but most roads conformed voluntarily, including the Milwaukee, which gave in to the standard body by 1943.[186]

It was another ten years before agreement was reached on basic interior arrangements. In 1950 the A.A.R.'s Mechanical Division adopted 19 basic floor plans: 4 coaches, 3 combines, 3 parlors, 1 diner, 1 diner lounge, and 7 varieties of baggage and combinations thereof. (No satisfactory scheme for sleepers could be agreed upon.) The plans were the result of a study conducted by the American Railway Car Institute, a New York–based trade association funded by the car manufacturers.

The A.A.R. Mechanical Division considered strengthening the standards for the floor plans, but it decided to stop short of more specific rules. Interior decoration, flooring, fixtures, seating, and other equipment were left to the discretion of the buyer. It was felt that any attempt to standardize car details more rigidly would disqualify some suppliers and in no way aid the railroads. Unlike the specifications for freight cars, no concrete standard drawings were ever adopted. This step might have come had passenger travel continued as part of normal railroad operations. In 1971 it seemed possible that Amtrak might take on the job the railroads never finished, but this did not happen. Amtrak has purchased both high- and single-level cars in recent years.

Late and Unusual Developments

Streamlining was credited with "saving" the railway passenger business in the 1930s. Twenty years later, railroad managers hoped for another miracle through the introduction of attention-getting passenger equipment. But by then the competition was stronger and the public was more highway-minded—in fact, the rules of the game had changed. Also, what the railroads had to offer in the way of innovations was unimpressive. Most of the schemes were simply old wine in new bottles. The extra-lightweight trains of the 1950s (for example, Train X) were a final and, in the opinion of many observers, a poorly conceived attempt to save the passenger train (their dismal story is covered in Chapter 8). Apparently a recurring theme in industrial history is that radical design changes proliferate both during the development stage of a technology and during its decline. Certainly it seems true that desperate measures are common to the last days of every dying trade.

During the 1950s a few alternatives to conventional designs were offered that were more plausible and successful than the extra-lightweight trains. These were the articulated, bilevel, and dome cars. Articulation is actually an ancient solution, traditionally put forward when all else fails. As in the case of the monorail, an impressive list of hypothetical advantages can be assembled:

1. A great weight and cost saving by the elimination of one truck.

2. An increase in passenger space by the elimination of one vestibule.
3. A shorter train (an advantage at short station platforms).
4. No body overhang at one end of the car.
5. An improved ride, because all cars are joined more solidly together.
6. Simplified end and vestibule construction.
7. A more streamlined exterior, because all cars are joined in one long, smooth tube.

The weakness of the case lies in the fact that these points, all so plausible-sounding, are mostly fallacious. In reality the motion of the cars is amplified through the central truck's center plate. Much of the movement of a conventional set of cars is damped by friction of the draft gears and vestibules.[187] The eminent railway historian and engineer Brian Reed has spoken of the disagreeable ride characteristic of articulated trains in Britain. As for the other real or fancied advantages of articulation, they are more than canceled by the fixed consist necessary with fully articulated trains. This feature has been wholly unacceptable in the operating routine of American railroads, where pickups and set-outs are the normal practice.

A convenient though little used compromise was found in the "marriage" of two or three cars as an articulated set. So far as can be determined, the scheme was first tried in Great Britain during the first decade of this century by N. H. Gresley of the Great Northern Railway.[188] Two- and five-car sets were operated, and by 1924 over 200 sets were in service. During the same year the London and North Eastern placed a three-car diner-kitchen-diner unit in service. Gresley's 1907 patent was controlled by the Leeds Forge Company, which in turn gave Commonwealth Steel an American license. There was considerable experimentation with articulation by American streetcar lines, and one of the Brooklyn subways obtained some three-car trains from Pressed Steel. So far as is known, however, no main-line units were tried by United States railroads until the following decade.

In 1936 Pullman produced an articulated pair of sleeping cars named *Advance* and *Progress*, and beginning in 1937, both the Union Pacific and the Southern Pacific acquired a number of articulated diners. But only the Southern Pacific championed articulated passenger cars: it ran a sizable articulated fleet that included many twin-unit coaches (see Figure 2.78).

While articulation was to remain an oddity in this country, bilevel and dome cars were more widely accepted. Somehow the split-level car has always been appealing, since it offers opportunities for greater architectural expression as well as added space. Morgan's land barge of 1828, the *Columbus* of 1831, and the Gothic cars of the 1830s were described in Chapter 1. In the 1890s the East Indian Railway built two triple-deckers—eight-wheel, iron-frame cars that seated 100 third-class passengers.[189] According to a contemporary report they were found unwieldy, and presumably no more were built.

The general scheme was not revived in this country until the 1928 patent of Albert E. Hutt.[190] In 1932 a sample car was

Figure 2.101   The Milwaukee Road was late in introducing gallery cars for commuter service when it acquired a double-decker from Budd in 1965. But this set of cars was among the most expensive of their type to be built. (Car Builders' Cyclopedia, 1965)

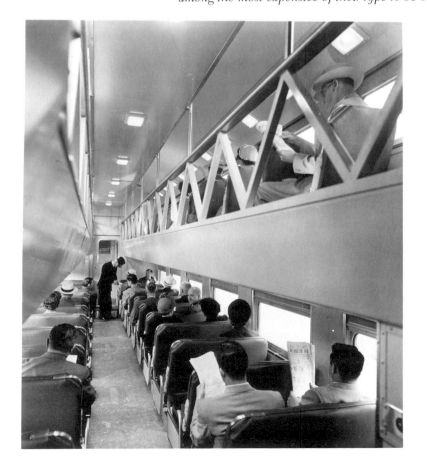

fabricated to this design at the Altoona shops for the Long Island Railroad. Hutt's design was aimed at double seating within the general limitations of standard clearance. This was an important consideration on the Long Island, where tunnel clearances allowed only a slight increase in overall height. With an ingenious staggered seating plan and only 5½ extra inches of roof structure, the 72-foot Long Island car (Number 200) offered seating for 120 passengers. Extensive use of aluminum held the total weight down to only 36 tons. Sixty-two more cars on this general design were built for the Long Island's electrified suburban service (see Chapter 8).

Although the Long Island cars had an impressive seating capacity, they were not well liked by passengers. In addition, they could not take full advantage of the bilevel's space potential because of the restrictions on height. The railroads offering commuter service in the Chicago area, however, were unencumbered by overhead clearances and decided to capitalize fully on the double-deckers (Figures 2.101 and 2.102). Undoubtedly the idea was suggested by the success of similar suburban cars on several European railways.[191] In 1950 the Burlington received thirty 85-foot double-decked "gallery cars" from Budd.[192] They towered 15 feet, 8 inches over the rails and weighed 61 tons. The

Figure 2.102   An interior view of a gallery car on the Chicago and North Western Railway. Several lines in the Chicago area are now operating similar cars. (Chicago and North Western)

Figure 2.103   *In 1954 the Santa Fe introduced a luxury bilevel car that has come to be accepted as the best form for long-distance travel. New equipment ordered by Amtrak in 1976 follows this general arrangement.* (The Budd Company)

upper seats and overhead floor were supported by brackets attached to the side walls and ceiling of the car, thus leaving the main floor open and free from posts and other encumbrances. Passengers mounted narrow, curving stairways placed near the center doors. End vestibule entrances were entirely eliminated by the central door so that more space was available for seating, which totaled a whopping 148. In the five cars that had toilets, seating was still an impressive 145. For short runs the facilities provided were satisfactory and the cramped, swaying ride was tolerable.

The Burlington acquired more cars of this type during the following year. With 124 gallery cars in service or on order by 1965, the road had converted its entire Chicago commuter fleet to bilevels.

It was obvious that the Burlington had come upon a good thing, and imitators quickly appeared. The Rock Island, the Milwaukee, and the North Western began acquiring similar cars for their Chicago suburban trains. The first group of North Western cars, which were made by St. Louis, had the largest seating capacity (169) of any car on this continent.[193] Within a few years the road had fifty of the cars and today uses equipment of the same design for almost its entire passenger service. Gallery cars were introduced on the West Coast by the Southern Pacific and did much to reduce operating costs on its San Francisco commuter trains, but none have been used on Northeastern railroads because of clearance problems.

The success of these cars suggested their application to mainline service—an idea which illustrates the tendency toward desperate innovation in the 1950s. The Santa Fe was not encum-

bered by a major suburban service, but its managers found the bilevel cars intriguing and decided that they might stimulate long-haul travel. More commodious accommodation would be necessary, of course, though the gallery plan was not really suited to this end. The road envisioned a double-floor or bilevel car with ample space and generous seating. Placing the daytime passenger compartment on the upper level, some 8 feet above the rails, would give travelers a better view of the countryside and remove them from the track noises. The lower level between the trucks could be used for crews' quarters, lounge areas, equipment rooms, or storage. The lower floor could be dropped between the trucks to reduce the overall height of the car and lower the center of gravity, which was held to 65 inches, or about the same as that of a diesel locomotive. The crawl space above the trucks at each end below the passenger-floor level would be filled with air-conditioning equipment, air brake apparatus, and the diesel generator.

Two experimental high-level coaches were made on this scheme and began service on the Santa Fe's deluxe coach train, El Capitan, in 1954.[194] The El Capitan was announced during the same week that Train X and the Talgos were ordered. In terms of weight per square foot of usable space, the Santa Fe decision looks remarkably wise. The results were so good that the road bought forty-seven more cars from Budd in 1956 at a cost of about $275,000 each. Thirty-five were center-door coaches seating 72 passengers and measuring 85 feet in length and 15 feet, 6 inches in height. They weighed 79½ tons (Figure 2.103). Six were double-deck lounge cars, which were heavier. Six were diners with kitchens on the lower deck, and these cars were so heavy

Figure 2.104   Interest in the underslung or possum-belly style of car has persisted since the 1830s. The Pennsylvania Railroad's Keystone of 1956 was a late example of a long series of failures to perfect a workable design.

that they required six-wheel trucks. All these cars ride as well as conventional equipment does, even at 90 miles per hour. Despite their great height, they have only a slight noticeable lean on curves.[195] The Santa Fe purchased twenty-four more of the coaches from Budd in 1964.

The only road to try gallery cars for long-distance travel was the North Western.[196] More workaday than the Santa Fe high-levels, these cars were put into service in the fall of 1958 between Chicago and Green Bay. Thirteen cars were produced at Pullman's Worcester, Massachusetts, plant: ten coaches, a parlor car, a parlor-lounge, and a coach-lounge. The North Western remodeled four existing cars, two diners, and two Railway Post Office baggage-lounge cars with false high roofs to conform to the new bilevel equipment. The coaches were 85 feet long and 15 feet, 10 inches high. They seated 96 passengers. The ride was described as rather bouncy, with a tendency to noticeable rocking on tight curves. The one parlor car was eventually converted to a coach and pressed into commuter service, but the other twelve cars remained unchanged. They were sold to Amtrak after the Green Bay train was discontinued.

It seems as if no idea ever becomes so discredited that someone will not give it another trial. Certainly the phenomenon of radical innovation in a dying industry is starkly demonstrated by the late revival of the old possum-belly car. This scheme, described in Chapter 1, originated in the early 1830s and was tried by both the B & O and Richard Imlay. It appealed to amateur inventors and

was repeatedly patented. During the 1870s the New York Elevated operated a number of drop-center cars, and some thirty years later the London subway acquired some steel cars in the style. Its great advantage was a low center of gravity, but this necessitated a complex and expensive drop-center floor frame. The short set of steps required to connect the two levels was probably even more hazardous than the stairways of the gallery or high-level cars, because it was less noticeable. Unsteady on their feet from the motion of the train, passengers often stumbled on them.

The misguided revival of the super-lightweight articulated motor trains in the 1950s undoubtedly caused designers to dust off the drop-center plan. It had one important advantage over the low-level lightweights: the ends of the car could be made the same level as that of standard-sized cars, so that drop-center cars could be integrated with the existing fleet. In addition, the danger of telescoping was minimized. The scheme was a compromise between the radical and the conventional.

Deciding to give the drop-center plan another trial, the Pennsylvania purchased eight cars from Budd for its New York–Washington service (Figure 2.104). The train, named the Keystone, began running in June 1956.[197] Budd had hoped to produce the cars for about $80,000 each, but the cost proved to be nearly double this figure. The coaches were standard in length (85 feet) and were carried on two four-wheel trucks. Couplers and end platforms were the regulation height. However, the

Figure 2.105   The dome car was introduced in 1945 to give passengers a good look at the scenery. Such cars became extremely popular in the West. (Railway Age, October 11, 1947)

overall height was 11 feet, 9 inches over the rails, and the coach floor was only 24 inches above the rails, or some 28 inches below standard floor level. Undersized 34-inch wheels further lowered the cars. The center of gravity was said to be 45 inches, or some 9 inches lower than the standard. The overall weight was held to 46½ tons. The interior was divided into three sections. One of the elevated ends over the trucks was outfitted as a smoking section, and in two cars this area was arranged as a buffet. The depressed center section had coach seating, as did the other elevated end section. It should be noted that even the elevated end sections were slightly below the regular floor level. The floor was connected to the vestibule level by a ramp on a 7-inch incline.

In actual service, the Keystone did not measure up to expectations. The ride was not smoother, or the noise level lower, than that of conventional cars, and the steps, as already noted, were a nuisance. The one innovation was the head-end power system, which eliminated generators, batteries, and considerable weight. It was necessary to run all the cars together as a unit, coupled to a conventional diner and parlor cars at the end opposite the power car. Budd received no more low-level orders, but the Keystone piled up a fantastic mileage record before it was retired in 1969 and sent to Altoona for disposition. Amtrak has since purchased it for possible rebuilding.

At the close of World War II there had been expansive talk of what the public might expect in the way of innovative postwar passenger equipment. Most of the prophecies had been promotional hot air, and most of the new trains featured cars not unlike their predecessors. The one spectacular exception that did see service on many U.S. lines after 1945 was the dome car. A glass enclosure and high-level seating offered a thrilling rooftop view of the passing countryside. Vista Dome, Astra Dome, blister, bubble, glass top—whatever name was applied in praise or criticism, these cars were effective traffic generators. Despite their enormous weight and cost, they were probably the most successful of the new designs put forward to save the passenger train. Unfortunately they were really only suitable for use on Western lines, where clearances were not a problem.

Perhaps the caboose cupola could be regarded as the forerunner of the dome, but there seems to be no direct thread of development between the two. Evidence does exist that a passenger car with a high-level observation compartment was in service on the Burlington in the early 1880s.[198] J. M. Forbes requested use of the "birdcage car" for a Western sightseeing trip in 1882. The president of the C B & Q was willing to lend the car, but he warned that it would be unable to clear the tunnels or snowsheds on the Central Pacific. No description of the car has been uncovered; it may have been a special model used for track inspections, or perhaps it was some form of private business car. The 1891

patent of a Canadian, T. J. McBride, might be cited as a more direct ancestor of the modern dome car. It was called an observation-sleeper by *Scientific American* and was pictured as a luxurious twelve-wheel car with three graceful observation domes.[199] No cars appear to have been built to the patent specifications. The Canadian Pacific's Montreal shops produced a boxy version of McBride's plan in 1902 for sightseeing service in the Rockies.[200] Three more were built in 1906. All were outfitted as parlor cars (described in Chapter 4).

These early efforts appear to have been forgotten by the railroad industry—as well they might, since only a handful were ever constructed. The idea seems to have been reinvented in July 1944. At least, there is a monument along the tracks of the D & R G W which states that the scheme occurred to Cy R. Osborn at this exact spot while riding in the cab of a diesel locomotive. The elevated cab windows offered a magnificent view of the Colorado Rockies, and Osborn, an official of General Motors Corporation, thought that it should be available to more people than the engine crew. The designers at General Motors agreed. They put drawings and models for the projected "Astra Liner" dome car on display in a spring 1945 exhibit on the future of railroad equipment that was especially created for the chief railroad executives. Apparently the only executive much taken with the idea was Ralph Budd, president of the Burlington. Because there was no hope of obtaining a new car, Budd ordered the road's Aurora shops to rebuild a modern stainless-steel coach into a dome car.[201] Certain compromises were necessary because of the existing car's design and the difficulties of obtaining materials. The plan called for a depressed floor below the dome section, but this was not possible without a major reconstruction of the frame. Since the ceiling height below the dome was sufficient for seated passengers, seats were installed. They were placed lengthwise—an awkward arrangement, but the best solution that the Aurora shops could devise. The dome itself was also a makeshift, being built up from flat sheets of glass rather than from the more graceful curved panes called for in the General Motors design (curved panes were not available in an industry geared for war production). Whatever the dome's aesthetic shortcomings, the Burlington aimed to make it as comfortable as possible. Double panes with an air space between them were used for insulation. The dome seated 24, while the lower level carried another 34 passengers. The remodeling was accomplished quickly, and the car, renamed the *Silver Dome*, entered service in July 1945.

The public was delighted with the *Silver Dome*; people felt that it was a preview of the next generation of railroad passenger cars. Within six months the Burlington had placed orders for forty new dome cars (Figure 2.105).

Not everyone shared the Burlington's enthusiasm, however.

Figure 2.106 General Motors' Train of Tomorrow, completed in 1947, featured several dome cars. Here enthusiastic models pose as passengers enjoying the sights. (General Motors Corp.)

During a symposium on postwar car design held just after the *Silver Dome* entered service, some doubts were expressed regarding the merits of the dome car and the desirability of exciting expectations for a revolution in passenger cars.[202] There was much talk of providing plain, comfortable new cars and not wasting energy and time on fanciful innovations. It was also obvious that only roads with generous clearances (up to 16 feet) could handle domes. This excluded virtually all the Northeastern roads. In addition, much of the dome car's seating was "nonrevenue." Passengers could not be expected to travel long distances in the dome; other space would have to be available. Domes might be traffic generators, but there was a price to pay—especially since the drop-center floor, the stairway, the extra air conditioning that was required, and the dome itself made the first cost well above that of an ordinary coach.

But the pressures for innovation were strong. A dramatic response to the competition was thought necessary, and improvements in equipment and service were the only course available. It was hoped that the dome car and similar improvements could hold a substantial share of the postwar passenger traffic market. And so other roads began to follow the Burlington's lead. Of course, the Western roads were the big buyers, not only because of their ample clearances but also because of their longer haul and the glorious scenery along the lines.

In 1947 the first domes came off the assembly line. That spring Pullman completed four dome cars for G.M.'s Train of Tomorrow.[203] All were 85 feet long and stainless-sheathed. The 30-foot domes were squared off, much like the Burlington's *Silver Dome*. Only one of the cars was a coach; it seated 72 and weighed 73½ tons (Figure 2.106). After making numerous press runs and being exhibited at the Chicago Railroad Fair in 1948 and 1949, the train was sold to the Union Pacific for service between Portland and Seattle.

The Burlington received its first commercially built domes from Budd in the fall of 1947.[204] These cars were used on the Twin Zephyrs running from Chicago to St. Paul–Minneapolis. They were the first fully streamlined domes with curved glass. The dome seated 24, the lower level 54. The washrooms were located below the dome (Figures 2.105 and 2.107). A narrow passageway to one side of these compartments allowed transit between the passenger sections on the main floor level. The floor was dropped at this section 19¾ inches for overhead clearance. The possumbelly floor frame and the opening in the roof for the dome required a heavier, more complex structure than a normal coach, but Budd's testing machine calibrated its strength as well above the A.A.R. minimum requirements.

Next came dome cars for the California Zephyr, a Chicago-to-Oakland train sponsored by the Burlington, D & R G W, and

Figure 2.107   The Silver Terrace was built for the Burlington in 1952 by Budd. (The Budd Company)

Figure 2.108   The Milwaukee Road's superdome illustrates a good idea that was developed too far. Ten of these 112-ton giants were produced by Pullman in 1952 and 1953. (Pullman Neg. 62835)

Western Pacific railroads. The initial train sets consisted of sixty-six cars, thirty of which were domes.[205] The Missouri Pacific also began to assemble a smaller fleet of domes for its Eagle trains.

The Pere Marquette introduced the dome east of the Mississippi River in 1948. To the surprise of many, the B & O became the next (and for many years the only) dome car operator in the East. Adjustments were necessary on a line notorious for its limited clearances, but the B & O was eager to maintain its pioneer image. In 1949 it added two brand-new dome coaches to the Baltimore–Chicago Columbian. A year later it purchased three

dome sleepers from the C & O, which was having second thoughts about its massive postwar equipment orders. About the only other Eastern line to attempt dome service was the N & W, which transferred its three Wabash Bluebird dome coaches to the Norfolk–Cincinnati run in the late 1960s. Perhaps the dome car category should be expanded to include the three sun-roof lounge sleepers built for the Seaboard Air Line by Pullman in 1956.[206] The roof lines of these cars were kept to the standard 13-foot, 6-inch clearance. Seating was at the same level. A unique effect was created by large side and roof windows for about one-third of the

Figure 2.109   Interior of a Milwaukee superdome car of 1952. (Pullman Neg. 62840)

car's length. But the high-level view so important to the dome car's success was not available, and no other cars of this type were built.

It was inevitable that the success of the dome concept would lead to the baroque extravagance of the superdomes. The largest land passenger vehicles built for everyday use, they are either the greatest achievement or the most flagrant excess of railway travel, depending on the point of view. Anyone standing beside the track as one of these ponderous cars rolls past is tempted to step back a few paces. The swaying "grand saloons" give the impression that they are ready to fall on some hapless bystander. The first superdome appeared late in 1952, a swollen, glass-topped 112-ton giant.[207] It was superlative in all ways except for its overall height—15 feet, 6 inches, compared with the *Silver Dome*'s 16 feet, 2 inches. Pullman built ten Sky Top domes for the Milwaukee Road at the staggering cost of $320,000 each (Figures 2.108 to 2.110). Seating was provided for 68 in the dome and 28 in the lower lounge. The top-heavy cars required special twelve-wheel trucks and huge outboard springs to support and stabilize them. The large glass area of the full-length blister (625 square feet)

required an air-conditioning plant of 16 tons' capacity with on-board diesel generator sets. Once in service, the Sky Tops were found less satisfactory than their smaller predecessors. The forward view was hampered by low seating and high end bulkheads, and the cars rode hard. Six of the superdomes ran on the Milwaukee's Olympian Hiawatha until the road abandoned its Chicago–West Coast service in 1961. These six were subsequently sold to the Canadian National. The other four remained in Chicago–Twin Cities service and were purchased by Amtrak.

In spite of its disadvantages, the superdome was attractive enough to be copied by other lines. The Santa Fe purchased fourteen similar cars from Budd in 1954, while the Great Northern bought six identical full-length domes in 1955. Stainless-steel construction held the weight to 96 tons. The standard or short dome was generally regarded as providing the best view. All the Santa Fe cars were sold to Auto Train in 1971.

By 1958, 234 dome cars were in service on American railroads.[208] Although they were not able to reverse the downward trend of the American passenger train, they remain the most outstanding of the postwar cars.

BODY ROLL DIAGRAM

FULL LENGTH DOME CAR
LOT 6908
CONDITION SHOWN IS FOR 70 M.P.H ON 4° CURVE
WITH 4.5" SUPER ELEVATION WITH FULL PASSENGER LOAD

Figure 2.110   This diagram illustrates the calculated stability of the superdomes. A hard ride and poor visibility were their chief defects.

CHAPTER THREE

First-Class Travel

IN THE EARLY DAYS of the republic, Americans warmly denied that any public institution would promote any form of privilege. All citizens belonged to one class, and all would enjoy (or endure) identical treatment. The October 2, 1852, *Scientific American* said proudly, "we have no second class cars for the inferior classes, because all our citizens rank as gentlemen and every man has his own coat of arms." A year later a trial lawyer at the Winans eight-wheel car case eulogized the single-compartment American car as "republican in its character because it allows persons to mingle freely," while he disparaged the European compartment cars that encouraged exclusiveness.[1]

In his *American Notes* of 1842 Charles Dickens mocked this boastful Yankee egalitarianism. He agreed that there were indeed no first- and second-class railway cars as in England, but said, "There is a gentlemen's car and a ladies' car; the main distinction between which is that in the first, everyone smokes; and in the second nobody does."

The philosophical satisfactions inherent in a single class of cars were bolstered by practical considerations. Using one simple style of car for all travelers, the capital-poor American railroads of the early nineteenth century were able to stock their lines cheaply and easily. A long, open-compartment eight-wheel day coach became the standard American car.

Yet from the very beginning of railway travel there were strong pressures for better accommodations than those of the standard coach. And comfort was not the only reason; in 1835, for example, a testy traveler on the Boston and Providence Railroad complained: "The rich and the poor, the educated and the ignorant, the polite and the vulgar, [are] all herded together in this modern improvement in travelling."[2] That passenger would surely have paid double fare for a parlor car seat had one been available, and within a few years special cars did begin to appear on American railways. In his railway inspection tour of this country in 1838 to 1840, the great Austrian engineer F.A. Ritter Von Gerstner reported on the surprising variety of passenger cars in service on American lines.[3] While details in his and other accounts are sparse, there is enough data to conclude that the much-glorified single-class ideal began to break down at an early date. Von Gerstner said, for example, that the Utica and Schenectady inaugurated second-class passenger cars in 1836 but gave up the service three years later because only emigrants would ride in them. The Mohawk and Hudson sold cheap tickets to people willing to ride in freight cars. And according to Von Gerstner, the three major lines running out of Boston were providing second-class cars. Several years later another Austrian railway engineer, Karl Ghega, reported that the Boston and Providence was operating second-class cars and that the New York and Harlem had mounted roof seats for economy-minded passengers.[4]

Moreover, Von Gerstner said that Negroes were often obliged to travel in the baggage car. In many cases head-end cars had a compartment fitted up with seats for Negroes, and half fare was generally charged. The Petersburg Railroad of Virginia, however, permitted Negroes to travel in the coaches at full fare. By 1850 Dionysius Lardner described the segregation of Negro travelers, including freedmen, as a general practice and said that they were obliged to ride in baggage cars.[5] Even after the Civil War, separation of the races continued in the South, where it was often enforced by state law. Cars that were politely called "partition coaches" were built as late as the 1950s. The emigrant car that was furnished for a poorer class of traveler had inferior accommodations, but this was not always true of the Jim Crow car. Particularly in later years, the doctrine of "separate but equal" facilities was obeyed.

Not only second-class accommodations but also first-class cars were in existence by the 1830s. Thus even during the pioneer period of railroads in this country, the American self-portrait of single-class travel was more mythology than fact. Strong demand both for cheap and for luxurious conveyances forced the railroads to develop specialized cars.

Early Sleeping Cars

Sleepers formed the largest group of the specialized first-class cars. They appear to have been the earliest luxury vehicles envisioned for railway use. It has been claimed that Napoleon's traveling coach suggested the railway sleeping car, but more direct antecedents can be found. In 1819 Benjamin Dearborn of Boston speculated that railway cars not only could provide a safe retreat from highway robbers, but also might be outfitted like an ocean packet, with facilities for dining and sleeping. Richard Morgan's proposed double-deck railway car of 1827 and its numerous "births" on the first level was pictured in Figure 1.7. These and other prerailroad projections cannot be dismissed as idle visions, for the traveling public had already experienced comfortable travel on the sea if not on the land. A year before Dearborn's suggestion, scheduled packets began to cross the Atlantic. Some offered the refinements of a good hotel, with bedchambers, dining halls, decorated saloons, and before long, bathing facilities. Canal packets began to emulate their oceangoing counterparts, so that by the time the railway movement began, prototypes for sleeping and feeding the traveler during the journey were well established.

But as Chapters 1 and 2 have pointed out, the great bulk of railway travel was in coaches, even after the development of the sleeping car. That the coach was also called the day car is apt. A trip of half a day in a coach could be a pleasure, but later the passing scenery began to lose its appeal. It seemed as if no comfortable position could be found by turning or shifting in the upright seat. The sway and bounce, the clatter and roar, and the dust, cinders, and wind pouring in through the windows began to seem intolerable. A few miles ahead there might be a transfer point where the travelers, festooned with luggage, rushed for seats and the empty victory of another foul coach, its floor a litter of crumpled papers, boot scrapings, and saliva. Naturally the car was already filled with coarse-faced occupants whose rude conversation was punctuated by the squall of teething babies. The prospect of the night's journey was grim for everyone.

Figure 3.1   The Cumberland Valley Railroad's Chambersburg, *1838c., is often credited as the first railway sleeping car. This drawing was prepared around 1881 for a patent suit.* (Pullman vs. Wagner, *1881*)

The compounded aggravations of a transcontinental trip were summarized by a contemporary traveler:[6]

You will be worn out with fatigue. You will be cramped and stiff with the confinement. You will turn blacker than the Ethiop with tan and cinders and be rasped like a nutmeg grater with alkali dust. You can never sleep a wink for the jarring and noise of the train, and never will be able to dress and undress and bathe yourselves like Christians. Above all, your nearest and dearest, under the influence of the fatigue and the monotony and the discomfort, will be ready to turn and rend you before you get down into the Sacramento Valley—and *you* will desire nothing better than to make a burnt offering of them and every one else insane enough to shut himself up seven days and nights in a railway car!

Travel is a necessity and it was the duty of the railway manager to make it as bearable as possible. The development of more comfortable cars is among the acts of public charity that make life more tolerable in small but deeply appreciated ways. The short-distance traveler and the poor were left to the coaches, but the wealthy and privileged demanded and would pay for some-

thing better. This demand led first to the introduction of sleeping cars and later to parlor, lounge, and private cars.

Tradition credits the Cumberland Valley Railroad with operating the first sleeping car, called the *Chambersburg*. Popular accounts say that it entered service as early as 1836 or 1837, but circumstantial evidence indicates that it began running sometime in 1838.[7] Although no contemporary accounts have been uncovered to establish the exact date, it is known that regular train service did not commence on the line between Chambersburg and Harrisburg until February 1838. In later years several old officials of the line said that the sleeping car was not placed in service until some time after the inauguration of the scheduled trains.[8] Stage passengers from Pittsburgh used the Cumberland Valley as a connection with Harrisburg, where trains of the Pennsylvania State Works took them on to Philadelphia. Many travelers complained about the late-evening connection; most were already exhausted by the long trans-Allegheny trip. Could not the railroad, as the marvel of the new age, offer something better than a low-backed coach seat?

mens comp.

mens comp.

ladies comp.

at night

in the day

← 76" →

34'

Figure 3.2 A reconstruction of the Chambersburg's general plan, based on a sketch by an employee of the Cumberland Valley Railroad in 1871. (Traced by John H. White, Jr.)

The railroad responded by ordering a special car from Richard Imlay of Philadelphia. Whether he supplied an existing car or built one to order is not recorded, but it may have been an existing car, since Imlay is known to have furnished sleepers for other Eastern lines at this time. A fairly complete description of the *Chambersburg* can be reconstructed from testimony given in two early sleeping-car patent suits.[9] Although the car itself had long since been dismantled by the time of the legal contests, the roof, a berth, and certain other parts were stored in one of the Cumberland Valley repair shops. Measurements were made by the master car builder, Jacob Shaffer, who had worked at the car shops since 1835.

The *Chambersburg* was an eight-wheel car with end platforms and an arched roof. The body measured 34 by 8 feet, with an inside clearance height of 6 feet, 5 inches.[10] The roof arch was given as 3⅝ inches. The exterior was painted brown and was picked out in shaded bronze or gold-leaf striping inside the body panels. The lower body panel was convex.[11] The interior was divided into four compartments, with a center aisle extending the length of the car (see Figures 3.1 and 3.2). At one end was a ladies' compartment with two upholstered benches running parallel to the central aisle. At night hinged edge flaps were raised to make a wide double berth. No upper berths were placed in this section of the car, but the two center sections had three levels of berths on each side of the aisle. They were convertible into side bench seats for daytime travel. The other end section was outfitted with ordinary cross seats. Total sleeping capacity was 16 persons.

The berths in the men's sleeping compartments measured 19 inches wide by 6 feet, 4 inches long.[12] Bedding consisted of some form of coverlet and round or bolster-type pillows covered in figured cotton calico; sheets were not offered. The mechanics of convertible berths were described by Jacob Shaffer:[13]

The first was a stationary seat with a loose cushion on it, the second one, which formed the back for the first one, was hung with hinges to the side of the car. In day travel this formed the back for the seat or bottom berth, and at night it was raised up horizontally and formed a middle berth. The top berth was hung to the partition at each end by a peculiar kind of an iron; as near as I can describe it it was formed in the shape of a sickle. The one end had a round boss on which part when screwed up to the bottom of this berth held the berth. There was a plate connected with each iron. These plates were let into the partition and the ends of these irons ran into a hole in the middle of the plate. These irons were on the back end of the berth. That berth, in day time, could be swung up towards the top of the car. It was held up by two girths. Those girths were fastened to the carlines or rafters of the roof by a wood strip running along the head-lining. The other ends of the girths had a bottom hole which hooked onto a button on the face of the berth and held it up. In the night it was unhooked from those girths and let down to a horizontal plane, where it was held by two sliding bolts, one at each end of the berth near the front lower edge. There was an opening or slot cut into the partitions into which those bolts entered. There was also two girths extending from the strip at the roof down. On the bottom of that berth there were two brass plates having hooks at the front end. The long girths had a square loop of brass which hooked onto the hooked plate and made a support for the centre of the berth.

Several witnesses in an early sleeping-car patent case said that the *Chambersburg* had an oilcloth ceiling, candle lighting, a water closet in the ladies' compartment, and a small, tapered iron stove for heating in the winter.

This first car proved so satisfactory that a second sleeper, the *Carlisle*, was produced in 1842 or 1843 at the Cumberland Valley shops by remodeling a day coach. The work was supervised by Jacob Shaffer, who patterned it exactly after the first sleeper.[14] In about 1850, however, the road suspended sleeping-car service. Both cars were too worn for continued first-class use, and most of the overnight traffic had been diverted to the Pennsylvania State Works railway, which by that time was offering through service between Pittsburgh and Philadelphia.

The earliest sleeping car for which a contemporary account has been discovered began running on the Philadelphia, Wilmington, and Baltimore Railroad in the fall of 1838 between Havre de Grace, Maryland, and Philadelphia.[15] Since the 24-passenger car was built by Richard Imlay, the probability is that it was very similar to the *Chambersburg*. The arrangement of the berths sounds nearly identical:[16]

During the day the bottoms of the upper tier of berths fold downward so as to form the back of the lower seat and render them as comfortable as can be. Whenever an upper berth is wanted it is only necessary to draw up the back to a longitudinal position where it is sustained by an iron catch at each end, take a pillow and blanket from the rack above . . . and there is formed at once a snug bed.

The Washington *Globe* of November 23, 1838, carried an advertisement for the new car that promised "a comfortable night's sleep . . . without the least disturbance of rest," and an early-morning arrival in good time for a train to New York. It also said that ladies could expect "commodious retiring rooms, attended by female servants." The 1838 P W & B annual report announced that the service was a success and that a second car was under construction. And then silence; there was no more comment by the road on sleepers until 1847, when a company report noted that some cars had been fitted with special seats for night travel. T. Wesley Bowers, a Wilmington car builder and onetime employee of the P W & B, claimed that the two Imlay sleepers were rebuilt as day coaches in 1844.[17] Bowers offered no reason for the suspension of sleeper operation; apparently the management of the road decided that chair cars were adequate for the run. The journey between Baltimore and Philadelphia was broken by the ferry crossing at the Susquehanna River, so that a through sleeper was not possible. In 1854 the road did make a slight improvement by installing reclining night seats.[18] No more sleepers appeared on the line until 1859, when two coaches were rebuilt for through service between Washington and Philadelphia.[19] By this time the Susquehanna ferry had apparently been remodeled to carry railroad cars. These company-operated sleepers were soon followed by Knight's patent cars, which are discussed in a later section.

At about the time that the P W & B's sleepers first appeared, Von Gerstner reported seeing sleeping cars on other Eastern railways. The Philadelphia and Columbia had one or more eight-wheel sleepers built by Imlay at a cost of $2,350.[20] Water closets and ladies' compartments were provided. The cars were said to accommodate 42 passengers. During an inspection of the Richmond, Fredericksburg, and Potomac Railroad in 1839, Von

Gerstner saw an eight-wheel car that was outfitted for night travel.[21]

Stephen W. Worden worked in the Richmond, Virginia, shop of James Bosher when some of the first cars were built for the R F & P in 1835.[22] Worden testified years later in a patent case that two of these cars were long eight-wheelers meant for through service. Four years later they were fitted as sleepers, with folding berths in three tiers. The lower berth could be folded out to form a double bed. A green curtain separated the ladies' and men's compartments. Hair mattresses and pillows covered in maroon were furnished for each passenger. The exterior was distinguished by a small three-paned Gothic sash over each window. Both cars were remodeled as coaches in 1849.

Von Gerstner mentioned one more Southern sleeping car dating from the first decade of American railroading. However, it was not meant for revenue passengers; it was a six-wheel track-workers' bunk car that the Austrian engineer saw on the New Orleans and Nashville Railroad.[23]

Clearly Richard Imlay played a large and perhaps a central part in the early introduction of sleeping cars. Three railroads had sleepers of his construction in service before 1840, and it is possible that there were others for which no records survive. Another scrap of evidence on Imlay's pioneering work in this area comes from accounts that were published in the 1890s of an antique car model owned by the Baldwin Locomotive Works.[24] It was a large wooden model of about ¾-inch scale that the firm had acquired, together with a lot of patterns, years before from Imlay, possibly at the time that his business was closed. His shop had been adjacent to Baldwin's in the Bush Hill section of Philadelphia. A label attached to the model read: "Made by Richard Imbrey [sic] and Jacob Dash, car builders, Bush Hill, Philadelphia, built in 1840."

Whether this model depicted a car that was actually built or not is unknown, but it does contain features Imlay is known to have employed. Note in Figure 3.3 the clerestory roof and longitudinal berths, which are arranged in the same way as those of the *Chambersburg*. Had a full-sized car been built from the plan of the model, it would have been unusually large for the period. Assuming that each berth space was equal to at least 6 feet, the car would have measured 42 feet over the body. If all spaces on both sides were fitted with berths, it could have slept 42 passengers. But according to the drawing in Figure 3.3, one side had only one level of berths, equaling some 28 berths. It is also likely that the blank end-panel section was set aside for a water closet. Perhaps the model builder simply did not bother to install the folding berths on the left side. Unfortunately the disposition of this extraordinary model is not known.

There were a few other sleepers in existence before 1850. In the summer of 1842 the New York and Erie began running two large eight-wheel cars, the *Erie* and the *Ontario*.[25] They were built by John Stephenson on his lattice or Diamond frame pattern. Because of the Erie's 6-foot gauge, it was possible to make the bodies 11 feet wide and to provide extra-large cross seats capable of accommodating three passengers abreast. The seats were made

Figure 3.3 *A model made around 1840 of an eight-wheel sleeping car built by Richard Imlay. The model was once in the possession of the Baldwin Locomotive Works, but it has since disappeared.* (Locomotive Engineering, *January 1898*)

from a plan patented by Nicholas McGraw on December 10, 1838 (No. 1030), for a convertible sofa bed. McGraw's design was intended for household use, but Stephenson adapted it for railway service. Even so, the basic concept of a convertible sofa bed was retained, and it is remarkable how similar the design is to that used generations later by Pullman (Figure 3.4). The likeness was so strong that Wagner and Pullman are reported to have settled one of their patent suits out of court to avoid publicizing the early origins of the convertible seat berth.[26] Apparently neither wished to emphasize that others had anticipated the basic idea of the American sleeper—a convertible day-night car—so many years before their own patents. Such a revelation would undercut their carefully constructed patent deterrents and possibly encourage other firms to enter the field.

The berths on the Erie cars were made up by pulling out two support bars fitted to the underside of the seat frame, thus forming a bridge between the seats. The loose back cushions were laid across the supports to complete the berth. Pillows, blankets, and a berth enclosure were apparently not offered, nor was any provision made for upper berths. In fact, the Erie had so little track when these cars were placed in service that there was no possibility of a long overnight trip. By the time more of the line was opened, the sleepers had been downgraded to slow mixed-train service. They remained in use until sometime in the late 1850s, when they were further downgraded as track-workers' bunk cars.

Competition with a steamship line that ran a parallel route compelled the Petersburg and Roanoke Railroad to build two sleeping cars at its own shops in 1846 and 1847.[27] By offering a good night's sleep, the steamers were taking all the traffic, and the railroad president was determined to recover some of it by providing equivalent service. The car bodies were 48 feet long. The roofs had a pronounced arch of about a foot and provided an inside clearance of 7 feet, 6 inches. Ample space was consequently available for three berth levels. Each car contained eight sections, or twenty-four berths. A ladies' compartment with a water closet was placed at one end of the car. At first the upper bunks were fixed and only the lower berth was convertible, but

the lack of headroom made it necessary to refit the upper berths so that they could be folded aside in daytime. For unexplained reasons sleeping service was suspended in 1853–1854, and the cars were remodeled as coaches.

During the same year that the Petersburg car was placed in operation, the Baltimore and Ohio Railroad decided to try a similar experiment. What inspired this move is not known; perhaps it was memories of Imlay cars on neighboring lines, or perhaps some now forgotten official wanted to improve the traveler's lot. Were it not for a French traveler's account, nothing more might be known of this car than the simple fact of its existence. Although it is listed in the B & O's 1846 annual report, no details are offered. However, the July 22, 1848, *L'Illustration* contained an article on a Frenchman's visit to North America in 1847–1848. One convenience of the journey which he particularly praised was the sleeping car. An illustration of the car's interior accompanied the article, but its barn-like proportions and the general lack of detail suggest that it was not a sketch from life but a dubious reconstruction based on the traveler's verbal description. The Frenchman wrote:

I wish to speak of the interior arrangement of these carriages (sleeping cars) destined to receive travelers for the night. They are actually houses where nothing, absolutely nothing is lacking for the necessity of life. They are divided into several compartments or sleeping rooms, some for men and some for women alone. Each of these rooms holds six beds, or rather couches, placed in three tiers along the sides. During the day the lower couches make excellent sofas. When the sleeping hour arrives, they take the trouble to raise the backs of the couches, and when they reach the horizontal position necessary for their use, strong iron braces moved by some interior mechanism, seize and hold them firmly. Three perpendicular straps guarantee the sleeper from falling. To tell you that these beds are perfectly comfortable would be to lie, but one is very thankful to find them, such as they are, and to be able, thanks to this precaution, to pass a pretty tolerable night. These six-bedded rooms are all connecting, so that one can walk, if necessary, from one end of the car to the other. Lanterns suspended from the ceiling light the interior, which presents a picturesque and novel spectacle.

Figure 3.4   A New York and Erie Railway sleeping car of 1842, built by John Stephenson of New York City. The drawing was prepared around 1881 for a patent suit. (Pullman vs. Wagner, 1881)

Like the other pioneers already mentioned, the B & O sleeper appears to have been withdrawn after only a few years of service. It disappeared from the official roster and was probably returned to work as a day coach. A hint that no more sleepers were operated on the B & O for another decade can be found in the published account of the opening of the line's Western connection to St. Louis in 1857.[28] On the first night out of Baltimore, it notes, the special train stopped overnight at Grafton, Virginia. The adult members of the party were put up in a hotel, while the youngsters spent the night in "comfortable sleeping-cars upon the side tracks of the Road." Presumably these were bunk cars of some variety.

A final sleeper for which there is a record before the middle of the century was reported in operation on the Georgia Railroad by *Disturnell's Railroad Guide.* In the scattered copies of the guide available for 1848 and 1849, stateroom cars for night travel were listed as running on the road between Augusta and Atlanta. The Georgia Railroad's annual reports do not mention these cars, and no other details or sources on them have so far been located.

To summarize, the available evidence, sleeping cars were operating on at least eight railroads before 1850. These sleepers were relatively sophisticated and embodied all the basic features of the convertible day-night car that was to become a resounding success after the Civil War. Why did the idea fail to take hold earlier? Certainly not from lack of technology; the mechanical details had been fairly well worked out before 1840. A principal reason for the railroads' early lack of enthusiasm was doubtless the small demand: few roads offered through travel for much

over 100 miles. Even at 20 or 25 miles per hour, the trip was over in a few hours. Of even greater importance, most trains operated in daylight. The few that traveled at night could offer a form of makeshift sleeper. Ladies' cars were relatively common by the mid-1840s, and they usually had a comfortable sofa where a woman could stretch out for a nap.[29] Since husbands could accompany their wives in the ladies' car, a few men enjoyed the same privilege. For the coach traveler, high-backed seats were in use on some roads by the 1830s. Reclining seats were not introduced until the middle of the century, but even standard coach seats were convertible into a poor sort of bed by means of hammocks, foot straps, and headrests fastened to the seat back or frame. Some of these devices were homemade gadgets brought along by the passengers; others were manufactured; and to some degree they satisfied the cheap overnight market.

In addition, sleeping cars were more expensive and heavier than regular coaches, and they carried fewer passengers. The bedding and internal fittings required more maintenance. The fairly limited demand and modest income potential of sleepers were probably all the reason any practical railroad manager needed to discontinue them.

It was during the 1850s that long-distance travel became a dramatic reality and precipitated demand for more luxurious accommodations. The Eastern lines were being rapidly extended into the heart of the continent. By 1851 the Erie reached Lake Erie; two years later the B & O touched the Ohio River; four years later a through rail connection opened between Baltimore and St. Louis, except for the two major river crossings. Passage between New York, Cleveland, and Chicago was soon available. But who could endure forty hours in a day car? Even though through cars were not run until several years later, and overnight stays at cities along the way were traditional, night travel became more common. For the commercial traveler anxious to get on with his business, the loss of time for rest stops was an intolerable waste. The most valuable part of the day was squandered by sitting on a rattling railway car. To travel at night would save time for the workday hours, and time was money.

All these pressures combined to bring about the sleeping car's renaissance. Some railroads produced their own cars. The Philadelphia, Wilmington, and Baltimore, which had been in and cut of the sleeping-car business since 1838, reentered the field in 1859 with two coaches remodeled as sleepers. The road offered through service between Philadelphia and Washington; the cars were moved on the southern leg of the trip by the B & O. During the following year the P W & B remodeled two more cars, and in 1863 it converted a fifth coach more elaborately, giving it a clerestory roof and an "improved interior."[30] But in the same year, the road decided to quit sleeper operation and contracted with the Central Transportation Company to take over the service.

Some years earlier the Illinois Central had made a serious bid to enter the sleeping-car business. It did not dabble in makeshift converted coaches but set out to provide luxurious railway travel long before Pullman contemplated the idea. In 1855 it placed an order with the Buffalo Car Company for some Gothic stateroom cars which were to measure 50 by 10 feet.[31] The first six cars were delivered in June 1856; a Detroit newspaper described their splendor:

LUXURIOUS RAILROAD CARS—The *Detroit Advertiser* says the cars of the Illinois Central Railroad, for comfort and convenience, excel those of any other road in the West. One of them contains six staterooms, each room having two seats with cushioned backs long

enough for a person to lie upon. The backs of the seats are hung with hinges at the upper edge, so that they may be turned up at pleasure, thus forming two single berths, one over the other, where persons may sleep with all the comfort imaginable. In one end of the car is a small wash room, with marble wash bowls, looking glasses, etc. On the opposite side of the car from the stateroom is a row of seats with revolving backs, similar to barbers' chairs, so arranged that the occupant may sit straight or recline in an easy attitude at pleasure. The other five cars each have two or three similar staterooms.

More sleepers were added to the fleet and the Illinois Central continued to operate its own cars until 1878, when a contract with Pullman finally ended the road's sleeping-car operations.[32]

Two years after the Illinois Central introduced its Gothic cars, the neighboring Burlington line altered two coaches for sleeping service.[33] In 1861 it converted a third car, but a few years later this road also decided to contract with Pullman for the specialized service.

As one railroad after another began to fall into line, their master car builders showed a new interest in the sleeping car. At the same time that inventors and private contractors were entering the field, several railroads encouraged their car superintendents to perfect their own designs. S. C. Case of the Michigan Central introduced a sleeper on his own plan in the late summer of 1858.[34] In June of that year Case received a patent (No. 20622) for his "Excelsior Plan for combining seats and couches." His three tier, folding-bunk arrangement offered 56 berths, which the inventor claimed could be converted by one attendant in three minutes. A lady traveler rhapsodized over Case's plan:[35]

Ah! how pleasant it looked, after leaving the other dusty cars—I could almost fancy I was at home, there was such a parlor-like air about it, and in less time than I can tell it, my seat was converted into a low spring bedstead, and a double one at that, and I and my weary

Figure 3.5   A Baltimore and Ohio sleeping car of about 1859 was pictured in a German text on American railways. The berth arrangement does not appear to be one of the many schemes patented during this period. (A. Bendel, Aufsätze Eisenbahnwesen in Nord-Amerika, Berlin, 1862, plate E)

little pet were soon wrapt in quiet slumber. At dawn of day I awoke very much refreshed, and on going the length of the apartment, I discovered a neat little dressing room, with all the necessary conveniences attached. I really disliked to leave this huge cradle which had so gently lulled us through the night, and I did so with a feeling that I will not try to suppress: it was gratitude to the inventor of so comfortable a conveyance, and I thank him in the name of mothers, who have hitherto found it so fatiguing to travel by night with little children.

Case's successor, J. B. Sutherland, continued to improve on the Michigan Central's sleepers. His patent (No. 29635) of August 14, 1860, for larger cars superseded Case's design. Using Sutherland's plan, the company's Detroit repair shops produced eight new twelve-wheel cars, which were described by the April 25, 1861, *American Railway Review* as the ultimate in luxury and convenience. The interiors were paneled in native woods; the clerestory roof had stained-glass windows; dressing rooms were located at each end of the car; Ruttan's heating and ventilating system was used. No less than $4,500 had been spent on each car. The operations of the Excelsior Line extended to the Michigan Southern, an arrangement probably facilitated by Vanderbilt control of both roads. The *Elkhart* was one of the many splendid cars

built at the Adrian, Michigan, shops by John Kirby. That and a compartment car, the *City of Toledo*, were both meant for the Excelsior Line's Chicago-Buffalo run.[36] The beautiful mahogany, maple, and gilt interior was lighted by gas. The Wilton carpet was in cherry and drab, berth curtains of brown terry were bordered with cherry velvet, and a flowered ceiling completed the interior's rich decor. Fully $20,000 had been poured into the *Elkhart*. By this time Pullman cars were operating on the two Michigan lines, and very soon Wagner would displace both Pullman and the railroad-operated Excelsior Line.

The Great Western Railway of Canada produced a sleeper at its Hamilton, Ontario, shops late in 1858.[37] Most early sleeping cars attempted to seat and sleep an equal number of passengers, but the Great Western car was designed to carry only 36 passengers and hence offered more space for each. Cross seats were provided at one end for day travel. The fact that the car had triple-tiered berths and a single washroom, however, indicates that it may not have been much superior to the maximum-capacity sleepers of the time. The berths were equipped with springs and carpet-covered horsehair mattresses. Pillows, quilts, and silk damask curtains were provided. The car had a black walnut interior, which suggests that it was nicely trimmed inside. This sleeper was built for the Albany-Chicago run via the suspension bridge.

In 1859 Draper Stone produced a sleeper for the La Crosse and Milwaukee Railroad. Meanwhile the Baltimore and Ohio had again become intrigued with sleeping cars, perhaps through memories of its 1847 sleeper, or perhaps because its exchange of through cars with the P W & B reintroduced sleepers to its tracks. The B & O remodeled four coaches for sleeper service on the main line to the west and began to operate other night cars on the north-south line.[38]

Again it was a foreign traveler who has left a record describing one of these cars.[39] The report of a German government official, A. Bendel, offers not only a detailed word picture (based on a spring 1859 journey) of a second-generation American sleeper, but also an excellent mechanical drawing (Figure 3.5). According to the scale on these sketches, the car's length was 54 feet over the end platform; its width was 9½ feet outside. The inside ceiling clearance was 7 feet. The berth and seat arrangement does not seem to follow any of the convertible car patents and was probably the idea of the B & O's own shop personnel. The lower berth was formed by a frame bridging the leg space and the loose seat-cushion backs. The upper berth was supported by two open wooden frames covered with woven cane. The frames were made in two sections for more compact storage against the roof during the day. The bolster pillows and upper-berth mattresses were rolled up for storage in the seat boxes. Sheets for both berths were rolled up like a spring-loaded window blind in the daytime. Stationary wooden partitions separated the berths. Passengers were expected to provide their own blankets.

To sleep 48 in a relatively small car, it was necessary to put two passengers in each berth—a 45- by 60-inch space. It was a cramped, low-ceilinged car like this one that justified Pullman's criticism of the early sleepers as being jingling old arks with macadamized mattresses and a greasy haircloth bolster for a headrest. Actually, the B & O car offered some comforts. A water closet with washbasin and a tank to one side was available at each end of the car. And Bendel reports that the interiors were nicely furnished with oilcloth or carpet floor coverings and wall paintings outlined in gilt and upholstery on the panels between the windows.

Halfway through the Civil War, the B & O shops were producing far more luxurious sleeping cars for the new through Washington–New York service, made possible by the cooperation of the B & O, the P W & B, and the Camden and Amboy railroads.[40] The cars offered one or two staterooms each, with berths like those aboard the best coastal steamers. The B & O soon turned its sleeper service over to contract operators, but in 1880 it once again began operating its own line of night cars.

Because of the soaring interest in sleepers just before and after the Civil War, sleeping cars were soon being operated in all sections of the country, with equipment ranging from the barely adequate to the luxurious. In perhaps a slight exaggeration, A. Bendel claimed that by the time of his 1859 visit, *all* U.S. railroads operating night trains were providing sleeping cars. In 1862

four major lines running out of Chicago advertised their sleeping cars as luxurious and elegant, while the Erie proclaimed its own design a "Splendid Ventilated Palace Sleeping Car!"[41] Contemporary evidence does demonstrate that the cars of this date were more than the "old arks" that popular history has called all pre-Pullman sleepers. While statistics for the period are not available, *Appleton's Railway Guide* gives a general idea of the extent of sleeping-car service. The 1866 *Guide* lists twenty-three railroads, operating in all regions from New England to points west of the Mississippi River, that furnished sleepers.[42] The Chicago and North Western had sleeping cars running to five cities from Chicago. The Illinois Central's night service from Chicago to New Orleans was probably the longest sleeper run in the country.

Another indication of the accelerating interest in sleeping cars was the growing number of mechanics who were attracted by the idea. The usual battalion of threadbare opportunists appeared to patent and reinvent countless old or obvious designs. Even the most talented seem to have offered only modest alternatives to the basic scheme worked out generations earlier by Richard Imlay. All had one goal—the perfection of a combination car that could be easily transformed from a day coach to a sleeping dormitory. The obvious advantage of the dual-purpose car was that it eliminated the need for two separate cars for the same number of passengers. The convertible car was invariably achieved by some form of folding coach seats and a hinged overhead bunk. The more closely the car resembled an ordinary coach during the day, the more skillful the design was considered. It was also desirable to seat and sleep as many passengers as a regular coach could carry, but no plan was ever devised that successfully achieved this goal within the limits of what came to be an acceptable berth size. Some of the earliest three-tier bunk cars did so, but with considerable discomfort for the passengers.

In the late 1850s mechanical trade journals like *Scientific American* were filled with descriptions and engravings of ideas for sleeping cars. The Patent Office suddenly faced a boom in applications in a nearly dormant subclass that had seen only a few additions since the first sleeper patent (No. 11699) was issued on September 19, 1854, to Henry B. Myer.[43] The rush began in 1858, and by the end of the following year over thirty patents had been issued. This record was never approached in later years, but patents continued to be granted even into the 1960s.[44]

Close behind the inventors came the entrepreneurs. Their enthusiasm for sleepers was sharpened by the understanding that the greater profit was in operating them, not in perfecting an ideal sleeping car suited to the needs of the railroads. The majority of the lines were soon weary of the troublesome special cars, with their bedding, porters, and all the fussy details of first-class travel. Since the investment was enormous and the management problems endless, it seemed best to turn sleeping-car operations over to an outside contractor. The railroad would then be free to concentrate on its regular affairs, while sharing in the profits of night cars. A pattern for contracting business out was available in the express and mail lines already in operation. A few railroads, notably the New Haven, the Great Northern, and the

Table 3.1   Railroad-Operated Sleeping Cars

DATE BEGUN	RAILROAD	DETAILS	DATE BEGUN	RAILROAD	DETAILS
1838?	Cumberland Valley	Service suspended 1848	1865c.	Boston & Albany	Its own cars and Wagner's
1838	Philadelphia, Wilmington, & Baltimore	Suspended 1844, reestablished 1858c.	1869c.	Central Pacific (Silver Palace)	Operated its own cars; 1883 contract given to Pullman
1838	Richmond, Fredericksburg, & Potomac	Service suspended 1849; Woodruff cars after 1866	1873	Wisconsin Central	Five cars by 1882. Possibly a Woodruff operation.
1839c.	Philadelphia & Columbia	Reported in service by Von Gerstner	1880	Baltimore & Ohio	Ended Pullman service, ran its own cars and those of Mann and Monarch; returned to Pullman 1888
1842	New York & Erie	Two cars retired about 1850			
1846	Petersburg & Roanoke	Service suspended 1853–1854	1890s	Florida East Coast	Cars by Jackson & Sharp; Pullman contract 1909
1846	Baltimore & Ohio (and see entries for 1859, 1880)	Service short-lived	1890	Chicago, Milwaukee, St. Paul, & Pacific	Contract with Pullman until 1890; Returned to Pullman 1927c.
1848c.	Georgia Railroad	Stateroom cars			
1856	Illinois Central	1878 contract given to Pullman			
1858	Michigan Central	First used Woodruff, Case, later Sutherland patents; Pullman 1865, Wagner 1875	1926	Gulf, Mobile, & Northern	Pullman cars were used before this time.
1858	Chicago, Burlington, & Quincy	1865c. contracts with Pullman	°	Central Georgia	Pullman contracts 1906 for some but not all service.
1858	Chicago, Rock Island, & Pacific	Myer & Furniss patents	°	Flint & Pere Marquette	Pullman contract 1904
1859	La Crosse Railroad	D. Stone's patent used	°	Detroit & Mackinac	Pullman contract 1911
1859c.	Rensselaer & Saratoga	Compartment Car	1890	Great Northern	Contracts with Pullman 1922; by 1925 all sleepers operated by Pullman
1859	Baltimore & Ohio	A second attempt to offer sleeping cars: its own, Knight's, and Central Transportation's			
1860c.	Michigan Southern & Northern Indiana	Excelsior Line	°	Soo Line (M St. P & S St. M)	Pullman contract on some through lines, 1934; railroad-operated sleepers until about 1953
1860c.	New York, New Haven, & Hartford	Several private lines as well as its own cars: Mann, Monarch, and Wagner at various times; contract given to Pullman in 1913			

° Dates service began cannot be supplied.

Note: Railway Age, Sept. 13, 1895, summarized railroad-operated sleeping-car service in the United States and Canada. It reported that of 134 individual lines and 154,224 route miles, 19 railroads representing 22,144 miles were independent. About one-half of these were Canadian.

Milwaukee, continued to operate their own sleepers into the twentieth century. But the real story of the sleeping car in America is that of a concession business (see Tables 3.1 and 3.2).

It was a confused tangle during the first decade as swarms of petty capitalists struggled to gain new sleeping-car contracts. Some were small proprietary operators who personally manned the one or two cars under their contract and owned nothing more than the furniture and bedding. In return for the fares collected, the operator might pay the railroad a fixed amount or a portion of the fares. In addition to this tithe, the concessionaire was usually obliged to pay a license fee to one of the patentees. Arrangements of this type were standard on the Rock Island and Wabash railroads.[45] Adding to the confusion was the fact that some railroads operated their own cars in addition to giving contracts to private concessionaires. The New York Central had Woodruff, Wagner, and Wheeler cars on its eastern lines and company-sponsored cars on its western subsidiaries. In addition, the New York Central was carrying Great Western sleepers over a portion of its road. The B & O permitted Knight, Central Transportation, and Woodruff cars to run among its own cars, apparently without restrictions. The chaos of these mixed operations bothered some far-

sighted men, who saw a great need and opportunity to consolidate the sleeping-car business. The potential profit of such a monopoly was obvious. Late in life Andrew Carnegie readily conceded that his fortune was started by an early investment in sleeping-car operation. There was genuine conviction in his epigram: "Blessed be the man who invented Sleep."

Woodruff's Sleeping-Car Enterprise

Theodore T. Woodruff (1811–1892) was the first major figure to emerge from the early concession operators. At the start he seemed destined to dominate the business and technical development of the sleeping car. He was among the first to operate cars on a concession basis on several lines, and before 1860 he had obtained such major contracts as the Pennsylvania Railroad's. His patents covered basic features of the convertible car. The historian of technology Siegfried Gideon thought that Woodruff's plans were the most inventive and clear-headed of all those among the pioneer patentees in this field.[46] But history has largely ignored his accomplishments; he is generally treated as just another pioneer who built a few cars and was then eclipsed by Pullman. George Pullman had reason to play Woodruff down, but it is harder to explain why Carnegie dismissed the inventor as little more than a country rustic, and why such scholars as Gideon, who studied the matter in some detail, did not give more prominence to Woodruff's contributions. Of course, Woodruff's early withdrawal from the sleeping-car business and his subsequent life, which reads like a Frank Norris novel, tend to place him in undeserved obscurity. Had it not been for his brother Jonah, who perpetuated the Woodruff name with his own line of sleeping cars, Theodore might have been an even shadowier figure than he is today.

Like Pullman and Wagner, the other two giants of the sleeping-car trade, Woodruff was born in New York State in the early years of the nineteenth century. He began life as a simple tradesman apprenticed to a Watertown wagonmaker.[47] Later he worked in a pattern shop and eventually became engaged in railroad car work. There are several conflicting stories about how he became interested in sleeping cars. He is said to have considered it as a very young man when the first sleepers began to appear in the 1830s, but he was too busy at the time to pursue the scheme.[48] Another version is that he was inspired to develop a more humane means of night travel after a grueling journey with his chronically ill wife.[49] Or according to an equally romantic story, Woodruff was literally hit on the head by the idea.[50] As a trainman riding atop a freight car in 1854, he was knocked off by a low bridge and while in a dazed condition began work on a sleeping-car model. He made this model in the office of James Tillinghast, superintendent of the Rome, Watertown, and Ogdensburg Railroad. Woodruff asked Tillinghast to help him finance the scheme and secure a patent, but the superintendent declined, seeing little future in the idea. Ironically, Tillinghast

Table 3.2 Sleeping-Car Companies

DATE FOUNDED	NAME	DETAILS	DATE FOUNDED	NAME	DETAILS
1857	T. T. Woodruff & Co.	Joined Central Trans. 1862	1870	Pullman Southern	Consolidated with Pullman Palace Car Co.
1858	Wagner Palace Car Co., also called New York Central Sleeping Car Co.	Taken over by Pullman 1899	1871	Woodruff Sleeping & Parlor Coach Co.	Formed by Jonah Woodruff; joined Union Palace 1888
1858	Case Sleeping Car Co. (Mich. Cent. R.R.)	Succeeded by Pullman 1865c.	1875c.	Leighton Sleeping Car Co.	Operated over New Haven; succeeded by Wagner?
1858	Eli Wheeler (N.Y.C.)	Patents taken over by Central Trans.?	1875	Lucas Sleeping & Parlor Car Co.	Cent. of Georgia and other Southern lines; sold to Woodruff 1878
1859	Knight Sleeping Car Co.	Joined Central Trans. 1862			
1859	Gates Sleeping Car Co. (L S & M S)	Succeeded 1873 by Wagner	1882	Flowers Sleeping Car Co.	No operation; office Bangor, Maine. William Flowers Patent, 1874.
1859	Pullman	(See Pullman Palace Car Co.)	1883	Mann Boudoir Car Co.	Began U.S. operation 1883; joined Union Palace 1888
1860c.	Paine-Harris Sleeping Car Co.	Succeeded by Anderson & Co.	1885	Monarch Parlor-Sleeping Car Co.	Operated 30 cars; gone by 1894
1861	Hapgood Sleeping Car Co.	New York to Boston via Springfield; succeeded by Wagner? 1868?	1887	Crescent Safety Parlor & Sleeping Car Co.	Jersey City—no operation
1862	Indianapolis & St. Louis Sleeping Coach Co.	Woodruff licensee; bought by Pullman 1871	1888	Union Palace Car Co.	Combine of Woodruff and Mann; purchased by Pullman 1889
1862	Central Transportation Co.	Successor to Woodruff, Knight, etc.; leased 1870 to Pullman	1889	Harris Palatial Car Co.	One car built; in receivership by 1892
1864	John B. Anderson & Co. (L & N)	Purchased 1872 by Pullman Southern	1887	W. J. Brashears	Baltimore—no operation
1865	Southern Transportation Co.	Controlled by Central Trans.	1889 or 1891	Jones Vestibule Sleeping Car Co.	Office in Denver; not certain if any cars built
1865?	Crescent City Sleeping Car Co.	(See Paine-Harris)	1890	Allen Hotel Car	Two cars—New Haven
1865?	Rip Van Winkle Line	(See Paine-Harris)	1892	Krehbiel Palace Sleeping Car Co.	Operated briefly on Milwaukee Road
1866	Silver Palace Car Co.	Associated with Central Trans.	1895	Williams Palace	Kansas City—no operation
1867	Pullman Palace Car Co.	Succeeded Field & Pullman	1895	L. F. Ruth	Pittsburgh—no operation
1868	Pullman Pacific Car Co.	Formed to operate cars over Central & Union Pacific, but C P withdrew in 1870c.	1896	Lockwood Stateroom	Cincinnati—no operation
			1903	Owensboro Sleeping Car Co.	Maine—no operation
1870	Erie & Atlantic Sleeping Car Co.	Used Woodruff patents; succeeded by Pullman 1875c.	1905	American Palace Car Co.	Several cars on Canadian Northern
			1907c.	National Sleeping Car Co.	Texas Central Railroad

Figure 3.6   The first patent received by Theodore T. Woodruff, the most important of the early sleeping-car pioneers, in 1856. (U.S. Patent Office)

was to become president of the Wagner Sleeping Car Company in later years. Woodruff then packed up his model in an old-fashioned bandanna handkerchief and took it to Thomas W. Wason, the car builder of Springfield, Massachusetts. As the story goes, Wason agreed to build a test car early in 1857, though his employees advised him not to become involved in it.

There appears to be some truth in the third story, except that Woodruff already had several substantial supporters before his visit to Wason. Around 1855 he was employed by the Terre Haute, Alton, and St. Louis Railroad as master car builder. The chief engineer of the line was Orville W. Childs (1803–1870), a fellow upstate New Yorker with a distinguished career in canal building. Childs liked Woodruff's scheme and encouraged the railroad's president, E. C. Litchfield, to draw up patent papers for Woodruff and advance him the money for the filing fees.[51] Childs encouraged the inventor to form T. T. Woodruff & Company, induced two local businessmen to invest in the concern, took a quarter interest for himself, and gave Woodruff a quarter share for his invention.[52] Woodruff received two patents (Nos. 16159 and 16160) on December 2, 1856 (Figure 3.6). A contemporary engraving shows their general plan (Figure 3.7).

Hence when Woodruff approached Wason early in 1857, he had two patents and the backing of three investors. The Springfield *Daily Republican* of October 7, 1857, describes the car:

We have been much interested in examining a new seat and couch car nearly perfected at the car factory of Wason & Co., in this city, the invention of Mr. Woodruff of Alton, Ill. The car is of the average length, containing upon each side 28 good seats, the whole car containing 56, or four less than the usual number.

These seats face each other in pairs, that is, two persons face two other persons always when the car is full. Everything is elegant, and one would not suspect at the first glance that the whole car could be thrown into couches in which every one of the whole 56 passengers could lay down at full length, yet such is the case, and it is so arranged that each one in a compartment may go to bed singly without depriving the others of their seats, and for the last one two seats will be left. In the first place there is an upper berth six feet long which can be swung down into location and be above the heads of all. Into this one may retire, drop the curtain and go to sleep; when the second wishes to retire, the backs of the seats next to the wall of the car are swung up and united to a stationary section of the berth that stands as a table between the two outside passengers, and thus the second couch is constructed, and over this the curtain drops.

WOODRUFF'S PATENT SEATS AND COUCHES FOR RAILROAD CARS.

Proprietors of Patent.—T. T. Woodruff, G. R. Dykeman, J. S. Miller, O. W. Childs, Syracuse, N. Y.

T. T. WOODRUFF & Co., title of firm.

VIEW OF CONVERTIBLE COUCHES.   VIEW OF CONVERTIBLE SEATS.

Figure 3.7 The interior of a Woodruff sleeping car of 1859 is shown in this engraving from a railroad guidebook of the period. The section on the right is arranged for daytime travel, while that on the left is made up for sleeping.

There are two passengers left sitting face to face next the passage-way; when the third wishes to retire, the two seats (which are double) under the berths and hung together by an adaptable hinge, are un-folded and just fill the space between the seats, thus making the third couch 6 feet long. There are then two unoccupied seats left, and in these, two men can sit, thus making five in each compartment, or the fourth can make his couch in the same manner the third did. We do not see but the thing is perfect. It certainly needs nothing but to understand it to make it practicable everywhere, and it is a very great improvement; these long night trips are made comfortable by it; it must at once attract the attention of railroad men, and passengers on long routes will give it their universal blessing whenever it may be brought into use.

The car is finished in Wason & Co.'s best style, and will be deemed at the West, whither it is bound, a good specimen of Yankee skill.

The owners, having no service contract, took the car on a tour in hopes of securing a contract for the use of their sleeper. The first trials were reportedly held on the New York Central between New York and Buffalo. The Central's management had little faith in the merits of the car and insisted that Woodruff, who person-ally acted as the conductor, pay his own fare—which absorbed nearly 70 percent of his earnings.[53] The cheese-paring parsimony of the Central was discouraging, but the other lines were more receptive. Woodruff was allowed to carry out free trial runs on the Michigan Central, the Chicago and Alton, Woodruff's own Terre Haute, Alton, and St. Louis, and other roads. Friendly as these lines were, few seemed willing to build or operate cars under licenses with Woodruff. It was probably at this juncture that Woodruff and his associates realized they must do more than offer designs and advice: they must take over the provision and operation of a full service package. Their investment of $5,000 to $6,000 in the first car was just the beginning.

One way to avoid direct investment was to induce subcon-tractors to operate sleepers, and in 1858 Woodruff was able to make several such arrangements. Some of these contracts, printed in *Bulletin* 59 of the Railway and Locomotive Historical Society, included agreements with Benjamin Field and Associates (April 5, 1858) to supply service on the Galena and Chicago Union and

the Illinois Central Railroads; with T. N. Parmelee, G. B. Gates, Morgan Gardner, and Webster Wagner (May 8, 1858) to operate Albany-Buffalo sleepers for the New York Central and with J. D. Morton (May 12, 1858) to run sleepers on the Michigan Central (with 20 percent of the net profits to go to the railroad). In the same year two large sixteen-wheel Woodruff sleepers built by Wason were placed on the Michigan Southern and the Northern Indiana.[54] They had fourteen sections and small bay windows. They were said to be owned by a Mr. Bates of Utica, presumably a Woodruff contractor who could not place his cars locally. Be-tween 1858 and about 1865 the Wabash contracted with several small concessionaires to operate Woodruff cars; the cars belonged to the railroad, but the operators owned the interiors.[55]

By the fall of 1858 Woodruff claimed to have twenty-one cars in service, and the 1857 sample car continued its tour in search of more customers.[56] Then T. T. Woodruff & Company secured a rich plum in the form of a contract with the Pennsylvania Railroad on September 15, 1858, to operate sleepers between Philadelphia and Pittsburgh.[57] The terms of the contract were very much like those of a twentieth-century Pullman agreement: the railroad was to maintain the cars except for the interiors; it was to transport the attendants (porters) without charge; it was to be free from pat-ent claims; and it would receive coach fare for each passenger. Woodruff was to collect a fee not exceeding 50 cents for each berth and must provide enough sleepers by January 1859 to ser-vice all the night trains operated over the road. Four new sleep-ers, numbered 22 to 25, were ordered from Murphy and Allison, car builders of Philadelphia. The 1857 Wason car was not good enough for this prestigious run; it was sent to end its days on the Ohio and Mississippi Railroad. The results of the Pennsylvania contract were described in September 1859 by Jonah Woodruff, who served as conductor aboard car C.[58] He reported that be-tween 10 and 50 passengers were carried daily, producing a total monthly revenue of $550. Deducting wages and supplies, the sleeping-car company netted a handsome return of $483.09 a month.

A fanciful story about the Pennsylvania contract is often quoted from Andrew Carnegie's *Triumphant Democracy*. The great steelmaker claimed that he was approached on a train one day by a "tall, spare farmer-looking kind of man [who] wished me to look at an invention he had made. With that he drew from a green bag . . . a small model of the sleeping berth for railway cars. He had not spoken a minute before like a flash the whole range of the discovery burst upon me. Yes, I said, that is something which this continent needs." At the time Carnegie was private secretary to the Pennsylvania's general superintendent, Thomas Scott. He claims to have pressed Woodruff's invention on Scott and thus brought "sleeping cars into the world."

The nonsense of this tale is dispelled in Joseph Wall's exhaustive biography of Carnegie, which quotes a letter written by Woodruff himself in 1886 stating the events as they happened. By 1858, Woodruff notes, he was already an established sleeping-car operator, and he was introduced to Scott by the road's president, J. Edgar Thomson. He was hardly the forlorn, "farmer-looking" person wandering about with a model in a green bag, as described in *Triumphant Democracy*. Carnegie knew better, but he liked the story. Without directly denying Woodruff's corrections, he refused to acknowledge his fabrication publicly. He did frankly admit his debt to Woodruff and in later years gave the old inventor a monthly pension. It was a well-deserved subsidy, for Carnegie acquired an eighth interest in the sleeping-car company for a cash outlay of just over $200. By 1860 he was receiving $5,000 a year in dividends.

THE CENTRAL TRANSPORTATION COMPANY

It was clear by 1862 that Woodruff's operation could no longer be effectively managed as a partnership. The firm had expanded wonderfully in five years. The partners were no longer riding the cars; they could afford to hire porters. But if they were to grow and meet the challenge of the new sleeping-car companies then emerging, they needed additional capital to buy out the competition and purchase more and better cars. T. T. Woodruff, who is described as having a rather poor head for business, did not see the need for such radical action, but his older brother Jonah saw that they must expand or expire.[59] Some of their licensees, like Gates and Wagner, had grown increasingly independent and were expanding quickly. Edward C. Knight, the fabulously wealthy Philadelphia sugar merchant, was introducing his cars on many lines and had the capital to develop a serious rival enterprise. It seemed vital that steps be taken to deal with the situation.

At last a compromise was negotiated: the consolidation of the Woodruff and Knight companies, together with the acquisition of the more important sleeping-car patents. The deal was engineered by Jonah Woodruff and O. W. Childs, one of the original partners, who had come to Philadelphia around 1860. The new firm, chartered on December 30, 1862, took the ambiguous corporate title of The Central Transportation Company. Capitalization was set at $200,000; the charter expired in 1891. Into the pot went the contracts, equipment, and patents of T. T. Woodruff & Company—the largest contributor to the assets of the new joint stock company. Of the 4,000 shares first issued, Woodruff received a controlling interest of 2,141.[60] The other patents acquired were those of E. C. Knight, H. B. Myer, W. A. Brown, P. B. Green, Eli Wheeler, J. B. Creighton, and Edward Burke. In this group only Knight and Wheeler appear to have operated any cars. The others, so far as is known, were paper tigers; but they held registered patents, and even if the designs were never used, they were effective tools for discouraging potential competitors. The deterrent effect of these twenty-five patents was emphasized in the company's advertisements and statements. Even the cars carried the patent message on their letter boards.

The inclusion of Knight in the Central Transportation Company was essential to its success. His three patents, issued between 1858 and 1860, were a minor menace, but Knight's wealth and influence (he was an officer and shareholder in several railroads) made him a real threat to Woodruff. It is reported that his interest in perfecting a night car was the result of a sleepless rail journey to New Orleans. Knight's first car was not produced until 1859 or 1860. One of his cars was engaged for the Prince of Wales's trip between Washington and Philadelphia in 1860.[61] This vehicle, built by Murphy and Allison, was described as substantial and elegant, with twelve sections and a total capacity of 60 passengers. It had a double floor insulated with cork chips. Within a year Knight cars were running on the Baltimore and Ohio, the Philadelphia, Wilmington and Baltimore, the Camden and Amboy, the Virginia and Tennessee, and the Cincinnati, Hamilton, and Dayton railroads. Had it not been for his merger with Central Transportation only two years after entering the industry, Knight might have become a major name in the sleeping-car world.

The Central Transportation Company was the strongest sleeping-car operator in the country. It controlled the principal Eastern routes and the basic sleeping-car patents. Its cars rolled over the B & O, the Pennsylvania, the Northern Central, the Central of New Jersey, and many other lines. Childs, the new president, was eager to expand. In 1863 the following advertisement appeared in the *American Railroad Journal* and the *U.S. Railroad and Mining Register*:

CENTRAL TRANSPORTATION COMPANY.

THIS COMPANY, a Corporation organized in pursuance of the provisions of a General Act of the Legislature of the State of Pennsylvania, having, by purchase, recently become the sole owner of

WOODRUFF'S, KNIGHT'S, MYERS'
AND OTHER PATENTS FOR
SEATS AND COUCHES IN SLEEPING CARS,

would respectfully give notice to all RAILROAD COMPANIES IN THE UNITED STATES, that may desire Sleeping Cars on their Roads, that this Company are now prepared to negotiate for placing wholly at its own expense, on such Railroads as may require them, their

SLEEPING CARS,

and operate them upon terms at once liberal and satisfactory to RAILROAD COMPANIES.

The Cars of this Company are constructed of great strength, and contain their late improved and patented plans of Seats and Couches, with STATE ROOMS AND BERTHS, finished in a style the most elaborate and tasteful, and are furnished with all of the modern conveniences and means of comfort usually found in the SLEEPING APARTMENTS OF OUR FIRST-CLASS HOTELS. The interior is lighted with gas; they are well ventilated, and at seasons of the year requiring it are warmed with pure heated air, regulated in degree at pleasure by registers in each State Room and section.

Where known, these Cars are not only regarded as indispensable on all through NIGHT TRAINS, but have become desirable, above all others, as DAY CARS.

Communications from the officers of Railway Companies desiring Sleeping Cars on their Roads, addressed to the Central Transportation Company at their office, No. 1347 BROWN STREET, PHILADELPHIA, will receive immediate attention.

Railway Companies using, or permitting to be used on their respective lines, Sleeping Cars or Couches, that infringe upon the patents owned by this Company, are respectfully notified that satisfactory arrangements for such infringement will be expected. Address as above, O. W. CHILDS, President.

New contracts, of course, created a need for more cars. Again the company advertised in the railroad press, this time for five cars by February 1864 and an additional twenty before the end of the year.[62] Builders were told that the cars must be patterned after the existing cars lettered Q, R, S, M, N, O, P, and T which were running on the various lines serviced by Central Transportation. It is likely that some of the new cars were built by Kimball and Gorton of Philadelphia. A few years before becoming president of the sleeping car company, Childs had been head of the Philadelphia car works. In 1862 Kimball and Gorton were succeeded by J. R. Bolton, and two years later George Dykeman, another of Woodruff's original partners, took a major interest in that firm. It is probable that a good number of Central Transportation cars were produced at the shop before it closed in 1867. This interesting liaison between the operating company and a car manufacturer, with overlapping officers and interests, was a forerunner of the gigantic combine assembled by Pullman in later years.

Two of the new cars appeared on the Pittsburgh, Fort Wayne, and Chicago Railway in June 1864. The *Chicago Tribune* described them as equal in elegance to the few first-class hotels in America.[63] The staterooms were fitted with large French plate mirrors. The cars had hot and cold running water, a hot-air furnace mounted under the rear platform, a perfect ventilation system, and "bedding of excellent quality," all of which guaranteed the traveler a comfortable night's rest. It might be noted that George Pullman was just starting work on his first luxurious sleeper, the *Pioneer*, in a nearby shed. Perhaps some inspiration was offered him by the two Woodruff cars.

Three years after its charter was granted, the company had expanded so much that it was necessary to increase its capitalization tenfold—to 2 million dollars.[64] The business had spread into the Midwest during the same period, and now that the Civil War was settled it was time to move into the South. The Southern Transportation Company was formed in the fall of 1865 by

Knight, Childs, and their associates. The controlling interest of 2,000 shares was assigned to the Central Transportation Company. The remaining 1,100 shares were held by Childs and other investors, the most prominent being Henry Plant, the Connecticut Yankee turned Southern railroad promoter. The parent company granted the use of its patents in return for a 15 percent share of the Southern's revenue, in addition to the controlling interest already owned by Central Transportation. During the following year the Southern negotiated contracts with seven Southeastern railroads, including the Virginia and Tennessee, the Alabama and Florida, and the Richmond, Fredericksburg, and Potomac. However, the enterprise did not flourish as well as had been hoped. Southern traffic was not comparable to that in the North, and other independent competitors operating out of Atlanta and Louisville took over many lines that Southern Transportation had hoped to acquire. The concern managed to pay an average dividend of 7.5 percent before being acquired by Pullman Southern in 1870.

Elsewhere, Central Transportation was doing well. In 1866 its eighty-eight cars were valued at $8,000 each.[65] During the summer of that year the company's vice president and general manager, Jonah Woodruff, inaugurated the first Silver Palace cars. A special train pulling twelve Silver Palaces was run from New York to Chicago via Philadelphia and Pittsburgh to promote the new luxury sleepers. The excursion party, made up of railroad men, the press, and their families, was said to have enjoyed a week of purest pleasure. A musician was commissioned to compose a march dedicated to the Silver Palace car.[66] A colored engraving of the Silver Palace car *Chicago* (Number 76) was printed on the cover of the sheet music. *Scientific American* heaped praise on the Silver Palace cars and their promoter, calling Jonah Woodruff "King of the Sleeping Cars" and declaring that he deserved the golden harvest he was now reaping.[67] It said that the cars were the "most brilliant vehicles that ever rolled on iron wheels." It described richly carved black-walnut interiors, enhanced by velvet window hangings, Brussels carpets, and silver-plated lamps and hardware. Similar cars are shown in Figures 3.8 and 3.9. There were stained-glass clerestory windows that "shed a flood of blue-tints upon the glittering silver pillars and fret work below." The builder of these cars was not named. A few years later a York, Pennsylvania, newspaper reported that two Silver Palace cars, Numbers 112 and 113, had just been finished at the Northern Central shops. The bodies were yellow, the platforms a salmon color, the trucks green. The window sash was varnished cherry. Each car cost $10,000. Late in 1866 the *American Railway Times* mentioned that the New Jersey Central had finished three Silver Palace cars in its shops and had three more underway that would be named *New York, Chicago,* and *Pittsburgh*.

In addition to the extravagant decorative treatment, the Silver Palace cars featured an improved berth design patented by Woodruff on November 19, 1867 (No. 71,258). The upper berth was double-hinged and built so that the outer leaf could be folded perpendicularly to the main inner leaf. The whole structure was

Figure 3.8   Woodruff began to introduce palace cars in 1864, at the very time that Pull-man was starting work on what is erroneously believed to be the first of the breed, the Pioneer. The Woodruff cars here were built around 1869 by Jackson and Sharp. (Hall of Records, Delaware Archives)

Figure 3.9   The exterior of the Woodruff Silver Palace car, the Empire *(Number 85), shown in Figure 3.8. (Hall of Records, Delaware Archives)*

L-shaped when closed and was fastened to the ceiling at a 45-degree angle (Figure 3.10). This allowed a double-wide berth to be folded up against the ceiling. Woodruff also included removable berth-divider panels and folding tables in his patented design. The silver-plated columns were more than decorative; they helped to support the roof and upper berth structure.

Silver Palace cars placed Central Transportation well ahead of its competition and gave the firm invaluable opportunities for further expansion. In the late 1860s the most prestigious upcoming contract was for the transcontinental railroad. Negotiations began in 1867, with both Pullman and Central determined to win

the prize. Actually, two contracts were involved because of the division of the line between the Union Pacific and the Central Pacific. It appears that Woodruff was successful in landing the Central Pacific contract. At least one builder's picture survives showing a Central Pacific car ascribed to Woodruff (Figure 3.11). However, some Western railroad historians believe that sleeping cars were never operated by Woodruff on the Central Pacific. All the early sleepers on this line carried Silver Palace lettering and seemed to feature the interior arrangement and decor of Woodruff's cars, but it is possible that nothing was signed except a licensing agreement for the name and design. It is also possible

Coupe longitudinale

Elévation de bout   Vue intérieure   Vue intérieure   Coupe transversale

Figure 3.10   A Silver Palace car built in 1869 for the Central Pacific by Harlan and Hollingsworth. This cutaway drawing appeared in a French work on American transportation. (E. Malezieux, Public Transport in the United States, 1873)

that the operating contract, if one did exist, was canceled at the time that the service was to begin because of Pullman's lease of Central Transportation early in 1870.

Whatever the facts about the contract, the Central Pacific's sleepers were operated under the name of Silver Palace. The first car was received in July 1869 from Harlan and Hollingsworth.[68] Other cars were furnished by Jackson and Sharp. By 1870 twenty-one were in operation; Pullman sleepers were not permitted on the line, and passengers were obliged to change cars. In 1875 Barney and Smith added eight Silver Palace cars to the fleet.[69] They had 55-foot-long bodies and six-wheel trucks, and they were painted a light canary yellow. Before it abandoned independent sleeper operations in 1883, the Central Pacific's fleet had grown to forty-one cars. The cars and service were not considered equal to Pullman's best, however; one waggish British traveler said: "I met with no silver whatever in the Silver Palace sleeping-cars. The fittings, lamps, . . . door handles, etc., are of the white metal called pinchbeck or Britannia metal."[70] Thus within a decade of its appearance, Woodruff's triumph of elegance was being called a cheap sham.

If Central Transportation could not secure a hold on the Far West, it was doing magnificently in the East. In 1869 it had 119 cars in service on sixteen major railroads and was by far the strongest sleeping-car company in the nation.[71] During the fiscal year 1869 it earned over $256,000—equal to a 17.8 percent return on invested capital.[72] A 20 percent profit was confidently ex-

pected for the following year. It seemed likely that the firm would enjoy a long and prosperous life.

The seeds of its unexpected demise were sown in the course of competition for the Union Pacific contract. It was during these negotiations early in 1867 that Andrew Carnegie met George M. Pullman. Carnegie, immediately impressed by Pullman's ambition and energy, saw that he was more than "a lion in the path"; he was a business genius likely to outrun his competitors. Carnegie decided to abandon his old associates and join Pullman. He devised a "compromise" that squeezed out the Central Transportation Company and gave the Union Pacific contract to a new company, the Pullman Pacific Car Company, with shares owned by Pullman, Carnegie, Thomas A. Scott, and the Union Pacific Railroad.

Both Carnegie and Scott were shareholders in Central Transportation. The officers of that company, particularly Childs, were annoyed by Carnegie's double-dealing and did not think that the $20,000 patent fee arranged by Carnegie for the Pullman Pacific was adequate. Carnegie finally quieted their complaints by selling his shares in Pullman Pacific to his Central Transportation associates—pocketing a $20 profit per share on the transaction. Although Childs believed that Pullman could be beaten in court with their arsenal of patents, he was finally persuaded to accept the compromise plan. But Pullman, hard-pressed for cash, was putting everything he could raise into new cars needed for the transcontinental service. His delays in paying the patent fee hardly sweetened relations with Childs; yet somehow Carnegie

WOODRUFF'S SILVER PALACE SLEEPING CAR,
Built by the Harlan & Hollingsworth Company, Wilmington, Delaware,
For the Central Pacific Rail Road of California.

Figure 3.11   The Central Pacific Railroad operated its own sleeping cars, but its first sleeper, the 1869 car shown here, was patterned on the patented designs of Woodruff. Note the Creamer spring brake on the left side of the end platform railing. (University of California, Bancroft Library)

managed an even more startling agreement between the two firms.

After months of quiet parley, Pullman leased the entire property of the Central Transportation Company. The lease, retroactively effective to January 1, 1870, was signed on February 17 and conveyed the rolling stock, patents, and contracts for 4,400 route-miles on sixteen railroads to Pullman.[73] An annual rental of $264,000 was to be paid during the 99-year life of the contract. At the same time, Pullman leased the Southern Transportation Company for a yearly rental of $20,000. Its property consisted of seventeen cars and five contracts covering 1,500 miles of line. Pullman had overtaken his chief rival and was on the highroad to realizing his dream of a national sleeping-car monopoly—all because of Andrew Carnegie. Since the early years of the Woodruff firm, Carnegie had had little regard for "the cautious old men in

Philadelphia" who ran its affairs. At the time of the lease Childs was sixty-eight and may have realized that his life was coming to an end; he died in September of that year. Knight was busy with other affairs. Theodore Woodruff had been out of the company since 1864. Only Jonah Woodruff protested, but he was unable to stop the Carnegie steamroller.

The Central Transportation management retreated to a small office and for the next several years did little more than record Pullman's rental payments. A dispute that arose in 1877 over Pullman's failure to pay the Southern Transportation rental marked the beginning of a long legal battle.[74] Central won the first suit and even threatened to reenter the sleeping-car business. A settlement was reached in 1878, when Pullman purchased the Southern Transportation Company outright for $90,000. A more serious battle developed in 1885 over the renewal of the Pennsylvania contract. The railroad demanded a more favorable agreement than the original one that Pullman had inherited from Central Transportation. Because this was one of the principal assets Pullman had taken over from the old company, he contended that the annual rental of $264,000 which he was paying to Central Transportation must be reduced to $66,000.[75] Pullman also maintained that most of the Central cars had been retired and many of its patents had expired. Central countered that Pullman's earnings were never less than $800,000 per year, or nearly three times the rental.[76] Pullman protested that his company's average income resulting from the lease was only $300,000. Central won a judgment, but it was thrown out by the U.S. Circuit Court, which ruled that the firm was *ultra vires*: acting beyond its corporate powers in transferring its operating contracts to another party. The bickering continued through a decade, benefitting only the lawyers.

At last the case reached the U.S. Supreme Court, where a lower court's judgment against Pullman for 4.2 million dollars was set aside in 1898.[77] Speculators who had paid around $40 a share in hopes of a fat settlement with Pullman faced the prospect of clearing little more than $29 a share. They now recognized that their failure to accept Pullman's 1885 offer to exchange Central shares at 4 to 1 was a golden opportunity gone by. The affair was finally settled in 1899, when a special liquidating dividend of $470,000 put to rest the tortuous life of the Central Transportation Company.[78]

WOODRUFF SLEEPING AND PARLOR COACH COMPANY

At the time the Pullman lease was negotiated, Jonah Woodruff was sixty years of age. Most men would have retired, but Woodruff thought that the lease was all wrong, that a sleeping-car monopoly was not a good thing, and that a Pullman monopoly was particularly objectionable. What is more, he thought that a better sleeping car could be built and set out immediately to do it. He had money, and he found a friendly workshop in one bay of the Harlan and Hollingsworth car works. His former patents, assigned to the Central Transportation Company, were now Pullman's property. But no matter; Woodruff had a new scheme in mind and obtained a patent in July, 1870 (No. 105,288), covering some novel ideas.

More striking was the general plan of what Woodruff came to call his rotunda car. It was the most original, if not the most practical, sleeping-car design since the commercial beginnings of sleepers. The upper berths were completely dismantled and stored away during the day. Inside, the car looked like an ordinary parlor car because it had none of the bulky upper-berth structure and paneling. All this gear, including the bedding, was stored in the recesses between and under the seats. Woodruff regarded the openness of his design as a decided improvement over the "burial casket aspect of the old sleepers."[79] He pointed out that it permitted better circulation of air and heat. The car's center of gravity was lowered, he said, and the possibility of

being accidentally shut inside the upper berth in the event of a wreck was eliminated.

Woodruff also claimed a weight saving of 4 to 6 tons, which was considerable for the chronically overweight sleeping car. But the greatest novelty of the plan was the oval lounges at both ends of the car. These lounges contained chairs and the water closets. One was usually assigned to ladies; the other was reserved as a gentlemen's smoker. Since the lounges were over the trucks, no berths had to be placed in this undesirable location. The lounges also served as pleasant observation or sitting rooms. The unique domed ceiling carried out the rotunda treatment of the oval rooms, which added a nice architectural touch and, Woodruff said, assisted in the ventilation.[80]

Most of these advantages that Woodruff announced to the press were fabrications, partial truths, and wishful thinking. But the design did have one merit: it was strikingly different, and as such it attracted the attention and support Woodruff needed in order to form a new sleeping-car company. By May 1871 he completed two sample cars, *Lady Washington* and *Lady Franklin*, at the Harlan and Hollingsworth plant. Both went on tour to woo contracts and investors. A few months later a charter was obtained from the state of Pennsylvania incorporating the Woodruff Sleeping and Parlor Coach Company, with Jonah Woodruff as president, August Trump as secretary and treasurer, and James Irvin as vice president.[81] A paid-in capital of $150,000 was reported. Pullman instituted a patent suit not long afterwards, but was forced to withdraw when it became clear that Woodruff's design was unique and could not be suppressed by the 100 car patents in Pullman's control.[82]

Freed of this entanglement late in 1871, Woodruff put his first cars in service on the Indianapolis, Peru, and Chicago Railroad. By mid-1873 eight sleepers and four parlor cars were running on the Chicago, Danville, and Vincennes, the New York, and Oswego Midland, the Camden and Atlantic, and other lines. In 1872 Woodruff took over the sleeping-car contract of Charles Doubleday on the important Cleveland, Columbus, Cincinnati, and Indianapolis. The operation was small but thriving. Woodruff's reputation and old contacts were undoubtedly helpful. Railroad travel and mileage were expanding, and the number of small and medium-sized lines made it difficult for Pullman to corner all the available contracts. In these circumstances small operators like Woodruff could prosper. By the third year his company had twenty-four cars in service (Figures 3.12 to 3.15).

Unfortunately, Woodruff's health appears to have declined in 1874 or 1875. He tried to recover it by going into semiretirement on his New Jersey farm, but to no avail. After his death in March 1876, the affairs of the company were taken over by several Pittsburgh businessmen headed by Frank Rahm. The new management continued to seek new lines. It found the South receptive and was soon providing service between Richmond and Atlanta.[83] It achieved a major advance in 1878 by purchasing the Lucas Sleeping Car Company of Atlanta, which had been formed three years before by Christian E. Lucas with the backing of W. M. Wadley, president of the Central Railroad of Georgia.[84] It

Figure 3.13   The Flora *was a Woodruff rotunda car built around 1875 by Jackson and Sharp.*

Figure 3.14   This Woodruff rotunda car, the Roberval *(Number 23) of about 1875, combined parlor and sleeping facilities. Not all Woodruff cars devoted so much floor space to parlor chairs.*

Figure 3.12   After Pullman's lease of the Central Transportation Company, Jonah Woodruff organized his own firm. In 1871 he introduced a unique style of car with round ends. The example shown here dates from about 1875 and appears to be named the Lottie.

is said that the line had been founded as a protest against Pullman, who began to admit Negroes as passengers on the formerly segregated sleepers. Lucas had only five cars, but he was a practical builder with several valuable sleeping-car patents, and he had gained the good will of other Southern lines (see Figure 3.16). Lucas was made head of Woodruff's Atlanta office. The Chesapeake and Ohio and the Virginia Midland quickly joined the Woodruff clients. In 1879 it was reported that Lucas patent cars were being completed for the Jacksonville, Pensacola, and Mobile. At the same time new Woodruff cars were being built for the Chicago and Eastern Illinois. Like most of the other early Woodruff cars, they were manufactured by Harlan and Hollingsworth

and were painted a rich wine-maroon color with green, blue, and gold striping.[85]

By 1880 Woodruff had 64 cars in service—a puny fleet compared with Pullman's 800-car armada. And the disparity in strength was even greater because the Woodruff operation was so scattered. Most of its lines were isolated, often with only two or three cars running between relatively minor cities. An example was the contract with the St. Louis, Keokuk, and North Western, which involved two cars that began service in 1877. By 1882 the cars needed heavy repairs, but the revenue they brought in barely covered operating expenses.[86] Meanwhile the railroad had come under the Burlington's control, and the new owner considered turning the operation over to Pullman rather than repairing the cars in accordance with the contract. It also had the option to buy them outright for $12,000. The Burlington decided to keep the Woodruff contract alive because its president saw an advantage in having two sleeping-car companies on the string. No matter how poor or weak Woodruff might be, it could be used as a threat against Pullman. A few years later when the contract came up for renewal, Woodruff offered to run the cars at 2 cents a mile and give 25 percent of the net earnings to the railroad. Pullman charged 3 cents a mile, and the Burlington wanted to pay only half that.

Marginal as many of Woodruff's operations might be, the company controlled several good lines. As early as 1877 it was running a through car between New York and St. Louis via the New York Central's Bee Line. This plum was taken away in 1881 by Wagner, but Woodruff received a payment of $1 per passenger for giving up the contract. At that time all New York Central

Figure 3.15 A Woodruff rotunda car of about 1875, rebuilt around 1890 as an office car for the Southern Railway. (Southern Railway)

Figure 3.16   Christian E. Lucas received six patents for sleeping cars between 1875 and 1880. Some cars built from his designs were used in the South. All became the property of Woodruff in 1878. (Hall of Records)

Figure 3.17   Interior photograph of a Woodruff sleeper of about 1875, showing three upper berths made up.

subsidiaries were under extreme pressure to deliver their sleeping-car contracts to the Vanderbilt-controlled Wagner company. Pullman was also busy trying to grab Woodruff's better contracts. Soon after the Wabash took over the Indianapolis, Peru, and Chicago line, Pullman disputed the I P & C contract (which had been Woodruff's first business deal). Pullman contended that its own contract with the Wabash superseded those made with the now-defunct Indiana line.[87] But for every contract that Woodruff lost,

the company found a new one to take its place. In addition to the routes already mentioned, Woodruff eventually secured the Wisconsin Central, the Evansville and Terre Haute, the Grand Rapids and Indiana, and the New York and Manhattan Beach.

During these years the rotunda car design was the mainstay of Woodruff's rolling stock. Jonah Woodruff had improved the stowaway upper berths by mechanisms covered in two patents issued in February 1874 (Nos. 147,538 and 147,539). Neat metal

Figure 3.18 This interior of a Woodruff sleeper displays an upper berth framework patented in 1874. The rotunda parlor can be seen through the end door. Note the hot-water heating pipes visible under the seat end frames.

Figure 3.19 When the Monongahela was delivered by Jackson and Sharp in 1885, the old-fashioned rotunda ends had long been abandoned. After 1889 the car was operated by Pullman under the name Melissa. (Lancaster County Historical Society, Pennsylvania)

frames, apparently of cast brass, took the place of the 1870 design. Two excellent interior photographs made at the Jackson and Sharp plant sometime in the 1870s clearly show the apparatus (Figures 3.17 and 3.18). One of the rotunda rooms can be seen through the doorway. By the early 1880s, however, the rotunda cars were considered out of date. The design had lost its novelty, and the extra expense of the rounded car ends made it desirable

to develop a new plan. In addition, the round ends were weak and not likely to withstand a telescoping accident. The complicated tinwork of the dome roof must also have been difficult to keep watertight. More conventional-appearing cars became the standard. Woodruff's collapsible berth was replaced by those of the folding type, and eventually only an expert could distinguish one of the late Woodruff sleepers from a Pullman (Figure 3.19).

Figure 3.20   W. D. Mann, who helped found Pullman's European counter-part, Wagon Lits, attempted to introduce compartment-style sleepers in America. An 1878 Wagon Lits car is shown here.

With few exceptions, the Woodruff cars were built by Jackson and Sharp or Harlan and Hollingsworth. Both companies had accepted stock in at least partial payment for car orders, and over the years both had become large shareholders in the firm. It is likely that their power brought about the downfall of Woodruff's president, Frank Rahm. Early in 1882 he took over an idle car plant at Swissvale, Pennsylvania, with the idea that Woodruff, like Pullman, could build its own rolling stock.[88] The shop's extra capacity could be used to produce equipment for outside buyers. Whether this venture was a success or failure has not been discovered, but within the year Rahm resigned and Job H. Jackson, president of Jackson and Sharp, took over the management.[89] Afterwards all Woodruff cars were produced at Jackson's establishment; the Swissvale plant closed and was sold a few years later to Westinghouse.

A second management shakeup occurred in January 1884, when Woodruff's longtime vice president and general manager, J. H. Irvin, died. The offices were shifted back to Philadelphia. A year or two later Jackson stepped aside as president in favor of Daniel C. Corbin, a New York financier.

Before 1877, Woodruff paid a steady 12 percent dividend, had no mortgages, and in general was doing rather well in the bad time following the Cooke financial panic.[90] The firm remained profitable after this date, but dividends were suspended for several years and all cash was put into new equipment. In 1881 it was necessary to take out a $300,000 mortage. To win contracts and hold the existing clientele, more new cars were needed. Between 1884 and 1888 fifty new cars were purchased at a cost of over $594,000. All this buying was highly profitable for Jackson and Sharp but proved a severe strain on Woodruff's increasingly shaky finances. Capitalization was set at $5 million, but less than $1½ million was paid in; the bonded debt was up to $122,000 and other notes were held privately. Bonds sold at a 25 percent discount, and little stock had been bought since the 1870s. Some shares had been traded at 75 percent below their par value. The company was not broke, but its obligations so exceeded its earnings and modest cash reserves that one bad year would surely precipitate a failure. Short of selling out to Pullman, the only solu-

tion appeared to be a merger with the Mann sleeping-car company.

The Mann Boudoir Car Company

William D'Alton Mann (1839–1920), an American who built a sleeping car that became popular in Europe, had a great ambition to repeat his success in his native land. The life of Mann reads like a caricature of a hero in one of Theodore Dreiser's novels. He was flamboyant, greedy, pursued by scandal, and one of the fascinating characters of American business history. The details of Colonel Mann's nefarious life have been recorded elsewhere.[91] It is enough to say here that after a few years of war profiteering, oil well speculation, carpetbagging, and railroad promotion in Mobile, Alabama, he turned his thoughts to perfecting a sleeping car.

In 1871 Mann settled on a not very original plan for a compartment sleeper in which each compartment had its own side door, like a European coach. It was not possible to pass through the car on the inside—an arrangement totally unacceptable on American lines. Far superior compartment cars with side aisles extending the full length of the car had been running in this country since the 1850s, but they were apparently unknown to Mann and the Patent Office examiners. (A floor plan of a drawing-room car dating from 1859 was shown earlier in Figure 3.5. This chapter has also discussed stateroom sleepers on the Illinois Central in 1856 as well as on other railroads.) Mann received a patent on January 9, 1872 (No. 122,622).

At this stage of his career he left Mobile, apparently for good reasons, and went to England with the idea that his sleeping car might be well received there. However, it turned out that the relatively short runs characteristic of British railway travel were not favorable to sleeping-car service. Later in 1872 Mann went to Brussels and built a prototype car.[92] During the following year he helped to form what was to become the International Sleeping Car Company, better known as Wagons-Lits. He also established a line between Munich and Vienna, and Mann cars were said to

Figure 3.21   *The daytime arrangement of a Mann boudoir car is shown in this advertisement of 1885. (Biographical Directory of Railway Officials, 1885)*

be coming into wide use in Germany by July 1873.[93] Within two years, Mann had fifty-one cars placed on a dozen railways and was transporting 3,000 passengers weekly between Berlin, Vienna, Paris, and Bucharest (Figure 3.20).[94] The London, Chatham, and Dover was also testing a Mann car on its line. In 1873 Mann had gained a valuable partner in Georges Nagelmackers, a member of the Belgian banking family. The company greatly expanded its service in the next several years. It introduced improved cars with side aisles, and Mann managed to obtain a U.S. patent covering the design on January 8, 1878 (No. 198,991).

Just as Wagons-Lits really began to prosper, Mann decided to return to the United States. He sold out to Nagelmackers for 2 million dollars and in March 1883 obtained a New York State charter incorporating the Mann Boudoir Car Company. One of the original investors was the political cartoonist Thomas Nast, an odd associate for an entrepreneur as gaudy as Mann.

Within a few years Mann had made the two worst decisions of his business career. The greatest blunder was to sell his interest in Wagons-Lits, which was about to become Pullman's European counterpart and would have made Mann wealthy beyond even his avaricious dreams. His second miscalculation was to start an independent sleeping-car operation in the face of Pullman's growing monopoly. When Jonah Woodruff had attempted the same thing a decade earlier, Pullman's hold on the industry had been much weaker. Within five years Mann was to lose his fortune in the attempt to invade the North American sleeping-car business. But at the beginning he was fully convinced that the technical superiority of the boudoir car was enough to win a large share of the American market. He plunged half of his savings into new cars, a number of which, produced by the Gilbert Car Company, were placed in service between Boston and New York via the New Haven and the Boston and Albany railroads in October 1883.[95]

The first group of Mann's sleepers included a luxurious private car named after the opera singer Adelina Patti. Modestly describing it as the most magnificent private car in the world, Mann made it available to the celebrated star for a national tour. The press printed countless stories about the $40,000 car, its decor, and its glamorous passenger. Mann also produced two other private cars: the *Etelka Gerstner* and the *Janisch*. All three were rented to private parties and served to publicize Mann's new enterprise.

Some years earlier Mann had learned that associating with celebrities was good publicity. In 1878 he had received much favorable press notice through a tenuous relationship with the Russian royal family, for whom he oversaw the construction of two private cars (presumably incorporating his Boudoir arrangement) at the Ashbury Carriage and Iron Company of Manchester, England. America had its own royalty of sorts—government officials, financial moguls, and stage personalities—and Mann was alert to opportunities for ingratiating himself with this elite whenever possible.

For his regular series of sleeping cars, Mann adopted the Wagon-Lits plan for American use. The overall size was enlarged,

and the cars were mounted on four- or six-wheel trucks. The bodies were stretched out to 64 feet, but the compartment or boudoir system was retained.[96] Each room had its convertible seat-bed and overhead berth (Figures 3.21 and 3.22). The oversize seat back was folded upward to form the upper berth; bedding was stored in the seat box. Note that there were single and double compartments which slept two and four passengers, respectively (Figure 3.23). The little fold-up seats in the aisle were for passengers who wished to see the view from the opposite side of the train. The interiors were quite different from the typical car of the period. In place of the usual varnished wood paneling, much of the wall surface was covered with embossed leather, which was laid over hair padding and provided a soft surface when passengers bumped against it. The arched roof and 42-inch, spoked Krupp wheels were other design novelties. Mann had devised his own heating and ventilating system. He claimed

Figure 3.22   At night the boudoir sofa became a double berth. The compartment door stands open to the side-aisle corridor, which ran the length of the car. (Mann Boudoir Car Company catalog, 1885c.)

Figure 3.23   Floor plan of the Mann boudoir car La Traviata, *built in 1883 by the Gilbert Car Works. (Mann Boudoir Car Company catalog, 1885c.)*

that his fan-ice-water apparatus collected all the dust and cinders and lowered the temperature by as much as 20 degrees. The effectiveness of the device was unquestionably exaggerated, but it was an early move toward air conditioning.

In short, Mann was offering the traveling public a radically different style of sleeper. It answered the most frequent complaint about the Pullman by furnishing greater privacy and more

room. Passengers no longer had to begin undressing in the aisle and struggle to complete the task inside the cramped berth behind the aisle curtains. Mann hoped that this basic advantage, together with the other original features of his cars, would create enough demand to displace the traditional open-section sleeper.

Several railroads were attracted by the boudoir cars soon after the first lot went into service. The Cincinnati Southern signed up

Figure 3.24 *Exterior of* La Traviata, *showing its arched roof. The raised portion of the roof was part of the air-washer ventilating system used by Mann. Note also the Miller couplers.* (Mann Boudoir Car Company catalog, 1885c.)

for six cars in 1884. Then came the Boston and Lowell, the Grand Trunk, the Wabash, and the B & O. The Boston and Albany asked Mann to operate its dining cars as well as some of its through sleepers. So many of his cars were running south out of Cincinnati that Mann opened a repair shop across the river in Ludlow, Kentucky. The Wabash substituted Mann cars for Wagner's on its Chicago–St. Louis trains.[97] Four Mann cars were exported to Australia in 1885 by the Gilbert Car Works; three of these were still in service in 1960, according to the railway historian G. M. Best.

By 1885 forty-three Mann cars were operating over seven railroads, which might be considered good progress in such a competitive field (Figure 3.24). The company was losing money, however, and management decided that acquiring additional cars and contracts was no guarantee of success. A more conservative policy was formulated: rather than striving for blind expansion, the company would seek contracts for more heavily traveled lines.[98]

To implement the new policy, a more experienced and conservative executive than Mann was sought to head the operation. In 1885 Thomas C. Purdy, former general manager of the Mexican National Railways, was named president. Mann was demoted to vice president and general manager. But efforts to consolidate the business's operations seemed impossible. Like Woodruff, Mann was left with the marginal lines that Pullman or Wagner did not control. Also like Woodruff, the company's activities were scattered over the Eastern, Southern, and Middle Western states. New contracts with the Georgia Pacific and other lines did little to aid the sagging fortunes of the Boudoir Company.

Other factors were working against Mann. The smaller capacity of the compartment car has always been the chief economic argument against its use. A Mann car slept 22, while an open-section car of the same size and weight accommodated 30. Because of the smaller capacity, Mann's fares were nearly double those for an open-section car. But Mann maintained that his cars remained popular and that on the New York–Boston trains, where compartment and section cars were both available, passengers always selected the boudoir.[99] His statement was confirmed by a report that travelers made the same choice on the Chicago–St. Paul sleepers, where 75 percent of the passengers preferred the Mann cars.[100] In the very long run Mann's contention was borne out by the success of the roomette, though it took Pullman sixty years to evolve it.

Other objections were made to the compartment car. A major one was that the partitions added weight and expense. Since the earliest days of the American railroad all practical railroad men had agreed on the disadvantages of the compartment arrangement. There was also an undercurrent of disapproval in the Victorian mind concerning the compartment sleeping car. The very name "boudoir" was suspect. The *National Car Builder* thought it too exotic and best "left to thrive in its native soil on the other side of the water, where the people know how to pronounce it properly."[101] There was the suspicion that unspeakable acts took place behind the locked doors of the boudoir compartments. Or some people viewed the compartments as dens harboring villains who were waiting to pounce upon the innocent. No such goings-on were possible in the section sleeper, where all passengers mingled in democratic openness under the watchful eye of the porter. Female passengers might complain about the mixing of the sexes with only curtains for privacy, but somehow this seemed less objectionable than the perfumed privacy of the boudoir. These fears were confirmed by the railroad trade press, which delighted in reporting the latest outrages that had occurred aboard European compartment cars. The *Police Gazette* ran a story called

"Love in a Palace Boudoir" which described how a wife discovered her husband and his sweetheart aboard an infamous boudoir sleeper. Mann's reputation thus suffered a mortal blow.[102] The boudoir was not only sinister and immoral; it was French!

By 1888 Mann was ready to admit defeat. He had lost his fortune trying to keep the company afloat, and now he was heavily in debt. In five years, only the last season had shown a profit, and that was too modest to encourage even the ebullient Mann. It had been a costly experiment; the time had come to end it.

The Union Palace Car Company

By the late 1880s it was clear to the creditors of both the Woodruff and Mann companies that desperate measures were necessary if either firm was to survive. As early as 1884 Woodruff was considering acquiring Mann's patents and was weighing the possibility of taking over his entire operation. But these discussions led nowhere, for Mann still expected to outrun the established sleeping-car firms with his exotic Continental dormitory cars. It took only four years to prove the folly of this dream. Meanwhile Woodruff had overextended itself through purchases of new equipment—a reasonable gamble, but one that did not pay off. It was in no position to buy Mann now that he was eager to sell. An outside agent was needed to salvage the faltering companies.

That the combination of two failures will produce a success seems to be an undying hope in the business world. Such marriages must occasionally work, but the main examples that come to mind seem to be negative, such as the Penn Central and the Erie Lackawanna. Was the Woodruff-Mann merger, which was proposed in 1888, a misguided but sincere effort to save them, or was it a cynical maneuver to extract a large price from Pullman to be rid of these nuisance competitors? On first glance the conspiracy theory might seem the most convincing, but actually the parties who brought about the merger had another motive entirely.

Job Jackson, president of Jackson and Sharp, led the effort to combine the two firms. He was encouraged by another large creditor, Harlan and Hollingsworth. Both of the Wilmington car builders were holding large quantities of Woodruff and Mann paper, and it was very much in their interest to keep Mann and Woodruff alive. If both companies failed, the rolling stock would most likely be sold at a discount, with both car builders taking the loss. More important, there would be no more sleeping-car orders at the Wilmington shops, for both Wagner and Pullman were producing their own stock. The two builders were anxious to maintain an independent sleeping-car operation, no matter how poor that firm's prospects might be.

And the prospects were indeed bleak. Both companies were financially distressed, but Mann was shockingly weak, with an income that rarely paid interest on the bonds. Both had large debts, poor earning records, and too many lightly traveled routes. To gain some contracts they had undercut Pullman to the point where those routes made no profit.

Jackson felt that only a new corporation could save the situation. Mann and Woodruff were both discredited, but fresh capital might be raised to support a new enterprise. Accordingly, in September 1888 the Union Palace Car Company was formed, with a capital of 3 million dollars.[103] Two-thirds of this amount would be needed to buy Woodruff and Mann; the remainder would be used to purchase new cars and expand the business. The scheme was poorly conceived, for even if the stock could be sold at full value, the outstanding liabilities of the old companies, plus the expenditures planned by Union, would result in a shortage of $823,000. From the beginning investors recognized this weakness, and the stock of the Union Palace Car Company sold at a 50 percent discount. But Jackson was happy enough, because he and the Harlan company had transferred their holdings in Union to a group in New York that put up 1.5 million dollars.[104] The sale was managed by Thomas Purdy, the former president of Mann and the new head of Union. The Wilmington car builders were also paid off with an order for thirty-four cars, most of which appear to have been patterned on Mann's design.

At the outset Union had Mann's 54 cars and 2,000 miles of route, together with Woodruff's 127 cars and 3,500 miles of route. Union also claimed to have an additional 4,000 miles through a new contract with the Richmond and West Point Terminal Company. There was talk of running cars on the New Jersey Central and the Reading.[105] During the same year an advertisement appeared in *Official Guide* claiming that the Union Palace Car Company's operations would cover 15,000 miles of railroad by January 1889. Even if Union reached this goal, it was far short of Pullman's 82,000-mile empire.

Shortly after the merger it was already clear that Union was not going to make it. After only three months of corporate life, Pullman purchased the Union Palace in January 1889 for just over 2½ million dollars.[106] *The Commercial and Financial Chronicle* contended that Pullman acquired the stock for 50 cents on the dollar. Yet even if Pullman paid the full price, the sleeping-car giant was satisfied to be freed of this rival at last. Curiously, the corporate identities of Woodruff, Mann, and Union were maintained for a number of years afterwards. Advertisements appeared in *Official Guide* that gave G. M. Pullman or Thomas Wickes (a Pullman official) as head of the firms. Their general offices were located in Pullman's Chicago office. Apparently Pullman saw some advantage in maintaining this scheme, or perhaps old contractual obligations made it necessary.

Monarch Parlor-Sleeping Car Company

A dozen or more insubstantial sleeping-car companies were organized after 1880. All were founded in a futile effort to upset Pullman's monopoly. Most were shadowy paper corporations. A few, like Harris (1889) and American (1905), actually built sample cars, but of these enterprises possibly only Monarch is worth more than passing notice.

The genesis of the Monarch Company appears to be a patent

Figure 3.25   *Gustave Leve's 1876 patent for sleeping cars was followed by the Quebec, Montreal, Ottawa, and Occidental Railway for several cars built at its own shops in 1880–1881. Leve's patent was also used by the Monarch and Mann sleeping car companies.* (Canadian Illustrated News, *February 19, 1881*)

(No. 181,857) issued September 5, 1876, to Gustave Leve of New York. The design's main objective was to provide a day-night car with parlor seats. The revolving chairs were collapsible, and the bedding was stored in narrow closets along the sides of the car. Curtains formed berth enclosures, as in the ordinary American sleeper. The mechanism and cabinetry to accomplish this trans-

formation was complex and probably not worth the trouble or expense.

Leve's design apparently lay dormant for four or five years; then in 1881 the *Canadian Illustrated News* described some cars of his design on the Quebec, Montreal, Ottawa, and Occidental Railway. The cars were built in the railroad's shops, with certain improvements added by the road's mechanical superintendent. (An engraving accompanying the article is reproduced in Figure 3.25.) In 1883 the Gilbert Car Works constructed two cars using Leve's patent for export to Australia.[107] Early the next year four more cars were in service on the Occidental Railway of Canada and the Florida Transit.[108] Leve had meanwhile acquired a partner named Alden and was engaged in the "excursion business," but little information can be found on the nature of this enterprise. Reports were circulated in July 1885 that the Monarch Parlor-Sleeping Car Company had been formed in New York to exploit Leve's patented car.[109] A. F. Higgs, owner of several refrigerator-car lines, was to head the 5-million-dollar organization. The cars would be named after ancient rulers, and soon sleepers called *Zenobia*, *Cleopatra*, and *Rhodope* were running under Monarch's letter board. Thirteen cars were in service by the following spring.[110] They were 75 feet long, with 66-foot bodies, and offered 24 rotating parlor seats by day and 12 double-berth sections by night. The 35-ton cars were carried on eight or twelve wheels and cost $22,000 each.

In the summer of 1886 the Gilbert Car Works produced an extraordinary parlor observation car for Monarch named *Ymir*. It offered service from New York City to the White Mountains via the Connecticut River Railroad and featured an open side aisle for viewing the scenery. (In keeping with its unusual floor plan, the car was decorated in an original scheme that is described in Chapter 4.)

In addition to its Canadian, New England, and Florida contracts, Monarch operated cars for a time between Baltimore and Pittsburgh, Indianapolis and Evansville, and Fort Smith and New Orleans. Like Mann, it had entered the field too late; and like Mann and Woodruff, its operations were too scattered and marginal to pay. The company appears to have struggled on until 1893 or 1894 and was said to have built a total of thirty cars.[111] The details of its disposition have not been uncovered, but it does not appear to have been taken over by Pullman. It is possible that the Gilbert Car Company had a large interest in the concern, after the fashion of the Wilmington shops' involvements with Mann and Woodruff. Since Gilbert is known to have built all but one or two of Monarch's cars, Gilbert's unexpected and complete failure in 1893 may explain Monarch's disappearance.

The Wagner Palace Car Company

By the late eighties Pullman had vanquished all its major competitors but one. The semi-independent sleeping-car line trading under the names of its founder, Webster Wagner (1817–1882), continued to flourish. Its cars ran principally over the New York Central and the other Vanderbilt-controlled lines. It carried a large, profitable traffic between the major cities of the industrial Northeastern–Great Lakes region of the country. The massive resources and influence of the Vanderbilts seemed to ensure its continued operation. Much to Pullman's frustration, there ap-

peared little hope that the Wagner Company would topple of its own weight or sell out as smaller competitors had done. Wagner was a formidable rival.

Webster Wagner was one of the earliest Woodruff concessionaires. In partnership with George B. Gates, T. N. Parmelee, and Morgan Gardner, he signed a contract with the New York Central Railroad on May 8, 1858, to operate sleeping cars between Albany and Buffalo.[112] Erastus Corning, president of the railroad, favored the scheme; possibly he saw the concession arrangement as a cheap way to introduce some measure of luxurious travel.[113] The hard times following the 1857 panic had left the road with little excess capital for frills. Four cars costing $3,200 each were said to be in service on this route by September 1858. Woodruff reportedly received a royalty payment of $500. Several months later *Leslie's* magazine published an interior view of one of the cars showing its three-tier berths made up for the evening.[114] The engraving, which has been frequently reproduced, is one of the earliest pictorial representations of an American sleeping car.

Just two years after the cars entered service, a British traveler recorded his experience aboard a sunflower-yellow car which carried the following legend in red lettering: "Albany and Buffalo Sleeping-Car." The charge for a single upper was 25 cents; the double lower cost 50 cents. The passenger wrote:[115]

When we were between Little Falls and Herkeimer the officer of the sleeping-car entered and called out, "Now then, misters, if you please, get up from your seats and allow me to make up the beds." Two by two we arose and with quick hand the nimble Yankee turned over every other seat so as to reverse the back and make two seats facing each other. He shut the windows and pulled up the shutters, leaving for ventilation the slip of perforated zinc open at the top of each. Then he stripped up the cushions and unfastened from beneath each seat a light cane-bottomed frame which he fastened to the side of the car at suitable heights and covered each with cushions and blankets.

I went out on the balcony to be out of the way and when I came back the whole place was transformed. There was no longer an aisle of double seats but the cabin of a small steamer with curtained berths and closed portholes. The bottom berths were wide and comfortable with room to roll and turn. There were two high berths to choose from; both wicker trays, ledged in, cushioned and rugged. The one was about a foot and a half higher than the other and I chose the top one as being nearer the zinc ventilator. Some of the passengers had turned in and were snoring. Others, like young cows balancing on spring boughs, were swinging their legs from the wicker trays and peeling off their stockings or struggling to get rid of their boots.

I clambered to my perch and found it was like lying on one's back on a narrow plank. If I turned my back to the car wall the motion of the train bumped me off my bed altogether and if I turned my face to the wall I felt a horrible sensation of being likely to roll backward into the aisle, so I lay on my back and settled the question. It was like trying to sleep on the back of a runaway horse. At each place the train stopped there was the clashing of the bell and if I peered through the zinc ventilator into the outer darkness a flying scud of sparks from the engine did not serve to divest my mind of all chances of being burnt alive. Then there were blazes of pine torches as we neared a station, more bell clamor and jumbling sounds of baggage, slamming doors and itinerant conductors.

At last I fell into a precarious and fragile sleep that lasted until daylight returned.

In later years statements attributed to Wagner about the origin of his company are in conflict with what appears to be the true story. Perhaps he was seriously misquoted or had an exceptionally poor memory, or he may have purposely altered the story to flatter his latter-day patron, Cornelius Vanderbilt. The Commodore, a man of no small ego, doubtless approved of the efforts of his faithful associate to keep the name of Vanderbilt central to the action. Wagner's accounts made no mention of Woodruff, Corning, or the original partners.

Wagner claimed that he was allowed to remodel a Hudson River Railroad coach early in 1858 to demonstrate his ideas for a sleeping car. The car was intended to compete with the palatial steamers on the Hudson River. With the help of William H. Vanderbilt, the Commodore's son, the gruff old capitalist was induced to inspect the car. Wagner expected to hear the worst, but instead Vanderbilt asked: "How many have you got of these things?" According to Wagner, "I told him I had only one." "Go ahead," said Vanderbilt, "and build more. It's a devilish good thing and you can't have too many of them." "My heart dropped back in its place and I knew that my fortune was made."[116]

In another version of their first meeting, the story goes that a sample car was attached to a train taking the Commodore through Albany.[117] A chance inspection led to an immediate partnership, with Vanderbilt providing the capital for a 75 percent interest. The Commodore made it clear that he would have no cars operating over his line which did not pay him a major share of the revenue, and that if Wagner did not like the terms he could take his cars elsewhere. The general drift of the story is probably true enough, but the details and chronology are badly distorted. Vanderbilt had little direct connection with railroads until the mid-1860s. He was a steamboat man at the time that Wagner was involved with his first cars, and would not have looked kindly on new ways to kill the overnight river trade. Wagner operated no cars south of Albany until June 1860. Nor is he known to have done more than simply use Woodruff's designs; at least, no sleeping-car patents were issued in Wagner's name.

The meeting with Vanderbilt most likely took place in 1868 rather than 1858, for it was not until late in 1868 that Vanderbilt gained control of the New York Central. It was probably only then that he gave Wagner notice to accept his terms or find a new client. Wagner wisely consented to Vanderbilt's proposition; it was essential to his survival against Pullman, and it made him a rich man. The Wagner Company was basically a New York Central subsidiary. Though it maintained a facade of autonomy, its officers were New York Central men, Vanderbilt agents, or Vanderbilt in-laws. Although stock was issued, it was a closed corporation. Unlike the Pullman Company, Wagner did not publish public annual reports, and considering the scope of Wagner's operation, the trade press gave its affairs little coverage.

Wagner's prospects for success before the alliance with Vanderbilt appeared modest. He had failed as a carriage maker and in 1843 had taken a job as a station agent for the New York Central.[118] In this work he discovered the need for sleeping cars and, as already discussed, became a small contractor for Woodruff in 1858. In 1860 he left his railroad post for the grain and produce trade. But the sleeping-car business continued to hold his interest; his original fleet of four cars grew to a dozen, and he decided to concentrate on expanding this line of business. In the spring of 1860 he contracted to run sleepers over the Hudson River Railroad.[119] By June Wagner and his partners were running cars from New York City to Buffalo; through cars were not possible, however, until the bridge at Albany was opened in 1866.[120]

Noticeable improvements were introduced at this time. Sheets and pillows were added to the bedding, and far grander cars were placed in service. The new Hudson River cars built by Eaton and Gilbert featured clerestory roofs, double windows, and three lamps with large reflectors.[121] (Wagner liked to claim the clerestory as his innovation and managed to obtain a patent for the raised roof, but as discussed in Chapter 1, the idea can be traced back to 1835.) A description of Wagner's improved cars was given in the May 31, 1860, *American Railway Review*:

Two new and improved sleeping cars yesterday went into operation on the Hudson River Railroad, which promise to inaugurate a more healthy and convenient mode of traveling than has hitherto been enjoyed by our locomotive community. The improvement consists in the thorough ventilation of the car, complete exclusion of dust, and an arrangement by which every four seats are converted into a stateroom, which renders a family of four almost as complete as if they were under the lock and key of their own stateroom on a steamboat. In fact, the car throughout is simply a steamboat saloon on wheels. The roof is raised in the centre, and in the sides of the elevated part are inserted twenty-eight ventilators, which preserve the air pure and sweet, and keep up a steady and healthful circulation—a desideratum which has long been desired, but which, until now, has never been supplied. The system of seats and berths is not unlike that already in use. By day the mattresses and pillows are tucked away beneath the cushions, the cane-bottomed berths rest on their ledges at the top of the car, and four seats are all that meet the eye. By night these are ingeniously brought together and turned into a double bed, while still another tier of horizontal accommodations mysteriously finds its way into existence three feet above, and offers its downy temptation to the unprovided voyager; the whole being inclosed by a curtain, which throws its protecting folds around the sleeping inmates within. In addition to these staterooms, which are paid for at the rate of fifty cents a berth, there is, at the end of the car, a ladies' dressing-room, where are all the conveniences of the toilet, and every thing necessary as a prelude to a breakfast but the meal. The floor is constructed of two layers, with sawdust between to deaden the sound of the wheels; and for a similar purpose, and to exclude the dust, the windows are likewise duplicated. The cost of each car is about $3,500. They are said to be the finest cars ever built in this country.

A few years later, increasing extravagance was evident in eight new cars that Wagner purchased for $8,000 to $9,000 each.[122] The ceilings were covered with green and gold frescoes. More

Figure 3.26   The Boston and Albany advertised Wagner sleeping cars in this broadside of July 20, 1869. The parlor car floor plans may also represent Wagner equipment. (Smithsonian Neg. 76336)

important than the costly decorations, however, were the improvements in passenger comfort. The old three-layer Woodruff berth arrangement gave way to the two-berth plan. Here again were spacious, well-appointed sleeping cars in advance of, or at least contemporaneous with, Pullman's *Pioneer*.

At the same time that Wagner was improving his night cars, he decided to introduce parlor cars for day travel. In September 1867 the first Wagner parlors entered service. A heavy first-class daytime traffic could be expected between New York City and the upstate cities. Wagner was eager to capture that traffic, and within a decade one-third of his fleet was made up of parlor cars (Figure 3.26).

By the end of the Civil War, Wagner was solidly in the sleeping-car business and ranked second only to the Central Transportation Company. Since the old partnership arrangement was no longer adequate to the firm's growing business, in January 1866 he formed the New York Central Sleeping Car Company.[123] Its capital was limited to $80,000. George Gates, one of

Figure 3.27   A floor plan for a Wagner hotel car of about 1870 is shown in this French engraving published in 1873. The legend reads: A. aisle, B. seats, B'. porter's seat, C. compartment 4 seats, C'. compartment 5 seats, C''. compartment 6 seats, D'. platform, F. smoking compartment, G. ice water fountain, L. wash stand, P. heating stove, Q. coal box, R. heating stove (special type), S. parlor seating area, U. urinal, V. kitchen, W. water closet. (Malezieux, Public Transport in the United States, 1873)

Wagner's original partners, was made vice president of the new firm. Gates held the Lake Shore and Michigan Southern sleeping-car contract and was also running sleepers between Buffalo and Erie and Rochester and Cleveland. It is likely that Wagner and Gates strengthened their alliance at this time to facilitate through service between New York and the west. Wagner took over the company, entirely succeeding Gates, in 1869 or 1873.[124]

When Vanderbilt consolidated the New York Central and Hudson River railroads, he issued an ultimatum for a more favorable contract. Wagner had little choice but to comply. In the usual sleeping-car agreement, the railroad was paid a percentage of the profits; or the road might even be obliged to pay the contractor a mileage fee if the cars ran at a loss. But in the 1869 Vanderbilt contract, Wagner was obliged to pay a percentage of *gross* revenues to the railroad. Vanderbilt placed a levy of 20

percent on drawing-room car revenues and a twenty-five percent charge on sleepers.[125] Wagner was also responsible for the upkeep of car exteriors and was solely liable in the event of their loss through collision or fire. Typically the sleeper concessionaire was obliged to maintain only the interiors and to provide new cars as required. Moreover, the Commodore demanded a major number of the shares in Wagner's company.

In return Vanderbilt offered capital, contracts, and the prospects of a golden future as his rail empire spread west. He was already buying into the North Western and other lines west. To meet Vanderbilt's demands for more and better cars, Wagner had to sell six stock issues in 1869–1870. Capitalization increased more than sevenfold to $600,000.[126] The business was developing too rapidly for the makeshift methods and the simple corporate structure of the existing firm. It was necessary to form a joint stock

Figure 3.28   Some impression of the elaborate maroon and gold exterior of a Wagner palace car is shown in this engraving made around 1875. (J. R. Asher and G. H. Adams, Railroad Atlas, *1879)*

Figure 3.29   A Wagner sleeper that was operating on the Lake Shore and Michigan Southern Railway around 1875. The interior arrangement differs little from a Pullman car of the same period. (J. R. Asher and G. H. Adams, Railroad Atlas, *1879)*

company in November 1875.[127] Wagner transferred his contracts to the new firm, and Vanderbilt's old gross-revenue tariff was abandoned. The new body retained the old corporate title of New York Central Sleeping Car Company but remained a closed corporation, with all shares held by Wagner, Vanderbilt, and a few trusted associates.[128]

Just months before the reorganization, Wagner cars replaced Pullman's on the Michigan Central. After Vanderbilt gained control of the road in the fall of 1875, one of his first acts was to force a new sleeping-car contract on the directors.[129] Pullman was outraged. It was one of his oldest and best contracts, and he offered new terms: 4 cents a mile, plus the first-class fare. Wagner agreed to provide service for the fare alone. The Michigan Central management would have preferred to do business with Pullman, a

Figure 3.30 New York State Senator Webster Wagner and one of his cars, pictured in the New York Graphic *January 17, 1876.*

Chicago man, but Vanderbilt told them that they could accept Wagner or face an end to through New York–Chicago service. The Commodore embargoed Pullman cars at Buffalo, which would funnel the first-class traffic to the other east-west lines offering through cars. Vanderbilt prevailed; Pullman was banished from the Michigan Central except for the Grand Rapids and Montreal (via the Grand Trunk) trains. Wagner ordered sixteen new cars from the Wilmington car builders. The New York Central–Michigan Central combine offered the smoothest and levelest between New York and Chicago, and it was now all Wagner's.

Pullman, embittered by his loss, became a furious enemy of the Vanderbilts. Since he could find no immediate way to even the score with the powerful family, he turned to their associate as a promising target. Before this incident he and Wagner had been merely business rivals, but now Pullman meant to make trouble. His patents were always a handy weapon for harassing the opposition. After acquiring the assets of the Central Transportation Company in 1870, Pullman had agreed to give Wagner a new contract licensing use of the basic sleeping-car patents now in his control. Wagner used the Woodruff-Pullman designs for the hinged upper berth and the other standard features of the American convertible day-night cars of that day. He had no special designs, and aside from compartment cars, his sleepers were undistinguishable from Pullman's (Figures 3.27 to 3.31). The November 1870 agreement allowed Wagner use of patents for a yearly fee of $5,000 on the New York Central cars and $500 for

Figure 3.31   Another engraving from the 1876 New York Graphic *showed a Wagner interior. The illustration was captioned "The luxuries of Modern Travel."*

the Shore Line cars (Boston–New Haven).[130] It was also agreed that Pullman could run four cars a day over the New York Central and Wagner could run the same number over the Michigan Central. Thus each could have through service between the major cities involved.

That agreement had been broken by the Michigan Central affair, and Pullman meant to prosecute for patent infringements. The first case opened the following spring in New York Federal Circuit Court, but it was withdrawn for some reason. In 1881 Pullman mounted a second attack, issuing a complaint against the New York Central Sleeping Car Company in the U.S. Circuit Court of Chicago. Damages of 1 million dollars were claimed. By this time Pullman controlled over 100 patents and felt that he could humble, if not ruin, his adversaries. Wagner, however, assembled an astonishing number of old car builders whose recollections of ancient sleeping cars cast a deep shadow of doubt on the priority of many of Pullman's patents. Most of the basic ideas had been in use by 1840. The case continued for several years and was then quietly settled out of court for $50,000.[131] However, Pullman took the Wagner company to court again in 1887 over the Sessions vestibule patent and enjoyed another minor triumph.

His attacks on Wagner were not confined to the courtroom. He took pleasure in running down his rival in press interviews, saying that Wagner's success was attributable to his backing by certain "rich men."[132] Pullman neglected to mention the support he himself had received from several Chicago millionaires. He went on to say that Wagner was grossly overcapitalized at 5 million dollars—that his 250 cars would then be worth $30,000 each, when they might not fetch more than $5,000 in a forced sale.

Pullman's effort to punish the Vanderbilts by joining in the West Shore railway scheme was conspicuously less successful. The Hudson River line paralleled the New York Central's east shore main trackage, but it hardly brought ruin to the Vanderbilts. In fact, they acquired it at the 1885 foreclosure sale. Pullman's vendetta had cost him dearly, and when the final victory over Wagner came in 1899, Pullman had been dead for nearly two years.

At the time of the Michigan Central contract, Vanderbilt was also promoting Wagner's fortunes elsewhere. The *New York Graphic* for January 17, 1876, said that Wagner's 125 sleepers and 60 drawing rooms were worth 1.5 million dollars. More of his fleet would soon be operating in the West. After the Vanderbilts had acquired a large interest in the Chicago and North Western, the road was ordered to use Wagner cars (though it was necessary to permit the transit of some Pullman cars to honor an agreement with the Union Pacific). The C C C & I began using Wagner cars in 1875 on its Cincinnati-Sandusky trains for the same reason.[133] In New England similar pressures were brought to bear. The Fitchburg and the Troy and Boston railroads were forced into line.[134] Wagner had had a few New York–Boston cars on the New Haven's Shore Line since 1870 or before, but the schedule was slow because of the New London ferry. Although the Vanderbilts pressed the Boston and Albany to adopt Wagner sleepers for its faster route via Springfield, the company was still independent enough to resist, at least partially, for several years. The road built and operated its own sleepers into the 1880s.[135]

During the seventies and eighties, Wagner attempted to improve his equipment (Figures 3.32 and 3.33). Gilbert produced some drawing-room cars in 1873 with very large windows for the time: 27 by 45 inches.[136] Two years later a few cars were outfitted with panoramic 36- by 54-inch windows. A fancy New York decorating firm was employed to create the finest interiors. More

Figure 3.32   The Brightwood was built for Wagner's New York–Boston service by the Wason Car Works around 1875 to 1879. (Car Builders' Dictionary, 1879)

Figure 3.33   The Toledo was built for Wagner in about 1879 by the Gilbert Car Works.

important, Wagner made an effort to get around the Pullman-controlled patents. In the late 1870s he tried a patent sleeping car on the plan of James T. Leighton (who died in 1892), the superintendent and part owner of the New Haven Car Company.[137] In 1873 and 1874 Leighton and his brother obtained four patents. Their scheme was similar to other sleeper patents of the time, being an effort to eliminate the hinged upper berth and devise a collapsible bed that could be stored under the seats. Leighton claimed that he had made substantial weight savings, noting that some sleepers weighed as much as 34 tons and that his car was only 19 tons. Seven cars were built early in 1876 for the Atlantic, Mississippi, and Ohio Railroad.[138] Wagner saw enough merit in the design to order several Leighton cars during the following year.[139] How far he pursued the plan is uncertain, but he does not appear to have given it more than passing attention.

Wagner experimented again with the same general idea in 1881, when he purchased three or more cars from the Wason Car Works.[140] The plant's superintendent, William H. Paige (1842–1885), had patented a sleeping-car design in 1876. Like the Leightons', the scheme was for a convertible day-night car free of the hinged upper berth. There seems to have been no large-scale acceptance of either the Leighton or the Paige designs, and Wagner stayed with the conventional Pullman open-section plan.

By the early 1880s Wagner was spending $16,000 to $17,000 a car. Some were built with body lengths of 70 feet, 6 inches and total car weights exceeding 32 tons. The following description offers other details on some Wagner cars produced by Gilbert in 1880.[141]

It is so complete in every way as to leave little or no margin for further improvement. Its entire length, exclusive of platforms, is 69 ft.

5 in. At one end, and occupying a little more than one-third of the length of the car, are a state room, smoking room, ladies' saloon, locker and heater, and at the other end a lavatory and saloon for gentlemen, occupying a much smaller space. In the intermediate space are 12 sleeping sections, and there is also another in the state room. The upper berth frames work with weighted chains, and also have spring beds for the mattresses. The entire finish of the interior is in solid mahogany, relieved with inlay work upon the section panels. The ceilings are decorated oak, and the seats are upholstered in scarlet plush. There are two windows to each section with spring-roller curtains; and also projecting dust-guards on the outside to intercept cinders. These guards are placed vertically on each side of the windows, and work with a hinge and spring catch, so each can be set alternately at a right angle with the car side, according to the direction in which the train is moving. Hitchcock lamps are used for lighting, and the hot-water pipes from the Baker heater run under each of the seats. There are small portable tables provided for each section, in case they are wanted for lunches, card-playing, and the like. Howard's nickel-plate curtain rods, brackets, etc., add to the effect of the interior fitting. As an example of the many little conveniences that are being added to the comforts and luxuries of railway travel, we noticed in the gentlemen's lavatory an elegant nickel-plated receptacle for collars and cuffs, attached to the lower end of the mirror frame. There are the usual supply and exhaust ventilators, and also some of the latter placed in the roof and working with the wind. They may be opened, closed or regulated from the inside.

The cars have 6-wheel trucks, with Allen paper 33-inch wheels, Miller platforms and couplers and Westinghouse automatic brakes. Altogether these beautiful coaches embody about all the luxuries consistent with railway traveling. What their average weight is we do not know, but considering their great length, it must be somewhat ponderous. The average proportion of paying weight hauled, as compared with the non-paying weight, must be something almost frightful to contemplate as it affects railway economy. But unfortunately no skill or cunning of builders can abate the evil without curtailing the luxuries, and these the traveling public will not consent to forego.

Wagner's death in the January 1882 rear-end collision at Spuyten Duyvil prompted rumors that Pullman would take over his firm. But the Vanderbilts moved quickly to silence these speculations. Augustus C. Schell (1812–1884), an old friend of Vanderbilt's and a prominent member of the Tweed ring, was named president. He was a director of several of the New York Central lines, and though familiar with the legal side of railroading, he had little practical knowledge of the sleeping-car business. He was one of the few privileged outsiders with a large interest in the firm. Together with his son-in-law, James D. Taylor, Schell managed the fortunes of the New York Central Sleeping Car Company for just under two years.[142]

After Schell's death in March 1884, James Tillinghast (1822–1898) was appointed president. He was another longtime Vanderbilt associate, but he lacked the financial and political influence of his predecessor. Tillinghast was a practical railroad manager who had been general superintendent of Central since 1869. Ironically, many years earlier he had declined to help Woodruff with his first car. He had, however, been a vice president of the New York Central Sleeping Car Company since 1881 and was Schell's logical replacement. Yet he held the office for less than a year, because the Vanderbilts decided that a younger, more aggressive man was needed to direct the enterprise. William H. Vanderbilt had just the man: his new son-in-law, William Seward Webb, M.D. (1851–1926). Webb had practiced medicine for several years, but he gave it up for Wall Street. He never returned to the care of the sick, but he did have an ailing corporation to treat and he was eager to prove his ability to Vanderbilt.

The Wagner line had fallen into eclipse since the death of its founder. Complaints of "antediluvian sleepers" and "cheese box" smokers were reaching the trade press.[143] A letter appeared in the 1885 New York State Railroad Commission Report protesting the poor condition of Wagner cars. According to one passenger, cold air poured through loose windows, making the car's name, *Arctic*, all too appropriate. The interim management had not concentrated on keeping up the property, but Webb meant to bring about an immediate and full recovery. New cars were purchased and old ones rebuilt. Six months after Webb took charge, the improvement in service was noticeable.[144] Webb had long-range plans for the company that included expansion outside of the Vanderbilt empire and direct competition with Pullman. But first it was necessary to improve the physical plant and consolidate the contracts that could easily be acquired. Webb secured full rights to the Bee Line's sleeping-car operation in 1885. The following spring a contract was signed with the Central Vermont.[145] There was even talk of a Woodruff takeover. Of course there were inevitable setbacks: the Wabash, for example, banished Wagner in favor of Mann, but Webb had the pleasure of driving out the last Pullmans from the Big Four (the C C C & I) a few years later.[146]

Webb felt that the firm's old name, the New York Central Sleeping Car Company, was too parochial. For railroads outside the Vanderbilt empire that Webb hoped to entice into leaving Pullman, a title featuring the Central might prove offensive. In December 1886 the directors adopted a new corporate name: the Wagner Palace Car Company.[147]

During the same year Webb decided to rationalize the entire process of repairing and procuring cars. For years Wagner had acquired his stock from a variety of builders. Most of the orders seemed to go to Gilbert. But Jones; Jackson and Sharp; Harlan and Hollingsworth; Osgood Bradley; and others were also represented. Repairs were contracted out to the builders or handled at the New York Central's Albany and Niagara Falls shops. In 1879 Wagner was said to have taken a large interest in the Gilbert Car Works, which may explain why so many new car orders went there.[148] In 1883 an effort was made to consolidate the repair facilities at Buffalo, but only cosmetic work was attempted at first; mechanical repairs were scattered around as before.

As a stopgap measure Webb leased the Jones Schenectady plant, but his long-term aim was a major upgrading of the Buffalo shops. He did not aspire to a plant like Pullman's, but to an industrial establishment that could construct and repair a large number of cars. Several large buildings were added to the 40-acre site in 1886.[149] Within a few years almost every square foot of the plot was covered with buildings. The idea was that 1,600 to 2,000 men would be employed in fabricating 80 and repairing 500 cars a year. One of the smartest car builders in the business was hired to head the plant: Thomas Bissell, superintendent of Barney and Smith and at one time head of Pullman's Detroit shops.

The new plant was not ready in time to meet Wagner's immediate needs. New cars were required on the West Shore and the Bee Line. Commercial car shops were apparently too crowded to take the work, because Webb turned to Pullman—an unlikely move if there had been any alternative. Between 1884 and 1886 Pullman's Chicago plant received eleven orders for a total of sixty-five cars.[150] During 1887 Bissell concentrated on overhauling the Wagner fleet; 500 cars were run through the shop, but only a dozen new ones were built.[151] During the following year at the Buffalo shops, a large new-car program was initiated in which 1 million dollars was spent on seventy cars (Figures 3.34 to 3.37).[152] About twenty-five of these were for three luxurious vestibule trains scheduled for the Michigan Central. Several private cars were produced for Webb, the Vanderbilts, and various other New York Central officials. Three new private cars for rental parties were built at about the same time. Once Wagner's needs were met, the shops would seek outside orders on a contract basis.

Bissell's work went on public display December 7, 1887, when the new Chicago express entered service.[153] The six-car consist included a buffet-smoker-library, two parlor cars, two sleepers, and a diner. Steam heating and vestibules were special features of what was described as "the most substantial and handsome lot of railway carriages ever constructed."[154] The interiors were designed by Louis Tiffany in an elegant but unobtrusive manner. In

Figure 3.34   The Indiana *featured narrow vestibules, Miller couplers, gaslights, and other improvements typical of the palace cars. It was built in 1888 by Wagner's Buffalo shops. (William Voss,* Railway Car Construction, *New York, 1892)*

Figure 3.35   An interior view of the Indiana, *1888. (William Voss,* Railway Car Construction, *1892)*

addition to the library, the first car had a barber shop and bathroom. Hot and cold water were drawn from a 400-gallon tank. Electric lights were installed in some of the cars a few months later—such a novelty that the *National Car Builder* suggested there might be a bowling alley and a billiard table next. The new train was to compete with the luxurious special put on the Pennsylvania a few months before by Pullman. Like the arms race of modern times, the competition for first-class travel fostered an early version of the industrial-military complex. In both cases the purchaser, fueled by fear, bought more and more elaborate hardware to counter the competition. The contest was an enormous benefit to the supplier as new trains rolled from the production line.

In another effort to combat Pullman, Webb decided to promote the compartment-style sleeping car. Pullman was opposed to the idea and, except for a few experiments, seemed determined to

stay with the open-section car. Yet the public's interest had been aroused by Mann's boudoir design, and there was no question that the room car was far more comfortable than the berth car. Wagner had been running such cars for many years, and now Bissell was directed to produce a more modern version. In 1886 some combined berth-compartment cars were put into service.[155] Berths and center aisles were placed at the ends, with six staterooms and a side-corridor passageway in the center. Two ladies' and three men's washrooms were provided. Three years later, four more compartment cars began operation on the New York–Chicago run.[156] The 45-ton, twelve-wheel cars were named *Lorraine, Magenta, Barcelona,* and *Normandy.* The 70-foot body of each was divided into ten rooms and had a staggered side aisle (Figure 3.38). It slept 24 passengers. The small end compartments for the use of all passengers were set aside as a men's and a ladies' lounge, respectively. A small library and a

STANDARD BODY FRAMING.

STANDARD FLOOR FRAMING.

Figure 3.36 Floor and body framing of a 76-foot-long Wagner sleeping car. (William Voss, Railway Car Construction, 1892)

CROSS SECTION.

END BRACE.

Figure 3.37   A cross-section drawing of a 76-foot-long Wagner sleeping car of about 1890. (William Voss, Railway Car Construction, 1892)

buffet were somehow worked into the floor plan. The library consisted of several triangular bookshelves in the center cross aisle.

The most interesting Wagner-built room cars were produced in 1890 on the plan of Elbridge G. Allen, superintendent of the New Haven, for the New York–Boston Shore Line trains. The cars had eight open, double-berth sections and eight single-berth staterooms laid out in a curious zigzag pattern (Figure 3.39). They slept 24 passengers and seemed to have no particular advantage aside from their novel floor plan. In 1893 Jackson and Sharp built several more Allen cars for New Haven's Old Colony division.[157] Apparently they were not Wagner-operated and were eventually rebuilt as parlor cars. They cost $15,000 to $20,000 each, which was not excessive for an elaborately finished car, but the limited capacity and great weight prevented their wider use.

At first Webb's strategy in expanding the Wagner line seemed successful. By the early 1890s Wagner was operating cars on 20,000 miles in eighteen states and Canada.[158] It owned about 600 cars worth 10.5 million dollars and employed 3,000 people. By the end of the decade its 725 cars were running on thirty railroads, with routes going as far west as California and as far south as the Gulf of Mexico.[159] Pullman was operating some 2,400 cars over 120,000 route miles; Wagner was carrying one-third the number of passengers on its much smaller system because of the densely populated territory it served.

Webb drove hard to take even more business away from Pullman. In 1896 he won a contract with the Chicago, St. Paul, Minneapolis, and Omaha Railroad.[160] He was unsuccessful in negotiations with the Northern Pacific and the D L & W, however, and competition with Pullman was proving very expensive. The cost of new cars to compete with Pullman's best was staggering. Wagner put together an extravagant train for the 1893 Columbian Exposition and produced new cars to handle the anticipated crowds (Figure 3.40).[161] In 1897 the company spent even more money on costly stock to reequip the Lake Shore Limited.[162]

Since the beginning of his administration, Webb had spent lavishly. The Vanderbilt resources were adequate, and the family had not been timid about investing in a good thing. When Webb took over in 1885, Wagner's capital worth was fixed at 6 million dollars and it was netting a comfortable $137,000 on a $537,000 gross revenue.[163] Three years later, the capital stock was doubled.[164] In 1891 it was boosted to 20 million dollars. At the same time, profits fell to $628,000 on a $3,297,000 gross—a disappointing record compared with that of six years before.[165] The regular 8 percent dividend (1.6 million dollars) continued to be paid out of the treasury's surplus, though actual earnings were rarely above 5 percent. And so Webb's presidency was a mixture of triumph and failure. He had built up the company and competed with Pullman's best trains. But he had won few new customers outside of the Vanderbilt circle, and the company was showing a poor return on so large an investment.

Because the firm remained a closed corporation to the end, few records are available on its internal workings. Speculation based on what evidence exists leads to a few conjectures about why

Wagner was sold to Pullman. In addition to Wagner's mediocre financial picture, it must have been obvious that Pullman had won the war, if not the battle. By 1890 it was clear that his monopoly was complete and that the other independents must eventually fall into line. That Wagner did not take over the Union Palace Car Company is one indication of Pullman's growing mastery. Possibly Webb did not even try for the Union Palace, but considering his earlier interest in the Woodruff merger, it seems logical that he would attempt to persuade the Vanderbilts to move before Pullman. It may be that they were already disenchanted with the sleeping-car business, or that they thought Union a poor investment—which indeed it was. Wagner really had nowhere to

WAGNER STATEROOM SLEEPING CAR "LORRAINE."

WAGNER SINGLE-BERTH SLEEPING CAR—BOSTON AND NEW YORK SHORE LINE.

Figure 3.38   Wagner was more receptive than Pullman to the idea of compartment-style sleeping cars. The Lorraine, *illustrated in the upper floor plan, was built in 1889 at a cost of $15,000. It weighed 45 tons and looked very much like the* Indiana, *shown in Figure 3.34.* (Railroad Gazette, *October 31, 1890*)

Figure 3.39   E. G. Allen (1850–1899) designed a bedroom-style car, the Mayflower, *that was in use on the Old Colony Railroad. Jackson and Sharp built the* Mayflower *in 1893. Wagner operated similars cars; see Figure 3.38.* (Railroad Car Journal, *October 1893*)

go; Pullman had the contract lines, and the few railroads that were running their own cars seemed content to remain independent. One sure sign of skepticism about Wagner's future was the sharp decline in new-car acquisition after 1893. Before then an average of fifty-five new cars was added annually, but between 1894 and 1899 the average fell to only seven cars a year.[166]

It is also possible that Webb himself was ready to quit. He had labored fifteen years to upgrade Wagner and could well have been weary of that uphill battle. He would soon be fifty, and possibly he felt that it was time to relax a bit and enjoy his fortune. Visitors to the Shelburne Museum, Webb's Vermont estate, can understand the good life that awaited him after retirement from the sleeping-car business.

When Wagner's earnings fell to a dismal $356,000 in fiscal 1898–1899, the shareholders saw another reason to sell.[167] The firm had no debts, but the earnings trend and the need for new cars meant that capital could be raised only through a mortgage. The company's prospects were just not that promising; it might be wise to sell before the present owners were saddled with a fixed debt of major proportions. The Vanderbilts were ready to unload.

The final and perhaps most convincing reason for the sale was the generous price that Pullman was ready to pay. Wagner's physical plant was valued at 8.1 million dollars. Its stock was said to be worth 20 million. Yet the purchase price was $36,667,000—a rather large premium, it would seem, to pay for good will.[168]

Figure 3.40 The Isabella *was part of the elaborate train that Wagner had specially built for exhibit at the Columbian Exposition of 1893. (Pullman Neg. 2269)*

According to an article published in *Pearson's Magazine* during the height of the muckraking era, the sale involved a conspiracy between the giants.[169] The article claimed that Pullman and Wagner had agreed to respect each other's territory and "went forward in harmony to pluck the public." Thus when the managers of Wagner saw that a sale was in the offing, they began to pump out new stock issues to inflate the price. Pullman was also accused of watering its stock to balance the exchange. The author of the piece failed to mention the bitter competition between the firms and the large investments in new equipment that justified their periodic increases in capitalization.

Another rumor circulated at the time was that the Vanderbilts were drawn into the sale by an offer of 5,000 shares of Boston and Maine stock from the Pullman estate.[170] That they coveted a larger share in the B & M was not questioned, but the story of the transfer was disproved by the news that the executors had sold the stock some time before the merger was consummated.

The negotiations were apparently prolonged, perhaps for several years. The first public notice that something was definitely underway was Wagner's move in July 1899 to register its entire stock holdings with the New York Stock Exchange's Unlisted Department. Yet before Wagner could go public, an agreement with Pullman was announced in October. The directors of both concerns agreed to a one-for-one exchange of shares. For 200,000 Pullman shares, Wagner would transfer its rolling stock, Buffalo

car plant, contracts, and other assets. Because of its large cash surplus, Pullman had to raise its capital structure by only 20 million dollars to complete the sale. William K. Vanderbilt and Dr. William Seward Webb were given seats on the Pullman board. The plan was ratified by the shareholders in December 1899, and on the first day of the new century, the Wagner Palace Car Company was no more.

The Pullman Company

Until recent years the Pullman Company figured large in the everyday life of Americans. It was a singularly public business. Except for the very poor, most travelers had first-hand experience with the sleeping car; it was the normal conveyance for long-distance travel. And to the average citizen a sleeping car was simply a Pullman. The name was part of the language, and it was the only name that the public readily associated with railroad car history. Pullman was popularly accepted as the inventor of the sleeping car. Overseas, where few true Pullman sleepers ever operated, the word was known to the educated classes, who associated it with luxury and excellence in travel.

The term is rapidly losing its meaning, however; many young people have never ridden on a train, much less spent the night behind the swaying curtains of an open-section sleeper. To such an audience one must explain Pullman as if it were some ancient kingdom—a mysterious, vast business empire once proud and rich, but like Byzantium, gone and nearly forgotten. And like Byzantium, it has a complex and romantic history. As shown statistically in Table 3.3, its rise and decline was epic. The life of its founder was the classic American success story—the poor cabinetmaker who became an industrial titan. Pullman formed a corporate giant that systematically took over all sleeping-car operations in North America, established branches in Great Britain and Europe, and constructed the largest model community and railroad car plant on earth. The resulting multi-million-dollar corporation was remarkably stable. It was celebrated for its profits, comfortable cash surplus, and absence of fixed debt. Its conservative, well-managed progress seemed imperious to the vicissitudes of time.

But the Pullman empire, like so many of its historic predecessors, was brought down unexpectedly. Internal decay may have played a small part, but it was the general decline in railroad passenger travel after 1945 that hurt Pullman irrevocably, for first-class travelers turned to the airways and highways. The separation of Pullman's manufacturing and sleeping-car divisions, which occurred at the same time, weakened the already troubled giant. The end came with merciful speed; on December 31, 1968, in its one hundred and first year, the Pullman Company suspended operations.

George Mortimer Pullman (1831–1897) did not set out to monopolize the sleeping-car business, or even to be a railroad man. His first ambition was to rise above the poverty that his family endured. The Pullmans were New York State farmers and

mechanics who had never enjoyed much in the way of comfort, luxury, or social distinction. George seemed destined to follow the same honest but colorless path. He had learned woodworking in his brother's cabinet shop, and at the age of twenty-two he became a contractor who moved buildings near the Erie Canal, which was being widened. He did well, but the work gave out after a few years. Around 1856 Pullman went to Chicago, where the general level of the city was being raised above Lake Michigan. Again Pullman's skill in house moving paid off, but again the work was temporary.

Pullman was soon casting about for a more stable career or some larger enterprise to occupy his considerable energy and ambition. Around 1858 he returned to Albion, New York, to see his old friends, among whom was Benjamin C. Field (1816–1876). Field, a man of influence and money, was active in state politics and railroad building.[171] He had helped Pullman in his first years as a contractor, and their association remained a friendly one. Field mentioned that he had received the rights for Woodruff's sleeping-car patent on certain Western railroads —presumably in return for his help in pushing favorable sleeping-car legislation through the state assembly. He needed a Western partner to help develop the franchise. Chicago was rapidly becoming a railroad center, Pullman knew the local men of affairs, he had some cash to invest, and he was looking to the future. Why not join Field and his brother in the scheme? The older man was persuasive. Many years later, Pullman also claimed that a ride in a sleeper out of Buffalo (presumably aboard one of Wagner's first Woodruff cars) convinced him of the need for better sleeping cars.[172]

In April 1858 Field signed contracts with the Illinois Central

Table 3.3   The Pullman Company, 1868–1968

YEAR	NO. CARS	ROUTE MILES	ASSETS	GROSS	PROFIT, LOSS	PASSEN-GERS
1868	50°		$ 1,470,800	$ 258,000	$ 169,700	
1870	300°		3,312,800	746,500	331,800	
1875	600°	30,000°	11,257,000	2,568,600	1,260,800	
1880	700°	60,000°	13,280,000	2,635,000	1,416,400	2,000,000°
1885	1,195	71,400	28,466,000	5,613,600	2,793,400	
1890	2,135	120,680	43,013,000	8,860,900	4,563,700	5,023,000
1895	2,556	126,660	62,792,000	8,547,600	4,290,300	4,788,000
1900	3,258	158,500	78,895,900	15,022,800	6,623,400	7,752,000
1905	4,138	184,100	96,151,900	26,922,000	13,038,000	14,969,000
1910	5,285	124,100	135,989,600	38,880,800	18,050,300	20,203,000
1915	7,287	215,800		41,512,800	20,580,100	24,252,000
1920	7,752	119,700	143,678,000	14,519,700	12,913,500	39,255,000
1925	8,776	130,300	329,148,900	81,490,300	16,779,000	35,526,000
1930	9,801	133,800	352,276,400	19,061,600	7,404,000	29,360,000
1935	8,007	115,400	276,275,300	50,063,400	Loss — 502,000	15,479,000
1940	6,910	109,500	304,469,800	60,095,500	2,411,800	14,765,000
1945	8,590	95,700	327,446,500	147,855,700	26,150,924	31,484,000
1950	6,226	102,700	83,131,400	103,756,600	1,325,200	15,606,000
1955	4,776	89,100	44,129,000	94,506,500	282,317	11,438,000
1960	2,650	67,400	28,000,000	55,876,900	286,046	4,484,000
1965	1,494	51,000	24,162,000	35,805,900	Loss — 21,513,000	2,507,000
1968	765	33,400	20,126,500	17,778,100	Loss — 22,000,800	1,073,000

° Estimates
Note: Figures before 1950 represent both sleeping-car and manufacturing divisions.
Source: Poor's Manual of Railroads and *I.C.C. Reports.*

and the Galena and Chicago Union to provide Woodruff-style sleepers.[173] Since Field signed for himself and his associates, it is not known whether Pullman was a partner at this time; it appears, however, that the deal somehow went wrong. Possibly Pullman was brought into Field's enterprise as the local agent after the failure of these first contracts, in order to prevent future fiascos. During the next year Field and Pullman convinced the Chicago and Alton to let them remodel two coaches for sleeping-car service.[174] The exact details of the sleeping furniture are not known, but if the arrangement of an 1897 replica can be believed, it differed from Woodruff's standard three-tier scheme.

The replica, presumably a fair representation of Pullman's first sleeping car, the Number 9, was built during his lifetime. The recollections of old employees were apparently not taken into account. The original conductor recalled eight sections, and one of the car builders involved in the remodeling says that oil lamps were used.[175] The replica, however, was made with twelve double-berth sections. It was 50 feet long over the platforms and 12 feet, 9 inches high, and it weighed 17.5 tons. The ceiling clearance was only 7 feet. Three candle center lamps lit the car. The upper berths were arranged with ropes and counterweights, much on the order of the Eli Wheeler and C. M. Mann 1858 patents. If this was the actual arrangement of the Number 9, it is possible that Pullman was attempting to avoid Woodruff's patents. According to the few records available, the original Number 9 was destroyed by fire.[176] The replica, made in 1897 for display at the Tennessee Centennial Exposition, was produced by rebuilding a Santa Fe tourist sleeper, Number 402—an 1870 product of Barney and Smith. A year or two later one end of the replica was removed, and the car was used to carry tour parties around the Pullman plant. In 1948 it was restored for the Railroad Fair in Chicago. After several years of outdoor storage it rotted badly and was destroyed in 1953 or 1954 at the Calumet shops. Countless interior and exterior photographs of the replica have been reproduced in school textbooks and popular histories, but it is almost never correctly identified as a replica.

The two cars, Numbers 9 and 19, were remodeled at the Chicago and Alton's Bloomington repair shops and ready for service in the fall of 1859.[177] The sleeping-car business was thus underway, but it was not promising enough to hold Pullman's full attention. Two sleeping cars connecting several frontier settlements did not add up to the exciting future that a young man with his ambition had envisioned. Traffic on most Western lines was relatively light, and the more heavily traveled routes were already taken. The Illinois Central chose to run its own cars. Woodruff, Wagner, and their associates were signing up the eastward lines. Field, however, was presumably happy with their progress and began buying more sleepers for the Alton. Barney and Smith produced a new car for him in 1860 or 1861.[178] Operations were extended to the Galena and Chicago Union Railroad. By July 1863 the Alton had four Field-Pullman cars in service and had contracted for four more.

Meanwhile, Pullman left Chicago for the Colorado gold fields. This move might be explained as the act of a restless youth seeking adventure, except that even as a young man Pullman was not impetuous. He was strong-willed, deliberate in his judgments, and brave, but not idly daring. It seems more likely that he felt disappointed in the limited prospects of his Chicago business. The Field brothers might be happy with the sleeping-car investment, but Pullman at twenty-nine was seeking a fortune. He was impatient with the dowdy bunk cars; they failed to pay large profits or excite his interest. Perhaps he decided that they had no future, or perhaps he recognized the need for more capital to build a better line of sleeping cars. It has also been suggested that during the early years of the Civil War, Federal government interference with operations on the Western lines damaged Pullman's business and caused him to invest his time elsewhere. But it appears more likely that Pullman had simply not yet decided to concentrate on the sleeping-car business. He had other investments that seemed more attractive at the moment.

As early as 1860 Pullman was buying land in the Pikes Peak area, and though he retained his interest in sleeping cars, he began to spend more time in Colorado. Stanley Buder claims that he stayed there from the spring of 1862 to the spring of 1863.[179] During that year he was involved in mining, storekeeping, and operating a wagon line. The idea that it was a pastoral interlude spent mooning about in a mountain cabin, dreaming of the perfect plan for sleeping cars, has been largely discredited.[180] Pullman was making money in a business-like way; provisions and transportation were necessities, and he was supplying them. He acquired an interest in a silver mine. He brought his partner in the still active Chicago-based house-moving firm of Pullman and Moore out to Colorado to help expand his growing business.[181] His visits to Chicago were frequent enough to keep him active in the sleeping-car business—and to permit eventual expansion into a shirt factory and into the Third National Bank of Chicago. In the spring of 1863, he returned to Chicago a rich man.

Possibly it was at this time that Pullman decided to concentrate on developing the sleeping-car business. Up to now the partnership with Field may have been just another side venture, like buying an interest in a hotel or a lumberyard. But things were changing; the prospects of the business were now quite different, or at least Pullman was seeing them differently. The possibility of building something grand was taking shape, and Pullman sensed an opportunity that few others saw or were bold enough to attempt. The East and South were blanketed with railways; the war had been decided; the Union would prevail. Work was about to start on a Pacific railroad, so that a transcontinental rail network would be a reality within a few years. Long-distance rail travel would flourish, and the sleeping-car trade with it. Whoever controlled that trade would become rich and famous. Perhaps Woodruff realized this too, but he was moving slowly. As Carnegie said, the Philadelphia-based company was run by "a few cautious old gentlemen." Pullman's vision may not have come until sometime after 1863, but his actions indicate that his course was set by that time.

Evidence of his renewed commitment was the splendid new sixteen-wheel palace car purchased by Field and Pullman in July

1863 for service on the Alton. Originally numbered simply the 40, it was renamed the *Springfield* in honor of the location of its builder, the Wason Car Company. A description of the car was given in the *Western Railroad Gazette* for July 18, 1863:

The car is 58 feet long and of the usual width. The interior is of black walnut finish, highly polished. Instead of the ordinary cramped seats, the berths are so arranged that in the daytime the car is really a pleasant *salon*, the sofas extending along the sides, with camp chairs for those who desire. The car contains fourteen sections, capable of accommodating four persons to a section, or fifty-six passengers in all. The curtains are of [mohair] plush, with curtains of volin finished lasting. To convert this into a sleeping car at night, all that is necessary is to slide out the sofas, and let down the cushion back, and the couch is complete. When the beds are made up, about three feet of space is left in the aisle, and the berths are screened by costly damask curtains, extending from the roof to the floor. The car is provided with double doors and windows to deaden the sound, and it is said the passengers can converse with as much relief from noise as though they were in their own parlors. In the raised ceiling are eight of Westlake's patent ventilators in the roof. They render ventilation most complete, while dust and cinders are entirely excluded. At either end of the car, by a very ingenious arrangement, is a pleasant and convenient state room, with doors, windows, a table for writing, &c., for the accommodation of a party of three or four persons; the whole stateroom so arranged that in the day time it slides back into the side of the car, and does not occupy more than a foot of room. At night the stateroom is drawn out, the bed is let down and the weary passenger has a soft bed upon which to pass the night, secluded from "the rest of mankind." The car is also provided with every convenience usually found in first-class hotels. It is brilliantly lighted with six large Kerosene lamps, suspended from the ceiling. One very desirable feature, not found in ordinary sleeping cars, the mattresses are of the very best material, and the berths provided with clean, white counterpanes, pillow cases and sheets, that are daily changed.

The body of the cars rests upon two double trucks, having in all sixteen wheels, hence the irregularities of the track are reduced by a compound lever lessening the motion, which, by passing through several sets of springs reduces it practically to nothing. The swing-beam is placed in the frame connecting each pair of short trucks. The cost of this railroad palace is $6,500.

Pullman and Field promised the Alton to provide even more luxurious cars than the *Springfield*. The two men were joining the race to supply the best in sleepers; no more provincial rebuilds or bare-bones bunk cars. Whatever the Central Transportation Company offered, they would attempt to match or beat. Pullman wanted to produce the ultimate palace car and decided to build it himself. His plan was not definite enough to give to a car builder; he would have to work it out in a shop. The Alton rented him a shed not far from his office, and he began work in the winter of 1864 under the direction of a practical car builder.

Pullman's car, completed in the spring of 1865, was called simply A; it is better known by its later name, the *Pioneer*. Over the years its importance became greatly exaggerated. It was credited with revolutionizing the car-building art and as exceeding anything before imagined "even in the wildest flights of fancy."[182] All previous sleepers were dismissed as primitive "rat-

tlers." Pullman himself appears to have started the myth. In a statement before the U.S. Senate in 1874, he claimed that the early sleepers were "crude and unsatisfactory." In later years his remarks became more extravagant, and soon sleeping-car history was divided into two great epics: before and after the *Pioneer*.

The *Pioneer* was Pullman's natural favorite, not because it was his first car but because it was the first that he could call his own. Its predecessors were rebuilt products or the handiwork of a contractor, while the *Pioneer* was like a love child: he had seen each plank and cushion put in place. Its success was a great personal triumph, and in the passing years sentiment and frailties of memory colored its victories ever brighter. It was not just an important advance in the art; it was the greatest milestone since the beginnings of the railway carriage; it was the biggest, the most advanced, and the most expensive car ever constructed to that time. The exaggerations began to feed upon themselves, and what had been slight overstatements grew in the retelling until fantasy overwhelmed fact. Pullman officials, perhaps with the best intentions, began to parrot and embellish the story. Repetition has made it a fixture of technical history. The old truism that history belongs to the victor is borne out in the story of the sleeping car. It is likely that if Wagner or Woodruff had surmounted their competition, we would hear little of Pullman or the *Pioneer*.

In fact, the *Pioneer* does not seem spectacular when compared with other first-class sleepers of its time. Both Wagner and Woodruff were operating magnificent cars before the *Pioneer* had left its shed. Pullman himself had purchased a comparable car two years before; according to the description of the *Springfield* quoted earlier, it was in no way inferior to the *Pioneer*. All the *Pioneer*'s supposedly remarkable features were already in use: sixteen-wheel trucks, clerestory roofs, under-the-floor heaters, and fanciful interior decoration.

In overall size the *Pioneer* was said to be a foot wider and 2½ feet higher than any other car ever built. The legend that station platforms had to be reconstructed and bridges raised for it may be true, as Joseph Husband and other writers on Pullman have stated, but it seems unlikely that so practical a man as Pullman would have intentionally built an oversized vehicle.[183] This over-construction has been explained by Pullman's supposed need to gain space for a better convertible day-night car—especially space for overhead capacity to accommodate the folding upper berth. The berth design was described in a patent issued to Field and Pullman in April 1864 and in a second patent granted in September of the following year. Both involved a hinged upper berth, but neither were original inventions. Imlay's sleepers of the 1830s used it, and both Woodruff and Green managed to patent it some twenty years later. Yet the idea is commonly credited to Pullman. The arrangement does not require any more space than the other clerestory roof cars then in service. Any increase in inside dimensions was made more for show than utility. And the *Pioneer* was unquestionably showy; its interior, outfitted like a gaudy ocean steamer, helped drive the total cost up to $20,000. This was said to be four or five times that of an ordinary coach,

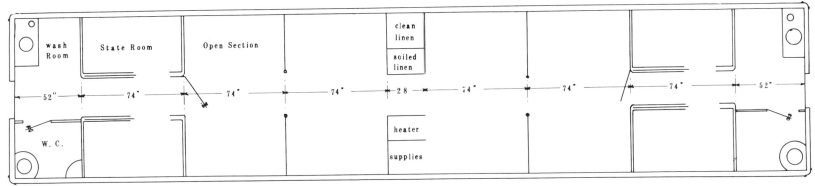

Figure 3.41   *Reconstruction floor plan, based on early newspaper accounts, of Pullman's palace car the* Pioneer. *Completed in 1865, the car was remodeled several times before its demolition in 1903. (Traced by John H. White, Jr.)*

Figure 3.42   *The Pullman hotel car* Viceroy *was completed at the Great Western Railway's Hamilton, Ontario, shops in 1865. It seated 56, weighed 37½ tons, and cost $20,000. (*Engineering, *June 12, 1868)*

and equal to the price of two locomotives. But the inflation of the Civil War had doubled and even tripled railroad equipment prices; at prewar figures the *Pioneer* would have cost little more than an ordinary first-class sleeper.

Pullman's invention of the hinged upper berth appears to be a fiction which developed after his lifetime, when the basis of his success was somehow ascribed to a fertile creativity rather than an extraordinary business sense. Pullman himself was frank to disclaim any talents as an inventor or even particular skill as a mechanic. He rightly considered himself a capitalist and was listed as such by contemporary biographical sketches. Skilled mechanics could be hired, and Pullman might generally oversee their work. But he left the details to his artisans; his time was better spent in directing the business.

The actual history of the *Pioneer* cannot be traced before 1893, but legend has it that the Chicago railroads agreed to rebuild platforms and raise bridges so that the car could be used in 1865 on Lincoln's funeral train. The inclusion of the *Pioneer* was a double coup for its owner. The first—the reconstruction undertaken by the Chicago lines—is obvious; the second was the prestige gained from its use by the Presidential party. No photographs of the *Pioneer* as built are known to exist, but it is possible that the car appeared in one of the Lincoln funeral train photographs. In

particular, a picture taken at Michigan City shows an unusually large clerestory roof car under the memorial archway.

Little information can be offered about the *Pioneer* beyond its legend. By sifting through the surviving newspaper stories, and by discounting such obvious errors as that it slept 100 passengers, the following details can be assembled.[184] The 54-foot body had twelve open sections. The end sections were fitted with sofas that faced the center aisle. The original floor plan was not satisfactory, and the dressing rooms were enlarged after the car had been in service a short time (Figure 3.41). Over the years it went through several remodelings, including a change to six-wheel and then to four-wheel trucks. It was retired in 1889, and photographs made two years later show it as a well-used relic in the boneyard. Its labors were not at an end, however; Pullman was hard-pressed for equipment during the Spanish-American War, and the *Pioneer* carried troops to San Francisco. The car was held in storage until August 1903, when the scrappers put an end to this famous, if overrated, railway veteran.

While the details of the *Pioneer* legend can be questioned, the essence of the story is true enough. Pullman recognized a widespread public desire for luxurious railway travel and a willingness to pay for it. He saw the sleeping-car trade as a public-service enterprise where taste, fashion, and publicity were essential in-

Figure 3.43   The Pullman sleeper Omaha *was built in 1866 at the Galena, Illinois, shops of the Chicago and North Western Railway.* (Engineering, *February 7, 1868*)

gredients for success. A first-rate restaurant must have more than nourishing food; it must offer a sophisticated atmosphere, excellent service, elegant tableware, and exotic dishes. The affluent classes have always demanded these accessories in the institutions they patronize. Pullman intended to put the premium hotel on wheels. If it meant fanciful cars with stained glass, polished brass, and veneer panels picked out in gilt, he would oblige. If that was the taste of the times, the eye would never lack for some new

detail, carving, or rich hanging to gaze upon. Pullman carried his belief in the merits of luxury into his personal life and spent lavishly on his family. Their great stone mansion on Chicago's Prairie Avenue was that of a merchant prince. In the mid-1870s a very special private car was produced for Pullman that was meant to impress, and it created this effect on all who saw it. Pullman believed in a grand scale of living, and so did the patrons of his sleeping cars.

Thus the *Springfield* and the *Pioneer* were only the beginning of a long line of overstuffed palace cars (Figures 3.42 to 3.45). The fleet expanded rapidly during the late sixties as Pullman acquired more contracts. Some idea of their appearance can be gained from a description in the *American Railway Times* of June 9, 1866:

The wood of the interior is of black walnut so worked as to bring the beautiful grain of the wood out to the best advantage. Exquisite carving adorns every place where it can be tastefully placed. Some of the principal pieces show the designers and executors masters of their art. In the day time, all portions of the furniture of the car, used for sleeping purposes, are carefully concealed from view by richly painted screens, which are carried to and kept in their places by a neatly arranged system of pulleys. The upholstery is of rich Brussels carpet, with curtains of heavy damask. The metal work is all silver plated—the mirrors (one, surrounded with rich carvings, to each section) of heavy French glass. The windows are double, and so arranged as to assist the ventilation; each sash has a newly invented fastening, working simply and effectively. Each car is heated by Westlake's patent heaters and ventilators, with registers in each section. The washing apparatus is beautiful and convenient. Each car has one state-room, accommodating four persons, so arranged as to be, at will, completely separated by sliding doors, from the remainder of the car. The seats are commodious and luxurious; each car is also provided with four sofas.

The exterior is as splendidly finished as the interior, having received the best effort of gilder, landscape painter and varnisher. They are throughout, most perfect specimens of car architecture. Each car is set upon two double eight-wheeled trucks, built upon principles peculiar to the Aurora shops, and so constructed that, when in motion, and even in turning curves, but the slightest jar is perceptible; in fact, when in rapid motion, one can write easily and legibly upon the tables in the car.

The "Omaha" car, built at the Chicago and Northwestern shops in this city, differs somewhat from the Aurora built cars, the seats being all sofas, leaving during the day, the central aisles wider, and the general appearance of the car lighter, and possibly more home-like. Extra seat room is provided by the introduction of camp chairs. The "Omaha" is further provided with an organ, set in the centre of the car, so that the passengers may indulge in music and singing. The "Omaha" cost $28,000.

The need for solid comfort was attended to as well. Pullman wanted to offer amenities beyond that of a good bed. Travel was boring, and a meal or drink would enliven the trip and make it a more social affair. The idea was an old one, but Pullman felt that it had been neglected on the railroads. In 1867 he introduced the hotel car to the American public, and in 1868 he inaugurated the full dining car.

Pullman also proved a master at devising celebrations and excursions that gained newspaper space. The *Pioneer* was attached to the first Alton train between Chicago and St. Louis just weeks after its Lincoln funeral-train journey. The following spring a special train of new Pullman cars carried five hundred Chicago notables to what was described as a "country dinner" in a picnic grove near Aurora. In reality, the meal was catered by one of Chicago's best restaurants. Printed silk menus indicated the sim-

Figure 3.44 A drawing room and a portion of the main compartment of the Pullman hotel car City of New York *(1866) are shown in these two engravings. The drawing room was available for an extra fare beyond the regular sleeping-car tariff. (Illustrated London News, October 2, 1869)*

Figure 3.45   The 1866 City of New York *was a product of the Burlington's Aurora shops.*
(Illustrated London News, *October 2, 1869*)

plicity of the affair, and the Chicago Opera Company presented the evening's rustic entertainment.[185]

Other excursions followed, usually to celebrate additional Pullman service or new Pullman cars. The most elaborate of all was the Boston Board of Trade's transcontinental trip in 1870. The mayor of Boston, Alvah Crocker, and other wealthy men were Pullman's guests. With his usual attention to detail, Pullman put a small printshop aboard one car so that a newspaper could be published. The trip produced reams of copy and a resolution of thanks from the passengers that included a prayer for "no delay in placing the elegant and homelike carriages upon the principal routes in the New England States." On the other hand, it did little to promote large-scale travel to the Pacific coast, nor was it effective in preserving Pullman's hold on the Central Pacific's sleeping-car trade. Certainly the excursion was useful in building a favorable corporate image and introducing Pullman's name to a wider audience. And now that he was a public figure, excursions were less necessary; Pullman's normal activities became newsworthy in themselves.

BUILDING THE PULLMAN EMPIRE

New cars, investment, and publicity brought new customers. Within a year after the *Pioneer*'s inaugural trip, Pullman had signed up every major line operating out of Chicago. The Michigan Central joined his system in October 1865; the Burlington came in a month later. At the end of 1866 Field and Pullman were running over forty cars on seven lines.[186] A dozen more cars were under construction, according to the credit reporter R. G. Dunn. And once the Great Western's third rail was laid down, Pullman cars would be running along the southern boundary of Canada.

But if the firm was to expand beyond the Middle West, Pullman knew that it must attract more capital and influential friends. Incorporation was essential. Apparently Field did not agree; per-

haps he was frightened by Pullman's drive to expand, or perhaps he was simply weary of the partnership. In all events, he retired sometime in 1866. In the fall of that year Pullman decided to incorporate. A charter for the Pullman Palace Car Company was granted on February 22, 1867, with capital set at an even 1 million dollars. By this time Pullman was an insider with the Chicago money barons, and John Crerar and Marshall Field were pleased to promote the first issue of stock. Their admiration for Pullman was genuine; he was respected as a clear-headed industrialist with a genius for organization and finance. Carnegie thought that he was one of the ablest men of affairs he had ever known.

Pullman's subsequent career confirmed these opinions. A practical visionary, he began to map out a plan to implement his dream of a national sleeping-car network. At the time of incorporation he had tied up much of the Midwest and established a Southern subsidiary in Altanta. An opportunity lay in the West, for the Union Pacific contract would soon be tendered. The greatest obstacles were in the East, where the Central Transportation Company and Wagner lines exercised control. There seemed little hope of upsetting these giants, and for the moment Pullman decided to press on into the South and West.

The last battle of the Civil War was hardly over before Central Transportation turned toward the South, hoping to take over its sleeping-car business. To counter this move Pullman formed a partnership in 1866 with Hannibal I. Kimball and Robert H. Ramsey of Atlanta.[187] A Southern banker, George Rice, was made head of the firm, but Kimball (1832–1895) appears to have been its driving force. A New Englander who went broke in the carriage business, Kimball moved to the South after the war, like many other bankrupts. He recouped his fortunes with wonderful speed through bank, hotel, and street-railway promotion. The very model of a successful carpetbagger, he became a good man to know in Atlanta. After two years Pullman, Kimball, and Ramsey had cars on ten Southern lines.

Pullman, however, remained dissatisfied with this share of the

Figure 3.46   Barney and Smith of Dayton, Ohio, built the Dayton in the 1870s for Pullman. Dimensions for a similar car are given in the diagram in Figure 3.49. (Arthur D. Dubin)

traffic. After acquiring the Southern Transportation Company early in 1870 through the Central lease, he became determined to consolidate the remaining independents of the region. He had the Southeast in hand, but to the west were the so-called Paine Lines. Enoch H. Paine of Louisville had begun assembling a small sleeping-car operation after the war.[188] By 1870 he controlled four companies, including the Crescent City and Rip Van Winkle lines. Negotations for some form of consolidation with the Pullman firm were underway late in the same year, although Kimball had soured on his Chicago partner by then. He wrote to Paine: "You know his [Pullman's] way of dealing: if he could buy us out at half what we were worth he might do it but don't think he would [do so] otherwise until he thought some one else wanted us."[189] An agreement was reached despite these misgivings, and in February 1871 the Pullman Southern Car Company was formed to consolidate the Pullman-controlled lines in the Cotton States. Paine and Kimball both were told to join up or face patent litigation.[190]

The consolidation promised greatly improved service, including through New Orleans–New York cars outfitted with changeable-gauge trucks.[191] The projected 58-hour schedule was declared "one of the grandest railroad arrangements of this fast age." But the new era that Pullman Southern promised was slow in coming. Traffic was light and service not of the best, and Paine managed to keep a bitter legal battle going with Pullman for many years. He accused Pullman of milking the operation, while Pullman claimed that he was subsidizing it. By the early 1880s, however, the affairs of Pullman Southern were much improved. The debt was down to $88,000 and a profit was visible for the first time in years.[192] By 1882 the subsidiary had served its purpose, and it was absorbed by the Pullman Palace Car Company after an exchange of stock.

If Pullman Southern was a mixed success, the Western auxiliary company must be considered a failure. Just months after the Palace Car Company was incorporated, the Pullman Pacific Car Company was also organized. As described in the earlier section on Woodruff and the Central Transportation Company, this new firm was formed at the suggestion of Andrew Carnegie, who had just met Pullman and became so impressed with him that he abandoned the Woodruff enterprises. Joining forces with Pullman, Carnegie helped the Pullman Pacific to beat the Central Transportation Company out of the coveted Union Pacific sleeping-

car contract. To sweeten the deal various officials of the railroad were taken in as stockholders, as was Carnegie for his help in securing the contract. Pullman was made president and general manager, but he controlled only 1,200 of the 5,000 shares. The state of New York issued the Pacific Car Company a charter in April 1868, with a capital limit of ½ million dollars.[193]

Pullman was delighted with his triumph over Central Transportation. His dream of a national sleeping-car company seemed ever more real, and he plunged into an extravagant car-building program. Even Carnegie, the supreme optimist, was taken aback by Pullman's enthusiasm and wondered if so many palace cars were needed to traverse 1,200 miles of desert.[194] After the transcontinental opening in May 1869, the future steelmaker's question was answered in the negative. Traffic in general was disappointing once the initial novelty of rail service to the West had faded. Both the Central Pacific and the Union Pacific seriously undercut Pullman's position. After July 1870 the Central Pacific refused to handle through Pullman cars, and passengers were forced to change at Ogden.[195] The C P had built up its own fleet of Silver Palace cars and wanted no competitors on its line. Meanwhile the Union Pacific cut into Pullman's revenues by offering tourist sleeper service. The road acquired twenty-three such cars at about the time that through service began. The glamour of the Pullman Pacific plummeted, and the Chicago headquarters began to buy up its shares in 1870.[196] In October of the following year, a new 12½-year contract was signed with the Union Pacific that ended operations of the Pullman Pacific Car Company. An Association agreement (a joint ownership scheme sometimes used in the sleeping-car business) was made giving equal shares to both Pullman and the railroad. When the agreement was renewed in 1884, the old Pullman Pacific corporation was officially liquidated.

During the first years of the Pullman Palace Car Company, George Pullman's general strategy did not seem to be going well. The Southern and Western ventures were costly disappointments. Then by a fortuitous accident, the largest sleeping-car company in the country came into Pullman's possession. Even he must have been surprised at his good fortune and the relative ease with which he took over the Central Transportation Company. What appeared to be growing into a decade-long battle was settled in a few months of negotiations. After February 1870 Pullman con-

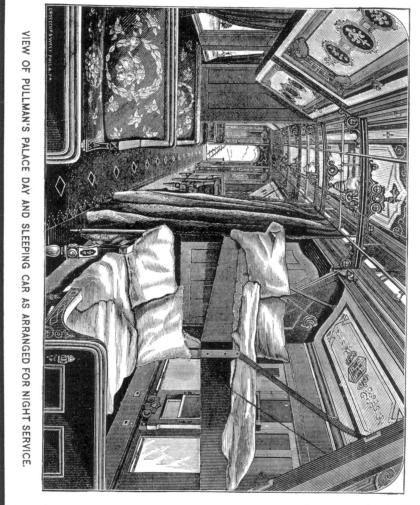

DOUBLE TRACK.

PULLMAN'S PALACE DAY AND SLEEPING CARS,
Via "*MANTUA JUNCTION,*" run as follows,
FROM NEW JERSEY RAIL ROAD DEPOT, FOOT OF COURTLANDT STREET:

LEAVE NEW YORK DAILY, except SUNDAY, at 9.30 A. M.	ARRIVE CHICAGO, 3.00 P. M.
LEAVE NEW YORK DAILY at 7.00 P. M.	ARRIVE CHICAGO, 6.00 A. M,
LEAVE NEW YORK DAILY, except SUNDAY, at 9.30 A. M.	ARRIVE CINCINNATI, 12.45 P. M.
LEAVE NEW YORK DAILY at 7.00 P. M.	ARRIVE CINCINNATI, 10.30 P. M.
LEAVE NEW YORK DAILY, except SATURDAY and SUNDAY, at 9.30 A. M.	ARRIVE LOUISVILLE, 5.45 P. M.
LEAVE NEW YORK DAILY at 7.00 P. M.	ARRIVE LOUISVILLE, 7.40 A. M.
LEAVE NEW YORK DAILY, except SUNDAY, at 9.30 A. M.	ARRIVE ST. LOUIS, 6.45 A. M.
LEAVE NEW YORK DAILY, except FRIDAY, at 7.00 P. M.	ARRIVE ST. LOUIS, 1.30 P. M.

RATES OF FARE ALWAYS AS LOW AS BY ANY OTHER ROUTE.

VIEW OF PULLMAN'S PALACE DAY AND SLEEPING CAR AS ARRANGED FOR NIGHT SERVICE.

FEWER CHANGES OF CARS THAN BY ANY OTHER ROUTE.

PULLMAN'S PALACE DAY AND SLEEPING CARS,
Via "*ALLENTOWN LINE,*" run as follows,
FROM DEPOT OF CENTRAL R. R. OF NEW JERSEY, FOOT OF LIBERTY ST.:

Leave NEW YORK DAILY AT 5.00 P. M.   Arrive CHICAGO, 6.00 A. M

COURTEOUS EMPLOYEES.

Figure 3.47   An Erie Railway timetable of 1870 included this engraving of a Pullman interior.

trolled the great majority of sleeping cars. His system was nationwide, and he held the basic patents.

Most men would have been satisfied to consolidate their position and enjoy their new fortune. Pullman did so, but he also pressed for more sleeping-car contracts and continued to acquire such desirable properties as the Erie and Atlantic Sleeping Car Company in 1873. Three years before, he had decided that he needed a car works of his own. He took over an old plant in Detroit and in less than two years transformed it into one of the nation's finest car shops.

By the early 1870s the Pullman organization had become incredibly complex. The Pullman Palace and its two subsidiaries had 400 cars, employed 2,000 people, and controlled some 30,000 route miles (Figures 3.46 to 3.49). Another 400 cars were jointly owned by Pullman and the individual railroads having Association contracts. On top of this was the car plant and Pullman's efforts to break into the general area of commercial car building.

But these activities were not enough to occupy his energy or satisfy his ambition. Now that the United States was conquered, Pullman looked overseas. His new goal was an international

Figure 3.48   A berth and headboard divider plan patented by Pullman and Field in 1865. The patent covered only the details of the hardware, not the idea of the folding upper berth itself. (Engineering, January 26, 1872)

Figure 3.49   This drawing, published in 1877 by the British journal Engineering, is believed to represent a design of about 1870. (See Figure 3.46). The car slept 24. (Engineering, November 2, 1877)

sleeping-car cartel. In 1872 James Allport, general manager of the Midland Railway of England, made a tour of American railways. Impressed by the superior accommodations of Pullman's equipment, Allport recommended their use on the Midland.[197] Pullman visited England early in the following year and returned to Chicago with an order for six cars and an operating contract. Not long afterwards, a group of English investors organized a company to promote the use of Pullman cars in Great Britain, Ireland,

Europe, and India.[198] Calling itself the Car Syndicate Limited, the firm tried to interest foreign railroads in the merits of American-style sleeping and parlor cars. But it encountered a general rejection of the idea: just as North Americans seemed to feel a revulsion toward compartments, Europeans saw little merit in the open American plan.

Pullman was not about to compromise his basic plan, and the cars he sent abroad were faithful copies of their American coun-

Figure 3.50   The Enterprise *was built in Pullman's Detroit shops in 1873 and was exported to England for service on the Midland Railway. Except for the wooden-center wheels, it appears to be a purely American design in all details. (*Engineering, *April 2, 1875)*

terparts. They were manufactured at the Detroit shops, knocked down, and shipped to Derby for reerection at the Midland's repair sheds. One of the sleepers was ready for service in January 1874. The *Midland* measured 58 feet long by 9 feet wide by 13 feet, 2 inches high. It weighed about 25.5 tons and slept 30. Within a few months *Leo, Enterprise, Excelsior, Britannia,* and *Victoria* were all in service. The last two were parlor cars. In all particulars except for their narrow width (a foot less than the usual 10 feet) and their wooden-centered wheels, these cars were standard Pullman sleepers of the time. Excellent drawings of the *Enterprise* were published by *Engineering* magazine in April 2, 1875 (see Figure 3.50).

Public reaction to the cars was favorable, and Pullman secured the good will of the press by running well-provisioned excursions. The resulting clippings were reprinted by the Pullman Company in a small booklet, *Official Correspondence with Reference to the Relations of the Pullman Palace Car Co. to Railway Companies* (Chicago, 1874). In its pages timid ladies were assured that their safety was more secure under the watchful eye of a Pullman porter than it would be if trusted to the confines of a lonely compartment. Pullman stock had all the modern improvements, and each car was a "museum of Patented appliances." Another newsman proclaimed that "the Pullman car is an absolute civilizer. It encourages the social amenities, . . . and, by its elegance of surrounding, and beauty of adornment, encourages aesthetic tastes and the refinements of life."

These and similar accolades failed to produce a mass conver-

sion to American-style cars; in fact, other British lines were rather slow to sign on. The need for sleeping cars was limited in the British Isles, because most trips were short and there were many excellent railway hotels. Pullman's open parlor cars had little to recommend them over the first-class British carriage. By 1882 only four lines were using Pullmans; the fleet was a meager fifty. In 1888 the Midland itself withdrew to operate its own cars. In 1907 British Pullman was sold to a group headed by Lord Dalziel. Under native management its affairs prospered greatly, perhaps a sign not only of the benefits of local control but also of the universal resentment felt toward foreign ownership.

Pullman had even less success on the Continent. In 1874 he sent the *Midland* on a three-year tour of European railways, an expedition which landed a contract in Italy but nowhere else.[199] Sleeping cars were definitely appropriate to European travel, but the local talent was already engaged in perfecting such a system. Seventy years before the Common Market, Europe was characterized by intense nationalism. Nations were jealous of their borders and disinclined to permit free entry of foreigners. Pullman faced no such problem in the United States, but in the Old World extraordinary skill and patience were needed to arrange for the interstate movement of railway cars. Georges Nagelmackers, who was Pullman's equal in talents for persuasion, finance, and management, had succeeded in winning such concessions for the Wagons-Lits. It was futile to challenge so powerful a competitor, and after Nagelmackers declined a suggested merger, Pullman had nowhere to go but home.

There was enough to keep him occupied in Chicago, since the day-to-day direction of the multilevel Palace Car Company was demanding even for a man of Pullman's ability. The business decline following the 1873 panic made it essential to devote more attention to the firm's management. Independent sleeping-car contractors such as Woodruff and Wagner remained a threat, and the competition for contracts as new railroads opened would continue briskly for another twenty years. Pullman also faced a new problem: his success and far-reaching efforts to "run everything" had earned many enemies, who lost no opportunity to abuse and ridicule Pullman and his business practices. In the first year of the Pullman Company, complaints against the sleeping-car monopoly were voiced by the *American Railway Times*, which said that the contract arrangement was against the best interests of both the traveling public and the railroads.[200] The cars were "tawdry, tapestried, cushioned and over-ornamented, receptacles of filth and fever." Could not the railroads run plain, economical cars to the benefit of the public and their own stockholders? According to the *Times* article, the sleeping-car companies were a ring of parasites that feed and fatten because railway managers are too inept or lazy to run their own affairs.

Protests against the alleged abuses of the sleeping-car monopoly came to a head in 1874. Maine's State Railroad Commission issued an angry statement about the gross excesses of the palace cars, declaring that they benefited only the privileged class of traveler and were subsidized by the general public.[201] Pullman could ignore the state of Maine, but not the U.S. Senate. That year Senator William Windom asked Pullman to report on its affairs. Rather than appear in person, Pullman prepared a written statement that was published in the Windom Committee's report.[202] Pullman gave a calm, reasoned defense of the sleeping-car monopoly. Its main justification was one of public convenience, safety, and comfort. Because individual railroads were not offering through service for long-distance travel, Pullman was acting as the public's advocate in perfecting a nationwide system. The railroads were relieved of a burdensome, specialized service. And a national system like Pullman's could afford to run cars on poorly patronized lines because of the revenues received from its better-paying runs. Only a national system could offer a uniform standard of service.

Pullman only hinted at the benefits of the car pool—a point that in later years became a favorite justification for a sleeping-car monopoly. According to this argument, only the Pullman Company could afford to maintain a reserve of cars for seasonal emergencies or special charter service. Most of the cars could be in constant use when deployed on a nationwide basis through a central organization. Cars for the Florida trade, for example, could be shifted from the Western lines at the end of the summer holidays. Cars used one week for a Shriners' convention could be used the next for delegates to a political conference.

In his final arguments Pullman noted that his company had developed a specialized conveyance that "by a number of ingenious devices, very greatly improved the comfort, safety and healthfulness of railway passenger-cars, lessening the fatigue of travel and making night-journeys convenient and easy." He pointed out that while the weight and even the decoration of these cars was questioned in some quarters, the standard sleeper weighed 28 tons, or about 2.5 tons more than a coach. He also said that the specialized expertise of the Pullman organization was evident in the discipline and training of its porters and conductors. In general the Windom Committee seemed to accept Pullman's reasoning, for no Federal control was imposed until the early years of the twentieth century.

While the Senate investigation was underway, the stockholders of the Pennsylvania Railroad conducted an investigation of the road's management in which the Pullman contract was a matter of particular concern.[203] The road's officers defended the contract, maintaining that operation of its sleepers was best handled by outside contractors and that no larger share of the revenues could be expected. In addition, Pullman had invested $2,550,000 in 170 cars on their lines, freeing this large amount of capital for other uses and thus effecting a saving to the shareholders. Neither side mentioned possible collusion between certain officers of the railroad who held an interest in the Pullman Company.

While Pullman seemed to enjoy general public approval, there were occasional complaints. It would be impossible for any organization to satisfy everyone, particularly when the majority of its customers were of the fussy, demanding middle and upper-middle classes. In the nineteenth and early twentieth centuries these groups had raised fault-finding and fastidiousness to a neurotic preoccupation. Criticism of toilet facilities and cramped dressing rooms was frequent. In 1870 the *National Car Builder* said that some cars had no washbasins and that the toilet was nothing more than a "dry closet."[204] During the same decade the journal remarked that the term palace car must be a whimsical stroke of American irony, for most sleepers were more like dungeons than palaces.[205] Their closed-in spaces, poor ventilation, and overheating were inexcusable. Worst of all, the evil-smelling oil lamps provided so little light that the passenger could not read and was forced to retire early, suffering the pangs of claustrophobia behind the curtains of a narrow berth that rolled like a cork in a tempest. After viewing Pullman's Centennial exhibit, the *Railroad Gazette* could not resist grumbling: "We always admire the Pullman decoration with a protest and regret that the money is not expended in ways which would give so much more pleasure to passengers."[206]

More substantial accusations were made about rates charged. In 1875 the Illinois Central Railroad contended that Pullman's $2 fee was too much and that a fair profit was possible at $1.50.[207] At the present rate, the road said, cars paid for themselves in only fifteen months. Some years later a complaint in the *Railroad Car Journal*, while admitting that a sleeper might not pay for itself in less than three years, took up the same theme: berth fees were too high and profits were excessive.[208] It said that a typical car might travel 300 miles a day, and even allowing for only 50 percent occupancy, it would earn $3,000 a year.

The various complaints about the Pullman company prompted more than an occasional editorial. Efforts to regulate and tax its operations were attempted by several states. Ohio considered taxing Pullman as early as 1875. The Illinois legislature conducted an investigation of the firm's affairs with an eye toward rate regulation.[209] Robert T. Lincoln, acting in Pullman's defense, contended that the company's profits were very small, that no one was compelled to use sleeping cars, and that it was foolish to go into the regulating business, for where would it all end? The classic statement made during these hearings was by another friend of Pullman, who said simply that "the company was not running cars for the fun of the thing." In the end Pullman's legal department seems to have been successful in fending off state regulation by arguing that the firm operated an interstate business.

Carping critics and suspicious state legislators were so many gnats compared with the economic problems that Pullman faced in the 1870s. During this decade the national economy suffered one of its most severe and protracted financial collapses. Thousands of businesses failed and others barely survived. But it was not the general depression that troubled Pullman; the company weathered the storm with relatively little trouble. It was, however, shaken by the loss of the Michigan Central contract. Its stock fell $20 at the news, but soon recovered to nearly its full par value.[210] Yet a certain doubt about the company's future was abroad, and whereas no shares had previously been available, few buyers could now be found.

Even so, the next year's earnings were down only $33,200, representing a decline of about 1.5 percent.[211] The Centennial traffic admittedly helped cushion the blow, but actual earnings on investment were an impressive 18.75 percent. The corporate structure was wonderfully solid as well, with assets exceeding obligations by 42 percent. A cash surplus of 2 million dollars was on tap to tide the company through difficulties. Pullman might be accused of overexpansion, of a foolhardy overseas adventure, and of extravagance in his personal life, but he was always the practical visionary who took care not to overextend.

Just before the Centennial year began, Pullman decided to retire part of the bonded debt by reducing the annual dividend from 12 to 8 percent.[212] He knew when to retrench. Sleeping-car revenues seemed to hold up well enough, but the Detroit car shops were running at a loss. Wages were cut back to stop this drain. The shop's deficit was too minor, however, to have any real effect on overall earnings. Another economic setback occurred in 1878, when the firm's secretary, Charles W. Angell, absconded with $115,000.[213] It was an embarrassing scandal, but the stockholders were assured that even so large a theft could not adversely affect the company's financial position. The hard times bottomed out in the last year of the decade. In 1879 revenues were down a solid $300,000 compared with average earnings of the previous five years.[214] Even so, profits were substantial enough to cover all expenses, pay the usual dividend, and leave a surplus of $337,000.

The next ten years were an era of growth and prosperity. In 1881 earnings were just short of 3 million dollars.[215] Profits were so large that an extra stock dividend was declared. Some idea of Pullman's expansion in this decade is given by the dramatic increase in capitalization. In 1880 the capital stock was valued at 6

million dollars. This figure increased several times over the next few years, and by 1889 it stood at 25 million.[216] During the same time, route mileage and the number of cars doubled, while assets tripled.

Pullman's progress was the result of more than good times and the general increase in railway mileage and travel. It was the reward of a many-sided effort to obtain new contracts and hold existing ones. One obvious tactic was to drive out or take over the competition, as Pullman did in the Union Palace and Wagner purchases. Another was to seek new markets. Pullman never entirely forgot about European possibilities; he reportedly even considered starting a transatlantic steamship line in the early 1880s. But he was realistic enough to leave this to the Cunards, just as he had prudently left the continental sleeper trade to Wagons-Lits. Mexican railways, however, were too tempting to be ignored. There were not many roads operating in that country when Pullman made a contract early in 1884 for service between El Paso and Mexico City.[217] Except for an interval during the revolution of 1914–1920, Pullman cars continued to run over Mexican lines until the company itself was dissolved.

But the techniques of takeovers and new-market development were less important than the simple maintenance and acquisition of contracts. Pullman saw that good contracts were more vital to his success than the most perfect equipment or the best-trained porters. Prudent contracts were essential to the company's profits and its relations with its landlord railroads. As a tenant he must satisfy a railroad's desires, yet somehow extract terms favorable for a profit. Just the right balance of concession and insistence was needed to forge such an arrangement.

The standard contract required each car to gross $7,500 a year (later this was raised to $9,000). Pullman agreed to furnish suitable cars in the required number; to maintain interiors, supply fresh linen, provide porters, and supervise their conduct; and to hold the railroad safe from patent claims. The railroad was to haul the cars, handle routine running-gear repairs, and provide heat and light. It was generally understood that the road would receive the coach-fare portion of the ticket while Pullman took the first-class portion. In the early years Pullman accepted a simple $2 berth fee. If revenue fell below the $7,500 base, the railroad would pay Pullman a mileage fee, usually 3 cents a mile. Contracts generally ran for fifteen years, though a few were for thirty.

Some lines demanded and received additional concessions. The Burlington required Pullman to give Burlington employees free passes and to pay all repair costs if revenues reached $9,000.[218] The Pere Marquette was granted a "scaled mileage" contract that provided for a graduated schedule of charges: 2 cents a mile if annual revenues were below $5,000; 1 cent a mile if revenues were between $5,000 and $6,000; no charge if revenues averaged $6,000 or over on specific lines; no charge on the entire system if revenues averaged $7,500.[219] The B & O received a special rate of 1 cent a mile for its tourist cars but no mileage fee for its regular sleepers. Another form of contract variation was the split-concession agreement with the Chicago and Northwestern that permitted the railroad to operate its own parlor cars. All such variations in terms aroused contention, jealousy, and suspicion among the roads. In an attempt to avoid quarrels, Pullman kept the contracts confidential until regulation by the I.C.C. made them a matter of public record.

When necessary, Pullman tried various means to placate difficult customers. In his earliest years he had new cars built in the shops of the contracting railroads. This gave the railroad a bit of extra revenue and won the loyalty of the presiding master car builder, who had the opportunity of showing off his skills. But the practice was costly, and it quickly disappeared after the Detroit shops opened in 1870.

A more successful method of winning railroad enthusiasm for Pullman service was the joint ownership scheme called an Association line or contract. This device had been used since the beginning of the sleeping car. In it the railroad and Pullman held a quarter, a half, or a three-quarter share in cars and their maintenance. A fifty-fifty split was the most common Association agreement. Revenues were shared accordingly. In this way the railroad was at least part master in its own house. At the same time, Pullman was relieved of providing capital for that portion of the cars and was thus free to divide his efforts among a larger number of operators. In effect, it gave him control of that much more mileage. Even if his ownership was not total, no competing firm could put its cars on the line. In some cases an Association contract might be the only type that a line run by an independent management would accept. It was a wedge, however, and Pullman could always hope to take over the service in time.

Several roads were drawn into Pullman's web through Association contracts. The Rock Island ran its own sleepers for nearly twenty years before joining Pullman in an Association agreement in 1880. Within eighteen years Pullman had full ownership of the line's sleeper service. The Northern Pacific wanted to run its own affairs, but was forced into Pullman's camp by its weak financial condition. Though it was operating seven cars at a good profit in 1880, it was being hard-pressed to finance its Western extension. It gave in to Pullman, knowing that funds would not be available for a full sleeper fleet once through service began. In 1881 it signed a fifty-fifty Association contract calling for sixty cars to be furnished within two years. To assure his hold on the Northern Pacific, Pullman purchased a large number of its shares.[220] An even more dramatic example of this technique was evident in the case of the Boston and Maine. In 1892 Pullman bought 5,000 of its shares; in the following year a twenty-five-year contract was signed.[221]

During the 1880s roughly one-third to one-half of Pullman's agreements were Association contracts. They included such major lines as the Pennsylvania, the Illinois Central, and the Santa Fe. A breakdown of the number of cars running under the two types of contract shows that Pullman gradually gained control:[222]

1881	502 Pullman	232 Association	=	734 total cars
1885	606 Pullman	509 Association	=	1195 total cars
1889	1066 Pullman	344 Association	=	1420 total cars

By 1890 Pullman and Association cars were almost equally divided—probably the result of the Union Palace takeover. By the late 1880s Pullman was clearly disenthralled with the Association

concept. In its early years the company did not have the capital to provide all the cars needed and was content to share investment and revenues with a partner. By 1890, however, its surplus was over 16 million dollars and its credit was unlimited. The monopoly was nearly complete; few lines seemed tempted to join Wagner, and even fewer were venturesome enough to run their own sleepers. There was a shift in Pullman policy toward discouraging new Association contracts and buying out the existing ones wherever possible. The chronically insolvent Erie was one of the first Associations to go in 1887. In the next year the Santa Fe and its subsidiary, the Atlantic and Pacific, gave way. In 1898 the Union Pacific sold its Association interest for 1½ million dollars. That nearly half of this figure was for "nonphysical assets," or good will, shows how eager Pullman was to end all its fourteen Association contracts.[223] In the early years of the new century, such major Association lines as the Southern Pacific, the Missouri Pacific, and the Atlantic Coast Line sold out. The final holdout, the Spokane, Portland, and Seattle, was a very small joint ownership involving only three cars. It was terminated in 1922.[224]

In general, the Association plan had been an effective provisional measure for holding service contracts until a complete takeover was possible. There were a few notable failures, however. The most dramatic breakaway was that of the Baltimore and Ohio. In 1880 the road's president, John W. Garrett, decided to follow a new policy of controlling all services connected with B & O operations. Beginning in the early 1870s he began to take over telegraph, rail supply, and printing services. He hoped to develop on-line passenger travel by providing resort hotels, and he intended to assume full control of the sleeping and parlor car operations once the Association contract with Pullman expired. The agreement itself was a hangover from the days of the Central Transportation Company. Before the contract ended, the road ordered new cars, and on the date of expiration, October 1, 1880, Garrett placed his stock in service and forbade the movement of Pullman cars over the B & O.[225]

Pullman immediately began legal action, asking for an injunction against operation of the railroad's sleeping cars and claiming patent infringements. He initiated a similar suit against Barney and Smith, who were supplying some of the cars to the B & O. Pullman further contended that the road's management was endeavoring to persuade other railroads to take up independent sleeper and parlor operations and was thus guilty of malicious conspiracy to ruin Pullman's business. The Baltimore and Ohio countered that it had every right to operate its own cars now that the contract had expired, and that it had decided to do so because Pullman had furnished it with inferior cars. It also alleged that because of the unfavorable terms in the old contract, interest and repair expenses had resulted in a loss to the railroad. The courts sided with the B & O on all issues except that of the patent claim.[226] The thirty-three cars owned by the Association were placed in the hands of a court-appointed receiver. At a public sale held in 1882 the equipment was sold for $5,000; Pullman claimed they were worth $18,000 each, and so a third legal battle began.[227] Litigation was pursued vigorously over the next few years. To Pullman the B & O represented not only a valuable contract, but a maverick who might tempt others to stray outside the syndicate he had so carefully constructed. If the other Association lines saw that malcontents could go unpunished, they might all develop wanderlust.

In the end the B & O returned to the fold, not because of the courtroom maneuverings but owing to a change in its management. The B & O's new president, Samuel Spenser, found the line in financial trouble and sought partial relief by eliminating its peripheral activities. He signed a new twenty-five year Pullman contract in June 1888.[228] An immediate settlement was made out of court, and the B & O remained forevermore a loyal Pullman subject.

No other Pullman defectors can be directly linked to the B & O's unsuccessful try for freedom, though at least one major Western line decided to go it alone during the same period. The Milwaukee Road had reluctantly formed an Association with Pullman in 1882.[229] They were so unwilling that Pullman was forced to agree to very liberal terms to win a contract. The Milwaukee would give Pullman only a 25 percent interest in the Association, and the contract could be terminated at five-, eight-, or eleven-year intervals, with a maximum lease of fifteen years. Pullman was to furnish forty-five sleeping cars and manage the road's diners. But the road did not consider even this weak Association agreement desirable, and late in 1890 it canceled the lease.[230] The rumors were that Wagner would take over the contract, but this was not so; the Milwaukee intended to run the 150-car sleeper, parlor, and diner fleet on its own. A suit developed over compensation for the cars furnished, but this litigation and other harassments did not persuade the Milwaukee to give up its independent course. Only when the road was near financial collapse in the 1920s did Pullman again secure a hold on its first-class car service.

These cases of contract disagreements and legal struggles should not be taken to mean that Pullman built and maintained its business through bullying. It resorted to open conflict only when no other measures were possible. The company recognized its subservient position with the railroads and the need to maintain harmonious relations with them. This was achieved by making contract concessions and by providing good service. Service was the long-term basis of Pullman's success. Throughout its corporate life there was a paramount emphasis on good equipment, polite employees, meticulous housekeeping, and gracious and well-controlled food service. A uniform and uncompromisingly high level of service could be expected not just on the Broadway Limited but on the meanest plug run to Muncie, Indiana. The car to Muncie might be older but it would be well maintained, and its crew would be every bit as professional as those assigned to the fanciest name trains. That so uniform a standard could be upheld in an operation as large and diverse as the Pullman Company speaks well for its management and its employees. Even in its years of decline, when all hope of reversing the slump in railway travel was painfully evident, the spirit of perfect service continued.

At the same time that the Associations were being discouraged, Pullman decided to deal with another menace to its affairs. The

tourist sleeper was the logical outgrowth of the old emigrant car. These primitive boarding cars handled a class of traffic that Pullman had no hope or ambition to attract. Emigrant cars had been used since the 1840s and were often little more than converted boxcars. But the Central Pacific began a new trend in the 1870s by introducing a more civilized emigrant car. It followed the general arrangement of the standard sleeping car, though it had no bedding or other frills. Within a few years, other Western roads were running improved emigrant cars that began to rival the comforts, if not the ostentatious luxury, of Pullman's finest.[231] The addition of bedding and porters upgraded the emigrant car to the tourist sleeper, which amounted to a very acceptable cut-rate overnight excursion car. The difference in fares was astonishing: a trip to the West Coast was $13 by Pullman, $3 by tourist car.[232] In the mid-1880s thousands of middle-class passengers were taking advantage of these low-priced sleepers.

Pullman became alarmed; if the scheme spread, it could ruin his business. The Western roads were not interested in abandoning the popular people's sleeper. In the late eighties, however, Pullman succeeded in taking over the operation of these cars. In 1890 he acquired the Northern Pacific's tourist cars and thereby gained control of a potentially serious rival. Now he had just over 250 second-class sleepers in service.[233] Much as Pullman may have disliked the whole idea, the tourist car was established, and many would continue in operation until after World War II.

Thus during the eighties Pullman was busy with the Associations, the rebellious B & O, the tourist cars, and the Union Palace merger. In general it was a decade of consolidation, growth, and enormous prosperity for the company. The car plant and model community south of Chicago was built in the early 1880s. The town of Pullman was to be its founder's fondest creation and the part of his enterprises in which he seemed to take inordinate joy. Ironically it proved his undoing, for the bitter 1894 strike at the car plant exposed Pullman to nationwide censure. Never before had he been subjected to ridicule on such a scale. How greatly it affected him can never be gauged, but it is thought to have contributed to his death in 1897. He died confused and embittered, under the ugly shadow of the strike. Perhaps fate's cruelest trick was that his death came just months before the company's takeover of Wagner. He was cheated of witnessing the ultimate victory, the fulfillment of a thirty-year struggle to create the great American sleeping-car cartel.

EXPANSION AFTER PULLMAN'S DEATH

Pullman's death left a vacancy that the firm's directors seemingly would not or could not bring themselves to fill. The company attorney and son of a former President, Robert Todd Lincoln (1843–1926), was named acting president. But even such a prestigious figure was not thought capable of assuming Pullman's office. Lincoln was not fully elevated to the presidency until after the Wagner merger was consummated.

Lincoln found himself in charge of a vast corporation, rich in property, rolling stock, contracts, and cash. It controlled the great bulk—nearly 90 percent—of the sleeping-car service in North

America, and operated the largest railroad car plant in the world. Its cash surplus was so gigantic that one of Lincoln's first acts was to pay out a 50 percent stock dividend amounting to 18 million dollars. And there were still ample resources to pay an inflated price for Wagner the following year. With the Wagner merger, the company decided to reorganize in January 1900. Capital was jumped to 74 million, and a new corporate title was adopted. The Pullman Palace Car Company seemed fussy and out of date even in the first year of the twentieth century. The firm became simply the Pullman Company; nothing more was required to describe this giant among American businesses.

Pullman flourished during the fifteen-year Lincoln administration, owing either to his stewardship or to the momentum of the company itself. However, it does seem evident that Lincoln was never ardently enthusiastic about his job. He seemed to prefer the study of law or the quiet pleasures of his hobby, which was astronomy.[234] But times were good, passenger traffic was up, and railways had no competition for long-distance land travel. Between 1900 and 1910 Pullman's traffic tripled, while its assets doubled. Net profits peaked in 1910 with a return of 14 percent, which was equal to over 19 million dollars.[235] During these years the stockholders received two spectacular extra stock dividends: a 36 percent dividend in 1906 and a 20 percent bonus in 1910. If the 1898 split is included, the shareholders were awarded extra dividends equal to 106 percent, or 99 million dollars.

John S. Runnells (1844–1929) succeeded Lincoln in 1911. An attorney with the firm since 1887, he had been Pullman's legal confidant for years. He was the last Pullman president to descend from the original organization, and like his predecessor, he inherited a healthy company. Some vexing problems were gathering, however. The Populist movement was adding momentum to the campaigns of the antimonopoly activists, and Pullman was an obvious target. The more radical Populist spokesmen charged that the company operated unsafe (that is, wooden), uncomfortable, unsanitary, and positively indecent cars.[236] They said that Pullman won the favor of judges, legislators, and other influential men by the distribution of free passes. They accused Pullman of reaping exorbitant profits. And finally, they declared that the cynical tolerance of those in power, coupled with public indifference, had permitted the growth of this monstrous swindler among corporations.

The cries of outrage were somewhat amplified by Pullman's takeover of one of the last major independent sleeping-car fleets. By 1912 the New Haven's president, Charles S. Mellen, had pushed the road into a desperate position through a reckless expansion program. In a last attempt to save his crumbling rail-steamer-trolley empire, Mellen agreed to sell the road's sleeping-parlor car business to Pullman for 3.3 million dollars.[237] The cash was used to shore up the tottering company. The transfer also relieved the New Haven of the enormous expense of reequipping with steel cars. Its safety record had been shocking in recent years, and a series of wrecks not only evoked a public outcry but convinced the New York Central that nothing but steel cars should be permitted to run into its Grand Central Terminal. The terminal was to serve as the New Haven's New York City station

under a rental agreement. The possibility of establishing a convenient separate terminal was out of the question for the nearly bankrupt New Haven. The opening of Grand Central in 1913 made action by the New Haven imperative. Since Pullman had the resources to provide steel cars, on January 1, 1913, the great sleeping-car monopoly became greater by 252 sleeping, parlor, and composite New Haven cars. After the New Haven sellout, the Milwaukee and Great Northern were the only large-scale non-Pullman operators left in this country.

The real and imagined abuses of Pullman were now so much a part of public debate that in 1906 the company was subjected to regulation by the Interstate Commerce Commission. This official gesture by Congress helped quiet the editorials, but it did not silence the Pullman legal staff, who argued that the Pullman Company was an "inn-keeping" operation and therefore not liable to I.C.C. control. For years Pullman lawyers had warded off state regulation by demonstrating that the business was interstate in nature. Now their own argument had been used to put Pullman under Federal control. While no one could agree on a precise definition of Pullman's operation, it was clearly interstate. In 1910 Pullman at last admitted defeat and began to report on its affairs to the I.C.C. However, the regulation was more in form than in substance; the Commission required uniform accounting procedures, but did little at first to govern Pullman's operations. Its corporate structure was in no way altered; it was not required to shed any part of its property, or to modify its contracts, or in general to do much more than report on its financial situation.

One exception to the I.C.C.'s permissiveness occurred during the early years of regulation. A St. Paul traveling man, George S. Loftus, had tried for years to force Pullman to reduce its upper-berth rate. Finally he took the case to the I.C.C., which decided that uppers were indeed inferior accommodation and that the charge for the top berth should come down by 20 percent.[238] This was hardly a crushing defeat for Pullman, although it did represent an annoying interference. Its cars generally ran half full, and most patrons preferred the lower berth and were willing to pay a small premium to avoid the ladder necessary to mount to the top bunk. In this decision a dissenting opinion was expressed by I.C.C. Chairman Martin A. Knapp. A true believer in Pullman as a public benefactor, Knapp's thinking may have represented the silent majority's view of that day. Knapp believed that the Pullman charge above regular coach fare was a bargain for a ride in an uncrowded, luxurious car, whose "occupants are usually persons of good appearance and unobjectionable manners."

In its own defense Pullman could contend that it was working for the public welfare by its crash conversion to steel cars. This program began in earnest in 1910. It involved the early retirement or downgrading of many first-class wooden cars—an action that increased the depreciation account to 6.8 million dollars, or double its normal rate.[239] The extraordinary expenditure for new steel cars cut Pullman's profits substantially. Millions were spent on retooling and building new shops. Repair facilities had to be remodeled as well. Pullman was genuinely devoted to a fast changeover to steel cars and achieved it long before the class-one railroads did. However, the benefits of the conversion far exceeded Pullman's investment. For its money Pullman acquired a durable, safe fleet, and its manufacturing division enjoyed its most profitable years during the wood-to-steel changeover.

In a more direct effort to polish its image, Pullman turned to the popular press for support. Apologists could be hired, while journalists sympathetic to the cause could be well treated. One of the industry's most faithful defenders was the journalist-historian Edward Hungerford. His 1911 book *The Modern Railroad* contained a flattering chapter on Pullman. Articles by many writers began to appear in national magazines, such as *Harper's Weekly*, explaining the necessity for Pullman's monopoly. The propaganda effort reached its high point in 1917 with Joseph Husband's *The Story of the Pullman Car*. Whether or not it was company-sponsored is unclear, but the author seems to have had the company's wholehearted cooperation and he produced a work filled with praise of Pullman. Despite its factual errors and partisan viewpoint, it became a standard reference and was for many years the only volume available on American railroad car history. A company-sponsored history was written in the late twenties, but did not progress beyond proof sheets.[240] Like Husband's work, it was a lavish tribute but a shallow and unsatisfactory business history. Afterwards Pullman seemed content to tell its story through giveaway booklets, exposition leaflets, and general news releases. Its best public relations series was published late in the twenties as "Pullman Facts." Subjects from the history of the sleeping car to the operation of the company laundries were covered in these informative booklets.

The Populist menace had no sooner faded than nationalization was visited on Pullman in the form of the U.S. Railroad Administration. This was a Federal agency organized to manage the railroads during the World War I transportation emergency. Pullman, now the twenty-fifth largest corporation in the land, lost its sleeping-car operation to the U.S.R.A. in July 1918. The car works remained in Pullman's control. The government paid a fixed annual compensation of $11,750,000 for the sleeping-car concession.[241] Pullman personnel operated the cars as before, and record-breaking numbers of passengers were moved. Federal operation was in reality quite short, ending in March 1920. But to Pullman's management, the entire episode was as unthinkable as it was inequitable. The company's claim against the government was one of the last to be settled.[242]

The postwar era opened the second period of growth for Pullman in the twentieth century. The trauma of Federal ownership was soon forgotten as the nation's prosperity seemed to promise an unending boom. Travel expanded with the good times. Salesmen seeking orders, vacationers exploring the Far West, speculators rushing to Florida—all were filling berth, bedroom, and drawing-room space. The leasing of private cars flourished as new millionaires sprang up. Everyone seemed on the move, and everyone seemed rich enough to go Pullman.

The excursion business was doing well too, so that the pool cars earned their keep. The fleet of extra cars was one of Pullman's greatest assets, and one of the major reasons why the railroads were amenable to the existence of the sleeping-car monopoly.

Pullman had about 1,500 cars in its active pool by the late 1920s. Some individual cars ran more than 150,000 miles in a year. Extra cars were kept at each of the company's forty-seven district yards, and they were transferred from one district to another to meet demand. A political or business convention, a major sports event, or some other special event might create a sudden need for several hundred extra cars, all on short notice and all for a limited engagement. No individual railroad could afford to keep such equipment on hand for occasional use, much less to maintain the cars in top condition for instant service. But Pullman could and did, because as a national system it could keep its pool constantly employed. In the railroads' view, the antimonopolists be damned; Pullman was a positive necessity.

The pool cars followed the tide of travel. In winter they made the second and third sections necessary to handle the holiday rush, or they ran south with the Florida traffic. In summer many of the same cars carried vacationers to Maine, Michigan, and the Western mountain and lake resorts. During the summer peak, for example, the Boston and Maine required 115 extra cars. Conversely, the Florida trains ran comfortably with 300 cars in normal times, but in the winter season they needed nearly three times that number.

Special events required skillful deployment of the pool cars, with all assignments coordinated by the Chicago office. A Knights Templar conclave in Denver might require 450 cars; the next week the same cars might be rushed east for anticipated World Series crowds or for a major political event (845 cars were used for F.D.R.'s first inaugural). The greatest single pool-car job was the 1926 Eucharistic Congress, which took 1,199 cars to carry the delegates to Chicago.

New cars were rushed through production during the 1920s to handle the crowds of travelers. In 1925, 25 million dollars was spent on 933 cars. During most of the decade, 300 to 400 new cars were put into service yearly. In this busy period the fleet grew to 9,800 cars; passengers numbered 30 to 35 million yearly (or roughly 55,000 a night); some 2,800 conductors and 12,000 porters were employed.[243] Except for the freak traffic surge created by World War II, the twenties were Pullman's busiest and best years.

It was also a time for expanding the monopoly. Washington was favorable to business, and Federal antitrust actions were rare. Pullman was free to pick up the few stray sleeping-car operations still running. In 1922 the Great Northern joined Pullman and within three years ended its sleeper operations. The sleeping-car historian Arthur Dubin was told by old Pullman officials that this concession was arranged because of the general indifference of James J. Hill's son to the affairs of the railroad. Whereas the father had been determined to control all the operations of the line, his son was less inclined to do so and permitted, if not welcomed, Pullman's management of the pesky sleeping-car trade.

Next to enter the Pullman system was the comparatively minor Central of Georgia.[244] Since 1880 the little Southern road had run its own sleeper and parlor car service, but in May 1925 it came to see the advantages of a national interchange via Pullman. In this way through traffic might be routed over its tracks. Independent operations limited first-class travel to on-line customers.

The final and most satisfying take-over was accomplished in 1927. For nearly forty years the Milwaukee had snubbed Pullman—owing, it has been said, to a personal feud between the road's president and George Pullman.[245] But now the Milwaukee was the weakest of the Western transcontinentals, with a financial structure that was sagging badly. It could not afford to reequip its name trains, and it needed new cars in order to hold even a small share of the traffic. Pullman, ever alert to ailing railroads, promised the Milwaukee to deliver 127 new or rebuilt cars ready for service by May 21, 1927.[246] Even so the road did not give in on a systemwide basis. Pullman moved in slowly, and as late as 1940 the Milwaukee was still running a few of its own cars. However, Pullman now controlled 96 or 97 percent of the sleeping-car operations in the United States.[247] Only 120 independent sleepers were in service, and these were mostly on unimportant runs. The Canadian Pacific–controlled Soo Line and a pitiful operation on the Gulf, Mobile, and Ohio represented the last of the independents. On a dollar revenue basis, Pullman must have held a 99.9 percent monopoly—an achievement rare in American business.

The consolidation of the sleeping-car industry was surely satisfying to Pullman's management, but it was more a filling in of details than a major advance. The company had, after all, dominated the trade since the turn of the century. Some greater challenge was needed, and expansion into car building seemed a logical avenue for investment. The Chicago shop showed excellent profits in good times—far better than the sleeping-car operation, and surely better than the government bonds in which the company had invested heavily. In hard times when car orders fell off, the sleeping-car division could be counted upon for a steady cash flow. Its business was stable, and depressions appeared to have little effect on it. Like a smooth-running flywheel, it always carried Pullman over the economic dead spots that the country seemed to experience every twenty years or so.

Thus Pullman decided to expand its car-building operations—not in the passenger car area, since the company had about all that business it could hope to win from American Car and Foundry Company, but in freight cars, which represented an even larger market. Pullman's own shop was obsolete, however. Purchasing an existing plant was simpler and faster than building a new set of shops. In nearby Michigan City was the nation's largest freight car manufacturer, Haskell and Barker. At its head was Edward F. Cary (1867–1929), one of the most dynamic railroad car builders in the nation. It seemed a perfect match.

For Pullman, there was more to the merger than an expansion into car building. Pullman's president, well into his seventies by 1920, was eager to retire, but finding a suitable replacement was not easy. Cary was asked to take the presidency, but he declined to do so except on his own terms. Finally a deal was made in 1922 that has been called a Haskell and Barker take-over of Pull-

man.[248] Pullman bought the Michigan City car works for 15 million dollars. Cary then became president of Pullman. His Haskell and Barker associates filled major Pullman positions: David A. Crawford and Charles A. Liddle were made vice presidents, Ellis Test became chief engineer, and Harold Dudley was named comptroller.

The new men were eager to revitalize the wealthy but stodgy old firm. In the summer of 1924 a major reorganization separated the manufacturing division from the Pullman Company. The parent corporation was called the Pullman Car and Manufacturing Company. The firm would now have a free hand in expanding its manufacturing interests and at the same time could remove this important area of its business from the scrutiny of the I.C.C. Cary engineered a second and more fundamental reorganization in June 1927.[249] A parent or holding company called Pullman Incorporated was chartered in Delaware. The charter was perpetual and liberal, allowing Pullman Inc. to hold the capital stock of both the Pullman Company and the Pullman Car and Manufacturing Company. In the process another major stock dividend was engineered. The reorganization enabled Cary and his associates to acquire new manufacturing properties for the holding company. Yet despite this freedom, they moved cautiously and stayed within the traditional Pullman borders. Negotiations were opened with several firms, including the General American Tank Car Company.

The most serious bargaining was with the Standard Steel Car Company, which had plants in Butler, Pennsylvania; Hammond, Indiana; Sagamore, Massachusetts; La Rochelle, France; and Rio de Janeiro. It also held a large interest in the Osgood Bradley car works.[250] The firm was controlled by the Mellons, a wealthy Pittsburgh banking family. Standard's founder and longtime manager, John Hansen, had retired in 1923 and the Mellons now found the car works a troublesome though profitable burden. They were interested in combining the property with an established concern whose management would direct the day-to-day operations. While the merger was under discussion in 1929, Cary was constructing a new freight car plant in Bessemer, Alabama. He was determined to capture a larger share of the car market by any means available. He may have been pushing himself too hard, for he died unexpectedly in April 1929. Daniel Crawford succeeded his old associate and pushed through the Standard Steel merger late in the same year. The price was 50 million dollars and a directorship for Richard K. Mellon. Pullman's board now included representatives of the largest financial interests in the country: the Morgans, the Vanderbilts, and the Mellons.

PULLMAN IN DECLINE

The great stock market crash had occurred that fall, but Pullman had weathered many a panic, and of course no one knew that a worldwide depression would follow. Nor were there many who foresaw how prolonged or catastrophic the business collapse would be. Pullman soon realized its huge blunder; new-car orders evaporated and its newly acquired workshops stood silent. As one railroad after another slipped into bankruptcy, Pullman's greatly expanded manufacturing branch became a colossal liability. In 1930 it showed a respectable profit, but in 1931 it lost over 2 million dollars.[251] The debt mounted alarmingly in 1932 and 1933, averaging nearly twice the loss of 1931.

The sleeping-car side of the business, believed to be depression-proof, was demolished by the Great Depression. The old dependable flywheel stopped revolving. For the first time in its history the sleeping-car operation showed a loss: 1.2 million dollars in 1932, 56,000 dollars in 1933, up again in 1935 to 1.64 million. Only its giant surplus saved the company, and some 50 million dollars from this fat kitty was paid out to cover losses and dividends during the thirties. Even at that, no dividends were paid in 1931–1933, and in 1934 Pullman reduced its capital stock for the first time in history from 120 to 108 million dollars.[252]

To survive, drastic economies were necessary. Personnel in the car plants that had once employed 16,000 were cut to 2,000 men. New-car orders were slashed; old sleepers were scrapped for what salvage money they would bring in. It seemed pointless to go after new outside car orders, since there were none to fight for. Pullman's only real hope was to rebuild as much of the sleeper trade as possible. By 1933 that traffic was down by 58 percent relative to 1929. Parlor car trade had nearly vanished; no one seemed too proud now to ride up front in the coaches. But it did appear possible to increase sleeper patronage, and rate cutting appeared to be the most direct means of doing so.

In the spring of 1930 rates for berths were cut by 22 percent.[253] A novel arrangement, worked out with the Pennsylvania to attract budget-minded travelers offered a combination rail-highway ticket between New York and St. Louis. During the day passengers traveled by bus; at night they boarded a connecting train and took an upper berth. The fare was equal to that of a coach ticket. In the following year, upper-berth rates were cut a full 50 percent on certain lines. To stimulate lower-berth patronage, the S.O.S.—single occupancy section—plan was adopted. For a premium of 40 percent over the normal lower rate, a passenger could enjoy the luxury of an entire berth section for himself. This eliminated the possibility of a noisy overhead companion. If he wished, the S.O.S. passenger could have a double mattress by using the one reserved for the upper berth. The scheme was well received, a fact that illustrated the growing disenchantment most Americans felt for the open-section sleeping car.

A more basic fare revision was made in 1932 when the 50 percent surcharge was removed, thus reducing Pullman fare to just 0.7 cents a mile above coach fare.[254] *Railway Age* over-optimistically claimed that the reduced fare would double Pullman's business; in reality the company did well to maintain the current levels of traffic and had little success in adding to them. A major deficit in 1935 shook Pullman's faith in reduced fares, and in 1936 the company petitioned the I.C.C. for a 10 percent hike. A weak recovery in the following years encouraged Pullman to install some new cars and upgrade the standard car fleet,

but the "little" depression of 1938 revived business pessimism, at least temporarily.

By 1940 things had come right again. The manufacturing divisions were busy with lightweight car orders, and the sleeping-car trade stabilized as the economy began moving again. But the Depression had badly shaken the company, and it was only a prelude to the graver problems that lay ahead. The history of Pullman took a downward turn after 1930, and the decades ahead would not be good ones.

The euphoria resulting from the economic recovery ended abruptly for Pullman on July 12, 1940, when the U.S. Justice Department filed a complaint in the Federal District Court at Philadelphia against Pullman Inc., its subsidiaries, and its officers. It was a civil rather than a criminal complaint, because Pullman affairs had been a matter of public record since the time of I.C.C. regulation and hence no secret conspiracy could be alleged. Yet the charges in Civil Action 994 were serious enough. Principally Pullman was accused of creating an unlawful monopoly in violation of the Sherman and Clayton antitrust acts.

The result of that monopoly was, according to the Justice Department, inferior sleeping-car service given the American public by a self-serving, miserly corporation which ran old-fashioned cars of its own manufacture. By its policies and restrictive contracts, Pullman had systematically stifled, thwarted, and restrained the railroads and other independent car builders from introducing modern sleeping cars. So long as Pullman enjoyed its monopoly, there would be no incentive to improve service; and so long as Pullman was at once the operator and builder of sleeping cars, little innovation could be expected. Through its operating contracts, some of them extending into perpetuity, the production of sleeping cars by non-Pullman subsidiaries had been blocked. In an eighty-page printed statement, the Justice Department outlined Pullman's minor and major sins. A separation between the sleeping-car and manufacturing divisions was sought in the public interest. And so began a legal battle that plodded through the courts for seven years.

Why had the Justice Department waited until 1940 to move against Pullman? As a very public corporation its monopoly had been obvious for years. True, the climate in Washington was no longer probusiness, and the White House was less inclined to discourage the Justice Department from implementing the antitrust laws. But the true source of the suit became evident as the case progressed. It appears that the Budd Company had promoted it because of Pullman's success in suppressing the purchase of sleeping cars from other builders, such as Budd. It seems more than coincidental that the case opened in Philadelphia, the home of Budd's car plant, rather than in Pullman's headquarters town of Chicago.

The quarrel with Budd had begun in 1935.[255] In January of that year Pullman's contract with the Burlington expired, and the railroad was not anxious to renew the agreement. Pullman continued to provide service for over a year while new terms were under negotiation. Meanwhile the Burlington decided to purchase some sleeping cars from the Budd Company to match its latest streamlined equipment. This was heresy to Pullman, which refused to consider operating sleepers that were not of its manufacture. In the ensuing argument Pullman claimed that the Budd cars were not up to standard and insisted on innumerable petty changes in bedding and accessories after the cars had been delivered. Pullman also stipulated that a "hold harmless" clause be added to the contract in the event of a structural failure—thus casting doubt on the soundness of Budd's Shotweld construction. It further insisted that the Burlington buy no more than ten sleeping cars from Budd. If the Burlington would not meet these demands, it could establish an independent sleeping-car operation. The railroad gave in to Pullman. A similar battle between Pullman and the Santa Fe developed two years later and again Pullman won, permitting Budd to furnish no more than five cars. After the antitrust suit opened, Budd claimed that it was impossible to secure sleeping-car orders because of Pullman's control.[256]

Pullman denied that Budd had been prevented from gaining a legitimate portion of the passenger car business and that Budd offered a product superior to that of the Pullman shops. To the contrary, Budd had openly threatened to instigate government proceedings if Pullman failed to purchase a license for Shotwelding and buy a percentage of new Budd cars. Thus Pullman accused Budd of blackmail, and to the various charges of the Justice Department, Pullman heatedly denied that it was a sedentary monopoly, indifferent to progress or the public's comfort. Pullman took particular umbrage to the criticisms of its rolling stock. Admittedly the bulk of the fleet was composed of heavyweight steel cars, but many of these were under fifteen years of age and all were well maintained and comfortable. Indeed, the company had spent some 30 million dollars on air-conditioning the standard cars during the past decade. While its 388 lightweight sleepers represented only 5.6 percent of its fleet, this figure compared rather well with the general passenger car stock of all the nation's railroads, in which only 3.7 percent were streamliners. Pullman contended that it was moving with all reasonable speed to improve its equipment: 71 lightweights were on order, and many old cars were being remodeled with room-style accommodations. Its lawyers also argued that Pullman was a benevolent monopoly, even if it was not a natural one like the postal and telephone systems, and should not be disbanded merely because it was a nationwide enterprise.

The final judgment, announced early in May 1944, ordered Pullman Inc. to separate its manufacturing and operating divisions. One or the other must be sold within a year; the company had ninety days to decide which one to keep. Meanwhile it must respond to any reasonable demand for sleeping-car service, whether covered by a contract or not, and it must seek competitive bids for all new sleepers—thus ending the in-house purchase of equipment.

The choice of which division of the business to retain was apparently not difficult. Even in the mid-1940s the declining fortunes of railway passenger travel were clear enough. Pullman decided that the operating division would go on the block. At the same time, management resolved not to appeal the case. Perhaps Pullman was secretly happy to have an excuse for unloading the sleeper business before it slipped into a major decline. The court-ordered sale price was based on an assessed value of 75 million dollars. At first no one seemed interested; the anticipated sale to the user railroads did not materialize.

As a device to lower the asking price, Pullman then agreed to sell its 600 streamlined cars to the railroads, thus trimming 35 million dollars from the valuation. By 1945 several customers were interested. The most active was Robert R. Young, (1897–1958), the controversial financier who had alienated the railroad industry through a publicity campaign about the poor quality of passenger service. His advertisements—"A hog can ride across the country without changing cars but you can't"—received wide attention. Young's simplistic solution for reversing the decline in railroad passenger travel was to install new equipment and offer coast-to-coast service. Through the investment bankers Otis & Company, he petitioned the courts for permission to buy Pullman on the same terms offered to the railroads.[257] He promised a grandiose modernization program costing 500 million dollars.

Suddenly the railroads were keen to negotiate with Pullman. The prospect of Young managing all sleeping-car service, or in any way having a wider field of influence, was anathema to many railroad managers. Pullman then sought and won a court order in January 1946 requiring that the sale be made to a consortium of railroads.[258] Possibly Pullman took this action to retain the favor of its future customers. In March the Justice Department appealed the sale to the Supreme Court, contending that railroad ownership would perpetuate the abuses of the old monopoly because the owner and lessor would be one and the same. Meanwhile the railroads sought I.C.C. approval of the sale, hoping to influence the Court's decision. Young threatened to organize his own independent sleeping-car company and thereby give Pullman a taste of competition it had not experienced since Wagner's demise. Nothing came of Young's plan; like his promise to reequip the Chesapeake and Ohio (a subsidiary of the Allegheny Corporation, which he headed), it just faded away.

Late in March 1947, almost three years after the sale had been ordered and seven years after the case had opened, the Supreme Court approved the sale of Pullman's sleeping-car property to the railroads. The purchase was completed on June 30, 1947, when 731,350 shares of stock were transferred to fifty-seven cooperating railroads. They had established the new Pullman Company for $40,202,482.[259] Carroll R. Harding (1888–1963), a civil engineer and former Southern Pacific executive, was named president of this major enterprise owning 6,000 heavyweight sleepers and employing 30,000 people. While passenger miles had declined 45 percent and revenues were down 39 percent since the wartime peak, its performance was still well above the average traffic of the Depression years.

The anticipated growth in the travel market gave Harding hope that Pullman could survive and prosper. There was some danger that the railroads might undercut Pullman by installing superior coach service, except that the roads could now expect to benefit directly from any profits Pullman might earn, and maintenance of the pool cars was as important as ever. Another advantage was the fact that the railroads were still a major means of intercity travel, and on the long hauls, passengers were willing to pay for sleeping accommodations. Americans wanted to travel and enjoy spending the money they had saved during the war, and Pullman was likely to share in that substantial cash flow.

By eliminating jobs and consolidating operations, Harding trimmed 10 million dollars from the 1948 budget. Railroad purchases of the new lightweight sleepers relieved Pullman of this major investment and built up the stock of attractive new cars. By 1949, 1,000 streamlined sleepers were running. Between 1947 and 1950 about 100 open-section standard cars were remodeled as room-style cars. It was hoped that all the heavyweights could soon be eliminated from regular trains. By retaining the traffic it already had, cutting costs, and upgrading equipment, Pullman seemed to have a future.

Yet the results for 1948 were not encouraging. Traffic continued to slip—not gravely, but enough to cause concern. Profits were a meager 1.7 percent. During the following year the forty-hour week drove wages up by 9.8 million dollars.[260] In the spring of 1949 air passenger travel exceeded Pullman's for the first time. Pullman cut service by 16 percent, eliminating most of the marginal runs. The streamlined cars, once regarded as the saviors of railway travel, were proving more costly to maintain than standard steel cars because of the added wiring, plumbing, and gadgetry. They also earned less because of their smaller payload.

In the early 1950s the decline accelerated. Traffic was below Depression levels; the airlines now moved twice as many passengers. In 1955 Pullman had to borrow 1.2 million dollars to meet the regular dividend payment. A major retrenchment program was necessary to cut losses. In 1953, 82 more lines were suspended. During the next few years all repair shops except the St. Louis and Chicago works were closed or sold.

A greater threat to Pullman's corporate life came in the late fifties, when some of its major users began to pull out. In 1957 the Pennsylvania took over its own parlor car operations. In July of the next year Pullman lost one of its major customers when the New York Central withdrew completely.[261] It was paying Pullman mileage fees of $814,000 in excess of the sleeper revenues. Few cars were in interchange service, and extra train movements were now so small that the pool cars were little used. The Pennsy threatened to follow the Central, but somehow Pullman managed to stall off wholesale withdrawals. Major cuts were necessary; nearly 3,000 employees were let go, so that the Pullman work force shrank to one-third of its 1947 size. Most of the heavyweights were retired or sent to government storage yards. Traffic had fallen off so badly that there was now even a surplus of lightweights.

The long postwar Indian summer that Pullman had enjoyed was at an end by 1960. There was no more talk of rebuilding traffic; all efforts were directed toward cutting costs and somehow keeping the company alive as long as possible. Barely sol-vent, superannuated like some ancient Civil War veteran, its only ambition now seemed to be to reach its centenary. Pullman's decline was irrevocably linked with the general collapse of railway passenger travel. By 1960 the great mass of intercity travelers were moving by private automobile. The 10 percent or so who still used commercial carriers could choose between buses, railways, or airlines. Bus schedules on new interstate roads now rivaled rail schedules, and buses were far cheaper. The airlines had the great advantage of speed. Businessmen, salesmen, field engineers—Pullman's bread and butter—abandoned overnight travel when they found that they could attend to their business in a single day by flying. Air fares also offered a considerable saving when compared with first-class rail tickets. Pullman began to price itself out of the market: as patronage declined, it raised fares. In 1956 and 1957 two increases jumped fares by 15 percent. Soon it cost twice as much to go by Pullman as it did to fly.

By 1965 the Pullman Company was heavily in debt. Its losses exceeded the value of its stock by 300 percent that year. Its traffic dropped, not by 5 or 10 percent compared with the previous year, but by nearly 50 percent. Its cars were now all railroad-owned except for 135 heavyweights in storage and 300 standard cars leased to Mexico. The Pennsylvania pullout in August 1967 was another blow to Pullman's sagging fortunes. By 1968 Pullman was operating on just twenty-seven railroads, and while every cost had been trimmed, its losses were a runaway 22 million dollars. After twenty-one years of railroad ownership, Pullman decided to quit. Late in 1968 it was announced that all operations except some minor housekeeping and maintenance service would cease on January 1, 1969.[262]

In its one hundred and first year, the sleeping-car company expired. But the Pullman name lives on with the manufacturing division, which has prospered since the 1947 separation.

Mechanical Development of Sleeping Cars

Through most of its history the development of sleeping car design and manufacture was dominated by Pullman. The first generation of Pullman cars began in 1865 with the *Pioneer*. Then came a series of twenty nearly identical sixteen-wheel giants.[263] Floor plans and details varied, but all appeared to follow a standard plan of berth arrangement. Next, many different builders and shops produced a largely mongrel assortment of cars. At the same time, Pullman was experimenting with car types, trying to decide whether to offer a straight sleeping car or some composite form which would include parlor, dining, and sleeping facilities. The diner car immediately proved to be a financial disaster, and Pullman hurriedly ended that service unless the railroad agreed to subsidize it. The hotel car was a more feasible scheme, and various combinations of sleeping-buffets and parlor-buffets became a part of Pullman's regular fleet. Light food service proved both popular and profitable. Luxury cars for daytime travel began to appear in Pullman consists during the early 1870s.

SECTION OF SLEEPING CAR,
showing Berths down and folded up.

Berths down.

SCALE

Berths folded up.

LONGITUDIN'L SECTION OF SLEEPING CAR, *showing two sections.*

Figure 3.51 The mechanics of the convertible lower and the folding upper berth are shown in two drawings first published in 1879. Note the bedding stored in the right upper berth. (Car Builders' Dictionary, 1879)

Pullman did not invent the parlor car, as is sometimes claimed. However, he recognized it as a natural adjunct to the luxury night trade and provided and ran such cars under contract for participating railroads.

One of Pullman's most important acts was his early decision to concentrate on the open-section sleeper. It made up the backbone of his fleet and remained the standard American sleeping car for three-quarters of a century. The open-section car was the clear favorite by 1870. Compartment sleepers were rejected again and again during the following decades, and while room cars were occasionally produced, Pullman remained firmly committed to the open-section sleeper until the late 1930s. No matter what its defects in passenger privacy or comfort, it was the cheapest, lightest form of sleeping car.

Coupled with Pullman's loyalty to the open berth was a strong allegiance to a single floor plan. The twelve-section (twenty-four berths), single drawing-room car was the favorite. Again so far as can be determined, Pullman settled upon this plan by the early 1870s. The construction and appearance of a standard Pullman sleeper is shown in many illustrations in this chapter (Figures 3.51 to 3.54, for example). The following description appeared in the July 1879 *National Car Builder*:

The "Keystone" and "Empire" are specimens of the present standard patterns of sleeping cars used by this company. Their length outside is 57 feet 3 inches. The interiors of the cars are finished with walnut near the floors with panels of prima vera, a California wood as above. The raised part of the panels is Circassian walnut, and all the mouldings are of amaranth, a beautiful dark-red wood. The panels on the under sides of the upper berths are decorated with inlaid work of prima vera and ebony. The ceilings are finished in oak panels with amaranth mouldings, and decorated with painted floral designs. The seats are covered with ponceau plush; the window curtains are made of Scotch tapestry of two shades of tan. The berth curtains are of entirely new design, and made in patterns expressly for these cars. The cars contain 24 berths. The ladies' dressing-room is at one end and the gentlemen's at the other. The latter is arranged with a state-room between it and the main saloon of the car, so that it is entirely shut off

Figure 3.52 The upper berth is down, ready for its nocturnal occupant, while the lower seats remain as they would appear during the day. The center table was removed and stored in a closet when the lower berth was made up. (Car Builders' Dictionary, 1879)

from the latter. This is the best arrangement for sleeping cars that has yet been devised.

Pullman followed the general practices of wooden car construction described in Chapter 1. Because sleepers tended to be the longest and heaviest cars in service, the more elaborate methods of side framing were employed. The earliest form of Pullman trussing for which drawings can be found is that of the *Enterprise* (Figure 3.50), presumably a typical Pullman design for the 1870s. The growth in complexity and length over the next

Figure 3.53 A berth is made up in the sleeping car Anton of about 1890. The curtains are not in place on the second section in this view. The section in the immediate foreground is shown ready for daytime traveling. (Pullman Neg. 2827)

Figure 3.54 A cross section of a Pullman sleeping car from around 1895 shows the upper berth in the closed position. (Car Builders' Dictionary, 1906)

twenty to thirty years is shown by the wooden body frame drawings in Figure 3.55.

Pullman fought the same losing battle against increasing car weight that faced all American car builders. The trend toward larger, more luxurious cars forced tonnage up, perhaps even more noticeably with sleepers because of their extra burden of bedding. In his 1874 statement to the Windom Committee, Pullman tried to dismiss his car weights of 28 tons as being only 2.5 tons heavier than a twelve-wheel coach. But a more representative coach weight for the period was 20 or 22 tons, making Pullman's cars closer to 6 or 8 tons overweight. And weight became an ever worse problem. By the mid-1880s a 70-foot Pullman car weighed 39 tons (Figures 3.56 and 3.57). By the first years of the next decade 50-ton Pullmans were common, and by the end of the wooden era sleepers had reached 60 tons.[264]

There seemed to be no way to avoid raising weight as overall size increased, as end construction became heavier, and as extra appliances were added. Eight-wheel trucks were abandoned by the late 1860s, but the six- and four-wheel trucks quickly gained in mass and weight as larger wheels, axles, and frames were required to support longer cars and sustain higher speeds. Lighting and heating methods were constantly improved. Gas lighting

called for tanks and piping, electric lighting for generators and batteries. Hot-water heating required a massive stove and a system of piping; steam heating did not reduce weight by much. Improved toilet facilities, demanded by the Pullman first-class clientele, were very heavy. Water tanks and pumps were common by the seventies. By the end of the following decade, air-activated water systems were installed. Vestibules were introduced during the same years, adding several tons to each end of the car. But Pullman had to feature the latest improvements, no matter what effect they had on total weight.

Adopting the newest auxiliaries was one way to upgrade the fleet; remodeling older cars was another. Such veterans as the *Pioneer* were kept acceptably modern by progressive alterations in the lighting and heating systems and by adding air brakes and automatic couplers. Roof lines, trucks, and exterior paint schemes were changed to follow the style. Older cars were reassigned to less important runs (Figure 3.58). But it was also desirable to retire obsolete cars, so that while the size of the Pullman fleet might appear relatively stable from one year to the next, the cars themselves were continually changing as new ones took the place of retired stock. In 1873, 120 new cars were placed in service.[265] Most of the old Central Transportation Company's cars were

Figure 3.55 Pullman sleeping-car wooden-framing plan as evolved by about 1895.
(Engineer, *January 22, 1897*)

Figure 3.56   A Pullman sleeping car of about 1880.

Figure 3.57   The Vancouver *was built for service on the Rock Island line in 1889.*
(Pullman Neg. 164)

Figure 3.58   Older cars were often remodeled with new interiors, but the Oconee *was downgraded (see the rattan seat coverings) for tourist service. The berths have been made up on the right side of the car except for the front section. (Pullman Neg. 2749)*

Figure 3.59   The Laconia *of 1898 had a wide vestibule end. (Pullman Neg. 4254)*

retired between 1879 and 1881, according to Pullman's annual reports. The reports note all additions and retirements—1883, for example, was a very busy year, with 121 new cars added and 50 old ones sold or scrapped.

To maintain the fleet on a nationwide basis, Pullman established regional repair shops between 1873 and 1889 in Wilmington, Delaware, Elmira, New York, St. Louis, Denver, and Buffalo. In later years repair facilities were opened in Atlanta, Georgia, Richmond, California, and Calumet, Illinois (the Calumet shop was adjacent to the main Pullman car plant south of Chicago). New-car construction was undertaken in Pullman facilities at Detroit (1870) and Chicago (1881). Repair work was done at both plants.

Since Pullman's maintenance of the everyday sleeper fleet was not particularly newsworthy, the press emphasized the premium cars that the company furnished for the name trains. There was only so much to be said about a standard twelve-and-one sleeper, but the special cars prompted extended comment. Two luxuriously furnished cars, the *Pilgrim* and the *Puritan*, were turned out for Boston–Montreal service in 1883. The *Railroad Gazette* of February 2 described them with admiration:

The woodwork is mainly dark mahogany, finely carved and highly polished. The upper berth panels and partitions are inlaid with satin, ebony and other fancy woods elaborately carved. The upper deck and

roof are lined with whitewood veneering, which is artistically hand-painted. The upholstery is in crimson plush, and the window and berth curtains are of raw silk. There are ten large centre lamps of beautiful design. These and the other trimmings in the car are of burnished brass or silver. The carpet is of a rich pattern, and the glass in the doors and partitions is of French plate, neatly embossed. The car is luxuriously finished and furnished. A spacious wash-room is located at one end of the car, fitted with bowls, mirrors and toilet utensils of the best make. Adjoining this is a cosy compartment luxuriously furnished for a smoking room. A passageway around this smokers' salon leads to a handsome buffet which faces the main compartment of the car. This little room is attractively fitted up with a side-board of fanciful design, which contains a set of handsome breakfast table-ware for serving out tea or coffee, oyster stews and other light refreshments at a nominal price to the passengers. A small safety stove and other utensils are arranged in a separate compartment, the whole being in charge of a competent steward. At the other end of the car is a handsome parlor or drawing room for ladies, with a commodious lavatory adjoining. The car is warmed by hot-air pipes under each seat. The exterior of the car is painted a dark olive green, and is richly ornamented in gold and colors. The 'Pilgrim' is similar in all respects to the 'Puritan,' and cost some $22,000.

By the end of the eighties whole trains were being produced to capture the luxury trade. One of the first was the Pennsylvania Special.[266] The five cars—the first Pullmans to be fitted with

Figure 3.60   Elaborate exterior gilt decorations were abandoned long before the Lucerne *entered service around 1900. (Smithsonian Neg. 48376)*

vestibules—were ready for public display by the spring of 1887, but they did not enter regular service until the following summer. Three of the cars were sleepers: the *America,* the *England,* and the *France.* The interiors were paneled in mahogany and satinwood. The berth seats were covered with a pale-blue glacé plush; the bedrooms were upholstered in a terra-cotta-colored plush.

Other railroads began to acquire whole trains of new stock. In 1888 the Golden Gate Special which began running in the West featured electric lights, steam heat, and an observation car with a recessed end platform and floor-length end windows.[267] Next year came the Montezuma Special, a splendid new train for the New Orleans–Mexico City run. Two of the cars, exceptions to Pullman's open-section rule, were on the compartment plan, with six drawing rooms each.[268]

In keeping with the eclectic taste of the times, Pullman interiors became more diverse in style and even more elaborately decorated as the final decade of the nineteenth century opened. Or rather, ornamentation increased in interior appointments; curiously, the exteriors became more chaste during the same time. The heavy overlay of gilt striping, panels, and baroque scrolls on the outsides of cars slowly gave way to a relatively simple gold border lining (Figures 3.59 and 3.60). The focusing of attention on interior furnishings appears to have been motivated by the desire to create something special for the coming Columbian Exposition. Pullman was a sponsor of the 1893 World's Fair and naturally wanted his own firm to be well represented. Wooden car technology was near its final limits; sleepers with 70-foot bodies were common. But the subtle improvements now possible in car construction would be lost on a general audience, whereas a major change in interior decor would be clear to everyone who had traveled in a sleeping car. Accordingly Pullman's chief draftsman, August Rapp, produced a design in 1892 that skillfully wove the rich Empire style into a form suitable for sleeping-car architecture.[269] The ceiling and berth fronts took on full curves like a Roman vault (Figure 3.61). The seat ends looked like an oversized copy of a Napoleonic bedroom set. In the flashy Empire manner, the moldings, torches, winged lions, and rosettes were finished in gold ormolu. The sleeper-observation *Isabella*

Figure 3.61   While exteriors became more chaste, Pullman designers put greater emphasis on interior decor. The Empire-inspired Columbian style of interior was introduced in 1892. (Car Builders' Dictionary, 1906)

was one of the special cars built for the 1893 Columbian Exposition. It set the style for the top-grade Pullman sleepers for nearly a decade. No basic improvement or change in general arrangement or comfort was incorporated in what came to be called the Columbian style of interior. The berth and seating mechanisms were identical with those of the 1870s.

With the coming of the twentieth century Pullman continued as one of the great forces in passenger car architecture. In 1904

Figure 3.62   The body plan for Pullman's experimental all-steel sleeper, the Jamestown of 1907. The car was considered too heavy and the design was not repeated. (Master Car Builders' Report, 1908)

Figure 3.63   The Carnegie *of 1910 was the first production model all-steel sleeper to enter service. The 68½-ton car was converted for tourist service in 1936. (Pullman Neg. 12112)*

the railroads abruptly dropped the amazing carvings and the North German Lloyd grandeur of the Columbian style for the plain lines of the "sanitary" interiors. The rustic mission style overthrew the luxuriant Moorish. Pullman was quick to accept the new style. It showed considerably less enthusiasm, however, for the shift to steel cars. The company was much more comfortable with the traditional materials of construction. Only after some major customers showed their clear intention of entering the steel era did Pullman begin serious investigation of metal car design. Sometime in 1906 the directors, including J. P. Morgan, met to consider the advisability of such a move. Richmond Dean, the chief engineer, described the historic meeting:[270]

The directors considered and disposed of some unimportant routine matters, and then Mr. Lincoln—Robert T. Lincoln, then president of the Company—raised the question of steel construction. Before he had uttered three sentences of his preliminary statement, Mr. Morgan interrupted:

"Can we build all-steel cars?"

Mr. Lincoln turned to me, passing the question along, and indicating that it was my time to talk. It had come more suddenly than I had expected. But I replied:

"Certainly we can;" and while I was pulling myself together for the detailed exposition, Mr. Morgan went on:

"Then I move that we proceed to do it."

The motion was adopted; nobody wanted to see my plans and estimates; nobody asked any more questions, and we just went back to the works and proceeded to build steel cars.

In truth mass production of steel cars was nearly four years away. Pullman did, however, begin work on an all-steel twelve-section sleeping car, which was completed in March 1907 for exhibition at the Master Car Builders' Convention. It was named the *Jamestown* in honor of the tricentenary of the first English settlement in North America. The *Jamestown* was not mechanically a notable success, although in general arrangement and appearance the car was acceptable enough (Figure 3.62). It was comfortable, pleasing (even if a bit severe in its internal appointments), and as good-looking as any standard sleeper of the day. It was a skillful steel reproduction of its wooden prototype, with rivets that were countersunk and puttied over, and queen-post truss rods.

Pullman apparently attempted to build the car from commercially available steel shapes, using as little special fabrication as possible. After all, the company had little metalworking equipment and was probably trying to see how good a car could be produced without extensive retooling. The center sill was formed from two standard 15-inch I beams joined by a ½-inch bottom plate. The side sills were 6- by 6-inch angles; the belt rail was a flat 3-inch bar. The cast-steel cross sill, platform, and bolster end piece was one of the few special items used in building the *Jamestown.*

The *Jamestown* was said to ride smoothly and quietly; none of the expected battleship rattle developed.[271] But the car was grossly overweight, so much so that its tonnage was kept something of a secret. Of the detailed reports that appeared in the American technical press, none mentioned total weight. The 1908 Master Car Builders' report on steel car construction estimated the weight at 80 tons. The Gutbrod Report of 1913 gave a figure of 81 tons. Presumably the latter was correct; probably the initial embarrassment of the *Jamestown*'s poundage had passed sufficiently by 1913 for the company to release full specifications.

Few engineers expected that a lightweight design was possible in view of the safety, comfort, size, and cost requirements. Nevertheless, 81 tons seemed excessive. Surely a compromise plan could produce a lighter car. A second steel sleeper was reportedly designed in 1907–1908 for service on the Union Pacific, but no record of it can be found except for incidental mention in the 1908 M.C.B. Report. Possibly the car was an improvement on the *Jamestown.* In 1908 Pullman built five steel combine cars for its own service. Again not enough information exists to determine whether they contained notable design advances. During the next year or two, nearly ninety steel passenger cars of various types were produced at the Pullman car works for the Harriman Lines and the New York Central. A new steel car plant was outfitted at the Chicago works in 1909 to facilitate the quantity production of such cars.[272]

Pullman designers were developing a feeling for steel construction. The new plant allowed greater freedom in the manufacture of special shapes, so that plans could be more sophisticated and reductions in weight more substantial. Pullman's experience in building steel cars commercially was of great benefit. In addition,

Figure 3.64   Details of body construction used in the Carnegie *and many other early steel Pullmans were published in 1910. (*American Railroad Journal, *October 1910)*

Figure 3.65   The Carnegie's *twelve open sections and single drawing room repeated* Pullman's *most favored floor plan.* (American Railroad Journal, *October 1910*)

Pullman as well as most other car builders profited from the achievement of the Pennsylvania Railroad designers who had perfected an improved design for passenger cars. It came to be recognized as the key to practical steel car design.

In 1909 Pullman began work on another prototype all-steel sleeper that would incorporate everything that the company had learned since producing the *Jamestown*. The new car was intended for service on the Pennsylvania. Completed in January 1910, it was named the *Carnegie* (Figures 3.63 to 3.65). The 73-foot, 6-inch body held twelve open sections and one drawing room.[273] The metal interior was grained to imitate mahogany paneling. The total weight was 68½ tons, a saving of nearly 12 tons over the *Jamestown*.

There was no question about the practicality of the *Carnegie*; it was a car that would serve long and well. Within a month Pullman was at work on an order for 500 steel cars. The company made a voluntary commitment to convert its sleeping cars to steel equipment on a nationwide basis as soon as possible. No new wooden sleepers were ordered after 1910. The better wooden cars were sent back to the shop and steel-underframed; by 1913 over 600 had been processed.[274] In the same year, 2,100 all-steel sleeping cars were running on American lines. Pullman claimed that one-third of its fleet was steel; 857 cars at a cost of $22,000 each were installed in 1913 alone. It was a remarkable record, a conversion program in which Pullman outpaced the railroads by decades. In 1926 three-quarters of Pullman cars were all-steel, and all but 4 percent of the remainder were steel-underframed. By the middle 1930s the last wooden-bodied cars were retired.[275]

The coming of the steel car gave Pullman an opportunity to standardize. Because the conversion to new stock was made quickly, it was possible to stay with a single basic design that was not likely to become obsolete within the period of the change-over. Of the 8,000 cars built for Pullman use in the quarter century of the heavyweight era (1910–1935), one-half were of a single uniform pattern.[276] The old standard floor plan of twelve open sections and one drawing room continued into the steel era. Based on the *Carnegie*, one lot after another of twelve-and-ones was manufactured. A new design was introduced in 1923 (referred to as plan 3410), but only an expert could see much difference in it. This is not to say that Pullman did not operate a great variety of cars. Sleepers were offered in all combinations of open-section and room arrangements. Some all-section cars were built; others were parlor-lounge, parlor-buffet, sleeper-lounge, parlor, observations, and all-room cars. Eventually there were 100 different floor plans and, if subclasses are counted, 457 additional styles.[277] It may not sound like such an effective job of standardization, but the cars themselves were manufactured from standard elements. The basic cars were identical shells. The frame, trucks, roof and body parts, and mechanical equipment were the same. Berths, seats, lighting fixtures, hardware, window frames, and doors were standard elements that could be com-

bined to produce a great variety of car types. Hence standardization was realized to a very large extent in the heavyweight era. In the faster-changing lightweight period, such uniformity was not possible.

Standardization, so beneficial to economical operations, also discouraged innovation. One can argue about what is useful innovation and what is merely change for its own sake, but except for air conditioning, Pullman could hardly name any major improvements made during the standard era. A list published in 1932 enumerates items such as legible bulkhead signs, bottle holders, and new window curtain fixtures.[278] Pullman's critics pointed to lists like this as signs of neglect and reaction, as the Justice Department did during the 1940 antitrust suit. The fairness of such criticism might be challenged, for Pullman was offering comfortable, safe accommodations and sincerely believed that little could be gained from any radical design changes.

The most unfortunate consequence of Pullman's conservatism was the decision to stay with the open-section sleeper (Figures 3.63 to 3.69). Criticism of this plan became more pronounced with the passing years, and Pullman came to regret that it had not shifted to room-style cars at the beginning of the steel era when such a change would have been possible. As a convertible day-night car, the open-section sleeper was a compromise, and like many compromises it did poorly in satisfying either requirement. It was an inherently makeshift arrangement. In the daytime the rigid seat was not very comfortable. It could not recline, and because of the face-to-face arrangement, half of the passengers were forced to ride backwards relative to the train's movement. The windows were necessarily small owing to the upper berth. The difficulties of undressing and of clambering into the upper berth have already been mentioned. One magazine spoke indignantly of the degrading spectacle of disrobing in a Pullman aisle: "It is indecent, primitive, rural and jay."[279] The limited toilet facilities were a second failing of the section cars. Ladies' and men's washrooms were located at opposite ends of the car. Going to the toilet in the middle of the night meant a stumbling journey down a narrow, curtained aisle to the far end of the car. In the morning it meant standing in line awaiting a turn at one of the community washbasins. As indoor plumbing became a common feature, the limited facilities offered by Pullman seemed ever more inadequate.

Bedtime entailed minor pandemonium as the porter began to assemble the berths. Passengers had to find seats elsewhere while the beds were unfolded; they could take a seat with a neighbor or retreat to another car. Some passengers wished to retire early, others late. As a result the car was partially made up for much of the evening and thus had a confused appearance that seemed to underscore its makeshift plan. In making up the berths the porter opened the upper berth, where the bedding, curtains, and mattresses were stored. The upper was already made up, but the lower had to be assembled from scratch (Figure 3.70). The seat cushions were first rearranged to form the bed, the mattress was

Figure 3.66   A standard Pullman open-section sleeper as it appeared in daytime, with all the bedding hidden inside the closed upper berths. The seats, unlike those in a coach, could not be reversed. The photograph was made in 1916. (Pullman Neg. 20672)

Figure 3.67   This cutaway drawing shows berths both open and closed. Notice the extra-large men's washroom, which also served as a smoking compartment.

Figure 3.68   The four-square appearance of a standard steel Pullman was well represented by the Lake Garfield of 1924. It had ten open sections, two compartments, and one drawing room. (Pullman Neg. 27713)

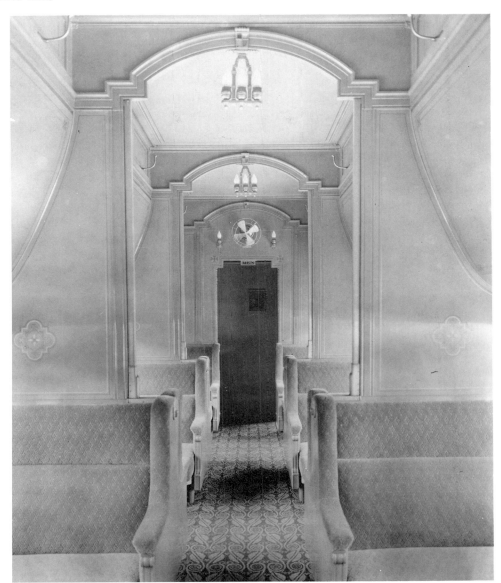

Figure 3.69   This fourteen open-section sleeper was built in 1927 for the Milwaukee Road's luxury train, the Pioneer Limited. *The built-in headboards and arches formed semiprivate spaces that helped eliminate the open coach-like appearance of most sleeping cars. (Pullman Neg. 31047)*

then put in place, and the bed was made up with sheets, blanket, and pillows. Next the curtains were hung. An experienced porter could do the entire job in just three minutes.

Basically the section cars did not offer the privacy or comforts that American passengers now demanded. And as more Americans traveled overseas, more were exposed to the conveniences of the all-room cars operated elsewhere in the world. Why was this country so far behind the times? Why did Pullman cling to its obsolete cars? *Scientific American* accused the Pullman management of "a blind unreasoning conservatism."[280]

Pullman's answer was the same as it had always been: the open-section sleeper offered the cheapest style of sleeping accommodation possible. Section cars carried 27 passengers; room cars could accommodate only 18.[281] By eliminating the bulkheads of the room car, section cars saved 7,000 pounds per passenger. Because they were lighter and could carry more people, section cars cost less to operate and thus could offer the most economical sleeper space. Pullman also claimed that compartment cars were a proven failure in America. The Mann Company had gone out of business, Pullman implied, because it room cars had been unpopular with the public.

In time, however, changing market conditions forced Pullman to accept the compartment car. Competing forms of transport, coupled with a growing demand for greater luxury, brought about a basic shift in policy, perhaps the greatest since the organization of the Pullman Company.

By the middle twenties Pullman was concentrating more on room- or combination-style sleeping cars that offered several rooms and a lounge or club section. The shift to this style of car in preference to the standard twelve-and-one sleeper was obvious by 1935, when 77 percent of the new cars were of the composite type.[282] Another indication of the open berth's declining popularity was passengers' growing reluctance to travel in the upper berth. By 1936, 80 percent of the riders were choosing the lower berth despite the 20 percent higher fare. Economy was no longer enough to attract passengers. If the uppers were to run empty, there was little point in keeping them. If passengers would accept only single-occupancy space, it was necessary to offer room cars. The space would be more costly, but it seemed to be what the public wanted, and satisfying that want would surely produce more revenue. And so began the search for an inexpensive, single-occupancy design of maximum capacity. These cars would

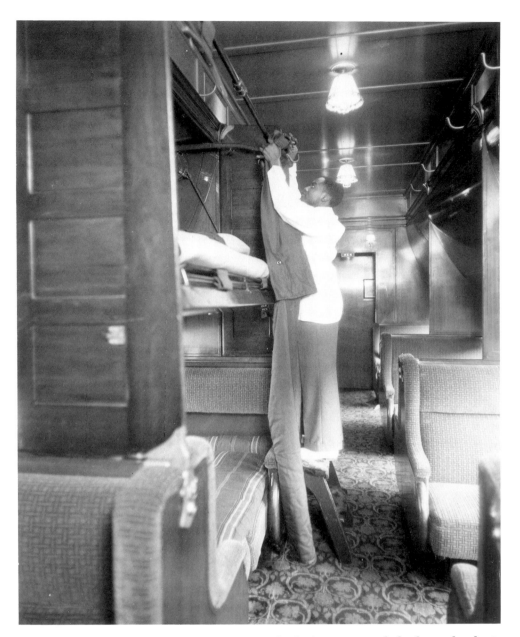

Figure 3.70   A skilled porter could assemble both the upper and the lower berths in three minutes. In this view the porter is hanging the curtains. (New York Central)

Figure 3.71   The Eventide was an experimental all-room car on the duplex plan. The 16-room model was rebuilt from a library-baggage car in 1933. (Pullman Neg. 37692)

be aimed at the traveling salesman and other business and professional men. Couples and wealthy riders would, as before, choose bedrooms or drawing rooms; what was needed was some ingenious scheme for a single-room sleeper.

Meanwhile the open section did not completely die. It is true

that few section cars were built after 1930, but Budd did produce six 16-section sleepers for the California Zephyr in 1948. Other roads purchased a few combination open-section and room cars during the lightweight era. As late as 1955 the New Haven bought some combination cars with six open sections to accom-

Figure 3.72   The Major Brooks *of 1938 was a later form of the duplex single-room scheme. The* Brooks *had several bedrooms as well. It operated on the Pennsylvania Railroad. (Pullman Neg. 41262)*

modate government employees, who were required by law to travel on the cheapest sleeper accommodation. This Federal regulation perpetuated the antique open section into the modern age. The cars remained on a Washington-Boston night train until sometime in 1968.

Room cars were not new to the Pullman Company. From the beginning the standard open-section sleeper contained at least one bedroom for patrons willing to pay for the luxury. By the eighties some all-room cars were being operated on trains catering to the affluent traveler. In 1927 came the first single-occupancy room cars, aimed at the lone traveler who wanted superior overnight accommodations enough to pay a rate set at 25 percent above rail fare plus two lower-berth charges.[283] It was not economy class; for his money the traveler received the privacy of a tiny but fully enclosed compartment, a small but stationary bed (32 by 78 inches), a toilet and folding washstand, and individual heat control. It occupied just over 33 square feet of space, but compact as the rooms might seem, only fourteen could be fitted into an 80-foot car. It is not likely that they were profitable, but they were popular enough so that forty-five were eventually placed in service.

These room cars, which Pullman titled the Night series, were hardly the model that the company was seeking. Something more imaginative than slicing a car up into fourteen narrow rooms was needed to produce a single-room car with enough capacity to replace the open section. Pullman designers experimented with the Bradish plan for remodeling open sections with stationary headboards and semipermanent curtains. But the scheme offered few tangible improvements over the existing plan.

Pullman was looking for a more complete departure from orthodox sleeping-car design. Few more radical plans had been devised than that patented by Albert E. Hutt on April 30, 1929 (No. 711,317). The inventor offered a split-level arrangement for day-night service, but it was initially dismissed by Pullman as too visionary. As Chapter 2 mentioned, Hutt's scheme was tried out by the Long Island for dual-level commuter cars. Pullman began

to think better of the general idea, if not of Hutt's specific plan, which was essentially a variation on the old open-section plan. Surely the best use of a given space was to deck it, and Pullman meant to cram as many single-room spaces as possible into the confines of a 10- by 75-foot space. Every other room was raised just a few feet above the car's main floor level. This provided enough free space for a sofa bed and, as the scheme was eventually developed, almost doubled the number of rooms over the one-level arrangement. The design was called the duplex.

To test the duplex idea, in 1931 two heavyweight steel cars were rebuilt at one end.[284] That part of each car was fitted with four duplex rooms and given an arched roof for greater headroom. Small, ungainly windows for the upper duplex rooms were punched into the high side of the body. The rooms were all on one side of the car, with a side aisle on the other. After running these two hybrids, which had been renamed *Wanderer* and *Voyager*, for a year on the Pennsylvania Railroad, Pullman decided to give the duplex a full-dress trial. Two heavyweight baggage-buffets were run through the shop in 1933. Rechristened *Eventide* and *Nocturne*, they emerged with arched roofs and sixteen duplex rooms each, again all on one side. Viewed from the aisle side, the cars appeared reasonably conventional, perhaps lacking the usual number of windows, but in general not conspicuous. But the room side of the car, with its awkward jumble of staggered windows, had the grace of an armor-plated hen house (Figure 3.71). The car's poor looks were not its greatest disadvantage. Pullman did not consider sixteen spaces sufficient for an economical single-room sleeper. The duplex, like all bilevel cars, also had an inherent safety defect: passengers had to mount three steps to reach the upper rooms. Step lights and handrails could not completely eliminate the hazard.

As the Depression worsened, Pullman dropped any plans it might have had for a large-scale car-building program. Duplex rooms were incorporated in a few new cars over the next years, and despite its obvious defects, it was an attractive enough room plan to remain in at least partial favor so long as Pullman stayed in business (Figures 3.72 and 3.73). It had something of a resur-

Figure 3.73   *This schematic drawing shows the duplex room's two-level arrangement. It was designed to replace the old-fashioned open-section plan. (Traced by E. Tone)*

gence in 1942 when a blending of duplex and roomette plans produced a car with twenty-four single rooms—a record at that time.[285] The floor space per unit was squeezed down to only 20.9 square feet, yet somehow each contained a toilet, a sink, and a rollaway bed. The payload was now large enough that a profit seemed possible even if the fare was only 10 percent above that of a lower berth. Some experts claim that the duplex cars built for the Pennsylvania in 1949 were the most comfortable single-room units ever placed in service. These cars, named the Creek series, had twelve duplex roomettes and four bedrooms. In the more usual arrangement there was another bedroom, but by dividing this space among the duplex rooms, enough space was gained for permanent rather than folding beds.

Meanwhile Pullman designers developed a better floor plan, which became known as the roomette. The rooms were arranged on one level on both sides of a center aisle, and they were placed longitudinally rather than transversely. Much of the hardware perfected for the duplex, such as the folding sink, was salvaged for the roomette. The roomette's advantages were that it was on one floor level and that two more rooms could be fitted into a standard 80-foot car. During the day the passenger occupied a wide seat with armrests. The bed was folded upright into a niche at the head of the compartment (Figures 3.74 to 3.77). When the bed was pulled down at night, by either the traveler himself or the porter, the seat and the toilet were covered. The tiny room was in fact almost completely occupied by the bed in its down position. A thin passenger could stand in the narrow space between the doorway and bed. A bit more space could be gained by opening the door, which slid from view between the hollow aisleway bulkhead. Privacy was possible even with the door open, because there were full-length doorway curtains that could be zipped shut. The roomette was a self-contained living unit with its own temperature control, ventilator, fan, sink, mirror, luggage rack, toilet, coat closet, and even such refinements as a

night light. The room was small—suffocatingly so, for some passengers—but a great many found it a cozy and pleasant retreat.

The first roomette car was completed in June 1937. The 67-ton, eighteen-room streamliner seemed destined to succeed the old twelve-and-one as the new standard sleeping car. It was hailed as a miniature hotel and as Pullman's best weapon in its battle for overnight travel. More roomettes were built; by 1941 sixty-eight were in service. During the same years (1935–1941) Pullman rebuilt over 300 heavyweights with room-style accommodations. The company anticipated that after the war more new roomettes would join the fleet and that the open section would vanish from the American scene. Indeed, another fifty-five roomettes were built, but the taste of the traveling public took an unexpected turn. Economy seemed less important than comfort, and passengers preferred bedrooms and drawing rooms to the cramped roomette. Fewer full-roomette cars were ordered. Instead the railroads settled on the ten-and-six (ten roomettes, six bedrooms) as the standard sleeping car. Thus after some sixty years the argument over the best form of sleeper was resolved: history had sided with W. D. Mann.

At the same time that Pullman was trying to devise a replacement for the open section, it became involved in a program to develop a lightweight sleeping car. Except for the diner, the sleeper was the heaviest of all passenger cars. Pullman's interest in the problem was more dispassionate than that of the railroads, but the company could not afford to show indifference if it wished to maintain its reputation as a leading car builder. The manufacturing division's investigations into the art of lightweight construction drew Pullman into the new age of aluminum, Cor-Ten, and stainless steel.

In 1931 Pullman produced some aluminum-bodied, high-speed interurban cars for the Indiana Railroad. The same year Pullman

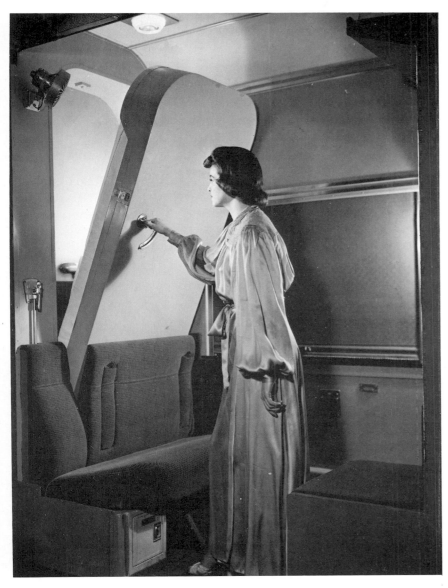

Figure 3.74 *The Roomette, introduced in 1937, offered a single-occupancy room on one floor level. This became Pullman's replacement for the open section. The front wall of the Roomette mock-up has been cut away to make the interior visible. (New York Central)*

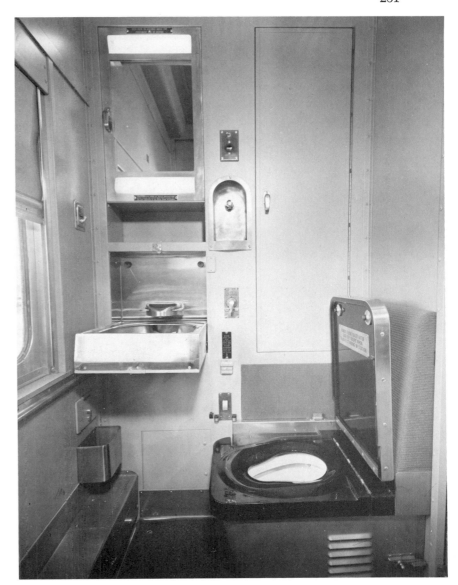

Figure 3.75 *An interior view of a Roomette taken from the seat. The toilet seat is closed and the sink folded back into the wall when the berth folds down from the opposite wall behind the seat. The oblong door above the toilet is a clothes closet. (American Car and Foundry Company Neg. 61245-Z)*

decided to build a lightweight sleeping car. The project languished, and work did not begin until 1932. Then the car was rushed through so that the experiment might be ready in time for the 1933 World's Fair.[286] Named the *G. M. Pullman*, the sleeper came as close to being an all-aluminum railway car as any vehicle built (Figure 3.78). The underframe, body, roof, and most of the hardware were of aluminum. The center sill, 20 inches deep, was on the fishbelly pattern and was fabricated from thick aluminum plates. More remarkable, the truck side frames were cast aluminum. Even the wheels had aluminum centers; steel tires provided the running surface. Structurally, the 84-foot-long car was a radical departure from the standard steel sleeper of its day. It featured a round, enclosed observation room, a buffet, and four bedrooms.

The total weight was proudly reported as 48.5 tons, or about one-half that of a steel sleeper. Yet the *G. M. Pullman* outweighed its sister aluminum observation coach, described in Chapter 2, by over 11 tons (Figure 2.61). This demonstrates how much extra weight was added by the bedding, partitions, and other gear necessary for a sleeping room car. The *Pullman* created the hoped-for notoriety during its exhibition. Afterwards it

ran on the Florida Arrow and the Super Chief. It proved to be a serviceable car, and after leaving the Pullman fleet in 1952, it was used for another twelve years on the Chicago Great Western. The first lightweight, streamlined sleeping car created little excitement within the ranks of the Pullman management. No more lightweights joined the sleeper fleet for another three years, except for the Union Pacific motor trains.

Again, novelty seemed uppermost in the minds of the designers who laid out the next streamlined sleepers. Aluminum was forsaken for Cor-Ten steel, and the established eight-wheel arrangement gave way to articulation. The two units of the resulting 150-foot, fourteen-wheel mammoth were named the *Advance* and the *Progress*. The rear unit, the *Progress*, was divided into an observation lounge, a buffet, and three bedrooms. The *Advance* was cut up into sixteen duplex rooms. The 112-ton pair hardly proved to be the wave of the future. They had not been long out of the shop when a more conventional lightweight sleeper appeared. This car, named the *Forward*, was ready late in 1936 (Figure 3.79). It was the first to incorporate Pullman's truss side frame and fluted stainless outer skin. The 81-foot, 6-inch, 55-ton car was divided into eight open sections, two com-

Figure 3.77   Some modern sleeping cars were a mixture of drawing rooms, bedrooms, and Roomettes. The car shown here was part of the 1949 California Zephyr.

Figure 3.76   Roomettes are placed on both sides of this narrow center aisle. Metal sliding doors, not visible, can be left open and the curtains closed (or opened), depending on the wishes of the occupant. (American Car and Foundry Company Neg. 61245-FF)

partments, and two double bedrooms. Structurally it proved a satisfactory model for a production lightweight sleeper (Figures 3.80 and 3.81). But Pullman was not ready to make any commitment for a wholesale conversion to streamline equipment.

The reasons were more economic than technical. After all, the existing fleet of heavyweights was adequate for Depression-level traffic. The cars themselves were solid, safe, and comfortable, albeit old-fashioned (Figure 3.82). The cost of a new car was calculated at $75,000 and up, whereas a standard sleeper could be built for $30,000.[287] But the railroads began to join the streamliner bandwagon, and streamliners proved to be profitable traffic generators. Then, as discussed in the previous section, Pullman too began to buy more lightweight cars.

In the same years that Pullman was working to perfect a practical room-style car and lightweight construction, it was also attempting to design a cheap sleeper. The tourist sleeper was still used in the West, but it never gained much of a hold elsewhere in the country. Room cars were driving sleeping-car rates up. Based on lower-berth rates, a roomette cost 140 percent; a compartment cost 200 percent, and a drawing room cost 250 percent. Commercial travelers and the affluent did not protest, and Pullman foresaw even greater profits as more open-section cars were retired. But there was still a market for the economy-class overnight traveler: students, retired people, and just plain poor people.

As early as 1925 Pullman tested a three-tier budget sleeper. In 1940 the coach-sleeper scheme wes revived.[288] Ten compartments with face-to-face seating were placed along one side of the car. A narrow aisle ran on the other side. The arrangement was like a European sleeper, except that the compartments had no ends. Curtains were used at night. The seat formed the lower berth; the middle berth was raised up against the top bunk during the day and pulled down at night; the top berth was stationary. Each unit had a washbasin, but no toilet. The first two cars that were converted to this plan could sleep 45 persons: five compartments slept three each, and the other five slept six each. A second pair of coach-sleepers, fitted in 1941, slept only 42. Because of their large capacity, it was necessary to charge only $5 above the regular coach fare on trips as long as Chicago to Seattle. The coach sleepers seemed successful and certainly appeared destined for wide use after the war. The famous troop

Figure 3.78   In 1933 Pullman produced its first lightweight car. The aluminum-bodied sleeper, buffet, and observation car was used mainly on the Santa Fe. It was scrapped in 1964 after standing idle for many years. (Arthur D. Dubin)

Figure 3.79   The Forward, 1936, was the first of what might be called a conventional lightweight sleeping car. Except for its room arrangement, it set a pattern for Pullman's lightweight fleet. The Forward was remodeled as a baggage-dormitory car in 1963. (Arthur D. Dubin)

Figure 3.80   Pullman built the Paria and six sister cars in 1938 for service on the Santa Fe. The car, which had seventeen Roomettes and a single open section for the porter, was remodeled into a coach in 1960. (Pullman Neg. 41575)

sleepers of World War II gave the country a large-scale demonstration of what basic overnight service might be like. Pullman constructed 1,200 of these boxcar-like sleepers on an emergency basis. The chief criticism of the coach-sleeper was the lack of privacy, and the troop sleepers had not only this failing but about every other discomfort that the discerning traveler tries to avoid.

The strong trend in passenger car design after the war was toward greater luxury. Even so, the major car builders came forward with fresh proposals for a coach sleeper.[289] Budd offered a 32-passenger design called the "Budgette," a poor man's duplex with double bedrooms and roomettes. The American Car and Foundry Company created a "Slumberliner," while Pullman came forward with a 30-passenger "Railotel." None of these de-

Figure 3.81   The Shenandoah *was originally built for the Chesapeake and Ohio in 1950. The six-bedroom, ten-Roomette sleeper was sold to the B & O in 1957.*

Figure 3.82   Pullman remodeled many of its heavyweight cars after the coming of the lightweight car. Some were remodeled into room cars, and all were air-conditioned. The Clover Bluff *was built in 1910 as the* Green Tree; *it was rebuilt and renamed in 1935. (Pullman Neg. 39275)*

Figure 3.83   The slumbercoach was an economy duplex-Roomette style of sleeping car introduced in 1958 that could accommodate 40 passengers. The Slumberland *was built by Budd in 1958.*

signs were carried beyond the mock-up stage, and the idea remained dormant for several years.

If Budd never controlled the passenger car-building trade, it did become the leading innovator. The company never let any attractive idea disappear without a trial. In 1953 Budd was ready to reintroduce the coach sleeper. The slumbercoach was a stripped-down duplex-roomette with twenty-four single and eight double compartments (Figure 3.83). It could sleep 40 passengers, surely an all-time record for an all-room car, and something of a benchmark in the history of space-saving design. The beds were small (24 by 73 inches), as were the windows. But the same general luxuries offered by the regular roomette were found in the Slumbercoach.

The first cars of this type entered Pullman service under the sponsorship of the Burlington in October 1956.[290] Cars were run on the Denver Zephyr and the City of Denver. A single room cost $7.50 more than coach fare, a double room $13.50. In 1958 and 1959 the Baltimore and Ohio, the Northern Pacific, and the New York Central bought another fourteen Slumbercoaches for

Pullman service. These cars represented final efforts both to perfect a cheap sleeper and to introduce a major design reform in sleeping-car arrangement. The technical problems were admirably solved, even to the costly and involved wiring, waste, and water lines needed to service so many individual rooms. It was the general collapse of long-distance railroad travel that cut short the Slumbercoach's wide acceptance.

The sleeping car's future is now a matter to be resolved by Amtrak. Initially the firm was forced to make do with existing equipment, but large capital-improvement funding (over 400 million dollars) was provided by Congress beginning in 1973. Orders were eventually placed with Budd and Pullman for nearly 800 new cars. The majority of these were coaches, but 70 new bilevel sleepers were included. The basic design follows the Santa Fe hi-level cars; however, no single-room accommodations are included. First-class and economy-fare double bedrooms are planned for the first new sleeping cars built for United States service in about twenty years. Appropriately, Pullman will build this equipment; deliveries are expected to begin late in 1977.

CHAPTER FOUR

First-Class Travel

PARLOR, DINING, AND PRIVATE CARS

Parlor Cars

THE LUXURY DAY CAR was an embarrassing phenomenon in democratic America, which claimed to have a single class of railway travel for a single class of citizen. This equalitarian boast became ever more hollow in the nineteenth century. Some apologists attempted to explain the parlor car as not really anti-republican—as merely a more comfortable style of day coach. But its patrons knew better; the parlor was intended for those who were willing to pay for exclusiveness. The timid and the sensitive found the common mixed-class cars filled with a boisterous crew who seemed to "drink whisky and eat onions simultaneously."[1] Besides those who were nervous of their fellow citizens, there were people who simply enjoyed marching into the svelte, club-like interior of a parlor car while a platform full of coach riders looked on. They could thus buy distinction, and in a nation of new-rich, many were buying.

The extra-fare passenger received a single armchair, a window, and the services of a porter. In terms of real comfort this was little more than that provided by a good coach. Certainly there was no comparison with the difference between a coach seat and a sleeping berth. Sitting up all night in a parlor car was much less comfortable than reclining in a bed, yet the price of a parlor seat was nearly equal to that of a berth. In the nineteenth century a parlor car ticket was generally a flat $1 above the coach price, while in more recent times it has been about 50 percent higher than the coach fare. However, many people have considered it a good buy. In the 1870s one British tourist wrote: "By paying an extra dollar for about every 200 miles, you can have a seat in a luxurious saloon, with sofas, armchairs, mirrors and washing rooms—and the inevitable spittoons."[2]

Parlors were not found on every train and every railroad. They were a relatively rare class of car even compared with sleepers or diners. They never represented much more than 2.5 percent of the total passenger car fleet. Most of them operated in what has come to be called the Northeast corridor, where Boston, New Haven, New York, Philadelphia, and Washington are the principal terminals in a chain of cities. Smaller parlor operations linked such close-lying centers as Chicago–Milwaukee and San Francisco–Los Angeles. State capitals with major satellites, like Chicago–Springfield or Albany–New York City, were connected in order to facilitate the comings and goings of politicians and lobbyists. Affluent resort areas could also support parlor operations, at least during the season. Florida's Jacksonville–Miami–Key West and New England's Boston–Bar Harbor are examples. Thus short runs, usually not over 300 miles in length, between major cities were characteristic of nearly all parlor car operations.

Although parlors were not a noticeable feature of American trains before 1875, many examples can be found. As early as 1845 the Eastern Railroad purchased several fancy cars in which "each seat is a separate arm chair, made to turn on a pivot."[3] Clearly this was parlor car seating, but the fact that each car accommodated 70 persons makes it doubtful that the cars were meant for the exclusive traveler, and it is not stated that an extra fare was charged. Late in the summer of 1853 the Hudson River Railroad began operating the first of several saloon cars built by Eaton and Gilbert.[4] The 45-foot-long body was divided into five staterooms, four of which were 8 feet square. Each had a sofa, chairs, a center table, and a splendid looking-glass. The paneling was ornamented with landscape paintings; the ceiling was hung in silk. At one end of the car was a washroom attended by a chambermaid. A side corridor formed a passageway. The sample car was said to be "a most magnificent fixture—combining sociability and comfort to an extent never before approached in railroad arrangements." The road planned to order another twenty saloon cars if the car became popular, but no record can be found of the success or failure of this experiment. Sometime in the same decade the Rensselaer and Saratoga Railroad introduced a similar car, which a German engineer saw during a visit in 1859. A sketch of the floor plan made by the engineer shows that the car had six compartments, face-to-face seating, tables, chairs, and a 48-foot body (Figure 4.1).[5] Several years later the *American Railroad Journal* noted that this car was very comfortable and had proved a great favorite with families journeying to the spa, but said that such cars were not widely patronized because "they had an appearance of exclusiveness."[6]

During the 1860s parlor and drawing-room cars began to appear on many railroads in the Northeastern and Great Lakes states. In November 1863 the B & O, the P W & B, and the New Jersey lines began to offer through service between New York and Washington. The P W & B provided three stateroom cars to inaugurate this service, but no details on them have been uncovered. A long description of the successors to the first lot of cars does exist, however. The last of four new parlors was portrayed by a writer in 1869:[7]

A description of this car will answer for all. It is fifty-six feet long and nine and half feet wide, the exterior painted and varnished light yellow, the color agreeably relieved by the bright hues of the lettering. The exterior of the car presents a handsome appearance by the graceful outlines of the mouldings on the sides and at the ovals of the windows. On entering the car at either end the passenger steps into an ante-room or vestibule, separated from the main portion by a dark walnut door, the upper half of which is of ground glass, worked in elegant patterns. Each vestibule contains four roomy arm chairs cushioned with dark velvet, marble washing basins with silver faucets, and the private closets. These apartments are intended for the use of smokers, who will appreciate the provision which enables them to smoke a cigar without being driven into what is usually the most uncomfortable and crowded car on a train. The main apartment of the car contains thirty-seven arm-chairs in double rows on either side of the central aisle, and by merely working a wheel and screw, set below the seat of the chair in the framework, the occupant can adjust it at any incline. When the back is adjusted vertically the chair is an ordinary arm-chair, but by turning the wheel back is lowered and from under the chair a rest for the feet is thrown out, it then becoming a cozy and luxurious arrangement for sleeping. The cushions are of crimson velvet, the wood work of dark walnut grained and polished, and the metal work heavily silver plated. The floor of the car is grained and polished in light yellow, the

Figure 4.1   The earliest known illustration of an American parlor-stateroom car is this Rensselaer and Saratoga Railroad floor plan of about 1859. (A. Bendel, Aufsätze Eisenbahnwesen in Nord-Amerika, Berlin, 1862, plate E)

Figure 4.2   The Boston and Providence operated an English-style parlor car in 1866 on some of its boat trains. Side-door cars remained a rarity in America. (Railway and Locomotive Historical Society)

aisle is carpeted with a neat pattern of velvet carpet, and the interior walls are of dark yellow walnut and light yellow satin wood, each polished in a beautiful manner. The ceiling is of a pure white material, worked out with tracery of blue and gold in graceful designs. The patent ventilators above the roof are of ground glass, and all the metallic finishings are plated with silver. The artistic combination and contrast of colors on the interior of the car, the dark walnut, the brilliant satin wood and crimson velvet, the white, blue and gold of the ceiling and the sparkle of the metal, all form a beautiful and attractive spectacle.

Meanwhile New England railroads were introducing their first parlors. No lusher market could be found than aristocratic Boston and its dignified citizens, who regularly took the boat trains to Stonington or Fall River for excursions to less civilized regions. In 1866 the Old Colony Railroad placed two English-style first-class cars in service on its Fall River line trains.[8] The idea was said to be that of the railroad president's son, who had just returned from Great Britain a confirmed Anglophile. He was convinced that compartment cars were the perfect form for first-class day service. Two were manufactured in the Old Colony shops. They were novel enough in appearance to be featured in the pages of *Leslie's* illustrated newspaper. Eventually two more were placed in service.

The competing Boston and Providence decided that it must have an English saloon car for its Stonington line. The road ordered a 52-foot, six-compartment car from the Salem Car Company, which produced a $14,000 round-end beauty that weighed

14 tons (Figure 4.2).[9] The English cars soon fell into disfavor, however. Passengers disliked being locked into the side-door compartments. The lack of toilets was a nuisance to passengers and crew alike. The conductors feared the hazardous outside running boards when they had to pass from one car to the next. One or more of the Old Colony cars were reduced to a mass of broken timbers in the Wollaston, Massachusetts, wreck of 1878.[10] The B & P car had a less tragic ending: after standing in storage for some years, it was sold for use as a seaside café in 1885.

Parlor car experiments continued in New England. In 1868 the Boston and New York express line introduced two handsome compartment drawing-room cars, the *City of Boston* and the *City of New York*.[11] Built by Wason, they had 45-foot bodies divided into six staterooms, each with accommodations for seven passengers. The seat fee was $1, or an entire compartment could be secured for $6 above the regular fare. The interior was paneled in walnut and cedar and outfitted in the lavish style of the period. Hot-water heating was a new luxury. The exterior was painted in dark green and drab, with maroon panels picked out in blue. The governor of Connecticut, railroad officials, and other prominent men were aboard for the first trip.

An awareness of luxury day cars seemed to spread slowly in some quarters. It is claimed that the New York Central did not consider using them until a vacationing director chanced upon one of the Old Colony English cars during a visit to Newport.[12] Apparently the 1853 Hudson River saloon cars were already for-

Figure 4.3   A conventional parlor interior, dating from about 1875, is shown in this photograph of a Wagner-operated car. (Railway and Locomotive Historical Society)

Figure 4.4   The Garden City, a progressive style of drawing-room car operated on the Lake Shore and Michigan Southern in 1867. It illustrates the advanced state of the palace cars not associated with Pullman. (New York Central Neg. 357)

gotten. In all events, operating officials of the Central were summoned to Boston to examine the latest marvel in passenger comfort. They acknowledged the merits of the scheme, but the Central's superintendent, James Tillinghast, objected to the compartment plan and insisted on an open, coach-like style of car. The project was given to Webster Wagner, who introduced his first parlor in August 1867.[13] Among his earliest cars in this style were the Catskill and the Highlander, which were described as having two staterooms and a large, open central compartment with twenty swivel armchairs. In addition to the usual end doors, they featured a center entrance.[14] Wagner was operating through prime parlor car territory. Albany to New York City was an ideal run for such equipment. Patronage developed quickly, and by 1876 sixty Wagner drawing rooms were in operation. The appearance of a Wagner car of this period was preserved on a stereoptican card, reproduced here as Figure 4.3.

Figure 4.5   A large advertising broadside issued by the Atlantic and Great Western around 1865 included this interior view of a drawing-room car. (Chicago Historical Society)

At the same time that the New York Central was experimenting with parlors, the Michigan Southern began introducing the idea in the West. The Illinois Central had acquired some sleeping cars with revolving parlor seats as early as 1856, but the hybrid scheme was never really intended as a luxury day car. The Lake Shore cars appear to be the earliest true parlors to run out of Chicago. The first two were built in 1867 by John Kirby at the road's Adrian, Michigan, shops.[15] The *Garden City* and the *Forest City* cost $17,500 each—an incredible sum for that day. They had two drawing rooms at each end, with a gaslit central saloon. The interior trappings were furnished by a New York decorator. By chance a photograph of the *Garden City* has survived (Figure 4.4). The picture has been reproduced before, though usually incorrectly dated 1878. By 1870 the Lake Shore was running five parlor–drawing rooms with such names as *Central Park* and *White Pigeon* between Chicago and Cleveland.

The Erie appeared next in the growing number of parlor car operators. Actually, its Western connection, the Atlantic and Great Western, appears to have pioneered parlor service as early as 1865. A broadside of that time (now in the collection of the Chicago Historical Society) illustrates the benefits of travel on the broad gauge. One panel, reproduced here as Figure 4.5, depicts the interior of a compartment-type parlor car. It was not long before the Erie far outstripped its subsidiary in luxury daytime travel. In 1871 the road introduced seven of the most extravagantly decorated parlors ever built. The first four were named *James Fisk Jr., Morning Star, Evening Star,* and *Jay*

Gould. The work was done in the Erie's own shops, but even so the cost came to $35,000 each. The *Railroad Gazette* of March 25, 1871, gave the following description:

The Jay Gould is 56 feet by 11 feet; will seat fifty-six passengers; is lighted by Ganster's gas machine, heated with Baker, Smith & Co.'s patent car steam-heating apparatus, with pipes under each seat and sofa; has two water-closets, two wash-rooms in white marble, black walnut, maple, rosewood and gilt; a ladies' lavatory; ice-water tanks and attachments; five reflecting gasaliers suspended from the ceiling, trimmed with silver, by Geo. H. Keitching & Co. The interior of the car is trimmed with silver, wherever metal is used, including arms, hinges, rails, knobs, hat-racks, clothes-hooks, brackets, window fastenings, door-slides, etc., by Newman & Capron, of New York.

The bridal chamber has velvet Brussels carpet; the furniture is covered with drab medalions of crimson scarlet, white and rose colors; drab silk curtains, intermixed with blue and gold, bought of Stewart; sofa beds with easy pillows and bolsters, chairs on revolutionary hinges; beds with improved spiral steel spring mattresses, covered with crimson reps; beds raised by automatic or spiral springs, so complete that a child can raise or lower them. The ceilings were frescoed by Sebastian Brodt.

The car is trimmed with French moquette, and green, drab, and gold crimson are the prevailing colors. The windows are of best French plate glass—one large or center one and two smaller or movable ones—the whole of etched glass by Downing, and in all the panels is the Goddess of Music, with a lyre etched in the glass. The wood used is black walnut, ash-bars, black walnut panels and rosewood mouldings, inlaid with gold, the two former coming from the West. There is a

Figure 4.6   In 1874 Pullman exported this American-style parlor car for service in England. The drawing is representative of standard Pullman design except for the wooden-center wheels. (Engineering, April 2, 1875)

mirror in each door and panel. The seats are peculiarly adapted for ease and comfort, being upholstered at the ends, back, and extra or lounging back with French moquette of green, drab, crimson and gold. The cushions are on the best of steel spiral springs and are adjustable; arms silvered. The upper berths are self-acting. The busts of Mr. Gould, inside, are carved of white holly, the Erie Company doing all the carving of the car at a cost of $4,000 each. The portraits in oil of Mr. Gould on each side of the outside of the car will attract much attention. The oil paintings in the inside or frescoed panels are representations of scenery along the line of the Erie Railway. Chairs are hung with silver slipjoint hinges. The ventilation is very complete and is formed by wire screens in the roof over each platform, and from this point is conducted through the car, being increased or lessened by the use of beautiful glass-paneled swivel ventilators in the roof of the car. The wire screens prevent cinders from going in. The electric or magnetic brake is used; Miller's platform buffer and bumper; six-wheel trucks with steel tires on wheels, the steel and iron being cast solid together; eight guard chains from trucks to car body, and elliptic and rubber springs combined. The several departments of the company's workshops are so complete that these cars in all their elegance are made therein, embracing cabinet work, upholstering, painting, gilding, body trucks, etc. The Jay Gould weighs but twenty-five and one-half tons on a wide track, whereas a Pullman narrow-gauge car weighs twenty-eight to thirty tons. Commerce, railroad and shipping are represented on the outside corners of the car. None but the best selected timber was used. Mr. Gould gives a personal supervision over the workshops and employes in the construction of these and other cars.

By the early 1870s the parlor car was an established part of the American railway fleet. In addition to the examples already men-

tioned, small roads such as the North Pennsylvania and the Terre Haute and Indianapolis were reported to be acquiring luxurious vehicles of this type.[16]

Where was Pullman during the upsurge of interest in expanding luxury service? Historians, including Joseph Husband and Stanley Buder, have claimed that Pullman did not introduce his first parlor until 1875. Actually, however, he was running parlors as early as 1871 on the Great Western Railway of Canada.[17] Two cars named *Leo* and *Mars* were produced by the Taunton Car Company for the run between Hamilton, Ontario, and Suspension Bridge. Each 62-foot body was divided into five compartments. The central section seated 20, the two staterooms accommodated parties of 6 each, and the two end rooms were used as smokers. Enthusiastic press reports said that the great 27-by 48-inch windows suggested a crystal palace, and that the large openings would provide easy egress in the event of a smashup. These cars were considered something of a bargain at the advertised rate of 3 cents a mile. Three years later Pullman began operating parlors with reclining seats between New York and Washington. He placed a similar car on the Michigan Central in the summer of 1875.[18] Some months before, Pullman shipped the first American-style open parlor to England for service on the Midland Railway. Drawings of the *Victoria* published not long after the car entered service present the most detailed plans available for a parlor of that time (Figure 4.6).[19] Records uncovered in recent years by Arthur Dubin indicate that the *Victoria* and its companion parlors sent over later by Pullman were not newly constructed; they were existing cars that had been

Figure 4.7   Even some relatively small roads, such as the R W & O, found it profitable to operate parlor car service. The St. Lawrence *began service in 1878. (New York Central Neg. 719-3)*

Figure 4.8   A parlor car interior from an advertising booklet issued in 1876 by the Pennsylvania Railroad to stimulate travel to the U.S. Centennial Exhibition.

refurbished especially for overseas service. The *Victoria* itself had an extraordinary service life, running until 1932.[20]

In 1876 Pullman placed six parlors on the Pennsylvania's best trains out of Philadelphia. Described as "triumphs of elegant comfort and costly ornament," they were named *Hebe, Ariadne, Andromeda, Proserpine, Daphne,* and *Queen.* The *Queen* formed

part of Pullman's exhibit at the U.S. Centennial.[21] The major portion of the 48-foot bodies was given over to an open compartment, with a small drawing room at one end and toilets at the other. This became the typical floor plan for an American parlor car. The seating arrangement was not so typical: the chairs were placed in double rows on one side and in a single row on the opposite side. A total seating of 31 was thus achieved, but at the sacrifice of spaciousness, comfort, and exclusiveness—the commodities that parlor car patrons most desired. However, the chairs with reclining seats built to Wood's patent design were intended to mollify the disgruntled first-class traveler. The windows were arranged in the patterns of three that were popular at the time, with a large center glass flanked by two narrow panes.

By the end of the 1870s the parlor car was a customary part of the better trains. Pullman, Wagner, and Woodruff ran a complement of such cars in their concession fleets. If theirs are counted along with the railroad-operated cars, it is probably safe to estimate that between two and three hundred parlors were in service on U.S. railroads by 1879 (see Figures 4.7 and 4.8). Curiously, the term "parlor car" was not widely accepted by this date, for the *Car Builders' Dictionary* of that year continued to list only the "drawing-room car."

As the parlor car proliferated during the next decade, so did the number of its critics. To some, parlors represented an unconscionable swindle that offered doubtful comforts at high prices. Others contended that it was all a conspiracy engineered by rapacious railroad tycoons who ran dingy coaches to force passengers to buy tickets for extra-fare cars. The first school argued that the parlor was nothing more than a common day coach with oversized windows and a single row of chairs.[22] The chairs were uncomfortable, with convex stuffed seats and backs and poorly placed armrests. The big windows were wasted, for half of the seats were aligned with posts. As one writer put it, windows are like toothbrushes; each passenger should have his own.

Figure 4.9 *The parlor car* Brandywine *was built in 1885 by the Baltimore and Ohio Railroad's Mt. Clare Shops in Baltimore.* (National Car Builder, *December 1885*)

Figure 4.10 *The* Lavinia *was produced for the B & O by Jackson and Sharp in 1886. It is very similar to the* Brandywine *and may have been constructed from drawings furnished by the railroad.* (Hall of Records, Delaware Archives)

This critic was annoyed by the absence of luggage and coat racks and totally enraged by the miserable parlor that he and several other passengers were put aboard during a change of cars in Philadelphia. Contributed by the New Haven for the cooperative through-car fleet, it was a dusty relic in deplorable condition. The window-sash latches were missing, it smelled worse than a fishing wharf, and the stove smoked. Perhaps, the writer suggested, the New Haven thought that the car was good enough for Yankee travelers; or perhaps the road thought it *not* good enough and had therefore sent it south.

Even when the New Haven put some new cars on its routes— and they were very much needed, according to the *Railroad Gazette*—the fault-finding continued. The bodies were duplicates of cars intended for coach service, so that while the big-window humbug was avoided, the seats and window openings did not correspond.[23] The seats, however, had been improved. The backs were not too high, and they were actually shaped to fit the human spine. The top was rolled back, offering a support for the head or neck. The seat cushion was tilted back so that the occupant could settle in. And the back was not too high, a common fault with most parlor chairs. The neighboring Boston and Albany's efforts at parlor car reform were mentioned in the same article, which praised the plainness and lightness of the cars. Car Number 208 seated 30 passengers, yet weighed only 45,275 pounds—some 10 tons less than a Wagner drawing-room car. The article described the rattan seats as a refreshing improvement for summer travel, though it said that the backs were too perpendicular.

During the years when it operated its own sleeping and parlor cars, the B & O made an effort to acquire rolling stock equal to

Figure 4.11   Interior of a parlor car, very likely the Lavinia, *built in 1886 by Jackson and Sharp for the B & O. (Hall of Records)*

Pullman's best. The parlor car *Brandywine* (Number 544), constructed at the Mt. Clare shops in 1885, had nineteen pedestal armchairs in the main saloon that were carefully arranged to correspond with the window openings (Figure 4.9).[24] The 35-ton car measured 65 feet in overall length. A heavy silver-plated handrail was placed in front of each large window. The small buffet was outfitted to serve coffee, tea, and other light refreshments. A very similar car, the *Lavinia*, was produced a year later by Jackson and Sharp. Surviving photographs of this car, coupled with the *Brandywine* drawing, give a good general impression of a parlor car of the time (Figures 4.10, 4.11, and 4.12).

A major attraction of daylight travel is the view. Watching the passing scenery is relaxing, amusing, and free. The designers of parlor cars were deeply concerned with ways to improve the view. By the eighties the big window had been thoroughly exploited, but bay windows, which were then popular in domestic architecture, seemed to offer another approach. In the world of fashion true solutions are not really necessary; novelty is often

enough. And so began the bay-window parlor car craze. Actually the term "bay window" was an exaggeration: the windows were set at a slight angle, say 15 or 20 degrees, so that the sash could be made relatively small and would not invade the car's floor space. Even so, the window enabled the passenger to look ahead much more comfortably than he could through a conventional window opening.

In 1884 the Pennsylvania built a sample bay-window parlor at its Altoona shops.[25] Number 901, which seated 37 passengers in its 62-foot body, was successful enough that more bay windows (called class CA) were produced at Altoona (Figures 4.13 and 4.14). The Boston and Albany caught the bay-window fever and ordered Wason to construct the finest car possible on this plan.[26] Bruce Price, the New York designer, and L. C. Hyde of Wason's drafting department took over the design commission. Price claimed the bay-window design in general, but Hyde's lunette window plan was used. The darkly elegant interior was considered remarkable—nothing less than "an art study." The

Figure 4.12   A Baltimore and Ohio parlor car of about 1885. Note the foot cushions and the white cloth arm- and headrest covers.

PARLOR CAR.

CA

SEATING CAPACITY 39 PERSONS
6 WHEEL TRUCK 13/00 LBS
WEIHGT 65300 LBS

Figure 4.13   A class CA parlor car built in 1884 by the Pennsylvania Railroad's Altoona shops.

Figure 4.14   A floor plan of a bay-window parlor car constructed by the Pennsylvania in 1884. (Railroad Gazette, August 29, 1884)

Railroad Gazette was sufficiently impressed to devote a half page to an engraving of the interior (Figure 4.15). The B & O and Pullman soon followed the fashion. The B & O put a bay-windowed parlor on its plush Royal Limited. Pullman's Golden Gate Special featured a combination observation-parlor-sleeper in which the parlor compartment had bay windows.

In the background of this activity was a patent case doggedly pushed by a poor trainman named W. K. Tubman, who had somehow purchased George S. Roberts's 1877 patent on railway car windows.[27] Tubman claimed that it was the basis for all the fashionable bay windows then in service. During the suit, the defense was able to show a long array of prior examples. Between 1852 and 1866 several patents had been issued that envisioned some form of bay window protruding outside of the car body.[28] The Erie, the New Haven, and a few Western lines were said to have tested one or more of these schemes. Tubman claimed that Roberts's patent was the first to have the bays *inside* a normal car body's outer limits. Roberts's plan, however, resembled the bay window then in use only in the most general sense, for his had a bold zigzag system of interior panels that would have reached 2 or more feet into the interior, whereas the windows in the current designs were set on modest angles that extended only a few inches inside the car. Tubman appears to have had little success with his case against the railroads, and the entire fad seems to have fallen from favor rather quickly. Pullman exhibited a parlor with bay windows at the 1893 Exposition, but that appears to have been the end of it.

In addition to the early examples presented during the course of the Tubman trial, there were some lantern-window cars operated over the New York and Harlem Railroad and the Boston–New York express line that the German engineer Bendel saw during his visit of 1859.[29] He was so impressed by the unique sheet-metal appendages that he included a sketch of them in his published report. They protruded outside the car body from the window openings, but because they were so small there was supposedly no danger of collision with opposing trains. To see out, it was necessary for the passenger to duck his head inside the lantern. The lanterns required constant attention; one or more panes had to be replaced every day. The New York–Boston cars were soon retired, and one of them served long afterwards as a tiny match factory at Saybrook Junction.[30]

During the bay-window era an even more radical parlor car plan was introduced by the Monarch Sleeping Car Company. It featured open galleries at the body's center, with observation rooms at each end. The scheme was reminiscent of the private car built in 1860 for the Caliph of Egypt by Wason. It also resembled the royal train of cars built for Napoleon III at about the same period. Monarch, however, enclosed the midsection so that passengers could choose between the breezy outside gallery and the snug central compartment. The drawings show the arrangement more clearly; note the half round end windows (Figures 4.16 and 4.17). More details on this unique sight-seeing parlor are given in an account from the *Troy Daily Times*, June 15, 1886:

There is being constructed at the Gilbert car works, Green Island, a novel and magnificent observation car, which is to be used between New York and the White Mountains. The car is constructed on a plan different from that of other observation cars. The coach is about sixty-five feet long, and it is supported upon trucks which have twelve large-sized wheels. The exterior of the car is of a "silver flake" color, trimmed in old gold. It will glitter and sparkle in the sunlight with a dazzling brilliancy. At each end of the car is a drawing-room, about fifteen feet long and the full width of the car. The ends of the car are rounded, and are of French plate glass. The sides are of glass also, which will enable the occupants to have an unobstructed view of the country through which the car will pass. The interior will be fitted up in magnificent style. The floor is of tile. The ceiling is covered with beautiful pearl-tinted Lincrusta-Walton. The small spaces between the windows are also covered with the same material. At the ends of the drawing-rooms are large French mirrors. The rooms will be supplied with easy-chairs. Between the drawing-rooms are three compartments, that will each accommodate four or five persons. These compartments are narrower than the drawing-rooms, which leaves a platform about two feet and a half wide on each side of the car, running from one drawing-room to the other. On the outer edge of this platform will be a silver railing, about four feet high, which alone will cost $1,000. Doors from the smaller rooms open upon the platforms, so that the excursionist can walk out upon the platforms at pleasure. Doors from each side of the drawing-rooms also open upon the platform. The small rooms are fitted with the same elegance as the drawing-rooms. The sides are almost entirely of plate glass. Large mirrors are arranged at each end of the rooms, while from the ceilings are suspended handsome silver chandeliers. There is also a room fitted up for a buffet and there are toilet rooms in the car. A galvanized-iron tube runs along the whole length of the car on top, furnishing ventilation by means of small openings from the ceilings of each room. The railings around the platforms at the ends of the car are all silver plated. The car is supplied with the latest improved platforms and air brakes. On each side of the drawing-rooms appears the name "Ymir" in old gold letters, and along the top of the car the words in old gold, "Monarch parlor-observation car." The coach will cost, when completed, about $16,000, and will be ready for use about July 4.

It is not known whether Gilbert built other cars like the *Ymir*, though it seems improbable. After Monarch suspended operations, the *Ymir* came into the possession of the Jackson and Sharp Car Company, where it was refitted as a demonstration car for

Figure 4.15   The Wason Car Works of Springfield, Massachusetts, produced this magnificent arched-roof parlor-drawing room car for the Boston and Albany in 1887. (Railroad Gazette, October 21, 1887)

Section and Side Elevation.

Plan.

Figure 4.16 The Gilbert Car Company of Troy, New York, built an unusual parlor-observation car for the Monarch sleeping car company in 1886. (National Car Builder, *September 1886*)

Figure 4.17 The Monarch parlor-observation car around 1895, when it was being used to demonstrate a new system of electric lighting. (Lancaster County Historical Society)

the Moskowitz electric car-lighting company and was occasion-ally used by Job H. Jackson for private excursions.

By the late 1880s the parlor car had developed a multiple personality. It was no longer an extra-fare day coach with swivel seats and a drawing room; it was now expected to serve as a lounge, a buffet, an observation car, a library, and even a gather-ing place for restless travelers who were weary of their assigned seats. Under the pressure of these demands the parlor took three lines of development. In one direction it evolved into the chair car—a supercoach that often, but not always, had a double row of reclining parlor-car-like seats. In another, the traditional par-

lor car continued as before. In the third direction the lounge car, whose space was not for sale, emerged as a special luxury car set aside for the free use of first-class passengers.

In the straight parlor, the basic plan remained the same. The car grew in size, changed in decor with changing fashions, and made successive shifts from wood to steel and finally to light-weight construction, but the standard plan of 1875 was identical with that of all later generations (Figure 4.18). The last word in parlor design can be seen in the Metroliner's Metroclub cars. The term "parlor" was considered too old-fashioned to use for them, since it might conjure up a mental picture of a fussy, overstuffed relic of the Victorian age. Yet the Metroclub is a traditional

Figure 4.18   The timeless appearance of the parlor car's interior is illustrated by this B & O streamliner of 1935. (American Car and Foundry Company)

parlor car, and its rows of high-backed armchairs on swivel bases are alternately comforting, disappointing, or amusing, depending on your viewpoint.

The chair car might be called the poor man's parlor. It falls somewhere between the better day coach and the extra-fare parlor. It is analogous to the tourist sleeper as a railroad effort to offer a cheap competitor to the luxury cars run by Pullman and Wagner. It represents an effort by certain railroads to take trade away from the concessionaires or to offer some attractive accommodation not available on a rival line. The chair car was aimed at the parlor car market and at those coach passengers who wanted something better than a stiff-backed double seat. The big attraction was added luxury at no extra charge. Each passenger had a reclining chair that could be turned toward the scenery or the other way to converse with a neighbor across the aisle. At night it could be dropped back like a sofa, and while it was not as good as a berth, it was adequate for a snooze.

To provide this added comfort without extra cost, it was necessary to double up the number of seats. Four rather than two rows were common in the chair car. The legroom and exclusiveness of the parlor was missing, but so was the added tariff. The reclining chair basic to this style of coach can be traced back to the early 1850s. The inventors of that time, however, were preoccupied with perfecting a sleeping chair, and it seems to have been viewed as such over the next twenty years. The shortcomings of

the reclining chair when compared with a sleeping-car berth, at least for long-distance travel, are obvious. But for day journeys or short trips it had merit, and in the 1870s the reclining chair was again brought forward. The Chicago and Alton was among the first to push it. In 1877 the road began to run some handsome cars with Horton reclining seats.[31] Regarded as the most scientifically designed of all the new chairs, they were the product of a Kansas City doctor, N. N. Horton. A somewhat stylized engraving of the interior of a Horton car was used in an Alton advertisement of the period (Figure 4.19). Excellent drawings of a later Horton car produced at the road's Bloomington shops are shown in Figures 4.20 and 4.21. In 1882 twelve were in service and two more were under construction.[32] The short bodies carried between 32 and 40 seats. Chair cars gained favor on many railroads in addition to the Alton. They were a compromise between day coach, parlor, and sleeper that seemed to satisfy the class of traveler who would not or could not go first class.

The most exotic of what might be called the parlor's offshoots was the lounge car. Its arrangement and functions were so diverse that it cannot be assigned exclusively to the parlor family, but it fits no other category better. The lounge was the last of the major passenger car types to appear. It developed with the fast luxury trains that were not introduced in this country until the 1880s. A few express trains were operating a decade earlier, and they did include some luxurious sleeping and dining cars. Ac-

INTERIOR VIEW OF RECLINING CHAIR CAR, (DR. HORTON'S PATENT).

These Luxurious Cars are run in All Through Trains, Day and Night, between

St. Louis and Kansas City, St. Louis and Chicago, and Kansas City and Chicago, Without Change,

—VIA—

Chicago & Alton R. R.

NO EXTRA CHARGE FOR SEATS IN THESE CARS.

Figure 4.19   Reclining chair cars offered parlor car luxury at coach fares. The illustration dates from 1882.

Figure 4.20   Chair car Number 103 was built at the Bloomington, Illinois shops of the Chicago and Alton in 1882 to the design of William Wilson, superintendent of machinery. (National Car Builder, *June 1882*)

cording to the *Railroad Gazette*, as early as 1882 the Pennsylvania was running a parlor-library car on its New York–Chicago express.[33] But such luxuries were rare; most express trains were a mixture of new and old, common and palace. Passengers were essentially isolated in one car. Because of the open platforms, it was dangerous to pass from car to car when the train was in motion. Most roads forbade passengers to do so, and some locked the doors to prevent it.

In 1887 came the luxury train, a unified series of cars designed to serve travelers' needs in the same way that a hotel or a transatlantic liner does.[34] And it was a single establishment, for the cars were linked together with vestibules. Passengers might walk safely through the entire train. If it was stuffy in a sleeper, they could go to another car. If the diner smelled of fish, they could retreat without waiting until the next station stop. The new trains were extra-fast, extra-plush, extra-exclusive, and extra-fare. They were successful as well, and by 1890 more than a dozen Flyers, Limiteds, and Expresses were in service. The name train had arrived, marking a turning point in American railway travel.

It became the function of the lounge car to provide an inviting destination for passengers who were walking around. Strolling through the other cars won only the cold stares of their occupants. The lounge, on the other hand, was a clubroom where anyone might sit (Figures 4.28 and 4.29). Its existence was the result of the vestibule train, and its attractions were the devices of the ages for people with time to kill. Comfortable armchairs, sofas, and side tables were the customary furnishings. If the scenery or a cigar proved dull, there was a small library of popular classics for diversion. A barber shop with a full-sized chair provided another service, which businessmen appreciated as an efficient way to save time that was going to waste anyway (Figure 4.30).

All these features were available aboard what was probably the first true lounge car, the *Esperanza*.[35] It was completed in April 1887 by Pullman for the Pennsylvania Special. Bathtubs became a stock feature the next year, and by the mid-1920s showers were being installed.[36] The Seaboard Air Line went all the way by converting a club car into a traveling gymnasium that

Framing.   Inside Finish.   Side Elevation.

Floor Framing.   Floor.   Roof.

Figure 4.21   A general arrangement of the chair car illustrated in the previous engraving.
(National Car Builder, June 1882)

Figure 4.22   Barney and Smith of Dayton, Ohio, built this parlor car for the Cleveland,
Columbus, Cincinnati, and St. Louis Railroad in 1883. The vestibule ends were added
around 1890. (Railway and Locomotive Historical Society)

CHAIR CAR,

MISSOURI PACIFIC RAILWAY COMPANY. FRANK REARDEN, SUPERINTENDENT CAR DEPARTMENT.

Figure 4.23 A reclining chair car produced by the Missouri Pacific shops in 1893. (Railroad Car Journal, April 1893)

Figure 4.24   The Santa Maria, *built by Pullman for display at the 1893 Columbian Exposition, was perhaps the most elaborately decorated parlor car ever constructed. (Arthur D. Dubin)*

featured mechanical horses, a punching bag, and a small swimming pool.[37] Radios were also installed, and in later years movies and television were available.

Because they were nonrevenue vehicles, full lounge cars were rare. The seats were not sold; they were free to all passengers who had bought space elsewhere on the train. The earliest lounges were usually combined with a baggage area. Later a portion of the car was given over to bedrooms, open berth sections, or buffets. Light food and beverage service was a natural and frequent combination.

The return of alcoholic drinks in 1933 led to the club car, where a bar took the place of the buffet. The library quiet of the original lounge car gave way to a more tavern-like conviviality. It started modestly at first, with little more than an extra shelf for liquor, but full bars were soon installed. Some became flamboyant horseshoe affairs capable of producing both the atmo-

sphere and the full range of drinks offered by a smart Manhattan cocktail lounge (Figure 4.31).

The observation platform was another attraction that became an expected part of the name train's consist. The polished brass railings, drum sign, and scalloped canopy awning added a festive ending to the train (Figure 4.32). Before the coming of the dome car, the observation platform offered the best view on the train. The panorama of the passing countryside was uninterrupted by window posts or curtains. There was something exciting about the ride as well—the clatter and roar, the swirling wind and dust, and the hypnotic sight of the tracks as they disappeared into a subtle triangle at the far horizon. And as the train passed through a small town, passengers could stand on the rear platform to be eyed by farm-bound locals who wished that they too were aboard the glamorous car flying toward Chicago.

The observation platform came to the lounge car after first

Figure 4.26   This arched-window parlor, a Pullman product of 1903, is representative of the last wooden cars built in the United States. (Pullman Neg. 7394)

Figure 4.25   The Chicago Great Western lounge parlor car built in 1898 by Pullman had one compartment outfitted as a private library. (Pullman Neg. 4182)

appearing on private or office cars as early as the 1850s. The idea was supposedly suggested to Pullman in 1869 by Joseph Becker, an artist on the staff of *Leslie's* magazine.[38] Becker spent many hours on the rear platform of the last car of a transcontinental train. The view was so impressive that he urged Pullman to outfit some cars with proper railings so that his regular patrons could enjoy the scenery in comfort. There is no evidence that such cars were actually built until sometime in the 1880s, perhaps because passengers could not walk through the train until after the introduction of the vestibule.

Apparently no observation-lounge cars were in service before the coming of the luxury trains. If this is true then the first observation car was probably the *Aladdin*, rebuilt in 1888 for the Golden Gate Special, a deluxe train operated over the Union and Central Pacific railroads.[39] In the following year Pullman replaced the *Aladdin* with a new car, the *Sybaris*, which featured a larger open platform that was stepped back a foot or so inside the car body. That is, the rear bulkhead was not flush with the end of the car's sides. This provided more space and shelter for the platform passenger. The end compartment of the car was arranged as a saloon, with easy chairs and large end windows. During bad weather it was still possible to look out from the comfort of this observation room. A few years later the view was improved by moving the door to one side and installing extra-large single-pane windows that extended almost from floor to ceiling. Some windows were 50 by 60 inches. The view was grand but the end framing was rather meager, especially for the rear car, and many roads preferred the less exciting but safer center-door plan.

In its first year the observation platform was only a daytime diversion, but methods were later devised to transform it into a nighttime attraction as well. In 1926 the Northern Pacific acquired ten new 85-ton observation-clubs from Pullman.[40] Each had a 250-watt searchlight mounted in the end of the roof. By means of an overhead wheel, the beam could be directed at

Figure 4.27   Pullman completed the parlor-cafe-observation Number 303 in 1911. (Pullman Neg. 13665)

Figure 4.28   Model passengers enjoy the view in this photograph made around 1925 aboard the B & O's Capitol Limited observation car Mountain Creek.

Figure 4.29 The solidity and dignity of a heavyweight-era lounge car is shown in a view of the B & O's Capital Bridge. *The photograph dates from 1925.*

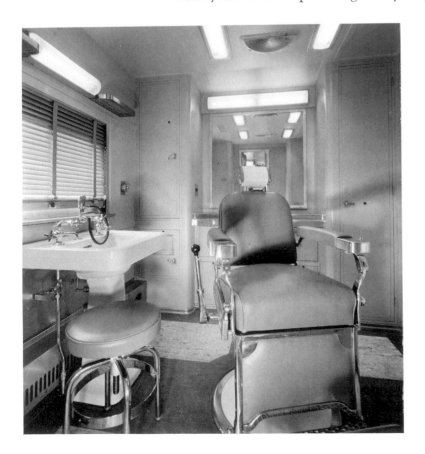

passing mountains and streams. In later years floodlights fixed under the rear platform effectively lighted the passing scene. The B & O also fixed a bank of lights on top of its dome cars for the pleasure of evening viewers.

Open-platform observations continued to be built until about 1930, but their popularity had begun to wane some years earlier. It was argued that they were uncomfortable and unsafe. No recorded instance of a passenger falling overboard can be found, but the hazard was real enough. The exposed platform was comfortable only in good weather, and even then it proved too windy or dusty for many patrons. Because every square inch of a passenger car represents precious floor space, the open platform was considered wasteful since it was so often unusable.

The solution was to enclose the end platform, and in 1909 the sun parlor or solarium car was born. The Burlington built several in its own shops and purchased more from Pullman.[41] The sun parlor did indeed increase passenger comfort and extend the usable floor space, but the view from it was rather limited. The Burlington liked the scheme, however, and acquired more sun-room observations. The Illinois Central became a convert in 1912. Pullman itself operated a number of sun-room cars that

Figure 4.30 Some first-class trains featured barber shops as early as the 1880s. This 1949 photograph shows such a facility incorporated in a Chicago and North Western lounge car. (American Car and Foundry Company)

Figure 4.31   Name-train lounge cars normally showed the latest thinking in interior design. This interior was created by Henry Dreyfuss in 1948 for the 20th Century Limited. (Henry Dreyfuss Associates)

Figure 4.32   The open observation platform offered a splendid vantage point for travelers. This scene was recorded on the Southern Pacific around 1925. (Southern Pacific)

Figure 4.33   A fully enclosed observation end, or sun-room, can be seen at the far end of this Burlington route lounge car. It was built in 1909 by Pullman. (Pullman Neg. 11790)

Figure 4.34   A sun-room observation car used on the B & O's National Limited in the 1920s.

Figure 4.35 Rounded-end observation cars were popular in the streamline era. The tavern-lounge car Franklin Roosevelt went into service in 1941. (New York Central Photo)

saw service on railroads in all parts of the country.[42] These boxed-in observations were not particularly attractive cars (Figures 4.33 and 4.34). They formed a blunt, inconclusive finish for a train, though this made them practical for runs where more cars were added en route.

In the style-conscious age of streamlining, the observation car's appearance was to undergo a radical change. The open-platform style was regarded as obsolete, while its blunt-end counterpart, the sun parlor, was discarded as just plain ugly. The observation car was the single form where the designer might readily display his talents. It is true that the car's exterior design was generally a source of frustration, for beyond smoothing its contours there was little opportunity to do much else. The observation end, however, could be molded into many shapes. The basic bullet or swallowtail configuration that came to be adopted after 1935 had been used some thirty-five years before by F. U. Adams. Adams's streamliner was a wooden prototype produced from a series of old B & O cars, but its general appearance was remarkably similar to those turned out by the aeroflow school. Appar-

ently Pullman was the first to resurrect Adams's plan, for use in the two aluminum observation cars which the company built for the 1933 World's Fair. The automotive lightweights produced by various designers a year or two afterward featured many strange ideas for the shape of the train's last car, but the graceful swallowtail plan won the day. The design that had seemed so startling at first soon became a cliché (Figure 4.35).

A notable exception to this prevailing style was the Skytop lounge cars on the Milwaukee Road. From the inside the passenger could not be sure if he were aboard a train or a spaceship. The windows swept not only around but above him: 90 percent of the car's end was transparent (Figures 4.36 and 4.37). Because of their great expanse of glass, the cars became known as "rolling greenhouses."[43] The imaginative design was created by the industrial designer Brooks Stevens. The Milwaukee shops built four parlor cars to his plan in 1948. Pullman built six in 1948 and 1949 which were nearly identical, except that they had bedrooms in addition to the observation lounge. These six were sold to the Canadian National in 1964.

Figure 4.36   The Milwaukee Road's Skytop cars of 1948–1949 were the most daring and original observations built in North America. (Association of American Railroads)

Figure 4.37   The observation parlor car Cedar Rapids was built by the Milwaukee Road's shops in 1948. (Association of American Railroads)

Figure 4.38 Burlington's midtrain or square-end observation parlor car Silver Tower *was built in 1952 by Budd. It is shown here in 1971 leaving Chicago on the Denver Zephyr. (Douglas Wornom)*

In time most railroads came to regret their investment in round-end observation cars, since they could function properly only at the end of the train. Cars coupled on were isolated because there was no vestibule; thus an extra switching move was necessary wherever cars were added to the train. For this reason few swallowtails were built after 1950, and those that were came equipped with end vestibules—an addition that did little for their appearance. Square-end observations, much on the order of the old sun-parlor cars, came into favor (Figure 4.38). These so-called mid-train observations could be placed anywhere in the consist. Older swallowtails were retired or rebuilt. The Santa Fe, for example, sent six of the Vista series back to Pullman for remodeling as square-end cars.[44]

In summary, the lounge car fulfilled the railways' promise of luxurious travel—a promise that had been spoken of before the first tracks were laid in this country. Although the common man could not afford its splendors, the lounge car did bring the posh comforts of the private car within the budget of the middle-class traveler.

Dining and Buffet Cars

The diner was one of the last major classes of passenger car to come into general use. Always a great favorite with the traveling public, it was viewed by the railroads as a costly liability. It was the heaviest and most expensive of all the cars in regular passenger service. It rarely made a profit and generally incurred appalling losses. The solution of unloading this service on a concessionaire, such as Pullman, proved vain; its red-ink reputation was too well known. Yet it was a service that first-class travelers expected and that the railroads were obliged to maintain.

At one time there were over 1,700 dining cars in this country.[45] Each was a marvel of compactness and efficiency. There was little unused space; cabinets and lockers occupied every corner, storing an array of tableware and provisions that included 1,000 pieces of crockery and glass, 900 tablecloths, 700 pieces of silverware, and food for 400 customers.[46] Since the car also carried a fully equipped kitchen, dining-room furniture, and heating, lighting, and ventilating fixtures, it is no surprise that it

Figure 4.39 *When this engraving appeared in 1871, only a few Western railroads were offering dining-car service.* (National Car Builder, *March 1871*)

weighed 80 tons and cost $50,000. The weight and cost increased again with the advent of air conditioning in the 1930s. The crew was equally large and expensive: seven to sixteen men were required, depending on the status of the train. On a top train such as the Broadway Limited, nothing less than a steward, an assistant steward, a chef, three cooks, and ten waiters could do the job. In 1925 10,000 men were employed in dining cars to serve 80,000 meals a day at a net annual loss of 10.5 million dollars.[47]

Like parlor cars, diners were primarily intended for first-class travelers. The great majority of railway travelers never patronized the diner. Among commuters there was little need for it; only long-distance travelers had to eat en route. Even then the thrifty person could bring his own box lunch, buy a sandwich at the station news counter before boarding, or pick up snacks at station stops along the way. At one time food could also be bought through the car windows from platform vendors. It is impossible to say how common this practice was, but it is docu-

mented in an engraving from a volume of American pictures published in 1872.[48] The scene, reminiscent of India or Guatemala, shows a crowded platform at Gordonsville, Virginia, where Negro women are selling coffee, hot cakes, and chicken from trays, some of which are balanced on their bandanna-covered heads. The newsbutcher offered another alternative, but his wares were meager, usually consisting more of smokes and cheap publications than food.

As for railroad station restaurants, they could be wonderful or terrible. The depot hotels at Altoona, Springfield, and Poughkeepsie had a national reputation for excellent fare,[49] and the Harvey Houses along the Santa Fe were recognized as model establishments from their beginning in 1876. But most depot eating houses seemed to fall below any conceivable standard of decency and were vigorously denounced in the nineteenth-century press. One account insisted that they were the most infamous of all the "pernicious institutions" in America.[50] It pictured the proprietors as dispensers of dyspepsia or descendants of the Borgias. It described piles of mouldering dishes and crumpled napkins amid swarming flies and an atmosphere ripe with the smell of yesterday's breakfast. It portrayed the help as unkempt, rude, and slow; the coffee as a black, gritty fluid; the fruit as knotty and decayed. It said that everything was dusted with coal soot, and that the prices would cause Delmonico's to blush.

Because of the time factor, meals at even the best depot restaurants were not always a pleasant experience. The train could afford to stop for no more than half an hour; fifteen or twenty minutes was more common. Two or three hundred passengers all wanting to be fed at the same time created an atmosphere that was not conducive to civilized dining. If the train were running late, the scheduled thirty-minute stop might be cut back by a conductor. The fear of being left behind ruined many a stomach.

The contrast between the depot restaurant and the dining car was remarkable. To the good food, good service, and pleasant surroundings of the diner was added the enchantment of the passing landscape. Few other places can induce the same feeling of serenity and well-being than a seat on a railway dining car. It was in many travelers' experience one of the most civilized pleasures created by Western society. J. E. Watkins captured the mood:[51]

The cry "Wilmington, fifteen minutes for refreshments!" is no longer heard, but instead the passenger enjoys his soup, his fish, joint, Cliquot, and coffee while the train runs fifty miles an hour. In fact, a dinner that is ordered at Newark is scarcely finished by the time Trenton is reached, and the stump of the cigar lighted shortly after completing it is often thrown away to the gamins at Broad Street Station, Philadelphia.

In the nineteenth century, small appetites were as suspect as six-course dinners were admired. A menu of the 1870s would include over eighty dishes.[52] To start, there would be Saddle Rock oysters on the shell or a choice of four soups. Among the entrees there might be broiled trout, stuffed loin of veal, and grilled mutton kidneys. Wild game like venison, quail, and prairie chicken were available as well, and so were a dozen vegetables and twice that many desserts. Such a meal was sold for 75 cents. The price was about equal to that of a good restaurant; it was not cheap, nor was it within the means of the average traveler, for it nearly equaled a day's wages for a workingman.

Only the best big-city restaurants of the time offered such a variety of dishes. In the menu just referred to, nearly everything was served with a sauce or was prepared in some special way. Even though some items were cooked at terminal commissaries, the assembly of such elaborate meals in a rolling, swaying kitchen that had little more than a coal stove seems miraculous. The kitchen of the 1875–1880 period was perhaps only two-thirds the size of modern diner kitchens. Curiously, as meals became simpler the size of the kitchen grew, though it remained a very busy place. In the mid-1920s, for example, the Broadway's galley produced twenty-four different dishes.[53] In its 50 square feet of open floor space (30 inches by 20 feet), four cooks simultaneously broiled steaks, made pies, cooked bacon, and baked potatoes. A dining-car cook had to be agile, well organized, and quick. He needed to know where each of a hundred pots and implements was stored and how to get at it without impeding the other kitchen workers. There was no room for waiters inside this busy area; the finished dishes were placed on a shelf for pickup and delivery.

In his 1819 memorial to the U.S. Congress, Benjamin Dearborn of Boston proposed that a system of railways be built and furnished with carriages offering meals to passengers during their journey. He cited the packet boat as an existing model for such service. A few years later as the railway slowly evolved into a public carrier, the same idea was proposed in England by William Chapman. Early railway managers had enough to do, however, to provide basic transportation. They could not achieve such luxuries as food service. Trips were short; let passengers dine before or afterwards.

Still, the idea of providing meals persisted, and in 1835 it was given an actual test. Imlay's eight-wheel car Victory, described in Chapter 1, went into service on the Philadelphia and Columbia Railroad. The fact that a serving counter and shelves occupied one end of the car indicates that food and beverages were available, though apparently no cooking was attempted. This was the earliest known instance of railway food service and represents the beginnings of the buffet car, if not of the diner. But like so many firsts, it seems to have been an isolated incident that led nowhere and was soon forgotten.

Yet food service was such an obvious travel requirement that it was always being discussed. In 1838 one of the Baltimore papers praised the new sleeping car installed on the P W & B and said that the only basic comfort required by the traveler was a dining car. This service, the paper predicted, would soon be offered. Five years later a passenger on the B & O tried to revive interest in the scheme. The B & O tracks reached Cumberland, Maryland, late in 1842, and in the next year Robert McLoed petitioned the directors of the company to place what he called a "refectory" car in service between Baltimore and Cumberland.[54] He suggested that an ordinary day coach be remodeled for the purpose, but apparently nothing came of his proposal.

Two special trains run over the B & O in January 1853 to celebrate the line's opening to Wheeling did contain refreshment cars.[55] Many notables were aboard, including the governors of Virginia and Maryland, state legislators, and judges. Food and drink, catered and served by the United States Hotel, were provided throughout the trip in new cars equipped with long center tables. However, it appears that the cars had been temporarily fitted up for the celebration; they did not offer regular food service.

It is claimed that a public dining car began running between New York and Chicago during the same year.[56] The reliability of this account can be questioned on several points, particularly the route and the date. It is doubtful that a through car could have

Figure 4.40   *The diminutive kitchens of Pullman's hotel cars were ca-*
pable of furnishing full-course meals. This engraving dates from 1869.
(Engineering, *July 17, 1869*)

that had been remodeled with a center partition. One-half of each was used as a smoking compartment; the other half housed an "eating bar." The food, which had been prepared at the depot kitchen, was kept warm in a steam box. On the other hand, an editor of the *New York Sun* described a diner of the same period in a 1924 memoir. In 1862, he said, he was traveling with an uncle to Washington. In the course of the journey they came upon an improvised diner, an old baggage car which showed signs of long service. The interior was bare except for an oblong counter in the middle that was surrounded by high stools. The center of the counter was cut away so that waiters could stand inside. The menu was limited to oyster stew, coffee, and strips of a deep-fried sweet cake known as crullers. Possibly both accounts are correct; the P W & B annual report for 1863 lists three refreshment cars, so that the third may have been the converted baggage car described in the editor's memoirs.

Information is available about several other dining cars of the Civil War period. On his trip to Washington in February 1861, President-elect Lincoln was served dinner between Buffalo and Albany aboard a car specially fitted for the purpose by the New York Central.[61] During a more solemn trip, Lincoln and his party were seated at a table in a baggage car for the noon meal while they traveled to Gettysburg.[62] A special train was run for the purpose, and although an office car was assigned to the President, it either did not have dining facilities or was not large enough to accommodate his entire staff.

During the war years the wounded were served food aboard hospital trains. Injured soldiers were at first carried in boxcars furnished with little more than straw or hammocks, but in January 1863 the P W & B refitted some cars with bunks and kitchens.[63] Hot soup and coffee were provided. In 1863 vastly superior hospital trains were assembled for the Army of the Cumberland by the U.S. Sanitary Commission. Full-scale kitchen cars were outfitted. At one end was the kitchen, with a range, cupboards, and a sink. A storeroom compartment was next to the kitchen. A narrow aisle to one side of the storeroom led to the dining area, which contained a long center table and benches. Although not elegant, these cars were neat and serviceable. They introduced the idea of dining on rails to many young men who probably had never before ridden a train.

Most of the early dining cars that have been described were makeshifts which created little interest outside of their own locale. None precipitated any movement toward the general adoption of dining cars, but they were the prelude to a new era in railway travel. In reality, the custom of eating while in motion had been established by canal packets and extended by river and coastal steamers. Rival lines competed through their cuisine. The great dining hall became the steamer's architectural showpiece. An attraction of travel was the promise of an excellent meal in a sumptuous cabin. Thus a public that was accustomed to eating and traveling considered it a nuisance to detrain every few hours for a snack, and not always a very good one. Yet curiously, the Eastern railroads, which competed most directly with the overnight steamers, did not lead the way in experimenting with din-

found its way between the two cities via the Northern Central, the Grand Trunk, and the Michigan Central, as the article claimed. The date of 1853 also seems a trifle early. But even if the story is as much as ten years off, it is early enough to be of interest. The diner was described as a remodeled coach, with tables and stools installed in place of the seats. The kitchen was located in one end of an adjoining baggage car. The waiters had some trouble carrying the food between the cars, but at 50 cents a portion, the service was very popular. Two stops were avoided, saving forty-one minutes. Who owned the car or initiated the service, and how long it remained in operation, are unknown.

During the 1860s, contemporary notices of dining cars began to appear. One journal reported a diner on a railroad in Pennsylvania in the fall of 1860.[57] Another spoke of a restaurant car on the Lehigh Valley Railroad in the same year.[58] That reporter was unimpressed: "We should require to be very hungry before we should desire to see another one. The 'restaurant' was a deserted entry to the baggage room; no one in attendance, and nothing to eat. We did not like its appearance—nor the odor with which it was pervaded." The same reporter, presumably Zerah Colburn, was in fact opposed to the entire idea of the dining car.[59] He questioned the wisdom of running diners because of their expense and weight. Since not everyone could be accommodated in a special car devoted to the purpose, it would be necessary to serve passengers at their seats. An army of waiters would be needed. Gravy, scraps of food, and dirty dishes would be everywhere. He insisted that the solution lay in upgrading the depot restaurants. Some railroads, like the B & O and the Central Pacific, did concentrate on improving their hotel-restaurant facilities, but others were willing to experiment with meals on wheels.

One such road, the P W & B, began diner service in the early 1860s. There are two contradictory accounts of these cars.[60] According to an old employee, they were placed in service in 1863 and ran for three years. They consisted of two 50-foot coaches

Figure 4.41   A British publication, Engineering, published four interior views of Pullman's hotel car City of Boston in July 1869. Portable tables and folding berths converted the car, as required, to function as a coach, sleeper, and diner.

ers. It was the lines running out of Chicago that introduced the dining car; Chicago was the place and Pullman was the man.

The beginnings of the commercial dining car were intertwined with Pullman's effort to establish a fleet of cars for luxury travel. Sleeping cars eliminated the most basic discomforts of long trips. Pullman thought that some form of food service would prove a

popular and profitable auxiliary. A small kitchen added to a palace sleeper resulted in the hotel car. In 1866 and 1867 several new cars were built on this plan for service on the Michigan Central, the Burlington, and the Great Western of Canada. A tiny kitchen was placed in the center of the car (Figure 4.40). It served the two ends, which were outfitted like a standard open-

Figure 4.42   Pullman's first full diner, the Delmonico *of 1868, is shown on an excursion
train operated over the Chicago and North Western Railway. (Chaney Neg. 9816)*

section sleeper except that small folding tables were installed
during the day between the seats. These could be used as writing
or dining tables; at night they were stored in a closet. Hotel cars
slept 30, but they could seat and feed as many as 40 passengers.
The food was prepared and served by the crew—generally four
to five men. The menu was surprisingly varied considering the
kitchen size, which in the case of the *Western World* was said to
be only 3 by 6 feet.[64] Built-in cabinets adjacent to the kitchen
stored the 133 items of food, the liquor, the 1,000 napkins, the
150 tablecloths, and the china and glassware needed for a long
journey.[65]

One of the first hotel cars, the *City of Boston*, was illustrated
and described by the British technical journal *Engineering* in the
year following its emergence from the Burlington's Aurora
shops.[66] The body, 60 feet by 10 feet, 8 inches, was divided into
four sections. At one end was a small drawing room; next came a
larger compartment with open sections; next the central kitchen;
and the far end was occupied by a compartment with open-
section berths. The illustrations accompanying the article mistakenly suggested that various sections of the car were permanently
set up for dining, sleeping, and day travel (Figure 4.41). Only
the drawing room, shown in the lower right of the four-part

engraving, remained the same day and night. The other three
areas shown are actually one of the open-section compartments
as it might be converted for dining, sleeping, or day travel.

The hotel car, though hailed as a great advance in railway
travel, had its faults. Travelers who suffered from motion sick-
ness found it disagreeable to be cooped up in a car where the
smell of food lingered. The price of the meals was high: an
entree with potatoes cost 50 to 60 cents, coffee and tea 25 cents a-
piece. The crew was too large for the number of patrons. The
investment required for a kitchen, linens, and the other accessories added to the economic risk of the hotel car.

Pullman hoped to solve all these defects by building a special
car that did nothing but prepare and serve meals. A complete
diner could provide more and better food for everyone aboard
the train. It would be cheaper, since it would leave more revenue
space in the sleeping cars and would surely cut labor costs. And
so only a year after the first hotel cars had appeared, work began
on a dining car.

That Pullman meant to build a magnificent rolling restaurant is
implied in the car's name, the *Delmonico* (Figure 4.42). No
restaurant in the New World enjoyed a greater reputation for
luxury in cuisine, service, and decor. Pullman planned to create a

Figure 4.43 A rare interior view of the Pullman dining car Cosmopolitan. The sixteen-wheel car was built in 1869 by the Burlington's Aurora, Illinois, shops. (Charles Clegg)

mobile eating carriage equal to its namesake. It would be no crude feedin' car with a steam table, or anything that could be criticized as greasy, untidy, or smelly.

Again the Aurora shops were assigned the job of construction. The car, which cost $20,000, was the same size as a hotel car—60 by 10 feet.[67] A kitchen 8 feet square was placed at the car's center and flanked by two dining rooms with twelve tables each, providing total seating for 48 persons. A large ice chest was mounted under the floor of the kitchen. Gaslights and a hot-water heating system were part of the up-to-date equipment. Two cooks and four waiters served 250 meals a day. The crew slept in the car at night by converting the Morocco leather seats into berths. The floor under the tables was covered with oilcloth for easy cleaning, while the aisle was carpeted. In the kitchen a side door to the outside permitted easy access for loading supplies. In April 1868 the *Delmonico* was ready for public viewing. After a test on the Illinois Central, the car was sent to New York and in June formed part of a special excursion train to celebrate Henry Keep's appointment as president of the Chicago and North Western. It was during this trip that the only surviving photograph of the historic car was made (Figure 4.42).

At the time of its introduction the press announced that the car would run regularly between New York and Chicago, but the lines involved thought otherwise. Either they were satisfied with

the hotel cars already in service or they found that the *Delmonico* did not pay. In the summer of 1868 it was leased to the Chicago and Alton. This and other disappointments led Pullman to move very slowly in the dining car trade; never in the firm's history were more than a handful of diners on its roster (see Figure 4.43). The hotel car remained in favor, however, demonstrating that there was a profit in light meals for first-class passengers. The kitchen was moved to one end of the car, less ambitious menus were attempted, and improved ventilation helped to dispel cooking odors. The railroads were also pleased with the hotel car, because it satisfied customer demand and relieved them of the necessity for providing dining-car service. Pullman managed the hotel cars with no trouble to the railroads, and if he could make a profit while doing so, all the better. The cars remained in vogue until about 1880. One was shown by Pullman at the Centennial, and a new hotel car named the *St. Nicholas* was described in 1879 as having twenty-four berths and a 10- by 6-foot kitchen-pantry at one end, all within its 63-foot, 10-inch body.[68] Within a few years, however, these cars fell from favor as the railroads reluctantly accepted the dining car. Pullman reduced the kitchen space, adopted the designation "buffet" in place of the obsolete term hotel car, and continued operating the cars much as before.

Following the *Delmonico*'s introduction only one railroad

Figure 4.44 *A cutaway engraving of a Rock Island dining car of about 1877. Note the men's bar at the right end.*

Figure 4.45 *A dining-car kitchen with its coal-fired range was depicted by an artist for* Leslie's Weekly Magazine *in 1877.*

seemed to have much faith in the full diner. In competing with the other lines connecting Chicago and St. Louis, the Chicago and Alton decided to try improving its food service. It encouraged Pullman to operate diners on its trains, presumably with a cost-plus guarantee. By 1872 the road was using five dining cars, which offered a bountiful meal at the fixed price of $1.[69] Trade picked up, and when the road's rivals countered with diners of their own, the Alton bought out Pullman's interest and reduced the meal price to 75 cents.[70]

By 1875 the dining-car fever had spread to the Michigan Central, and within a year or two the Burlington, the Rock Island, and the Wabash were also running diners.[71] All the Chicago railroads had decided that the dining car was a competitive necessity, no matter how great the losses. Increased patronage was expected to offset the out-of-pocket costs. A Rock Island

timetable of this period described the dining car as a democratic experiment which, unlike the hotel cars, was open to all travelers, be they millionaires or miners. The Rock Island also intended to have the best equipment possible, and in 1877 four cars, *Oriental, Overland, Occidental,* and *Australia,* were manufactured at its Chicago shops.[72] Each of these twelve-wheelers had a body 57 feet long. A 27-foot dining room seated 16 at four tables. A 10-foot kitchen with a gleaming copper floor was in the center, and a 7-foot lunch counter occupied the far end of the car. Light meals, snacks, and drinks were served at the counter. Formal meals—*table d'hôte* at 75 cents or *à la carte*—were available in the dining room. Royal Dresden china, elegant silverware, rare wines, and cigars were all part of the service. The car's exterior was painted a wine color. In addition to the usual gilt striping there were some "elaborately executed game pieces indicating the purpose of the car and the nature of the larder." A cutaway engraving of one of the cars is reproduced in Figure 4.44.

By the early 1880s the dining car was an established fact in the Midwest. The last major holdouts in the Chicago area were the North Western and the Milwaukee roads, and they fell into line in 1881 and 1882 (see Figures 4.45 to 4.47). An excellent contemporary account describes one of the C & N W's first cars:[73]

Each car is 69 ft. in length over all, the bodies being 62½ ft. long by 10 ft. wide. They have 6-wheel trucks, 33-inch Allen paper wheels, and well adjusted springs for easy riding. The main or dining saloon is 32 ft. in length, and is divided into five sections on a side, with a table and seats for four persons in each. Large plate glass mirrors cover the end partitions, and also the wide window panels, the effect of which is to make the size of the room seem much larger than it is. At the end of each double seat is a large window, the panels over the seat backs are narrow and those over the tables wide, with three alcoves for tumblers, salt, spices, etc. The seats are covered with maroon leather, and the floor laid with 3-inch alternate strips of walnut and maple. The seat ends and frames and the inside finish are cherry trimmed with mahogany; five Adams & Westlake 2-light chandeliers with argand burners hang over the aisle—one to each pair of tables. These, and all the metal trimmings of the car are gilt. The head-linings are decorated oak, paneled with mahogany moldings. Those below the clear-story have paintings of game, fruit and flowers. Hartshorn roller shades are used instead of blinds, and roll up out of sight under box-moldings. The windows are all double, the outside ones of the kitchen having game and fish designs ground in the center. The kitchen is 13 × 8 ft., and built of solid walnut, with all requisite utensils and conveniences. Water is carried in a long cylindrical reservoir in the clear-story overhead. Between the kitchen and dining saloon is a pantry 4½ × 8 ft. In the end corner of the car is an ice closet with glass-bottomed drawers opening into the kitchen, also a closet for small stores. At the opposite end is a Baker heater, linen and wine closets,

Figure 4.46 The Illinois was an 1881 product of the Chicago and North Western Railway's West Chicago shops. The car was painted wine color with gilt embellishments. (National Car Builder, July 1881)

Figure 4.47 Harlan and Hollingsworth of Wilmington, Delaware, completed this car in 1882. (Smithsonian Neg. 47984-G)

lavatory and ice chest. Without particularizing further, it may be said that these cars are believed to embody all the improvements pertaining to their class, and are unsurpassed in their luxurious accomodations and handsome finish by any others in the country. The outside is painted a dark wine color relieved with gilt tracery and ornamental designs representing storks, passionflowers, cornucopias, etc.

The demand for dining cars was also mounting in the East, though railroad managers there hoped that the entire diner system would quietly die. Certainly there was talk among the Chicago lines of mutual abandonment of food service because of the continued losses, but this never came to pass.[74] A few lines, believing that diners paid for themselves through increased traffic, would not join the abandonment collusion. The Eastern trunk lines were at last forced to admit that the diner would prevail. To the surprise of many, the conservative B & O was the first to give in. Early in 1881 it placed five new restaurant cars in service.[75] The Pennsylvania, deciding that it must follow suit, became mildly enthusiastic once the change in policy was made, and the Altoona shops were ordered to produce the best dining car possible.[76] The designer was Theodore N. Ely, head of the mechanical department. Four cars, which replaced hotel cars, were ready for the New York–Chicago express in the spring of 1882. A special feature was that each car carried a 125-bottle wine rack.

The New York Central was now forced to act. For years it had been content to run on a leisurely 36-hour schedule between New York and Chicago, providing a breakfast stop at Rochester, lunch at Erie, and dinner at Toledo.[77] Its more experimental Western subsidiaries, the Michigan Central and Canada Southern, had been using diners for some time. Finally in the autumn of 1883 the Vanderbilts decided that it was time to join the ranks of dining-car railroads.

No diners were seen in New England until the following year. The New Haven cautiously put two cars in service between Boston and New York in the summer of 1884.[78] The same pair was expected to double back to New Haven for the noon train to Worcester. Now the New Haven, like other Eastern lines, came to champion the diner, and it was one of the few railroads that realized a profit from food service.

By the mid-eighties the dining car was no longer an unwanted stepchild but the pride of the passenger car family. Every major railroad had at least a few restaurant cars, and they were common on all first-class express trains. Even the Southern railways, usually accused of lagging a decade behind, began to adopt diners almost as early as the Northern roads.[79]

It was the Far Western roads that resisted the dining car longest. The Union Pacific had no diners until almost 1890,[80] nor did the Central Pacific. Both companies had permitted Pullman to operate diners over their lines until the early 1870s, but then they returned to the old-fashioned system of stopping at station restaurants. The Central Pacific was known for the excellence of its eating houses, and it saw little reason to reverse this policy until the demands of faster schedules forced it to do so. For similar reasons the Santa Fe put off running diners for many years. Their contract with Fred Harvey had proved an excellent one. Beginning in 1876, the English immigrant who had turned station restaurateur set out to overcome the notorious reputation of such establishments. Within fifteen years his chain of dining rooms was serving 5,000 meals a day, and its reputation for good food was worldwide. But as other Western lines began to adopt dining cars, the Santa Fe felt obliged to do likewise. Friction developed between Harvey and the railroad, which ignored his protests. Harvey took his patron to court, where in 1891 he obtained an injunction restraining the operation of dining cars on the Santa Fe west of the Missouri River.[81] The railroad was forced to recognize its contract, but within a year Harvey was persuaded that both their interests might best be served if he were to operate the road's dining-car service.

Why did it take nearly a quarter century before passengers could expect to dine aboard a train in any region of the United States? The diner's poor earning performance was to blame, but why was the diner an inveterate loser? Even well-patronized trains could not seem to make money on their restaurant cars. High fixed costs and a limited market were the basic reasons for their fiscal difficulties. A sizable crew, generally ten men, was necessary to serve thirty to forty patrons efficiently. Even in the days of cheap labor, and even though the same crew served several seatings, this was a disproportionate ratio. The crew had to be fed and housed throughout the trip, and at least some of them had to be carried back to their homes as deadhead passengers. Unlike a restaurant, the railroad could not use part-time help. Free meals to other trainmen created an additional new expense. A dishonest crew could also add to the losses through pilferage of food, linens, and silverware, while a dishonest steward could steal supplies in wholesale lots and juggle his accounts to escape with cash as well.

Unused or spoiled food was another large expense. Cars were usually overstocked so that no customer would be disappointed, and what was not used could not always be salvaged. The old-fashioned icebox was not entirely effective in storing perishables.

Moreover, the clientele was limited to those aboard the train. No matter how good or cheap the meal, the dining car could not draw upon the population at large. And many travelers brought their own food; others, being country folk intimidated by the posh diner and its crew, skipped their meals. Only the first-class passengers could be depended on to patronize the diner.

The investment required for a diner could usually be applied more profitably elsewhere. In the 1880s such a car cost $12,000, or nearly three times the amount for a coach.[82] In 1905 the price was up to $17,000, plus $12,000 for dishes, linens, and other necessary gear. By the 1930s the average price was $50,000, and by the 1960s it was nearly four times that. Total investment on roads with large dining-car fleets was substantial. In 1930 the New York Central had 7.8 million dollars tied up in its 156 restaurant cars. Repairs and upkeep were higher than those for a coach or even a sleeper. In addition to renewing the tableware and linen, the kitchen required continual repairs. In 1925 a typical diner needed a major overhaul every eighteen months at a cost of $5,000 to $7,000.[83]

There was no effort to set prices in accordance with costs. No one would pay, say, $10 for a meal, or whatever figure the accountants might estimate was necessary to cover expenses. In the early years, 75 cents in the West and $1 in the East became established traditions. The dollar dinner had gradually disappeared by the time of World War I except on the New Haven, which continued to honor the old standard.[84] The *table d'hôte* vanished during the same period; Pullman and Wagner learned that *à la carte* meals offered the only hope of turning a profit in the food trade.

In 1887 it was estimated that each diner could be expected to lose $100 to $600 a month.[85] Some railroads were losing as much as $21,000 annually. A few, like the Pennsylvania, were just barely in the black—its diners showed a pitiful $29 profit in 1886. The Milwaukee had been losing $16,000 a year, but it succeeded in reversing this to a $4,000 profit by hiring a new manager. It was a temporary victory, however; many years later the same railroad reported an annual loss of $60,000 on just one train.[86]

These losses were justified as a necessary business expense—a cheap, direct form of advertising. Good meals increased traffic and won friends. Railroads became famous not only for the general quality of their food but for individual dishes. The B & O was known for its corn bread and country sausage breakfasts, the Union Pacific for steaks. The Northern Pacific was the great baked-potato route; a giant 20-ounce spud was the minimum size served.[87] This road also offered fresh fish in season. Its homemade fruitcake, which won a grand prize at the 1889 Paris Exposition, was produced in 15-ton lots to satisfy the demand

Figure 4.48 Plans for a Michigan Central 37-ton, 24-passenger dining car were published by the Railroad Gazette *in 1887.*

from its dining-car and on-line patrons. As losses mounted and general passenger travel declined, however, the good will created by dining cars was no longer enough to justify their operation.

By the early 1880s the dining car's arrangement was fixed in the general pattern shown in Figures 4.48 to 4.50. The kitchen was placed at one end, with a narrow side aisle connecting the vestibule and the dining compartment. Fixed cross tables and bench seats, much like those used in Pullman open-section sleepers, were at first favored. But sliding in and out of the seats was not easy, particularly for the overweight or elderly, and in the early eighties some lines began to install lift-up seats like those in theaters.[88] Apparently loose chairs were not used for almost another decade. This was, of course, the best form of seating, because it allowed easy entrance to the table and because the chair could be positioned to suit the comfort of a tall or short diner. By the middle eighties the table arrangement was altered so that one side of the car had the regular four-place tables while the other had smaller tables for only two persons.[89] The smaller tables were preferred by couples and single diners. The four-and-two table pattern created a wider aisle, improving the work space for the waiters. It also reduced the car's capacity, a result which was appreciated by both the crew and the diners. Forty seats in a 60-foot car were too many; twenty-four to thirty places came to be recognized as more realistic in view of the kitchen's capabilities and the general space needs of all who inhabited the car.

Even when car sizes grew to 70 and 80 feet in the 1890s there was no effort to crowd more seating into the cars, except on some Eastern lines where patronage was unusually heavy[90] (see Figures 4.51 and 4.52). During these years car designers allotted more space to the kitchen, so that the cramped 10-foot kitchens of the 1870s were expanded to as much as 27 square feet.[91] More space permitted a greater work area and larger ranges. In 1895 a full-sized coal range was 6 feet long, 34 inches deep, and 5 feet, 6 inches high, and it weighed 1,250 pounds. These substantial furnaces were also connected to a heater that supplied hot water for cooking and dishwashing.

During the mid-eighties a distinctive architectural feature emerged that was to remain in favor for many generations. This was the sideboard or breakfront, which was placed at one end of the dining room between the pantry and the rest of the car. It was a decorative focus, usually a handsome piece of cabinetry, and it served to screen off the working section of the car. China or silver plate was exhibited behind its glass doors, and a bowl of flowers was often placed on it.

As in the parlor car, large windows became a feature of the dining car (Figure 4.53). They were used for the same reason: a fine view added to the pleasure of the trip. The kitchen end of the car normally had half windows, which in some cases were made of opaque art glass. Its distinctive fenestration customarily set the diner apart from the other cars in the train. In 1893, however, the Delaware and Hudson broke with this tradition by purchasing a diner with a coach-like body.[92] All the windows

Sectional Half Side Elevation.

Plan of Kitchen, Pantry and Buffet.

Side View of Kitchen

Side Elevation of Kitchen.   Refrigerator.   Buffet.

Figure 4.49   The general arrangement and furnishings of a Chicago, Burlington, and Quincy dining car of about 1890 are given in this drawing. Other details are shown in Figure 4.50. (Car Builders' Dictionary, 1895)

were uniform in size, and the table and chairs were loose. The idea behind it was that these furnishings had only to be removed and chairs or seats installed to convert the car into a buffet-parlor or coach. The standard coach body was much cheaper than a regular diner, but sensible as the scheme may have been, it was never widely adopted.

Changes over time in dining-car structure and size followed the same general development of all passenger equipment. In 1886 a good-sized diner had a 64-foot body and weighed 41 tons.

By the end of the wooden age, there were some cars with 72-foot bodies and total weights exceeding 76 tons. An example of these large wooden-bodied diners was the New York Central's Number 405, rebuilt at the West Albany shops in 1910 for the Lake Shore Limited.[93] Its overall length was 80 feet, 8 inches. A crew of 13 was required to serve only 30 diners. A diagram of the car is shown in Figure 4.54.

The steel era brought no dramatic growth in size; few cars exceeded 80 feet in overall length, but their weight crept upward

Part Side Elevation.

Transverse Section showing Kitchen.

Transverse Section showing Bulkhead.

Plan of Dining Room and Conductor's Quarters.

Wine Closet.

Side Elevation of Conductor's Quarters.

Side Elevation of Conductor's Quarters.

Figure 4.50   A Chicago, Burlington, and Quincy dining car of about 1890. (Car Builders' Dictionary, 1895)

until it was between 80 and 85 tons (see Figures 4.55 to 4.60). The general floor plan did not change during this period except for a single major alteration: one or both vestibules were eliminated, which increased usable space by as much as 8 feet. An early example was the 1910 Lake Shore Limited car in Figure 4.54, which had no vestibule at the kitchen end. In 1914 the Burlington acquired a steel diner with no vestibules at either end,[94] and by the early 1920s nonvestibule diners were relatively common. These entranceways were considered nonessential, since most passengers did not board the train through the diner in any event. End doors and diaphragms were necessary, of course, for passage to and from the diner.

Beyond eliminating the vestibule platforms, most railroads seemed content with the standard floor plan in which the kitchen was at one end, the dining room occupied the other, and a side passageway ran alongside the kitchen. There were a few interesting exceptions, however. The earliest was a center-table plan tried out by the Erie in 1917 to provide greater privacy.[95]

Figure 4.51   Pullman's diner La Rabida *was an extra-elaborate display piece created for the Columbian Exposition in 1893. (Charles Clegg)*

Shoulder-high semicircular partitions in the car's center created a series of private dining alcoves. Each had a round table with four or six chairs. A center panel split some of these areas into small tables for two. Total seating was for 36. The tables could be reached from aisles on both sides of the car. The kitchen was placed at one end, as in a conventional car. So far as can be determined, the two sample cars built for the Erie by Barney and Smith were the only examples of this style.

In 1925 the Santa Fe, which was forced to operate two diners on certain heavily patronized trains, was seeking ways to cut costs and improve food service.[96] The road purchased a special two-car unit of which one was a full diner and its mate was a combination club car and open-section sleeper. The cars ran as a pair, with the club end of the first car always coupled to the diner end of the second car. The head end of the diner was blind and was attached to the baggage car. Because no one needed to pass beyond this point, there was no side passage, and the

kitchen was made full-width. The space gained was used to expand the dining-room seating from the usual 30 or 36 to 42. When the tables were filled, passengers could wait comfortably in the adjacent club lounge rather than stand in the vestibule or aisle, as was necessary with an ordinary diner. The club car was outfitted with a soda fountain which sold cigars, candy, and magazines in addition to ice cream and soft drinks. Service in the diner itself was speeded by an additional employee, who checked the trays against orders and operated a cash register. The scheme worked so well that the road bought a second pair of the cars from Pullman in 1927.

It may have been the Santa Fe's successful experiment that led to the development of twin diners. About thirty-five of these units were built between 1937 and 1953. Some were articulated, but most were fairly conventional cars operated as a married pair. The twin units were large-capacity food-service vehicles, with one car serving as the dining room and the second as a

Figure 4.52 *A photograph of the B & O's Royal Limited dining car, presumably taken in 1898 when a new set of cars was placed in service. (Smithsonian Neg. 48382)*

Figure 4.53 *The wooden diner Number 452 was outshopped by Pullman in 1909 for the C & O. (Pullman Neg. 11559)*

Figure 4.54   A diagram drawing for one of the diners used on the New York Central's train, the Lake Shore Limited, in 1910.

Figure 4.55   Santa Fe's Pullman-built diner of 1914 was equipped with indirect lighting. A battery of fans was required for comfort in the high temperatures of the Southwestern route. (Pullman Neg. 17610)

Figure 4.56   Passengers aboard an arched-roof steel diner on the Southern Pacific around 1925. (Southern Pacific)

kitchen and the crew's quarters. The Union Pacific had been satisfied with the articulated cars on its motor trains and had decided to use this plan of carriage for some full-sized passenger cars. The result was the twin-unit diners built for use on the Challenger trains by Pullman in 1937.[97] Each 140-foot pair had eight-wheel trucks at either end and a single six-wheel truck at the center connection of the two bodies. Aluminum construction held the total weight to just under 104 tons. The diner unit had a pantry at the center end toward the kitchen, while the remainder of the car was devoted to seating and tables for 68 passengers. In the second car a large, fully equipped kitchen occupied more than half the floor space. The rear portion was divided into four rooms for the twenty-man crew. Most of these bedrooms had two three-tier bunk beds. The public was delighted with the twin-unit giants, but the crews showed little enthusiasm. The long walk from the kitchen to the end tables created difficulties and extra work. Apparently the railroad management was not enthusiastic either, for it ordered no more twin diners until 1953. These new cars were nonarticulated, indicating that part of the design had not proved satisfactory.

In 1940 the Pennsylvania transformed eighteen heavyweight diners into nine streamlined, twin-unit ones for service on the Trail Blazer, the Jeffersonian, and the Congressional.[98] The cars were not articulated; four pairs of roller-bearing six-wheel trucks carried the 176-ton couples. The dining-room halves of these

units were identical and seated 68 passengers each, but the kitchen cars varied. Those intended for New York–Washington runs had lunch counters, while those meant for overnight service to the West had dormitories built into the same space. Following World War II, the Pennsylvania decided to expand its twin-diner fleet. Seventeen double units were produced in 1948 and 1949 by the Altoona shops, American Car and Foundry, and Budd.[99] They cost $335,000 a pair and measured 165 feet in length. They could run through to Chicago on the most heavily patronized trains without restocking food or supplies. They provided storage for 2,016 pieces of linen and 2,721 other items.

If twin units were good, triple units would be better, or so reasoned the Southern Pacific in 1939–1940 when it acquired two sets of three-car diners for its popular San Francisco–Los Angeles train, the Daylight. Formerly it had operated separate coffee shop and dining cars, each with its own kitchen. A substantial economy could be realized by placing a single kitchen between the two food-service vehicles. To ensure maximum efficiency, a full-sized mock-up of the 201-foot-long triplex was constructed at the Sacramento shops.[100] The fourteen-man crew was invited aboard to collaborate with the design staff, and the interior working space was shifted around until an optimum layout was achieved. More cars of this general design were purchased in 1941 and 1949 for other Southern Pacific trains (Figures 4.61 and 4.62). In the later sets, the kitchen dormitory

Figure 4.57 The general arrangement of a heavyweight Pennsylvania steel diner of 1922. (Railway Mechanical Engineer, December 1922)

AERIAL PERSPECTIVE DRAWING OF DINING CAR

NOTE!
ROOF AND UPPER HALF OF
NEAR SIDE OF CAR REMOVED
TO SHOW INTERIOR ARRANGEMENT

BUILT BY PULLMAN CO.—1930
FOR NEW YORK CENTRAL LINES
OVERALL LENGTH — 83 FEET

Ⓑ BUFFET   Ⓡ REFRIGERATOR
Ⓛ LINEN LOCKER Ⓕ FRUIT REFGTR.
Ⓔ SWITCH BOARD Ⓑ BEVERAGE REFGTR.

Figure 4.58 A perspective drawing of a heavyweight steel diner used on the New York Central's 20th Century Limited. (Edward Hungerford, Run of the 20th Century, 1930)

Figure 4.59   The Virginia Dare, a standard diner delivered in 1927 by Pullman. It was remodeled in 1938 as a streamliner.

Figure 4.60   Union Pacific dining car Number 383 was delivered by Pullman in November 1923. (Pullman Neg. 27544)

was put at one end. The diner and tavern-lounge sections were coupled at the point of articulation by a "hidden" joint, so that the two units had the appearance of one long room.

More exotic than multiple-unit diners were the dome and hi-level dining cars that appeared in the 1950s. The Santa Fe bought six hi-level diners from Budd in 1956 for its plush El Capitan.[101] The upper-level dining room seated 80; the kitchen was below. These were the largest single-unit diners ever built: 85 feet long and 15 feet, 6 inches high, with a weight of 96.7 tons. The food was sent up to the dining room on a dumbwaiter. The system and general arrangement were so new that service was

slow and orders were often confused at first, but the crew soon caught on.

A year before the hi-levels entered service, the Union Pacific offered its passengers the luxury of dining in a dome car. Competition for the 18 dome seats was keen. Another 28 passengers could be served in the lower dining room. Food orders were telephoned to the kitchen and the meals were sent up by dumbwaiter. Ten aluminum cars of this type were built for the U P at A.C.F.'s St. Charles plant in 1955.[102]

The dome cars on the Union Pacific represent the final development of the full diner. Few dining cars were built after that

Figure 4.61   A Southern Pacific diner publicity photograph made around 1955. (Smithsonian Neg. 73-739)

Figure 4.62   A triple-unit dining car built by Pullman in 1949. (Pullman Neg. 60431)

Figure 4.63   On shorter runs where full meals were not required, trains had café cars, such as this 1905 Pullman product. (Pullman Neg. 7881)

Figure 4.64   The Buffalo, Rochester, and Pittsburgh's café-observation cars Number 206 and 207 offered light food service for first-class passengers. Pullman completed the steel pair in 1913.

date, although a large number of half diners, variously called buffet, snack, grill, club, and café cars, performed an important on-train food service (see Figures 4.63 to 4.65). These cars were in many ways more significant, if less glamorous, than the full diners, since they probably fed more travelers.

The buffet car served drinks and snacks and might consist entirely of a closet-like kitchen and a serving bar. The remainder of the car was devoted to parlor or sleeping space. On the other hand, the café car might be a true half diner—that is, a small kitchen and dining room serving cooked meals, but without the variety or grandeur of a full restaurant car. Cafés were useful both on express trains whose runs were too short to serve many

Figure 4.65   Buffet cars were often combined with baggage cars. The Number 1336 was a Pullman product of 1914. (Pullman Neg. 17713)

Figure 4.66   The diminutive kitchen of this B & O buffet car, produced about 1925, is visible through the doorway.

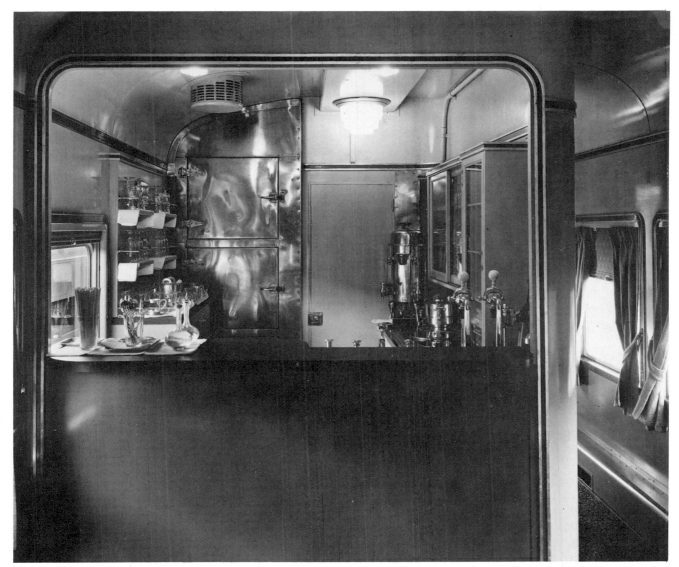

Figure 4.67   In many modern buffet cars, the kitchen became more of an open serving bar. The example shown here dates from 1935.

full-course dinners and on less well patronized first-class trains that had to offer proper if somewhat limited food service. And then came a mongrel group of second-class diners that were difficult to catalog. In some the entire car was devoted to food service, but in others part of the floor space was given to alternate uses. Many were rebuilt from old cars that were not necessarily diners. In design and finish they ranged from the professional to the amateur, and they went by many names befitting their uncertain origin and station: grill car, snack car, sandwich car, and lunch-counter car. Still another class of diner that developed in the years before Amtrak took over passenger train operations was the Automat car. Whether it can be accurately described as a third- or fourth-class accommodation is a matter of opinion.

The buffet car has been credited to Pullman; it is said that the idea came to him during a trip through Italy, where the need for a small snack such as a cup of coffee or tea, a sandwich, or some bouillon might "relieve faintness" between meals.[103] On a train a full diner's menu was too heavy for such needs, and the car would probably be closed during off-hours. How nice it would be if a pleasant sideboard were available at any time. In 1883 Pullman ordered some cars equipped in this manner for the West Shore Railway. A buffet measuring 3.5 by 8 feet was placed between the smoking and drawing rooms of a new car. This innovation seems to be more a natural outgrowth of the hotel car, which was then passing out of favor, than a new idea. There was also a precedent in the buffet aboard the *Victory* back in 1835.

Whatever its historical forerunners, the significance of the buffet car is that it permanently reestablished Pullman in the food trade. From that time forward, light food was sold on many Pullman-operated cars. The commissaries, stores, and personnel necessary to offer this service seemed to lead Pullman ever farther into the dining-car trade. Some roads, like the Milwaukee, bought diners from rival manufacturers but asked Pullman to operate them.[104] Full diners were rare in the Pullman fleet; there were usually no more than twenty, and these were in pool service for lease to heavy seasonal or special convention trains. Most Pullman food service cars were of the buffet or café class. In 1933, for example, Pullman's commissary department was operating 64 restaurant cars on nineteen different railroads.[105] It also had 171 buffet cars and 137 other composite sleeper, parlor, or lounge cars that were combined with a soda fountain, a sandwich bar, or a broiler buffet capable of serving simple grills. As the operation of independent food service grew more expensive, more railroads began to contract with Pullman's commissary department. In 1950 Pullman was catering aboard 400 cars.[106] The service ranged from full-course meals to beverages only.

Figure 4.68   The oaken interior of this Burlington café car was paneled with antique timbers imported from England. (Pullman Neg. 8255)

Twenty-six commissaries and 1,200 employees were required to run this part of the business.

Buffets were almost always first-class in their appointments (Figures 4.66 and 4.67). They were often combination cars, with a baggage compartment ahead and a parlor or sleeping section behind. Some buffets were little more than a counter with a gas-fired hot plate. Others, like those Pullman produced for the Burlington in 1906, had a kitchen, two small dining compartments for private parties, a public dining room seating 18, and a café smoking section seating 12.[107] The interior of these cars was decorated in high style, like a fine German restaurant, somber and rich in dark, old wood (Figure 4.68). The paneling was in fact sawed from some ancient English oaks that Pullman managed to buy from the Rockingham Park estate. When the planks were cut, a bullet dating back to Cromwell's time was discovered embedded in one of them.

During the same year that the Burlington's elegantly cabineted buffets were put in service, the American Palace Car Company was trying to revive the hotel car in the East. For some years the company had been operating a combination sleeper-parlor-diner built by Wason in 1890 on the patents of L. J. Harris.[108] In 1907 it sold four cars of the same general design to the Canadian Northern. Barney and Smith produced these 20-seat passenger cars. Meals were cooked in a buffet-sized kitchen at one end, and folding tables were set up for the meal. It was ingeniously worked out, but since the plan could not overcome the inherent defects of the old hotel cars, little more was heard from the venture.

Simpler forms of dining cars rather than more complicated ones were now receiving attention. The common man's eating car came into being in the early years of the new century. Fast, cheap food service was the purpose of the lunch-counter car (Figure 4.69). These utilitarian vehicles were often remodeled coaches. Out came the seats; in went a long stand-up counter, a grill, some shelves, and if necessary, a row of stools. The Pere Marquette put such a car in service in 1904 to accommodate the weekend crowds traveling from Chicago to the Michigan resorts.[109] It was designed for 60 people, but 150 to 200 at a time

Dining Car, C., B. & Q.

Dining Car. Pullman Co., Builders.

Café Parlor Car, El Paso & Southwestern. Pullman Co., Builders.

Observation Café Car, Chicago Great Western. Pullman Co., Builders.

Café Coach. New York Central. Barney & Smith Car Co., Builders.

Café Parlor Car, C., B. & Q. Pullman Co., Builders.

Lunch Counter Car, Père Marquette.

Figure 4.69 The variety of food service cars is illustrated by this series of floor plans published in 1906. (Car Builders' Dictionary, 1906)

Figure 4.70   Lunch-counter cars offered fast and economical service to passengers who did not want the complete meals offered in the regular diner.

would crowd in on a busy day. The Illinois Central operated a similar car in about the same period; the Pennsylvania produced an updated version in 1914. Gradually lunch-counter cars spread. On the Union Pacific they were called sandwich cars. More elegant versions, known as coffee shop cars, were built new for the Baltimore and Ohio, the Pennsylvania, and the Southern Pacific in the 1930s and 1940s (Figure 4.70).

A direct descendant of the lunch-counter diner was the cafeteria car, which served more elaborate hot meals. After passing the steam table, the passenger would carry his tray to a table. In 1945 Pullman offered to produce cars on this plan, which could serve two or three times as many people as a conventional diner.[110] Self-service would greatly reduce the labor overhead, and because only a small staff was necessary, it would be economical to keep the car open in off-hours. It seemed an attractive plan, but few railroads were much impressed by it. Perhaps passengers, inexperienced in carrying food trays on a moving train, too often found the task beyond their skill. The Union Pacific did, however, remodel four of its lunch-counter cars for cafeteria service in 1960.

The economics of the dining-car business had a direct effect on the design of the rolling stock and the general style of service eral ways: they offered more attractive service to increase vol-offered. Managers tried to reduce their continuing losses in sev-ume and revenue, they installed more efficient equipment to increase productivity, and they experimented with streamlined operating practices to lower out-of-pocket costs. Second-class diners, like the lunch-counter and cafeteria cars, answered the needs of the coach traveler, but more effort was expended on the luxury traveler. Sumptuous decor, food, and service continued to be pushed through the 1950s and is still offered on a few Far Western trains (Figure 4.71). But nothing remains today that is comparable to the New Haven's Yankee Clipper deluxe diner of 1930.[111] The waiters were in smart white uniforms and shoes; the steward wore formal attire (Figure 4.72). Only the choicest cuts were served, and all vegetables and fruits were fresh—no canned foods were used. Patrons might even choose between eggs with white or brown shells. When the C & O's George Washington was inaugurated, a colonial dining car was part of the consist (Figure 4.73). It featured colonial decor as well as

Figure 4.71 Western lines continued to purchase first class dining cars for the affluent long-distance traveler. This example was produced by A.C.F. in 1949. (American Car and Foundry Company, Neg. 61236-B)

Figure 4.72 The fact that dining cars were labor-intensive is amply documented by this B & O photograph of the 1920s.

Figure 4.73   Beginning in the mid-1920s, both the B & O and the C & O railroads operated cars with colonial interiors on their better trains.

colonial dishes. Fresh-cut flowers were common on all the best diners (Figure 4.74). The railroads' concern for their first-class diner is evident in the fact that the first passenger cars to be air-conditioned were always the restaurant cars.

These efforts to appeal to the luxury traveler undoubtedly attracted more business, but the increased revenues rarely covered costs. Cutting costs seemed to offer a more certain solution to the dining-car dilemma; how to minimize losses became the chief concern of the dining-car department. One of the most successful methods was through wholesale buying. Large railroads required prodigious quantities of food, and they could turn this to the same advantage enjoyed by a grocery store chain. The Southern Pacific, for example, consolidated its dining-car, hotel restaurant, and maintenance-of-way kitchen car purchasing.[112] Every effort was made to buy a one-year supply of basic commodities when they were in season, thus securing the cheapest price and best quality. Since the railroad had its own refrigerator cars and

warehouses, transport and storage presented no problem. Vast stores were kept in reserve, since the road's food consumption in a single day included 1.5 tons of potatoes, 8,400 eggs, 500 pounds of butter, 300 to 400 pounds of coffee, and 350 pounds of chicken.

Food preparation was another area in which costs could be cut. Most large railroads and the Pullman Company maintained commissaries at major terminal points—kitchens that made pies, ice cream, puddings, and soups. There potatoes were machine-peeled on a production basis, and meats were cut, trimmed, and wrapped. The commissaries saved time and work for dining-car cooks, as well as ensuring uniform portions and quality. When frozen food became a commercial reality after World War II, railroads quickly adopted it. The Rock Island, in fact, began to serve frozen precooked dinners in the early 1950s.[113]

Idle equipment makes no money, and deadheading is equally profitless. Careful scheduling achieved maximum utilization of each dining car. A diner sent through from New York to Chicago

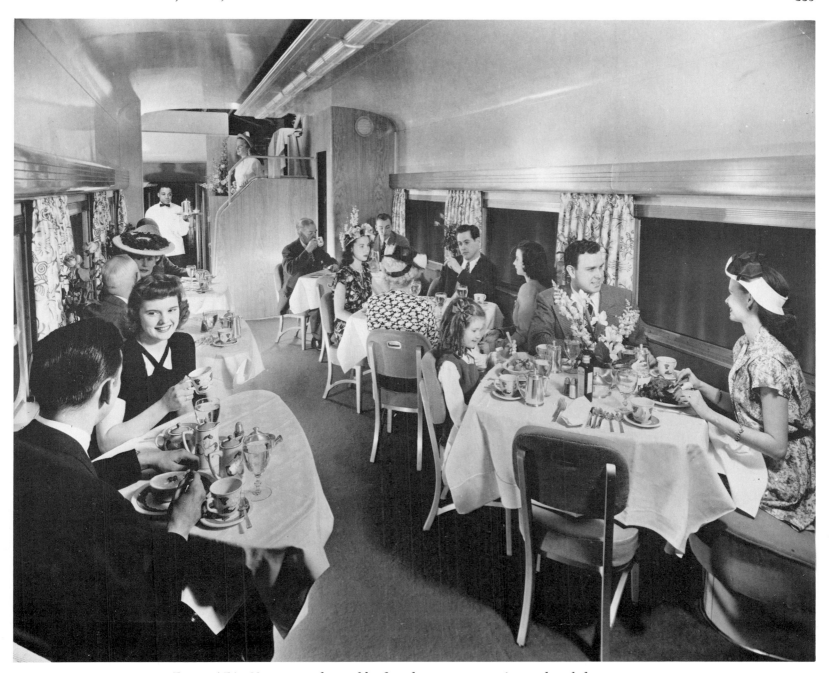

Figure 4.74   Name-train diners, like first-class restaurants, featured such luxurious accessories as linen tablecloths and cut flowers. (General Motors)

would spend long, empty hours rattling through the night. Hence it was common to cut out the diner along the line after the last evening meal was served. The car would be cleaned and restocked so that it was ready to serve breakfast on a train that picked it up the next morning. Cars were worked back and forth over a large system, serving meals on several trains before returning to their home terminal. Like the cars, crews would be picked up and dropped off en route. This method of operation was employed from the earliest days of dining-car service. The New Haven's first diner introduced in 1884, for example, ran on the noon train between Worcester and New Haven. It was then deadheaded to Boston to provide dinner on the evening express to New York.

Technical improvements in the dining-car kitchens themselves could also effect savings (see Figures 4.75 to 4.77). Simple changes like adding a side-door entrance to the kitchen from the outside speeded the loading of supplies. It was also an efficient way of ventilating. When business was slow, the side door pro-

vided a breezy view for a lounging cook. Improving the stove was a more complex process. Coal-burning iron ranges were traditional. It was also a tradition that the junior cook had to come aboard hours before the scheduled departure to light the giant black monster. Charcoal was sometimes used, but coal was the standard fuel and remained so until very recent years, when the older cars went into retirement. More convenient fuels were tried out as they were developed, but diners were relatively slow to put them into general use. Steam was widely available after 1900, because vapor heating became common. It was not considered useful for cooking, but it did serve to keep dishes hot. Gas appears to have been the first practical alternative to coal. Nearly all first-class cars had gas lighting by the 1890s, and the large reservoir of this combustible fuel under each car naturally suggested its possibilities as a cooking fuel. Pintsch gas was used aboard buffet cars during the period. To a limited extent, bottled gas was employed to fire full-sized ranges in the 1920s. By the 1940s propane was installed on some of the better new diners,

Figure 4.75   Dining-car kitchens were arranged like a ship's galley. They had the utilitarian look of a miniature food-processing factory. (Pullman Neg. 19955)

such as those assigned to the Daylight. Before that, the Southern Pacific had been using Presto-logs.[114]

Electric ranges held the greatest promise for dining-car service. This type of stove was installed in a few diner kitchens in the twenties and thirties, but the limited power supply on board train and the old standard of low-voltage direct current made it impractical. Abundant high-voltage power became common with the advent of air conditioning, thus opening a way for the electric range. The Pennsylvania's twin diners of 1940 had combination cooking: electric for regular dishes and a charcoal grill for broiling.[115] An electric coffee urn was a marvel of efficiency, producing several gallons of fresh brew in just two minutes. In 1946 the Illinois Central began work on an all-electric kitchen at its Burnside shops.[116] Basically the plan was an adaptation of a design perfected for submarine galleys. Postwar shortages prevented completion of the first car (Number 3987) until early in 1949. Besides the range and refrigerators, the car had a 3-gallon coffee maker and an electric deep-fry kettle capable of producing 100 pounds of fried potatoes, chicken, or fish an hour. Mounted on the car's underside were two 50-kilowatt diesel generators that supplied the power. Not many other all-electric kitchens were installed, but electric ranges became relatively common after

1950. During this period experiments with super-fast infrared and microwave ovens were also underway.[117]

Other gadgetry entered the twentieth-century dining-car kitchen. In 1940 came dishwashers, garbage disposals, and automatic door openers that eased the way of tray-laden waiters into and out of the pantry.

The electric refrigerator was adopted before these niceties and proved to be a more basic improvement. A well-insulated icebox could keep food for as long as three days without servicing, but not all boxes were this well constructed, and losses from spoilage grew less tolerable as food prices continued their historical rise. Another disadvantage of the icebox system was the space taken by the great blocks of ice. There seemed little enough room to stow the 1,200 to 1,500 pounds of food needed for a fully stocked car without giving up space to this fast-melting refrigerant. Around 1925 the Milwaukee Road began to experiment with electromechanical refrigerators.[118] The technology was already well advanced for home and commercial use, but it had yet to find acceptance on the dining car. The Milwaukee's installation was a success, and within a few years seven cars had mechanical refrigerated main meat lockers, although their other units remained iceboxes. In September 1927, however, the road con-

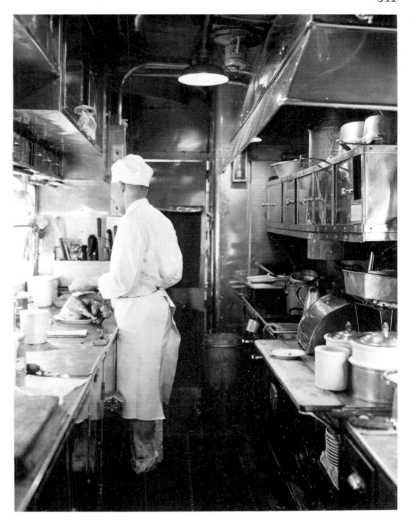

Figure 4.76 A chef at work in a B & O dining car in February 1936. Coal-fired stoves and ice-box refrigerators remained standard equipment.

verted all the boxes on diner Number 5138 to mechanical refrigeration. In 1930 the New Haven followed the Milwaukee's lead.[119] Other roads gradually joined them, although the old-fashioned icebox remained a part of the dining-car picture for many years. A few roads, such as the Union Pacific, regarded dry ice as the preferable cooling medium.

Labor was the largest single operating cost, and historically it rose faster than any other. In the twelve years between 1937 and 1949 the wages for cooks doubled, while those for waiters tripled.[120] So far as the full dining car was concerned, wages were the most difficult item to keep under control. The diner was labor-intensive by its very nature. The only way to reduce the work force was to change over to lunch-counter cars or self-service grills, and this was not an acceptable alternative on first-class trains. As trade unions became stronger during the twentieth century, it became more difficult to close out jobs. Thus a dramatic drop in patronage was not necessarily followed by a comparable decline in the number of chefs, stewards, and waiters. In 1925 when railway passenger travel was in full flower, some 10,000 people worked in dining-car service. Yet in 1961 after passenger traffic had dropped by more than 75 percent, 9,600 people were still employed.[121] Some managers hoped that waitresses would offer a way out of the rising labor cost spiral. Women hostesses were hired by the New Haven and the C & O, but most roads stayed with all-male crews. The only effective way found to control labor costs was to eliminate the full diner.

When the railroads were hauling the vast majority of intercity travelers before 1950, they were content to accept dining-car deficits as a necessary business expense. The 8- to 10-million-dollar annual loss was charged off to advertising and good will.[122] For every dollar of revenue in 1950, it was necessary to spend another 40 cents to cover dining-car expenses.[123] The C & O, which had a rather small passenger train operation, reported a $700,000 dining-car loss during the same year. A few years later the industry as a whole claimed a loss of 29 million on diner operations.[124] Individual large systems, like the Pennsylvania, were losing as much as 12 million dollars a year; cutbacks were inevitable. Some lines, like the Cotton Belt and the Long Island, simply abandoned diners altogether. Others, like the Lehigh Valley, cut back on personnel and menus. Cooked-to-order breakfasts gave way to a Continental morning menu of sweet rolls, dry cereal, and coffee. One man did the job formerly requiring three. In 1961 the Pennsylvania pulled off all full diners running between New York and Philadelphia. In their place mobile food carts were pushed through the train by a single vendor, who offered a sad variety of packaged sandwiches, cakes, and soft drinks.

Many of the genteel refinements of the dining car were already gone by the time the food carts took over. The New York Central once spent $2,000 a month for fresh-cut flowers. By the early fifties only passengers on the Twentieth Century Limited could expect to find flowers on the table. On other lines, linen gave way to paper napkins and paper tablecloths. Ever more rigorous economies soon sliced away the last vestiges of the elegance once

common to all American diners. Only a few Western lines attempted to maintain the old standards. These lines did not include the Southern Pacific, which set out to cut its losses when dining-car patronage took a nose dive in the late fifties. The first step was not drastic; the triple-unit diners were replaced with single-unit cars.[125] But trade had fallen off too much to support a conventional car, and in 1961 the road decided to further downgrade the food service by substituting Automat cars. Eleven old sleeping cars were remodeled to dispense soft drinks, candy, and hot dishes through coin-operated food machines. Tables and chairs were available at one end. An attendant made change and tidied up. This scheme became the subject of so much criticism that the cars were withdrawn from service after Amtrak assumed passenger operations on the Southern Pacific.

The lesson of the Automat was not lost on passenger train designers. Yet patrons expecting a return to the palmy days of the traditional dining car were disappointed by what the Metroliner and Turbo Train have to offer. On Metroclub (parlor) cars and the Turbo Train, airplane-style meals are served at the seats. The food is prepared in advance and kept warm in a heated tray box at one end of the car. The meals are acceptable and not overly expensive, but the special ambiance that the experienced traveler had come to associate with dining on the rails is not present. Now the passenger must crouch over a mechanized meal which is perched on a low, folding table in front of the seat. The Metroliner coach riders feed in even less comfort at snack bars located in the center of every other car, where pastry, sandwiches, and coffee are served from a stand-up counter. A single attendant dispenses these modest items, which passengers may consume at their seats or at a counter opposite the serving bar.

Figure 4.77   American Car and Foundry equipped this Kansas City Southern dining car kitchen (1948) with a gas stove and electric refrigerator. (American Car and Foundry Company, Neg. 61229-R)

Such a dining area is objectionable on at least three counts: standing to eat is essentially a fueling operation, trainmen and passengers block the aisles at station stops, and the absence of windows in this section of the car is oppressive. The passenger who chooses to take the food to his seat must balance it on his lap and ignore the looks of his neighbors as he crumbles and spills his way through the meal. Ironically, the very line on which commercial dining cars first appeared has come full circle back to the stand-up lunch counter of the 1860s.

Private Cars

The private car was the least representative of the general passenger car fleet; the average traveler rarely saw one, much less rode in it. Their numbers were small even when business cars were included in the total. In 1930, which appears to be their peak year, only about 900 were in operation. Wayner lists approximately 750 in his *Car Names and Consists*. In the whole history of American railroads it is doubtful if more than 2,000 private cars were built for North American service, and again

this figure is considerably inflated by the inclusion of business cars.

Yet no other form of railroad conveyance has so completely captured public attention. A treatise on the all-important day coach has little popular appeal; the story of the sleeping car interests a small group; but the private car appears to stir enthusiasm among all sorts of readers. Fascination and curiosity about the very rich, their lives and possessions, seem indigenous to all societies and eras. Romance and glamour have developed around the excesses of kings and princes, and in America we look to our merchant princes to do their duty and spend royally on such baubles as private railroad cars. Lucius Beebe insisted that the private car added another dimension to an age of wealth, well-being, and conspicuous consumption (Figure 4.78).

Some private car owners, like A. A. McLoed, went to considerable lengths to obtain the showiest car possible. The more extravagant and costly, the greater the publicity. Like Jim Brady's diamond ring, a flashy railroad car immediately established credit and social position.

Car manufacturers were equally interested in the private car as an image builder. An order for a thousand boxcars or even a

Figure 4.78   The private car was a symbol of luxury and prestige in the nineteenth century. This unidentified group posed with obvious pride beside their private car sometime in the 1880s. (Bar Harbor Historical Society)

dozen coaches rarely gained much newspaper space, but a new rail palace for Senator William Sharon was news. The purchaser was a big name; hence any major acquisition was news in itself. The price of $50,000 was news, and the splendors of its interior furnishings made exciting copy for journalists. The handsome paneling, rich hangings, and overstuffed furnishings were standard, but what about the dozen master wood-carvers who worked on it, or the Italian artist imported expressly to do the ceilings and the cluster of golden angels ringing the transom frieze? A dedicated reporter could even find magic in the kitchen: "The fire glowed in the range and shone on polished pots, pans

and porcelain. . . . Pudding and jelly molds, skewers, steamers and saucepans as bright as silver hang on the hooks of the dresser. Stores of delicate china are nestled in the snug closets, crystal and silverware crown the oaken buffet."[126] The journalists were encouraged with luncheons in their honor and special predelivery trips in the car. Their descriptions were inspired as much by free champagne as by the car itself, according to *Railway Age*.[127] In an effort to gain free advertising some builders, like Ohio Falls, were claiming prices as great as $30,000 when the car actually sold for $9,000. They would also attempt to spread gaudy rumors —say, that the car would be used by President Garfield.

Figure 4.79   A special eight-wheel car was furnished the Duke of Wellington and his party for the opening of the Liverpool and Manchester Railway in 1830. (Detail of Smithsonian Neg. 8953)

In the strictest definition, a private car was the personal property of an individual owner. Such cars did exist, but the greater number of cars commonly referred to as private were actually railroad-owned. Technically they were the railroad's office or business cars that were intended to transport its officials over the line in the performance of their duties. The Association of American Railroads came to classify both private and business cars by the letters PV. It is often difficult to differentiate between private and official ownership, since many private car owners were directors or major shareholders in railroad corporations. In some cases a railroad-owned car was reserved for their exclusive use, yet it might be carried on the company's books as an official car. Who inside the ranks would argue the point with a Vanderbilt or a Gould? In other instances the car might be purchased by an individual, but as a director of a railroad he could expect free storage, maintenance, and carriage. Interchange privileges with other lines were a welcome extra dividend. Again, who would challenge a financial tycoon to prove that his itinerary was official or personal? In 1930 884 PVs were reported in service. The Pullman Company estimated that less than 50 of these were privately owned, and even in that group many were undoubtedly somewhere in the shadowland between private and railroad ownership. The semantics of the subject are also significant. Railroads became very concerned about emphasizing the utilitarian nature of their PVs through the designation of "business" or "office" car. The intent was to dispel criticism of wanton extravagance and to demonstrate that the stockholders' investment was being used for the affairs of the railroad. Yet in most office cars there was little space devoted to desks, file cabinets, or typewriters.

The abuses associated with private cars came to silence the high-flown reports of their luxuriousness that were once considered such good publicity. To obscure the size of the private car fleet, some major roads began to buy cars under the name of a

subsidiary. In 1895 the B & O charged a new PV to the account of the moribund Sandusky, Mansfield, and Newark. In this way it could be made to appear that the parent corporation had only six office cars, when in fact it might own twice that number. Shareholders' complaints about profligate or self-indulgent officers led to suppression of news about private cars. It is conceivable that the paucity of data available for the historian on the earliest PVs is due to this self-imposed censorship.

The proliferation of business cars had reached scandalous proportions by the early 1890s.[128] On the Union Pacific, for example, every employee above a flagman seemed to have his own car. Railroad officers would have better understood the problems of the average traveler if they had taken passage in coaches rather than in the splendid isolation of PVs. Money always seemed available for office cars; small roads, as well as larger ones on the verge of receivership, spent their funds on PVs. Junkets to fashionable watering places were commonplace. On a single day in February 1891, twenty-one private cars were counted on a siding in Jacksonville.

The exchange of privileges was another costly abuse indirectly borne by the shareholders and the public.[129] An official of the smallest line might travel free of charge over the entire national network. For example, during the 1930s the president of a minor Southern line rode a mere 167 miles on his own property and 61,000 miles on other systems. At one point so many private cars were awaiting free carriage east that the New York Central was obliged to run a special train. Eventually the I.C.C. felt obliged to limit these abuses. In 1928 it ended the free carriage of private cars except for on-line movements.

The ancestor of the private car was an elaborate vehicle for dignitaries who attended the opening of the Liverpool and Manchester Railway on September 15, 1830. The Duke of Wellington,

Figure 4.80   The Western Railroad (Massachusetts) produced a private car for its directors in 1839. The drawing is a reconstruction based on early photographs and verbal descriptions by employees of the Western Railroad. (Tracing by John H. White, Jr.)

hero of the Napoleonic wars and now Prime Minister, was the principal guest. He and his party rode in the imposing eight-wheel car shown in Figure 4.79. The following description is taken from a newspaper of that day:[130]

> Built by Messrs Edmundson's of Liverpool, the floor is 32 feet long by 8 wide, and is supported from 8 large iron wheels. The sides are beautifully ornamented, superb Grecian scrolls and balustrades, richly gilt, supporting a massy hand rail all round the carriage, along the whole centre of which an ottoman will be the seat for the company. A grand canopy 24 feet long is placed aloft upon gilded pillars, and is so contrived as to be lowered for passing through the tunnel. The drapery is of rich crimson cloth, and the whole is surmounted by the ducal coronet.

The ducal car was rarely used and appears to have excited no demand for more private cars in Great Britain. The Dowager Queen Adelaide acquired what is the next recorded instance of such a rail vehicle in 1842, and Victoria's first royal car appeared two years after that.

In theory America's First Citizen was expected to sit among the common folk, yet in 1841 when President Harrison was in transit between Baltimore and Washington, the superintendent asked if he should provide a "distinct car." Two years later President Tyler journeyed to Boston for the Bunker Hill Monument dedication. The Camden and Amboy felt obliged to offer a separate car, and since the road had no official car, it provided its coach Number 19 fresh from the shops after a complete rebuilding.[131] The coach had been fitted with six-wheel trucks, a variety of seats including several styles of reclining chairs, and a privy on one platform. It was regarded as very special and hence suitable for so honored a patron, and after the trip it was renamed for Tyler.

After leaving the Camden and Amboy, it is possible that the President continued his journey in an actual private car, for two New England lines did own them. In 1841 or 1842 the New York and Hartford completed a PV for its chief executive, Samuel R. Brooks.[132] And the Western Railroad had acquired a special car for its directors as early as 1839. This is the first American private car for which a record can be found, and because it figured in

two Pullman sleeping-car patent suits, some details of its construction have been preserved.[133] It was a boxy, four-wheel car that hardly seems like a grand beginning for the private car in America, but it may have been luxurious in its day. The body plan has been reconstructed from dimensions supplied by old employees during the lawsuits (Figure 4.80). An engraving published by *Scribner's* magazine shows the body as it appeared in the 1880s. The car was divided into three compartments. It had face-to-face seats that could be converted into sofas by placing a plank between them and rearranging the cushions. The car was not a favorite with the directors, however; either it was too small or there was not enough call for it. In addition, the introduction of the double-truck car probably made it almost immediately obsolete. The car was reportedly operated as a regular coach between 1840 and 1845. Later it was used on mixed trains and provided the crew with a deluxe caboose. By the late 1850s it was set off the tracks near West Springfield as a watchman's shed, and a few years later a car inspector moved the body to his garden for a toolhouse. There it stood until rot and neglect obliterated the last vestige of America's first mansion on rails.

So long as railways remained short, there was no real need for private cars. A brief excursion could be tolerated in the meanest car. But as the system was extended, railroad officials, rich people, and prominent people who were obliged to travel yearned for a better class of car. During Jenny Lind's tour of North America in 1850–1852, for example, she rode thousands of miles. Added to the normal fatigue of travel was the adulation of her public, which assembled in oppressive numbers. The problem was magnified by her marriage in Boston toward the end of the tour. So that the honeymoon couple could travel to Pittsfield in peace, a coach was quickly remodeled for them.[134] The seats were pulled out and replaced with luxurious household furnishings. Some months earlier, Miss Lind was reported to have made a portion of her Southern tour in a coach converted into a private drawing room. In later years other stage celebrities found private cars a necessity.

As the B & O reached out toward the Middle West, office cars for its officials were quietly added to its stock. No PVs are listed until the 1854 annual report, which records two cars. One of

Figure 4.81   The two floor plans of the Baltimore and Ohio Railroad's director's car of about 1855 are from A. Bendel's 1862 book on North American railways and Isabella S. Trotter's travel book of 1859.

them was described as "new," indicating that the other had been placed in service some time before. Three years later the line had three office cars. In 1857 the Ohio and Mississippi built a 64-foot director's car with four compartments and 30-inch-wide passageway along one side.[135] The interior paneling was Gothic in style and painted white with gilt highlights. A Brussels carpet and crimson-velvet-covered furnishings added to the richness of the interior. Two of the sofas could be converted into double berths. The washroom had a "toilet table." A patented ventilating system was effective in cooling and cleaning the air. The car was said to be something "new in the history of railroad traveling," and if this statement is accurate it would indicate that private cars were still a rarity.

The most complete data on the second generation of American business cars is found in the published accounts of two foreign visitors.[136] Both described B & O cars observed in 1858–1859, and both include floor plans. As shown in Figure 4.81, the drawings differ slightly although the general arrangement and the overall size—roughly 8 by 40 feet—agree almost exactly. The three compartments offered an observation-sitting room at one end and a central stateroom, with a side aisle leading to the pantry–crew's quarters. The observation room sofas probably doubled as night roosts for junior members of the party. The absence of a cooking stove indicates that only light meals were served. By far the most interesting feature is the rear observation platform; the drawings are documentary evidence that this basic feature of the private car was in existence before the Civil War. And it may be that the car (or cars) shown in Figure 4.81 was used by Lincoln on his trip to Gettysburg. He was known to have borrowed a PV from the B & O for this historic journey.

In the fall of 1858 a young English gentlewoman, Isabella Stuart Trotter, toured America. One of the floor plan drawings comes from her travel account, which also gives the following description of life aboard a private car in that distant time. Luxurious railway travel, at least for powerful railway men and their favorites, was a reality long before George Pullman's crea-

tion of civilized train-board life. Mr. Ralph Greenhill of Toronto, Canada, kindly brought Miss Trotter's memoir to my attention and provided the following excerpt:

Though Mr. Garrett talked of the directors' car, we presumed it was only a common carriage such as we had been accustomed to, but appropriated to their use; instead we found a beautiful car, forty feet long by eight wide, of which the accompanying diagram shows a plan drawn to scale. Outside: painted maroon, highly varnished with Canada balsam; the panels picked out with dark blue. Inside: painted pure white, also varnished. Ceiling the same, divided into small narrow panels, with excellent ventilators at each end. Round the car there were twenty-two windows, not shown in the plan, and three brilliant lamps in the sitting-room and hall, and one in the bed-room; these were lighted when passing through the tunnels. There were three hooks in the wall serving for hat pegs, and at the same time to support two flags for signals. A large map of the mountain pass from Cumberland to Wheeling hung over the sofa opposite the table. The table was covered with green baize stretched tightly over it. On the table were placed a large blotting-book, ink, and pens, three or four daily newspapers which were changed each day, the yearly report of the railway, a peculiar timetable book, containing rules for the guidance of the station men, times of freight and passenger trains meeting and passing each other, &c. Papa has these. The sofas are covered with a pretty green Brussels carpet (small pattern) quilted like a mattress with green buttons, chairs covered with corded woolen stuff, not a speck or spot of ink or smut on anything. A neat carpet, not a speck or spot on it, a sheet of tin under and all round the stove. Pantry cupboard containing knives and forks, spoons, and mugs. Bed-room berths much higher and wider than in a ship. Red coloured cotton quilts, with a shawl pattern, two pillows to each bed, pillowcases of brilliant whiteness, sofa bed larger and longer than a German bed. White Venetian blinds occupied the places usually filled by the door panels and window shutters. Green brussels carpet like the cover of the sofa; three chairs to match. The windows in the sitting-room had grey holland curtains running on wires with very neat little narrow strips of leather, and a black button to fasten them, and a button and well made buttonhole below to keep them from blowing about when the window is open. Looking-glass in neat gilt frame, hung over a

Figure 4.82   This extraordinary sixteen-wheel car was produced for the directors of the Chicago and North Western Railway in 1867 by the road's Fond du Lac Shops. (Chicago and North Western Railway)

semicircular console in the bed-room, another near the washhandstand, where a towel also hangs. Two drawers for clothes, &c. under berths. Table-cloth for meals, light drab varnished cloth, imitating leather, very clean and pretty, china plates, and two metal plates in case of breakages. Luncheon consisted of excellent cold corned beef, tongue, bread and butter, Bass's ale, beer, whiskey, champagne, all Mr. Tyson's. We supplied cold fowls, bread and claret. The door at the end opens on a sort of platform or balcony, surrounded by a strong high iron railing, with the rails wide enough apart to admit a man to climb up between them into the car, which the workmen always do to speak to Mr. Tyson. Usual step entrance at the other end. The platform can hold three arm chairs easily, and we three sat there yesterday evening, talking and admiring the view.

During the 1860s, accounts of official cars became more common. The *American Railway Review* described four during the first years of that decade.[137] The Michigan Southern and Northern Indiana provided an elegant car for its directors modeled after a special car that the Michigan Central had fitted out for the Prince of Wales. In 1862 the La Crosse and Milwaukee produced a directors' car with such features as a clerestory roof, bay

windows, and built-in furnishings. Native Wisconsin woods were used for the interior paneling. During the following year, T. W. Kennard, chief engineer of the Atlantic and Great Western, acquired an "elegant little dwelling house" on wheels. Officials lower on the corporate ladder were now beginning to enjoy the comforts of the private car. Kennard's car, however, represented more than an executive status symbol to be used for an occasional inspection. Construction of the road was in full swing, and the chief engineer was traveling over 1,200 miles a week. His rolling office was well justified and well used. The 48-foot body contained a parlor, a bedroom, a washroom, and a kitchen.

The grandest private car of this early age was built in 1866–1867 at the Fond du Lac shops of the Chicago and Northwestern Railway (Figures 4.82 and 4.83). A contemporary newspaper described the sixteen-wheel giant as more like an "ocean steamer than a car."[138] It might be properly called the *Great Eastern*'s counterpart on rails. The 70-foot-long Gothic structure came complete with a 24-foot sunken central parlor. The parlor floor was dropped nearly 2 feet through the use of a shadbelly frame, making the ceiling 15 feet high. Here was the car with every-

Figure 4.83   A portion of the interior of the Chicago and North Western Railway's directors car of 1867. (Chicago and North Western Railway)

thing: mirrors, a frescoed ceiling, Brussels carpets, ottomans, sofas, easy chairs, a piano, crown-glass windows, and oak, cherry, walnut, and ash panels—in short, every trick known to the car builder's trade. At each end were 12-foot-long parlors; bedrooms and closets took up the remaining free space. Since there was no room for a kitchen, a small commissary car was attached for long runs. In later years the once-splendid directors' car was given to a lowly division chief. It ended its days as a station house somewhere in South Dakota, and sometime before 1910 was destroyed by fire.[139]

In 1870 the directors of the Lake Shore and Michigan Southern acquired their own car, the *Northern Crown*.[140] It was a mere 60 feet long, its central drawing room a paltry 17 feet. But it did have extra-large windows for the time (37 by 47 inches), and its chrome-yellow exterior was decorated with "well executed landscapes representing the great industries of the age." The iron handrails had silver mountings. Still, the car was no match for the hulking magnificence of the North Western's.

By 1870 the business car was well established: most trunk lines owned one or more. The opening of the Pacific railroad had provided an even more plausible argument for expanding the PV fleet. Only the most heartless shareholder would begrudge the comforts of a clean bed, a writing table, and good meals to any railway official forced out along the line in the Far West. Thomas Durant had a fine car as early as 1866 for his end-of-the-line inspections on the Union Pacific.[141] During the same year he purchased a special car constructed by the government for President Lincoln. This second car was used by guests or government officials traveling over the line. Meanwhile the Central Pacific shops built a large private car for Leland Stanford that could sleep ten and was paired with a provision car on long journeys.[142]

Eastern managers' need for private cars was less convincing, but there the Directors' car was already a well-established institution and it was not difficult to justify individual cars for the top brass. John W. Garrett accepted the car *Maryland*, produced at the Mt. Clare shops in 1872, as a necessary fixture for the office of the president of the Baltimore and Ohio. This historic car was magnificently engineered and furnished:[143]

Some months ago, a car was built at these shops designed for the special use of the president of the road, John W. Garrett, Esq. The best skill and resources of the car department were brought into requisition to produce a vehicle creditable to the builders and suited to the service for which it was designed. It has already made several trips to New-

York, and has elicited the admiration of all who have examined its structure and finish. When not in use it is kept in a special house at the Camden station, in Baltimore, where we had an opportunity to inspect it, and for which we are indebted to the politeness of Mr. Schryack, of the Mount Clare works.

The car is named the "Maryland," and is 51 feet long by 10 wide outside the body, runs on six-wheeled trucks with strong check-chains attached, and with Dinsmore springs on the equalizers. Iron body-transoms are used of the kind represented in the foregoing cut. The outside of the car is painted a light yellow, with nothing in the way of external ornamentation to attract special attention or indicate the quality of the interior fitting-up, which is in the best Pullman-coach style. The car is designed to run always in the rear of a train, so as to afford a view of the track from the end windows. A glance at the arrangement of the interior indicates at once that the vehicle is designed for but one principal occupant with his traveling suite. There are four distinct compartments, a porter's room, state-room, a sleeping and toilet-room with side passage and closets, and the parlor or drawing-room. The porter's room occupies about ten feet of the forward end of the car, with a closet for table-ware on one side of the door, and a Baker & Smith heater on the other. Next to this is the state-room with an upper sleeping-berth on each side, and seats for eight persons underneath, and which can be transformed into two lower berths. Next is the central compartment, occupying a space of about 17 feet in length, and containing the principal sleeping and toilet-room, and corridor, with a door and two windows in the partition between. There are also included in this division a water-closet and linen-locker, communicating with the sleeping-room. This room is sumptuously furnished. Across one end is a curtained lounge. The bedstead is an elaborate piece of cabinet-work in French walnut, surrounded with heavy double damask crimson and green curtains. The window-curtains are of the same description, and hung, like the others, on silver-plated rods. Two large mirrors occupy the spaces between the windows. The remaining portion of the car, comprising a space of about 15 feet and extending to the rear end, constitutes the drawing-room, which is entered by a door from the side passage above mentioned. Its furniture consists of a large and elegant sofa-lounge, an oblong black-walnut centre-table with marble top, two easy-chairs of the Pompadour style and two others of a different pattern. This apartment has five windows on each side and three in the end, these last affording a fine view of the track. The spaces between the side windows are occupied with mirrors, and the curtains are of the same kind as those in the sleeping-room. The floors have Brussels carpets. The interior finishing throughout is in solid black-walnut, with elaborate raised paneling of French walnut "burl," with semi circular tops. The general effect is somewhat sombre, but is relieved by light gilt mouldings above the windows. The artificial lighting is done by four of Williams, Page & Co.'s improved plated lamps. Carroll ventilators are used, and all the windows have double sashes. In the central passage-way is a wash-stand with a reservoir underneath holding a barrel of water, which is raised by a small force-pump. The inside door windows are beautiful specimens of embossed glass.

During Garrett's lifetime the car was used by many dignitaries, including several U.S. Presidents.[144] In 1908 it was sold to the Quebec Central and was reported still in operation in 1949.

Thomas Scott of the Pennsylvania Railroad acquired a personal car the year before Garrett's *Maryland* was completed. An experienced operating man, Scott believed in frequent over-the-line inspections. Several years later he was partially paralyzed by a stroke, so that a special car was even more of a necessity. In addition he was attempting to build a second railway empire in the Southwest, a venture that called for regular journeys to Texas. Scott's car was one of the first PVs to be numbered rather than named. The simple designation 120 indicated its utilitarian purpose, and the same number was used by P R R presidents until recent years. The first Number 120 was a product of Altoona. The 63-foot twelve-wheeler was completed in 1871.[145] In addition to its pointed Gothic windows, red body, brown trucks, and green wheels, the 120 is thought to be the first American railway car equipped with a bathtub.

Cornelius Vanderbilt was a greater railroad chieftan than either Scott or Garrett, and as a poor boy become suddenly rich, he was quick in acclimatizing to the luxuries of this world. In the spring of 1871 reports began to appear about a private car nearing completion for the Commodore.[146] It was shorter than a coach but ample enough to accommodate a 25-foot drawing room, a parlor, a bedroom, and a pantry. The triple style of windows was used, with the large center pane measuring 42 by 36 inches. The exterior was painted cream with buff trim, but no gaudy ornamentation was permitted. Nor would its owner allow a public showing; he wanted no more publicity. The car builders wanted to install gaslights, but this scheme was also overruled by the old man, who was content with candles. According to Charles Sweet, the car was not a favorite with its owner.[147] For his personal jaunts Vanderbilt is said to have preferred the Wagner parlor car *Dutchess*, with some of the parlor seats pulled out, other furnishings installed temporarily in their place, and a portable zinc floor and kitchen added at one end. Some years after the Commodore's death, his private car could be seen in service between New York and Chicago. Wagner was running it as a parlor car under the name *Iroquois*.

The early history of the private car revolves almost exclusively around vehicles built to serve railroad officials. It is difficult to identify the first individually owned passenger car, particularly in view of the tangle between railroad and private ownership. For example, the cars just described may actually have been true private cars, since Durant and Vanderbilt may have paid directly for theirs even though they were railway officials. The car built for President Lincoln between 1863 and 1865 would seem to be a clear-cut example of a private car, because it was surely not built for a railroad official. Yet it was never used by Lincoln during his life, and after the funeral trip it was sold to the Union Pacific as a business car. The personal private car does not appear to have come into being until sometime in the 1870s.

Some historians have suggested that P. T. Barnum was the first private citizen to own his own car. The date of his purchase, said to be in the seventies, is uncertain, and like so many facts ascribed to the great showman, remains a source of continuing speculation.[148] According to Lucius Beebe, the Western banking-mining king Darius O. Mills bought a private car in 1870 or 1872. However, a contemporary description puts the date ahead to 1875 or 1876. Early or late, Mills's car set the pattern for other

Figure 4.84   Darius O. Mills named his private car the D.O.M. *It was built in 1875 or 1876 by Harlan and Hollingsworth.* (National Car Builder, *January 1876*)

Figure 4.85   Mills's D.O.M. *stands ready for delivery at the factory's transfer table.* (G. M. Best)

Western moguls to follow. He commissioned Harlan and Hollingsworth to fabricate a first-class vehicle; there was to be no scrimping on price and no Vanderbilt-like repression of ornament.[149] The exterior was in fact a festival of color. The body was canary yellow with brown, red, and dark-green trim. The trucks were orange with green wheels. The body measured 53 feet, 10 inches by 9 feet, 8.5 inches. The car, named *D.O.M.* after the initials of its owner, cost $15,000 (Figures 4.84 and 4.85). Peter Donahue, a West Coast capitalist, hoped to eclipse the *D.O.M.* by having a PV built locally from native wood by the Kimball car works of San Francisco in 1876–1877.[150] Leland Stanford, realizing that his Sacramento-built car was outclassed by these late arrivals, placed an order with Harlan and Hollingsworth for a car equal to Mills's or Donahue's. According to the Pacific Coast railway authority G. M. Best, the *California* was delivered to Stanford in 1877.

Of all the PVs built during the seventies, George M. Pullman's own car was easily the most famous. The car was called the *P.P.C.* for the Pullman Palace Car Company, and the letters were interwoven to form an elaborate monogram on the car's side. The *P.P.C.* was thus not only a conveyance for Pullman and his family but a subtle rolling advertisement for the sleeping-car business. It was put to other uses as well, for Pullman delighted in lending the car to traveling dignitaries. Presidents, South American emperors, European royalty, great men of industry and finance—all enjoyed complimentary excursions aboard the *P.P.C.* As James W. Holden said, "Its deep luxurious carpets cushioned the tread of a long array of celebrities."

Despite its normally glamorous entourage, the *P.P.C.* was actually built as an excursion car for rental parties. A newspaper account of the day and the order book listing confirm that it was in fact leased out at $85 a day when not otherwise engaged.[151] Pullman thus had the best of all reasons for building a PV: it would be a revenue producer and not a cash-draining parasite

Figure 4.86   George Pullman's own private car, the P.P.C., was remodeled in 1892, but the interior retains much of the look of the 1870s. (Arthur D. Dubin)

like other private cars. As late as 1892 the availability of the car on a rental basis was reported in the New York press. Although the *P.P.C.* was technically one of the company's rental cars, it has always been known as Pullman's personal pet. Little else would explain the car's long service record and the special attention it received in later years.

The *P.P.C.* was completed at the Pullman Detroit shops in June 1877. Mounted on twelve wheels, its body housed an observation room, several bedrooms, a large central parlor-dining room, and a kitchen-pantry at the forward end. The parlor-dining room featured a reed organ. The car could sleep 10. After the fashion of the day the interior was finished in the Eastlake style, and the side walls were pierced with triple-window sets laid out with a broad center pane that was flanked by narrow sidelights. Fifteen expert wood-carvers worked on exquisite geometric turnings, panels, and moldings of the most opulent cabinet woods obtainable. The ceilings were painted to represent such scenes as the "bulrushes—with young Moses omitted; another of fuchsias and humming-birds."[152] The lamps and other metal fittings were gold-plated. A lilliputian bathtub completed the car's special equipment. Exaggerated estimates on the cost have been circulated ever since the car was produced; $25,000 was quoted in 1877, while in later years the figure was doubled.[153]

Within a decade the *P.P.C.* was just another veteran private car, yet Pullman exhibited a strong affection for it. Although he could easily have afforded a replacement, and no man on earth was in a better position to obtain one, he would not give the car up. In 1887 electric lights were installed in it, and in 1892 the body was lengthened to 66 feet, 10 inches and remodeled on the outside with a more modern style of clerestory roof.[154] A drawing and a few photographs survive from this period (Figures

Figure 4.87   The P.P.C. was built in Pullman's Detroit shops in 1877. This drawing shows it as remodeled in 1892. (Pullman, Inc.)

Figure 4.88   The Ellsmere was used by W. S. Webb, President of the Wagner Palace Car Company. It was built in 1888 at Wagner's Buffalo shops. (Railroad Car Journal, August 1897)

Figure 4.89   The Manhattan was produced for Austin Corbin, president of the Long Island Railroad, in 1885 by Jackson and Sharp. (Hall of Records)

Figure 4.90   Observation-room interior of the Manhattan, *1885.* (*Hall of Records*)

4.86 and 4.87). Mrs. Pullman shared her husband's affection for the old *P.P.C.*, and after his death in 1897 she kept the car, renaming it *Monitor*. When Mrs. Pullman died in 1920 the aging wooden relic, now totally unfit for service among the heavyweight steel equipment of the time, was put in storage at the Calumet shops. Unfortunately no one in the Pullman Company had the sentiment or foresight to preserve this historic memento. It was unceremoniously broken up the next year.

During the decade of the eighties the private car came to its full development. It was no longer a rare plaything to be enjoyed by a few top executives and financial leaders. Its use now spread to a wide circle of lesser officials. In addition, actors, actresses, and opera stars—people who were viewed with some suspicion by the polite strata of society—were now arrogantly flying about the country in fabulous private cars, which were often the gift of a patron or lover. On a less affluent level were circus, medicine-show, and theatrical groups, which found comfort and convenience in private car travel. Frequently these troupes were content to take over some shopworn veteran or remodel an existing car for their use. The Robbins iron passenger car, discussed in Chapter 2, is an example (see Figure 2.12).

During the eighties, with the notable exception of Pullman, the wealthiest car owners developed the habit of trading a relatively recent purchase in on a new one. The Vanderbilt sons switched cars frequently and soon discovered that one or two could not satisfy the needs of the entire family. In 1882 W. K. Vanderbilt had a 74-foot yellow giant built by the New York Central shops.[155] Scenes painted on its sides depicted the major improvements on the line—such as Grand Central Depot and the West Albany bridge—and such scenic spectacles as Niagara Falls. A Wagner sleeping car was connected to the *Vanderbilt*, with some form of vestibule fixed between the two for the safe passage of travelers. Within a few years, however, the railroad

king tired of his twin cars and had the Wagner shops fabricate a single-unit successor, the *Idlehour*.[156] His brother Cornelius decided it was time to trade cars as well, and in the following year he acquired a new one from Wagner done up in the Louis XVI style.[157] But it was a brother-in-law, W. S. Webb, who showed the family how to go through cars. In 1885 he obtained a handsome one from Barney and Smith.[158] Two years later a replacement was under construction at the Wagner shops.[159] At his direction a number of private cars were produced for Wagner's private rental fleet and placed at the Vanderbilts' disposal. Of this group Webb favored the *Ellsmere*, with its 18-foot dining room, Italian Renaissance decor, and embossed leather ceiling (Figure 4.88).

Jay Gould and his family were also steady patrons of the PV. Gould's delicate health and secretive personality made a private car a positive necessity. At first he used existing official cars; a photograph shows him standing on the rear platform of the Atlantic and Great Western's Number 200, which was produced at the road's Kent, Ohio, shops in 1876 to replace the old *Atlantic*.[160] After acquiring control of the Missouri Pacific, he was content to use Commodore Garrison's old car, the *Convoy*.[161] In 1886 he is said to have purchased a set of twin cars from Pullman named *Penola* and *Bedford*. The cars were built under lot number 1346, but no other record of their existence can be found. During the following year he acquired a new car named in honor of his yacht, the *Atalanta*.[162] This 70-foot, 78,000-pound, twelve-wheeler was produced at Pullman's Detroit shops at a cost just over $20,000. It was described as rich but not garish. The *Atalanta* served out Gould's remaining years, but both his son and daughter preferred to travel in their own cars. The family could easily afford it.

Other wealthy users of PVs included the Stanfords, husband and wife, who had separate cars, and C. P. Huntington, who

Figure 4.91   The Alexander, *a Pullman creation of 1892, was one of the most lavish private cars ever produced. It was made to the order of A. A. McLeod. (Pullman Neg. 2108)*

roughed it for several years with a single car.[163] Huntington found that his guests, servants, and luggage were always underfoot, however, and bought a second car in 1895 for the overflow.

Other money barons less well known today than the Vanderbilts and Goulds also had sumptuous private cars. Although many were Pullman products, other builders too were busy satisfying this small yet prestigious market. Jackson and Sharp constructed some of the best PVs of the period, but because the cars were ordered by relatively obscure figures they did not attract publicity. In 1885 the New York real estate and railroad entrepreneur Austin Corbin bought from Jackson and Sharp what some authorities regard as one of the best cars ever built up to that time.[164] He called it the *Manhattan*. The Munich-lake-color sides were emblazoned with rich gilt decorations. The windows were glazed with clear French plate and color-leaded glass. Sage-colored carpets, Mexican mahogany, dark brass hardware, and silk tapestry in subtle hues created a somber but rich interior. One unusual decorating note was the sheepskin mats in the master bedroom. Other features included the bathtub that was hidden under a sofa, the five water tanks storing 300 gallons, and the air and vacuum brakes that were part of the mechanical equipage. The builder's photographs in Figures 4.89 and 4.90 give a better idea of the car's appearance.

The financier Angus A. McLeod was at the height of his fortune during the early 1890s, when he managed to assemble several railroad and coal properties, using the unstable Reading for an anchor. During those brief peak years McLeod purchased two

private cars from Pullman. The first, the *Alexander*, delivered in 1890, did not satisfy McLeod's patrician taste. In 1892 he ordered Pullman to produce a truly deluxe car, something that would please a Hapsburg prince. According to a Pullman official, the resulting vehicle was "without a doubt, one of the finest ever built."[165] Surely few cars were more elaborately finished, from its polished brass end rails to its Empire dome ceiling. The 70-foot body was literally crammed with costly hangings, carvings, and leaded glass. Because of the car's vulnerable position at the rear of the train, McLeod had insisted on safety as well as splendor, and the second *Alexander* was solidly built. Its frame was in fact unique—formed of a solid row of 4- by 8-inch timbers (Figures 4.91 and 4.92). Canvas and flooring were nailed over this deck. The body was heavily reinforced with steel plates. No tally of the car's total weight has been found, though with such a massive floor frame it must have set a record. Sometime after McLeod's business failure in 1893, it was sold to the Southern Pacific.

McLeod's *Alexander* was in many ways the apex of the private car during its gaudiest period. Other cars imitated and perhaps a few even surpassed the *Alexander*, but most private purchasers backed away from such excesses. Henry M. Flagler, the developer of Florida who had been an early partner of John D. Rockefeller, ordered a new PV that would be rich-looking but in an Empire style less ostentatious than McLeod's. The new car, unobtrusively numbered the 90, was produced by Jackson and Sharp in 1898.[166] The body measured 71 feet, 6 inches, and the

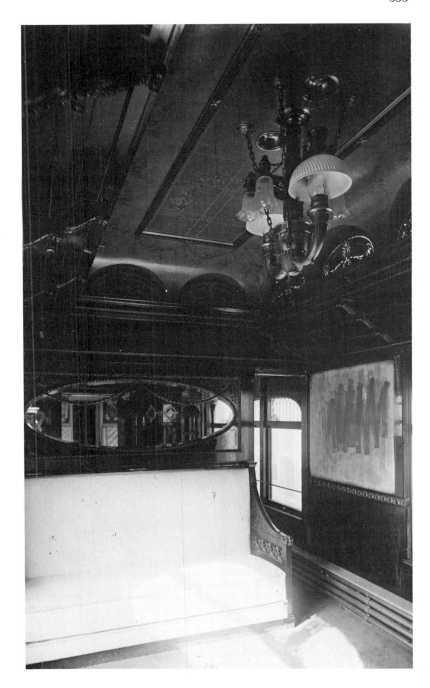

Figure 4.92   Interior of the Alexander's *drawing room, 1892. (Pullman Neg. 2111)*

platforms added another 7 feet, 2 inches. The interior was paneled in white mahogany and satinwood and was lighted with Pintsch gas fixtures. The plan and appearance of the car are shown in Figures 4.93 and 4.94.

Much of the private car's glamour can be traced to the association of stage personalities with railway travel. Although the pattern was established in 1850 by Jenny Lind, it does not appear to have taken hold for another thirty years. And then in the 1880s came three exciting women whose American tours made their every movement a matter of public interest. The first two were opera stars: Etelka Gerster and Adelina Patti. The third was a beautiful socialite turned actress: the legendary Lily Langtry. All three crossed the country aboard Mann-styled private cars.

The singers traveled in cars bearing their own names—both rented from the Mann Company. Patti's car was completed by the Gilbert Car Works late in 1883.[167] Its dark embossed-leather interior was decorated with morning glories picked out in gold and silver leaf. The 55-foot interior was divided into three compartments, with furnishings that included a piano and a pink boudoir. Beebe summarized the decor: "Looped, fringed, frizzled and ferned, its rococco splendors were the tinsel stuff of grand opera itself."[168] Within a few years the *Adelina Patti* was taken over by Pullman and renamed the *Coronet*. Late in 1901 it was sold to a used-equipment dealer who specialized in restocking circus trains, and no more tales of its Byzantine splendors were circulated to the press. In its day Patti's impressario, Colonel Mapleson, had boasted of a solid silver bathtub, 18-carat-gold keys, and other such nonsense for the entertainment page.

The *Etelka Gerster* toured the country several times before Gerster's retirement in 1896. This 1884 product of Jackson and Sharp lingered in Pullman's private car rental fleet for several years and was then sold to the Bangor and Aroostook Railroad.[169] Between 1899 and 1924 it was B & A business car Number 97, after which it was moved to Stockton Springs, Maine, for use as a summer cottage.

Mrs. Langtry rode in rental cars until a wealthy admirer presented her with her own PV, the *Lalee*, built by Harlan and Hollingsworth in 1888 for a reported cost of $65,000. The bright Jersey blue *Lalee* was compared to Cleopatra's barge, but the dazzling car received less attention than it might have because it was overshadowed by the private life of its owner, particularly her rumored dalliance with the Prince of Wales.[170]

The private car was not the exclusive property of bizarre financiers or ladies of flamboyant reputation. Some business cars were actually used for business travel, and after the initial extravagance of the first-generation "directors' cars," their plan and finish seems to have become more practical (Figure 4.95). Although the elaborate cars of the very rich received most of the attention, records do exist that provide details on a few office cars. For example, during the 1870s the general superintendent of the Pennsylvania Railroad traveled over the line in a small, spartan, 46-foot-long car. A floor plan, together with a brief description in James Dredge's 1879 report on the Pennsylvania, supply these facts.[171] The end saloon was outfitted as an office,

with long desk-like tables on either side. The central compartment could be used as an auxiliary office by day. A collapsible table converted it into a dining room, and at night, folding berths made it into a sleeping room. At the rear of the car was the general superintendent's tiny stateroom and kitchen (Figure 4.96).

Another office car of the same period was described in a French report published in 1882.[172] Owned by the Louisville and Nashville Railroad, it was a combination pay and track-inspection car. The paymaster's compartment at the forward end contained tables, seats, a cage, a safe, and a signal bell. The right-of-way superintendent occupied the far end of the car. The end bulkhead had extra-large windows for a clear view of the tracks. The stove, linen closets, and toilet were located here. At the center of the car was a bunk room shared by both crews. Double-glass windows were an interesting feature. The framing plan in Figure 4.97 shows other structural arrangements.

The cost of business cars was well below the reported prices of the more widely publicized PVs. In 1880 the Northern Pacific paid the Ohio Falls Car Company only $4,339.13 for its office car

FRAMING PLAN   INSIDE VIEW OF CAR   OUTSIDE ELEVATION

71 ft 6 in. × 9 ft 8 in. over Frame   FLOOR PLAN OF CAR

FLOOR PLAN OF KITCHEN END OF CAR   FRONT VIEW OF CAR   SKETCH OF FINISH IN PARLOR.

PRIVATE CAR NO. 90. BUILT BY
THE JACKSON & SHARP CO. WIL. DEL.

Figure 4.93   Jackson and Sharp produced the Number 90 for Henry M. Flagler, president of the Florida East Coast Railway, in 1898. (Railroad Car Journal, *August 1898*)

A-3.[173] Some years later it spent $2,481.75 to produce the A-8 in its own shops. The figures of $30,000 to $50,000 for the fancy private cars of the time were admittedly exaggerated; but even if the actual prices were only one-third of those amounts, the cars were a silly luxury. The nineteenth-century business car offered the same space and comfort at relatively tiny prices by providing a plain interior and simple furnishings (Figure 4.98).

Because PVs were so often custom-made to suit individual needs, they were the class of car least amenable to standardization. In the first rush of enthusiasm, according to a Pullman official, the customer overwhelms the designer with novel concepts; his car must be radically different and much better than all that have preceded it. "But when he comes face to face with certain blue printed mechanical and engineering limitations that are imposed on the car builder, the enthusiast finds that much of the striking originality in his scheme will have to be eliminated."[174] Yet within the tubular confines of a standard car body, decoration, furnishings, and even floor plans to some degree, were matters of idiosyncratic taste. In 1901 Charles M. Schwab, who spent $40,000 on his *Loretto*, chose an enclosed sunroom in place of the traditional open observation platform. Four years later Louis Hill had his second PV fitted out as a mobile garage so that he could travel with his favorite automobile. Max Fleischmann took an active hand in the design of his *Edgewood*, which is remembered for its heavy underframing, noiseproof construction, and elaborate filtered air system. The Woolworth heiress, Mrs. J. P. Donahue, insisted upon gold fixtures and other

extravagances for her *Japauldon*, built by American Car and Foundry. Fairfax Harrison, the aristocratic president of the Southern Railway, required a pair of cars for his travels.

Try as they would, however, designers were confined to what a 10- by 8- by 80-foot box could contain. It was a foolish shape for a "house," since all the rooms were to be private, a hallway was necessary, but then the rooms or the hall had to be made impossibly narrow. The designer was forced to produce a compromise floor plan. Excluding the freaks, there appear to be three basic plans. The most common had an observation platform at one end, a parlor or observation room, and then a series of three or four bedrooms, bypassed by a narrow side aisle which led to a dining room. Both the parlor and dining room were full-width. Behind the dining room were the pantry, kitchen, and crew's quarters (Figure 4.99). This arrangement was in full vogue by 1900. In the next most popular plan the observation platforms were fixed to both ends of the car, the dining room was at the forward end (in place of the kitchen and crew's quarters), and the kitchen was toward the center of the car (Figure 4.98). The third plan, which seems to have gone out of favor by the 1890s, had a large central, full-width room that doubled as parlor and dining room (Pullman's *P.P.C.* was an example).

The first steel PV was Robbins's steel car of 1889. It was not until 1907, however, that another such vehicle was constructed. In that year Pullman's Buffalo (formerly the Wagner) shop produced an all-steel business car for J. M. Schoonmaker, vice president of the Pittsburgh and Lake Erie.[175] Early in 1910 the Al-

Figure 4.94   The dining room of Flagler's Number 90, 1898. (Hall of Records)

toona shops were building a similar car for an executive of the Pennsylvania.[176]

It is unlikely that many new wooden private cars were produced after this date for domestic service (see Figure 4.100). However, steel underframes and steel side plating rejuvenated many an aging wooden PV for long years of use. An example is Austin Corbin's *Oriental*, which Pullman built in 1890. Corbin died in 1896, and the *Oriental* was taken over by the Louisville and Nashville for its board chairman, August Belmont.[177] A steel underframe was subsequently added. Metal sheathing, applied in 1947, so disguised the car that few persons would have guessed its true age. In 1958, after almost seventy years of service, the *Oriental* was retired. It is now exhibited in the Adirondack Museum at Blue Mountain Lake, New York.

Car builders produced new private cars rapidly during the great prosperity of the 1920s. By this time a new PV took up to four years to construct and cost between $70,000 and $100,000.[178] The price was figured at cost plus a 15 percent profit plus an

allowance of 175 percent of direct labor for overhead. Furnishings, china, silver, linen, and kitchen utensils added another $7,850. Railroad officials and wealthy private individuals were not the only customers for the cars; corporations such as U.S. Steel, Maryland Oil, Fruit Growers Express, Anaconda Copper, and even the automotive giant General Motors also became PV owners. The crest was reached in 1927, when builders turned out 108 new private cars.[179] During the next year a total of 886 PVs was reported in service, but the Pullman Company estimated that no more than 40 of these were actually privately owned.

The economic chaos following the 1929 collapse abruptly ended the private car era. In 1930 only one new PV was produced. A handful of business cars were constructed after that time, but for all practical purposes the PV was obsolete. Many owners sold or stored their cars during the Depression years. Those who could afford to keep them could not always afford to operate them. *Fortune* magazine published the following table of annual operating charges in its July 1930 issue.

SALARIES (per year)
Steward	$2,400	
Chef	1,800	
Waiter	1,500	
		$5,700

RAILROAD CHARGES

(Depending, of course, upon the car's use. Here it is figured that the car's habitat is Chicago, that it makes one trip to the Kentucky Derby, one trip to the Yale-Harvard boat races, one trip to Pasadena, three trips between Chicago and New York, two trips to Washington, one trip to Hot Springs, one trip to Quebec, and one trip to Jaspar Park. The remainder of the time it is paying parking charges in the yards.)

Tickets	$20,219	
(Twenty-five for each trip)		
Surcharge	2,021.90	
(10 percent)		
Parking	681.60	
($3.60 per day for first seven days; $12 a day thereafter)		
		$22,922.50

FOOD

Passengers	$750	
(average five)		
Allowance for crew		

(Depending again upon whether the owner maintains a full crew, or takes on the chef and waiter when needed. In general, however, it will be found more satisfactory to keep a full crew. At $3 per man per day, crew's food will cost . . .)

	3,285	
		$4,035
		$576

PARKING CHARGES FOR PERMANENT STORAGE
(Usually at $2.40 per day. After trips are subtracted, total days in Chicago storage are 240.)

TIPS		$500

(Station masters, yard masters, and their assistants must be tipped.)

TOTAL	$33,733.50

By 1940 the private car fleet was down to 572, a one-third decline in a single decade. Furthermore, economic recovery did not stimulate PV construction, for other negative factors were at work. The very rich were beginning to fly or drive; the private railroad car had acquired a stodgy, Victorian aura. And by late 1941, conservative travelers who would have continued using PVs could not because of the war effort. There was no place for luxury pleasure travel; the railroads handled essential business movements, but the PVs were otherwise sidetracked. Many private car owners sold out. Franklyn Hutton's *Curley Hut* was

bought by the C & O for use as an office car. Barbara Hutton's car went to the Western Pacific, while Henry Ford's fabled *Fairlane* was sold to the Cotton Belt.[180]

At the war's end the PV did not return to favor (see Figures 4.106 and 4.107). Between 1949 and 1955 only fourteen lightweight business cars were produced.[181] Some roads found it cheaper to rebuild existing cars. The business car fleet soon stabilized at its prewar level, but privately owned cars virtually disappeared. Benjamin Fairless, president of U.S. Steel, was an exception; he commissioned Pullman to produce the *Laurel Ridge* in 1949 at a cost of $200,000.[182] After his retirement in 1955 his company's use of the car declined, and in 1962 the maroon and silver streamliner was quietly sold to the Santa Fe. Except for the Anheuser-Busch Brewing Company's *Adolphus* (1954), the *Laurel Ridge* was the last of the species.

By 1971 there were only 179 business cars in operation. A new generation of private car owners has emerged in recent years, but they are like an exclusive club of antique yachtsmen. The railroads' top management has largely abandoned the office car for the speed and prestige of the executive jet. The general manager or division superintendent still finds the business car useful, but for railroad presidents it is too slow and cumbersome for everyday business or that "fast trip to the Coast." With the coming of Amtrak and a skeleton system of intercity passenger trains, it is no longer possible to travel between many major cities. The inspection trip, once a major justification for the business car, is increasingly made in luxurious automobile-rail cars. In 1954 the New York Central, for example, remodeled a Chrysler four-door sedan for the use of its chief executives.[183] What was once the railway passenger car's most prestigious form has gradually faded away.

Rental Cars

For generations aspiring plutocrats have rented jewelry, seaside mansions, and limousines. In an earlier time they rented private cars. Some of our wealthiest citizens preferred to rent (or borrow) rather than own a PV. J. P. Morgan, who could well afford to buy every private car in North America, declined to own any but was willing to charter an entire train as the occasion arose.[184]

From the beginning of the railroad era it has been possible to charter a private car. Special parties and regular commuter clubs have taken advantage of this service since the 1830s.[185] So far as is known, the early renters made short trips and used ordinary day coaches. Apparently it was that genius of railway travel, George M. Pullman, who first recognized the potential profit in rentals, particularly in luxury rental cars for long-distance travel. After the transcontinental railway was opened in 1869, the scenic wonders of the West might be seen from the comfort of a passenger car. How much better it would be if one's family, friends, or close business associates could enjoy the trip together in the home-like privacy of their own car. In 1870 Pullman offered to

Figure 4.95   A Baltimore and Ohio office car, possibly Number 41. Its plain appearance suggests that it was a division superintendent's car. The photograph dates from 1858.

Figure 4.96   G. C. Gardner, superintendent of the Pennsylvania Railroad from 1871 to 1878, used this office car. Such cars were considerably less elaborate than the private cars previously shown. (James Dredge, The Pennsylvania R.R., London, 1879)

Figure 4.97   A Louisville and Nashville combination superintendent and paymaster car of about 1875. The engraving was published in a French text on U.S. railways based on data collected in 1876. (E. Lavoinne and E. Pontzen, Les Chemins de Fer en Amerique, Paris, 1882)

Side Elevation

End Elevation.

Sectional Plan of Floor.

Figure 4.98   A Chicago, Burlington, and Quincy business car of about 1890. (Car Builders' Dictionary, 1895)

PRIVATE CAR "IDLEHOUR" PLAN 2052.

Figure 4.99   W. K. Vanderbilt's Idlehour was a Pullman product of 1905.

Figure 4.100   Business cars generally carried low numbers. The Number 100 was completed by Jackson and Sharp in 1911. (Hall of Records)

Figure 4.101   James C. Brady, "Diamond Jim," was a successful railway supply salesman who could afford his own car. This early steel PV dates from 1914. (Pullman Neg. 17014)

Figure 4.102   The Santa Fe purchased some very short 52-foot office cars in 1924 for its division superintendents. (Pullman Neg. 27959)

Figure 4.103   Jackson and Sharp was a division of A.C.F. when it produced this office car in September 1925. (Hall of Records)

Figure 4.104   Top officials of major trunk lines suffered few privations even when rail-road business kept them away from headquarters.

rent a hotel car, with its crew, for $85 a day.[186] Food would be supplied for an additional $2 a day per passenger. Within two years Pullman was leasing sleeping, drawing-room, and hotel cars. A prospectus dated April 1, 1872, offered the following terms:[187]

For a regular SLEEPING CAR, containing twelve open sections of two double berths each, and two staterooms of two double berths each, (in all twenty-eight berths), with Conductor and Porter, SEVENTY-FIVE DOLLARS per day.

For a DRAWING-ROOM CAR, containing two drawing-rooms, having each a sofa and two large easy chairs by day, and making up at night into two double and two single berths each, three state-rooms, having each two double berths and six open sections of two double berths each, (in all twenty-six berths), with Conductor and Porter, SEVENTY-FIVE DOLLARS per day.

For a HOTEL CAR, containing one drawing-room, as above described, one state-room, having two double berths, and ten open sections of two double berths each, (in all twenty-six berths), and having also, in one end, a kitchen, fully equipped with everything necessary for cooking and serving meals, with Conductor, Cook and two waiters, EIGHTY-FIVE DOLLARS per day.

The Conductor, if desired, will make all arrangements for the excursionists with the railroads for procuring transportation of the car, and in the case of their taking a Hotel Car, will also act as Steward, purchasing for them the requisite provisions for the table.

The number composing the party will, of course, make no difference to us, the car being chartered, with its attendants, at a certain rate PER DAY, from the time it is taken until we receive it back again.

So far as the Railroad Companies are concerned, they require eighteen (18) fares for a car-load. If this number of fares is furnished, there will be no EXTRA charge on the part of the Railroad Companies for hauling the car.

In 1877 Pullman produced the *P.P.C.*, a private car equal to the most deluxe standards of the day, for rental parties. As mentioned earlier, Pullman came to use the *P.P.C.* more for himself and less for rentals as time went on. But it appears to be the first true private car built for the rental market. Around 1880 Pullman added two special cars, the *Davy Crockett* and the *Izaak Walton*, to the fleet for hunting and fishing parties.[188] Sportsmen could hire one of them, plus a cook and a waiter, at the bargain rate of $35 a day. These cars were less highly finished than the typical PV and thus appealed to parties wishing to "rough it" in comfort.

Figure 4.105   The interior of the B & O office car 97 as shown in a photograph of 1930. (The car's dining compartment is shown in Figure 4.104.)

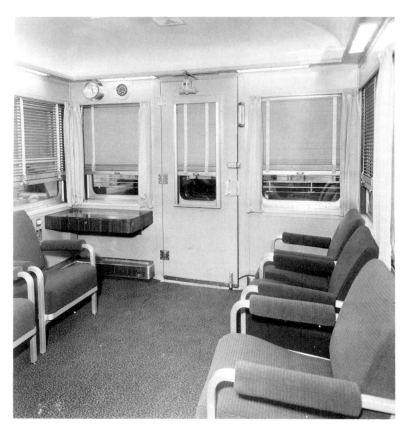

Figure 4.106   Business cars lost their club-like interiors with the coming of the streamline era. This Great Northern office car, the A-28, was completed in 1947. (Pullman Neg. 52052)

They carried eight to ten persons and were outfitted with a kennel for hunting dogs.

By the late 1880s Pullman's rental fleet had grown to nineteen cars—thirteen private cars and the rest hunting and hotel cars.[189] They were available for single trips or extended tours. Pullman offered to stock the cars with provisions and, if desired, to furnish "polite and skilful attendants."

Pullman's rental business was apparently profitable, because the company's competitors began to seek a share of it. Among the first was Jerome Marble (1824–1906) of Worcester, Massachusetts, a banker and man of affairs who returned from a private car hunting trip to the Dakotas convinced that luxury excursion cars would prove a good investment.[190] In 1878 he had Jackson and Sharp remodel the car *Delaware*, which the firm had built two years before for display at the U.S. Centennial Exhibition.[191] The 52-foot body was refitted as a private car, an observation platform was added to one end, and four-wheel trucks were installed in place of the original six-wheel Ashbel Welch–style trucks. Renamed the *City of Worcester*, the transformed coach began making cross-country tours with parties as large as twenty (see Figures 4.108 and 4.109). Marble thought so well of the scheme that he organized the Worcester Excursion Car Company in July 1878.[192] In the following year the *National Car Builder* said that a trip to California could be made cheaper aboard Marble's car than on a first-class train. Fourteen passengers took the *City of Worcester* to the Dakotas—a 78-day, 4,000-mile journey—at a cost of only $203 each.[193]

Marble bought two more cars and in 1883 acquired a fourth

Figure 4.107 *As rail executives turned to private airplanes, the business car became something of an anachronism. This late example was produced by Budd in 1957 for the Santa Fe. (Robert Wayner)*

Figure 4.108 *The Worcester Excursion Car Company rented private cars, mainly for Western travel. The* City of Worcester *was originally built in 1876 as a luxury coach but was remodeled as shown here for rental service. (Hall of Records)*

Figure 4.109   The City of Worcester's *interior as remodeled for rental service. (Arthur D. Dubin)*

car from Jackson and Sharp that he named the *Edwin Forrest*, after the actor.[194] The *Edwin Forrest* featured a central 24-foot-long grand saloon. This car, and perhaps the earlier Worcester cars, used William Paige's patented collapsible lower berths. The metal-framed, canvas-covered beds were put away during the day, leaving the floor clear for easy chairs. Wagner-style folding berths were used overhead. Through these arrangements Worcester hoped to avoid Pullman's patent trust. Another interesting feature of the Worcester cars was a fireproof safe for valuables. The company purchased more cars and by the late 1880s was operating nine, including the luxurious *Railway Age*, which it had leased from the publisher of that journal. Among Worcester's customers were Sarah Bernhardt, Edwin Booth, and Theodore Thomas. Its main business continued to be Western hunting parties. When the American frontier vanished in 1890 the plentiful supply of wild game disappeared with it, and according to Lucius Beebe this was a major reason for the Worcester Excursion Car Company's suspension around 1895.[195]

Worcester was unique in concentrating on private car rentals, but it was not Pullman's sole rental competitor. Mann, Woodruff, and Wagner were all engaged in the rental trade to one degree or another, though only Wagner proved to be much of a rival for Pullman. Wagner was rather late in entering the field; it was after Seward Webb took charge that the firm acquired its first rental car. The *Marquita*, produced at Buffalo in 1887, was soon followed by the *Riva*, the *Grassmere*, and the *Wanderer*. But before Wagner could invade this sphere on a large scale, the Vanderbilts at last accepted Pullman's natural monopoly, and

Wagner's rental cars became part of the Pullman fleet after the 1899 consolidation of the two firms.

Pullman was now without rivals, even in the rental field. In 1903 private car leasing was a fast-growing trade; Pullman regularly had twenty-four cars engaged at one time.[196] Wealthy commuters, particularly in the New York and Philadelphia suburbs, often banded together and leased a parlor car. These arrangements were not made exclusively with Pullman. In a few instances the car was jointly owned by the commuters themselves, as when a group of New Orleans cotton brokers purchased the *Club-on-Wheels* in 1886.[197] A similar private parlor-style car was produced in 1893 for a group of Plainfield, New Jersey, businessmen by Harlan and Hollingsworth.[198] The *Plainfield's* many comforts included lounge chairs, card tables, and a smoking and nonsmoking section.

Pullman rentals rose and fell in concert with the stock market. In 1931, despite the deepening gloom of the Depression, there was enough demand to keep twenty-three cars employed.[199] Daily rates were adjusted for the length of the lease. One- to two-day rentals cost $175 per day, with a graduated scale that declined to $50 a day when ninety or more days were involved. These prices included a cook and two attendants. Food was supplied at cost plus 25 percent. In 1938 Pullman was still operating twenty-three rental cars, but in 1939 the number was cut back to seventeen.[200] It had been a bad year, with financial reverses that prompted many business cutbacks throughout the economy. Pullman's rental division not only reflected the downturn by reducing the number of cars in the fleet, but also added a poor people's private car. This was a restaurant-sleeping car with open-section berths for 16 passengers at one end, a central kitchen, and a parlor-buffet compartment at the opposite end.

World War II put a stop to PV rentals, just as it closed down private car traffic in general. Pullman disbanded this portion of its operations in order to concentrate on more essential classes of passenger transport. Several cars were sold during the war to serve as business cars. The Central of Georgia thus acquired the *Marco Polo*, while the Seaboard Air Line picked up the *Pioneer*. During the same years a car similar to these two was extensively remodeled for President Roosevelt. At the war's end, the Pullman Company was embroiled in its legal contest with the Justice Department (described in Chapter 3) and was apparently too preoccupied to revive its rental car pool.

Presidential Cars

Just how to accommodate the President of the United States was a dilemma in the early days of the republic. In a society self-consciously opposed to privilege, the head of state was in theory no better than any other man, and on a train he might be expected to take his seat among coach passengers. Yet he was a very special citizen holding an office that seemed to call for some mark of distinction when he traveled. Monarchs routinely acquired royal railway carriages with as little fuss as they enlarged their collection of golden state coaches. Pope Pius IX casually accepted an incredibly ornate baroque train from Napoleon III. The nineteenth-century rulers of Mexico showed a special fondness for private rail cars and were inclined to go through one or more sets during their dictatorships. But the U.S President was expected to be more restrained. As temporary head of state, he had the use of a mansion, one or more carriages, and a yacht. Ironically, during the great age of railway travel there was no Presidential private car. If the President wanted one, he had to seek the favor of a railroad executive.

A questionable tradition was thus established that encouraged a measure of obligation toward anyone who provided a car for the President and his party. The practice persisted until 1942, although it seems to have caused little concern among the usually vocal critics of those holding political office. The watchdogs apparently saw nothing amiss when Tom Scott or John Garrett regularly offered the President free use of a private car. Garrett's *Maryland* was probably the most frequently occupied by Chief Executives during the Gilded Age. It was so simple to run the car down from Baltimore, and Garrett was so ready to oblige, that Hayes, Harrison, Cleveland, and McKinley came to think of it as their very own.[201] Cleveland, in fact, borrowed the *Maryland* in 1886 for his honeymoon trip. Pullman, always sensitive to the value of publicity, was delighted to lend the *P.P.C.* to traveling Presidents. Theodore Roosevelt, who came to feel uneasy about this practice and its potential for breeding scandal, was dismayed to learn that during his first term he had unwittingly accepted well over $100,000 in free rail travel.[202] In 1906 Congress added a travel allowance to the White House budget, but it was only for $25,000. Presidents continued to depend upon the generosity of the railroads until the private jet became the Chief Executive's mode of long-distance travel.

Plans for a government-owned private car were made during Lincoln's administration, when the exigencies of the Civil War required him to travel. In May 1862 he went to Fredericksburg, occupying a baggage car for part of the trip.[203] A few weeks later he journeyed to West Point by train. Late in the following year, when it was necessary to make a fast trip to Gettysburg for the battlefield dedication, the B & O lent its directors' car. It seemed clear, however, that a special car should be available to the President. Both his prestige and his security demanded it, and both would best be served by providing a car that was under the control of the War Department.

It is not entirely clear who authorized construction of the Presidential car. The Lincoln scholar Victor Searcher suggests that Secretary of War Edwin M. Stanton initiated the scheme.[204] Stanton was solicitous of the President and took a great interest in his safety and comfort. Another account attributes the idea to B. P. Lamason, superintendent of the U.S. Military Railroad's car shops in Alexandria, Virginia. If it was Lamason's idea, he would surely have needed approval from higher authority before beginning the project. A preliminary search of the U.S.M.R.R. papers at the National Archives failed to yield any information on this subject. Like so many incidents in Lincoln's life, the story of the Presidential car is rich in mythology but short on docu-

mented facts. There are, in fact, several foolish tales about the car. The most persistent is that it was armor-plated. In later years, however, two veteran car builders who actually assisted in the car's construction heatedly denied that boiler plate was inserted between the inner walls to form a bulletproof shield. The following narrative is drawn largely from the recollections of these two mechanics, who appear to have recorded the most dependable information available on the Lincoln car.[205]

Construction is said to have begun in November 1863. Lamason based his general design on a Pennsylvania Railroad coach of the period. The car's 42-foot body was divided into three compartments: a parlor and drawing room at either end, with a side aisle and stateroom in the center. The stateroom was planned as the President's compartment. A small washroom was located in the drawing room. The absence of a kitchen and crew's quarters suggests that extended journeys were not envisioned, although the sofas were convertible into berths for overnight travel. In reality the car was outfitted more as an elegant drawing room–parlor than as a self-sufficient private car. The interior woodwork was oak and black walnut. The upper panels were upholstered with a corded crimson silk which had a tufted pattern for added richness. The headcloth, also of crimson silk, was gathered at the center of each panel to form a rosette. The furniture was covered in dark-green plush, and the curtains were light-green silk. The interior of the upper deck, or clerestory, was painted zinc white and decorated with the seals of state in full color.

The exterior was painted a rich chocolate brown, rubbed out to a fine polish with oil, rotten stone, and the bare hand. The national coat of arms was painted on the oval sheet-metal panels. In the window panel directly above, the words "United States" were lettered in gilt. The body and clerestory were highlighted in a restrained manner by fine gold striping. The decorative polished-brass washers for the bolster tie rods are visible below the third window at each end of the car (Figures 4.110 and 4.111).

Lamason's original plan called for conventional four-wheel trucks, but during the course of construction he decided to mount the car on sixteen wheels. He adopted Ambrose Ward's patented bolster. Each truck had a short wheelbase of only 4 feet, 10 inches. There were no equalizers. Broad-tread, 33-inch cast-iron wheels were used so that the car could travel on both standard and Southern gauge tracks. The axle box pedestals were very ornamental; Lamason was said to have devoted weeks to their design.

According to W. H. H. Price, the car was not finished until February 1865 because the Alexandria shops were busy with emergency repairs to keep trains running to the front. Another employee in the U.S.M.R.R. shops claimed that the car was actually finished in May 1864, but that Lincoln then refused to set foot on it. Articles in a hostile New York newspaper had severely criticized the pretentious private car being completed for a head of state who so earnestly proclaimed his homespun, plain-folks origins.[206] The great commoner could not afford to jeopardize his public image, and so the car sat in readiness for its inaugural trip, a trip that was not made by Lincoln during his lifetime.

Figure 4.110   The U.S. Military Railroad shops in Alexandria, Virginia, completed this private car for President Lincoln in 1865. (Library of Congress)

Figure 4.111   Lincoln never rode in the car during his lifetime, but his body was conveyed in it from Washington to Springfield, Illinois. Afterwards it was sold to the Union Pacific Railroad. (Library of Congress)

Figure 4.112 This narrow-gauge car was reportedly used occasionally by President Cleveland. It was built in 1880 by Billmeyer and Small of York, Pennsylvania, as a coach for the Bradford, Bordell, and Kinzua Railroad. It was sold in 1907 to the East Broad Top Railroad. (Edmund Collins, Jr.)

After his assassination on April 15, 1865, his body was carried to Springfield, Illinois, aboard this car. The 1,700-mile journey, which meandered among the major Northeastern cities, took thirteen days. After the funeral the car was returned to Washington; late in June it carried Mrs. W. H. Seward's body to Auburn, New York.

Now that the war was over, the military railway began to wind up its affairs. Among the surplus rolling stock put on the block was the Presidential car. Even while the funeral train was on its somber journey an advertisement appeared in a Washington newspaper offering to sell the car to "some enterprising man."[207] A rival paper characterized the scheme as "humiliating," but crass or not, the car was put up for sale. One account claims that Lincoln's old law partner bought it to keep it out of the hands of exhibitors.[208] Others say that it was sold at auction, after spirited bidding, to T. C. Durant, vice president of the Union Pacific.

Whatever the circumstances of the sale, it is certain that the car came into the possession of the Union Pacific sometime in 1866.[209] Durant was proud of his new acquisition and enjoyed carrying excursion parties in it over the U P tracks, which were then being laid. The Lincoln car formed part of the special train that ran in October 1866 to celebrate completion of the line to the one-hundredth meridian. The car's glamour, however, could

not really make up for its notoriously rough ride, ascribed by some critics to its cumbersome sixteen-wheel undercarriage by others to its broad-tread wheels. Durant abandoned the Presidential car late in the same year for the greater comfort of a Pullman.[210] During the next three years it served as an office car for government inspectors and various railroad officials. It was then converted into an emigrant car. Some of the elegant interior furnishings were retained by the Union Pacific's president, Sidney Dillon.[211] Later it was sold to the Colorado Central, but ultimately it returned to U P ownership, for the Union Pacific took over the Colorado line.

By 1886 the car was stationed on the Marysville branch of the U P—"a relic of faded gentility."[212] Now it was fitted with rough bunks to shelter section hands; few would have recognized the former palace on rails. And Lincoln's ghost could settle comfortably into the shabby interior with no fear of censure, for it was as humble as the meanest cabin car. It was stored in the North Platte yard by the early 1890s, a weather-beaten hulk with its windows boarded over.[213] There was talk of restoring it for the 1893 Columbian Exposition, but the necessary $5,000 could not be found. After another five years of neglect it was refurbished for exhibit in Omaha at the Trans-Mississippi Exposition, where souvenir hunters picked at its rotting flanks. The remains

Figure 4.113   The Ferdinand Magellan *was a private rental car placed in service in 1927. It was remodeled for President Roosevelt in 1942. (Pullman Neg. 33771)*

Figure 4.114   The Ferdinand Magellan, *as remodeled for the use of President Roosevelt. Presidents Truman and Eisenhower also traveled on it. (National Model Railroad Association)*

were again placed in storage. In 1903 came the final humiliation: a showman named Franklyn B. Snow decided that "the most sacred relic in the United States" should not be hidden from the American public.[214] He persuaded the Union Pacific to sell the car and arranged to exhibit it at the 1904 St. Louis World's Fair. The cross-country carnival following the close of the fair so horrified the growing band of Lincoln admirers that a group was organized to take the relic away from Snow. In the fall of 1905 Thomas Lowry, president of the Soo Line, purchased the car with the intention of placing it in a museum.[215] He presented it to the Minneapolis Park Board, which found that it could not furnish the imposing museum Lowry had envisioned. The board erected a temporary wooden shed for it in a park in the northeastern section of the city. In March 1911 while plans were underway to move the car to a state museum, a group of boys built a bonfire in a grassy field surrounding the shelter.[216] The fresh spring wind spread the flames to the car, and the damage was so severe that the remains were broken up and given away for souvenirs. From its initial rejection to its fiery cremation, the proud creation of the Alexandria shops had suffered a series of degradations. Today a handful of relics preserved at the Union Pacific Museum in Omaha is all that remains of the historic Lincoln car.

The Lincoln car was already a venerated relic before another effort was organized to provide the President with his own car (see Figure 4.112). It galled some railroad men that the Chief Executive should have a purely ceremonial conveyance like a yacht while he lacked the practical benefits of a railroad car. In February 1897 the *Railroad Car Journal* announced a scheme to put matters right.[217] The editor had formed a committee of prominent master car builders headed by Fitch D. Adams and was busy persuading the supply industry to donate the necessary materials. By April the committee had completed its preliminary design for a married pair of eight-wheel cars. The matter was discussed in the editorial pages at some length, but not with such gravity that it did not print several humorous designs incorporating the outlines of the capitol and the White House. The supply trade showed little enthusiasm, however, and the Master Car Builders Association refused to endorse the plan. At this point the project was tabled.

A special car for the President's use became an issue again during World War II.[218] Since the dangers of sabotage and assassination were more acute than an peacetime, there was no opposition in 1942 to the idea that the President should have a secure car to travel in. The Pullman Company had been chauffeuring Franklin Roosevelt even before he became President, for in his years as governor of New York he often used Pullman's rental car, the *Marco Polo*. It was coincidental that the Association of American Railroads, in cooperation with Pullman, selected a sister car, the *Ferdinand Magellan*, for remodeling into a mobile security vault (Figure 4.113). Protection against attack was the first priority; the car was not notably luxurious, and its exterior was far from handsome. The *Magellan* was one of six rental cars built in 1928 at Pullman's Chicago shops. The remodeling was carried out at the nearby Calumet repair shops in 1942. One of the five bedrooms was eliminated so that the observation and dining rooms could be enlarged. The roof and sides were sheathed in ⅝-inch-thick armor plate, and 3-inch-thick bulletproof glass was installed in the windows. The rear platform door was like a bank vault's; it weighed 1,800 pounds. The roof was redone as an ungainly turtleback casement. The undercarriage became a clutter of boxes, pumps, and extra mechanical gear. The total effect was that of a monstrous caterpillar. The swollen giant's weight of 142½ tons made it the heaviest passenger car ever to operate in this country. One engineer said that it "drags like a brickbat on the tail of a kite.[219]

Roosevelt made many journeys in the *Magellan* after its metamorphosis was completed in December 1942 (Figure 4.114). He traveled 50,000 miles aboard it in the following two years. After his death in April 1945, the Presidential car was part of the funeral train. However, it did not carry the President's body, as is often assumed.

In 1946 the A.A.R. sold the car to the U.S. government for the sum of one dollar. No figures have ever been released on the cost of the remodeling. President Truman was a devoted railroad traveler and ran the car throughout the United States during his tumultuous election campaign of 1948. During this time the car lost its identity. In the war years it had been lettered "Pullman" in an effort to avert suspicion that it was something special, but it had carried no name or number. After the war even the Pullman lettering disappeared, and the only marking that distinguished the car was a dark-bronze Presidential seal affixed to the observation railing.

Truman's successor showed less interest in rail travel and rarely used the car. It was Mrs. Eisenhower, in fact, who made the last official trip in it in 1954. Four years later it was declared surplus property. Early in 1959 it was given to the Florida Development Commission, which in turn presented it to the University of Miami.[220] The university in turn passed the car on to a group of railroad enthusiasts who operate the Gold Coast Railroad in Fort Lauderdale, Florida. Today it is tourists and not Presidents who tread its thickly carpeted corridors.

CHAPTER FIVE

Passenger Car Accommodations

FROM DECORATION TO VESTIBULES

OF ALL THE COMMODITIES CARRIED BY RAILWAYS, none is more fragile, sensitive, or quarrelsome than human cargo. A lump of coal or a stick of lumber can be trundled about in the simplest sort of car; it requires no shelter, heat, seating, or washrooms. Schedules, rough track, and rude trainmen are of no consequence. But these and countless other matters are of paramount concern to the railway passenger. The degree of heat, the mechanics of the sunshade, the angle of the seat back, the texture of the cushions, and even their color form the basis for endless complaints, comments, and recommendations. To answer the demand for greater comfort, the railroad passenger car became an increasingly complicated structure with a profusion of subsystems and fancy auxiliary apparatus. Year by year the addition of some new improvement for convenience, sanitation, safety, or decorative effect increased weight and cost. The trend was irreversible and explains more than any other single factor the car builders' failure to reduce dead weight.

Thus the history of the passenger car cannot be told simply by tracing its increase in size or the shift from wood to steel construction. The story must also cover the progressive development of its fittings and the evolution of a simple shelter on flanged wheels into a complex vehicle interwoven with pipes, ducts, wiring, and gadgets.

Seating

To the first railway passengers, land travel was a hardship to be borne. They expected little more than a hard seat, a leather curtain, and perhaps a hand strap. Lighting came from the sun and ventilation from chance breezes, while heating was self-generated. The ride was wonderfully smooth compared with highway travel, and since the journey was short, pioneer lines gave minimal attention to most interior fixtures.

Seating was an exception, however. A good seat is a basic comfort and was recognized as such from the beginning. The earliest four-wheel cars had high-backed upholstered seats arranged face-to-face, highway coach fashion. When the large, open, eight-wheel cars came into fashion, this time-honored plan was abandoned, because Americans had developed a phobia about riding backwards. Apparently the syndrome was peculiar to this country; British travelers seemed content with the old system, and indeed it is universally followed today on the Continent. Of necessity Pullman reinstated the stationary-back, face-to-face seat in his sleeping cars, but it was considered a drawback. Coach passengers would accept nothing less than a reversible seat back, and until very recent times the swingback or walkover seat was the standard for American coaches.

Conduce Gatch, car builder for the B & O, claimed to have introduced the reversible seat around 1833.[1] He said that it was used on several experimental eight-wheel cars built at the time. The idea was an old one; in fact little was new in the area of furniture design by the opening of the last century. (An attractive bench with a reversible back appears in a fifteenth-century painting of Saint Barbara by Robert Campin.) The Camden and Amboy adopted the design for its eight-wheel coaches of 1836. The B & O itself employed both metal-frame swingback seats, as recorded in the Austrian Karl Gegha's 1844 report, and a high-backed wooden Empire-style seat which seems to have had a platform rocker base. The Empire chair appears in a B & O drawing (1835c., shown in Figure 1.70) and the engraving of a B & O coach in Douglas Galton's 1857 report on American railways (Figure 1.82).

British visitors were well pleased with American seating. A correspondent for the *Illustrated London News* wrote glowingly of the standard of comfort available to travelers in North America and the absence of the mean wooden benches in third-class carriages.[2] Except for emigrants, all passengers could expect a seat that was "well stuffed, and covered with a fine plush, the arms of the seats being made of polished black walnut or mahogany." The reporter was equally impressed by the generous seat size—18 inches deep by 39 to 42 inches long—and the deep, comfortable cushions.

American travelers of the period were not so well pleased. Compliments were few, complaints legion. An advertisement for a luxury seat in the 1851 *American Railway Guide* claimed that even after a short time in a common car seat, the passenger felt as if his bones were pounded in a bag. Another critic wrote that the experienced traveler "will breathe dust and sparks, and submit to any amount of jolting with entire serenity," but that his language would become irreligious when faced with a car seat that lacked such conveniences as a simple footboard.[3]

A dozen solutions were devised to resolve the car seat dilemma. Because no one design could comfortably accommodate all passenger sizes, shapes, and weights, some form of adjustable seat was regarded as essential. August Mencken and other historians have claimed that the first adjustable or reclining railroad car seat was introduced in 1852. This is the year when the first U.S. patent was issued for such a device, but as Chapter 4 described, similar seats were being tested on the Camden and Amboy more than a decade earlier. Reclining seats were also used by the Eastern and P W & B railroads in the mid-1840s. It is true, however, that no wide-scale interest in reclining seats developed until the middle of the following decade. Although the inventions which flooded the Patent Office at that time aimed generally at a more comfortable seat, their specific purpose was to provide a recumbent sleeping-car seat. Bailey, Kasson, Smith, and Hammitt seats were rushed onto the market (see Figures 5.1 and 5.2). The usual exaggerated claims were made for them. Hammitt announced that his seat "will render railroad traveling more pleasant than staying at home."[4] A number of railroads adopted these new designs for at least some of their better cars, and the introduction of parlor and chair cars encouraged a new wave of reclining-seat inventions. Adjustable seats were expensive and heavy, however. They required more space and hence reduced the total seating capacity, and a fee had to be paid for the patent. More important, they failed to supplant the sleeping car. In testimony during the Pullman-Wagner legal battle of

Figure 5.1   American passenger car seats typical of the 1850–1870 period. The reclining seat on the left was patented by C. P. Bailey in 1854. (A. Bendel, Aufsätze Eisenbahnwesen in Nord-Amerika, *Berlin, 1862, plate E)*

ALBERT M. SMITH'S
Patent Premium Reclining and Self-Adjusting
CAR SEAT.

For a Night and Day High or Low Back Seat combined in One.

PATENTED AUGUST 21st, 1855.

It was awarded two first premiums, a Gold Medal, at the great "Fair of the American Institute," held at New York, and a Diploma at the State Fair, held at Elmira, N. Y., 1855.

This valuable improvement is adapted and can be applied at a very trifling expense, to the ordinary Seat now in use, without impairing its present qualifications as a day Seat, and a new Seat provided with it, and made in single seats, costs no more than the ordinary seat.

By an arrangement that is very simple and not liable to get out of order, the back is so hung at points, varying from the centre, that it can be converted into a High Back Night Seat, by pushing against the upper part of the Back, which disconnects the lower part, and allows the sides to be reversed, the outside placed in, which is the natural form and shape of the person, and raised high enough to support the head, this slides the seat forward on a curve, so as to be in conformity with the angle of the back, and it is then self-adjusting to any position of the person, and cannot be moved from it by the sudden motion of the Cars, making a seat as perfectly adapted to its intended use, (as a High or Low Back), as if made specially for it, and no other.

The Seat can be seen and examined, and orders will be received for the improvement to apply to old Seats or for new Seats at the office of

ALBERT M. SMITH,
Patentee and Manufacturer,
13 NORTH St. PAUL STREET, ROCHESTER, N. Y

or **TAULMAN & LOW,** *Agents,*
157 BROADWAY, NEW YORK.

Figure 5.2   Smith's reclining seat, from an advertisement in the Railroad Advocate *of December 15, 1855.*

Figure 5.3 Numerous inventors attempted to produce headrests to convert ordinary coach seats for sleeping purposes. (American Railway Review, April 16, 1860)

1881, W. N. Bragg, who had patented a reclining seat in 1857, admitted that he could never sleep in one of his own chairs.[5]

An elaborate car seat would be worth the money if it eliminated the need for sleeping cars, but because it could not, most railroads were content to give this trade to Pullman. The less affluent traveler was forced to make do with a low-backed, one-position coach seat. American inventors offered the coach passenger a wonderful variety of headrests, stirrups, and hammocks (see Figure 5.3). Their advertisements claimed that each individual willing to buy and carry his own apparatus might convert an ordinary seat into a commodious chair or bed. The earliest device of this nature seems to be a headrest patented in 1860 by Amos Gore.[6] One of the wildest, patented in 1889 by Herbert M. Small, was a small canvas hammock, fitted with hooks, to be mounted between two seat backs. The sleeper's legs hung down between the supporting straps, and the feet were held by a pair of looped rope stirrups. The hammock must have had approximately the stability of a trapeze in a hurricane. It required the cooperation of the people in the forward seat, since three of Small's patented hooks protruded over their seat backs. The experienced traveler avoided such gadgets and paid the extra fare to go Pullman.

For almost the entire history of railway passenger service in America, the typical car seat was a small, relatively uncomfortable two-passenger bench (Figures 5.4 and 5.5). The low one-piece back was reversible. The seat cushion was flat or convex. To withstand heavy use, both the back and seat cushions were hard, with stiff, unyielding springs. The armrests were often unpadded; they were commonly of wood, though in some instances nickel-plated cast iron was considered good enough. No adjustment was possible except that the position of the seat back could be reversed. The tall, the short, the fat, and the lean were expected to adapt themselves to the shape of the seat. Footboards were often dispensed with because they made cleaning more difficult. (A fast sweeping was a necessity at many terminals, where the cars were sent out on another train almost as soon as they were emptied.) The typical coach seat was not really a full sofa, though in size it appears to duplicate a love seat. A pair of legs was commonly omitted, for one end of the seat was attached to the inside wall of the car. This saved weight and the cost of the legs. It also simplified cleaning under the seat.

In the universal American-style coach, double passenger seats were placed on either side of the aisle. A rare exception was the rows of double and single seats used by narrow-gauge lines since the 1870s and adopted many years later for some deluxe standard-gauge cars. Suburban cars were occasionally equipped with three-abreast seats on one side of the aisle and conventional two-abreast seats on the other. The Southern Pacific adopted this plan as early as 1912 on some of its Oakland commuter cars, and the New Haven followed suit several years later.[7] The scheme was never popular, however, and when it was proposed in recent years for the new Erie-Lackawanna suburban service, it was bitterly opposed by the riders.

Before 1850, car seat frames appeared to be fabricated almost entirely of wood. After that time metal began to play a greater role. Cast-iron end frames in fact became very popular by mid-century. This style of seat framing has been credited to George Buntin, foreman of the Eastern Railroad's paint shop.[8] Although the claim cannot be verified, Buntin did patent a peculiar style of reversible seat that employed a cast-iron frame. The principal feature of the device was its cradle-like body; the seat and the back were identical, so that the seat was the back when it was tipped one way, and vice versa. It was called the poor man's reclining chair, because its central pivot did offer a variation in angle. Of course, both occupants of the seat had to agree on the same position. A number of New England railroads used the Buntin seat, and it was being produced at least until the late 1880s. A variation on the design is shown in Figure 5.6.

Others tried to improve the standard car seat at a realistic cost for the average coach patron. Reclining seats were simply too expensive for wide application; a coach seat mechanism had to be simple. One of the best of these schemes was the rocker seat patented in May 1879 by C. C. Mason of Altoona, Pennsylvania.[9] The loose seat cushion rested on an elliptical cradle and could thus be shifted and inclined fore or aft according to the position of the seat back (see Figure 5.7). This simple design made it possible to slope the seat cushion so that the passenger slid back against the seat back. In 1883 the inventor claimed that over 20,000 of his seats were in use. They were a great favorite on the Pennsylvania Railroad.

The comfort of the day coach rider was the abiding concern of Mathias N. Forney, locomotive designer, editor of the Railroad Gazette, and longtime secretary of the Master Car Builders Association.[10] Forney was appalled at the lack of science displayed in car seat design. The car seat manufacturer's aim was to produce the cheapest form of chair, add a few fistfuls of stuffing, and let the human shape conform as best it could. Forney understood the need for a cheap, simple seat, but felt that economy should be carried just so far. In analyzing the problem, he concluded that no one had bothered to study human anatomy or the basic elements of physiology. He set out to design an organic seat that suited the human form.

Forney's study led him to stipulate five conditions that were necessary for a comfortable seat:

1. A large bearing or seating surface for maximum support.
2. A concave seat cushion to receive and support the human posterior.

Gardner's New Reversible Car Seat, No. 8,
Patented December 6th, 1881.

THE WAKEFIELD RATTAN CO., BOSTON,

HORTON'S RECLINING CHAIR FOR CARS.
Solid Comfort in Riding Found at Last.

A real Sleeping Chair for night and a delightful Parlor or Reading Chair for day riding.
It has an automatic head-rest or pillow, an adjustable leg-rest, also an adjustable
foot-rest. Turns on a pivot, which avoids all difficulty in reversing the seat in cars
and is so simple and easy in its operations that a child
can adjust it to any of its many positions. It is a *very
great comfort* to passengers riding at night, and will
greatly increase *local night travel*. It will
be used by nine-tenths of the traveling public, and tend to *increase
travel. Send for Circular
and further information*
N. N. HORTON,
Kansas City, Mo.
It has been adopted by the
Chicago & Alton, Chicago,
Burlington & Q'y; Wabash; Han. & St J.;
Mo. Pac.; St L., K. C. & N.; Kan. Pac.;
Ky. Cen.; Col. Cen.; D.& R. G.; Den.
& South P.; Detroit. Grand H. & M

PATENTED FEB. 20, 1877.

*Figure 5.4   The variety of seat styles available in 1870–1890 is suggested by this selec-
tion of advertising cuts. The Gardner bench had a pierced wood veneer seat and back
rather than upholstered cushions. (Miscellaneous trade journals)*

3. A seat cushion of the proper depth and height, with a soft
front edge that would not dig into the legs.
4. A seat cushion inclined toward the back, to relieve strain
and cradle the passenger more securely in his chair.
5. A high and properly shaped seat back that supported both
the shoulder blades and the small of the back.

The common car seat was deficient on all five points. On
August 25, 1885, Forney patented a seat (No. 324,825) that satisfied his conditions. A clever arrangement of levers, which he
called cross links, permitted the seat back to reverse (Figures 5.8
and 5.9). Forney promoted his design in the *Railroad Gazette* and
by personal appearances before such groups as the New England
Railroad Club. Several railroads, including the New York Central
and the Long Island, adopted his seat. How widely it was used
cannot be determined, but it did enjoy a long life (see Figure
5.10). The Scarritt Car Seat Company of St. Louis came to control the patents, and as late as 1928 its successor, Hale and

Kilburn, continued to manufacture the Forney seat under the
trade name of Scarritt.[11]

Except for the Forney seat and the relief offered by an occasional chair car, the coach passenger seemed condemned to suffer
forever in the walkover seat (Figures 5.11 and 5.12). By 1930,
however, luxurious long-haul motor buses unexpectedly began to
cut into railroad passenger traffic.[12] Since a notable feature of
the buses was their comfortable bucket seats, the venerable
arguments of railroad managements about maximum capacity
and the undesirability of squandering cash on fancy coach seating were silenced. More legroom and soft, well-formed seats became the new order. Of course, it was not until the lightweight
era that coach passengers escaped the old walkovers, but the
railroads were at least committed now to improving the lot of the
coach patron (see Figure 5.13).

The bucket seats introduced in 1930 had a double set of
springs. The upper decks were encased in nearly airtight upholstery to form an air cushion. In 1936 molded foam-rubber

Figure 5.5 Reversible or walkover seat backs were standard in American coaches. Both wooden and cast-iron seat ends were common. (Pullman Neg. 2776)

Figure 5.6 This cast-iron seat, patented by T. J. Close in 1860, is similar to the popular Buntin seat except for the wooden-slat seat and back.

cushions were introduced.[13] Gone was the walkover back; the seats had individual reclining backs (Figure 5.14). To reverse, the entire seat rotated on its base. On some cars it was possible to turn a seat 90 degrees and view the scenery in comfort. Some manufacturers used a simple pedestal base, but one of the largest car seat makers, Heywood-Wakefield, devised a platform-mounted center plate. The seat could thus be placed tight against the interior wall, ensuring maximum aisle width. It was pulled away from the wall along the track of its platform and then swung around on the center plate. Even with such an ingenious mechanism, more generous spacing was needed for the seat. It was possible to keep it within a few inches of the old 38-inch center spacing, but most roads were willing to give their patrons even more legroom. The Illinois Central changed to a 52-inch spacing that cut a standard-sized car's capacity to a mere 48 passengers.[14]

The success of the deluxe bucket seat inspired further investigation into this subject by the Association of American Railroads' Mechanical Division. In 1945 with a grant from the Heywood-Wakefield Company, the group hired Professor E. A. Hooton of the Harvard anthropology department to conduct research on car seat design.[15] Hooton, an expert in the field of human anatomy, set up test chairs in major Boston and Chicago stations to gather the needed measurements. Data collected from the 3,867 adults who volunteered for measurement were used to prepare a standard profile guaranteed to satisfy the greatest number of coach passengers (Figures 5.15 and 5.16). The most critical single dimension was the seat height of 16¾ inches.

Hooton's recommendations led to the development of the famous Sleepy Hollow car seat that became very nearly the industry standard in the years after World War II. Heywood-Wakefield offered the seat in three basic styles: a night chair with reclining leg rest, a day chair without the leg rest, and a lightweight model with a tubular frame. By the time the Sleepy Hollow line appeared, its maker all but controlled the car seat business. (Through a predecessor firm it claimed an establishment date of 1826. In 1897 the Heywood Brothers and the Wakefield Rattan Company had combined, and during the next half century they absorbed or outlived their major competitors, including their old rival, the Hale and Kilburn Car Seat Company of Philadelphia.)

Deluxe car seating was expensive not only because the seats themselves cost $3,000 to $4,000 but also because they added dead weight.[16] For generations seat designers had tried to trim weight. Early in the century pressed steel was substituted for cast-iron parts, but with the introduction of oversized deluxe seats the weight per seat jumped to 175 to 200 pounds each. Even if a car had only thirty seats, they weighed a total of 3 tons. Designers hoped to cut that figure in half, and one manufacturer actually succeeded in producing a 105-pound coach seat by using a welded tubular frame.[17] However, the overwhelming popularity of the Sleepy Hollow seat ended all hope of cutting overall car weight by creating lightweight furniture.

A great variety of materials was used for seat coverings. Bare wooden seats appeared only in the poorest class of cars. Emigrant cars, commuter cars, and special-accommodation cars for railroad shop employees might be outfitted with slat-bottomed seats. So far as comfort was concerned, the perforated veneer seats manufactured by Gardner and Company of New York probably ranked one level above the slat seat. In them thin sheets of veneer, steam-molded to the desired contour, were perforated with a geometric pattern of holes for ventilation and extra springiness. Even so they made a poor seat and had only

Fig. 1123. *Elevation.*
MASON ROCKER CAR-SEAT, *Pennsylvania Railroad Standard.*
Fig. 1124. *Cross Section.*

Fig. 1125.
With lock.
SEAT-STOP. (4 5-16x⅝ *in*)

Fig. 1126.
Without lock.

Fig. 1127.   Fig. 1128.
MASON ROCKER CAR-SEAT.
Perspective Views.

Fig. 1129.
SEAT-ARM PLATE.
(2 7-16x1⅞ *in.*)

Fig. 1130.
SEAT-ARM PIVOT.
(*Head, ¾ in. diameter.*)

Figure 5.7   It was possible to tilt the seat cushion of the Mason seat by adjusting its position on the rocker-like base. The design was patented in 1879. (Car Builders' Dictionary, 1888, figs. 1123–1130)

limited use in steam railroad coaches. Gardner's major market was in the rapid-transit field.

From the beginning, coach passengers could expect some form of upholstered seat. Before 1880 the covering might be cloth, leather, rattan, velvet, or even woven wire.[18] Of these options leather appeared to be the best choice, since it is handsome, durable, stain-proof, and readily available. But it is expensive and slippery, and in the winter months passengers complained that it was cold. Even so, leather has had a steady if not universal application throughout the history of passenger cars.

Almost no information exists on seat coverings in the early period, though it seems that velvet was favored for the superior class of cars. In 1849 the Hudson River Railroad's best cars were upholstered with the finest plush velvet.[19] Several years later the Erie draped its cars with 300 feet of the same material at $5 a yard.[20] Red hues, from crimson to dark maroon, were most often chosen, though in the mid-1870s Pullman was said to favor Utrecht velvet, whose olive bronze was such a subtle shade that only an expert could identify it.[21]

Around the mid-century a miraculous material was introduced

that was to become the standard railroad car seat covering: bristly, bright mohair. It could endure heavy wear and retain an attractive appearance after years of use. Two and one-half times as strong as wool, mohair was the longest-wearing natural fiber known.[22] It could be counted on to last twelve to fifteen years, and some seats were still good after twenty-five years. Its stiff pile tended to hold the passenger in place, unlike a slick, soft-pile velvet. Mohair was also a sanitary covering, since it did not attract dirt and would not hold it. Because the oil was inside the individual hairs rather than on the outside, as with wool, normal processing did not destroy its silky luster.

Mohair reportedly came to the United States in 1839. At that time the first Angora goats were sent to a South Carolina planter by the Turkish sultan in gratitude for the American's help in showing the Turks how to grow cotton.[23] For more than forty years, however, plush made from the fleece of the Angora goats was imported. It was an expensive handmade textile until 1882 when, after three years of experimentation, the Sanford Mills in Maine perfected a practical mohair loom.[24] The domestic machine-made textile soon captured the market. It had few seri-

Figure 5.8   M. N. Forney, editor of the Railroad Gazette, *patented a seat in 1885 which became widely used on many American railroads.* (Railroad Gazette, *June 4, 1886*)

Figure 5.9   Forney designed his seat so that the back and seat conformed to the human anatomy in either their forward or their reverse positions. (Railroad Gazette, *June 4, 1886*)

ous rivals for railroad car seat upholstery until the introduction of cheap, soil-resistant fabrics during the 1940s.

Heating

During most of the nineteenth century, passenger cars were heated in the winter months by the common iron stove. It was the least expensive and simplest form of artificial heat available at the time. In a land of cheap firewood and coal these fuel-hungry ovens were not considered inefficient, and the brakeman's work in stoking them was regarded as time well spent. By modern standards, however, the stove is second only to the open fire in inefficient heating, because it warms by direct radiation. This is adequate for a small enclosure, but in the long tube of an American passenger car, a stove heated only a portion of the

structure. To equalize the effect, it became common during the heating season to remove several seats on one side of the aisle and place the stove in the middle of the car. After about 1870 this practice seems to have been abandoned and a permanently fixed stove was installed at both ends. But neither method was really effective; passengers next to the stove roasted, while those far away from it froze. The car was difficult to heat not just because of its shape, but because the side walls were thin and generally uninsulated, while the windows and doors were loose-fitting. The doors seemed to be open more than closed as trainmen walked through the car, passengers boarded, and porters propped the door open to load luggage. Added to the inefficiency of the iron stove was the very real fire hazard. On every count the car stove was unsatisfactory, but for many generations it was accepted as preferable to no heat at all.

Figure 5.10   A Forney-style seat was still being produced by Scarritt-Comstock in 1915. (Railway Mechanical Engineer, March 1915)

Besides the stove, there were other major car-heating systems over the years. Hot-air systems used a stove as the source of heat but broadcast a stream of warm air through the car, usually by baseboard registers. Hot-water systems again used the stove but gained better distribution through a system of pipe radiators. Steam, usually drawn from the locomotive, was common in this country from about 1890 to 1910. Vacuum, a low-pressure steam system, was seldom chosen for car heating. Vapor, a popular extremely low-pressure steam system, became standard after about 1910.

There were, of course, countless experimental systems. An idea that was regularly proposed was the storage heater, for which cannon balls, bricks, or iron slugs would be heated in a central plant and deposited in the cars. Such schemes may have been plausible for city transit cars but not for long-distance trains. Electric, alcohol, and even gasoline heaters were proposed at an early date. The most original system, developed by a Swede in 1870, aimed at heating the train with a central furnace in the baggage car—not by circulating hot water or steam, but by propelling heated sand from car to car.[25] The inventor felt confident that he could make sand circulate through a labyrinth of iron pipes, flexible couplings, elbows, and tees the length of the train and then return to the head-end furnace for reheating. Such an invention could only convince practical railroad managers that stoves must be the best system after all.

In the very first trains there were no car heaters, and the passengers complained not about the poor heating but about the lack of all heat. Travelers bundled up and tried a variety of foot and lap warmers fueled by hot bricks, steaming water, or live coals. These methods were a holdover from coaching days, and as long as railway cars were divided into small coach-like compartments, the era of self-heating prevailed. No railway could be expected to install a stove in each compartment. But the early emergence of the large single-compartment eight-wheel car quickly changed the heating possibilities. First came the single space heater, a modest enough accommodation. In December 1835 a Boston newspaper reported the advent of anthracite coal-burning stoves on the lines then running out of the city.[26] A month later it was announced that the B & O was using a small, under-the-floor heater patented by Alexander McWilliams of New York (Figure 5.17).[27] Other Northern railroads began to install stoves at about the same time, according to reports in accounts by David Stevenson and Von Gerstner.[28] The Austrian engineer said that the Utica and Schenectady had mounted a wood-fired stove outside of the car body and run a pipe under the seats to distribute the heat.[29] A large Stephenson-built, double-truck car purchased by the Auburn and Syracuse had a similar system, according to a note in the *Rochester Democrat* of August 28, 1839. But a few roads, such as the New Haven–Hartford line, were running unheated cars as late as the winter of 1839.[30]

Almost no specific information appeared about the early railway heaters; it can only be assumed that the stoves used were so ordinary that no description was thought necessary (see Figure

Figure 5.11   Cast-iron-frame, mohair-covered seats such as these were the industry standard throughout the heavyweight era. (Standard Steel Car Company)

Figure 5.12   The standard walkover seat was given modern ends and high individual seat backs, but basically it was unchanged from the traditional design. (American Car and Foundry Company)

Figure 5.14   An aluminum-frame reclining seat on a swivel base made by Karpen and Brother in 1937.

Figure 5.13   Lightweight tubular frame seats made for the New Haven in about 1937 by Heywood-Wakefield. (Railroadians of America)

5.18). One exception was the Littlefield car heater, a double-walled coal burner patented on March 4, 1856.[31] Manufactured by Erastus Corning of Albany, it was put into use on the New York Central Railroad, of which Corning happened to be president. The self-feeding central hopper, filled with some 50 pounds of fuel, was good for the entire run between Albany and Buffalo. It operated at one-tenth the cost of a wood-fired stove and was the only heater on the train that could keep frost from the windows when the temperature dropped to zero (Figure 5.19). Passengers, particularly ladies and children, were said to crowd into the cars fitted with Littlefield's miracle, leaving the other coaches nearly empty. Even the trainmen, who rarely had a civilized word to offer on any subject, eloquently praised the heater because it required no attention throughout the long trip. No record exists on how widespread the Littlefield stove became. Its introduction suggests, however, that a general conversion from wood to coal firing began at about the same time that a similar shift in locomotive fuels took place. The shift from wood to coal represents a significant advance in railroad car heating.

Another special form of car heater that gained some attention was the cone disc stove manufactured by Gardner Chilson of Mansfield, Massachusetts (Figure 5.20). It was a common cast-iron stove of the 1870s except for the large-diameter disc that deflected the heat, preventing the immediate rise of hot air to the clerestory and circulating some heat to the floor where it was most needed. As a safety feature in the event of a wreck, long

Figure 5.15 *After measuring 3,867 adults, Professor E. A. Hooton of Harvard University proposed this seat contour as a standard for railroad car seating. The design was completed in 1945.* (Railway Mechanical Engineer, *August 1945*)

Figure 5.16 *The Sleepy Hollow seat, based on Hooton's profile, provided luxury seating for coach passengers.* (Burlington Lines)

Figure 5.17 *McWilliams's coach heater, originally proposed for carriages, was used by the Baltimore and Ohio in the 1830s. C is the stove; d the heat pipe; b the register, and e the exhaust.* (U.S. Patent Office)

bolts secured the stove to the floor, and the fire doors could be locked as another precaution against the scattering of live coals.

The basic defects of the car stove, however, were not solved by Littlefield, Chilson, or a host of other inventors. The most prevalent complaint was that the stove was too hot for passengers unlucky enough to be sitting near it. The experienced traveler made a dash for the intermediate seats—not too close, not too distant. But for the people pressed near its simmering iron walls, the atmosphere was dry and baking as a kiln, so that they dozed and then woke sodden with perspiration. A giant 350-pound heater crammed with a quarter cord of wood could "burn the air and blister the varnish."[32] The villain here was the oafish brakeman whose job it was to tend the stove and who invariably went to extremes of overstoking or understoking. Overstoking was the more common; it seemed to be the pleasure of this barbaric race to burn up as much of the company's fuel as possible.

During the dark age of the car stove, many mechanics were busy devising improved heating systems. The most promising, the hot-air heater, was both simple and cheap enough for large-scale application. The basis of all such plans was to direct a stream of air over the stove to be heated and then to distribute it through the car by registers or ducts. The forward motion of the train created the air current, which was gathered by a roof-mounted hood or scoop. Axle-driven fans were also proposed, but most inventors wisely chose to avoid this complication. One of the first forced-air systems was patented on July 24, 1846 (No. 4654) by Tappen Townsend of Albany. It was a wind scoop mounted on the locomotive. After being heated in a furnace, the air was directed back along the train by ducts, with flexible joints between the cars. As far as is known, Townsend's plan was never tested. A few years later a Canadian, Henry Jones Ruttan, began to promote a forced-air system that received many trials during the 1850s and 1860s. Ruttan's chief concern was the purification of air for car ventilation (covered in the following section), but he incorporated a hot-air heater as part of the system.

In 1862 William Westlake (1831–1900), better known in later years as a partner in the giant Chicago railway supply house of Adams and Westlake, devised a forced-air heating system.[33] As well as the usual roof-mounted scoops, it had an under-the-floor heater, and registers running along the floor on both sides of the car assured a more even distribution of hot air than other systems provided. Westlake's heater was used by the La Crosse and Mil-

Figure 5.18   The common wood or coal stove is shown in this coach interior engraving.
(Illustrated London News, *April 6, 1861*)

waukee Railroad. Pullman was an early champion of the West-lake design but abandoned it within a few years in favor of the Baker hot-water heater.

The most popular of all hot-air stoves, patented on July 7, 1857 (No. 17,756), by James Spear of Philadelphia, was a cast-iron potbelly enclosed in a sheet-metal circular casing (Figures 5.21 and 5.22). The radiating surface was increased by a similar casing around the smoke pipe. Spear mounted a wind scoop on the roof that directed the air to the base of the stove and thence out along the length of the car through a baseboard register. The heater was so effective that it was adopted as standard by the Pennsylvania Railroad. This in itself guaranteed the success of Spear's patent, and soon such imitators as A. P. Winslow of Cleveland and F. S. Bissell of Pittsburgh were competing for a portion of the trade. A major attribute of the Spear plan was its low first cost—$45, about one-tenth the cost of a hot-water system.[34] However, the wind scoop heaters were not faultless. When the train was stationary, the air flow was reversed and the heat was exhausted out of the stovepipe.[35] The clerestory vents were too large, creating an imbalance that tended to draw cold air in and around the windows, doors, and other openings. The same imbalance was said to produce crosscurrents that inter-

fered with the lamps, causing them to smoke and blacken the chimneys.

Another class of car heater closely related to these hot-air systems was the suspended, or pendant, stove. It was mounted outside the car under the body, thereby leaving valuable space for more passengers. Suspended stoves were said to pay for themselves by the revenue from the extra seats. Optimistic supporters believed that the stove would be less dangerous in a collision because it would be thrown clear of the train.[36] And since the stove was not in the car, passengers were not burned, bruised, and blistered by bumping against it when the train lurched or abruptly stopped. Also, the stove's dust, ash, and smoke were not produced inside the car. The scheme had so many apparent advantages that it was tried at an early date. (McWilliams's patented heater of 1832 shown in Figure 5.17 was of this type.) The Old Colony tried some form of under-the-floor furnace in 1846.[37] Westlake pushed the suspended stove during the 1860s.

The chief advocate of the pendant heater was the Reading Railroad. Its partiality for this system was explained by an unusual circumstance. The road was double-tracked at such an early time that the clearances between the tracks, tunnels, and cuts were made extremely narrow.[38] Although adequate for their

Figure 5.20 *Chilson's car stove, patented in 1876 and 1878, featured a large cast-iron disc top that was intended to deflect the heat to the floor. Note the tie rods securely mounting the stove to the floor.* (Poor's Manual of Railroads, 1875–1876)

Figure 5.19 Littlefield's self-feeding car stove was patented in 1856. Coal-fired car stoves gradually replaced wood stoves after 1860. (New York State Railroad Report, 1856)

day, they proved burdensome as larger and wider rolling stock came into favor. The cars passed so closely that passengers indiscreet enough to poke a head or hand out of the window were in danger of being hit. Enough damage claims resulted that wire grills were mounted over the windows. And now that the passengers were effectively caged inside the cars, the management felt compelled to do as much as possible to lessen the danger of a human barbecue. Gaslights and under-the-floor heaters were installed beginning around 1860.

The ascendancy of the pendant stove was assured on the Reading when John E. Wootten, the chief mechanical officer, took an interest in developing the idea. On September 29, 1868, Wootten was granted Patent No. 82,672 for a pendant-style heater. Since the general scheme was an ancient one, Wootten could claim little more in his patent application than the invention of a

reversible air vane, or scoop, mounted on the side of the furnace. Within ten years it was reported that all the Reading's cars would soon be outfitted with under-the-floor units, despite the fact that successive rebuilding of the line had corrected its clearance problems and the window screens might soon be phased out.[39] A better understanding of the Wootten stove, its radiator, and the stovepipe arrangement can be gained from Figure 5.23. These drawings appeared in the 1888 *Car Builders' Dictionary*. The major parts are designated by the letters A through H. A, B, and C are not shown, but the furnace is clearly represented. D is the cold-air supply pipe, E the hot-air flue, F the smoke pipe, G the hot-air delivery pipe, and H the coal box.

Soon after Wootten's retirement in 1886, the Reading's enchantment with the suspended heater evaporated. It was referred to as "old-fashioned," and though 400 cars were still so equipped, no new cars had been outfitted with it for some time.[40] The Wootten stoves were said to consume enormous quantities of fuel. Because they were hung out in the cold, much of the heat was lost simply in heating the car's underside rather than its interior. They required special attendants and fueling carts at every major stop. And of course, it was impossible to tend their fires while the train was in motion. The pendant idea was not entirely rejected in the industry, however; both W. C. Baker and George Westinghouse promoted a steam heater of this type.

Hot water was an attractive alternative to the methods just described. It is recognized as one of the best heating systems for small and intermediate-sized structures like railway cars. It distributes heat evenly from a central source and retains its warmth after the fire dies. It is relatively safe—certainly when compared with steam. A major drawback, however, is its high first cost: it requires a boiler, an expansion tank, piping, and radiators, which are much more costly than an equivalent hot-air system.

HEALTH!   COMFORT!   SAFETY!

Figure 5.21   James Spear manufactured a popular style of hot-air stove based on his patent of 1857. (National Car Builder, *October 1881*)

Curiously, hot-water heating for railroad cars seems to have received little attention until the middle 1860s. And even then it was pushed by a single man, William C. Baker (1828–1901) of Dexter, Maine. The Baker heater became the best-known, if not the most widely used, single make of car stove, and its inventor's name is one of the few that can be recognized by those with only a casual knowledge of railroad car history. Baker's interest in central heating began soon after he moved to New York City around 1850. Apparently he was drawn to the subject through a lecture delivered by Professor Benjamin Silliman, Jr., a popular scientific speaker and engineering gadfly of that day. In 1851

Baker helped Silliman to install the first steam-heating system in New York City.[41] Baker's interest gradually turned from heating buildings to heating railroad cars. Around 1865 he was ready to test the car heating system he had developed; during the next year he received the first of forty patents he would eventually acquire. The first Baker heater was demonstrated aboard the private car of the New Haven Railroad's president. It proved so successful that it was tried on some of the road's coaches. Then the New York–Boston Through Line adopted it for sleeping and parlor cars. In an advertisement of 1868 Baker listed seven railroads, including the Erie and the Michigan Southern, that were

Figure 5.22   Spear's hot-air stove had floor-level registers and a roof-mounted air scoop.
(Journal of the Franklin Institute, *1897*)

using his apparatus.[42] Within a few years the Baker heater was found on nearly every first-class car in this country. But few roads felt that they could afford to spend $500 each to install the heaters in their coaches when $2 to $10 coal stoves did an adequate job.[43]

Baker's invention was not an original contribution; in the main he was simply the first person to apply hot-water heating to railroad cars. However, his system was well arranged and executed. The source of heat was a small coal-burning stove shrouded with a sheet-metal jacket placed in one corner of the car (Figures 5.24 and 5.25). An iron pipe coil was mounted inside the stove.[44] The lower end of the coil passed out of the stove and looped its way around the floor under each seat, thus forming a continuous radiator. The pipe returned to the area of the heater and rose to terminate in the cistern, or expansion tank, mounted on the car roof. The upper end of the coil ended in the same tank. A filler plug and a safety valve were attached to the

expansion tank, which was a hollow, melon-shaped iron casting. The pipe used was 1- and ¼-inch, with 11 running feet allowed for each passenger. No pump was necessary; the natural currents set up by the fire at one end of the pipe were enough to circulate the water.

A basic problem was that the water would freeze in cold weather unless the fire was kept burning, but Baker thought of an ingenious solution. He devised a simple antifreeze by saturating the water with as much fine, dry salt as it would dissolve. The strong brine solution, which any mechanic could make, would not freeze until the temperature dropped to one degree above zero. In these extreme conditions, of course, the system would have to be drained or the fire kept going. Because the system was airtight, the salt did not attack the pipes—a fact that Baker claimed was proved by long years of service.

The salt water in turn caused a problem that Baker confessed had been difficult to work out. If the safety valve were to open,

Figure 5.23 Under-the-floor stoves were adopted by some lines to reduce the fire hazard. J. E. Wootten's patented design shown here was used on the Reading between 1870 and 1890. (Car Builders' Dictionary, 1888, figs. 1529–1532)

WARMING AND VENTILATING
Railroad Cars
BY HOT WATER.

BAKER'S PATENT CAR WARMER.—One way of Applying it.

A very simple, safe and efficient plan for

Warming Railway Carriages !
— BY —
HOT WATER PIPES,

Which Radiates the Heat Directly at the Feet of Each Passenger without the Necessity of Going to the Stove to Get Warmed !

☞ All the finest Drawing-Room and Sleeping Cars in the United States have it, or are adopting it. Full descriptive Pamphlets furnished on application

Baker, Smith & Co.,
Cor. Greene and Houston Sts., N. Y., and 127 Dearborn St., Chicago.

Figure 5.24 Baker's hot-water car heater was superior to ordinary car stoves because it could distribute heat more uniformly through the car. (Railroad Gazette, August 5, 1871)

salt crystals would form on the seat, preventing it from closing tightly once the pressure was relieved. His solution was a rubber-ball safety valve, which was used for many years until even Baker came to admit that it was a menace, for reasons that will be explained later. The valve was set for 150 pounds of pressure. Care was needed to fill the system and expel all the air. During this operation the system was subjected to a 400-pound hydrostatic test to expel the air and test all joints for absolute tightness. Theoretically, once this was done no more water need ever be added, but in practice an occasional topping off was found necessary. A pressure gauge attached above the heater, on one end of the coil, indicated a blockage in the system if it registered more than a trifling pressure. Once heated up, a dull fire could normally maintain a comfortable temperature. Two pecks of coal were sufficient for a twenty-four-hour period.

The Baker heater offered the great advantage of a steady, even warming of the entire car. It took up less floor space; one Baker could easily do the job of two stoves. It saved fuel. Unlike the Spear and most other hot-air systems, its operation was not dependent on the motion of the train. The brakeman, since he was not forever having to attend the heater, was free to "answer the call of the whistle." In a serious smashup, Baker cheerfully suggested the pipes would rupture and put out the fire. Apparently the scalding brine was not supposed to touch the passengers.

One of the Baker heater's failings has already been mentioned: the first cost discouraged its use for anything but palace cars. Also, it did not eliminate the fire peril; as long as live coals were carried inside a car, that danger was present. And it was said to be a noisy system, given to fits of pounding.[45] Far worse, however, was the danger of explosion. Like any closed system with a heat source, it could burst, with terrible results. During the 1890s several explosions were blamed on Baker heaters.[46] One of the worst, which occurred on the Santa Fe, wrecked one end of a smoking car and injured five passengers, two of them seriously. Baker blamed the safety valves; he said that trainmen would screw down the rubber ball to stop them from leaking. To solve

Figure 5.25   A schematic drawing of Baker's hot-water car-heating system. (Institution of Civil Engineers, *vol. 53, 1878, plate 4*)

the problem he devised a new throwaway style of cast-iron valve.

The popularity of Baker's apparatus encouraged others to produce variations on the same theme. A Cincinnati inventor, J. Q. C. Searle, introduced a hot-water system in 1870 that broke Baker's monopoly and was eventually manufactured by the Union Brass Company of Chicago. It was good enough for use on some of Pullman's deluxe cars. A second, less popular hot-water system was patented by J. Johnson in 1881. The Pennsylvania Railroad tested its own hot-water system in 1879.[47] It differed from the Baker system in that it had a single boiler, placed in a baggage car, for the entire train. A steam pump circulated the water. The device did not succeed, but it was one of the first instances of continuous train heating in this country.

In 1882 Baker, anxious to outdistance the competition, introduced an entirely new car heater: a pendant, low-pressure steam cooker, to be produced by a new company formed in Pittsburgh with the help of George Westinghouse.[48] The cup-shaped boiler held 8 gallons of water and measured roughly 2 feet in diameter and 1 foot deep. Two inclined chutes to the firebox from either side of the car served as hoppers which automatically fed in coal through a shaking action induced by the motion of the car. The hoppers carried 175 pounds of coal—enough for twenty-two hours of fire. Within two years the business was reorganized as the Standard Car Heating and Ventilating Company, with Westinghouse as president.

In a few years Baker was back in business for himself, again as

a champion of hot water. Steadily growing car sizes called for larger heaters, and Baker increased the coil length from 16 to 30 linear feet.[49] The profile of his new model was squatter, and to lessen the danger of the stove's shattering in the event of an accident, the iron castings were replaced by pressed steel or malleable iron. A safe-like wrought-iron box offered further security against fire (Figure 5.26). This stove, if its fire was properly laid with 75 pounds of hard coal, could burn without attention for twenty-four hours. Baker introduced a heater of even greater capacity in 1892. It featured two coils, both made of tightly wound, tapered pipes for a maximum heating surface.[50]

In 1890 some 7,000 cars (or approximately one-quarter of all passenger cars) were heated with hot water; the majority of these were most likely of Baker's manufacture.[51] An 1891 Baker catalog in the Library of Congress claims that 30,000 Baker heaters had been sold since 1868. In 1893 Baker's business was so prosperous that he built a new plant in Hoboken, New Jersey, but after 1900 all forms of heating were superseded by steam for main-line passenger cars. Baker heaters gradually disappeared, although they continued in limited use for cars that required some independent form of heating. This was especially true for private cars that were set out at isolated stations which had no steam. Some self-propelled rail cars also used Baker-style heaters into the 1920s.

All the early forms of car heating depended on a live fire, and this meant that the danger of incineration was an everyday hazard. But the fire peril was an accepted fact of life at home or

Figure 5.26   To lines willing to pay the extra costs, Baker offered iron safes as a precaution against fire. (Railroad Gazette, *March 11, 1887*)

lished. "Not long since," one correspondent said, "the writer was riding with a friend on a train which passed through a short tunnel. The register of the heater was in such a position as to allow the fire to shine out of it through two apertures about two inches apart. When the car became dark the fire glared out of these openings like the eyes of some demon. This caused the friend to remark: 'There is death, staring us in the face'."[53]

The combination of stoves and wooden cars was unquestionably dangerous. Tinder-dry wood heavily coated with varnish, oilcloth head linings, upholstered seats, lamp reservoirs sloshing with kerosene—it was a rich flammable mixture waiting only for the stove to topple over and spew out its cherry-red coals. As if there were not enough kindling aboard already, car builders discovered that wood shavings were a cheap insulating material and blithely packed the spaces between the upper and lower floorboards with this highly flammable material. In time the wood shavings were replaced by mineral wool, a material that was adopted by the Pennsylvania Railroad in 1882.[54]

Other measures were taken to lessen the likelihood of fire. Stoves were, of course, set on metal sheets, often zinc, to protect the floor from sparks. The doubtful precaution of pendant heaters has already been mentioned. Stoves were bolted to the floor sills in the hope that they would not fly loose and break open during a wreck. The B & O locked the stove doors to prevent the fire from spilling out in case of an upset. Each brakeman had to carry a key to open them. Often a water box was built into the stove's base so that if the stove tipped over, the water would spill and theoretically extinguish the fire (Figure 5.27). In 1868 Ruttan's hot-air stove was reported as having a water bottom containing five pails of water.[55] Many other stove manufacturers, such as Winslow and Bissell, offered similar features if desired. How effective these tanks were remains a question. If they were widely used, which is uncertain, they do not appear to have solved the fire problem.

More elaborate extinguishers were also developed. In 1887 the Pittsburgh, Chicago, and St. Louis tested a water sprinkler that would be automatically set off by a derailment.[56] Even if there was no fire, the passengers would receive a hosing. The Dyer Railway Automatic Fire Extinguishing Company of Chicago advertised a device that promised to kill any stove or lamp fire. Apparently it was a simple chemical (soda) extinguisher with tubes running to every open flame in the car. A more practical short-term solution was to install a better class of stove: cheap, brittle, cast-iron models were easily shattered, while as early as 1875 it was noted that wrought-iron stoves could withstand all but the most disastrous shocks.[57] Most railroads did not bother to make this simple substitution for car safety, however, and *The Railroad Gazette* announced that they had foolishly ignored the public interest. "And now the day of stove reform is past, and the people demand a root and branch revolution in [the] methods of heating. Who can blame them."[58]

Even if the railroads were not ardently interested in passenger safety, dozens of valuable cars were destroyed yearly by fire. The most publicized losses were the result of wrecks. Trains that were

abroad, since all heating and lighting depended on an open flame. Houses, barns, sheds, steamships, and churches regularly burned to the ground, and there were few communities that had not experienced at least one courthouse fire. Consequently there was no general outcry from critics of the railroads about the dangers of stove heating until rather late in the century. Beginning in the early 1880s, however, a storm of indignation suddenly seemed to explode. True, there had been a number of disasters in which passengers were burned in the wreckage, and some were surely burned alive. Although this was not a new development in railway travel, and although a minute number of people died by fire relative to the millions of passengers carried, it was an unspeakably gruesome prospect. Perhaps public wrath was stirred by the culmination of years of railroad accidents. Certainly the press also played a major role in the antistove crusade. The illustrated national weeklies reported all gory details of railway calamities. *Harper's* editorialized in a number of cartoons, the most stirring of which, entitled "The Modern Altar of Sacrifice," pictured a glowing car stove standing in the midst of a flaming wreck. Draped on top of the stove was a prostrate female figure. The wooden railway car and its demon heater were likened to the Prophet Elijah's fiery chariot.

The railroad trade press joined the cause. In June 1882 the *American Railroad Journal* pointed out that with the coming of warm weather, "the season of railroad barbecues is over, the annual sacrifice of baked and roasted victims has been offered up." Another journal described stove-heated cars as "perambulating crematories."[52] Dramatic letters to the editor were pub-

Figure 5.27 Water-bottom stoves were another way to lessen the fire peril from coal stoves. Bissell's stove was a product of the 1870s. (Poor's Manual of Railroads, 1879)

not badly damaged by the accident itself could be swept by fire, and wrecks had a way of happening in isolated places where no fire equipment was handy. Cars standing idle between runs were also known to go up in smoke, frequently because of an unattended stove. Between 1887 and 1890, ninety passenger cars were lost by fire.[59] Claims against the railroads for lost baggage, express shipments, and mail were a secondary expense. In these same fires twenty passengers died, and while not every death can be attributed exclusively to fire, the claims that the roads normally paid amounted to $5,000 each. Hence the liability loss itself was over $100,000, much of which could have been saved by eliminating the stove menace. After generations of stubborn resistance, even the railroads were beginning to see the need for some better way to heat their cars.

STEAM HEATING

The answer to the fire peril was steam heating. It was a clean, wonderfully efficient method that had become standard in factories, hospitals, apartment buildings, and other large institutions following its introduction in the 1780s by James Watt. It was superior even to hot water as a medium of heat transfer. Like hot-water heating, steam also required elaborate plumbing that made it an expensive installation. The initial investment was justified by the uniform heating it provided and by the savings in fuel it made possible.

In addition to its inherent advantages, steam appeared to be a natural for railway service. The locomotive at the head of every train was a ready source of steam; the hot vapor had only to be piped back to the cars. No stoves, hence no fire in any car of the train, put an end to the danger of fire. The logic of the plan must have been obvious to the earliest car builders, yet nothing can be found to show that any form of train-board steam heating was tried before 1843. Around then Jacob Perkins (1766–1849), who was described as an American mechanical genius with a head that fairly rattled with ideas, settled in England and pursued various projects to employ high-pressure steam.[60] A sickly offspring of his studies was a steam heater that Perkins fastened beneath the Queen's private carriage on the London and Birmingham Railway. The oil-lamp–fired boiler fed steam to pipes running between the double floor. Like so many of Perkins's efforts, it was prophetic of what would one day be an important innovation.

The first reported attempt at steam heating in the United States was on the Old Colony in 1851.[61] A few years later a test was made on the New York Central.[62] Although details are lacking, it can be assumed that nothing practical was developed, because the next news in the field came from Europe. In 1860 several German railways began serious tests of steam car heating. Within eight years another four German lines were using steam heat. In 1865 the Prussian railways were trying a low-pressure steam system fed by a separate boiler placed in the baggage car.[63] Sweden began to experiment in the early 1870s, and the

Alsace-Lorraine railways attempted to heat their cars with steam drawn from the locomotive in 1877.[64]

News of these experiments seem to have inspired fresh interest in America. A system patented by H. R. Robbins in 1868, which involved a small vertical boiler placed in the baggage car, was tested on the B & O. In 1875 a low-pressure scheme invented by William M. Fuller was tested on an unnamed railroad in the New York City area.[65] Separate stoves fitted under the floor supplied the steam. (It will be recalled that Baker revived the same idea in 1882.) A few years after Fuller, the Camden and Atlantic Railroad tried out a design patented by Wood, Scull, and Snyder.[66] All these pioneering efforts were failures, owing generally to the resistance of the railroads, defects in the individual systems, and the lack of prestige of the inventors themselves.

In contrast, circumstances were favorable for William C. Baker, who unveiled a steam heating system on the New York Elevated late in March 1878.[67] Steam was drawn from the locomotive and distributed through the cars by 1¼-inch pipes. Rubber hoses made a flexible connection between the cars. Leakage at the hose connections eliminated the condensate without the need of elaborate traps. While it is generally thought that steam heating was not adopted on the Elevated until 1883, *Locomotive Engineering* claimed that the El's first 700 cars were fitted with Baker's system. This inventor's success can be ascribed to the utility of his design, the receptiveness of the Elevated, and his own reputation as a leader in his field. The Elevated's desperate need for a compact, efficient heating system may indeed have been the chief reason that Baker was given his opportunity. The cars were so crowded that there was little space for customers, much less stoves, and the crews were much too busy to tend fires. Stops were frequent—the doors were open as often as they were closed—and only the most efficient system could keep the cars even tolerably warm. Baker's direct use of the locomotive rather than an auxiliary boiler as the heat source was especially significant, because it prefigured the future of steam heating in America.

Main-line managers discounted Baker's system as a novelty suitable only for rapid transit. They disparaged the devices of other inventors as foolish gadgetry, and one official stated that any more tests would prove "a waste of money simply to gratify the whim of a lot of cranks."[68] Railroad managements maintained that steam heating would tax the locomotive's strength and delay schedules; that separate boilers were too expensive; that the cars would be too hot. They argued that the fire hazard

Figure 5.28   Steam heating was introduced in the 1880s to eliminate car heating stoves.

could never be eliminated so long as there was a single open flame on the train, including cigars, lamps, and cooking ranges. And what of the locomotive itself? Had not the embers from the engine's firebox caused the Ashtabula holocaust?[69] Moreover, the wreck on the Maine Central had fried two mail clerks in a steam-heated car because the cooking stove had toppled over, setting the wreckage ablaze.[70] The bogeyman of scalding steam was another convenient argument against the new system; the rail-road establishment claimed that the choice was between being roasted by the stove or parboiled by the steam pipe. In addition, the rear cars of steam-heated trains were so chilly that the danger of pneumonia was greater than the danger of fire. Should these statements fail to convince the traveler that steam heating was visionary nonsense, a final stopper was thrown out: "The locomo-tive is meant to haul the train, not heat it."[71]

The inventors persisted in the face of official hostility. Not all railway managers were unsympathetic; some felt that steam heat-ing must be accepted in time and actually encouraged experi-

mentation. The problem was to perfect a workable system, and this could be done only in the field. The best laboratory was a train in everyday service, operated by everyday trainmen and patronized by everyday passengers. In the early 1880s half a dozen railroads cooperated in holding road tests that led to the real beginning of steam car heating in this country.

The proving ground was New England. The participants were the smaller trunk lines, such as the Connecticut River Railroad, which permitted James Emerson to outfit a train in 1881.[72] Emerson's separate boiler overheated, endangering the baggage car, so that he next tried taking steam from the locomotive with separate heaters under the cars to assist in warming the train. Since Emerson, like many others, believed that a separate boiler was preferable as the source of steam, he felt that an iron head-end car might solve the problem. Emerson's system worked well enough to remain in use for at least six years.

Lieutenant James W. Graydon of the U.S Navy was another proponent of the auxiliary baggage-car boiler. In the winter of

1881–1882 he tested apparatus on the Troy and Boston Railroad with fair success, but the interchange of cars with another line that was not equipped for the new form of heating ended Graydon's work after only a single season. At the same time the Maine Central allowed D. D. and J. H. Sewall of Portland, Maine, to try out their patented design. A notable feature was their snap coupling head for the steam hoses (Figure 5.28). The Sewall system was eventually adopted by the Maine Central and received some use on other New England lines and the Michigan Central.

William Martin of Dunkirk, New York, was even more fortunate with his design than the Sewalls and has been credited with the first commercial success in this specialty.[73] His initial patent was granted in 1882, and following a few small test installations Martin broke into the big time when the Boston and Albany adopted his plan (Figure 5.29). In 1884–1885 Martin outfitted 200 of the road's cars. By the spring of 1887 three railroads were using his equipment; ten months later twenty-one roads had signed up.[74] By the summer of 1889 Martin apparatus had been installed in over 2,600 cars.[75]

Martin had many rivals: Wilder, McElroy, Westinghouse, and others. None, however, posed a major threat except Edward E. Gold (1847–1931) of New York. He was the nephew of Stephen J. Gold, an expert in the heating of large buildings.[76] Gold had been raised in the business and came to the problem with great experience in the intricacies of steam heating. He was also a friend of Russell Sage, a Wall Street raider who bought and sold railroads together with his ally, Jay Gould. Sage encouraged Gold to pursue his interest in car heating; there might be big money in it one day if the government forced its adoption. Gold obtained a few patents (the first in 1881) while Sage, a major shareholder in the New York Elevated, smoothed the way for a test. In 1883 Gold equipped the first cars with his heaters, and the El's general manager displayed an unusual enthusiasm for the system.[77] Within five years all Elevated stock was remodeled with Gold apparatus. Because the line's small engines had to make the frequent starts necessary to rapid-transit operations, they were not capable of supplying steam for the cars. Direct steam heating was therefore not practical. In the compromise worked out by Gold, steam was sent back to hot-water heaters in the cars after the train was underway and the need for maximum steam was temporarily at an end (Figure 5.30). Each car had four cylindrical, floor-mounted radiators.[78] These double-tube reservoirs measured 4 inches in diameter by 10 feet, 10 inches long. The close-fitting inner pipe was filled seven-eighths full with salt water, which was heated by a steam jacket formed by the outer tube. The radiators could retain enough heat to warm the car for two hours. Gold calculated that 32 pounds of coal would heat an eight-car train for one hour.

The dramatic success of the Gold system on the nation's largest single passenger-carrying railroad prompted main-line managers to consider its adoption. The Elevated had grown into a major enterprise since the time of Baker's steam trials, and it was now managed by experienced main-line railroad executives. If they

Figure 5.29 *The Martin steam heating system, as advertised in the 1886 Railway Review.*

were satisfied with the Gold plan, it must be good. The Boston and Providence, the Long Island, and the Staten Island quickly fell into line. Next came the main-line roads like the B & O. By 1906 Gold boasted of being the largest car heating supplier in the nation, with apparatus on 40,000 cars.[79] By this time he was offering all types of heating apparatus—direct-steam, vapor, and electric—but his principal system was similar to that perfected for the Elevated. In addition to hot-water pipe radiators, he had devised a simple plan to convert existing Baker-type systems to steam. He replaced the original coil with a spiral pipe that had a smaller-diameter steam line inside. Thus the existing piping and heater could stay in place, while steam rather than a stove fire heated the water. In an emergency the fire could always be lighted.

By and large, government action to speed the cause of steam heating was the familiar story of too little, too late. Few laws

NO MORE PASSENGERS BURNED ALIVE.
THE DANGEROUS STOVE SUPERSEDED.
THE EDWARD E. GOLD SYSTEM.

Heating and Ventilating Steam Railway Cars, Horse Railway Cars, Cable Cars, Stations, Steamboats, Ferryboats, etc.   Safety, Economy, Comfort, Convenience, Uniformity, Pure Air for Breathing, and Increased Seating Capacity Secured by the Edward E. Gold System of Heating Cars without Stoves or Fire.

Sufficient heat is stored in each car to heat it for two hours when detached from the engine.

Vertical Longitudinal Section of a Passenger Car of the Long Island Railroad, with Gold Heating Apparatus. (Scale 5-32″ = 1 foot.)

ADVANTAGES.—Perfectly safe. Entirely efficient. Takes up no room available for passengers. Utilizes waste steam from locomotive for heating. No fire is carried by cars. Always ready. Easily managed. Very durable. Floors of cars heated. Uniform heat thoroughly diffused through the vehicle. Excels anything yet devised for car heating. Send for Illustrated Catalogue to

EDWARD E. GOLD & CO., N. E. Corner Frankfort and Cliff Streets, New York.

Proof of Value: 800 cars on Manhattan Elevated Railway are heated by this system. Adopted also by Long Island R. R.; Prov., W. & B. R. R.; Suburban Rapid Transit R. R.; Staten Island R. R.; Hoboken Cable Cars and Ferryboats; Interstate Rap. Trans. Co. (Wyandotte, Kansas).

Figure 5.30   The Gold steam heating apparatus, introduced in 1883, was one of the first commercially successful systems in America. (Railroad Gazette, March 4, 1887)

were enacted until the late 1880s, by which time the railroads were voluntarily, albeit slowly, installing the necessary apparatus. Actually the legislators had been talking about the stove menace for decades before the first bill was passed. A railway safety bill was introduced in Congress in 1868 that specifically called for safer car heating and the conversion to iron cars.[80] More bills were introduced, but none passed during the next years. The first measure enacted that had any effect at all was a rider on the 1881 Post Office appropriation specifying a reduction of 10 percent in payments to any railroad that did not provide mail cars with safe lighting and heating. Apparently the language and enforcement of the bill was too weak to have any major influence, but it did give the railroads notice that Congress was concerned about the problem.

Ohio was one of the first state legislatures to act on the matter of railroad passenger safety. It passed a law in May 1869 specifying that all car heaters be self-extinguishing in the event of a derailment.[81] The act was apparently ineffective, however, because Ohio found it necessary to pass an antistove law in 1892.[82]

In 1887 a number of other states began to hold hearings on antistove bills. The fire-peril hysteria was at its zenith. Several workable, if not entirely perfect, steam systems were on the market, and antistove groups hoped that a little pressure from the state capitols might inspire a more expeditious conversion. Albany appears to have been the most aggressive in pushing through a steam law. The New York State legislature announced in August 1887 that stoves must be removed from trains by May 1, 1888.[83] An extension was granted until November 1888, but as late as 1894 the New Haven was still operating stove-heated coaches in New York.[84] In fairness it must be mentioned that the state railroad commissions took the case through the courts and a $7,000 fine was upheld. Other state antistove laws were passed, some as late as Maryland's in 1894, but in fact the railroads appear to have converted at their own pace.

There was some justification for the measured rate of the conversion. The industry was already obliged to finance the installation of air brakes and automatic couplers on several million

freight cars in order to comply with the recently enacted Safety Appliance Act. Steam apparatus cost about $200 a car, so that 6.8 million dollars would be needed to change the whole fleet. To this must be added the cost of boilers at terminals to warm idle cars, as well as the investment lost by premature scrapping of the stoves. Yet one state legislator could not understand the fuss about expense.[85] Didn't stoves consume $35 to $50 worth of coal a season? Wouldn't this all be saved by using "free" steam from the locomotive? From his vantage point, steam heating would quickly pay for itself.

The progress of steam heating is difficult to outline precisely. Few overall figures appear to exist. Most early statistics are from the Master Car Builders reports, and the roads which answered the questionnaires on which they were based usually represented only a portion of the total passenger car fleet. The fragmentary information available reveals the following facts. In 1887, 150 railroads which owned one-third of the nation's passenger fleet (representing about 8,500 cars) reported that 392 were steam-heated.[86] The figure climbed to 800 in the following year, but only 24 roads reported. Individual manufacturers, such as Martin and Gold, claimed to have installed far more systems than that. In 1889 with only 45 railroads returning the questionnaire, they reported 1,572 steam-heated cars in service. In 1891 the first industry-wide total was offered.[87] According to the *Railroad Gazette*, 9,017 cars, or 26.7 percent of the total, were now steam-heated.

A better idea of the transition to steam can be gained from the reports of individual railroads. In 1893 the Pennsylvania had 41 percent of the cars on its Lines East under steam, and another 23 percent were being fitted out.[88] Its western lines reported that 56 percent of their cars were already converted. Early in 1895 the B & O had fitted all its 700 cars with steam, the majority of them with apparatus furnished by Gold.[89] It is probably safe to assume that by the turn of the century all first-class cars were heated by direct steam or by steam-actuated Baker heaters. However, some commuter and branch-line cars never gave up stove heating, and some railroads maintained stove-heated cars

Figure 5.31   Gibbs's steam hose coupling was marketed by the Safety Car Heating and Lighting Company. By 1892 it was used on more than 13,000 cars. (Railway Review, *December 5, 1892*)

for secondary or excursion trains years after the general conversion was in effect. This policy resulted in at least one major tragedy long after the fire peril was no longer an issue. On December 5, 1921, a head-end collision occurred on the Reading in a deep cut near Philadelphia.[90] The train, composed of aging wooden coaches, piled up and caught fire. Fire-fighting equipment could not reach the blaze because of the location, and twenty-seven people died. Most of the fatalities were blamed on the fire; once again the car stove had gathered in new victims.

Why weren't the railroads more enthusiastic about adopting steam in the middle 1880s? Some reasons have already been given, but a number of less obvious technical and operational problems were also involved. Temperature regulation was the major difficulty, and overheating was the chief complaint. The head-end cars received steam fresh from the boiler, and even if the pressure was reduced to 5 pounds or less, the temperature was well over 200 degrees. In mild weather the front cars were insufferably hot, like a fireless cooker. But because of radiation, steam leaks, and the insistence that train steam-line pressures be kept as low as possible for safety reasons, live pressure dropped to nothing by the time it reached the rear cars. No pressure, no heat. Automatic regulation could at least partially solve the problem, but the railroads complained about the expensive controls, so complicated that no ordinary trainman could learn to operate them in a lifetime.[91] Anyhow, the jarring of the trains upset most of the thermostats that were tried.[92] As early as 1888 the Milwaukee Road tested electric-pneumatic controls designed by George Gibbs. They had some success, but not enough to win many converts.

Traps were one automatic feature that could not be avoided. There had to be some way to discharge the steam once it condensed, or the system would fill with water. For maximum heat economy, the traps should be set to flush only after the water had cooled to, say, 120 degrees. But because traps froze with maddening regularity, it was common to adjust them well above this temperature. Once frozen, the system was blocked and all heating stopped at the point of blockage. Premature flushing made the direct steam system very uneconomic. In a test conducted in 1895, the difference between the costs of stove and steam heating were negligible, when theoretically steam heat should have been substantially cheaper.[93] Some lines attempted to avoid traps altogether, eliminating the condensate through leakage at the hose couplings. This was not only sloppy but self-defeating, because steam escaped as well as water.

The flexible connection between the cars was a fruitful source of mischief. Martin advocated a flexible pipe connection articulated through ball and sliding joints. Most heating suppliers favored rubber hoses as a cheaper method, and during the first half century of steam heating this was the standard connection. Manufacturing a durable rubber hose was not easy, however. Some lines received as little as two months' service from their hoses. Wire reinforcing actually shortened hose life, since it acted as a heating element. In 1894 the Peerless Rubber Company produced a durable hose from the finest Para rubber which was

guaranteed for at least one year.[94] In 1912 the Master Car Builders Association voted to issue very detailed specifications for steam hose manufacture in an effort to assure longer life.

The diversity in hose coupling heads was also a continuing problem. Each manufacturer had his own design, though all followed the basic air brake hose coupling devised years before by Westinghouse. This snap-lock coupling was made fast by a quarter turn and would automatically pull apart if the cars separated. Some modest advances in standardization were reached by 1892, when thirty-two railroads were said to be using the hose head patented in 1887 by George Gibbs (Figure 5.31).[95] These roads operated 13,301 cars, or about one-third of the entire fleet. But as late as 1912 the Master Car Builders were still struggling with the problem of nonstandard hose couplers.[96]

Steam heating placed a considerable drain on the locomotive boiler. In frigid weather the engine needed all the steam it could produce just to move the train. Snow and ice on the tracks, as well as journal oil that became stiff as grease, all impeded the train's progress. In 1887 it was calculated that a fourteen-car train consumed 22.6 horsepower of steam for heating, which was equal to 4.5 percent of the total engine power.[97] Computations made some years later showed that 100 pounds of coal an hour were needed to heat a ten-car train by steam.[98] Any way it was figured, steam heating robbed the locomotive of fuel and power. Larger locomotives were a necessity if schedules were to be maintained.

A possible solution to the steam-drain problem was an auxiliary head-end boiler. Many of the earliest steam heating promoters advocated this scheme, but few were able to sell it. Only while George Gibbs was on the Milwaukee Road's engineering staff did a heating engineer have any success in promoting the head-end system. He had initially favored drawing steam from the locomotive, but by 1890 he had come to believe that there should be an independent source of steam. On his recommendation the road built two 35-foot-long light and heat tenders— short, 32½-ton baggage cars that were encased with ¼-inch iron plate as a fire precaution.[99] A large horizontal boiler was justified, because it supplied both the engine generator for electric lighting and the steam for heating needs. The locomotive was thus entirely relieved of auxiliary requirements. The tenders worked well, especially in the bitter cold of the Northwest, but few other roads were ready to adopt Gibbs's plan until it was revived many years later.

A final difficulty in the adoption of steam was the problem of transition, particularly for cars involved in interchange service. For Pullman this was a major obstacle. Unless all lines interchanging cars were equipped for steam, it was pointless for any

=DIRECT STEAM HEATING SYSTEM=

=FOR=

=PASSENGER CARS=

THE SAFETY CAR HEATING & LIGHTING CO.

160 BROADWAY, NEW YORK, N.Y.

SCALE

DRAWN BY — CORRECT — APPROVED

JUNE 3RD 1900. — ENG.

SECTION AB

LONGITUDINAL SECTION

2" STANDARD PIPE — 1½" TRAIN PIPE — TRAP 833 — STANDARD PIPE

STEAM TRAIN PIPE — AIR BRAKE PIPE — SIGNAL PIPE — TOP OF RAIL

Figure 5.32 This direct steam heating system of 1901 used steam, at a reduced pressure, directly from the locomotive's boiler. (Safety Car Heating and Lighting Company catalog, 1901)

Figure 5.33 In the indirect system, steam from the locomotive heated water in the car heater. The water was then circulated through the internal heating system of the car. (Safety Car Heating and Lighting Company catalog, 1901)

Figure 5.34   After 1910, vapor heating replaced both the direct and indirect steam systems shown in Figures 5.32 and 5.33. The reducing valve pictured here was the heart of the system. (International Textbook Company, 1912)

one participant to convert (Figure 5.32). The problem could be partially solved by the use of dummy steam lines, so that the vapor could be passed the length of the train to the cars that were already fitted. A better solution was a combination steam and hot-water system, in which the stove could be lit if no steam connection was available. In reality this problem was not widespread, because the interchange of passenger cars was relatively small except for the Pullman fleet. Many lines ran trains that remained an integral package for weeks on end, moving back and forth between the same points. In these cases it was quite practical to have several different plans of steam heating on a single railroad, since the cars of the Golden Gate Special were rarely if ever mixed with, say, those of the Sacramento Limited.

A full solution to the problems of direct steam heating seemed impossible, and what developed instead was a major compromise: the union of steam and hot water. Edward Gold achieved this alliance in the early 1880s. Others, including Westinghouse, came to accept it as a blessing—the best of both schemes combined. Steam was introduced into a tube running inside the coil of a Baker-style heater. Hot water circulated through the pipe radiators. Because live steam was not flowing near the passengers, it was possible to boost pressure from 5 to 30 or even 60 pounds. The rear cars were thus guaranteed a lively supply of heat, and the danger of line freezes was reduced. Efficiency was materially improved, because the hot water fully utilized the latent heat rather than dumping it overboard, as in the direct steam system. More even temperatures were possible by regulating the hot-water flow. And overheating was less likely because no live steam was introduced into the radiators.

Railroads tended to favor the indirect method of heating, as it came to be called, because it solved what had seemed to be the unsolvable regulating problem (Figure 5.33). It also offered a way to salvage the existing investment in Baker heaters. A simple conversion kit would do the job. And it was still possible to operate the heating independently by stoking up a fire; no need

to worry about cold cars on nonsteam interchanges, or isolated terminals devoid of steam. It was also reassuring to have a fail-safe heating system for emergencies, like stalled or snowbound trains. If the locomotive's fuel ran out, the individual car heaters could warm the train. As late as 1905 the authority George W. Fowler, editor of the *Railroad Gazette*, said that the indirect steam system was the American standard.[100]

Yet for all its advantages, the indirect system was a makeshift that could be roughly compared to a plan for using a steam engine to pump water uphill to turn a waterwheel. Some more direct way to exploit steam must be found. The vacuum system was suggested, but it involved a supply-and-return line and a pump.[101] A far simpler scheme was introduced around 1902 by Egbert H. Gold of Chicago, who proposed using extremely low-pressure steam—no more than a few ounces above atmospheric pressure.[102] Heating engineers knew of this method as the vapor system, but no one had thought of applying it to railroad cars.[103] In Gold's modified version, pressure steam of 150 to 170 pounds filled the train line. At this pressure the rear cars would have plenty of steam, but its temperature would be 375 degrees, far too hot to introduce into the radiators. A reducing valve dropped the pressure to less than 1 pound, so that the steam entering the radiator pipes was at a manageable 212 degrees (Figures 5.34 and 5.35). Overheating of the cars was less likely, and regardless of the variations in the train-line steam pressure, the reducing valve kept the vapor pressure at a uniform level. Vapor also proved more economical than the old direct systems because steam was metered into the car as needed.

In 1904 the Chicago Car Heating Company was organized to exploit Gold's patents, and the business flourished as railroads switched over to vapor heating. Established suppliers like Edward E. Gold were forced to develop rival systems. Others fell by the wayside, while a few firms like the old Standard Heating and Ventilating Company were absorbed by the Chicago Car Heating Company, which in March 1917 became the Vapor Heating Company.[104]

Vapor was the greatest revolution in car heating since the beginning of steam in the 1880s, but like steam, vapor did not entirely solve the heating problem. The passenger car is inherently difficult to heat.[105] It is exposed on all six sides. Steel cars were said to require 20 percent more steam than wooden cars. Its rapid movement through cold air increases its radiation losses. It may pass through highly variable temperature extremes in a matter of hours—from boiling deserts to snow-covered mountains. It has a large window area, again causing radiation losses. Frequent opening of the end doors constantly admits cold air. The heating system must fit into a confined space yet remain accessible, and it must be able to perform for long periods with minimum servicing. Indeed, it is surprising that any system could be devised which partially meets these requirements.

Insulation, double-sash windows, and in later years Thermopane glass all helped to reduce radiation losses. But even a well-insulated car consumed 250 to 300 pounds of steam an hour.[106] Train pipe size and pressure were increased to satisfy the de-

Figure 5.35   A typical vapor steam heating system of the 1920s. Electrical controls regulated the degree of heating. (Car Builders' Cyclopedia, 1928, fig. 1917)

Figure 5.36   Fin-style radiators helped improve the efficiency of vapor heating. (Car Builders' Cyclopedia, 1953, fig. 918)

mand. For years 1½-inch pipe had been considered adequate, but in 1903 the Master Car Builders recommended an increase to 2 inches, and in 1911 the association made this size standard practice.[107] By 1948 the pipe size was increased by another ½ inch. Pressure followed a similar upward course, from 30 to 60 to 130 to full boiler pressures of 250 to 300 psi.[108] After pressures exceeded 130 psi, rubber hoses were no longer suitable. By the late 1920s flexible pipe connections between the cars were becoming more common. Steam hoses were made obsolete in January 1937 when they were barred from interchange service by the Association of American Railroads.

Other refinements were adopted in the 1930s. Electric thermostats had been reintroduced in 1921 and began to gain favor within a few years.[109] With the advent of air conditioning, thermostats became standard equipment. At this time the heating, ventilating, and cooling apparatus were integrated as a single system. Vapor continued to furnish the heat, but it was now distributed by overhead forced-air ducts and floor-mounted finned pipe radiators (Figure 5.36). Some cars had liquid-filled floor radiators reminiscent of the old indirect steam systems.[110]

The abandonment of steam locomotive operations did not bring an end to vapor heating. Small upright boilers stationed within the diesel-electric locomotive's body took over this duty. Similar oil-fired units had been used almost since the beginning of electric traction. Because main lines electrified only short portions of their tracks, through cars did most of their traveling behind a steamer. When it was clear that diesels were soon to be the dominant form of power, the design of steam heating generators was refined. Since the late 1920s the Otis style of boiler manufactured by the Vapor Heating Company was the best package unit available.[111] An even more efficient boiler was introduced around 1933 by an Englishman named Alic Clarkson. It was tried on the Union Pacific motor train streamliners with great success and soon became the favored style of steam unit. On long trains, Clarkson boilers were worked without respite at 100 percent of capacity. A typical steam generator burned diesel fuel and produced 4,500 pounds of steam an hour.

According to William D. Edson, the former manager of cars for Amtrak, head-end power plants—a modern version of Gibbs's 1890 boiler tender—began to replace small Clarkson-type gen-

Figure 5.37   Salisbury's dusters, a series of under-floor side curtains, created a tunnel
under the train to carry the dust away from the cars. The scene was photogaphed on the
Michigan Southern in 1868. (Pennsylvania State Railroad Museum)

erators in the late 1950s.[112] A head-end car could provide ample
room for a heating boiler and generator capable of servicing the
growing power needs of modern trains. The decline in passenger
revenues at this time ended hopes for the new investment necessary to produce such cars. A few experimental units were created,
however, such as the separate power car for the Pennsylvania's
Keystone. According to Edson, vapor heating has definitely been
abandoned in present practice. Electric heating was installed on
most cars after about 1965. Diesel-electric locomotives are being
fitted with alternators for head-end power. In some cases the
main traction motor generator is given enough extra capacity
to meet this requirement. Some older cars that have independent
engine generator sets, like the Santa Fe hi-level cars, will be
equipped with larger (60-kilowatt) units to handle electric
heating.

Ventilating and Air Conditioning

Ventilating involves nothing more than drawing the good air in
and forcing the bad air out. On first consideration it seems a
simple matter of throwing open the windows and letting in all
the fresh air in the universe. But in its wake come dust, smoke,
cinders, and the partially condensed steam that billows along the
train's smoky course. Just how to balance the conflicting requirements for a large volume of clean, pure air, without drafts, while
maintaining a comfortable temperature on cold days, was a problem that was never resolved until the advent of air conditioning
in the 1930s. Before that time ventilating was at best a sorry
compromise.

At first the problem did not receive much attention; passengers
seeking more air were advised to open the windows. For short
journeys the inconveniences of this method could be tolerated,
but as distances increased, it became a major hardship. The first
concession to improved ventilation appears to have been a row
of small openings above the top of the window sashes. The main
windows could remain closed while the open transom sashes
drew in fresh air above the passengers' heads. The oxygen supply

could be replenished without blowing the air and its disagreeable
load of soot into people's faces. An early example of this arrangement is found in the 1836 Camden and Amboy coach shown in
Chapter 1 (Figure 1.71). By the 1850s the scheme had been
refined: very small square openings were cut into the letterboard,
with either a simple wooden-center hinged, turnover sash, or
with a more elaborate sheet-metal, clamshell closure.[113] Wire
screen or pierced sheet-metal guards were sometimes placed over
the openings to keep out the cinders. Of greater importance, the
bad air was given an avenue of escape through roof-mounted
ventilators, which were usually short stovepipes with raised caps
to keep out the rain. Three or four of them were mounted along
the center line of the roof. Examples of cars with this primitive
form of ventilation are pictured in Chapter 1.

The combination of intake and exhaust vents provided a natural circulation absolutely essential to good ventilation. In theory
the stale, hot air would rise and flow out of the roof vents. In
practice, however, both the inlet and exit openings were too
small to move much air, particularly in the summer, when
sweltering passengers would have to throw open the windows
and suffer smoke and cinders for a little cooling air.

Passengers rarely suffered in silence; travelers are a complaining race whose voices are amplified by the grumbling of journalists.[114] One early-day editor compared the dust storm that
followed his train to a heavy Ohio River fog, as thick as gruel,
but dry and choking rather than cool and moist. He said that it
enveloped passengers and their garments with a heavy, dry coating. Early travelers on the Michigan Southern were sprinkled
with a shower of fine sand along the lake shore that changed to a
powdery clay dust near Cleveland—a dust so fine and dry that it
stuck in the throat like flour. According to another traveler, the
car interiors became so charged with dust that it was impossible
to see from one end to the other through the twilight haze.
Adding to the natural misery of a summer rail excursion was the
fresh-air fiend, a first cousin to the notorious seat hog. The fresh-
air addict would occupy a front position in the car with his
window raised high to admit a hurricane of smoke and dust that
blanketed everyone seated behind.[115] These worst of all health

cranks had "the constitution of sawlogs and the manners of oxen."[116] Trainmen were too indifferent or timid to restrain them, and less hardy travelers could only insist at the journey's end that something must be done about the dust.

One possible solution was to suppress the dust at its source. Ballast was a rare commodity on early railways; most tracks were simply set on bare ground. In dry weather a passing train naturally raised a minor dust storm throughout its journey. Because most lines insisted that they could not afford stone ballast, the only alternative was to sprinkle the tracks. In 1851 the Hudson River Railroad ran a sprinkler car in conjunction with its passenger trains.[117] The improvement was so remarkable that the Boston and Providence and the Stonington railroads followed suit.[118] These lines abandoned the practice, however, after only three or four years; apparently the cost was too great to continue it. In fact the service may have been initiated only to silence agitation for a state law requiring gravel or sod banks along the tracks as a dust shield.[119] The general practice was probably revived from time to time, for as late as 1898 the Long Island Railroad was reported to be sprinkling its tracks with oil.[120] The procedure not only increased passenger comfort by cutting down the dust, but reduced running-gear wear and allowed the use of cheaper forms of ballast such as sand or cinders.

A simpler solution to the dust problem was offered in 1855 by Elam C. Salisbury. He proposed directing it under the train by forming a wind tunnel of side curtains hung along the underside of each car. Rather than raising a cloud around the side of the train, the dust would then be pushed harmlessly out behind the last car. The side curtains, also called petticoats, could be made of canvas, boards, or sheet metal. They would serve secondary functions in deadening the noise of the running gear and preventing pedestrians from falling under the moving train. The scheme was attractive because it was low in first cost and involved no complicated maintenance problems. A year after Salisbury's patent, the Michigan Central was testing his invention.[121] The New York and Harlem also gave it a trial, but Salisbury's plan received the warmest welcome in the Midwest. The Michigan Southern was a strong supporter of the petticoat duster. A photograph, dated 1868, in the collection of the late Walter Lucas shows a full train of cars equipped with it (Figure 5.37). Yet within two years the scheme was abandoned, for the road discovered that a disastrous side effect of concentrating the dust under the train was excessive wear on the brake gears and journal bearings.[122] The mechanical department, deciding that Salisbury's invention was a luxury it could not afford, once again let the passengers fend off dust for themselves.

At the time that the Michigan Southern was stripping off its petticoat dusters, the Missouri inventor W. M. K. Thornton offered an improvement on Salisbury's scheme.[123] Rather than side curtains, he proposed a horizontal canvas curtain fitted to the truck frames below the axles, just 6 inches off the rails. Thornton thus hoped to keep the dirt below the journals and the passenger compartment. A secondary curtain was suspended between the trucks to form a continuous shield. After a trial in the West, nothing more was heard of Thornton's patent. A few years later, however, the Central Railroad of New Jersey tested a simpler form of duster that consisted of leather curtains stretched between and below the platforms. (The 1879 *Car Builders' Dictionary* shows the device in its figure numbered 216). Around 1880 the Pennsylvania Railroad put some fashion of duster on its Cape May express trains to help exclude the fine sand on that route.[124] Another style of duster was employed in 1930 by the Missouri Pacific on its observation cars.[125] A pipe frame covered with canvas extended out from below the rear platform and helped prevent the dust from billowing back to the passengers.

More elaborate antidust systems were proposed and tested during the 1850s. The most sophisticated involved air washers to cleanse the air. Some included ice cooling and might therefore be counted among the ancestors of modern air conditioning. In the various designs dust was collected by a water spray, a water bath, wet sponges, excelsior (which was often nothing more than wood shavings), or a combination of these methods.

One of the first air washers tried was patented in July 1854 by George F. Foote. A roof air scoop pulled air to a tank beneath the car, where a belt-driven rotary pump produced a spray. The air was then returned to the car's interior by floor ducts. August Mencken claims that Foote assigned his patent to the Michigan Southern shortly after it was issued.[126] The road installed the device at great expense in twenty-four new coaches received from Eaton and Gilbert that same summer.[127] A tank 18 feet long, 8 feet wide, and about 18 inches deep was fastened under each car, and water was added to a depth of 4 inches. In operation the system proved to be an effective dust killer—whenever it was in working order. Passengers did complain about the high humidity, but they rarely had to suffer long, because the pump or the spray jets quickly clogged with cinders, or else the belt slipped off one of the pulleys. And even if the system was working, the flow of air stopped when the train did. The windows were sealed, and passengers were left to endure mounting temperatures until the train got underway. During the following winter one of the Foote-equipped cars caught fire and was totally destroyed. The ventilating system was blamed for this calamity. A second series of trials was made during the next summer, and then the entire scheme was written off as a failure.

Foote's apparatus was also tested on the C C & C and the Erie. The Erie's mechanical personnel made some improvements to Foote's original designs.[128] The pump impeller was fitted directly to one of the axles or driven by a friction wheel. The water spray jets were attached to a standpipe reaching up into the intake air duct. A glass window was placed in one side of the duct so that the sprinkler action could be observed. Drawings of this apparatus were published in Douglas Galton's 1857 report on U.S. railways (see Figure 1.85). The service history of Foote's ventilator on the Erie and other lines where it may have been tested has not been uncovered. Presumably it was a short-lived phenomenon that promised much but delivered little.

The most talked-about air-washing system was introduced by

Figure 5.38   Henry Ruttan of Canada invented a ventilating and heating system that was used on a few U.S. lines between 1855 and 1875. Shallow water tanks below the floor helped trap cinders and dust. (Ventilation & Warming of Buildings, 1862)

Figure 5.39   In Ruttan's ventilating system, B–R was the roof air scoops; A directed the air to the water pans, M–P; B directed the cleansed air to the interior of the car via the register, F. (Ventilation & Warming of Buildings, 1862)

Henry Jones Ruttan (1792–1871) of Cobourg, Ontario.[129] Ruttan, a person of many occupations and interests, spent much of his life in the army, served as sheriff of his home town for thirty years, and became a representative in the House of Assembly. His avocation was heating and ventilating—an interest dating from the early 1840s that resulted in seven patents (1846–1858) and a major textbook on the subject published in 1862. In the course of his investigations Ruttan became interested in improving railroad car ventilating. In some respects his plan was very similar to Foote's, except that Ruttan dispensed with the pump and water spray and depended entirely on the water bath to collect dust and cinders (Figures 5.38 and 5.39). Tests of his system apparently began in May 1854 on the New York Central, and in the fall of the same year the Erie was ready to give his apparatus a trial.[130]

In Ruttan's device a 12-inch-deep tank under the floor provided 200 square feet of water surface. The bath was 3 inches deep. Baffles made the air take a meandering course to ensure maximum dirt collection. After passing over the bath, the clean air was directed into standing ducts that delivered it just above the passengers' heads. The intake was the familiar roof air scoop. No fans were employed; the current was dependent on the motion of the train. In winter one of the interior air ducts was replaced by a stove. The entire system, less the stove, cost a modest $100.

Ruttan was able to convince several railroads to install or at least test his apparatus during the following fifteen or so years. Few other inventors found so many openings, which says something for the system and its zealous originator. In 1857 the Grand Trunk was trying out an improved version.[131] Two years later the Boston and Lowell was also conducting road tests.[132] In 1863 the P W & B was expected to outfit thirty cars with Ruttan's ventilating system.[133] A year earlier all cars on the Michigan Central were reported to be so equipped.[134] By 1868 Ruttan announced that 300 cars were using his apparatus—a number far above the handful of cars that rival inventors could claim.[135]

After Ruttan's death, enthusiasm for the system declined. In 1873 the Burlington was still using it, but the road admitted that the apparatus had to be carefully managed in order to function properly, and that the task was too complex for the average brakeman.[136] Sometime before 1877 the Michigan Central, the largest single user, abandoned the Ruttan system.[137] It worked well enough on short trips, but after 50 miles the tank became choked with cinders, rendering it ineffective. Small leaks, which were unavoidable in the large sheet-metal tank, rotted out the floor sills. It is also likely that the car superintendent and his staff grew weary of attending to repairs on the complicated ventilator. Popular though it may have been with the traveling public, these men presented a good enough case about high maintenance costs to justify its removal.

The slow disappearance of Ruttan's system did not discourage other inventors from attempting to market similar devices. The railroad trade papers mentioned these efforts, but nothing of consequence occurred until the sleeping-car operator W. D.

Mann took an interest in the subject. Mann was seeking ways to discredit Pullman's service and style of car, and a major complaint against the open-section car was the poor ventilation. Passengers said that once they were closed inside the curtains, particularly in the upper berth, they felt as if they might suffocate. Mann thus found a ready-made issue and a ready-made answer in the ideas of Foote, Ruttan, and other early ventilation engineers. In Mann's version fresh air was drawn in through a low, wide hood that straddled nearly the full girth of the arched roof at one end of the car.[138] The air was directed into a closet, where it passed through a filter made of dampened mattress packing (which must have produced a conspicuous odor). In the summer the air was cooled by passing it over 350 to 600 pounds of block ice. In winter it was heated by steam pipes, though a small block of ice was still used to keep the excelsior moist with its melting water. In this way the need for a pump was avoided. The air was broadcast through the car by a register. Roof vents maintained the circulation.

With his usual bravado Mann made extravagant claims for the system, asserting that it achieved a full change of air every five minutes and that in summer it lowered temperature by 12 to 20 degrees. His claims were naturally challenged by the sober-minded *Railroad Gazette*, while the *Car Builders' Dictionary* conceded that the Mann system could effect temperature drops of about 10 degrees.[139] Opinions on the healthfulness of the air varied considerably between Colonel Mann and his critics.

Mann obtained a patent in September 1885, but he had already applied his system to some of his first North American cars in 1883. How many cars he equipped with it is uncertain, but as late as August 1887 the *American Railroad Journal* reported that the Mann sleeping car *Martha* had an ice-cooled ventilating system. Just how long he would have stood by his pet scheme will never be known, for the Pullman takeover in 1889 put an end to such unorthodox experiments for many years.

The dust problem simply remained unsolved and largely ignored. Some practical form of air-washer system could have been developed in the nineteenth century if the railroads had been willing to keep trying. The difficulty was not so much a lack of technology as a lack of commitment. An apologist for the industry might claim that the problem was indirectly solved by improved roadbeds, with crushed-rock ballast and mechanized ballast cleaners. Yet even today on European roads, despite their reputation for good track maintenance, major and minor clouds of dust are kicked up by trains. This powdery debris pours into non-air-conditioned compartments, coating passengers and their belongings with grime. One reason for the dust is that the leavings of the ballast cleaner are dumped close to the trackside, where passing trains are certain to stir it up. Whatever the cause, the dust problem lives on.

During the 1870s a new concern over ventilation arose in medical circles and spread to the general public. The purity, chemical composition, and quantity of the air people breathe became the subject of intense discussion. Railroad cars were singled out for

particular criticism. As early as 1840, when stove heating had become common practice, the atmosphere of the tightly closed cars was so foul that it made passengers delirious.[140] As the general ventilation craze grew, everyone attacked the railroads. A lady passenger complained of insolent train crews who prevented heat loss by refusing to allow passengers to crack a window sash.[141] To avoid the ill effects of "second-hand breath," she said that she had to spend part of the trip standing on the platform.

The members of the Master Car Builders Association, while determined not to pamper the public, tended to agree that railroad cars were the most poorly ventilated of all public gathering places. Their air space was extremely limited; even a prisoner was furnished twenty times more air then a coach passenger had.[142] The total volume of air in a railway car was adequate for no more than 4 people rather than the 50 to 75 who were crammed into a full car. Moreover, the air was hazardously contaminated with carbonic acid. Fitch Adams, the crusty master car builder of the Boston and Albany, expressed a minority opinion: "This matter of ventilation is a great hobby with some people. I knew a man who said more people died every year from breathing impure air than were carried off by whisky. The anti-whisky people say whisky kills $\frac{7}{8}$ of the people in the world, and if impure air kills more, where are they to be found?"[143]

Various guesstimates on the quantity of fresh air required per railway passenger were projected. In 1840, 240 cubic feet was given as the amount needed per hour. This figure was repeated in the 1874 Master Car Builders Report; twenty years later, however, nothing less than 3,000 cubic feet per hour was considered safe for the health and comfort of each individual.[144] At the same time, it was contended that the real villain was the high level of carbonic acid, now commonly called carbon dioxide or CO_2. Samples from railroad cars showed that the air contained 0.23 percent CO_2, nearly double the amount measured in theaters and other poorly ventilated places.[145] Carbon dioxide, it was claimed, was the cause of the lassitude, delirium, uneasiness, and dull headaches associated with long railroad trips. This notion was countered in 1910. In a paper read before the American Public Health Association, Dr. Thomas R. Crowder admitted finding high levels of CO_2 caused by normal respiration of passengers in such confined spaces as railroad cars.[146] But he contended that the passenger discomfort was produced primarily by lack of oxygen, excessive heat, and too much or too little humidity.

Throughout the pure-air discussion another basic theme was the need to coordinate heating and ventilating. The problem was that while it was possible to pull in enough fresh air to satisfy the most ardent CO_2 crank, it was not possible to heat the car unless the exchange was limited. In the days of stove heating, only a very small amount of the cold outside air could be taken in. Even with steam heating, only about 1,000 cubic feet per passenger per hour could be admitted if a 70-degree temperature was to be maintained.[147] This was one-third of the volume needed for a fresh interior atmosphere. The 1,000-cubic-feet compromise having been accepted, it was now necessary to find the best way to

NAMES OF PARTS, "EUREKA" VEN-
TILATOR; *Figs.* 1564–1565.

A. *Air-passage.*
B. *Air-inlet.*
D. *Deflectors.*
E. *Hood.*

Fig. 1565. *Section.*
"EUREKA" VENTILATOR.
(1-6 full size; for openings, 5x7" and
8x8".)

Fig. 1564. *Inside View.*
"EUREKA" VENTILATOR.

NAMES OF PARTS, CREAMER
AND "STANDARD" VENTILA-
TOR; *Figs.* 1566 *and* 1568.

AA. *Deflector-springs.*
BB. *Deflector.*
CC. *Casing.*
DD. *Deflector (in action).*
EE. *Hood.*

Fig. 1566.
CREAMER'S AUTOMATIC VENTILATOR.
(For openings, 3¾x7½" and 8x8". Outside appearance
similar to Fig. 1569, *or, without hood, to Fig.* 1567.)

Fig. 1567.
ROOF VENTILATOR.

Fig. 1568.
"STANDARD" AUTOMATIC VENTILATOR.

Fig. 1569.
VENTILATOR-HOOD.
1. *Deflector.*   2. *Hood.*

Figure 5.40 Creamer's clerestory-mounted ventilator and several other such devices
popular between about 1870 and 1890 are shown in this series of engravings. (Car Builders' Dictionary, 1888, fig. 1566)

exchange the air. The exhaust was no problem; with the clere-story roof that had been introduced in the 1860s, warm, foul air rose to the ceiling and flowed out the deck windows or vents. But the clerestory was not efficient in introducing new air into the car. Psychologically, open clerestory windows created a breezy effect, but most of the air that blew in blew back out before benefiting the passengers.[148] Only when the train was moving slowly did much of the outside air entering the clerestory drop down into the car. The crosscurrents set up by open deck win-dows often confounded the exhaust or made the lamps smoke, fouling the atmosphere and blackening the ceiling.

One way to improve the clerestory's effectiveness was to hinge the deck windows at the middle so that they could be swiveled partially open *away* from the direction of travel. The air blowing past the open end created a small vacuum, thus helping to suck out the stale air inside. The idea worked well enough if the brakeman positioned the deck windows correctly, but the brakemen had become an indifferent lot since the introduction of air brakes and steam heating.[149] In earlier days they had manned the brakes, tended the stove, trimmed the lamps, re-versed seats, and adjusted the clerestory windows, but now they would do nothing more than call out the stations.

A railway supplier, William G. Creamer, decided to solve the problem with a ventilator that automatically reversed itself. The idea came to him about 1860 when he was riding in a new car on the Hudson River Railroad, one of the first lines to adopt the newfangled clerestory.[150] Cinders rained in upon him intermit-tently because the slats in the deck window opening were facing ahead, thus acting as a scoop to pull in the smoke as it wafted over the top of the cars. The brakeman had failed to reverse the slats at the end of the run. On the spot Creamer thought of a simple solution: he would mount a series of small sheet-metal boxes, open at both ends, along the side of the clerestory. A hole, say 4 by 7 inches, would be cut through. A butterfly valve set inside the box would swing to a position opposite the direction of the train's travel because of the breeze acting on the other end of the valve (Figure 5.40). Cinders and smoke would be deflected, and no one need tend the valve. Light springs would hold it at a center or fully open position when the train was stationary to ensure an exhaust while standing at a station or siding. Naturally the deck windows would be kept closed. Creamer's invention was widely used and spawned many imitations.

Getting the bad air out was relatively easy; getting the good air in was more difficult. The problem was not lack of power,

Longitudinal Section of Car Showing Air Ducts and Intake.

Cross-Section of Car.

Hood and Downtake.

Fig. 4. *Fig. 5.*

Exhaust Ventilators.

Figure 5.41   The Pennsylvania used this roof-mounted, air-scoop ventilating and heating system between 1900 and 1935. (Railroad Gazette, September 30, 1904)

since the train's motion created a draft strong enough to ventilate the deepest mine or the stalest dungeon, to paraphrase Creamer.[151] But where could the new air be drawn in without producing too strong a draft or admitting clouds of dust and smoke? The side windows were out on both counts, and the clerestory could not do the job. That left the ends of the car body, where the end windows were a logical choice except that the wind tended to blow back directly on the passengers. It occurred to someone to make an opening, called the transom or hood sash, above the doors at each end of the car. Opening out under the platform roof or hood, it was protected from the worst of the smoke and cinders, which ideally blew over the top. The transom was relatively small and high, so that it was not likely to admit drafts or a great rush of air even when it was open. And it brought the air in at about the right level—even with the passengers' heads. Evidence on the origins of this scheme is tenuous, but Figure 1.79 in Chapter 1 shows that some of the earliest eight-wheel cars had small transom vents over the end windows, if not over the doors. The Boston and Albany had adopted the hood transom sometime before 1876. The sash win-

dow had a ground glass lettered to read "If You Want to Breathe Pure Air Keep This Ventilator Open."[152]

The history of early railroad car ventilation is a tale of indifference and half measures. Few prominent car builders took the subject seriously, and those who did seemed to push it aside after a few months of frustrating failures. One of the few railroad engineering figures of any prominence to give ventilation more than passing notice was Charles B. Dudley (1842–1909). For many years he headed the Pennsylvania Railroad research laboratory under the simple title of chemist. Sometime in the mid-1890s Dudley decided to investigate car ventilation, and he was shocked to find that the air was as foul as its reputation.[153] Passenger complaints were actually a reasonable demand for reform. Nevertheless, he said later, during twenty years in railroad engineering he had never faced such a baffling problem. The primary conflict was between heating and the fresh-air supply. It was Dudley who concluded that an acceptable temperature could be maintained if no more than 1,000 cubic feet of air per passenger per hour was brought in. Although not ideal, this com-

Figure 5.42   Garland's ventilator, introduced around 1905, was an extremely efficient aspirator that became standard with Pullman. (Car Builders' Cyclopedia, 1928, figs. 2094–2095)

promise was far better than the old ration of 250 or 300 cubic feet.

Dudley also hoped to salvage what he could of existing heating-ventilating technology. He sought no radical solution but saw merit in the P R R's existing standard, the Spear stove and hot-air register. The main defects of the Spear system arose because the source of heat (the fire) was higher than the outlet registers. This was easily remedied with steam heating, since the pipes could be placed low, near the floor (Figure 5.41). Dudley retained the old air scoop, with modifications. The outside air pulled in by the scoop was directed to an 8- by 14-inch sheet-metal box running under the floor for the full length of the car. Holes cut into the floorboards let the air rise to circulate the heat. The air scoop and its trunk, which were much enlarged compared with the old Spear plan, measured 14 inches square. Scoops were set at diagonal corners of the roof, and a vane inside each scoop could be flipped fore or aft, depending on the direction of travel. A butterfly valve, placed lower in the trunk, could be set to vary the flow of air. It could also be fully closed in tunnels, to prevent the scoop from sucking in smoke, and at terminals, to prevent the heat from escaping through a back draft. Globe-style exhaust vents were installed on top of the clerestory roof. This raised the escape opening 2 feet above the air intake.

Dudley and his associates tinkered with the project for several years, until by the late 1890s they felt that it was just right. Their belief was substantiated when a main-line train was stalled for more than four hours in a snow blockade near Harrisburg. The winds of the blizzard roared against the marooned train, but temperatures in the cars equipped with Dudley's apparatus never fell below 70 degrees. Soon the railroad embarked on a massive conversion, so that Dudley's system became standard on what

was sometimes called the standard railroad of the world. True, there was some grumbling about the necessity of cutting away the cross sills to make room for the ducts below the floor. Iron straps spanned the ruptures. By 1901 over 200 cars had been converted; three years later the total was more than 1,000. Even the B & O was adopting the scheme. Dudley's plan prevailed on the Pennsylvania until the age of air conditioning.

Dudley's efforts substantially increased the comfort of coach passengers, who were normally last in line for any technological improvements. Ironically, it was now first-class passengers who felt that they were being shortchanged. The sleeping car had the reputation of being the worst-ventilated of all railway conveyances, and Pullman was blamed not for its poor system but for the total lack of system.[154] Why pay extra fare to be closeted in an airless box that was upholstered with more suffocating hangings than can be found in a small hotel? For years Pullman smarted under this criticism, weakly contending that the clerestory did the job well enough. Finally in 1905 the company hired a Chicago physician, Thomas R. Crowder, to head its newly established sanitation department.[155] One of his responsibilities was to explore the subject of ventilation. Pullman's action might be explained as the belated recognition of a problem, or a step to quiet the public criticism stirred up by the reform-minded progressive movement. As described earlier, Crowder's investigations led to a dismissal of the old CO_2 bugbear. He also concluded that many complaints about sleeping cars were psychological in nature; claustrophobics inside a confined berth often experienced shortness of breath. Crowder admitted, however, that sleeping-car ventilation was far from perfect.

In the year that Crowder joined the Pullman company, a new device came on the market that promised to revolutionize car ventilating. It was developed by Thomas H. Garland, who had studied the problems of car ventilation while he was superintendent of the Burlington's refrigerator car business. His concern had been fruits and vegetables, not passengers. All previous ventilating systems had placed great emphasis on both the entrance and the exit of air, but Garland concluded that a strong exhaust was the key to good ventilation. Because no car body was perfectly tight, fresh air would find its way through the cracks, voids, holes, and crevices inherent in any man-made structure. Garland's roof-mounted, double-sided air scoop ventilator was based upon the aspirator principle (Figure 5.42):[156]

The theory is sound and well applied. Large openings in the ventilator face toward either end of the car, and these form the ducts through which the air is given most of its application to the purpose of exhaustion. The motion of the car forces the air in quantities into one opening or the other, according to the direction in which the car is moving; and the right angle bend in the duct conveys the air outward and away, leaving the ventilator by the openings at the outer edge of the car roof. In traversing the passage to its exit, the air goes from the intake duct into the transverse duct, and at this point a strong suction is effected which draws the air from the car through the opening provided for that purpose. But this exhausting action is effective also in other ways. Where the air from the intake leaves the ventilator it is drawn backward by the motion of the car and the current of the outside air passing over the mouth of the central duct, making that also effective in ventilating. In fact it is claimed that every passage in the ventilator contributes its share to the work, and the action of the whole is very definite and certain. It makes no difference in which direction the car is moving.

The sheet-metal housing conformed to the arch of the roof, draping over it like an anteater's snout. The suction created was astonishing: at 30 mph each vent pulled out 15,000 cubic feet of air. With ten vents to a car, a complete change of air was effected in less than two minutes.[157]

Pullman was convinced that Garland's was the ventilator it had been seeking. By June 1908 nearly one-half of the Pullman fleet was outfitted with this invention. Clerestory windows were soon a memory, while thousands of Creamer ventilators went on the scrap heap. Garland joined Burton W. Mudge & Company of Chicago, which acquired manufacturing rights to his invention.

Garland's device, though widely successful, had several faults. Like all wind scoop ventilators, it did not work unless the train was underway. Its powerful exhaust tended to waste heat, and in extreme cases where there were too many vents or the train was moving at high speed, it created an interior vacuum strong enough to make door closing difficult and even to cause toilets to overflow when they were flushed.[158]

Since the 1850s various inventors had proposed fans to induce artificial drafts beneficial to ventilation and cooling. Most of them had regarded fans as a means of stirring up the existing inside air for cooling purposes. In 1881 a Philadelphian named Isaac Fridenberg proposed a wind-powered turbine scheme.[159] The roof-mounted turbine was belted to a line shaft running the length of the car within the clerestory. Paddles were fastened to the shaft at intervals. Nearly twenty years after Fridenberg's patent the Reading was reported to be testing a similar plan.[160] In this case, each fan had its own wind turbine. Both systems depended on the train's motion, which meant that when the fans were least required, they were spinning to their utmost. Fans were most needed when the train was stationary or moving slowly. Thus some independent source of power must be supplied for fans not propelled by the train's motion.

The earliest attempt at such a plan for which a record can be found took place on the Milwaukee Road in 1891.[161] In that year George Gibbs, already deeply involved in steam heating and electric lighting experiments for the Milwaukee, placed electric fans aboard several dining cars: one fan in the kitchen, two in the dining compartment. Gibbs proposed that all dining, parlor, and sleeping cars be similarly equipped. The moving air must have seemed like a blessing to the occupants of a sweltering dining car standing in a sun-baked terminal. The idea was gradually adopted by other lines. In 1898 the B & O's Southwestern division was using electric fans to cool sleepers at layover points.[162] Passengers were commonly accepted on board several hours before a late-night departure so that they could bed down at a conventional hour, but the cars were unbearably hot and stuffy after standing in the terminal. Electric fans, plugged into the station power source, provided some relief until departure time.

Within ten years most parlor, buffet, and dining cars had electric fans.[163] As electric lighting was gradually adopted for coaches, electric fans began to become more common on non-first-class cars. By 1920 there were usually small fans at either end of a car, or in better installations a series of ceiling fans. Although they were an improvement, they mostly agitated the stale interior air rather than producing an effective exchange of air. In the mid-1920s Pullman was using powerful clerestory-mounted exhaust fans that created some circulation in stationary cars, provided that the porter was conscientious enough to turn them on. If it had a large enough capacity, an exhaust fan could do much to reduce stratification and equalize temperatures between the ceiling and floor. This was extremely important to the comfort of passengers in both upper and lower berths. A forced-air system, however, was considered too expensive for general use. Blowers of modest size were widely installed where they were most needed—in dining-car kitchens, sleeping cars, and smoking cars—but a large system for a coach was regarded as much too costly.[134] Moreover, a fan large enough to exhaust an entire car would require a substantial power source, whereas most generator-battery systems were barely adequate for electric lighting. It was also contended that a large blower system would be unbearably noisy. Even so, the gradual progress of electric fans was a prelude to the next major advance in car ventilation: air conditioning.

Air conditioning began a new era in railroad travel. It involved more than just lowering the temperature during summer heat.

Air conditioning was "manufactured weather," in which mechanized systems controlled the humidity, the purity, and the volume of air and regulated the temperature in all seasons of the year. It eliminated the filth associated with railway travel in the steam age. Because air conditioning was introduced during the same years as streamlining, it was overshadowed by the glamorous polished-aluminum and stainless-steel trains of the 1930s. Yet air conditioning was a major revolution in passenger comfort, a tangible improvement in place of the tin-deep eye appeal of the streamliners.

The origins of air conditioning in railway cars can be traced to a patent (No. 12,851) granted Job R. Barry of Philadelphia on May 15, 1855. Barry envisioned an ice system with axle-driven fans (Figure 5.43). It does not appear that his scheme was given a test, but as mentioned earlier, W. D. Mann employed a similar ice system on many of his sleeping cars. The most persistent effort made during the nineteenth century to perfect some form of car cooling appears to have taken place in India. There an ancient method of cooling was the tatty, a dampened grass mat that was hung in a doorway or window to lower the temperature of the air passing into the dwelling by evaporation of the water. This primitive scheme, improved with a semiautomatic method of wetting the mats, was applied to the Viceroy's private car in 1872–1873.[165] An extended series of articles explaining the system and its use on the Great Indian Peninsular Railway appeared in the British magazine *Engineering* in 1874. Fifteen years later the tatty system was still reported in service, but it was criticized as being ineffective when the train was idle.[166] Ice coolers and better insulation were proposed. In 1914 the G I P tested what is probably the earliest non-ice system used for passenger car cooling.[167] A carbon dioxide refrigerating unit was placed in the baggage car. Cold brine pumped back in pipes through the train was said to lower temperatures by 20 degrees. Because of the humid climate, condensation was a problem. Some passengers also complained that the unnaturally cool temperature was not healthy and petitioned for removal of the system. The outbreak of World War I drew the entire empire to the aid of Britain and was said to offer the Indian railways a convenient excuse for abandoning the experiment.

Meanwhile the Santa Fe hoped to perfect some method of car cooling to make its Southwestern desert route more appealing to transcontinental passengers. Since daytime temperatures often

exceeded 100 degrees, experienced travelers generally chose a more northern route. Late in 1911 the Santa Fe inaugurated a fancy first-class train named the De Luxe. The dining car was outfitted with an ice-cooled ventilating system.[168] The air was cleaned by a jet spray before being directed into the dining compartment. The apparatus was reminiscent of Mann's old system, but with an important difference: a large electric fan pumped 90,000 cubic feet of air through the compartment every hour. As many as thirty-three Santa Fe diners were similarly equipped.[169] As an air cooler, however, the system turned out to be relatively ineffective, and by 1926 it was at last abandoned.

A year or two after the Santa Fe's first ice-cooled diners entered service, the Pullman Company asked Willis H. Carrier, a pioneer in commercial air conditioning, if he could provide a mechanical refrigeration unit for sleeping cars.[170] After some study Carrier concluded that the state of the art was not ready for such an installation, and there the matter rested for more than a decade. In 1927 Pullman decided to reopen the question, and in April of that year the company installed a Servel electromechanical air conditioner aboard the sleeping car *Jacksonville*.[171] The unit, an apparent failure, was removed in August. Pullman next tried an ice system on the *McNair* in September 1929. Other companies were now awakening to the idea of passenger car air conditioning, and experiments began in rapid order on several railroads.

In the summer of 1929 Willis Carrier was asked to test his apparatus aboard a Baltimore and Ohio coach.[172] The experiment was successful enough to warrant a second installation, this time on a B & O dining car, the *Martha Washington*.[173] A 7½-horsepower, 110-volt electric motor worked the ammonia refrigeration compressor. A gear-driven 10-kilowatt generator powered the motor. The refrigeration unit cooled water, which was pumped to coils in roof ducts. A fan blew 2,500 cubic feet of air per minute over the coils into the dining room. (The kitchen was not cooled.) In a test held in April 1930, the steam heat drove the temperature up to 93 degrees, and Carrier's air-conditioning unit lowered it to 73 degrees in just twenty minutes. The car entered regular service a month later and was not retired until 1965. Carrier's original equipment was found unsatisfactory in some way. It is likely that the electrical power setup was defective.

Whatever the reason, the B & O turned almost immediately to the York Ice Machine Corporation for thirty-six gasoline-engine-powered, 4-ton units.[174] Electricity was needed only for the pump that circulated the brine coolant to ceiling coils. In May 1931 the B & O had equipped enough cars to make up the Columbian, which was said to be the first complete air-conditioned train in the United States. By the next spring the road had over 100 air-conditioned cars in service and several air-conditioned name trains.[175]

Air conditioning was introduced in the Far West just a few months after the B & O's first air-conditioned diner entered service. The Santa Fe seized upon the new technology to better its competitive position. In July 1930 it began running a new dining car, Number 1418, that was fitted with a Carrier electromechanical unit.[176] Unlike the apparatus that Carrier produced for the *Martha Washington*, conventional car-lighting current of 32 volts DC was employed. Two direct-drive 7½-kilowatt generators supplied the power at speeds above 15 mph. Batteries took over when the train was running slowly and at station stops. A special coolant was devised by Carrier to replace brine water. With the kitchen at full heat and the desert simmering at 104 degrees, the air conditioning maintained the inside temperature at 72 degrees. Passengers tended to linger at the table; the 1418 was the most popular car on the Chief. Yet the success of Carrier's electromechanical unit won no repeat orders from the Santa Fe; instead the line adopted a steam ejector system devised by Carrier and manufactured by the Safety Car Heating and Lighting Company. Early in 1932 the Santa Fe had ten diners cooled by Carrier's steam apparatus.[177]

Few innovations benefited passengers so directly as air conditioning. It brought immediate and dramatic relief from the worst discomforts of summer railway travel. Lines which did not offer direct routes between two cities found that they were winning business from lines which did, simply because they had installed air conditioning. Patronage on the Chesapeake and Ohio's George Washington jumped 25 percent because of air conditioning.[178] Passengers who had not ridden the train for as much as twenty years decided to take the roundabout way between Washington and Cincinnati just to escape the heat. People were equally delighted by the relief from dust. On May 10, 1934, the B & O's Capitol Limited journeyed through a nationwide dust storm, and during its 17½-hour trip the air conditioning kept the interior pure and dust free.[179] It was reported that the train's intake box had accumulated dust equal to that normally taken in during a two-week period. The spread of such wondrous tales created an air-conditioning bandwagon. All major lines were signing up. A market nonexistent two years before was being served by eight manufacturers in 1932. Minor lines like the M K T pushed in ahead of their slower-moving, big-system competitors. Conservative New England lines such as the Boston and Maine, which could hardly complain of Sahara-like summers, rushed ahead in 1931 with ice-cooled cars.[180]

No one showed more enthusiasm than the Pullman Company. Air conditioning offered a dramatic luxury that might bolster its declining sleeping-car business at a cost far below the cost of new cars. With the Depression, outside orders for new equipment had evaporated. The idle shops and men might as well be put to work, so in 1935 and 1936 Pullman began a massive air-conditioning program. A year beforehand the company had settled on a direct-drive, mechanical system after many experiments. The compressor was driven by a shaft geared to one of the wheel axles (Figure 5.44). An electric motor plugged into station power acted as a standby, but during the journey the cooling plant was dependent on the motion of the train. The blowers were independently powered by electric motors. The shaft drive system became Pullman's standard and was used on most of its heavyweight cars.[181]

Now that satisfactory equipment was on hand, Pullman lost no time in applying it. In 1936 there were 5,800 air-conditioned passenger cars running in this country, about 40 percent of them Pullman-owned.[182] One year later the company had 3,300 air-conditioned cars, on which it had spent 25.5 million dollars.[183] Before World War II Pullman had all but completed its air-conditioning program; only its oldest cars lacked the new equipment. Progress for the industry as a whole is summarized in Table 5.1. (The percentage of passenger cars that were converted to air conditioning is greater than the totals in the table might indicate, since baggage and express cars are included in the total passenger car figures.)

Table 5.1   Air-conditioned Cars in Service Compared with Total Passenger Cars

Year	A-C Cars	Passenger Cars°
1933	2,375	56,100
1935	5,809	50,400
1940	12,204	45,200
1945	12,598	47,200
1950	17,350	43,500
1955	16,084	36,900
1960	12,450	28,300

° Figures include passenger, baggage, and express cars.
Source: Railroad Car Facts, 1961, published by the American Railway Car Institute. Figures given here for air-conditioned cars in the earliest years are considerably greater than those reported elsewhere. The 1936 *Master Car Builders Report* claims that only 658 air-conditioned cars were running in 1933.

Air conditioning had many of the same problems associated with car heating. The passenger car was as difficult to cool as it was to heat because of the large window area, frequently opened doors, short air throws, restricted power supply, and limited space for ducts and machinery. And because cold air will not rise, it had to enter the car through the ceiling, thus necessitating a secondary duct system. The ceiling ducts were used to good advantage in the winter to distribute warm air via an auxiliary heater coil. In summer, railway passengers generated far more heat than that found in most air-conditioned areas because so many warm bodies were crowded into a relatively small structure. As a general rule 60,000 to 80,000 Btu must be absorbed hourly to maintain a comfortable temperature in a passenger car.[184] This required a 7.5-ton unit, and in a mechanical system 10 to 13 horsepower was necessary. Depending on the type of equipment used, weights varied from 4 to 7½ tons. Installation costs in the 1940s ranged from $4,000 to $10,000.[185] In an effort to hold down the power drain and the overall size of the plant, 75 percent of the old air was recycled; the remainder was made up of fresh air.

Three basic systems were adopted for railroad car air conditioning: ice, steam ejector, and mechanical. Ice was the simplest and least expensive in first cost. Its modest power needs were perhaps its greatest attraction. Ice provided the coolant independently of power-hungry compressors. Existing electrical power supplies were often adequate to power the fans and circulating

9. Strainer
10. High and low pres-
    sure switches and
    gauges
3. Speed control
4. Standby motor   11. Standby starter
5. Compressor   switch
6. Condensers (2)   12. Standby plug-in
7. Condenser fan   receptacle
1. Belt drive   8. Receiver tank   13. Fresh air intake
2. Drive shaft

14. Recirculated air intake
15. Filters (2)
16. Blower fan
17. Evaporator
18. Holdover cooling
    coil (when specified)
19. Suction line
20. Liquid line
21. Heating coil
22. Steam drain
23. Steam inlet
24. Air outlet
25. Heating thermostat
26. Cooling thermostat
27. Air ducts

The general design of the Pullman Air Conditioning System comes under the claims of the Carry
patent No. 2,000,477; patents No. 1,948,473 and 1,984,912, and other patents allowed and pending.

PULLMAN-STANDARD CAR MANUFACTURING COMPANY

*Figure 5.44   Pullman's axle-drive air-conditioning system was fully developed by 1932.
A standby electric motor drove the compressor when the train was not moving. (Pullman,
Inc.)*

pumps. If not, a modest auxiliary generator and battery set could
handle the extra load. Basic rewiring could thus be avoided. The
first cost of an ice system was calculated at 60 percent of a
comparable mechanical air-conditioner.[186] The maintenance
cost was only one-third that of a mechanical unit. It was all so
simple: ice melted in a bunker below the car, the cold water was
pumped to a radiator in the ceiling, and a fan blew air over the
radiator into the passenger compartment, lowering the temperature as much as 14 degrees. The water returned to the bunker for
recooling, and the warm surplus was discharged overboard.

One of the first railroads to adopt the ice system was the
Boston and Maine. It was drawn into the experiment through the
enlightened self-interest of the Metropolitan Ice Company of
Boston, which sponsored the tests.[187] For an average New England summer—short and relatively mild—the ice system made
sense. The investment was small, and the trains did not have to
haul the dead weight of a mechanical system. In cool months the
ice bunkers were empty, thus eliminating the greatest portion of
the system's weight.

Yet while ice seemed suitable mostly for the Northeast, there
were other large users of this most primitive form of air conditioning: Pullman, the Southern Pacific, and the Pennsylvania. By
1939 the country's railroads had nearly 3,800 ice-cooled cars.

After this time, however, the defects of the system were recognized as overwhelming. Ice was never really effective for sustained hot weather, such as that encountered in the South,
Midwest, and Southwest. It was expensive to operate: at $5.29
per 1,000 miles, it cost five times more to use than mechanical air
conditioning.[188] This cost was directly related to that seemingly
cheap commodity, ice. It was not the price per pound but the
quantity consumed and the labor required to handle it at the
storage depots en route. A railroad car consumed approximately
500 pounds of ice per hour.[189] Fifteen 300-pound blocks equaled
4,500 pounds of ice—enough for nine hours, or 360 miles. Some
roads installed giant 8,000-pound-capacity boxes on their cars,
but the basic disadvantages of the ice system remained unsolved.
After 1945 almost no new ice-cooled cars entered service, and
those already running were gradually converted or retired.

Willis Carrier, who had at first championed electromechanical
air conditioning for railway service, decided later that the power
problem could best be solved by switching to a system requiring
minimal electric input.[190] Steam, after all, was in plentiful supply. Borrowing a technique from elementary physics, where hot
water is made to boil at a low temperature in a vacuum chamber
and thus to give up much of its heat, Carrier felt that this simplest of all heat exchangers was perfect for car air conditioning.

Piping diagram of type RE steam ejector equipment.

The Safety Car Heating and Lighting Company, Inc.

Figure 5.45   Steam air conditioning was popular on some U.S. roads because it required only modest electrical service. But it proved so wasteful of steam that it was replaced by electromechanical units before 1970. (Car Builders' Cyclopedia, 1953, p. 709)

After two years of experimentation, he placed a commercial steam ejector air-conditioning unit rated at 5 tons' capacity aboard a Santa Fe car. Carrier worked jointly with the Safety Car Lighting and Heating Company and the Vapor Heating Corporation in perfecting the details of the 5,000-pound apparatus.[191] Its sponsors claimed that its steam consumption (206 pounds per hour) was less than that required for heating. Standard car-lighting power (32 volts DC) was adequate to run the small auxiliary fans and pumps.

Many railroads accepted the arguments in favor of steam ejectors, and it became a popular method of air conditioning. By 1937 nearly 1,100 cars were being cooled by steam; some lines had a few experimental installations, while others, like the Santa Fe, were enthusiastic converts. However, steam air conditioning had many drawbacks. It was deceptively simple in principle, but like many schemes with "only one moving part," it was bulky, inefficient, and dependent on countless auxiliaries (Figure 5.45). At $8,475 a car, it was the most expensive of the three major types of car air-conditioners.[192] While it scored well on maintenance, it was tricky to operate. And in spite of manufacturers' claims, it turned out to be a steam hog, consuming one-third more steam than that required for heating.[193] As steam locomotives vanished from passenger service during the 1950s, the ready supply of power for ejector air conditioning also disappeared. Some lines converted immediately to mechanical air conditioning; others, like the Seaboard, struggled along with head-end flash boilers until the late 1960s, when they too converted to electromechanical units.[194]

Mechanical air conditioning was recognized as the best method for cooling cars (Figure 5.46). But modest train-board

power and inefficient refrigerants inhibited its rise to prominence. Ammonia, successfully employed for generations in artificial ice plants, was the best refrigerant available. However, it was not well suited to compact cooling units because it required a bulky cooling tower and because brine was used as the coolant, since it was not considered safe to introduce compressed ammonia directly into the passenger compartment. A safe artificial coolant called Freon that came on the market around 1932 effectively solved this problem. The power problem was solved by spending money: additional batteries, larger generators, or gasoline-powered auxiliary units were costly but necessary for electromechanical compressors of sufficient capacity (Figures 5.47 and 5.48).

The Waukesha Motor Company produced one of the most popular independent engine-driven AC units. Chunky four-cylinder engines powered 7-ton-capacity ice engines. By 1943 the *Car Builders' Cyclopedia* claimed that twenty-five railroads were using these under-floor units. The Milwaukee Road, the Frisco, the Illinois Central, and the Southern Pacific were all enthusiastic Waukesha customers. Some of these lines also installed a second motor-driven power package to furnish current for lighting and other electrical auxiliaries. After about 1940 Waukesha pioneered the introduction of propane-fueled engines. They were cleaner and easier to maintain than equivalent gasoline or diesel engines, as well as being more convenient to refuel at station platforms. Instead of risking fuel spills in a congested passenger area, it was possible to exchange fuel tanks. But Eastern roads, particularly those entering Manhattan terminals, banned propane-powered units and required removal of the tanks before a car so equipped could enter the approach tunnels. According to W. D. Edson,

Figure 5.46   Railroad car air-conditioning systems normally used an overhead duct system. The cooling tower was eliminated when Freon replaced ammonia as the refrigerant. (Railway Mechanical Engineer, December 1930)

who supplied much of the data in this paragraph, few Eastern lines employed independent under-floor AC or generator units.

Pullman avoided the electric supply problem by devising a direct mechanical drive for the refrigerant compressor.[195] A gear box was fitted to one of the wheel axles, with a shaft to the compressor. A governor protected the compressor from overspeed. The system worked well and was adopted by some railroads for non-Pullman equipment. Although it was not efficient where there were frequent or prolonged stops, for short stops the residual cold air, together with the standby electric-powered blower, acted to maintain the temperature at a comfortable level (Figure 5.44).

Railroad air conditioning entered a final stage following World War II. The trend was toward developing more efficient large-capacity units, for 6- to 8-ton-capacity installations were no longer thought adequate.[196] Engineers now felt that 10-ton units were needed, and compressors of this size could not be economically driven from the 32- or even 64-volt DC service then common. These low-voltage systems were favored because of the train's dependence on batteries. As a general rule, the higher the voltage, the more expensive the battery. But there is a trade-off in favor of 110-volt circuits in terms of the lighter wiring possible; low-voltage systems require very heavy and hence expensive wiring if they are to service power-hungry units such as air conditioners. Thus 32 volts was satisfactory for lighting but not for motor currents. In addition, 110-volt AC systems could make use of conventional and cheaper motors, switches, and other

Figure 5.47   After air conditioning, the size and complexity of the electrical system required a closet in place of the small cabinet shown in Figure 5.48. (Trains magazine collection)

Figure 5.48   Before air conditioning, train-board electrical service was essentially a low-voltage lighting system. The modest wire cabinet here was typical of the 1920s. (Pullman Neg. 28894)

components. Most AC cars were equipped to run on 220-volt AC at stations, so that it was possible during layovers to plug standby power into a jack fastened on the underside of the car.

Westinghouse placed a 220-volt AC three-phase unit in service during the spring of 1947. With a more potent power package, overall weight was cut to just over 1 ton—a remarkable improvement over conventional equipment. To remove drag on the locomotive imposed by axle-driven generators, gasoline- and diesel-powered units became more popular in the postwar years. The trend today is toward head-end power.

Lighting

Some form of artificial lighting has always been necessary in the Northern United States, where daylight is short in the winter season. As night travel by railway increased, so did the need for lighting. It was provided at first in niggardly amounts: one or two candles were considered adequate to ensure safe passage in or out of a car. Common road coaches of the day offered nothing better, and in the 1830s even domestic lighting was provisional. In fact travelers from the backwoods had never enjoyed this luxury and were usually content to sleep on board until the journey's end.

In urban areas, however, artificial illumination was more advanced. Parlors, theaters, and stores were well lit; candles and oil lamps were plentiful. By mid-century, central gas systems were common in most large cities, and the affluent city dweller was appalled by the night train's "gloomy cage."[197] It was like stepping into some rude country tavern where thievery flourished in the shadows. Reading was impossible, as was conversation or any other civilized pastime. The educated traveler, who was usually age. In 1846 a traveler wrote: "A solitary lamp burned at one end of the car and its sickly light fell upon the faces of the sleepers."[198] Almost forty years later another traveler spoke of "the dim religious light" that was sufficient only to delineate the extent of the darkness.[199] Newsboys offered papers and magazines as if in "playful irony." The only diversion was to watch the orange flame of the lamp swimming in its glass globe like a smoky goldfish.

The technology for good lighting was available. Even candles could do an excellent job if enough were provided. Oil lamps were better, but this apparatus was expensive because it required eight burners instead of one or two fixtures. And as long as they could, most railroads kept their purses tightly closed.

Cost was an important consideration to railroad managers. A few cents an hour could sway the decision for or against better lighting. Trivial as such economies might seem, small savings could mount up with a large number of cars. Discussions of lighting in the trade press and the Master Car Builders Reports always emphasized cost. In one case an official discussing the whole matter of passenger comfort agreed that all reasonable efforts should be made to provide better accommodations, such as lighting, but candidly said that he did not feel a responsibility to indulge the public.[200]

It was not until after 1880, when the public outcry had grown loud enough, that the industry was willing to provide good lighting.[201] It is likely that first-class cars had been outfitted with proper lighting a decade earlier, because travelers who paid a premium felt entitled to the small luxury of an extra lamp or candle.

By the time railways became a public carrier, the art of artificial lighting had already passed through many stages and, like an ancient tree, possessed many branches and roots. At one time or another every conceivable burnable material, from yak fat to rape seed oil, had fueled the world's lamps. In the western nations by the early nineteenth century, candles were the most familiar source of lighting. Simple, smokeless, and relatively safe, they were the first source of light aboard railroad cars and enjoyed a long reign—from 1830 to about 1875. Tallow candles were the most common, but they were soft and tended to droop in warm weather and in direct sunlight. They were also eaten by rats and mice. Spermaceti candles, made from a rich wax extracted from the sperm whale, were the finest available, although they were not widely used because of the cost. After about 1860 paraffin wax manufactured from petroleum produced a cheap grade of candle that avoided the worst defects of tallow.

Originally two candles per car was considered adequate. Precise information is unavailable, but the 1836 Camden and Amboy coach may be taken as representative of the period (Figure 1.72) It has a small candle lamp built into the end bulkheads at diagonal ends of the body. At about the same time, tin side lamps were used (Figure 5.49). A flat sheet of tin-plated iron formed the reflector. The lamps were fastened to the interior wall panels between the windows. Lamps of this type were remembered by George Fowler, who said that there were rarely more

Fountain Car Lamp.
$30 per dozen.

Candlestick Lantern for Passenger Cars.
1 inch candle, $40 per dozen.
1¼ " "   44 "   "
Candles per pound 36 cents.

Springfield Pattern for Passenger Cars.
$30 per dozen.

Solar Side Lamp for Passenger Cars.
$20 per dozen.

Figure 5.50   By 1860, candle and oil side lamps were used for car lighting. Center lamps came into favor only after the clerestory roof elevated the ceiling. (Bowles Railway Supply Catalog, *1860c., p. 141*)

not content to bed down at sunset, was obliged to sit and stare out of a dark window because the railroad was too cheap to provide decent light.

Trainmen also cursed the dark cars as they stumbled through the aisles. Their dim lanterns were more than signal lamps; the conductor would raise his to read a ticket or decipher a timetable, and the brakeman needed his to move from one swaying platform to the next or to guide a passenger down the steps at a dark station.

Complaints began early and continued well into the railway

Figure 5.51 This 1874 advertisement testified to the continuing popularity of candle illumination. The center lamp was mistakenly printed upside down. (Poor's Manual of Railroads, 1874)

Figure 5.51 This 1874 advertisement testified to the continuing popularity of candle illumination. The center lamp was mistakenly printed upside down. (Poor's Manual of Railroads, 1874)

THIS CUT REPRESENTS ONE STYLE OF CENTRE LAMPS.

THIS CUT REPRESENTS THE SIDE CAR LAMP FOR CANDLE.

DANE, WESTLAKE & COVERT,

MANUFACTURERS OF

WESTLAKE'S

Loose Globe Car Lamps & Lanterns.

Send for Price List and Circulars.

228 LAKE STREET, CHICAGO, ILL.

than six to a car.[202] Large, 2-inch-diameter candles were used, but the light was too dim for reading. Around the middle of the century, open candle lighting gave way to more elaborate fixtures. Candles were placed in tubular spring-loaded holders. The cap of the tube was rolled over, leaving a small circular opening, and the spring pushed the candle up against the opening as it burned. This kept the flame in one position, making more efficient use of the reflectors. Glass globes or chimneys protected the flame from drafts—a necessity in a vehicle as breezy as the railroad passenger car. Side candle lamps were used in low-ceilinged, arched-roof cars (Figure 5.50). After 1860 when clerestories came into fashion, more elaborate center lamps were used. They provided aisle lighting, and the better class of car also had side lamps for reading.

After 1875 candles were considered obsolete for car lighting, although of course they did not disappear entirely.[203] In Ohio, candles were in fact the only legal form of car lighting after 1869, but the 1877 *Ohio Railroad Report* admits that the law was not actively enforced and that most lines employed oil lighting. Candles remained a secondary form of emergency lighting, much as they do today in most households. As late as 1942 the *Car Builders' Cyclopedia* showed a single candle side lamp for emergency lighting.

Historically, oil lamps predate the candle, but they do not appear to have been much used in railroad cars before the 1850s. They were more expensive and demanded more attention: filling the reservoirs and trimming the wicks was a chore. They gradually gained in popularity, particularly in areas having a good supply of oil, because a good oil burner gave more light at less cost than candles. Before the age of petroleum, however, there was no cheap lamp oil. A great variety of natural vegetable and animal oils—rape seed, whale oil, refined lard, and even Colza, a light oil made from a variety of cabbage—found their way into railroad car lamps. In the 1840s the Boston and Albany was using whale oil, which was said to be smoky.[204] Other New England roads employed the more expensive sperm oil with good success, and one line as late as 1891 thought it was superior to gas lighting.

One of the first artificial fuels introduced was camphene—a flammable mixture of redistilled turpentine and high-proof alcohol. It was first used in 1820 and grew in popularity as the price of whale oil increased. Just how early it was adopted for car lighting is uncertain; the *Illustrated London News* for April 10, 1852, mentions camphene lamps on a New York railroad. An engraving of the car's interior accompanying the article shows end corner lamps, indicating that even though candles were no longer used, the car had only two lamps. Although camphene was cheap and gave good light, it was explosive. It was said that "the friends of temperance will not object to the burning of alcohol," but camphene became notorious as a cause of fires and explosions.[205]

A safer artificial lamp oil was introduced commercially by a Canadian inventor in 1846. The oil was distilled from coal and

was commonly called coal oil, but its inventor preferred the trade name Kerosene. Commercial production began in 1856 at a cost of $1 per gallon. At the time whale oil cost $3 a gallon, lard oil $4, and sperm oil $7. Within four years 30,000 gallons of coal oil were being produced every day.

Coal oil was no sooner established than it was displaced by an even cheaper lamp fuel. Petroleum, known since ancient times as a curiosity or medical curative, was found in great subterranean pools in western Pennsylvania. Oil wells began to pump kerosene-rich crude in 1859. Petroleum was a far better source of kerosene than coal; a gallon of crude would yield nearly 3 quarts of kerosene. By 1865 improved refining methods brought the cost of production down to 5 cents a gallon. Its abundance and cheapness prompted a massive conversion to kerosene lighting. The *Western Railroad Gazette* of July 18, 1863, mentioned the sunlight glow provided by six kerosene center lamps on one of Pullman's new cars. The spread of this form of lighting in later years was attributable not only to its improved illumination but to its economy. A large kerosene fixture produced 24.5 candlepower at a cost just under 1 cent per hour.[206] Five candles could produce an equal amount of light, but at a cost of 12½ cents per hour.

Eighty years before kerosene was available, oil lamp technology had made some major advances. Possibly the most important were achieved by a Swiss, Aime Argand (1750–1803). In 1780 Argand devised an adjustable tubular wick holder with a center

WESTLAKE'S IMPROVED VENTILATING CAR LAMP WITH BELGIAN BURNER.

Figure 5.52   Oil lamps became a major form of car lighting after 1860. The fumes were exhausted through rooftop ventilators. (A. M. Wellington, W. B. D. Penniman, and C. W. Baker, The Comparative Merits of Various Systems of Car Lighting, *1892)*

Figure 5.53   A large oil-burning side lamp made by Post and Company of Cincinnati in 1882. (National Car Builder, 1882 advertisement)

tube for a good air supply. A glass chimney protected the flame from drafts and increased the area of combustion without cutting off the light. The gain in illuminating power was spectacular. All subsequent advances in oil lamp technology were really variations on Argand's original innovations. Beneficial as his design proved, however, oil lamps were less than perfect. The light was always too high or too low, depending on the quality of the oil and the condition of the wick. And air currents inside the car caused the lamps to smoke, although this defect was partially remedied when roof vents were installed above the chimneys (Figure 5.52).

The railroads' stinginess about the quantity of light continued to be a problem. Even at operating costs of a penny an hour, managers were reluctant to install more lamps than were absolutely necessary. In 1878 the Illinois Central began replacing candles with oil lamps on its suburban cars.[207] Three center lamps cost approximately $100 installed and produced approximately 72 candlepower. It was calculated that 120 candlepower was necessary for good illumination, but the railroad was not about to squander an extra lamp per car so that commuters might read their papers.

Actually there appears to have been more concern over lamp safety than over the quantity or quality of illumination. The fire-peril hysteria included oil lights as well as car stoves. Throughout the kerosene lamp era there was a widespread belief that such fixtures added materially to the dangers of railway travel. In reviewing the history of car fires in 1890 after nearly thirty years of oil lighting, *Engineering News* concluded that the hazard was a popular myth.[208] All major disasters involving fires could be

traced to stoves or to the locomotive's firebox. In only one instance could a fatal car fire be blamed on a lighting fixture. This accident occurred on the Great Western Railway of Canada in February 1874. A small lamp in the water closet broke loose and fell to the floor, starting a fire that went undiscovered. Before the train was stopped, eight passengers had been suffocated or burned to death; another thirteen had been injured. The car was totally destroyed. Yet this toll was minor when compared with such horrors as the 1867 Angola tragedy, when forty-two people died. Newspaper accounts blamed the kerosene lamps for the ensuing fire that claimed so many lives, but later reports corrected the error. They said that the wreck had occurred in daytime, and that candles were the only form of lighting on the Lake Shore cars.[209] Nevertheless, the image of flaming oil lamps was fixed in the public's mind; corrections had no effect. A few months earlier a locomotive plowed into the rear of a train on the Cincinnati, Hamilton, and Dayton Railroad. Its kerosene headlamps ruptured, setting the telescoped rear car ablaze. Not long afterward the palace car *City of Chicago* went up in smoke, and

Figure 5.54   An oil-burning center lamp manufactured by Howard and Company of Boston around 1876. (A. A. Merrilees)

again the blame was initially placed on the lamps. Later newspaper stories reported that the stove was the actual source of the fire, but they were lost among the back pages.

When kerosene was first introduced, railroads were foolhardy enough to use a household grade. Some bought from unscrupulous dealers who added naptha, thus lowering the cost but making it a volatile fuel too dangerous for lighting purposes.[210] After a few lamp explosions, railroads adopted less flammable, heavy oils sold under such reassuring trade names as Mineral Sperm Oil. They were free of naptha and were guaranteed to produce no vapor until heated to 300 degrees F.[211] They would not burn at less than 600 degrees. In addition to the built-in safety of high fire test oils, the railroads contended that the sudden movement of a derailment or collision would extinguish the lamp's flame.

Granting that the fire peril was greatly exaggerated, oil lamps did have many disadvantages. For all their ingenious burners and chimneys, they were smoky and evil-smelling. Many travelers objected to the pungent odor of kerosene, and more complained about dripping oil caused by leaking reservoirs, loose caps, or indifferent filling. Oil not only dripped on carpets, seats, and window hangings, but stained the passengers' clothing and luggage, creating a steady flow of work for the claims department.

Gas lighting was the best method of artificial illumination available in the nineteenth century. However, it was generally limited to wealthy homes, public buildings, and commercial establishments, because it was complex and expensive. Attempts had been made in the eighteenth century to use it for domestic lighting, but little real progress was achieved in this field until after 1800. In 1812 a city gasworks was established in London; four years later Baltimore opened a plant; and by 1850 over fifty American cities had gas. The number quadrupled in a decade, so

that gas lighting was common in most U.S. cities by the beginning of the Civil War.

The earliest attempt to use gas for car lighting that can be discovered was reported in 1834.[212] An unnamed British railway was said to be experimenting with a portable gas system. Many years passed before any similar tests were made on this side of the Atlantic. In 1851 the Hudson River Railroad began outfitting its cars with gasometers of sufficient capacity for a trip between New York City and Albany. The apparatus was more than a temporary experiment, since it stayed in service for at least five years.[213] A number of other railroads, including the Galena and Chicago Union, the Troy and Boston, and the Camden and Amboy, also used gas for car lighting during the fifties.[214] Many of these lines adopted a system perfected by Charles Starr of New York (Figure 5.55). Star compressed ordinary city coal gas to a pressure of 200 pounds in tanks with a capacity of 120 cubic feet. The meter was a clockworks device. The entire apparatus cost about 100 dollars per car. Starr's system was also installed with good results on several New York Harbor ferryboats.

The Troy and Boston's system, devised around 1856 by Marcus P. Norton of Vermont, employed a single gasometer in the baggage car for the entire train rather than individual tanks under each car. Norton's plan required some method of flexible coupling between cars, which undoubtedly proved troublesome.

In 1860 the Philadelphia and Reading began what was probably the first large-scale, long-term use of gas for car lighting. This move was encouraged by the success of the New Jersey Railroad and Transportation Company, which had replaced its stearin candles with compressed-coal-gas lamps.[215] Even considering the high cost of city gas—$3 per 1,000 cubic feet—the New Jersey line claimed a cost saving of 30 percent over candle lighting. A railroad could cut the cost of gas by making its own, which both the New Jersey and Reading lines elected to do.

Figure 5.55 Gas lighting apparatus offered by the New York Car and Steamboat Gas Company in 1860. B marks the gas tank, C the regulator, and E the lamps.

Within a few years the Reading decided to upgrade its lighting by converting to a better type of gas. In 1874 the road built a new plant in Reading to manufacture gas from petroleum for lighting both its cars and its repair shops.[216] The gas was far richer than the coal gas previously used; one tank of oil gas would last five nights, whereas an equal charge of compressed city gas was consumed in a single evening.

In 1860 or 1861 the Pennsylvania began compressing city coal gas at its Altoona car shops.[217] Tanks measuring 7 feet, 6 inches long by 14 inches in diameter held enough gas for sixteen to eighteen hours of service. The gas was compressed to the extraordinary pressure of 600 psi. Apparently the system was not notably successful; coal gas that was compressed to such a high pressure lost most of its illuminating power. It was also claimed that an accident associated with filling the tanks had started a fire that destroyed several cars.[218] The Pennsylvania temporarily abandoned gas lighting, but by 1871 the road was recommitted to it.[219] Thirty-five cars were already fitted with gas apparatus, and plans were underway to convert the entire fleet. The gas was made from petroleum. Each car contained five burners rated at 17 candlepower each and a tank carrying a sixty-hour supply of gas. By 1876 the line had compressor plants at Altoona and Jersey City.[220] In addition to the gaslights in a large number of its cars, the Pennsylvania's ferryboats were also being lit with compressed gas, though it was stated that city gas was again being used in these installations.

During the decade of the seventies other roads, such as the Chicago and Alton and the Erie, were using gas to some degree. It appeared likely that gas would become a general form of car lighting. Had it not been for the introduction of cheap kerosene a few years earlier, American railroads might have passed swiftly from candles to gas. But oil lamps were so cheap and gas so complex and costly that most roads compromised on kerosene. The Erie, for example, decided against expanding its gas-lighting program when its calculations showed that gas cost three times more than oil.[221] The price of kerosene lamps was half that of gas apparatus, mainly because the lamps did not require tanks or piping.

In addition to cost, early gas lighting was not entirely satisfactory. Coal gas could not stand compression without a substantial loss of illuminating properties, yet it had to be greatly reduced in volume or a reasonable supply could not be carried aboard a railroad car. In addition, it gave off carbon that fogged the lamp globes.

A gas rich in hydrocarbons could undergo considerable compression with no major reduction in its illuminating powers. Many engineers knew that such a gas could be produced from heating petroleum in a retort. As already noted, both the Pennsylvania and the Reading attempted to produce gas in this form during the 1870s. Why it was not successful is unclear. Late in the same decade an independent mechanic, J. M. Foster of Philadelphia, tried to work out a petroleum gas system and had at least moderate success.[222] His gas, manufactured from naptha or crude oil, was compressed to 40 atmospheres. With twenty burn-

ers he could produce 260 candlepower. The lamps were ignited by an electric spark. Foster's system was used both on the New Haven Railroad and in Mann's sleeping cars before an accident in the compressor plant around 1890 put an end to Foster's American Lighting Company.[223]

Just as European railways showed more concern than U.S. lines for improving car heating, so too they led the way in developing better forms of car lighting. While America tinkered with gas, Europeans methodically moved toward its large-scale application. In 1861 a Belgian railway began using enriched gas made from rock shale oil.[224] Among the other Continental inventors drawn to this field was an obscure tinsmith and lamp maker from Berlin, Julius Pintsch (1815–1884).[225] He began to develop a system of gas lighting specifically suited to railway service. Pintsch was not the first to explore the subject or to recognize the defects of coal gas; nor did he invent petroleum gas. But he does seem to have been more single-minded in his effort to perfect every detail both of the gas production and of the lamps and fixtures for use on railway cars. This concentration of effort resulted in the most widely adopted system of gas car lighting in the world. The name Pintsch became synonymous with gas lighting.

Pintsch began his experiments in 1867 and within two years was installing lights in Prussian cars. Between 1871 and 1878 the Baden State, Lower Silesian, and Alsacean railways adopted his system. Several British lines, including the London underground, converted to Pintsch lights at the same time. Steamships and buoys began to use Pintsch gas. By 1880, 6,000 railway cars were lit with lamps manufactured by the Berlin tinsmith.[226] In 1881 Pintsch decided to invade the American market and opened an office in New York.

The Erie appears to have been the first customer, contracting with Pintsch for 100 sets of lights.[227] In July 1882 a special train ran over the Erie with lighting so brilliant that it was actually possible to read in the cars. This was considered miraculous. Each car had a 344-cubic-foot cylindrical gas tank mounted under the floor (Figures 5.56 and 5.57). The gas, compressed to 127 psi., passed through a regulator valve box that was also under the floor and then through ¼-inch-diameter gas pipes to the lamps. The gas was reduced to less than 1-ounce pressure by the regulator before entering each of the thirteen burners. Every "sun burner" was rated at 17 candlepower. The lamps burned 8½ feet of gas an hour at a cost of 1.7 cents.

In 1883 the West Shore contracted for Pintsch installations on eighty-two cars, the New York, Providence, and Boston for installations on five cars, and Pullman for an indeterminate number.[228] A major obstacle to Pintsch in expanding the business was that the company had a lone gas plant in New York. Only lines operating in that locality could be serviced. Progress was slow in the American branch; by 1888 just over 400 cars had Pintsch lighting in the United States. In Europe meanwhile, 27,000 cars carried Pintsch lights, and Count de Rothschild was a major backer.

In 1887 several American capitalists decided that the Pintsch

Figure 5.56   Pintsch produced the most popular gas lighting system for railroad cars in the world. H is the gas tank, R the regulator, N and M the lamp vents, and V the filler valve. (International Textbook Company, 1912)

system was a good thing which needed some capital and an aggressive management.[229] George Pullman, Sidney Dillon, and T. C. Platt provided the cash. They hired Arthur W. Soper, a minor but respected railroad manager (and at one time general manager of Wagner), to run the business. The firm was reorganized as the Safety Car Heating and Lighting Company. In 1888 only four railroads were using Pintsch lamps. There were now six compressor plants, all in the Northeast except for a lone outpost in Marion, Ohio.[230] Soper began to push. It was easy to sign Pullman, who had a direct stake in the firm's success. Wagner was forced to follow, because its management could find no superior form of lighting available. By January 1891 twenty-nine railroads with 2,000 cars were using Pintsch lamps. Fourteen compressor plants were in operation in all regions of the United States. From this time on, the firm was a runaway success.

As line after line signed on for Pintsch lights, gas became a major form of car lighting in America. For the first time cars were being well lit, and lit cheaply enough so that gas could be used not just for the luxury sleeper and parlor trade but also for the workaday coach. In 1893 the Lackawanna began gas-lighting its suburban cars. It was reported that passengers were willing to stand in Pintsch-equipped cars rather than sit in kerosene-lit coaches, because at least they could read the evening paper while they were standing. Oil lamps were declared (prematurely, to be sure) "a relic of a past age."[231] The Chicago and North Western found that gas lighting reduced labor costs; the men who had been needed to clean and maintain the oil lamps were let go, and car interiors required less frequent cleaning now that the smoky oil lamps were gone.[232]

The progress of Pintsch lighting was recorded in advertise-

Figure 5.57   A Pintsch gas center lamp of about 1895. Pintsch lighting was a major system on American railroads between 1885 and 1910.

ments and news accounts fed to the trade press. By 1897 the Safety Car Heating and Lighting Company claimed that 10,809 cars carried its Pintsch equipment. Two years later its near-monopoly of gas car lighting was strengthened when the Reading Company closed its own gas plant to adopt Pintsch.[233] Foster was already out of business, and only a few other competitors were still feebly stirring. In February 1900 the New York Central announced plans to convert entirely: all 1,500 cars would glow with Pintsch gas, and all oil lamps would be scrapped.[234] Four years later it was reported that on a worldwide basis, 120,000 railway cars used Pintsch lighting. Of this total 20,000 were on U.S. railroads. The number of American compressing plants had grown to seventy. In 1908, probably its peak year, the Safety Company claimed that its system was installed aboard 32,455 cars in the United States and Canada.[235] This represented approximately 60 percent of all passenger cars in service.

Pintsch gas maintained its leadership in the field by offering a good product and by buying or squeezing out rivals. After the Foster company's collapse, Pintsch had few serious competitors. A minor threat was posed by the Frost "dry carburetter" system, which appeared on the market in 1884 (Figure 5.58). It was a gasoline vapor light especially designed for safe use on railway cars. To avoid the hazards of fluid gasoline, the reservoir or tank was packed with wicking, which absorbed the fuel and thus produced the so-called dry carburetor. The gas or vapor was made by passing compressed air, drawn from the air brake system, through the carburetor at a very low pressure.

In the spring of 1885 the Frost system was given a series of tests on the Pennsylvania Railroad.[236] The gas tank was placed inside the car. In later installations, however, individual carburetors were made for each lamp and were placed on top of the roof. This not only removed the gasoline from the passenger compartment but also placed the tank above the lamp chimney in order to heat the reservoir and improve vaporization in cold weather. The new style of carburetor, 30 inches in diameter and 4 inches deep, was like a flat cake pan with a hole in the center to ventilate the lamp. It was made of copper, so that it might crumple but would not burst in the event of a wreck. Each carburetor held 4 gallons of gasoline—enough for forty to fifty hours of lighting.[237] After tests made in 1889, the New York State Board of Underwriters declared it safe "beyond the suspicion of danger."

The Pennsylvania was satisfied not only with its safety but with the brilliance of its light and the economies it made possible. Frost burners were rated at 40 candlepower—more than twice that of Pintsch lights.[238] While they cost 3 cents an hour to operate, which was more expensive than Pintsch, the Frost system required no gas plant. It produced its own from an explosive liquid that distilleries could not sell for lamp oil and usually dumped in the nearest creek.

By 1892 the Pennsylvania, always the main user of Frost lamps, had installed them on 776 cars.[239] The road announced plans to equip all 3,700 cars on its associated lines with the lamps at a cost of 1 million dollars. Although it claimed that all the

Frost system's problems had been solved, some complications apparently developed, because the conversion never took place. Several other railroads, such as the Norfolk and Western, the Reading, the Burlington, and the Wisconsin Central, also used Frost lamps. Estimates made in 1900 were that 1,000 cars had this form of gas lighting—not a significant proportion of the market.[240] Among the drawbacks of the Frost system was that it did not give uniform light. Impure gasoline could foul the apparatus, and in hot weather the mixture tended to be overly rich, causing the flame to glow orange and smoky. In 1900 it was reported that Frost had been absorbed by a rival firm (presumably the Pintsch system's Safety Car Heating and Lighting Company), which was withdrawing the product from the market.[241]

Shortly before Frost was safely out of the way, another gas system was introduced to disturb the Safety Company's comfortable monopoly. Around 1898 the Chicago railway supply firm of Adams and Westlake began to produce acetylene lamps adapted for car lighting.[242] Derived from carbide of calcium and water, the gas had been known to science since 1836 and was one of the most potent illuminants available. The Texas Midland and the Santa Fe railroads tried the acetylene lamp system with some success. At one time the Safety Company even considered mixing 20 percent acetylene with Pintsch gas to double the candlepower. But beyond this mild interest, acetylene made little progress in car lighting. It was considered dangerous, it gave off an offensive odor, and it was highly corrosive.

During the peak years of gas lighting, 1890 to 1910, the only serious rival to Pintsch gas was electricity. But the early electric train-lighting apparatus was handicapped by mechanical defects and high costs. In addition, technical improvements within the field of gas lighting itself prolonged its domination of the market. In 1885 Carl von Welsbach of Vienna patented a mantle made of cotton thread and the oxides of metals called rare earth. The gas flame heated the mantle, making it incandescent and thus vastly increasing the amount of light produced. Mantle fixtures cut gas consumption by two-thirds, while improved burner design boosted efficiency by 300 percent.[243] Thus the cost of gas lighting fell dramatically even as its power of illumination soared. In 1909 a comparison of gas and electric train lighting underlined the superiority of the Pintsch system.[244] With a base of 400 candlepower at six hours, Pintsch cost 12 cents while electricity cost $1.42.

Although Welsbach's mantle was adopted for domestic lighting during the 1890s, it came into favor more slowly in other areas. Just why this was true, considering its remarkable benefits, is not understood. Yet the *Railway Review* of June 8, 1907, noted that the ordinary flat flame jet continued to prevail, that only 2,000 cars were fitted with the mantle-style burners, and that all these had been installed within the past two years.

But the resurgence of gas lighting was a temporary victory; electric lighting was the coming technology, and railroads used it on luxury trains despite the higher cost. The Pintsch system also had its faults, the chief of which was the terrific expense of the

Figure 5.58 The Frost lighting system ran on gasoline and was used mainly by the Pennsylvania Railroad and its subsidiary lines. (Railroad Gazette, January 2, 1891)

gas-producing plants. To service its relatively small fleet, the Milwaukee Road had to maintain seven compressor stations. The first cost represented an investment that railroads would have preferred to apply elsewhere. In addition, a minor but continuing annoyance was that the apparatus itself was manufactured in Germany, and the foreign standards, machine threads, and measurements complicated repairs.[245]

The real downfall of the Pintsch system was, of course, the perfection of electric lighting. After 1910 it began to grow in popularity, and its progress inevitably meant Pintsch's decline. The Safety Company now rarely reported hard facts about its business, so that statistics on the number of cars with Pintsch lighting are elusive. C. Malcolm Watkins of the Smithsonian Institution, a frequent commuter on the Boston and Maine in years past, recalls that the line began removing Pintsch lamps in 1928 and completed the process two years later, with the exception of a few head-end cars. He also remembers the faint odor of petroleum when one entered Pintsch-lighted cars, and the intense, slightly blue-green flame of the jets.

In the 1928 edition of the Car Builders' Cyclopedia the Safety Company advertised that the Pintsch apparatus was still available. In the 1931 issue there was no such advertisement, although the apparatus was illustrated in the technical portion of the text. By 1937 it was described in the text with a closing statement: "This system is now obsolete on American railroads." Nevertheless, W. D. Edson, identified earlier in the chapter, recalls gaslights on Milwaukee commuter cars in the 1940s.

Electricity was the first source of flameless lighting. In 1881, only a year or two after its commercial introduction, electric lamps were tried out on railroad cars. Gas had only begun to dominate car lighting, and there was reason to believe that electricity might overtake it and all other established forms of illumination. But this did not happen; instead, electricity waited in the wings for over thirty years while the old stars, somewhat beyond their prime, burned brightly on center stage.

To a limited extent, electric arc lighting had been employed for street and factory lighting since the late 1870s. It was far too intense for domestic service, however, and just how to harness this amazing source of illumination eluded inventors until Thomas Edison and Joseph Swan almost simultaneously perfected a practical incandescent light-bulb. The breakthrough was hailed by some as the greatest wonder since the invention of the locomotive itself. Others were more cautious, suggesting that the new light was dangerous to the human eye and might provoke "a fearful catalogue of nervous diseases."[246] The London, Brighton, and South Coast Railway chose to ignore such gloomy predictions and in October 1881 installed twelve Edison lamps aboard the Pullman parlor car Beatrice.[247] Current was furnished by thirty-two batteries. Late the same year the Brighton line had four cars with lights powered by batteries that were recharged at the end of each run.[248]

In the summer of 1882 the Pennsylvania Railroad decided to emulate the British experiment.[249] Batteries were imported from France, and the Edison Company assisted in wiring coach Number 397, the first car in the United States to have electric lights. Seven lamps with two 8-candlepower bulbs each were powered by a set of batteries weighing 1,200 pounds, which were rated sufficient for sixteen hours' service. In August 1882 the car ran between Jersey City and Newark. In 1884 the Pennsy's chemist, Charles Dudley, began a second series of experiments using Brush batteries. During the next year eight parlor cars with electric lights were placed in regular service. In 1887 two of the road's best trains, the Florida Special and the Chicago Limited, were outfitted with electric lights.[250] Steam-driven generators, stationed in the baggage cars, provided the power. From that time on the Pennsylvania persisted in operating at least a few electrically lighted trains, despite the many problems of this technology during its formative years.

Within a few years of the Pennsylvania's initial trial, other American lines began to test electric lighting. The Boston and Albany, the Connecticut River Railroad, and both Wagner and Pullman all had some electrically lit cars by 1886–1887. Although batteries were used in most of their installations, the Connecticut River line sought a practical way to manufacture current independently of accumulators.[251] Since the generator selected for this experiment was too bulky to be mounted on the trucks, it was placed inside the baggage car with a belt passing through an opening in the floor to a pulley fastened on one of the axles. The belt failed; so did the rope and chain drives that the line tried next. A 7-horsepower steam engine that drew steam from the locomotive was then placed in the car and belted to the dynamo. The system worked well enough to light a six-car train, but not well enough to encourage wide adoption of the head-end system.

Batteries, at first regarded as the simplest and best source of power for electric car lighting, proved to be a dismal failure. In November 1889 after a 2½-year trial, the Boston and Albany abandoned battery lighting for Pintsch gas.[252] In England the Midland and the Brighton railways decided to give up electric illumination after equipping nearly 1,000 cars and spending five years in the effort to develop a workable system.[253] The weak link was the wet cell battery, an expensive, short-lived mechanism that was at once heavy and fragile. Rough handling in the process of recharging, and agitation of the acid by the motion of the train, both ruined the delicate pasted grid plates. Batteries with a projected life of 7 hours rarely lasted 5½, and they took 15 to 20 hours to recharge.[254] Few weighed less than 200 pounds each, and with six to a coach, the workmen were hard-pressed to keep a continuous caravan of wet cells moving to and from the recharging shed. A few roads—for example, the C & O and to a lesser degree, the Pennsylvania—remained loyal to the straight battery system for many years, but most lines abandoned electric lighting after a few months of experimentation.

Small steam-driven generators were at first regarded as the best alternative method. Most of these installations, known as head-end plants, involved a single engine-generator set in the baggage car, with steam supplied from the engine or an independent boiler. In 1886 a Philadelphia inventor, H. E. Freese,

Section Showing Right Side.

Figure 5.59   The Milwaukee Road's head-end car of 1890 provided steam for heating and electric power for lighting. (Railroad Gazette, June 13, 1890)

promoted a steam generator set for each car.[255] Freese proposed using a "package" engine with a kerosene-fired boiler. The 2-horsepower unit could be placed under the car or in a closet. The scheme eliminated batteries and axle belt drives and gave each car an independent power source. But apparently the cost and maintenance expenses of so many individual engine-boiler units discouraged the adoption of the Freese plan.

Inspired by the personal enthusiasm of its electrician, George Gibbs, the Milwaukee Road became a champion of the head-end system (Figures 5.59 and 5.60). During the 1890s, when most of the industry adopted a wait-and-see attitude, the Milwaukee continued to operate electric lights. The road first tested batteries but quickly discarded them in 1888 for a head-end steam plant.[256] Two years later Gibbs designed special cars (described earlier in the section on steam heating) with boilers, high-speed engines, and generators to replace the makeshift baggage car generator units. When the lighting and heating tenders entered service, the line had forty-five cars wired for electricity. They ran primarily between Chicago and Minneapolis, and during the thirty-six-hour journey there was no need to worry about discharging batteries. In 1892 the Milwaukee had eighty-two cars wired and ran five electrically lit trains a day. The system was expensive both in first cost and in operating costs. It was necessary to have an electrician aboard the tender, and his monthly

salary of $85 was considered extraordinary at the time.[257] Hourly operating costs were figured at 35.5 cents an hour, or ten times the cost of gas. Occasionally in extremely cold weather it was necessary to shut down the generator's engine so that all steam could be used for heating. Despite these drawbacks, the Milwaukee stayed with electric lighting. By 1904 it had electricity on 300 cars—more than any other U.S. railroad.[258] It still favored the head-end steam tenders, although seventeen of the cars in commuter service had battery lights. Not long after this, however, the steam head-end generators were probably replaced by axle dynamos, which had been perfected in the meantime.

Electric lighting remained a novelty on American railroads until about 1910. The railroad trade press continually referred to it as a rarity found only on a few luxury trains. And even in these instances combination gas and electric fixtures were used, because the new system could not be depended on. In 1896 the C & O frankly admitted that it retained electric lights on its first-class cars simply for their advertising value.[259] Electricity was popular with the passengers, but it was not practical or economic. Meanwhile greater progress was being made in Europe: by 1898 an estimated 6,000 cars were electrically lighted.[260] In the United States it is doubtful if more than 150 to 200 were so equipped.

The situation was not unlike the period of transition from

Figure 5.60   The Milwaukee Road was one of the first American lines to make commercial use of electric lighting for passenger trains. Lamps and wiring details are shown in this drawing. (Railroad Gazette, June 13, 1890)

Figure 5.61   A Moskowitz belt-drive generator of 1896 led to the standard system of passenger car electric-lighting service in this country. (Railroad Gazette, October 30, 1896)

*Figure 5.62   Large flat belts driven by an axle-mounted pulley powered the car-lighting
generator. This photograph was made around 1910. (Pullman Neg. 17796)*

wooden to steel cars. The advantages of the conversion were obvious and much discussed. The technical problems were solvable, but the industry left the task to outsiders, arguing that the old form was proven, satisfactory, and economical. Cash requirements, inertia, and existing investments in gas and oil lighting reduced the railroads' interest in switching to electricity. Meanwhile, several private firms and individuals worked to produce a more acceptable system of electric lighting.

Morris Moskowitz devoted himself to developing the axle light system, so called because it involved an axle-mounted pulley that was belted to a generator. When the train was in motion, the generator powered the lights. When the car was standing, or when it was moving at very low speeds that did not produce enough current, batteries would take over the power needs. The system seems so simple, yet its perfection baffled mechanical engineers for many years.

The basic problem was how to maintain a constant voltage when the generator's rotation fluctuated with the speed of the train. In the normal range of speeds (zero to 70 mph), the generator's revolutions per minute varied from zero to 800. In addi-

tion, the generator's maximum wattage was reached at intermediate speeds of 22 to 25 mph. A third problem involved reversal of polarity when the train backed up or the car was pulled in the opposite direction. Moskowitz solved the voltage problem by a specially wound field magnet. By passing a current through a secondary large wire coil, the magnetic field was progressively weakened as the armature rotated faster, hence lowering the voltage so as to maintain the desired pressure of 32 volts.

The scheme was tried in Germany as early as 1882, and the first recorded North American test was made two years later.[261] This was almost a decade before Moskowitz became involved in axle lighting. In 1893 he placed a small belt-driven generator inside a car on the Central Railroad of New Jersey.[262] A chain drive from one of the axles powered a countershaft. Within three years Moskowitz had greatly improved the design by mounting the generator on the truck frame (Figure 5.61).[263] The Pennsylvania, Erie, and Santa Fe tested the apparatus, and a Southern Pacific private car traveled across the country demonstrating the reliability of Moskowitz's rig. In the spring of 1897 he was ready

Figure 5.63   Belt drives remained in favor throughout the heavyweight period, but after about 1915 the generator was mounted to the car rather than to the truck's frame.

to make an extended test of the axle light system and installed it on fifty cars.[264] Two years later his apparatus was on one hundred Santa Fe cars. Camel-hair belts were used at first, then friction drives, and finally belts were reinstated. By the summer of 1900 the Santa Fe decided that the Moskowitz system, although promising, was not ready for large-scale application.[265] All the original gear was removed, and a contract was signed for new tests on a dozen cars.

The axle light system thus suffered a temporary setback, but Moskowitz was clearly moving in the right direction. The Franklin Institute gave him an award that year. Other firms and mechanics, notably William L. Bliss, now recognized that axle lighting was the coming method of car illumination (Figure 5.62). The United States Light and Heating Company was formed to take over the Moskowitz patents. In 1902 the Twentieth Century Limited adopted axle lighting, and two years later the Chicago and Alton announced plans to convert all its passenger rolling stock to axle dynamos.[266]

While American railroads had now settled on axle lighting, massive conversion was still in the future. The familiar objections to the high cost and excessive weight of electrical systems were repeated as late as July 1909 in the influential *Railroad Gazette*. According to detractors of electric lighting, it was so expensive that railroad officials suppressed the actual figures to keep shareholders from accusing management of squandering company resources. Equipping a car with axle lighting cost $1,673. It added nearly 2½ tons of dead weight to the car. And annual operating expenses were $631.[267] Pintsch gas fixtures, even figuring in a prorated share of the compressor plant, came to only $550 a car, while operating costs were calculated at $55 to $65 a year.

But other factors began to overwhelm plain economics. When the Pennsylvania and New York Central railroads banned wooden cars from their New York City terminals, they simul-

taneously forbade stoves and open-flame lamps. This ruling suddenly rendered a great deal of serviceable equipment obsolete, and the effect on car lighting was dramatic. In 1902 the United States Light and Heating Company estimated that there were only 1,000 electrically lighted cars on American steam railroads. Within nine years this figure grew to 11,000.[268] In a year another 2,700 were wired for electricity. If cars with combination electric-gas fixtures were counted, nearly 30 percent of the entire passenger fleet now had electric lighting. By 1912 electricity was so clearly coming into its own that the Master Car Builders Association prepared an elaborate series of recommended designs and standards.

Lack of statistics prevents a more complete chronicle of electric car lighting in later years. Its slow progress can be traced for one representative railroad, because the Baltimore and Ohio published an annual booklet, *Summary of Equipment*, that included data on car lighting. The accompanying table provides some figures.

Year	Total B & O Cars	Electrically Lighted B & O Cars
1917	1,207	187
1924	1,400	473
1932	1,636	1,027
1938	1,217	1,049
1946	1,126	1,014

The progress of axle lighting is attributable in large part to the slow maturation which all mechanical devices undergo between the time of their invention to the time that they are finally ready for everyday service. This metamorphosis involves minor and unspectacular alterations in design and material. A slight increase in bearing area or the change from cast iron to malleable iron can make all the difference. This is an unpopular explanation of

Figure 5.64 When power requirements and generator size increased after 1930, shaft-drive generators replaced belt-driven units. (The Budd Company)

mechanical progress, however. And in the case of electric car lighting, there was in fact one dramatic breakthrough that corresponds to the popular notion of progress as something which is achieved in astonishing leaps: in 1907 a method was discovered for producing ductile tungsten lamp filaments. By 1911 bulbs of this type were being manufactured for commercial sale. Tungsten did for electricity what the Welsbach mantle did for gas: it radically reduced the power consumption for a given amount of light. This reduction was critically important to car lighting because it meant smaller generators and batteries—cutting down weight, investment, and operating costs.

In 1912, according to the electrical engineer C. W. T. Stuart, ball bearings replaced plain bronze bearings, making generators more dependable and easier to maintain.[269] A few years later the old standard of truck-mounted generators began to give way to an underbody fastening much on the order of the European plan (Figure 5.63). The underbody mounting had several advantages. It removed that much unsprung weight from the trucks and raised the generator and its belt above snow and flying ballast. Throughout the 1920s the old-fashioned flat belt drives remained in favor. Their defects were well established—rapid

wear, slippage, and complete loss of the belt if the splice failed—yet they were simple and cheap. So long as small 2- to 4-kilowatt generators were common, the flat belt remained secure. By the early twenties, however, there was enough dissatisfaction with the old system to encourage tests with shaft and chain drives. Of the many designs tried, the Spicer drive, developed by the Safety Car Heating and Lighting Company, was recognized as the best (Figure 5.64). The drive gear and case were designed so that the unit could be fitted to a rough-turned axle without dismounting the wheels or performing any machining. A safety clutch would automatically disengage in the event of any undue resistance from the generator. The clutch would also disengage when axle speed fell below the full load speed. It was possible to detach the shaft so that the generator could be manually turned, or "motored," during inspections without fully disassembling the drive mechanism. Spicer drives became increasingly popular during the thirties as the larger generators necessary for air conditioning came into use. By 1942 it was recognized as an industry-wide standard.[270] Flat belt drives continued on non-air-conditioned cars, and some were still in service in recent years.

While axle lighting was the orthodox system in this country,

Figure 5.65 A combination gas and electric center lamp of about 1910. (Smithsonian Neg. 49749)

some railroads chose to use head-end power plants. The case of the Milwaukee has already been described. Several other lines were tempted to adopt head-end power in 1907 when General Electric placed a Curtis steam turbine generator set on the market.[271] It was offered in 15-, 20-, and 25-kilowatt sizes and was designed for direct mounting on the locomotive boiler or as an auxiliary in the baggage car. The turbine overcame many of the objections to earlier steam head-end units. The vibration problem was eliminated, and because the sets were self-contained, no attendant was necessary. Head-end power was ideal for trains running point to point and not subject to re-marshaling en route. Through limiteds and commuter trains were then the province of steam turbine lighting. It was widely used on the suburban trains of the Boston and Maine and the Central Railroad of New Jersey. The latter road had 20 locomotives and 100 cars outfitted with turbine lighting by 1944.[272] A year later it had converted all 60 locomotives and 300 cars in suburban service. A 125-volt, 12-kilowatt generator was mounted on the tender. Ironically, these roads gave the very reasons that had once been used for justifying axle lighting to explain why they opposed it; they believed that an independent electrical system for each car was wasteful and unnecessary. A central power source removed 2 tons of dead weight per car. The tremendous "drag" of axle generators was figured at 33 horsepower a car, and on a fifteen-car train that amounted to 500 horsepower—enough to ruin a good schedule and tie up the line.

The motive power drain of axle lighting was also cited by advocates of head-end and under-the-floor gasoline or diesel engine generator units. The early Union Pacific motor trains employed head-end internal combustion plants. By 1941 the U P decided to place self-contained propane gas engine generators of 7½ kilowatts on each car of the new train, City of Los Angeles.[273] The Southern Pacific and the Milwaukee Road also used engine generator sets of this type. Most of the individual sets were 110 volts AC, but on cars not so equipped it was necessary to add small auxiliary 110-volt alternators to power electric shavers and water coolers. Rectifiers were tried out by some lines, but they were found unreliable and prone to burn out.

During the same year that axle lighting was being challenged by other methods of power generation, electric light fixtures were undergoing changes. The earliest lamps, particularly the combination gas-electric fixtures, were traditional in appearance (Figure 5.65). Electric lamp design evolved into a more functional form that was only partially obscured with a decorative overlay in classical, Adam, or mission motifs (Figures 5.66 and 5.67). In 1911 the Santa Fe tried indirect lighting, possibly for the first time on a railway car.[274] Hanging fixtures reflected the light to the ceiling, where it was diffused throughout the car. A few years later the Northern Pacific purchased some cars from Pullman with the lamps built into the ceiling along the line of the clere-

Figure 5.66 An electric side lamp of about 1925. Such lamps were used in parlor and dining cars.

Figure 5.67   This 1928 Kansas City Southern business car had dome center lights and ceiling fans. (Pullman Neg. 32916)

story.[275] They were spaced at 2-foot centers, and no center lamps were used.

The last major change in passenger car lighting was the introduction of fluorescent lamps. Railroads showed a progressive spirit in adopting this form of lighting only about a year after it was placed on the market. In its October 18, 1938, issue *Railway Age* noted that the New York Central had fitted a coach with some of these remarkable tubular lights. The low power consumption of mercury-vapor lamps made them ideally suited for railway car illumination. With only a small increase in wattage, the illumination was nearly quadrupled. A small motor-generator converted the power to AC. In June of the following year the Lehigh Valley began operation of a number of remodeled cars with fluorescent lamps on the John Wilkes.[276] Within a few years the fluorescent system was widely applied to car lighting. Special 60-volt tubes were used where normal 110-volt current was not available.

Toilet and Drinking-Water Facilities

Of all the excruciating discomforts of travel, none can cause greater desperation than the absence of a toilet. To be trapped aboard a crowded public conveyance without hope of relief until the next stop is a torment that seems to have touched even the stony heart of the most unfeeling general manager. And the extra cost of such accommodations was justified by more than compassion, for if each car had its own toilet and drinking water, passengers would not be tempted to try the next car. This would discourage the dangerous passage across the open platforms. It would also discourage through passengers from leaving the train at intermediate stations and either stranding themselves or delaying the train's departure.

In what may have been a coy effort to obscure its existence, the toilet was given such esoteric labels as saloon, lavatory, dressing room, washroom, water closet, and retiring room. In a price quotation for new cars dated May 10, 1855, the Buffalo Car Company delicately referred to it as a "loafing room."[277] Such a proliferation of titles might suggest something grand, but the earliest washrooms were less than sumptuous. Until rather late in the nineteenth century they were airless, cramped closets barely 3 feet square. Although contemporary evidence is sparse, the drawings reproduced in earlier chapters (for example, in Figures 1.81 and 1.82) show that the toilet itself was a primitive dry hopper—really nothing more than a wooden box with a circular hole cut in the top for a seat. It was much like an outhouse, except that the refuse dropped out directly on the tracks rather than into a privy. The first mention of facilities even so elementary as these is found in reference to cars built for the Philadelphia and Columbia Railroad in the middle and late 1830s.[278] It is possible that the special car *Victory* of 1835 had a toilet. The surviving contemporary model pictured in Figure 1.68 shows a small anteroom with two commodes side by side. One is quite low, indicating that it was for children.

How quickly such facilities spread is difficult to document; probably they were rather common by the early years of the following decade and became increasingly so as the length of rail travel increased. Passengers could no longer be expected to tend to their needs at the beginning or end of the journey. In 1842 the luxury of the washbasin was introduced.[279] Perhaps earlier cases could be found, but the pictorial evidence available indicates that few coach passengers could perform their ablutions on board the train much before the 1870s or 1880s. Such niceties were available to first-class passengers at an earlier date, of course, but even this favored group was obliged to wait in line during the morning and evening rush for a turn at the sink. Once inside the sanctuary, a passenger was obliged to make his toilet with as many companions as the tiny 3- by 6-foot compartment could hold. The commode itself was in a separate closet, often on the opposite side of the aisle. To draw water, one worked a plunger pump that pulled water from a tank (Figure 5.68). A public hairbrush and comb, tethered to a cord, were available for all who cared to use them. The towel was equally democratic. This practice persisted as late as 1914.[280]

Separate washrooms for men and women were relatively common on the first generation of sleeping cars, but apparently not on coaches until the 1880s. By then some luxury cars had bathtubs. One inventor envisioned a three-compartment bathing car modeled on a fully equipped Turkish bath.[281] The average traveler, however, felt well satisfied if water and a clean towel were available.

Air-activated water systems helped to bring the railway car bathroom out of the dark ages. The idea was patented in 1879 (No. 221,680) by David H. Jones of Belleview, Pennsylvania. A similar scheme was tried on the Fitchburg Railroad three years later.[282] Now that the water was under pressure, the old-fashioned sink pump could be retired. Even more important was the introduction of flush toilets and urinals. The aesthetic as well as sanitary benefits were substantial. Pullman began to use the air water system in 1888 and within five years was moving rapidly to install it on all cars.[283] At that point, however, serious technical problems were encountered, because the air water apparatus tapped into the brake system for the air supply.[284] A toilet that

Figure 5.68   A spacious men's washroom on a Pullman sleeping car of about 1885. Note the hand pumps and massive marble-top sink. (Car Builder's Dictionary, 1895, fig. 3468)

was flushed at the wrong moment could upset the pneumatic balance of the Westinghouse system enough to cause an unexpected release or application of the brakes. The leakage of water into the brake system was another troublesome side effect. The complexity of the system, if not its actual workings, is indicated by Figure 5.69.

Washroom furnishings had also improved substantially by the 1880s. The rude box commode was replaced by elegant ceramic bowls. Handsome paneled washstands were capped with white marble tops and nickel-plated fixtures. The disreputable bare wooden floors were now often covered with zinc or copper. The Michigan Central found slate an even more durable and sanitary flooring.[285] Linoleum and plastic tile came into favor in more recent times. The abandonment of marble-top sinks was caused by more than changing fashion, for these heavy slabs became lethal flying platters during a wreck. There is one reported incident of a Pullman porter being decapitated by a loose sink top.[286] In and around 1910, lightweight sink bowls of nickel silver were introduced (Figure 5.70). Vitreous ware or porcelain enamel also came into use (Figure 5.71). Stainless steel fixtures typified the modern period.

Despite all these improvements, the disposal of human excreta is no more advanced now than it was in 1835. Raw human waste continues to be dumped out of railroad cars onto the tracks. In 1947 the Association of American Railroads voluntarily investigated the problem, but no reforms were instituted. Around 1970 the consumer advocate Ralph Nader took an interest in the problem, and a few press statements appeared. Greater issues soon overtook this quixotic crusade, however, and the railroads were allowed to continue as before.

The American mania for ice water was endlessly recorded by British visitors during the nineteenth century. This passion was based on more than a local fancy; baking heat and billowing dust created a thirst best satisfied by cold water. As the duration of journeys increased with the expanding rail network, so did the need for drinking water aboard the cars.

In 1847 a Virginia railroad specified 4-gallon water coolers for its cars.[287] It was more common, however, for a boy to dispense water to thirsty passengers. Throughout the journey he walked the train from end to end, carrying glass tumblers and a brass or tin can made much like a tea kettle (Figure 5.72). On the Boston and Worcester, according to an 1853 account, the water boy used a silver urn and goblet.[288] Water boys were reported on trains as late as 1898.[289] No matter what system each road used, the industry as a whole was credited with liberal arrangements for providing drinking water.[290]

Ice coolers placed in each car allowed passengers to help themselves rather than await the water boy's next passage. But

Figure 5.69 Air-activated water systems eliminated hand pumps. The air was bled from the braking system. (International Textbook Company, 1912)

Figure 5.70   The men's washroom aboard a 1928 sleeping car was furnished with leather sofas and nickel-silver sinks. (Baltimore and Ohio Railroad)

the democratic cup remained, and everyone was expected to use it. When new, it was a polished, inviting vessel, but time reduced its bright surface to a grimy patina. A repulsive green crust formed around the lip, making it as appealing as a soiled napkin.

Public sanitation was a burning cause among the reformers of the Progressive era. Elimination of the public drinking cup was viewed as a necessity. The evils of the common glass were explored in the liberal press, and state legislatures were persuaded to outlaw its use. There was merit in this reform, but as in the case of Prohibition, not all of the public was pleased by the efforts of others to protect its well-being. *Railway and Locomotive Engineering* of October 1912 summarized the opposition's sentiments:

The cranks whose senseless agitation has eliminated the public drinking cup, even in Pullman cars, have inflicted much discomfort upon ordinary people and have largely increased the business of saloon keepers. The victims that deserve the greatest pity are the poor immigrant children who have need to make long journeys in hot dusty cars. When this pernicious movement against free water drinking has passed, people will wonder at the blindness which enabled cranks and saloon keepers to spread so much discomfort among honest people.

At last disposable paper cups eliminated the health hazard of the public glass (Figure 5.73). Just when they were first placed aboard railway cars has not been determined, but an early user was surely the Lackawanna, which began installing paper cups in 1909.[291] The scheme spread as local and state statutes began to outlaw the common cup. By 1917 it was reported that even where no such laws existed, the public water glass had vanished from railroad cars.[292]

To further assure the sanitation of its drinking water, Pullman made a double tank for ice and drinking water so that the ice which cooled the water did not come into direct contact with it. In the older style of tank where ice and water were mixed together, employees had simply been admonished not to handle the ice with their bare hands. With the advent of air conditioning, drinking water was cooled by small refrigerating units. After about 1940 Pullman provided almost all room-type cars with circulating ice-water systems that delivered drinking water to each compartment. The Pennsylvania continued the more traditional method of placing individual thermos bottles in each room. These heavy, silver hotelware bottles were kept filled by the sleeping-car porters. They presented an unexpected health hazard, because some patrons unaccountably mistook the water carafes for miniature urinals.

Decoration and Painting

The passenger car is the traveler's home. If it is neat and attractive, the pleasant surroundings can do much to ameliorate the tedium of the journey. Passengers have traditionally expected more than a weatherproof box with adequate lighting and heating; the interior must also reflect the taste of the times. Car builders responded to this demand, often with an exuberance that drew more censure than praise. While the industry could point to noted architects and interior designers who served as consultants, much of the planning was actually left to the chief draftsman. These solid, practical men turned to existing examples of the decorative arts for inspiration. Some translated successfully from classical forms, but the less talented car designers inadvertently transformed coach interiors into perambulating

Figure 5.71   A washroom and toilet of about 1950. (B & O)

museums of ancient decorative styles, from the Tuscan to the Hindu.

The more ostentatious embellishments were reserved for first-class cars. Car builders answered the popular demand, which seemed insatiable by the last decade of the nineteenth century, for ever more adornment. Their migratory products were boastfully described as "an active messenger of fame for the place of its fabrication."[293] The best minds of the day saw little to admire in the pretentious finery of the palace cars. Less discriminating critics rode the whirlwind of changing fashion, one season praising the design that they declared banal in the next.

To domesticate the new technology of steam railroads for the traveling public, measures were taken to introduce the familiar and traditional. The adoption of old-fashioned road coaches may not have been solely for the car builder's convenience; their appearance was also reassuring to the first railway travelers. So too the orthodox decorating motifs—the reproductions of old masters and scenes along the line—disguised the raw mechanism of the cars. It was reassuring to look at "The Wounded Tiger," "Napoleon Crossing the Alps," or "Washington at Trenton" rather than at a barren wooden box with a smoking metal steed at its head. The interior of the railroad car was also brightly colored, creating such a cheerful appearance that a German visitor wrote in 1859:[294]

The equipment of the cars must be called elegant in all cases, and sometimes even luxurious. Floor and ceiling are covered with oil-cloth or carpets, the walls are decorated with paintings and gold mouldings; in some cases, the pillars between the windows are upholstered. Good woods are used for the window frames, Venetian blinds, and seats; they are neatly polished; the upholstery is covered with red or blue plush. The outside of the cars is painted in a light color and varnished; above the windows, they show the names of the railroad to which they belong. All this, together with the spaciousness and good lighting of the interiors, gives them a friendly, inviting look.

The very highest style of exterior painting was evident by 1835, when John Baker of Lancaster, Pennsylvania, produced a car whose ends were embellished with scenes based on Byron's poem *Mazeppa* (appropriately, the verse revolved about a tale of fast running).[295] On the sides were finely executed paintings representing the seasons. During the same period a car built for the Tonawanda Railroad was pleasingly decorated in the Grecian idiom. Stephenson produced a car for the Auburn and Syracuse "gorgeously flowered and varnished." Heroic battle scenes, landscapes, panoramas, all painted in rainbow hues and as festive as a new shop window, were studied with tireless interest by city and country folk alike. In New England, scenes of everyday life were featured on the New London and Northern cars of the 1850s.[296] An old seaman would be pleased to step aboard a coach featuring an exterior painting of "Homeward Bound," "Cutting in a Whale," or "Storm on the Whaling Grounds." The most grandiloquent of the exterior paint jobs, and one for which a complete description has survived, concerns some cars produced by Eaton and Gilbert in 1835 for the Rensselaer and Saratoga Railroad:[297]

The outside of the cars is painted of a beautiful fawn colour, with buff shading, painted in 'picture panels,' with rose, pink, and gold borders, and deep lake shading, the small mouldings of delicate stripes of vermilion and opaque black. Within the panels are 'transferred' some of the most splendid productions of the ancient and modern masters, among which are copies from 'Leonardo da Vinci,' 'Horace Vernet,' 'David,' (the celebrated painter to Napoleon,) 'Stuart,' and many more of the modern school. The whole number of the subjects of the twenty-four cars, cannot fall far short of two hundred, as each car averages from six to ten subjects; among which may be enumerated, several copies from the antique, Napoleon crossing the Alps, the two splendid scenes in Byron's Mazeppa, the Hospital Mount St. Bernard, portraits of most of the distinguished men of our own country, among whom Washington (from Stuart's original) stands conspicuous, the wounded tiger, the avalanche, portraits of distinguished women, views of several of our popular steamboats, the railroad bridge near Philadelphia, and several views in the south. The 'tout ensemble,' is more like a moveable gallery of the fine arts, than like a train of railroad cars.

As late as 1848 this tradition of railroad car art remained very much alive. The South Boston *Gazette* of December 30, 1848, praised the work of local craftsmen:[298]

For some time past the painters employed by the Old Colony Railroad have been engaged in painting a passenger car. We had the

Figure 5.72   Water boys once patrolled American trains to dispense ice water to thirsty passengers. This scene dates from about 1880. (Dayton Rubber Company)

pleasure of visiting the shop a few days ago and examining this car, and pronounce it the most splendid specimen of car painting ever produced. It is of a vermilion hue, shaded with brown and black, while either corner is ornamented with gold scrolls done in a superior manner. But by far the greater attractions are the pictures upon the sides. These were executed by Mr. Kelley, a young artist of very promising talents, and are pronounced by excellent judges worthy of a place in any drawing room. Four in number, they represent scenes in such a manner as to make it appear as if the events portrayed were transpiring before us. A lion springing upon a gazelle as it is drinking from a brook is the subject of one of the pictures. A lady riding upon a powerful charger and accompanied by a dog is also portrayed, while on the other side is a fine view of a buffalo hunt. The fourth, and by far the most beautiful, is a painting descriptive of the inhuman mother and the wolves. The almost fiendish look of the mother, the agonizing features of the children, the appearance of the hungry wolves as they surround the sleigh, and the noble bearing of the horse as he exerts himself to the utmost to escape the ferocious animals are all portrayed in a most beautiful manner. The pictures are surrounded by gilt frames, and as if to make the delusion more perfect, these frames appear to be hung by small cords upon the sides of the car. The painter of the vehicle is Mr. James Hazeltine, who has acquired an enviable reputation as a car painter.

The talent required to execute such paintings did not come cheaply. John Stephenson paid his artists $30 a week in the 1840s. One of these men, James Walker, made his way into more respectable art circles in later years and is remembered for the heroic battle scenes in the National Capitol building's ro-

tunda.[299] The most highly paid artisans at Wason's car shop drew $8,000 a year—an extraordinary wage for a workingman in the 1850s.[300] William R. Tyler, Eaton and Gilbert's top-ranking artist since 1833, retired in 1858 when individual-portraiture styles of fancy paint jobs were replaced by simpler methods of ornament.[301] He opened a studio and lived to see his works exhibited in many public galleries. Possibly a more typical career was that of Jacob Scheller.[302] A native of Zurich, Scheller painted china, carriages, signs, and frescoes before leaving his native land during the 1848 revolution. Two years later he was working in the repair shops of the P W & B. At first his talents were well employed, but as simpler forms of car painting gained favor, Scheller's work became routine and he indulged his artistic bent as a Sunday painter.

From 1830 almost to 1860, car ornamentation was generally limited to what could be laid on by the brush (Figure 5.74). The bodies were plain boxy structures, largely devoid of any architectural ornament. The work was solid carpentry rather than fancy cabinetry. Stripped of their elaborate paint jobs, American cars of this period were as unimposing as a shoe box. The few surviving photographs of early cars almost all date from the 1880s or later, after the original ornate styles had been discarded for very plain painting. The plainness of the unadorned arched-roof car is clearly evident from Figures 1.24 and 1.26 in Chapter 1.

Probably it was very early in the history of car design that the costly mantle of exterior scenes gave way to a more serviceable exterior finish. The surviving Harlan and Hollingsworth draw-

Figure 5.73   Paper cups began to replace the common glass or cup around 1910. A Dixie cup dispenser is visible on the right side of the car. Note the water tank mounted in the ceiling over the washroom. (B & O)

ings, several of which are reproduced in Figures 1.88, 1.89, and 1.92, show that at least a few roads continued to honor the old ways into the mid-1860s by painting vignettes in at least a few panels. By this time most roads seemed content with the effect created by striping, lettering, and scrolls on the corner and door posts. It was a very rich finish: the lettering was double-shaded, and every visible nut and washer, even on the trucks, was high-lighted by striping.[303] All surfaces were varnished and rubbed out to a mirror-like finish.

The basic body color before 1870 was yellow—described as a very pale, creamy yellow, much like linen or straw.[304] Other colors that were occasionally used included light blue, pea green, gray green, drab, and Indian red. In the first decades of the American passenger car, bodies were sometimes painted dark

Figure 5.74   This panel from a head cloth of 1892 is representative of the decorative painting style common in American passenger cars forty years earlier. (Hall of Records, Dover, Delaware)

colors, but again pale yellow seems to have been the general favorite. It served as an effective ground color for accent decorations in vermilion, ultramarine, fine lakes, and gold and silver leaf. Yellow was also basic camouflage for the clay dust churned up by the train; a yellow car coated with yellow dust presented a tolerable appearance.

Interiors were done in much the same fashion and colors. Artwork there made more sense because it was inside, protected from the weather and placed where passengers could study and enjoy such primitive masterpieces. There was little opportunity to admire the artist's product while standing in the rush and confusion of a station platform.

The artist's main canvases were placed between the window panels and end bulkheads, but his greatest opportunities were in the large uncluttered ceiling. The roof ribs were covered with a head cloth. An extraordinary variety of materials was used for this purpose. Before 1880, colorful oilcloth or painted cotton duck were the most popular fabrics. Silk and German damask were more elegant, but they were difficult to clean and easily soiled by minor roof leaks until enameled cloth was developed. Common cotton duck was filled with sizing and then painted in bright colors, often in geometric patterns, that were hand-printed with wooden blocks. Head cloths for the best class of cars were decorated freehand at a cost of $125 each.[305] In 1850 Charles Stodder of Boston was said to have manufactured painted head linings for three years that were now extensively used by Northern railroads.[306] An Englishman, George Holt, settled in Worcester, Massachusetts, and became the major producer of "tasty and elegant" painted head cloths.[307] His business grew to such proportions by 1867 that he had to open a new factory. Within a decade, however, the chromo head cloth was out of fashion.

The brightly painted head cloth was in fact the last token of the old order; a new and more somber mode of decoration came into fashion at the end of the Civil War. The cheerful painted gaiety of the thirties, forties, and fifties gave way to what Lewis Mumford has named the brown decades. He explains this sudden change in taste as a gloomy reflection of a tragic war. In England, Victoria began a period of mourning for her late consort that was to continue for generations. Yet funereal as these times might appear, there was such richness in the dark walnut and heavily draped windows that the era was also called the gilded age.

Even before 1865, which Mumford designates as the beginning of the brown decades, there was evidence of discontent with garish car painting. In 1856, for example, the *Railroad Advocate* pleaded for simpler paint schemes within the capacity of an ordinary house painter.[308] It pointed out that once-beautiful cars became eyesores when their elaborate festoons were redone by the amateur decorators employed by most repair shops. And the best cars, according to the *Advocate*, were used on night trains, where fancy paint schemes go unnoticed. By 1870 strong condemnations of intemperate color schemes were a standard theme in the railroad press.[309] Cars decked out in showy garments were ridiculed as artistically and morally wrong. The "toy shop mixture" of colors was discordant; it offended the eye of people with taste. The *Manufacturer and Builder* stated that we judge cars, as we do men, by their dress, and an outfit wanting in taste shows that "some of the moral chords" are out of tune. Railroads were advised that the $500 wasted on showy paint jobs could be put to better use.

Pale yellow was replaced by olive green, chocolate brown, plum, and Munich lake. Car painters now argued that the darker hues were more practical.[310] Only three coats were needed for a well-finished job, compared with four to five coats of light color.

Figure 5.75   A car painter at work during the 1880s in Pullman's Detroit shop. Note the stencil in the extreme right foreground. (Chicago Historical Society)

Dark shades were more durable, and they helped to hide the black soot belched out by the coal-burning locomotives which rapidly came into favor after 1860.

Since polychrome decoration was not appropriate for the deep exterior colors, highlighting was accomplished with gilt and silver ornament, often magnificently shaded and outlined. Side and end panels, corner and door posts were embellished with luxurious scrolls and line work. The ebullience of the bright-color era was replaced by the opulent contrast of gold leaf on a dark base. The design was laid out by a stencil, but much hand work was required.[311] Six men might work on a car at one time, each devoting fifty to sixty hours to the job (Figures 5.75 and 5.76). In the 1880s the cost of ornamenting a car was calculated at $120 (less the lettering). Of this figure $50 went for the seventy-five or more "books" of gold leaf required. A cheaper job could be done with transfers or decals, and in 1878 Wason was using this method with a notable saving in time.[312] The idea was credited to the sewing machine industry, which had eliminated much hand work through the employment of decals.

The darkness of the exterior was repeated in the car's interior.

Figure 5.76   An exterior decorative scheme for a car produced by Gilbert in 1885.
(National Car Builder, *October 1885*)

Rather than somber painted surfaces, however, interiors were being finished in varnished hardwood panels: mahogany, walnut, rosewood, ebony, and many other rare cabinet woods. Ornament consisted of moldings, raised panels, carvings, inlays, and veneers. Contrasting wood tones and grains added to the effect—and to the cost of joinery and finishing. Cabinetmakers and woodcarvers replaced painters as the most skilled men in the shop. Often this elite corps was composed of Germans who sang songs of the old country or exchanged studio jokes with comrades from a Berlin academy while carving an antique head for a doorway.[313] Such artisans would spend up to 140 hours executing an end door panel.[314] Much carving and all moldings, traceries, and beadings were machine-made, yet the fitting of these elements required expert woodworkers.

Marquetry offered another opportunity for the decorative display of varnished wood. Simple straight-line inlays created attractive borders in contrasting colors. Rosewood, tiger maple, tulip, amboyna, and teak offered gorgeous hues and variegated patterns. Where nature failed, craftsmen would dye veneers to the tone desired. To imitate the crumple of a rose leaf, a minute piece of satinwood was lightly scorched in a bed of hot sand.[315] This and other tricks created the special effects of flowers, flying dragons, and allegorical representations of "Music" or "Night and Morning" that were emblazoned on berth fronts, bulkheads, doors, and transoms (Figure 5.77). No blank space was left to suggest that something might be lacking.

The beauty of natural wood was displayed on the outside of the car by naturally finished window sashes and doors, often of cherry wood. A very few U.S. railroads sheathed the entire exterior in naturally finished mahogany, and this scheme was enthusiastically followed by Canadian lines.

There were other minor revolts against the uniform solemnity of dark olive green. Surely during the Centennial year, it was argued, something more cheerful should be adopted, even if it were red, white, and blue.[316] A train of darkly finished cars was too much like a funeral cortege or a black-coated, black-hatted Hibernian procession on St. Patrick's Day. Some years later the Reading decided to lift the gloom by switching to brighter colors.[317] At an earlier period its cars had been red; then it had changed over to Pullman green. In 1888 the road adopted lemon yellow, with the letter board in a light orange. But this sunburst was short-lived, for within four years the Reading returned to the standard dark green.[318] The Pennsylvania was one of the few mavericks strong enough to follow an independent course, using Tuscan red rather than the standard green.

Varnished wooden interiors remained in high fashion until the advent of steel cars, but within that forty-year period—roughly 1865 to 1905—at least one major change occurred. In the early 1880s the fancy veneers, with their thin moldings painted in black, cream, and gilt, gave way to solid mahogany and oak paneling.[319] No contrasting moldings were used. Considering the times, the carving and other decorations were severe. Although to us the coach interiors of 1875 and 1885 may look equally elaborate, at the time the shift to honest, plain, solid paneling was regarded as a commendable return to basics. Pullman, however, did not adopt this reform, and first-class equip-

Figure 5.77   Rare woods highlighted by inlays typified the first-class cars of the 1870s to 1900. The berth fronts of this sleeper were decorated with the popular pineapple pattern of inlay. (Pullman Neg. 2137)

SCALE

12 0 1 2 3 4 5 6 7 8 9 10 11 12 13 14 15 FT.

Figure 5.78   The Pennsylvania Railroad championed what were extremely plain interiors for the times. The example here, dating from about 1880, was one of the Pennsylvania's more elaborate designs which featured carved panels between the windows.

Figure 5.79   A cross-section drawing of a Pennsylvania coach, published in 1881, shows the shallow beading used to decorate the oak interior. (Railroad Gazette, *August 19, 1881*)

taste-maker Charles L. Eastlake (1836–1906). In 1868 Eastlake had published a modest work, *Hints on Household Taste in Furniture, Upholstery and Other Details,* which called for more directness in design. He took up Ruskin's precept to ornament the construction and not construct the ornament. Eastlake especially disliked the French Renaissance furnishings that were popular at the time, with their rounded lines and deep carving. He suggested simple, straight lines to show clearly how the parts of the furniture were put together. For carving, he recommended shallow grooves or beadings that followed the lines of each part. Eastlake wanted to eliminate the pretentiousness of machine-made furniture. Somehow his little book sparked a passionate response, and the so-called Eastlake style soon raged like an epidemic throughout England. It crossed the Atlantic and took hold of American decorators, architects, and furniture makers. Car builders became equally intoxicated with the new fashion. But Eastlake's notion of purity in design was soon perverted as manufacturers tooled up for mass production, and it became a cliché by 1884.[321] Car interiors decorated in this style were criticized for being as angular as a carpenter's square. The Reading's round-end cars, with their angular, beaded, chamfered, and notched interiors, were compared to "a Swiss cottage on wheels."[322]

Despite such wholesale disparagement, elements that can be classified as Eastlake continued to appear in such praiseworthy efforts as the oak-interior cars of the Pennsylvania Railroad. Production of this series of coaches was begun in 1878 under the classification PD. The interior decoration was a conscious effort on the part of the car department's management, according to the *Railroad Gazette,* to "abandon the old styles of barbaric finish."[323] There was to be no jugglery—no wooden puzzles, no veneers, no fancy imported woods, and no superfluous molding. American oak or ash was the basic paneling material used. There was a clean, republican honesty about the selection of a native wood. It was common, yet cool and elegant. Its strong grain and light color made it right for a coach interior. Mahogany was considered handsomer, but it was a relatively soft wood not well suited to the rough-and-tumble service of coach traffic. Mahogany darkened with age, a characteristic disliked by some Victorian car builders, such as William Voss, yet even its critics admitted that it remained the most popular wood for parlor and sleeping-car interiors.

The Queen Anne style, an outgrowth of Eastlake's teachings, was the basic architectural form followed in class PD cars. Shallow, machine-cut beads, rosettes, and diapered work called for little hand finishing except in the window panels, where floral patterns were modeled on delicate reproductions of oak leaves, bulrushes, and mountain laurel (Figures 5.78 and 5.79). The patterns varied from one group of cars to the next, and patterns were alternated from panel to panel to relieve the repetition. Yet each piece of cabinetry was standard in size—a necessity because of the hundreds and perhaps thousands of coaches (classes PD, PE, and PF) built on this general plan by the Pennsylvania between 1878 and 1891.

ment in general became ever more baroque until the wooden car itself passed away.

The revolt against flashy burls and overdecorated details condemned the painted head cloths in particular calling them "common and vulgar in their gaudy paint and large figures to spoil the simplicity of the wood finish."[320] Plain, thin, wooden head linings were praised as chaste and harmonious. Beautiful as the new form was considered, some difficulties were experienced. Wooden head linings were built up in thin veneers to prevent splitting, but a leaking roof would loosen the glue and destroy the ceiling.

Ironically, one victim of the reform movement was the English

Figure 5.80 Luggage racks fastened above the seats were generally made of brass or bronze. This example dates from 1870. (Lyles Railway Manual, 1870–1871)

West Broadway Wire Works.
SPARK NETTING,
For Locomotives, Passenger Car Racks, etc.

The discipline of the designers was relaxed enough to allow the mixture of two other contrasting woods in class PD coaches. The window sash and blind frame were made of cherry, and the slats of the blinds were of selected bird's-eye maple. The head cloths were painted in unobtrusive patterns of deep red, drab, or grayish brown. The seats were covered in golden brown or cherry red. The hardware was polished bronze. For the times, these cars were considered models of restraint and good taste.

Much if not all of the actual interior design for coaches was left to the anonymous draftsmen or cabinetmakers. The decoration of first-class cars, however, was thought too important to be trusted entirely to car shop employees and their uninspired layouts, which were copied from a standard handbook of ornament or from architectural style manuals. By the 1880s the best palace cars had interiors created by the top-ranking decorators and architects. Consultants with famous names were undoubtedly costly, but the publicity received was considered a fair return on the money. In 1881 the Wagner Sleeping Car Company was having its interiors styled by the prestigious New York decorating firm of Pottier and Stymus.[324] A few years later Wagner hired Louis C. Tiffany, "an artist unsurpassed in this line," to decorate its new Chicago Limited; "proof that the Company is determined to turn out a train unequaled in appointments by any in the world."[325] Pullman's consultant was the architect Solon S. Beman, whose monthly retainer amounted to $5,000 by 1892.[326]

Railroads and car builders also sought expert advice. The Boston and Albany turned to the architect H. H. Richardson, the master of Romanesque revival, not only for stations but also for suburban car designs. His contemporary Bruce Price (1845–1903), who was famous for his churches and hotels, also devoted a great deal of time to car architecture. In 1884 he developed the bay-window parlor car for the Pennsylvania.[327] In addition, he worked with the Boston and Albany and with such car builders as Wason and Barney and Smith.[328] His dark, Byzantine interiors, while considered refined and restfully elegant (rather like a brown study), were regarded as too melancholy for the average American traveler. The Canadian Pacific obtained a more cheerful design from an imported French decorator, who installed flamboyant interiors in the Louis XV style.[329]

The educated taste of professional architects and decorators was meaningless without the encouragement of industry leaders. During the nineteenth century such a patron was found in George M. Pullman. In his own words:[330]

I have always held that people are greatly influenced by their physical surroundings. Take the roughest man, a man whose lines have brought him into the coarsest and poorest surroundings, and bring him into a room elegantly carpeted and finished, and the effect upon his bearing is immediate. The more artistic and refined the external surroundings, the better and more refined the man.

Pullman's interpretation of elegance resulted in a superabundance of ornament, for he began with the assumption that more was better. As a former cabinetmaker he had a love of beautiful woods and elaborate marquetry. While others pointed the way to

chastened oak interiors, Pullman remained loyal to the styles of the early 1870s. In fact he favored still more elaboration, and his influence was considerable because he oversaw the construction of the great majority of first-class cars in America.

That Pullman correctly assessed the public's liking for ostentatious finery seems undeniable. Many years later, H. L. Mencken said that no one ever lost money by underestimating American taste. And for generations leaders of the American literary and art worlds heaped ridicule on the Pullman palace car. It symbolized all that they despised about American life in the gilded age. The novelist William Dean Howells gave it a heavy broadside in his *Letters of An Altrurian Traveler* (1893–1894). Musing over moral and physical ugliness, he wrote:

But, in fact, do not the two kinds of ugliness go together? I asked myself the question as I looked about me in the ridiculous sleeping-car I had taken passage in from Chicago. Money had been lavished upon its appointments, as if it had been designed for the state progress of some barbarous prince through his dominions, instead of the conveyance of simple republican citizens from one place to another, on business. It was as expensively upholstered as the bad taste of its designer could contrive, and a rich carpet under foot caught and kept whatever disease-germs were thrown off by the slumbering occupants in their long journey; on the floor, at every seat, a silver-plated spittoon ministered to the filthy national habit. The interior was of costly foreign wood, which was everywhere covered with a foolish and meaningless carving; mirrors framed into the panels reflected the spendthrift absurdity through the whole length of the saloon.

Edward Bok, editor of the *Ladies' Home Journal*, used the Pullman car as an example of how not to decorate the home.[331] He crusaded to cleanse America of its penchant for filigree, turrets, figured carpets, and all other machine-made gewgaws (Figures 5.80 and 5.81). He wanted to raise the level of taste above frippery and the carnival arts, and he believed that railroad cars should be no exception. A train ride, he said, is about the only time that the average person finds free for meditation, and it should not be ruined by the absurdly busy trappings of the average parlor car.

By 1900 the relentless criticism of Howells, Bok, and other American Brahmins was having an effect. Popular humorists like George Ade described palace cars as a "chambermaid's dream of Heaven." They were part steamboat, part vestry, and part cuckoo clock (Figure 5.82). Even the trade press joined the attack. In its obituary notice of George Pullman, the *Railroad Gazette* praised his life and work but deplored the misapplica-

No. 5.

No. 1, Back	. . . .	5 inches high, 5 inches wide.
No. 3, "	. . . .	3½ " " 7 " "
No. 5, "	. . . .	5 " " 7⅞ " "

Figure 5.81   Late-nineteenth-century hardware was generally highly decorated cast bronze. The device shown here is a comb and brush rack. (Dayton Manufacturing Company catalog, 1888c.)

tion of money and labor in car decoration and Pullman's bad influence on public taste for so many years.[332]

The reaction against Pullman's excesses was in full swing by 1905, when the reformers who set out to cleanse the spirit of American decorating were poetically advocating the mission style (Figure 5.83). The railway industry was swept along with this popular movement. An uncompromising austerity banished all unnecessary detail. Plain, unadorned furnishings and bland colors were combined to create an understated elegance. Fusty damask curtains and ancient wooden slat blinds were replaced by cloth roller blinds.[333]

Very plain wooden paneling came into vogue at the same time. Steel car interiors were painted in muted shades of grey, green, and cream. Pullman favored wood graining to soften the stark metal interiors. The wood patterns reproduced—a process that required fifteen days—were subtle and without fanciful highlights or imitation inlays.

For the next twenty years car decoration remained spartan. By the middle twenties, however, car designers and the traveling public seemed a trifle bored with the monotonous appearance of the steel cars, inside and out. Perhaps there had been an overreaction to the excesses of the Victorian era, and the railroads should devise some more imaginative scheme of decoration (Figure 5.84). Early in 1924 the Baltimore and Ohio introduced colonial interiors for some new dining cars.[334] Hepplewhite chairs, fanlights over the windows, and Georgian pilasters added a comfortable element of pleasure to railway travel. Woodcarvers were again in demand as the Chesapeake and Ohio outfitted an entire train in the same fashion. Pullman happily returned to its traditional ways, though with considerably less abandon than in former times.[335]

The final era in railroad car decoration began in 1933 with the two aluminum cars shown by Pullman at the Century of Progress Exposition. They looked like no other cars seen before by the American public, and their interiors were designed in a startling new fashion that was called "modern Empire" for lack of a better name.[336] The seat frames were of square, polished aluminum tubes; the upholstery was jade green. Indirect lighting came from a gutter-like trough above the windows. Polished strips of aluminum were used for trim, and the floor was covered with marbleized rubber tile. In a complete break with the classical models which had been the basis of all decorating styles, streamlining thus came to the railroad car.

Streamlining as a decorative style was an outgrowth of art deco, which itself was introduced at a 1925 Paris exposition on the decorative arts. The intent was to unite art with industry. Hard lines, geometric patterns, and flowing curves characterized the radical new style (Figure 5.85). It was meant to be fresh, pure, and metallic. Futuristic materials like plastic, aluminum, and stainless steel emphasized its break with the past.

The railroads' enthusiastic acceptance of streamlining has been described in Chapter 2. Their decision to hire the best talent available indicates the depth of this enthusiasm, particularly during the formative years of the streamlined style.[337] The leading industrial designers—Henry Dreyfuss, Raymond Loewy, Walter D. Teague, and Paul P. Cret—were engaged to work out the shape, color, and decorative treatment of passenger cars. Some worked directly for the railroads, while others were engaged by car builders. The stylish trains created by these men produced revenues that confirmed the importance of appearance: good car decoration was clearly not a frivolous indulgence.

In the decades of the 1940s and 1950s streamlining became the convention. As more modern trains entered service, the streamlined effect became less dramatic and the railroads' interest in devising new designs seemed to dwindle. Except for top-ranking name trains, design was entrusted to lesser talents already on the payroll of the major car builders. The cessation of large-scale new passenger car orders in and around 1955 ended the incentive for more work in this area. In fact, it would be difficult to name any striking advance made in passenger car decoration during the past thirty years. A coach of 1940 is not materially different in appearance from coaches commonly seen in service today.

Passenger car painting during most of the nineteenth century required an extraordinary amount of labor and time to produce a durable, high-gloss finish. In the late 1860s a full ninety days, some of them devoted to the fancy decorative work, were considered necessary for a first-class job.[338] By the early 1880s sixty days were required for quality results.[339] Some slipshod painters turned out cars in as little as fifteen days, but most car builders agreed that good work could not be done in this time. Experienced men knew that haste guaranteed poor results; there were no shortcuts. A proper painting schedule included the following steps:[340]

Prime bare wood with raw linseed oil—allow to dry for at least 1 week.

Putty holes.
Begin with one light coat of color—dry for 12 to 24 hours.
Put on two coats of "rough stuff" (filler).
Apply one light coat of color—dry for 12 to 24 hours.
Add three coats of thin color—dry for 12 to 24 hours between each coat.
Stripe and letter.
Apply one coat of rubbing varnish.
Conclude with two coats of finishing varnish—drying time, 96 hours.

Figure 5.82   Pullman interiors were noted for their superfluity of decoration. In 1898 one critic spoke of their "Barbaric splendor and gloomy magnificence." (Pullman Neg. 4690)

It was necessary to sand and rub down (with pumice or rotten stone) between each coat. The drying time between the successive coats listed above was figured at 288 hours.[341] This horrendous labor was justified by the slick, waterproof finish produced. Beauty and elegance was required in a passenger car's paint job. In addition, a hard, smooth surface sheds dirt, water, and cinders and is easier to wipe clean than a rough or porous surface.

After the car had left the shop, it was returned once a year for a fresh coat of varnish. Faded or scratched areas were "cut in"

with new color, and the entire surface was rubbed out and varnished. Every seven to ten years it was necessary to burn off the old finish to the bare wood and then repeat the steps listed above.[342]

Car painters argued about the life of dark- and light-colored finishes. Some claimed that dark color attracted the sun's heat, causing the varnish to crack and alligator, and that it came off with the varnish during the annual rubdown.[343] Light color could be made from lead-base pigments, the most durable paint then available. But the enemies of light color said that it must be

Figure 5.83   Plain mission-style interiors came into fashion after 1905 in reaction to the elaborate palace car decor of former years. (Pullman Neg. 9254)

applied thickly and tended to crack. They also argued that lead paints "suck up" the varnish.[344] Fashion rather than durability, however, was the main factor in color selection.

The painter's workbench was the conservative stronghold in every car shop. No one seemed more resistant to new methods than the master car painter. He looked on the products of commercial suppliers with contempt; such materials were possibly fit for the household, but had no place in his domain. He boiled his own linseed oil and japan, ground his own pigments on a screeching dry mill, and hand powdered the fancy colors for lettering on a stone slab. The paint shop "was a Tartarus of hideous smokes and smells, and possessed many dark secrets over which the boss painter was the presiding alchemist."[345]

Slowly the old prejudices broke down. Until the end of the Civil War few car painters would accept anything but English

varnish, but they grudgingly began to use American varnish and by 1881 it had become the favored top coating.[346] A few years earlier commercial primers began to find acceptance, because they saved time and much of the heavy sanding necessary with "rough stuff."[347] The old guard insisted, however, that this shortcut produced a less durable finish. During the 1890s experiments were conducted with colored varnish and enamel. The Burlington switched to a four-coat enamel paint job without the usual varnish topcoat, a method which was heatedly criticized in the car painters' column of the *Railroad Car Journal* in April 1896.

A far greater heresy was announced the next year: William P. Appleyard, master car builder for the New Haven, proposed a plan for an outer sheathing of sheet copper that would entirely eliminate the need for exterior painting.[348] Thin copper (no. 30

Figure 5.84   The severity of the mission era gave way to more sprightly painted interiors during the 1920s. This car is decorated in the Grecian manner. (Pullman Neg. 28991)

gauge) was wrapped around each window post, sheathing plank, letter board, window frame, and every other exterior stick of wood. The sheets were carefully rolled around the pieces so that no edge was left exposed. It took only ten days to sheathe a car—a great saving in time compared with the traditional hand-painting method. The new system required 1,000 pounds of copper, which represented no serious increase in dead weight, since 275 to 300 pounds of paint and varnish were needed to do the same job. (Thirteen years earlier the Southeastern Railway in England had anticipated Appleyard's plan by paneling a car in sheet copper.[349] Its silver-plated moulding and lettering made it the only true "silver palace" car in existence. But apparently the

British experiment did not spawn a large family, nor did it employ Appleyard's wraparound technique.)

Appleyard claimed other advantages for his scheme. Copper was a durable, nearly permanent material proved by centuries of use for gutters, flashing, and so forth. It was maintenance-free. No undercoating or preparation was needed. Cheaper grades of wood could be used. It had a high scrap value, recoverable when the car was retired. A handsome, glossy, brown-green surface could be produced by skillful anodizing. Because annual varnishing and periodic repainting were eliminated, cars would spend more time in revenue service.

Early in 1897 the first copper-sheathed New Haven coach,

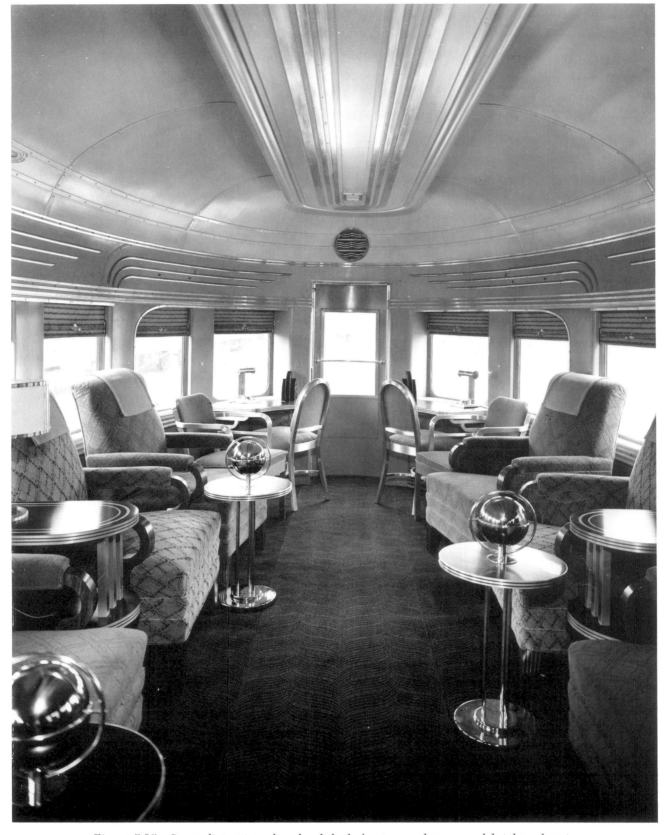

Figure 5.85 Streamlining introduced polished aluminum, chrome, and bright colors in the 1930s. (Pullman Neg. 39615)

Number 365, entered service.[350] After a year's test, two more New Haven cars were sheathed. The neighboring Fitchburg Railroad encased several of its cars in Appleyard's copper mantle. Predictions that it would wrinkle like a rhinoceros hide as wood expanded and contracted did not discourage the Boston and Maine, the Erie, and even the Southern Pacific from trying Appleyard's plan. New York's Interborough subway purchased

500 wooden cars with Appleyard's sheathing, but diverted them to elevated service after an accident in 1905 proved that they were not fireproof.[351] The New Haven itself had 800 to 900 copper-covered cars, but with Appleyard's death in 1905 the scheme lost its chief promoter. Some of this equipment remained in service until about 1940.[352]

After the initial excitement had subsided, old-line painters

Figure 5.86   Waterbury and Atwood patented a vestibule in 1854 that was actually used on several Eastern railroads. (American Railroad Journal, *July 1, 1854*)

were pleased to discount copper plating as a "red herring"—not at all the important improvement that had been claimed. However, the traditionalists soon faced a competing method that was not so easily put to rest. Compressed air was finding many uses around the repair shops. Already it powered riveting hammers, chipping guns, and wrenches; next some genius would think of a way to paint pneumatically. The earliest case of spray painting applied to railway work has been described by the prominent railway journalist Angus Sinclair, who visited a Southern Pacific engine house near San Francisco late in the 1880s that was in the process of spray painting.[353] He was told that the scheme would next be applied to freight cars. The use of air guns for painting some of the Columbian Exposition buildings in 1893 gave the technique even greater exposure in the mechanical trades. Many railroads adopted it for freight car finishing. But although spray painting came to be considered suitable for rough equipment, maybe even for passenger car trucks, it was certainly not regarded as a technique to use on the superstructure. The spray gun was only slowly accepted for passenger equipment; the brush and varnish pail were to remain the principal implements of the painter for many years.

Other innovations were adopted more readily. The most important was baked enamel finishing, which was first tried in January 1913 by the Pennsylvania Railroad.[354] A normal varnish job required attention every eighteen months, but the baked car still looked good after forty months of service. Painting time was also reduced from fourteen to six days. The baking time was three hours at 150 to 160 degrees F. The following year the B & O began testing baked finishes, but the road insisted that a good job could not be done in less than twenty-three days and that two coats of varnish were a necessity.[355] All agreed that the faster-drying oven jobs were not only more durable but smoother, because dust had less opportunity to settle on the surface.

Spray painting for large objects was facilitated by the introduction of the canopy exhaust hood. Powerful fans removed overspray and dust, improving the quality of the paint job and at the same time lessening health and fire hazards. The Milwaukee Road installed such a unit in 1931 and achieved estimated savings of $12,000 to $15,000 yearly through reduced labor costs.[356] Spraying consumed nearly 10 percent more paint than brushing, but the work went five times faster. However, the Milwaukee remained loyal to oil paint and varnish finishes. Pullman, after a fleeting enthusiasm for enamel and lacquer, also stood by oleoresinous paint topped with varnish.[357] For hard outdoor service no better finish could be found, in the opinion of the Pullman maintenance force. With a solid, red-lead base, seven finish coats, and fourteen days' labor, a Pullman finish would last for two years, or about 360,000 miles of running through sun, rain, snow, salt air, and alkali dust.

Synthetic lacquers were soon to overthrow old-fashioned varnish finishes. The automotive industry had been using the tough, fast-drying synthetics with great success since the 1920s, but it took the railroads almost twenty more years to fully accept the new technique.

Vestibules

Vestibules form an enclosed hallway between the cars. The accordian-like appendages, called diaphragms, at each end of the vestibules unite the individual cars of the train into a single unit, just as the couplers and drawbars hold it together physically. Passengers and crewmen alike can pass from one end of the train to the other, safeguarded against falling under the wheels and fully protected from the elements and the train's own smoky trail.

Figure 5.87 One of Pullman's engineers, Henry Sessions, reintroduced the vestibule in 1887 and was mistakenly given credit for inventing this device. (Car Builders' Dictionary, 1895)

Figure 5.88   Narrow vestibules prevailed until 1893. Most featured decorative end hand-rails. (Hall of Records, Dover, Delaware)

Pullman is popularly credited with inventing the vestibule, but like so many widely held beliefs about inventors, the facts do not bear this out. The experiments and applications that were made before Pullman form a long list. In 1836 W. Bridges Adams, an English inventor with many schemes for both highway and railway vehicles, patented a connecting passageway between cars.[358] Eleven years later Sir Henry Bessemer developed a vestibule in connection with experiments on high-speed trains. The device was actually intended to reduce wind resistance by filling in the space between the cars and making the train a reasonably uniform tube. S. R. Calthrop obtained a U.S. patent in 1865 with a similar purpose in mind, but his vestibule was also designed to accommodate human traffic between the pleated folds of the enclosure.

Ideas and patent papers, however, can be partly discounted as speculative rather than practical; the actual testing of vestibules on full-sized trains seems to have been the work of two New England railway employees, Charles Atwood and Charles Waterbury. Waterbury, the superintendent of the Naugatuck Railroad, obtained a patent in 1852 for platform canopies. The main purpose of the invention, particularly as explained in a second patent of 1855 taken out in conjunction with Atwood, was actually ventilation rather than passage between cars. The vestibules made the train into something of a wind tunnel, the air being scooped in at the open front of the baggage car (Figure 5.86). However, one of the patent claims specifically mentions safe transit between cars. The device was tested on the Naugatuck Railroad in the summer of 1853.[359] Over the next year or two it was tried on the Housatonic, the Erie, and the New Jersey Railroad and Transportation Company.[360] For all its originality, the plan did not work out well because of unequal platform heights and slack couplers. Also, the cars were said to become insuffer-

Figure 5.89   Full vestibules offered more room for loading passengers and luggage. (Pullman Neg. 31680)

The Wagner Company, which could not afford to be outclassed by Pullman, inaugurated a new vestibule train for the Lake Shore route with much fanfare late in 1887.[366] A vestibule craze swept the industry, stimulated by the enthusiastic public response. By June 1887 the Pennsylvania had three vestibule trains in service, with more on order. The Santa Fe, the Chicago and Alton, and the Rock Island were quick to respond. In August 1888 the B & O announced that all its best cars would be vestibuled within sixty days.[367] Apparently, however, industry-wide conversion of name-train equipment was not completed for another decade.[368]

Vestibules were primarily installed in order to improve accommodations for first-class passengers. Supporters of the vestibule also claimed that the friction of the diaphragm face plates steadied the ride by retarding the swaying of the car. The B & O considered this a very good reason to press ahead rapidly with vestibules because of its serpentine line through the Alleghenies. The vestibule was also said to be a safeguard against telescopes. But the claim that it was effective up to speeds of 40 mph was undoubtedly an exaggeration.[369]

The vestibule had critics as well. Some roads found it troublesome and expensive: cost estimates ranged from $400 to $1,800 per car, depending on the materials, ornament, and patent fees involved.[370] While end platform enclosures were useful in keeping heat in and cold air and dust out, they also blocked the free flow of air when it was most needed during the summertime. Some managers considered the stabilizing effect questionable; they did not believe that the friction created by springs and relatively small face plates had any substantial effect on the sway of a 30-ton car.

During the vestibule's period of introduction a legal battle developed between Pullman and Wagner. Only days after the first Wagner train entered service, Pullman began suit for patent infringement.[371] After hearing the evidence, the U.S. District Court in Chicago ruled that Pullman could claim rights only to the equalized tension spring covered in Sessions's patent of April 29, 1887 (No. 373,098)—not to the vestibule concept in general. Wagner was enjoined to remove the spring apparatus and did so willingly, noting that such devices were not really necessary on the smooth "Water Level Route," because the vestibule ends stayed together of their own accord. Unsatisfied, Pullman instituted two more suits in 1889 and 1890 and in the latter case succeeded in gaining a broader reading of its other patents. Wagner was given forty days to remove the vestibules from the 205 cars it had already fitted.[372]

To counteract this monopoly by preparing alternate designs, Wagner's chief mechanic, T. A. Bissell, devised a vestibule diaphragm with built-in springs that created the necessary end pressure. The steel face plates were replaced with soft, horseshoe-shaped pads. In 1892 he obtained a third vestibule patent that used a lever or pendulum face-plate pressure device.[373] This design was manufactured by the Gould Coupler Company of Depew, New York, and was widely used on all New York Central properties, including the Wagner line of cars. Meanwhile

ably hot when the train was stationary. After a few years the experiment was abandoned.[361]

The vestibule reappeared occasionally over the succeeding years. Between 1868 and 1877 the Burlington is reported to have used some form of car-to-car passageway.[362] The Michigan Central stretched canvas roof canopies between its cars, but the sides remained open. In 1875 the Lake Shore's fast mail train had a primitive sort of vestibule arrangement to enable the clerk to go from car to car.[363] These wooden partitions also discouraged tramps and robbers from riding the platforms. In 1882 Vanderbilt's twin private cars were joined by enclosed platforms. During the same year Pullman was reported to be fabricating a train for the Nickel Plate Road with enclosed platforms, "thus making it practically one continuous car."[364]

In about 1885 Henry H. Sessions, superintendent of Pullman's Chicago plant, began to investigate the vestibule problem. The most casual search would have revealed that the basic idea was an old one; thirty-four U.S. and British patents existed, in addition to the actual tests already described. But when the first cars to be fitted with Sessions's vestibules left the Pullman shops in April 1887, the scheme was proclaimed an innovation in railway travel (Figures 5.87 and 5.88).[365] The cars were operated by the Pennsylvania Railroad on a New York–Chicago express after a trial run on the Illinois Central. And while the claims for originality were unfounded, the vestibule train created the desired sensation. It became the talk of the traveling public, and for a time the name "vestibule" was used to indicate an extra-deluxe train.

Figure 5.90   At stops, it is necessary to raise the trap door (visible at right) to gain access to the steps. (B & O)

Figure 5.91   Full-width diaphragms improved the appearance of the train and had some effect on wind resistance, but they were expensive to install and maintain. (B & O)

Wagner appealed the 1890 decision, and early in 1892 a Chicago judge set aside Pullman's short-lived vestibule monopoly.

A year later Pullman produced an entirely new vestibule design.[374] Instead of the narrow aisle enclosed by double-hinged doors, the new wide vestibule extended the width of the platform. Trap doors covered the stairwells. Large side doors offered easier access, and the additional floor space made room for extra luggage and a place for a polite trainman to step aside for passengers. The vestibule became less of an appendage and more of an integral part of the car body. The narrow diaphragm passageway, however, remained unchanged. Essentially the wide vestibule patented on June 5, 1893, by Pullman is the same as that used on modern cars (Figures 5.89 and 5.90).

In later vestibules parts that were formerly wooden became

steel, but the rubberized fabric of the first Sessions diaphragm continued to be used for generations. On some classes of cars, notably dining, baggage, and mail cars, the vestibule and steps were eliminated from one or both ends. Diaphragms and end doors were retained so that passage between the cars was not interrupted (Figure 5.91). By the mid-1930s designers recognized the practicality of saving space by eliminating one vestibule in coaches, and Colonel E. J. W. Ragsdale of the Budd Company estimated that it saved costs of $4,000 a car.[375] He also opposed the outer diaphragm covers that were popular with some designers because they hid the unsightly end space of the exposed diaphragm. While they did unite the train of cars as a smooth tube, the covers were such a maintenance headache that Ragsdale started an "anti–outer diaphragm society."[376]

CHAPTER SIX

Head-End Cars

BAGGAGE EXPRESS, COMBINATION, EMIGRANT, AND MAIL CARS

BECAUSE BAGGAGE, MAIL, AND EXPRESS CARS are commonly placed at the front of the train, they are called head-end cars. They are the least elegant members of the passenger car family, so humble that in some way they resemble boxcars. Their proletarian connections are confirmed by their freight car–like side framing: since they have no continuous row of windows, they are built with full trusses and, like stocky peasants, tend to be shorter than the more aristocratic coaches, sleepers, and parlors.

Consistent with their utilitarian purpose, head-end cars are often plainer in finish, less influenced by changing fashion, and longer in service life than other cars in the passenger train. The trade press has traditionally regarded them as unnewsworthy and rarely refers to them, except for an occasional drawing or photograph accompanied by a caption that gives the basic dimensions. Because these cars are so poorly documented, it is necessary to depend almost entirely on surviving illustrations to trace their history. The evidence is especially weak in the early period, and since photographs are variously identified, hard-and-fast dates are not always available. All these pictures cannot be reproduced here, but most of them are at least cited in the notes to this chapter.

Baggage Express Cars

Baggage is integral to the whole business of passenger transport. Few travelers go on a journey without a bag or at least a package; even the commuter has his briefcase. Since the beginning of public transportation, carriers have been expected to make some provision for baggage. The earliest stagecoach lines set a limit of 14 pounds per passenger and charged a fee for baggage over that weight.[1] Steam railroads, however, could afford to carry extra baggage, and to win passengers from the highway, they offered to carry it free. Here was a service unique to America, and one that drew praise from European travelers. Passengers took their small grips on board with them, but larger pieces were placed in the baggage car. The traveler was given a brass check in receipt, and to the astonishment of most foreigners, the railroad assumed responsibility for the luggage in addition to transporting it without charge. To hold this traffic within some bounds, most lines agreed only to the free carriage of clothing; otherwise their lucrative express trade might disappear.[2] It was also necessary to place a top limit on weight—usually 100 pounds per passenger. In 1855 the General Ticket Agents Association began to suggest that 100 pounds be made the uniform free poundage limit, which apparently most railroads were willing to accept as a national standard.[3] By 1870 the limit was raised to 150 pounds and remained at that level for many years.

Generations of railroad managers must have looked upon baggage service as a costly nuisance that was unfortunately necessary in a competitive business. Large quantities of baggage during seasonal rush periods disrupted the smooth flow of train operations, whereas mail and express brought in extra revenue.

In the palmy days of the passenger train this additional income may have seemed incidental, but as passenger traffic itself declined after 1945, head-end earnings became a major source of revenue. Before a Federal subsidy was granted in 1971, mail and express business was vital to American passenger train operations.

Perhaps the earliest published reference to an American baggage car appeared in the January 19, 1833, *American Railroad Journal*. On the front page a paragraph reported a fire, presumably started by a spark from the locomotive, that consumed a Newcastle and Frenchtown baggage car and its contents, including $60,000 in greenbacks. The paragraph did not describe the car itself. Later that year the Baltimore and Ohio Railroad's annual report carried the following statement:

The conveyance of baggage on the tops of passenger-cars has been attended with so much labor and inconvenience in loading, unloading and shifting the baggage at so great a height, added to the manifest injury done the cars in the hurry and bustle of these operations, that the necessity of some improvement on the present plan is obviously indispensable. It is proposed to construct some cars and tenders to the passenger-cars for the exclusive purpose of carrying baggage and in constructing them to make two distinct apartments, one for the baggage going to the ultimate point of destination, which will remain undisturbed for the whole trip, the other for the accommodation of way passengers.

The Philadelphia and Columbia tried to overcome the problem of storing baggage by building possum-belly compartments under the floors of cars.[4] These compartments had side doors and, being close to the ground, were convenient to load and unload. In 1838 it was reported that the Morris and Essex Railroad had a similar car in service.[5] Most railroads, however, concluded that a separate car for baggage was more practical; it had greater capacity and was undoubtedly cheaper than fitting luggage compartments to every coach.

Little is known about the earliest baggage cars. Railroad annual reports list them but rarely give particulars beyond the fact that they had four or eight wheels. The Austrian engineer F. A. Ritter Von Gerstner named a number of lines that had eight-wheel baggage cars when he traveled in America from 1839 to 1840, but again with few details. Possibly the best picture of a first-generation head-end car is the engraving in Figure 6.1. It might be dismissed as just another fanciful printer's engraving, except that the locomotive is a faithful drawing of a Locks and Canals locomotive delivered to the Boston and Worcester Railroad in 1839. Since the entire engraving seems to be the work of the same artist, the rest of the train can be accepted as reasonably accurate. Another printer's train cut showing a larger four-wheel baggage car with outside bracing is reproduced on page 132 of August Mencken's book *The Railroad Passenger Car*. Judging by its mechanical details, the locomotive shown in the engraving would date from 1850, certainly no later than 1855. Yet the engraving is printed on a broadside dated November, 1858, and cuts of the same vintage were reproduced much later

Figure 6.1   The earliest style of four-wheel American baggage car is illustrated in an engraving of a Boston and Worcester train of about 1839. (Smithsonian Neg. 60600)

Figure 6.2   The first eight-wheel baggage cars were little more than boxcars with end platforms. This Old Colony Railroad train was photographed about 1855. (Chaney Neg. 5911)

in the nineteenth century. Caution must therefore be used when citing such illustrations as evidence of dates.

One of the oldest actual photographs of a baggage car is shown in Figure 6.2, which pictures a train on the Old Colony Railroad around 1855. The locomotive *Randolph* was built in 1846 by Hinkley, and although the age of the baggage car cannot be determined, it may have been no newer than the locomotive. An 1842 engraving of a Western Railroad train reproduced back in Figure 1.19 was made from a daguerreotype. The arched-roof, eight-wheel baggage car is similar to the Fall River car except for the lack of end platforms and doors. Another double-truck mid-century baggage car, shown in a lithograph of a scene on the Georgia Railroad, has end platforms, a central side door, four side windows, and the lettering on either side of the center door: "U.S. Mail" and "Baggage."[6]

Figures 1.21 and 1.22 in Chapter 1 provide two other pictures of baggage cars from the 1850s. And Figure 6.3, reproduced from A. Bendel's 1862 volume on American railroads, furnishes a scale drawing of a New York and Harlem car. It measured only 46 feet

over the platform ends; notice the extremely short 4-foot wheel-base of the trucks. During his visit of 1859 Bendel, an official sent by the German government to study American trains, was impressed by the popularity of combination express and baggage cars. He saw few full baggage cars in service, reporting that most were divided into at least two compartments.[7]

While conventional baggage cars served the needs of most railroads, lines that had steamship or ferry connections sought a better system. The transfer of luggage at such terminals was laborious and time-consuming. As early as 1839 (and perhaps even earlier) several railroads began to pack baggage in containers to expedite the transfer. Trunks, parcels, hatboxes, and bags were placed in crates measuring roughly 7 feet square by 5 feet high. The crates or luggage boxes were mounted on small wheels. As many as five could be carried on a flatcar. A few lines labeled individual crates for major stations and picked them up or dropped them off en route.[8] There was no waiting for the head-end crews to find and unload each bag; the crate was quickly deposited on the platform, the local agent distributed the

Figure 6.3 A New York and Harlem baggage car drawing made by a visiting German engineer in 1859. (A. Bendel, Aufsätze Eisenbahnwesen in Nord-Amerika, *Berlin, 1862, plate 15)*

Figure 6.4 By the late 1860s, baggage cars took on a more finished appearance. The car here was produced by Jackson and Sharp around 1867 for the Richmond, Fredericksburg, and Potomac Railroad. (Hall of Records, Delaware Archives)

individual pieces, and the empty crate was ready to be filled for the next train. The system was even more useful for through passengers on lines that had a rail-water portage. To ride the Erie, New York passengers were obliged to take a steamer up-river to Piermont. Evidence that the crate system was employed is available in a contract made with Davenport and Bridges in September 1840 for the necessary equipment.[9] The Camden and Amboy used baggage crates on its land-sea New York–Philadelphia route, as did the Philadelphia, Wilmington, and Baltimore Railroad at its Susquehanna River ferry crossing. The Stonington Line undoubtedly carried baggage in crates longer than any other American railroad. A line of fast steamers took passengers from New York to Stonington, Connecticut, where they transferred to the New York, Providence, and Boston Railroad. During his visit of 1839, Von Gerstner was impressed by the efficiency of the baggage crate system at Stonington. The boxes, fitted with 15-inch-diameter wheels, were easily managed by the deckhands. The competing Fall River Line also used the container system. In time both the Fall River and the Stonington lines were taken over by the Old Colony Railroad, whose management thought so highly of the scheme that a baggage crate car was included in the engravings on its stock certificate.[10] Reference was also made to the system in the railroad's annual re-

port of 1853. By 1896 the scheme was abandoned because of a change in operations.[11]

While the crate system was prophetic of what would one day revolutionize merchandise transport, it was not to become the prevailing system of handling baggage in America. Open-compartment baggage cars were found more suitable for this work, and their development closely parallels that of other passenger equipment. By the late 1860s the pictorial evidence shows more elaborate head-end cars, with clerestory roofs, arched windows, and doors (Figures 6.4 and 6.5). They seemed less alien to passenger service than their roughly finished ancestors, for the rustic outside body framing had been abandoned. Architecturally the head-end cars blended with the rest of the train, forming a suitable introduction to the elegant palace cars trailing behind. They were shorter and had fewer windows, but in elegance of form they duplicated passenger cars. In 1880 the Santa Fe brought out a new standard baggage car of graceful pattern.[12] The body measured 50 feet, 1⅝ inches by 9 feet, 9⅝ inches. The floor frame was composed of six 5- by 8-inch sills. The fine-looking Tudor arched doors offered a 6-foot, 2-inch by 4-foot, 4-inch opening (Figure 6.6). The normally utilitarian function of such rolling stock was occasionally upgraded for a special occasion, as when the Milwaukee Road outfitted a baggage car

Figure 6.5   The Cleveland, Columbus, and Cincinnati Railway built this clerestory-roof car in its shops in 1865. (Railway and Locomotive Historical Society)

as a ballroom car to enliven an excursion.[13] Carpets were laid and folding chairs were stationed around the side walls, which were hung with pictures and garlands of evergreens and roses. A grand piano and an eight piece orchestra played in one corner for the dancing excursion party.

A modern counterpart of the ballroom car can be found in the ski-train bar cars operated by such lines as the Denver and Rio Grande Western. In 1953 this road refitted an aging baggage car with ice lockers, stand-up counters, and other basic equipment for a primitive saloon. Holiday spirits were in plentiful supply.

In the final decades of the nineteenth century a material growth was evident in the baggage car's size and weight (Figures 6.7 and 6.8). The Pennsylvania's standard class BA baggage car, adopted in 1881, measured 46 feet long and weighed just over 17 tons. Eleven years later, the new class BC design measured 60 feet and weighed 33½ tons. After the turn of the century 60-foot cars remained popular, but heavier framing and six-wheel trucks jumped the weights to as much as 41 tons.[14] Eliminating the end platforms gained space and saved weight; in 1899 the B & O was reportedly removing them from all its baggage, express, and mail cars.[15] In general 60-foot cars were regarded as the optimum size, although 50-foot cars were acceptable for less heavily patronized trains, and 70-footers were considered necessary on runs involving regular movements of theater scenery.[16] Four doors, two near each end, were also favored because they made it possible to separate baggage and express handling, while both commodities could be piled in the large central space.

The shift from wooden to metal baggage cars occurred at the same time as the conversion of other passenger equipment. In the early years of the twentieth century, steel and steel-under-frame baggage, mail, and combine cars began to appear on American railways. This change had been anticipated in earlier decades by various patented schemes for burglar-proof express and mail cars. Except for the Green and Murison car of 1889 (Figure 2.16), none appear to have reached the full-size proto-type stage. Safety was the most potent argument for fireproof, collision-proof passenger cars ("safety" meaning the safety of paying customers, not of employees). The loss of mail and express shipments, however, entailed major claims—as did lawsuits initiated by widows of railway clerks. Furthermore, no train was fireproof if the head-end cars were not. The Pennsylvania Railroad did not exclude head-end equipment when it banned wooden rolling stock from the tunnels leading to Penn Station. A burning baggage car might not ignite a steel coach, but it could suffocate everyone on the train if the fire occurred inside one of the tubes. The railroads had not only a public duty but also a financial interest in upgrading every type of car. The mixture of wood and steel cars was a lethal blend. The temptation was to put "perfectly serviceable" wooden cars ahead of the coaches because they carried only nonrevenue passengers, but in practice it proved so dangerous that the railway mail clerks eventually mustered Congressional action against it.

As it happened, at the beginning of the steel car era when the first steel coaches were only inventors' dreams, a full-sized steel baggage car was already in service. In 1904 the Standard Steel

Side Elevation and Section.

Bottom of Floor Frame.   Floor and Roof.

End Elevation and Section.

Figure 6.6   A graceful four-door baggage car designed by the Santa Fe's mechanical department in 1880. The body measured 50 feet, 1⅝ inches long by 9 feet, 9¾ inches wide. Miller couplers and paper wheels were used. (National Car Builder, January 1881)

Car Company delivered an express-baggage car, Number 599, to the Erie.[17] The twelve-wheel, 60-foot car weighed 53 tons. The side wall sheets measured ³⁄₁₆ inch thick, and the roof plates were ⅛ inch thick. The car was built with a wooden lining, which was later removed to make it truly fireproof. However, crews complained so much about the lack of insulation that a fire-resistant composite board was installed.

Two years passed before another steel baggage car appeared, this time operated by the Pennsylvania Railroad. But in 1908, the same line had over twenty steel baggage cars in service and was publicly committed to restock its main-line fleet with them as rapidly as possible. In the Middle West the St. Louis and San Francisco Railroad decided to compromise by purchasing composite steel and wooden cars. In 1908 Pullman produced twenty-five baggage cars with distinctive fishbelly underframes (Figure 6.9). All four sills, including the visible outside members, were

deep (28-inch) plate girders. The effect was that of an old wooden body dropped onto a gigantic twelve-wheel steel flatcar.[18] Actually the body was steel-framed, and the wooden sheathing was laminated with sheet metal. The composite cars complemented existing wooden equipment, cost less than all-metal fabrication, and saved 5 tons in dead weight. The Santa Fe purchased a quantity of nearly identical cars from American Car and Foundry Company, but few other lines chose to buy composite cars. New all-metal or steel underframes for existing wooden cars were the most common direction for head-end equipment in the standard era.

Like coaches and sleepers, the new metal head-end stock simply perpetuated orthodox designs in a different material. Four doors and 60- to 70-foot lengths prevailed as before, though weight, strength, durability, and cost all increased. Individual designs continued to flourish after the introduction of steel con-

Figure 6.7   The Pennsylvania Railroad class Bb was introduced in 1887. Large cars like the Bb were needed for well-patronized through trains.

Figure 6.8   The K C S baggage-express car is representative of the final all-wooden cars. Pullman built it in 1901. (Pullman Neg. 5834)

struction. The variations were modest, but each railroad had its own plan. For a brief time the Harriman Lines agreed upon a common standard, and in the years just before World War I it was possible to see new baggage cars of a single design on the Union Pacific, the Illinois Central, and other Harriman roads. The U.S. Railroad Administration developed a series of standard

passenger car designs, but only its plan for a 64-foot baggage car seems to have been fully developed or produced.

The lightweight era represents the final phase in the development of head-end equipment. Low, arched-roof baggage cars were built as part of the dazzling motor trains of the mid-1930s. Full-sized lightweights of aluminum, Cor-Ten, and stainless steel

Figure 6.9   *The Frisco bought twenty-five steel-frame, wooden-sheathed baggage cars from Pullman in 1908. They measured 60 feet, 9 inches long and weighed 48.2 tons.* (Railroad Gazette, June 18, 1908)

Figure 6.10   *The Southern Pacific was a leader in adopting all-steel cars. The Number 6400 was produced by Pullman in May 1911.*

Figure 6.11   *The massiveness of standard-era equipment is illustrated by this Central Railroad of New Jersey baggage car. It was built by A.C.F. in 1923.*

Figure 6.12   *Standard Steel Car Company built the New York Central's Number 3361 around 1925.*

Figure 6.13 Baggage car interiors were open and spartan. The Atlantic Coast Line's 668 was a Pullman product of 1917. (Pullman Neg. 21096)

Figure 6.14 A Pennsylvania Railroad all-steel baggage-express car from about 1925. Cars of this design weighed 43.8 tons and carried 20 tons. (Car Builders' Cyclopedia, 1928)

Figure 6.15   A lightweight baggage-express car built in September 1950 by A.C.F.'s St. Charles, Missouri, plant. (Santa Fe Railway)

soon followed, as discussed in Chapter 2. Style dictated the need for new baggage cars, because nothing ruined the appearance of a streamliner more than hulking, clerestory-roof heavyweights at the head end. In later years some railroads reduced first costs by economizing on the materials in new head-end construction. By using conventional rather than specialty steels, heavier but cheaper cars could be produced. In 1956 the Santa Fe began to acquire "semilight" express-baggage cars.[19] To cut corners, Pullman fabricated the basic body while the Santa Fe finished the cars for service. Several hundred were produced between 1956 and 1964. Other Western lines continued to buy new baggage cars into the 1960s, and in surprisingly large numbers, considering the downward plunge of passenger traffic. Now much of this relatively new rolling stock has been converted to maintenance service, sold, or scrapped.

Combination Cars

No other passenger cars are more difficult to classify than combines. While relatively few in number, they were highly dissimilar in floor plan. Their taxonomy is a frustrating scramble of variations, exceptions, and special cases. The industry was satisfied to define them as passenger cars divided into two or more compartments to accommodate different classes of traffic. Officially they were termed as combined, combination, or occasionally composite cars. Colloquially they were called combines.

Cars that might technically be designated as combines include coach-baggage-mail, lounge-dormitory, café-coach, coach-buffet-parlor, baggage-dormitory-boiler, and coach-baggage cars. It could even be claimed that Pullman was largely an operator of combination cars, for its fleet was laden with buffet-parlors, bedroom-observations, coach-sleepers, and parlor-sleepers. In theory, all the variations of composite mail, baggage, and express cars were also combines. Yet what was commonly considered a combine was a simple coach-baggage car. A small baggage compartment occupied one end, with the larger part of the interior space given over to the passenger compartment. For the most part, this style of car was found on branch lines or accommodation trains. It was a serviceable maid-of-all-work vehicle suited to short hauls where few passengers or little baggage could be expected. On such routes combines were the last car on the mixed trains, which typically consisted of a dozen or more freight cars and a single combine. Combines also facilitated branch-line–main-line interchanges. A composite car could be switched directly onto a connecting main-line train, thus avoiding the transfer of passengers, baggage, and express. And on the main line itself, a combine was often part of the best name trains; high-class combines were found on such specials as the Royal Blue and the Empire State Express.

The passenger compartment of such a combine was normally set aside for smokers. It became a male retreat—a hideaway from wife and family where a man could sit down to enjoy a good cigar in solitude or in conversation with other men. In 1859 a German traveler wrote: "In these cars, smoking, which is considered bad manners in America just as it is in England, and

Figure 6.16   A combination car on a Providence and Worcester train around 1870. The car itself dates from about 1855 to 1860. (Thomas Norrell Collection)

Figure 6.17   The Lake Superior and Mississippi combine Number 2 was built by Harlan and Hollingsworth in 1869. (Traced by John H. White, Jr.)

which is always restricted to rooms specially designated for it, is permitted. This room contains plain wooden seats and is similar to our third class; it is easy to do without a smoke after looking into that room. Wherever such a room is not available, Negroes and smokers have to be satisfied to stay in the baggage room."[20]

The origins of the combine may indeed be found in a desire to separate Negroes from the regular coach passengers. Von Gerstner observed this form of segregation on several American lines during his journey of 1839–1840. He mentions baggage car passenger compartments that were set aside for Negroes on the Petersburg, Baltimore and Susquehanna, and Richmond, Fredericksburg, and Potomac railroads.[21] On the Petersburg Railroad Negroes might ride in the baggage car for half fare but could travel in the coaches if they paid full fare—a surprisingly liberal attitude compared with the policy of Northern railroads, where no blacks were permitted in the coaches at that time. Ironically, the antislavery sentiments so widespread in the North just before the Civil War did nothing to end segregation on board trains.

No illustrations of the first combines have been found; the earliest are a few photographs showing arched-roof cars of the 1850s. In the photograph in Figure 6.16, which appears to date from 1860 to 1870, the rear section was clearly for express, the center may have been for baggage or mail, and the forward end was for second-class passengers. A slightly later but remarkably similar combine is shown in the Harlan and Hollingsworth drawing of the late 1860s in Figure 6.17. This plain car was built for the Lake Superior and Mississippi Railroad. Unfortunately the complete date is missing from the drawing. The L S & M (which eventually became part of the Northern Pacific) started construction of its track in 1868 and opened the first section in the following year. Because the car is numbered 2, it must date from 1868 to 1869.

The most interesting style of combine—the open-side, corridor, or gallery car—also comes from this period. It was a clever scheme to create a separate room on one side of the car and yet allow free passage, so that trainmen and passengers could walk through the car without entering the room itself. Such security was desirable for express, mail, and even ordinary baggage. Photographs show that gallery cars were used on the New Haven, the Central Pacific, and the California Pacific railroads.[22]

Figure 6.18   This Cumberland Valley combine car was built in the road's own shops around 1855. The car is now exhibited by the Pennsylvania State Railroad Museum, Strasburg, Pennsylvania. (Traced by John H. White, Jr.)

Figure 6.19   Combines, like baggage cars, tended to be more elaborately finished after about 1865. The example here was completed in 1872 by Jackson and Sharp. (Hall of Records)

The Cumberland Valley Railroad also had a gallery car, which is now in the Pennsylvania State Railroad Museum in Strasburg, Pennsylvania. It was reportedly built by the Cumberland Valley in its repair shops around 1855. The passenger compartment seats 24, and the total weight is given as 31,500 pounds. The principal dimensions are shown in Figure 6.18.

Conventional cars could provide equal security without the space lost to the gallery simply by facing the baggage end toward the locomotive. Passengers had no need to pass beyond

this point, and trainmen communicated with the locomotive by signals. Thus conventional body styles became the standard for combine cars. In general appearance there was little to distinguish them from other passenger cars, except for the side door. Unlike baggage and mail cars, combines were usually built full-length. In finish they matched any coach, and for special service they were decorated grandly in the palace car style.

An example of a richly scrolled and outlined job is the Sodus Point and Southern's Number 1, shown in Figure 6.19. Two West

Figure 6.20   Virginia and Truckee combine cars Numbers 15 and 16 were built by the Detroit Car Company in 1874.

Figure 6.21   Combines often did more than carry passengers and baggage; some had mail or buffet compartments as well. The example here dates from about 1885.

Coast contemporaries now devoid of their original ornamentation have survived to document the combine of the 1870s: Numbers 15 and 16 of the Virginia and Truckee Railroad. The first is held by the Railway and Locomotive Historical Society's Pacific Coast Chapter awaiting the opening of the railway museum in Sacramento. The second car was sold to Paramount Pictures in 1938, and its future is uncertain. Both were built in 1874 by the Detroit Car Company. A diagram made from measurements supplied by Gerald M. Best is shown in Figure 6.20.

What might be labeled a typical combine of the next decade is the 50-foot car shown in Figure 6.21. It was built by Jackson and Sharp for an unspecified customer. The small mail apartment and the common stove straddling the partition between the mail and baggage rooms are of particular interest. The car was probably intended for a branch line; such roads were partial to small multipurpose combines. One of the most compact combines, and

something of a parody on main-line practice, was a coach-baggage-mail car produced in 1917 by A.C.F.'s Jackson and Sharp plant for the East Tennessee and Western North Carolina Railroad. All three working areas were crammed into the ridiculously short 40-foot body (Figure 6.22). The 3-foot track gauge further restricted the interior space. The car, antique in appearance even for 1917, had steel floor and body framing. The roof, sides, and interior were wooden. It should also be noted that the car is posed in the photograph on standard-gauge shop trucks for shifting in and around the maker's yard trackage.

A remarkable contrast to the stumpy, narrow-gauge triplex can be found in the long, elegant combines built for main-line service. A 60-passenger Erie Railway combine constructed in 1898 for through service is shown in Figure 6.23.[23] Only 14 feet, 6 inches of the 60-foot body was given over to baggage space, because the car was used on first-class trains carrying little lug-

Figure 6.22   The East Tennessee and Western North Carolina's diminutive narrow-gauge combine offered space for passengers, mail, and baggage. The Number 15, produced in 1917 by A.C.F.'s Jackson and Sharp plant, is shown on standard-gauge shop dollies. (Hall of Records)

gage. The conventional coach style of body framing and the freight car–like full trussing at the baggage end are clearly shown in the engraving. An even more elegant example is the New Haven stateroom car Number 2263 (Figure 6.24). Obviously this grand conveyance, manufactured by Pullman in 1900, was no country cousin intended for branch-line runs. (More on this level of combine is given in Chapter 4.)

The combine began as a lowly accommodation for emigrants, smokers, and Negroes. By 1890 it was occasionally adopted for most elite classes of service, and like its fancy relations toward the rear of the train, it evolved in the twentieth century from heavyweight steel to streamlined sleekness (Figures 6.25 and 6.26).

Emigrant Cars

Although emigrants were usually carried in broken-down coaches, some lines built special cars for emigrant traffic. Consequently it is impossible to characterize emigrant rolling stock exactly, but it shows more similarity to head-end cars than to any other general passenger car grouping.

As early as 1836 the Utica and Schenectady Railroad was running half-fare second-class cars which were said to be so inferior that only emigrants would ride in them.[24] After three profitless years they were withdrawn from service. During the next decade occasional mention was made of emigrant cars; the B & O's 1847 annual report, for example, listed two in the road's possession. As the tempo of Western settlement increased, the need for equipment that was adapted to the lowest possible fares became evident. A happy solution was found in specially outfitted boxcars.

The Illinois Central ordered twenty in 1855 at just $200 above the normal boxcar unit price.[25] Windows were placed on either side of the center doors, end doors were cut in, and seats were installed. In the same year the Michigan Central reported that it had fifty cars on the same general plan.[26] End platforms were added and the seats were removable, so that emigrants could be carried west and freight east. Here was passenger transport at the most basic level: it was not so much what the traffic would bear, but what the traffic could afford.

On the Grand Trunk Railroad emigrant accommodations reached a new low, according to a report published by *Railway World* in July 1867. An estimated 800 to 900 Germans were crammed into ten boxcars that had not been adapted to passenger service except for boards nailed partway up at the side door openings and a small hole cut into the upper rear corner for ventilation. No seats, windows, stoves, water coolers, or toilets were provided. The wretched assembly had endured these grim quarters for three days when they passed through Guelph, Ontario, and were seen by a local reporter. They had many more miles to travel before reaching their destination in western Canada. As the train stopped at Guelph, several pails of water were passed up into the cars, but the train pulled out before more could be provided. The reporter called it the most inhumane and shameful scene he had ever witnessed. A *Railroad Magazine* article written in the 1950s claimed that some emigrant families were content to load themselves, their livestock, and their baggage and tools into a boxcar for the journey west, and that such ragtag assemblies were called "Zulus" by contemptuous trainmen.[27] No descriptions of Zulu cars written by eyewitnesses have been uncovered so far, but the literature on the history of emigration has yet to be thoroughly researched.

Figure 6.23   An Erie Railway combine car, constructed in 1898, for first-class trains. The car was carried on six-wheel trucks.   (Locomotive Engineering, July 1898)

Figure 6.24   Combines are usually associated with branch-line or secondary trains, but some, such as this New Haven stateroom car, were built for deluxe runs that did not service large quantities of baggage. The Number 2263 was completed by Pullman in 1900. (Pullman Neg. 5208)

Figure 6.25   The Standard Steel Car Company produced this steel combine for Chicago suburban service in 1929.

Figure 6.26   Combines survived into the lightweight era. The Number 281, which seated 48 passengers, was manufactured by A.C.F. in 1947. (New York Central)

The essential meanness of the emigrant coach is shown in Figure 6.27's startlingly clear photograph of the Central Pacific emigrant car Number 319. This is hard evidence that such equipment was stark and comfortless, in an age that furnished opulence and plush ease for the traveler who could pay. Compared with the Central Pacific, the Pennsylvania Railroad pampered its steerage customers with padded seats, a clerestory, a stove, and a water cooler (Figure 6.28). But the severe interior and the presence of only two lamps indicate that there were no frivolous indulgences. A further sign of the Pennsylvania's emigrant policy was the reassignment of its class PC coaches (for example, see

Figure 1.44), built as cheap excursion cars for the 1876 Centennial Exposition, to emigrant service after the fair had closed. (Some of these cars were also used to carry miners in the anthracite coalfields.) Of course, the poorest grade of car could be justified because most emigrants could afford only the cheapest accommodations. Many of them were accustomed to such poverty that the coaches seemed no worse than their homes in the Old Country, or the stinking vessel that had landed them at Castle Garden on Manhattan.

The need for better emigrant cars became more obvious after 1870. For the next quarter century immigration reached its peak,

Figure 6.27   This emigrant car was completed around 1869 for service on the newly opened Transcontinental Railroad. Note the Creamer spring brake on the end platform railings. (Hall of Records)

Figure 6.28   A rare engraving of an emigrant car interior from a Pennsylvania timetable of about 1870.

Figure 6.29 The Western railroads began to furnish plainly finished emigrant sleeping cars in 1879. The example here was designed around 1885 by the Union Pacific. (Car Builders' Dictionary, 1888)

Figure 6.30   The Pullman tourist sleeper Number 260, from a photograph made in 1897. The car was very similar to the emigrant sleepers in the preceding drawings. (Pullman Neg. 3835)

with over 10 million persons landing on our shores. Those who chose to leave the Atlantic seaports and go west faced a wearisome journey. At least one Western railway official, Alban N. Towne (1829–1895), was sympathetic to their plight. Perhaps he remembered his poor New England boyhood and the discomforts that humble people face daily. As general manager of the Central Pacific Railroad he was in a position to reform passenger service on all levels. He chose to improve the emigrants' lot by providing supereconomy sleeping cars. In its basic design the emigrant sleeper was a duplicate of an open-section Pullman, but it was wholly devoid of ornament and frills (Figures 6.29 to 6.31). Plain oak-plank interiors and board seats with cast-iron ends made up the interior furnishings. A common cooking stove was provided, along with toilet facilities and water. Passengers were expected to furnish their own bedding, including mattresses. In April 1879 the road's Sacramento shops were busy fitting up twenty-five emigrant sleepers.[28] The Central Pacific's annual report for the year noted that sixty-eight second-class coaches had been converted into rudimentary sleepers.

Naturally the other Western lines had to match the improvements provided by Towne. The Santa Fe, the Rio Grande, the Union Pacific, and the Northern Pacific railroads all had emi-

Figure 6.31   Details of the Union Pacific emigrant sleeper shown in Figure 6.29 (Car Builders' Dictionary, 1888)

grant sleepers running before 1885. The desire to surpass the Central Pacific's standard led to the development of the tourist sleeper, in which the ideal of ultimate economy was abandoned for plate-glass windows, shutters, fancy lamps, paper wheels, and ice chests. The Denver and Rio Grande built six fancy tourist sleepers at its Burnham shops in 1883.[29] Within a few years the Santa Fe added steam heat, electric lights, and bedding. In 1889 Pullman succeeded in taking over tourist car operations on most major Western lines.[30] By this time they were distressingly similar to the palace sleepers, even including the services of a porter, and Pullman viewed them as a serious rival. Although tourist fares of 50 cents a night for the four- to five-day trip seemed wonderfully cheap to middle-class travelers, they were beyond the means of the typical emigrant, who was content with boarding cars and welcomed the largesse of such roads as the Santa Fe, which offered a mattress for the entire journey for an extra 75 cents.[31]

The primitive emigrant sleepers appear to have disappeared rapidly after 1900, for by 1915 the I.C.C. listed only 50 in service. The tourist cars remained, however; in fact Pullman had nearly 300. The Santa Fe promoted economical family travel via the tourist car in an elaborate color brochure.[32] Not only was the fare low, but meals could be prepared in the car's diminutive kitchen, thus avoiding the costly dining car. During the Depression, the Great Northern reintroduced the tourist sleeper in an effort to compete with the automobile and the bus.[33] The fare, including the berth, was only 2 cents a mile. By October 1933 such cars were running on eight trains.

The tourist sleeper fleet was expanded temporarily in 1943–1944 to handle troop movements. Even Pullman's large reserve of heavyweight open-section cars could not accommodate the traffic. The Federal government agreed to purchase 1,200 cars if Pullman would operate them. These boxcar-like carriers were extremely utilitarian in appearance and appointments. They measured 51 feet long and weighed 37½ tons. They had no vestibules; a center door was provided, as were twelve small side windows. Diaphragms and end doors provided a passageway to other cars in the train. The open, three-tiered berths were set across the car. Beds for 29 soldiers were provided. In addition to the sleepers, 400 kitchen cars built on the same general lines were produced to operate in conjunction with the troop trains. Both groups of cars were carried on a short-wheelbase, express-car style of truck that offered a safe (if not notably smooth) ride. After the war these cars were sold to various railroads for conversion to mail, express, or mail storage service. Others were retained as built and served admirably as bunk and kitchen cars on work trains.

None of the former troop-train sleepers were used for regular passengers, because the tourist sleeper went into a decline in the postwar period. During the war, rates had been boosted to discourage travel. Before 1941 tourist fare was normally one-half that of first class, but by 1942 it had jumped to 76.6 percent of Pullman fare.[34] In 1947 the Union Pacific and many other Western lines decided to eliminate tourist car service (such cars had

rarely, if ever, been run in the East at any period).[35] The roads were losing money on the tourist car, and they claimed that it was no longer necessary because new coaches with reclining seats offered the same comfort and carried larger revenue loads. On the other hand, the Great Northern decided to continue operation of its old-fashioned tourist cars, while the Milwaukee built new streamline equipment of the same class. Since both of these lines were weak contenders in the competition for Western rail travel, they found it necessary to offer special concessions. For all practical purposes, however, the tourist sleeper died in 1947.

Mail Cars

Because mail cars are the most complicated and interesting members of the head-end family, and because their link with government ensures documentation, their history has been very well recorded. Although their purpose was to carry goods and not people, they were intimately involved with passenger train history. Since mail cars were normally placed at the head of passenger trains, they were designed to resemble passenger cars in size, shape, and construction. Postal traffic represented a sizable volume of business and revenue.[36] During its peak years mail was carried on 9,000 trains over some 200,000 route miles; 30,000 clerks were employed; and gross annual revenues topped 50 million dollars. Some trains made up exclusively of mail cars carried 300 tons of mail daily between major cities such as Philadelphia and Pittsburgh. The Fast Mail was the most important train on the railroad, with precedence over everything on the line, including deluxe expresses. Every trunk line hoped to become known one day as the route of the Fast Mail.

For over a century the Post Office shared the industry's enthusiasm for moving mail by rail. Even after the coming of the airplane, the railroad remained unbeatable for speed, cost, and reliability. The mail service literally followed the railway; the two were interdependent, until eventually one of the partners faltered and was replaced by a more efficient rival.

The beginnings of the partnership were unglamorous. The railroad was viewed by the Post Office as another carrier to move letters in bags, even as it transported crates of shoes and baskets of vegetables in one corner of the baggage car. Mail was shipped as a bulk commodity from one large central post office to another for redistribution to smaller hamlets and substations within a hundred-mile radius. It is claimed that mail was first carried by rail in this country on the South Carolina Railroad in 1831. According to one source, the first official contract for moving the mail was granted to the Baltimore and Ohio in November 1834.[37] Another source claims that no contract was signed until August of the following year.[38] There does seem to be general agreement on the legal starting date of July 7, 1838, when Congress designated all railroads as official postal routes. Within seven years 3,900 of the 4,600 miles of American railroads then in operation were carrying mail.

Nevertheless, although mail was being transported on the majority of lines, the new technology of moving the mail by steam transit was poorly exploited in America. It was in Britain that the best coordination was achieved between railway and postal service. Early in 1838 the Grand Junction Railway tested an imaginative scheme for sorting the mail en route.[39] Clerks, stationed inside a car fitted with tables and pigeonhole boxes, prepared it for delivery as the Flying Mail traveled between Birmingham and Liverpool. Thus the mail was not being simply hauled but actually processed during the journey, so that it arrived ready for immediate distribution. At the same time a net permitted mailbags to be picked up at stations along the line without stopping the train, and pouches were dropped off through a side hatch. The premiere "traveling post office" thus had all the features of the modern mail car. The idea was adopted throughout Great Britain and in many parts of the Empire. A British journal's description in 1849 of clerks dexterously filing letters into thirty-five pigeonholes of a London–Glasgow mail car reads like an account written a century later.[40]

According to British authorities, the American Post Office waited twenty-five years before introducing this practical system, and even then viewed it with skepticism. At least, that is the story told by the published histories of the railway mail service, including those written by former officials of that agency. But it seems to be something of a half-truth, since there is evidence that the British system was being practiced in this country, at least in an elementary form, in the very year that the first Grand Junction car entered service. The following paragraphs from the *Niles Register* of May 18, 1838, make it clear that the mail was being sorted, picked up, and dropped off en route.

New mail arrangement. Mail cars, constructed under the directions of the post office department, are now running on the railroads between Washington and Philadelphia. They contain two apartments; one appropriated to the accommodation of the great mails, and the other to the way mails, and a post office agent. The latter apartment is fitted up with boxes, labelled with the names of all the small offices on or near the railroad lines. It has also a letter-box in front, into which letters may be put up to the moment of starting the cars, and any where on the road.

The agent of the post office department attends the mail from the post offices at the ends of the route, and sees it safely deposited in his car. As soon as the cars start, he opens the letter-box and takes out all the letters, marking them so as to designate the place where they are put in. He then opens the way mail bag, and distributes its contents into the several boxes. As the cars approach a post office, the agent takes out the contents of the proper box and puts them into a pouch. The engineer slackens the speed of the train, and the agent hands the pouch to the postmaster, or carrier, who stands beside the track to take it, receiving from him, at the same time, another pouch, with the matter to be sent from that office. This the agent immediately opens, and distributes its contents into the proper boxes. Having supplied thus all the way offices, the agent, when arrived at the end of the route, sees the mail safely delivered into the post office.

The mail leaves Washington at 6 o'clock, A.M. All letter-writers should, if possible, put their letters into the post office the preceding evening. The letter box at the cars will not be open until after 5 o'clock in the morning. It is not *a post office*; and letters will not be received at it while the post office is open. Its object is to allow all persons, until the last moment before departure, to send letters by mail, even although the mail has been closed at the post office, and even placed in the car.

The letter box will be open after the mail closes in the post office at Baltimore, and all the intermediate points. Philadelphia cannot at present enjoy its advantages fully, because the cars do not run into the city.

In this arrangement, the post office department has sought to give the greatest security to the mails, and to afford the community the best possible accommodation. Well executed, the plan must be almost the perfection of mail arrangements. It is intended, when it can be conveniently done, to extend a similar arrangement through to New York.

A year or so after the *Niles* article, Von Gerstner mentions seeing four mail cars with heated offices that collected, sorted, and delivered letters during the journey.[41] There were other reports of such service on the Boston and Albany in 1852 and the Richmond and Terre Haute in 1855.[42] Although their documentation is not conclusive, certainly the *Niles* article and Von Gerstner's account are indisputable. Furthermore, the Post Master General's report for 1840 speaks enthusiastically of the British system and provides a paper on the subject prepared by agents sent overseas to make a study of European methods.[43] Thus the Post Office was not unaware of the system. In addition, Canada began a traveling post office service in 1854, and in his report for 1859 the Post Master General praised its effects on cost and speed of delivery.

The facts seem to be that mail was sorted on some routes in this country, but that the system was not vigorously developed by the Post Office Department. Many routes were served by apartment cars such as the mail-baggage combine shown in Figures 6.32 and 6.33. The illustrations date from the 1860s and 1870s, but they might equally well represent cars of the previous decades, when even smaller, more poorly furnished compartments were the rule. In 1855 the Post Office began to press for better facilities so that its route agents could work more efficiently.[44] Their jobs encompassed what is now regarded as the function of a railway post office car. Letters were received, stamped, sorted, and dropped off (though exactly how much sorting was done during the trip is not clear). Pouches were picked up and dispatched en route. The Post Office asked for two rooms. The agents' room was to be no less than 7 by 12 feet, with a counter, boxes with sliding lids, a window and a door on each side, and locks on all doors. The second room, to be no less than 7 by 13 feet, was for through mail in locked pouches or canvas bags. In response to these requirements it is known that at least one railroad, the Lake Shore line, fitted three baggage cars in 1856 with a 7½- by 17-foot compartment for the Toledo-Chicago run.[45] By 1860 traffic had developed to the point where the road furnished cars with even larger compartments. In 1866 it was necessary to provide 35-foot-long cars devoted exclusively to mail.

Figure 6.32 No illustrations of first-generation mail cars have survived, but this drawing is a good representation of them. The antique-looking combine was built in 1866 by Harlan and Hollingsworth. (Traced by John H. White, Jr.)

Figure 6.33 A Denver Pacific arched-roof combine mail-express-baggage car, constructed around 1870. (Hall of Records)

Figure 6.34   The B & O's Mt. Clare shops built this elaborate mail car in 1864. It was one of the finest produced up to that time. (Leslie's Illustrated Magazine, October 8, 1864)

The Civil War created an unexpected volume of mail that overwhelmed the postal service. Official correspondence was only part of it. Letters between soldiers and their families and sweethearts created an avalanche of paper that the existing system could not handle. As letters stacked up, the public grew more impatient, and its complaints were forcefully passed on by Congress. Radical reform was needed.

Postal officials were ready to try any scheme to expedite the mails, so that William A. Davis, assistant postmaster in St. Joseph, Missouri, had little trouble in obtaining permission to introduce train-board classification of the Western mails passing through his home town in such volume. By sorting aboard the cars en route to the California Pony Express connection, considerable time would be saved. The Hannibal and St. Joseph Railroad agreed to fit two baggage cars with tables and pigeonhole filing boxes. The service began in late July 1862, but was discontinued after a year or two because of a shift in the California mail route. Many years later some of Davis's friends and relations—a son-in-law was particularly active—insisted that he had originated the railway post office scheme.[46] It was a foolish claim in light of the much earlier British and American examples already noted, but Davis's supporters went so far as to fake a photograph by grafting the image of a circa 1900 mail car to a picture of a wood-burning locomotive from the 1860s and labeling the fraud as the first railway post office car. Davis himself, it should be noted, never claimed the honor.[47] He was a lifelong postal employee and knew better.

It was George B. Armstrong (1822–1871) who might better be called the father of the railway mail service in America He envisioned, and around 1864 proposed in writing, a plan for a national system of railway mail service which would fully utilize the British concept of traveling post offices. A supervisor in the Chicago post office, Armstrong was appalled by the backlog of

mails created by the war. In cooperation with Eastern officials of the Post Office Department, he experimented with mail cars in August 1864. The first car, worked by two men, was run between Chicago and Clinton, Iowa. It was a 40-foot-long route agent's car with blind ends. It had been remodeled with a bank of seventy-seven pigeonholes and a cluster of newspaper boxes.[48]

On almost the same day that the Armstrong car began its first trip, a similar experiment was begun in the East. Actually, A. N. Zevely, third assistant post master general, had conducted tests since the spring of 1864. In a remodeled baggage car his crews had been able to sort 5,000 letters on the New York–Washington run. Zevely wanted a first-class Railway Post Office built to order and persuaded the B & O to fabricate such a unit at its Mt. Clare shops. The result was a magnificent eight-wheel car with a clerestory roof and bold lettering which read "Through Line between New York & Washington."[49] The Number 49 was ready for service by August 31, 1864. The inside body measurements were 9 feet, 4 inches by 45 feet, 5 inches. The interior furnishings included a 400-module pigeonhole box, tables, a stove, and space for newspapers and packages. A second car was under construction at the Camden and Amboy shops. The B & O car attracted so much attention that *Leslie's Magazine* published engravings in its October 8, 1864, issue (Figures 6.34 and 6.35). The car represented a remarkable advance, but the schedule of 12½ hours left something to be desired in the way of speedy service.

The successes of Armstrong and Zevely encouraged the mass application of the Railway Post Office idea. In 1865 such cars began to appear on the Burlington, the Alton, the Illinois Central, and the Lake Shore railroads.[50] Fewer R.P.O. cars were found in the East, where service was limited to the original New York–Washington route and two longer hauls between Albany and Buffalo and New York and Erie, Pennsylvania.[51] During that final year of the Civil War, sixty-four clerks were working

Figure 6.35   The B & O's 1864 mail car had 400 pigeonholes so that letters might be sorted en route. (Leslie's Illustrated Magazine, *October 8, 1864*)

R.P.O.s over 1,041 miles of track.[52] Little progress was made during the next few years, but the plan was firmly established in spite of heated criticism from the district postal supervisors, who feared that it would eliminate their patronage if not their very jobs. In 1869 the Railway Mail Service was officially inaugurated, with Armstrong as its head.[53]

After Armstrong's death in 1871, George S. Bangs took charge of the agency and set out to expand it. At the end of his first fiscal year he was in charge of 57 lines (covering 14,117 miles) and 649 clerks.[54] Traffic was very heavy on some routes, such as the Buffalo-Chicago line, which daily carried 50,000 to 60,000 letters each way and 3 to 10 tons of newspapers, magazines, and parcels.

More and better cars were now being run. Full mail cars were used on all major lines; the once standard apartment car (the mail-baggage combine) had been downgraded to lesser runs. A crook-arm mail catcher patented by L. F. Ward of Elyria, Ohio, on January 29, 1867 (No. 61,584), was now widely used.[55] In 1870 the Michigan Central turned out several finely finished cars with 48-foot bodies, 450 letter boxes, and 60 newspaper cases placed in a semicircular arrangement.[56] In addition to providing sorting tables and unusually good lighting, the Michigan Central

added two velvet-covered sofas so that the clerks could take quick naps. One-third of the space was open for pouches. The exterior was painted a buff color. Two years later the Central Pacific shops turned out a palace mail car that featured a parlor-dining-kitchen compartment in the center of the body.[57] It had a cooking stove, a table, chairs, a washstand, and several beds. More typical of the 1870s, however, is the utilitarian Lake Shore car depicted in Figure 6.36. In 1877 the Milwaukee produced a car after the design of J. E. White, later named head of the Railway Mail Service.[58] The 50-foot body was carried on six-wheel trucks and cost about $5,000. White's interior plan was designed to speed distribution. He dispensed with the semicircular box case and used a long case on one side of the car. Bags were held by iron racks resting on the floor. White claimed that his floor plan boosted capacity by 50 percent, but ultimately the three-sided end case prevailed.

These technical refinements were pale compared with the most sensational achievement of the R.M.S. in its early years: the Fast Mail. Rapid sorting aboard the cars meant little when the fastest train between New York and Chicago lumbered along at around 25 mph.[59] The 979-mile journey required just four minutes under thirty-seven hours. No wonder that beds were placed aboard the

Figure 6.36   The British trade magazine Engineering published this detailed drawing in its July 28, 1876, issue. The brief description did not mention the designer or builder.

Figure 6.37   The Lake Shore and Michigan Southern formed an important link in the New York Central's New York–Chicago fast mail train of 1875. (Chaney Neg. 13666)

Figure 6.38   The Number 2 is believed to be one of the Philadelphia–St. Louis fast mail cars produced in 1875. (Smithsonian Neg. 71262)

Figure 6.39   The Pennsylvania's class Ma mail car was introduced in 1876.

R. P. O.s. The average speed of passenger trains was not the only problem, for mail movements were tied to existing schedules. Departures were usually not geared to the next morning's deliveries, and connections were poorly scheduled for mail transfers. Bangs recognized that no material improvement in deliveries could be made until extra-fast mail trains ran independently of regular passenger service. He dreamed of a twenty-four-hour train between Manhattan and Chicago, and railroad men agreed that the existing equipment was fully capable of such a performance. But special train service and the twenty new cars that Bangs envisioned would be expensive. The R.M.S. had no funds and no prospect of obtaining any from Congress, which was even then quarreling with the railroads over a fair rate of compensation for carrying the mails.

In 1874 William H. Vanderbilt heard of Bangs's idea and became enthralled with it, refusing to be dissuaded by his father's caustic opinions on dealing with those rascals in Washing-

ton.[60] Support for the Fast Mail was explained as a public-spirited gesture on the part of the younger Vanderbilt. Privately, he was convinced that the publicity value of the project would be invaluable and that if it did improve service, public opinion would force Congress to fund it properly. Bangs was apparently willing to accept the circus aspects of Vanderbilt's promotion to see life breathed into his pet, and Vanderbilt for his part cut no corners. The New York Central and the Lake Shore shops at Albany, Cleveland, and Adrian, Michigan, were ordered to build a fleet of the finest mail cars ever produced and to do it quickly.[61] Three styles were devised: R.P.O. letter cars, newspaper cars, and bulk storage cars for through mail that was not to be sorted enroute (Figure 6.37). Other valuable details are given in a notice from a *New York Graphic* dated Sept. 15, 1875:

The letter-distributing cars are 50 feet in length, while those designed for the newspaper mail are 10 feet longer. All are uniform in width, 9 feet 8 inches, and 6 feet 9 inches high in the clear. The finish of the exterior does not differ, all of them being painted white, with cream-colored borderings and gilt ornamentation, highly varnished within and without. Midway on the outside and below the windows of each car is a large oval gilt-finished frame, within which is painted the name of the car, with the words "United States Post-Office" above and below. Along the upper edge and center are painted the words, in large gilt letters, "The Fast Mail," while on a line with these words, at either end, in a square, are the words, in like lettering, "New York Central" and "Lake Shore." The frieze and minute trimmings around the windows are also of gilt finish. At the lower sides and ends of the cars are ovals corresponding to those on which the names are painted, and inclosed at one end a painted landscape scene background, and in the relief an all-seeing eye, beneath which is a pyramid inscribed with gilt Roman figures, "MDCCCLXXV," and the motto "Novus ordo seclorum." At the opposite end, in the same colors, on a blue background, is the United States coat-of-arms.

In running appointments, as well as all others, the most important patents have been selected. The platforms are inclosed on either side by swinging doors, which can be fastened open at will to prevent disturbances by wind when the mail matter is being trucked from the tenders to the distributing cars, and also as a guard against any danger of its falling off. The cars are supplied with patent air-brakes and patent spring-brake, and the end doors of the cars are protected by an inclosed platform. On either side of each car are two doors supplied with patent mail-bag catchers for securing the mail along the route. The 60-foot cars are mounted on 6-wheel trucks, giving in all twelve wheels to each car, while the 50-foot cars are on 4-wheel trucks.

The article reported that the cars were named for state governors, such as Todd, Dix, Allen, and Hayes. The letter cars had a room for the chief clerk outfitted with a folding bunk, desk, and closet. The distributing room was furnished with the usual counters and pigeonhole boxes. An iron safe was provided for valuables. Washstands, water coolers, and hot-water heating were also available. The letter cars cost $4,200 each; the newspaper cars $3,300.

Every effort was made to draw public attention to the inaugural run. One hundred railroad officials, politicians, members of the press, and dignitaries such as the U.S. Vice President were aboard for the first trip on September 16, 1875. An equally impressive load of mail was placed on the cars—so much, in fact, that some of it overflowed into the V.I.P.s' drawing-room car attached to the rear of the four mail cars. The contents totalled 33 tons, a burden increased by another 150 letter pouches picked up at Albany. The first lap of the trip to Buffalo was made at the astonishing average rate of 40.9 mph (including stops), which equaled express-train speeds in England.[62] Once on the Lake Shore portion of the run, the average speed dropped to 33 mph. Even so, the train completed its run in just twenty-seven hours and fifteen minutes, a saving of nearly ten hours over the fastest regularly scheduled express train. The lightning run was to be more than a onetime stunt; this schedule would be followed on a daily basis.

The president of the Pennsylvania did not want to be outdone by the Vanderbilts, and just weeks before the first run of the Fast Mail, Tom Scott initiated a crash program to produce a competing train.[63] The cars were built in ten days (Figures 6.38 and 6.39). The Limited Mail was ready on September 13, 1875, but its inaugural trip from Jersey City to St. Louis was less well publicized than the New York Central's and its remarkable record of thirty-three hours received relatively little attention. Even so, Bangs's contention that faster train speeds were practical was again demonstrated.

What should have been the beginning of faster mail service, however, proved to be something of a false start. Congress refused to support such a luxury, just as the old Commodore had predicted. In fact, the two Houses reduced the overall rate of compensation to railroads by 10 percent in July 1876. Vanderbilt was infuriated and withdrew the Fast Mail service in a matter of days.[64] The Pennsylvania followed suit. The panic of 1873 was bottoming out, and the volume of mail declined as business slowed down. Congress reacted again in June 1878 with another 5 percent cut. This action reopened the long-standing battle between Congress and the railroads over a fair rate compensation for carrying the mails. In the early 1870s the industry protested against the niggardly rates paid, claiming that they were less than second-class freight rates.[65] Yet the cars required by the R.M.S. were expensive, highly specialized equipment, which spent as much time in the station as on the road. At one point the railroads threatened to withdraw all postal cars unless relief was granted. Refusing to yield, Congress hinted that it would take equally drastic action if the railroads interfered with the transport of mail. The battle continued, in various degrees of Sturm und Drang, throughout the history of the Railway Mail Service.

As the economy recovered, the volume of mail began to surpass old records. Congress relented somewhat, and the Fast Mail concept was restored to life. In 1881 the New York–Chicago train was revived, and a fast train from New York to Washington, Atlanta, and New Orleans brought the wonders of rapid mail delivery to the South.[66] In the same year a total of twenty-four hours was slashed from the transcontinental mail train schedules. In 1882 the B & O produced five elegant white cars, trimmed in ultramarine, for the Baltimore–St. Louis Fast Mail.[67] After the wonder of the new service had faded slightly, humorists satirized the Fast Mails as moving so rapidly that cows along the line were given timecards so that they could withdraw from the tracks in good time.[68] It was also declared necessary to tie down light stations, anchor handcar shanties, and use extra-large nails in fences because of the trains' terrific suction.

The R.M.S. was entering a period of enormous growth. In 1881 nearly 3,200 clerks and 1,892 cars ran over 90,200 route miles.[69] In 1888 another 1,900 clerks were traveling over an additional 53,000 route miles. By 1894, 3,059 mail cars were in service.[70] This growth in personnel, equipment, and mileage paralleled a similar expansion in the railroad industry. And even as the railroad business reached its peak in 1915, so the R.M.S. extended to its widest range in that year, when some 20,000 clerks and 4,000 to 6,000 cars (depending on which figures one cares to believe) moved the mails over 216,000 route miles.[71]

Figure 6.40   Jackson and Sharp built the Georgia *in 1879. It ran between Richmond and New Orleans over several Southern lines. (Hall of Records)*

Figure 6.41   Interior of the Georgia *of 1879. (Hall of Records)*

During these decades of growth the mail car followed the general development in size and materials of other passenger equipment. Full R.P.O.s, like baggage cars, remained on the short side, but combination mail and passenger cars were usually made full length. Until sometime in the early 1890s exteriors were painted white, with buff, red, or blue trim (Figures 6.40 and

6.41). The color scheme set mail cars apart from the rest of the train, but in later years colors became drabber and more standardized. A special mail car shown at the 1893 Columbian Exposition was one of the last known to have been painted in the old tradition. A more important development was the elimination of the end platforms. Many of the earliest R.P.O.s were built with blind ends; platform steps and handrails were retained for the convenience of the brakeman, but there were no end doors (Figures 6.42 and 6.43). However, the platforms were an inviting place for thieves and hoboes, some of whom would actually break through the blank end wall in search of plunder.[72] Around 1875 the Louisville and Nashville began to build head-end cars without end platforms. A drawing of such a car was reproduced in a French report on American railroads published in 1880 and 1882 (Figure 6.44).[73] Some of the data were gathered as early as 1876. By the mid-1880s nonvestibule mail cars were the general rule.

Representative postal cars for the period include the Michigan Central's *Indiana*, built in 1880, which measured 55 feet in length and weighed 26 tons.[74] Later in the decade the Milwaukee Road produced a 69-foot car weighing 33.5 tons.[75] At the 1893 Exposition two fine mail cars were shown, both featuring 60-foot bodies, which the Post Office fancied for many years as the optimum length for a full apartment mail car. A drawing of the 45-ton Pullman-built entry is shown in Figure 6.45.

Mail car furniture was also standardized during the final decades of the century. The most notable single development in this area was iron rack frames. In the mid-1870s wooden frames or racks with hooks replaced the circular cases formerly used to hold full, unsorted letter pouches.[76] The pouches hung on the

Figure 6.42   An early Railway Post Office car stands behind the locomotive's tender in this 1867 scene. It was photographed at Rockford, Illinois, on the Chicago and North Western Railway. (Smithsonian Neg. 73-403)

Figure 6.43   The N O J & G N's first Railway Post Office car was built by Jackson and Sharp around 1875. Note the blind ends.

racks were more accessible to the clerks. Gas-pipe racks were tested as a possible improvement over the wooden frames. On seeing the pipe racks, Charles R. Harrison, a R.M.S. employee, was inspired to devise a cheaper, more flexible design. Around 1879 Harrison worked out a plan for hinged, cast-iron mail car furniture (Figures 6.46 to 6.48). Pouches, racks, and sorting table could be set up, moved, or folded away as necessary, giving the individual car great flexibility.[77] It was even possible to clear the car for baggage or express without removing the R.P.O. fixtures. Costs were reduced, since the standard fittings were cheaper to make and replace. Harrison also claimed that his equipment was safer, because cast iron would not splinter and

Figure 6.44   The drawings for a Louisville and Nashville mail car of about 1875 were published in a French work on American railroads. The center table and rack was probably for newspapers. (E. Lavoinne and E. Pontzen, Les Chemins de Fer en Amerique, Paris, 1882, plate 6)

burn. Harrison opened a factory in Fond Du Lac, Wisconsin, in 1881.[78] Within a few years a number of railroads were using his furniture and by 1910 it was declared to be almost universally employed in American mail cars. In later years iron furniture manufactured under the trade name of Otis superseded the Harrison fixtures.

Although most car design was the exclusive province of railroad and supply trade engineers, mail cars were a minor exception to this normally inviolate rule. Since the beginning of the R.M.S., Post Office managers took a direct hand in the interior arrangement of the cars. It is true that frame, body, and running gear designs were left to the railroads, but even this fundamental arrangement was to change. Eventually the government was insisting that its standards and general specifications be carefully followed. In 1885 came what appears to be the first effort to achieve a system-wide set of standard floor plans.[79] Charles W. Vickery, head of the R.M.S., approved detailed drawings for a 40- and a 50-foot mail car. A U-shaped letter box at one end was to have 450 openings, 4 inches by 4 inches by 7½ inches deep. A sixteen-box letter case was also specified, as were particular door openings and inside dimensions. A number of railroads were said to be constructing new cars in accordance with Vickery's plan.

The Post Office also sought to improve productivity by requiring better facilities for clerks. Few jobs involved conditions of such stress. A man was expected to have unusual powers of concentration, dexterity, and precision, as well as the ability to work rapidly while standing for hours in a smoky, swaying railway car (Figures 6.49 and 6.50). To qualify he had to have the eyes of a hawk, the nervous system of a surgeon, and the balance of a sailor. The combination of exacting yet physically demanding labor was more than the average man could long endure. Hence the demands for better lighting, ventilation, and heating were not unreasonable.

Lighting was one reform that received early attention. Excel-

lent lights were necessary to read handwritten envelopes, and by the early 1880s special mail car lamps with fantastic umbrella-like reflectors had been developed (Figure 6.51). By the mid-1890s the Post Office was calling for an end to oil lamps, more for reasons of safety than of improved illumination.[80] It was contended that thirty fires aboard mail cars in a single year had been caused by kerosene lamps—a claim that industry called patently untrue. Even so, by 1897 one-third of the R.P.O.s had gas lighting, and the Post Office, looking beyond gas to electricity as a final solution, sponsored a test of electrical lighting in 1912 that ran for several weeks.[81] It was found that the style of reflector used could affect lighting efficiency by as much as 55 percent. A published report and new lighting specifications followed the tests.

Owing to pressure from top management, the clerks' union, and interested members of Congress, the Post Office exhibited increasing concern for the safety of its employees. In 1877 statistics on deaths and injuries began to appear in the annual reports of the Post Master General. During that year only 2 clerks were killed in accidents; 10 were seriously injured.[82] By 1884 7 were killed and 28 were seriously wounded. The total accidents jumped from 27 to 154. Increased train speeds and the growing number of mail cars caused the toll to rise. Between 1890 and 1900 mail cars were involved in 5,000 accidents that resulted in 75 deaths.[83] In 1897 alone, 14 clerks lost their lives. By 1907 a total of 210 clerks had been killed since 1875. During the next eight years an average of 15 names were added annually to the death rolls, but after 1915 fatalities fell off to a uniform average of 2 a year.[84] The improvement can be credited to more rigorous safety measures, the growth of automatic signaling, and the introduction of steel mail cars.

In the 1870s and 1880s the Post Office established standard floor plans and pressed for better lighting and heating. Yet the actual car structure was left to the judgment of the individual

Figure 6.45 Pullman built a mail car as part of its 1893 Columbian Exposition train. The car was fitted with the popular Harrison folding tables and fixtures. (Pullman's Souvenir Booklet, Columbian Exposition, 1893)

master car builders, most of whom were satisfied with a stock baggage car design.[85] Thus mail cars were generally the most lightly built of all passenger cars and, being placed at the head of the train, were the most likely to be crushed in an accident (Figures 6.52 and 6.53). Finally in September 1891, the R.M.S. formulated its own minimum standards for floor framing.[86] The specifications included more and larger sills as well as metal plate reinforcements and iron bolsters. In 1904 the R.M.S. requested even stronger framing, with end platform frames made from steel I beams, and larger-diameter axles.[87] Car builders contended that these changes would increase car weight by 14 tons. Even so, the R.M.S. pointed to a wreck of a Fast Mail on the Southern Railway the year before in which five clerks were killed. The existing framing was clearly not adequate, but the R.M.S. specifications were only recommended practice and no railroad could be forced to adopt them.

With the introduction of the steel passenger car, better wooden framing became a forgotten issue. Agitation began for the universal adoption of steel construction for postal cars. It was argued that because a Federal agency paid the fare and its employees were carried upon equipment furnished by the railroads, the government was justified in insisting upon the best and safest equipment available. In March 1911 a law requiring steel or steel-framed mail cars was passed by Congress (Figure 6.54).[88] The original bill had called for immediate suspension of payment to any line running wooden mail cars, but the final version was much weakened and allowed the use of wooden cars so long as they were not adjacent to steel cars. After July 1, 1916, the Post Master was to withhold payment to any line running wooden cars in trains having a majority of steel equipment in their consist. Railroads were also encouraged to adopt all-steel R.P.O.s as soon as possible. The statute was ineffectual compared with the strong legislation desired by the mail clerks, but it was a beginning.

In the fall of 1911 the Post Office sought the advice of the best talent in the car-building industry in preparing exact specifications.[89] Barney and Smith, A.C.F., Pressed Steel, and Pullman sent their chief engineers to Washington. Several leading rail-

Figure 6.46   A mail car of 1886 equipped with interior furnishings supplied by the Harrison Postal Bag Rack Company, Fond du Lac, Wisconsin. (Railroad Gazette, January 29, 1886)

roads also lent key men. Charles A. Seley, mechanical chief of the Rock Island, was named chairman of the committee. The engineers decided to formulate guidelines that would outline desired results rather than to compose ironclad specifications tied to particular designs. The committee continued to meet for nearly three years. Its most important action was the establishment of the 400,000-pound impact load as the minimum standard for floor and end framing. Until all-steel cars could be acquired, the committee recommended several short-term safety reforms. Only all-steel, steel-underframe, or steel-plate-reinforced wooden mail cars should be in main-line service. Wooden cars with weak ends (only six to eight posts) should be limited to branch-line service on trains of no more than four cars and at speeds no greater than 21 mph. A cheap end post reinforcement was suggested as well: old steel rails, four to an end, could be bolted on vertically to provide a very effective end bumper. The cost was figured at only $100 to $160 per car.

In 1912 the law was modified so that beginning in June 1913, one-quarter of the wooden mail car fleet would be replaced annually by steel.[90] Thus the conversion would be accomplished in

four years. It seemed that the steel postal car law at last had some real teeth and that the changeover was assured. The industry was cooperating, after all; long before either bill had been passed, some railroads had voluntarily acquired all-steel mail cars. The Erie was well in the lead. In 1905 its president, F. D. Underwood, ordered an all-steel, 68-foot R.P.O. from the Pressed Steel Car Company (Figure 6.55). The Number 699 weighed 70 tons and was soon running with two sister cars. During the same year the Santa Fe purchased thirty-nine steel-underframe R.P.O.s from A.C.F. The 60-footers had unusual four-girder fish-belly underframes. The outside sills were visible, making the cars look as though they were mounted on a long steel flatcar. But all-steel construction was obviously superior. Within a year after the Santa Fe composite cars entered service, the Pennsylvania and the Southern Pacific developed long-range plans to retire all wooden equipment. In 1906–1907 two of the Harriman Lines, the U P and the S P, produced sample all-steel R.P.O.s.[91] By June 1907 they were ready to order the cars for regular use (Figures 6.56 and 6.57). Within five years they had twenty-eight in service and another sixty-nine on order. The Pennsylvania's first steel

Figure 6.47   Details of the 1886 Harrison-equipped mail car. Note the trackside mail crane in the lower end-elevation drawing. (Railroad Gazette, January 29, 1886)

Letter Case with Stationary Table.

Details of Folding Racks and Tables.

R.P.O., Number 6546, built at a cost of $15,200, was ready for service in March 1907.[92] Nineteen more were running by 1909.

Other roads, however, showed less enthusiasm. Of the 5,866 mail cars in service by December 1914, 1,587 were all-steel and 677 were steel-frame, wooden-bodied cars.[93] Thus nearly two-thirds of the postal car fleet was still wooden. The railroads looked better when the statistics were limited to full R.P.O.s: of the 1,566 nonapartment mail cars, only 461 were wooden. Even so, no rush was being made to comply with the law. In 1922 the number of full R.P.O.s was down to 1,087, yet 67 steel-reinforced wooden-frame and 4 all-wooden cars were still running.[94] The ratio was much worse for the 4,074 apartment cars then in service: 1,104 were steel, 641 were steel-framed, 1,947 had reinforced wooden frames, and 382 were all-wood. It was not just the impoverished short lines that were reluctant to convert to steel; some wealthy trunk lines were remarkably indifferent to the Congressional mandate. The Santa Fe, for example, purchased no steel R.P.O.s until 1924.[95] The road was convinced that the best mail cars were still the steel-frame, wooden-body equipment purchased in 1905.

In the spring of 1926 the House Post Office Committee opened hearings on the situation.[96] Congressmen expressed considerable frustration over the industry's languid reaction to past legislation and introduced a tough bill calling for an immediate substitution of steel cars. Curiously, the Post Office Department now sided with the railroads in opposing the measure. Both parties denied the need for any legislation. They said that an instant conversion would amount to confiscation of valuable private property, because the railroads would be forced to retire serviceable cars. According to their argument, the cost involved would be unreasonable even if the cars could be produced immediately, which was impossible. Also, mail service would be disrupted in the interim. Post Office officials contended that they had not been negligent in carrying out the wishes of Congress, that the 1911 and 1912 bills applied exclusively to full R.P.O.s and not to apartment cars, or so the Post Office had interpreted their wording. After all, most full R.P.O.s were steel or steel-underframe, and all-steel cars were used on the heavy, fast trains likely to be most dangerous to life and limb. Thus 14,000 of the 18,000 postal clerks were riding in safe cars—that is, steel or steel-framed cars. Most wooden cars were in slow-speed, branch-line service. The new floor plans that had been introduced around 1915 did much to safeguard the clerks. The letter case was now placed toward the center of the car rather than at one end as before. This rearrangement, according to the Post Office, removed the clerks from the fatal end position.

The railroad spokesmen dwelled on the immense cost involved. The price of a new mail car was $24,000, and enough cars to satisfy the pending law would cost over $67,000,000. This investment in the face of Post Office cutbacks (for more mail was being presorted at central mail depots) made the investment in new cars unwarranted. Many were already in storage. And steel cars in themselves did not end the perils of railway travel. The great decline in fatalities was credited to safer operations that resulted in fewer wrecks. In 1906, mail cars were involved in 2,588 accidents; by 1922 the number fell to 757. The vulnerability of steel cars was underscored by the fact that of the 8 clerks killed between 1923 and 1925, 7 died in all-steel equipment.

The counterarguments centered on the railroads' unwillingness to abide by the law and the dangers that thousands of clerks faced daily aboard unsafe cars. Supporters of the tougher bill pointed to the suffering of the families of clerks who had been

Figure 6.48   Interior of a New York Central Line's mail car with Harrison racks. The car was built by Pullman in 1909. (Pullman Neg. 11926)

Figure 6.49   In 1873 an artist depicted the frenzied actions of clerks sorting mail en route. (Scribner's Magazine, June 1873)

Figure 6.50   This posed photograph made aboard a B & O mail car in 1938 fails to capture the rapid work of the clerks. (Baltimore and Ohio Railroad)

killed or maimed in the service. They also emphasized the fact that for each death, the government paid claims averaging $20,000—enough to buy a new steel car. However, they could not muster the votes to pass the bill.

Nevertheless, the hearings had some effect. The railroads added a large number of steel mail cars, including over 250 apartment cars, during the next two years.[97] In 1927 the Post Master General prohibited the operation of wooden or steel-frame cars in trains with a majority of all-steel cars. Steel-frame cars could no longer be operated between cars of all-steel con-

struction. Railway Mail Service specifications now contained sixty-one headings and filled thirteen pages of fine print.[98] The Post Master's authority was recognized as covering both full and apartment cars. In new Congressional hearings in 1928 the railroads repeated the arguments they had made two years earlier, but they could not claim that they had shown any better faith in carrying out the wishes of Congress. Again nothing was done; Congress decided that the old laws were strict enough. And then came the Depression, when Capitol Hill was swamped by problems greater than the existence of a few antique mail cars.

Figure 6.52   Mail cars, often built without end platforms to discourage thieves, tended to look unusually short. The Number 19 was produced by Jackson and Sharp in 1898. (Hall of Records)

Wooden cars were left to disappear by attrition at an exceedingly slow pace. As late as 1950, thirty-seven steel-frame and eight wooden R.P.O.s were reported in service, although none were on main lines.[99]

The interior arrangement of mail cars changed very little during the standard heavyweight period. Size was fixed by the Post Office decision in 1916 to pay for service by the floor space of the compartment. Units of 15, 30, and 60 feet became standard. Full-sized cars were set at 60 feet and were occasionally placed within an 80-foot car, the remaining space being used for baggage. Technically such a unit could be classified as an apartment car, yet the Post Office would insist that it was a full R.P.O. Mail cars followed the improvements introduced in the lightweight era (Figures 6.60 and 6.61). An 85-foot apartment car of the postwar period registered only 62 tons while a 74-footer of the standard era weighed 72 tons.[100] In 1940 mail clerks were first given the benefits of air conditioning.[101] A few years later the Pennsylvania produced a so-called dream car, with such luxuries as an enclosed lavatory, an electric coffee maker, and a refrigerator (Figure 6.62).

In a way, the superficial improvements just mentioned foretold the coming disappearance of the Railway Mail car. It had reached its apogee generations before, and had become an increasingly expensive and obsolete method for mail distribution in the postwar period. The basic problem was its operating cost: it was labor-intensive. A single car would have a crew of six to fourteen men. The upward spiral of wages, coupled with the forty-hour work week, soon made the mail car an economic impossibility. Machine sorting of letters eliminated much expensive personnel. This could best be done at a central post office, where

Figure 6.51   Before the time of electric lighting, mail cars were lighted with massive oil or gas lamps. The engraving here dates from 1883. (Engineering, August 31, 1883)

Figure 6.53   A Boston and Albany mail car designed by Fitch D. Adams in 1894. (Railroad Car Journal, June 1894)

Figure 6.54   *The Railway Mail Service pressed the railroads to provide safe cars to protect its clerks. Mail cars were among the first to be built with steel frames. The Number 20 was constructed in 1912. (Hall of Records)*

Figure 6.55   *The Erie Railway's mail car Number 699 was among the very first all-steel cars in America. (Traced by John H. White, Jr.)*

Figure 6.56   The Harriman Lines' all-steel R.P.O. car, designed in 1906. The first two sample cars, illustrated by this drawing, were produced in the following year. (Railway Review, June 15, 1907)

Figure 6.57   Interior arrangement of the Harriman Lines steel R.P.O. (Railway Review, June 15, 1907)

sorted mail was pouched and sent as a bulk commodity. The system of mail distribution had reverted to methods used before Armstrong introduced the R.P.O.

Actually, the decline of the R.P.O. began well before 1950. Twenty years earlier the Post Office had begun to turn to motor trucks, in which clerks sorted the mail just as they did in railroad cars.[102] Between 1929 and 1932 the railroads cut out 650 trains, forcing the Post Office to seek another medium of transit, which in turn forced the road to cut more trains; the trend was unending.

By 1951 long-distance truck transport had begun. A few years later first-class mail began to move by air on a space-available basis. By 1955 the volume of mail that was carried by rail was one-half of the peak wartime volume.[103] The railroads were hauling bulk mail, which produced the smallest margin of profit. Passenger train abandonments accelerated the decline of the R.M.S. In 1951, 700 routes were in service; within ten years only 262 remained.[104] As the Post Office shifted more of its traffic, the railroads were compelled to take off more trains. The Pennsylvania claimed that mail revenues amounted to 11½ percent of all

Figure 6.58   Because mail service was such an important source of revenue, a few lines named cars for Post Office officials. The W. Irving Glover *dates from about 1925.*

Figure 6.59   Pullman built this combine mail-baggage car in 1927. (Pullman Neg. 31681)

passenger train–related income, forming a vital subsidy to the continued operation of many trains.[105] In 1960 the Northern Pacific pulled off two trains when it was deprived of a $145,000 mail contract. The curtailment of such service had a devastating effect on overall railroad revenue figures. In 1961, the railroads' gross mail income was 341 million dollars; by 1973 this total had dwindled to 83 million dollars.[106] Following the introduction of the ZIP code in July 1963 came the mass abandonment of the surviving R.P.O.s. The last R.P.O. route—the New York–Washington run—was abandoned on July 1, 1977.

A few railroads that were slow to recognize how soon the R.P.O. would disappear unwisely bought new cars. In 1963 the

Union Pacific purchased nine new cars from Budd, and a year later the Santa Fe bought a dozen similar units.[107] Within three years the Santa Fe had thirty lightweight mail cars that were useless for any other service. Few were more than twenty years old—modern for an American passenger car. Yet because of the small side doors and window configuration, it was not practical to rebuild them for any other service. All were either scrapped or sold to Mexico.

Some other lines, such as the Missouri Pacific, avoided the mistake of investing in new R.P.O.s and instead bought some extremely long storage cars (Figure 6.63). Fifty cars were produced for the M P by the St. Louis Car Company in 1964. Here

Figure 6.60 The interiors of lightweight mail cars did not differ materially from the standard steel cars of twenty years before. This Great Northern car was built by A.C.F. in 1950. (Great Northern Railway)

Figure 6.61 The lightweight R.P.O. Number 4907 was part of the New York Central's reequipment program after World War II. (New York Central)

Figure 6.62   Before the lightweight era, mail clerks enjoyed few luxuries. Enclosed toilets and electric lighting came after the Number 21 entered service in 1907. (Pullman Neg. 9489)

is a good example of a car design influenced by the method of payment: the ceilings were kept low because the price was based on linear footage rather than weight or volume. An ordinary high-ceiling, 60-foot baggage car would probably carry the same load as these specially contrived 80-footers.

Railway mail service today is a high-speed freight operation with few ties to the passenger side of the business. In the early 1920s the New York Central experimented with containerized mail cars. The idea was revived in 1958 and has been expanded since that time. Some mail trains run with as many as forty cars, usually in a mixed consist of piggybacks and standard baggage car–like mail storage units. Attracted by these economical methods, some first-class mail has returned to the rails in recent years.

Figure 6.63 *Because the Post Office paid by the linear foot rather than by cubic area, some railroads bought long, low-ceiling cars to improve revenues. The St. Louis Car Company built this bulk-mail car for the Missouri Pacific in 1964. (William D. Edson Collection)*

CHAPTER SEVEN

Running Gears

PREVIOUS CHAPTERS have examined the obvious, eye-level portions of the passenger car that are most visible to the traveler; this chapter describes everything below the floor. In the shadows of the undercarriage lie many mechanisms which, although they arouse little interest in most passengers, are vital to their comfort and safety.

Trucks

Next to the body, the largest and most important subassembly of a railroad car is the trucks. Each truck is in effect a small independent car, consisting of a frame, wheels, axles, journal bearings, springs, and brake rigging. Not only do the trucks support the body, but because they can move laterally and turn on a center pin, they also materially reduce the car's resistance when entering a curve. The truck is expected to cushion the ride, and since no track is in perfect level or alignment, lateral and transverse irregularities must be absorbed by the springs and swing bolsters.

Spiral springs capable of a large deflection and hence a low frequency of vibration were favored throughout the history of the railroad passenger car. To dampen the bounciness of the "soft" coil springs, stiffer leaf or elliptical springs were arranged, usually through equalizing levers, to act in concert with the spirals. In modern times hydraulic, automotive-style shock absorbers were used as well. Side bearings were employed in yet another effort to control the car's roll, especially when it entered a steeply banked curve. Rubbing plates or roller-type bearings served this function. Side bearings could also be expected to avert derailments. If a wheel were to break, the side bearing would support the car and allow it to run on three wheels until the train could be stopped. In the days of cast-iron wheels this was an important duty of the side bearings.

With so many functions to perform, truck design presented the car builder with a major challenge. Advances in car weight and train speed compounded the difficulties. Faster schedules, particularly those requiring speeds above 85 miles per hour, dramatically increased noise and oscillation. Yet the car builder was expected to produce a safe, smooth-riding, lightweight truck for a reasonable cost. Ingenious plans were devised to satisfy this need, some of which will be outlined in the next pages. But no truck, no matter how well designed, could perform satisfactorily if not properly maintained. Hard-riding cars were usually the result of poorly maintained trucks. The most common ailments are listed below.[1]

Improper side bearing clearance—puts dead weight on bolster springs, which transmit vibrations to the body most noticeably when transversing a curve.

Weak elliptical springs—normally attributable to age, but also to improper tempering by the manufacturer.

Loose center plate—causes a rotating sensation.

Coil spring clearance—2½ inches are required. If clearance is not allowed, the equalizer will strike the truck frame a sharp blow and cause the car to vibrate for miles after the point of contact.

Worn equalizer ends—cause journal boxes to tip, creating a rocking motion.

Flat or out-of-round wheels.

Throughout their history American passenger car trucks have remained true to form: the four-wheel truck has always been the standard. Less consistency is found in the details of design, and less agreement on precisely what features produce a smooth-riding truck. This is normal, of course, in any active area of technology. Truck size has grown since the early days of the railroad. In a single century the wheelbase length increased from a scant 3 feet to 9 feet, the truck weight from less than 1 ton to nearly 10 tons.

The earliest trucks were indeed primitive affairs. Wooden frames built up from 4- by 6-inch timbers were tied together by a few light iron rods. Slender axles and wheels that would be considered barely sufficient for a modern handcar were expected to safely carry a carload of passengers. The wheels were set as close together as possible to minimize resistance in traversing curves. Leaf springs over each journal box provided the only suspension; no equalizers were used. Drawings of these pioneers are given in Figures 1.18, 1.19, 1.74, and other illustrations in Chapter 1.

In the 1840s the basic wooden-beam truck underwent gradual refinement. The so-called square truck came into favor during this decade.[2] It was thought that if the wheelbase equaled the track gauge, the truck could not fail to tram precisely because the geometry was so beautifully balanced. This myth prevailed for many years, and even after the steady-riding qualities of the spread- or long-wheelbase truck were established by locomotive designers, some conservative car builders held fast to the chattering square truck.

Two fundamental advances in truck design were introduced during the 1840s. The first and most important was the swing bolster. In an ordinary truck the bolster, the central transverse beam, is rigidly bolted to the "wheel pieces," the truck's side-frame members. The truck has no lateral flexibility, and each sideways shock the wheels receive from a poorly aligned truck or from their passage through switches or crossovers is transmitted directly to the body. The springs are effective only in absorbing vertical shocks. Charles Davenport of the Cambridge, Massachusetts, car-building firm Davenport and Bridges devised a simple but effective scheme to overcome this problem (Figure 7.1). The bolster was not fastened directly to the wheel pieces but tied to the truck frame with swing links connected to a secondary bolster, called the spring plank. The spring plank in turn was the lower support for springs fastened to the wheel pieces. The bolster was thus free to swing as much as an inch. This free motion, together with the semi-independent spring mounting, absorbed the lateral shocks encountered.

Davenport patented the swing bolster on May 4, 1841 (No. 2071). Some years later during the eight-wheel car case (covered in Chapter 1), Henry Waterman claimed that he had used swing-motion trucks as early as 1839.[3] Twenty years later a patent

Figure 7.1 Davenport and Bridges patented the swing bolster truck in 1841. It provided a fundamental improvement in the riding quality of railroad cars by dissipating lateral motion. (U.S. Patent Office)

Figure 7.2   Wooden-beam, four-wheel passenger car trucks changed little during the nineteenth century except for an elongation in wheelbase. This truck was designed in 1856. (Douglas Galton, Report on U.S. Railways, London, 1857–1858)

examiner overlooked both Waterman and Davenport's inventions and awarded a patent for the same idea to Henry Kipple and Jacob Bullock of Philadelphia (No. 26,502, December 20, 1859). The landmark invention of the swing bolster has been mistakenly credited to Kipple and Bullock in at least one modern account.[4]

While car builders came to accept the swing bolster as a necessary feature of passenger car trucks, it was less widely used for freight cars because of the additional cost. A few car builders even complained that it was not really effective for passenger service unless long links were employed.[5] The trend to short links actually worsened the ride, imparting a series of uneasy

jerks when the wheels struck a curve. Rapid flange and journal wear was a secondary result, because these heavy side thrusts pushed the truck out of square.

The second reform in truck design to appear in the 1840s was equalization. A lever or equalizer connected the axles, so that a bump or shock received by one wheel set was transferred to and partially absorbed by the other. The scheme had been used by American locomotive designers since 1838, but apparently it was not applied to cars much before the mid-forties. Early in 1845 a British engineering journal published a drawing of an eight-wheel American car with an elementary form of equalized trucks.

SCALE

Figure 7.3   A Pennsylvania passenger car truck of 1866 incorporated such improvements as equalizing levers and swing bolsters. The rubber springs shown were obsolete by that date, however.

The drawing together with the citation is reproduced in Figure 1.78. Two leaf springs over each axle were linked to a common central spring mounted below the wheel piece. Surviving drawings indicate that the equalizer was rarely applied before the middle or even late 1860s; just why remains a mystery, since it was universally installed on American locomotives generations earlier. Perhaps car builders felt that it was not necessary for very light cars traveling at low speeds. After the Civil War cars rapidly became heavier and speeds faster, which may account for the late adoption of equalizers. Once accepted they became universal, and they remain a basic part of modern truck design.

However, the merits of the equalizers have been challenged by critics who point to its ineffectiveness at very high speeds. They claim that it is actually counterproductive because the lever occasionally acts against, rather than in concert with, the springs and causes the truck to jump.[6] In the course of braking, the equalizer makes the truck tilt. Moreover, equalizers are heavy, adding to the car's dead weight; they break frequently, and they block access to the brake rigging. Yet most car designers seem to feel that the equalizer's merits outweigh its shortcomings and that it is necessary for a smooth ride.

Few new ideas gained acceptance in basic truck design after the introduction of the equalizer and the swing bolster. From this

time forward, the truck's development was a process of gradual growth in size within the rigid confines of a single pattern. By the late 1860s an archetypical design had evolved that was to remain in favor until the present day. Even the conversion from wood to steel did little to affect the basic plan. This universal truck is well represented by Figures 7.2 to 7.4. Its general features included open-jaw pedestals fastened to a rectangular frame and cross timbers with a floating swing bolster between them. A shallow U-shaped equalizer rests on each journal box, with a pair of coil springs mounted between it and the truck's main frame. The bolster is cushioned with elliptic springs independent of the equalizer. In appearance, wooden-frame four-wheel trucks twenty years apart in age look like duplicates. It is only when the actual dimensions are compared that the increasing size becomes obvious.

Six-wheel trucks are the second most common style found on American railways (Figures 7.5 and 7.6). They appeared as early as 1845 and, as mentioned in Chapter 1, a number were adopted by several railroads, including the Michigan Central. They were not widely used by the industry as a whole until after the introduction of heavyweight palace cars in the late 1860s. As sleeping, parlor, and dining cars became more numerous, six-wheel trucks grew in popularity. The extra axle at either end of the car was

Figure 7.4 Barney and Smith, car manufacturers of Dayton, Ohio, produced this wooden-frame truck around 1880.

Figure 7.5 Barney and Smith constructed this typical six-wheel truck around 1880. Note the decorative striping.

Table 7.1   Growth in Truck Size, 1867–1887

	1867	1887
Wheelbase	6 feet	7 feet
Wheel piece (frame top rail)	9 × 4½ inches	9 × 4½ inches
Equalizer springs	Rubber cylinders	Steel coil
Bolster springs	Leaf steel	Leaf steel
Brake shoes	4	8
Wheels	33-inch cast iron	36-inch cast iron
Axle (diameter at center)	3 15/16 inches	4 ⅛ inches

Source: Railroad Gazette, December 9, 1887, p. 792.

justified chiefly because of the car's great weight. Concentrating the load on eight bearings was considered undesirable for several reasons. It promoted hotboxes and axle and wheel failures.[7] In a time of cast-iron wheels, small journals, and faggoted wrought-iron axles, there was good reason for concern. The track was generally not up to supporting such a load safely; light ballast, slender rails, and widely spaced ties provided a shaky foundation. It was necessary to spread the load out as far as possible and divide it over not eight but twelve bearings. The longer wheelbase had the added advantage of imparting less motion to the car and thus in theory providing a smoother ride. The center axle had a generous amount of end play to help it ease around curves. Because brake shoes could be mounted to twelve rather than eight wheels, six-wheel-truck cars were considered safer. It was also claimed that brake shoe life was extended because the wear was more widely spread. In the wooden age, any car weighing over 25 tons was given six-wheel trucks.[8] After steel cars came into favor, cars weighing 42.5 tons and over were normally fitted with twelve wheels.[9]

At an early date critics gathered in opposition to the twelve-wheel truck. Their arguments were convincing enough: they cited the greater first cost and the increased weight, both of which easily exceeded the cost and weight of the conventional

Figure 7.6   An iron-plated, wooden-frame, six-wheel truck with paper wheels made by the Chicago, Burlington, and Quincy in 1886. (Railroad Gazette, April 2, 1886)

four-wheel variety by one-third. In 1890 a four-wheel truck weighed 5 tons, a six-wheeler 7 to 8 tons. Six-wheel framing was much more complicated, because the center or kingpin plate was directly over the middle axle. It was necessary to bridge over this wheel set and hang the truck on two auxiliary transoms or bolster beams. The bridge bolster was costly, complicated, and weak. The extra wheel bearings amounted to that much more drag on the train. The 10- to 11-foot wheelbase impeded the passage of curves, increasing friction and damaging the track. The supposed steadier ride resulting from the long wheelbase was a myth, according to tests made on the Milwaukee Road.[10] In this test the center wheel set was removed. The truck rode no better than a 9-foot-wheelbase truck and was considerably noisier. The critics claimed that long-wheelbase, four-wheel trucks with large journals could match any six-wheeler. But the majority of railroad men seemed to regard the six-wheel truck as the best form for passenger cars. Such an old-line executive as Daniel Willard of the B & O was seen walking along station platforms with eyes intent on the undercarriage of the cars. He would pick out a twelve-wheeler and climb aboard. A car man explained: "He won't ride in an eight wheeled job, no matter how many fancy stripes they paint on the sides."[11]

During the heavyweight era, twelve-wheel cars reached their peak of popularity. Six-wheel trucks were largely rejected for lightweight cars. They were rarely used after 1935 and then only for very special cars, such as the Milwaukee Road's superdomes (1952). An exception to this general rule was the New Haven, which continued to order twelve-wheelers and bought some twelve-wheel dining cars as late as 1949.

There was a short-lived interest in eight-wheel trucks, and a number of centipede-like sixteen-wheel cars were constructed in this country. Mechanics are occasionally inspired to indulge in excesses—reasoning, for example, that if twelve wheels are better than eight, then sixteen must be better yet. But in fact the belief that more wheels would ensure a smoother ride was ill-founded. They ground through curves and occasionally upset, and with so many axles, springs, and levers, they seemed to exaggerate rather than dampen the defects of a poor track. Lincoln's sixteen-wheel private car was abandoned by its second owner because of its galloping ride. The journalist Zerah Colburn said that traveling in a sixteen-wheeler "is more like sailing than rolling."[12]

It has been reported that such cars first appeared in 1839. In 1858 Wason built some sixteen-wheel sleeping cars for the Michigan Southern. Two years later the same firm produced an elaborate private car similarly mounted for the Viceroy of Egypt.[13] The 20-ton, 67-foot-long car had an open center section covered by a fanciful canopy and surrounded by scrolled iron railings. The special iron-frame trucks are illustrated in Figure 7.7. The car, minus most of its trimmings, is still occasionally used by the Egyptian government.

Pullman approved of the sixteen-wheel car and had twenty-three in his fleet.[14] His acceptance was doubtless inspired by his mechanical chief, C. F. Allen, who received a patent (No. 65,788) for a sixteen-wheel design on June 18, 1867 (Figures 7.8 and 7.9). Allen had built such cars for Pullman since 1863 and apparently had also been successful in converting his former employer, the Chicago, Burlington, and Quincy. Long after 1869, when Pullman had abandoned them, the Burlington continued to

Figure 7.7   An iron-frame eight-wheel truck built by Wason for the Viceroy of Egypt's private car, 1859–1860. (American Railway Review, January 19, 1860)

Figure 7.8   An eight-wheel truck patented by Pullman's mechanical chief, C. F. Allen, in 1867. (National Archives)

produce the sixteen-wheel centipedes. In 1876, for example, the road's Aurora shops were reported to have completed a new sixteen-wheel diner.[15] Travel brochures spoke of the splendid sixteen-wheel giants available to patrons traveling west. And then unexpectedly, the railroad reversed its policy in 1879 and condemned sixteen-wheel cars as a foolish mistake.[16] It rebuilt all its sixteen-wheelers with conventional trucks, ending this curious misadventure in passenger car suspension.

Late in the last century, cars grew to such proportions that wooden-beam trucks were no longer practical. To build a truck with a wooden frame of the requisite strength would call for such massive timbers that the structure would be absurd. The same compromise that was found so effective in body construction was also employed for the undercarriage parts. Iron plates and rods stiffened the wheel pieces and transom members. Flitch plates covering the backs and fronts of the major truck-frame timbers became common by the late 1880s. The outer plate was normally ¼-inch thick, while the inner plate measured ⅜ inch. Button-headed, ½-inch bolts secured the plates. By the 1890s composite construction was standard, except for the very lightest passenger cars.

Composite construction naturally suggests the possibility of all-metal trucks, but even before the coming of the commercial steel car, many attempts were made to introduce trucks of this type. Probably the earliest builder to try it was Ross Winans. Around 1835 he produced an all-metal truck with a large leaf spring that

Figure 7.9 A plan view of Allen's eight-wheel truck patented in 1867. (National Archives)

Figure 7.10 Winans's live spring truck, similar to the type that was used on B & O passenger cars in the 1830s and 1840s. The example here is a much later model that was in service on the Georgia Railroad's freight cars. (National Car Builder, January 1891)

served as both the side frame and the suspension. A stirrup bracket at the spring's center held the leaves together and formed a connection with the forged wrought-iron bolster. The cylindrical ends of the bolster were free to swivel in the bracket, giving the truck great flexibility. A Winans spring or live truck is illustrated in Figure 7.10. The B & O continued to use the live truck for its passenger cars until at least 1840. The truck's serious failing was its tendency to run out of square because there was no cross framing. This caused excessive flange wear and sus-

ceptibility to derailments. The constant flexing of the spring-side member led to metal fatigue, resulting in broken leaves and ultimately in total failure. It was practical only for very short wheelbases, a characteristic originally thought advantageous, but one that was soon recognized as a liability. The fact that it was impossible to mount brake shoes on trucks of this design was a defect so serious that it is a wonder live trucks were ever built at all. Yet they were simple, cheap, and well suited to light, slow-moving cars. In 1860 several Southern lines were still using

Fig. 1.

Fig. 2.

DAVENPORT & BRIDGES' IMPROVED PATENT IRON TRUCK FOR RAILROAD CARS, is presented above, and the attention of Railroad Companies is respectfully invited to the following description of their justly-celebrated invention:

Fig. 1 of the drawing above represents a top view or plan of our Improved Railroad Truck. Fig. 2 is a central, longitudinal, and vertical section. C, Fig. 1 and 2, represents the arched bars of the side trusses: they consist of two long bars of plate iron (about three inches wide by seven-eighths of an inch thick,) bent into the shape as seen in drawing 2. Each of them is placed directly over a flat and straight tie bar, A, which extends from one end to the other, as seen in Fig. 2. These parts, so arranged, receive between their ends the ends of diagonal cross bars or braces, B, which are united at their centres y being clasped and welded, as seen in Fig. 1. The bars so composing what may be considered as side trusses and diagonal cross braces, rest at their ends upon four pedestals, F, F, which receive the bearings or boxes for the axles to run on. Another flat tie bar, P, extends from the under side of one of the pedestals to that of the other, on the same side of the frame, and the whole is secured together by eight bolts, J, J, passing down through the ends of the several ars, A, B, C, and the pedestals, and on each side of the journals of the axles, O, O, in the positions represented in the drawings. From the above it will be seen that there are two bolts to each pedestal, and that this number is all that is requisite for the full security of the bars and pedestals together. The body rests and moves upon two sectional supports, D D, arranged on the sides of the truss frames, as seen in Fig. 2; they extend somewhat, or a sufficient distance above the truss frames, and are jointed at their lower ends by means of a bolt, L, which rests upon the top of the lower spring, M, which spring rests upon a bolt passing through the lower part of the inverted strap, E, which strap passes over and rests upon the top part of the upper spring, M, which is placed within the truss frame, and rests upon the top of the bar, A.

Two bands, N, N, are passed entirely around the central part of each truss frame the object of the same being to transfer the strain, or a portion thereof, of the spring, from the tie bar, A, to the arched bar, C.

These Trucks are adapted as well for eight-wheeled passenger cars as for baggage and freight cars, giving to each a more agreeable and easy motion than any other Truck heretofore constructed or in use. They are simple in their construction, combining strength and great durability, although weighing at least twelve hundred pounds less than the common Trucks. Besides these excellences, by reason of the elasticity of the braces, B, B, B, B, as seen in the drawing, and the other peculiarities of construction, the weight is *equalized* upon all the wheels, and yet any one may be raised so as to pass any inequality on the rails without lifting either of the other wheels from the track, thus rendering it almost impossible to run a car off. Being bound, and having as it were but four joinings, they are protected from injury by lateral strains, and in case of damage are easily repaired.

These excellences have been fully tested by use, for a long time, on the Eastern, the Fitchburg and Long Island railroads; and for proof of the above stated superiority of these Trucks over all others, we refer to the experience of those who have used and run them.

CAMBRIDGEPORT, April 1, 1845.   DAVENPORT & BRIDGES.

Figure 7.11   An inside-bearing, all-metal truck patented by Davenport and Bridges in 1844. (American Railroad Journal, July 24, 1845)

Winans's trucks on their freight cars.[17] The side-spring weight was given as 250 pounds each. More than thirty years later, the Georgia Railroad still had about 200 cars running on the old-fashioned live truck.[18]

In the 1840s other builders were promoting the cause of the iron truck. Charles Davenport patented such a carriage fabricated from iron straps on August 10, 1844 (No. 3697). Four double elliptic springs cushioned the ride (Figure 7.11). The inventor claimed savings of $100 in cost and 1,200 pounds in weight over a comparable wooden truck.[19] The device was promoted by Davenport's huge car plant—at the time probably the largest in the country—and received tests upon the Eastern, Fitchburg, and Long Island railroads. A similar design was patented by Fowler M. Ray on March 21, 1845 (No. 3962). Ray claimed that his limber truck could pass over obstacles such as logs or large stones without derailing.[20] He said that each truck weighed a mere 1,000 pounds, yet produced "delightful riding cars." Advertisements running in the *American Railroad Journal* throughout 1847 carried testimonials from several prominent lines—including the Philadelphia and Reading, which stated that it was making extensive use of Ray's miraculous invention. At the time that Davenport and Ray were promoting iron trucks, Eaton and Gilbert placed a competing design on the market. It was more conventional in appearance and, because of its relatively long wheelbase, must have offered a superior ride. How extensively it was used is not known, but a fine drawing of it is available in the plates of the American-style car built for Germany by Eaton and Gilbert (see Figure 1.79).

The initial flurry of interest in iron trucks produced few lasting results. Wood remained in favor, although mechanics did continue to experiment. In addition to the Wason and Allen eight-wheel iron-frame trucks of the 1860s., Allen patented four- and six-wheel trucks with strap iron side frames. He continued to use wooden transoms, however. Many years later, in 1884, the Philadelphia and Reading produced an all-iron six-wheel truck with a 10-foot wheelbase.[21] The trucks, with their 6-inch channel wheel pieces, were used under one or more parlor cars. A few years later the Lehigh Valley Railroad adapted a six-wheel tender truck design of 1884 for passenger car service. The frame was forged very much like a locomotive bar frame. The top rail measured 1¾ by 4 inches. Unlike a locomotive's, however, the frame was placed outside of the wheels. The success or failure of these two designs is unrecorded, but apparently they were not widely copied. Even so, all-metal trucks were coming into broader use at this very time, at least in the area of freight cars. Pressed-steel trucks were being introduced with some success in the late eighties. In 1889 Sampson Fox, a leading British manufacturer of such trucks, made a set of 7-foot-wheelbase bogies for passenger car service.[22] Praised by the railroad press as simple and handsome, they were surprisingly modern in appearance. The Pennsylvania and perhaps other American lines ran tests with Fox's experimental truck.

A far more durable and solid truck frame could be made from cast steel. A one-piece casting assured rigidity, guaranteeing that the frame would be square and that the axles and wheels would always run true. As early as 1896 Chicago's Lake Street Elevated was purchasing cast, malleable iron truck side frames.[23] Two years later the same line switched to cast steel, with bolsters made from 10-inch channels.[24] Steam railroads adopted cast-steel trucks somewhat later for freight cars. The earliest case that can be found of a cast-steel frame for a passenger car was in 1904 on the Big Four, a subsidiary of the New York Central. Cast-steel

Figure 7.12   A six-wheel, cast-steel truck with bolt-on pedestals, made in 1918. (Pullman Neg. 23662)

Figure 7.13   This cast-steel truck of 1929 had integral pedestals. Note its similarity to the wooden-frame, six-wheel trucks shown in Figures 7.5 and 7.6. (Pullman Neg. 32230)

four-wheel trucks weighing 13,000 pounds each were placed under a baggage car.[25] They proved so successful that within two years sixteen baggage cars had similar trucks. They replaced ordinary six-wheel trucks, with a weight saving of 9,500 pounds per car.[26] With the coming of steel cars, steel trucks were mandatory. The adjectives strong, massive, and heavy describe these ponderous subassemblies. In the heavyweight era four-wheel trucks weighed 7½ tons each, while a six-wheeler weighed 10½ tons (Figures 7.12 and 7.13). Designers would continue to agonize over the ratio between truck weight and total car weight. It was discouraging to find that even in the lightweight era a set

of trucks weighed nearly 10 tons each and equaled one-third of the car's total weight.

The age of steel revolutionized the materials of construction but did little to alter basic design. The steel truck was a literal facsimile in metal of what formerly had been made of wood. There were, of course, some exceptions, but only one variant design had any widespread use. Ironically, this radical truck style was chosen by the standard railroad of the world: the Pennsylvania. At the outset of the steel car age the Pennsylvania abandoned the conventional pattern for both four- and six-wheel trucks. The four-wheel-truck design that it adopted had several

Figure 7.14   The Pennsylvania's distinctive style of fabricated steel truck used on the first series of P-70 coaches. (Railway Review, June 8, 1907)

unorthodox features. No equalizer was used; a triple nest of helical springs was mounted over each journal box (Figure 7.14). Unlike traditional wooden-frame construction, the bolster and center plate were placed very low to provide clearance for the center sill. The brake dead levers were anchored to the car frame rather than to the trucks, eliminating the tilt and resulting jerky motion produced when the brake levers use the truck as a fulcrum. The Pennsylvania's six-wheel truck exhibited similar peculiarities (Figure 7.15). The first group of trucks was fabricated from pressed shapes or mill stock, but by 1915 one-piece cast-steel pedestals and end brackets were substituted.[27] Within the next decade one-piece cast-steel frames were adopted, and the equalizer was readopted.[28]

It is not surprising that efforts were made to reform truck

design during the early stages of the lightweight era. Critics of the railroad engineering establishment contended that no new ideas had been seriously considered for generations. One of the most vocal opponents of conventional truck design was the Milwaukee Road's chief mechanical officer, Karl F. Nystrom. Nystrom tried every reasonable alternative to the conventional patterns. He tested very long wheelbases, experimented with placing springs in the immediate vicinity of the boxes (in theory the ideal location), and studied such European innovations as the triple bolster. Yet after years of testing, he was forced to admit that the conventional pattern rode about as well under normal operating conditions and was cheaper than the exotic schemes.[29] It was one thing to produce a good riding truck for speeds up to 85 mph, but for travel above this rate the answers

Figure 7.15   This Pennsylvania six-wheel, all-steel truck of 1907 featured a massive plate-steel bridge bolster. (Railway Age, June 7, 1907)

came very hard. One of Nystrom's assistants summarized the lessons learned after ten years of testing:[30]

(1) High-speed passenger car operation begins above 85 m.p.h. Below this speed, ground wheels and good track maintenance on conventional four-wheel truck will give satisfactory results.

(2) No high-speed truck of any design will function properly with wheels that are eccentric more than .020 in.

(3) The vertical riding qualities of a truck are a function of spring deflection. The bolster spring of a high-speed truck must have a sensitivity ratio (spring rate divided by the load carried) of not more than .15, preferably .10.

(4) The truck must have two spring systems, namely, primary and secondary systems. The equalizer springs system must not have a sensitivity ratio lower than .25, or spring surging will result. Springs of high deflection cannot be set down on completely unsprung parts without spring surging. A rule of thumb is that the equalizer spring deflection should be not more than ½ that of the bolster.

(5) The truck must have equalizers, or truck frame gallop will result if operated above 100 m.p.h.

(6) Friction surfaces should be reduced to a minimum.

(7) Car-body-roll stability is a function of the square of the lateral spread of the bolster spring.

(8) The value of swing hangers, especially long ones, is badly overrated in producing a good lateral ride.

(9) Vibration dampers and insulators are necessary to give a quiet car body.

(10) Bolster springs should be placed as high as possible to reduce instability.

(11) Wheel base is not a function of good riding either vertical or lateral.

(12) Wheel shimmy is caused by the taper of the worn wheel metal and is not a function of the shape of the tread when new. Wheel shimmy cannot be controlled by truck design, snubber or other devices. What causes a wheel to wear a critical shimmy tread is not known. The Milwaukee has one truck design that is not plagued with this phenomenon while all the others at times will develop wheel shimmy. There are so many differences between the truck designs that it would be impossible at this time to state the cause.

(13) High-speed passenger car trucks have excessive vertical amplitudes on rough branch line tracks when operated at low speeds. When designing branch line or suburban cars, high speed truck design should not be followed.

The triple-bolster idea that Nystrom studied was modified for American requirements by the Pullman Company.[31] These trucks were distinguished by wing-like pedestals with coil springs at each end, quite near the axle boxes (Figures 7.16 and 7.17). The heavy equalizer was eliminated, and with it a large part of the objectionable unsprung weight of the conventional truck. The bolster, although carried by full elliptic springs, was not pivoted to the truck frame but rather supported by intermediate bolsters hung from the frame on short hangers and coil springs.

Figure 7.16   A cast-steel, triple-bolster truck with roller bearings, dating from about 1937. (Car Builders' Cyclopedia, 1937)

Figure 7.17   A triple-bolster, cast-steel truck of 1936 that was used on the Illinois Central's motor train, the Green Diamond. Note the truck-mounted air brake cylinders. (Pullman Neg. 39521)

The triple bolsters offered a soft ride; however, they were also prone to weave or teeter, and around 1940 they fell from favor.

The railroad supply industry played an important part in modern truck design, and in fact began to dominate this area after the passing of the wooden car. Many railroads and car builders were content to accept the stock models, especially for freight cars. Even in the passenger car field, a large proportion of the American fleet rode for many years on trucks manufactured by

the Commonwealth Steel Company of Granite City, Illinois (Figure 7.18). This firm, later part of the General Steel Casting Corporation, continued to play a leading role in passenger truck design until the market for such equipment expired. Around 1940 Commonwealth introduced the floating bolster, a scheme that freed the transom from the restrictions and wear of guide pedestals.[32] The bolster was held in place by a heavy tie rod, one end of which was fastened to the truck side frame and the other to a

Figure 7.18   A double-equalizer truck with inside swing hangers and roll stabilizers, used on a 1939 Budd-built Seaboard Air Line coach. The cantilevered air brake cylinder proved a poor arrangement, and side-mounted cylinders became the favored position in later years. (Seaboard Air Line)

bracket on one end of the bolster (Figures 7.19 and 7.20). Rubber discs at both unions allowed the necessary play. This became a common feature on four-wheel trucks. Another less widely used but promising innovation by Commonwealth was the outside swing hanger introduced in 1948 on a group of cars produced for the New Haven.[33] The bolster extended well beyond the side frames, so that the swing links were outside rather than inside (Figure 7.21). Because they were so far from the center pin, these trucks provided great stability and did much to prevent unwanted rolling. Only coil springs were used; the manufacturer claimed that they offered high static deflection. The trucks produced a soft ride, but their full potential was never realized because it was necessary to restrict the free swing of links to stay within existing side clearances.

Springs

As indicated in the previous section, various styles of coil and leaf springs were the basic forms used for railway car trucks. Since about 1870 they have been combined with great success, for the elliptics serve as an effective damper against the harmonic vibrations characteristic of the coil springs. Tempered steel was the only material used for metal railway car springs.

It was assumed by some early engineers that railways would furnish such a perfect pathway that no springs would be necessary. Smooth wheels running over an equally smooth permanent way should result in a faultless ride. But the motion of heavy trains, combined with the action of rain, sun, and frost, soon added unwanted kinks and buckles to what was once straight and true. In its annual report for 1833 the Baltimore and Ohio confessed:

Experience has demonstrated that the use of springs on all the carriages, whether for the transportation of passengers, or of merchandize, would contribute greatly to the preservation, both of the cars and of the road. Such an arrangement would assist in regulating the motion, and serve in some measure to alleviate the sharp jar inseparable from the sudden and violent collision of two hard bodies and no mitigating medium, and thus save both the cars and the road from that racking concussion which no perfection of construction, nor any reasonable care and foresight to prevent every accident, can entirely obviate.

In like fashion the Philadelphia and Columbia was rueful over the miserable performance of its initial lot of unsprung cars, most of which had pounded to pieces after little more than a year.[34] It was agreed at a very early date that springs were no idle luxury.

Figure 7.19   General Steel Casting Corporation made this floating-bolster truck for the New York Central in 1942. The horizontal stabilizer or drag link, visible above the right set of coil springs, held the bolster in line but allowed it to move vertically. (Pullman Neg. 45955)

Figure 7.20   A truck very similar to the cast-steel, all-coil-spring truck shown in Figure 7.19. Elimination of elliptical springs reduced maintenance costs, because the coil springs did not wear. (Pullman Neg. 48061)

Figure 7.21   Cross section of an outside swing hanger truck manufactured and introduced in 1948 by the General Steel Casting Corporation. (Car Builders' Cyclopedia, 1953)

In the pre-Bessemer period, steel was a scarce commodity. It was difficult and expensive to manufacture, and it could be exceedingly unreliable unless it was made by a warranted master. Since the American iron industry was generally unable to produce such rarified metals as spring steel, most orders went to overseas suppliers. There was a great need for some alternative—some new, resilient material that was cheap and within the capabilities of domestic manufacturers. Both practical and foolish schemes were proposed. Like alchemists of an earlier age, Yankee mechanics seemed ready to try any combination of materials or methods in the hope that one would give the desired results. They tinkered with rubber, felt, wood, and compressed-air cylinders.

Rubber was the most promising and widely used nonmetallic spring adopted for railway work. It is said to have been suggested as early as 1825 by Thomas Tredgold. Six years later Conduce Gatch claims to have used india-rubber blocks under the bolster of the eight-wheel car *Columbus*.[35] In 1835 a British patent was granted covering the application of such springs to vehicles.[36] The idea was obvious enough to have occurred to many men with an inventive turn of mind, but the difficulty lay in the material itself. Natural rubber was suitable to only a few uses because of its extreme reaction to changes in the temperature. It melted in the summer and froze in the winter. In 1839 Charles Goodyear perfected a practical method of stabilizing it through the so-called vulcanizing process. The method was not patented until 1844, but news of it appears to have spread quickly to those interested in developing a rubber car spring.

Fowler M. Ray of New York began experimenting with india-rubber springs in 1843.[37] In the following year his tests were transferred to the Osgood Bradley car plant in Worcester, Massachusetts. At that time Ray adopted vulcanized rubber, with results so good that actual road tests were made by the New Jersey Railroad and Transportation Company in the winter of 1845–1846. Rubber springs worked well in the light cars of the day. Ray convinced other railroads of the utility of his "invention" and secured a patent (No. 5696) on August 9, 1848. He claimed that rubber springs saved 444 pounds and $59.06 per truck compared with steel springs.[38] Ray's New England Car

Spring Company announced that it had sold 1½ million pounds of springs worth over 1 million dollars between 1847 and 1852.[39] During these years of expansion Ray quarreled with an English inventor, W. C. Fuller, who claimed exclusive rights to the manufacture of rubber springs. In 1850 the U.S. Patent Office decided in Ray's favor, but Fuller continued to protest the decision through the courts. Within a few years, however, the controversy became an empty battle, because the rubber spring fell from favor.

The declining fortunes of the gutta-percha spring were announced as early as 1857, when the *American Railroad Journal* said that "vegetable gum" was not the thing for car springs.[40] When it was new it was too lively and made the car dance, yet after little more than a year it crushed down into an inelastic mass. And in the winter when the tracks were rigid with frost and good springs were most needed, the gum block had frozen solid. In the same year the Illinois Central annual report announced its intention of reconverting to steel springs, and within two years all the rubber springs had been removed from its passenger cars. Some other roads remained loyal to rubber springs, and the fact that the New England Car Spring Company continued to advertise in the *American Railroad Journal* into the 1860s indicates that the market had not entirely disappeared (Figure 7.22). Rubber also continued to be used in place of coil springs on some lines into the 1870s, but the bolster (and hence the main weight of the car) was supported by leaf springs. An example of this form of spring was shown in 1879 in the first *Car Builders' Dictionary*.[41]

Air and wooden springs were tried briefly at the time that rubber was being accepted into railway service, but they never achieved any wide application. The attempts to use them indicate the extent and the intensity of the search for a substitute for steel springs. The first serious effort to introduce air springs was made by Levi Bissell of New York City, an inventor best remembered for his locomotive safety truck. Bissell secured a patent (No. 2307) on October 11, 1841. Each spring consisted of a cylinder 6 inches in diameter by 10 inches long. Leather rings and a 2-inch reservoir of oil and white lead formed the seal.[42] In the following year the Schenectady and Troy Railroad was re-

New England Car Spring Co.

SOLE MANUFACTURERS

OF THE

IMPROVED BEARING SPRING

BEARING SPRING

BUFFER SPRING

IMPROVED BUFFER

India Rubber Car Springs.

OFFICE, 61 CHAMBERS STREET,

NEW YORK.

Figure 7.22   Rubber springs were widely used by American car builders between 1845 and 1860, though they did not replace steel springs (American Railroad Journal, *February 8, 1862*)

ported testing "atmospheric springs" on some of its coaches.

Baldwin also experimented with the Bissell air cushion. Several years later the Boston and Maine was said to be applying air springs to new cars under construction at the Andover shops.[43] In 1861 the Philadelphia, Wilmington, and Baltimore claimed that it was still getting good results from such springs after four years of service.[44] Although the air spring never really challenged tempered steel, it remained a favorite with inventive minds and has been periodically reintroduced as the solution to the problems of truck suspension. As recently as 1969 General Steel Industries was promoting the idea for rapid-transit cars, and since that time air springs have been used more widely.[45]

Alert designers are always hopeful of employing some common material to solve a persistent mechanical problem. An example was the effort of James Millholland, at that time mechanical superintendent of the Baltimore and Susquehanna, to perfect a wooden car spring.[46] He reasoned that ash, being tough and resilient, would make a cheap, serviceable spring. Two pieces 8 feet long and 2 by 6 inches in cross section formed the spring for the six-wheel cars used on the B & S line. At least one passenger car was similarly equipped. The scheme was never widely used for passenger cars, but thousands of four-wheel coal jimmies were built on Millholland's scheme. The idea was patented on September 23, 1843 (No. 3276).

In the mainstream of car spring development, the most impor-

tant event during the second half of the nineteenth century was the general introduction of coil springs. They were rarely applied to railway cars much before 1870; the veteran car builder John Kirby stated that elliptic steel springs were about the only type used earlier.[47] Coil springs were too limited in capacity or were too high in price. Coil or "worm" springs were tried in carriage suspensions at least as early as the 1790s, but they exhibited a fatal weakness not found in leaf springs: a single break rendered the coil spring useless. In an elliptic, on the other hand, one or more leaves might fracture and the suspension would still be operative, although reduced in effectiveness.

Spring making, like all metal goods production in the early period, was an art. The design and manufacture was entrusted to an experienced blacksmith, "who looks wise and mysterious as he draws hieroglyphics on the dirt floor of his shop, or as he plunges the heated plates into a vessel of inscrutable liquid to give them the requisite temper."[48] The need for more rational analysis and less guesswork in this important area was felt by many railroad managers. Dependable springs were essential. European steelmakers were making advances in spring manufacture at the time of the U.S. Centennial. And a year or two before the Centennial, the Watertown Arsenal conducted a series of spring tests for the Pennsylvania Railroad. The trust in wise blacksmiths was disappearing; spring making became an increasingly specialized process carried on by large commercial producers.

Fig. 138.

SECTION OF WHEEL, AXLE,
JOURNAL-BOX, ETC.

Fig. 139.

Section through Journal.   Front View.

WHEEL, AXLE, JOURNAL-BOX, ETC.

A. *Centre, of Axle.*	9. *Stop-key Journal-bearing.*	15. *Dust-guard and Dust-guard Chamber.*	23. *Rim, of Wheel.*
B. *Neck, of Axle.*		16. *Dust-collar.*	24. *Face, of Rim.*
C. *Wheel-seat.*	10. *Journal-box.*	17. *Equalizing-bar Seat.*	25. *Tread of Wheel, or Wheel-tread.*
D. *Dust-guard Bearing.*	11. *Journal-box Cover.*	18. *Pedestal.*	
E. *Journal.*	12. *Journal-box-cover Hinge-pin.*	19. *Pedestal-horns.*	26. *Wheel-flange.*
F. *Axle-collar.*		20. *Pedestal-jaw.*	27. *Journal-bearing Stop-key.*
3. *Stop-plate.*	13. *Journal-box-cover Spring.*	21. *Hub, of Wheel.*	28. *Oil-cellar.*
7. *Journal-bearing.*		22. *Wheel-plate.*	29. *Stop Journal-bearing.*
8. *Journal-bearing Key.*	14. *Journal-packing.*		

Figure 7.23   A typical journal box, bearing, and axle arrangement that has changed little
in its general plan since about 1850. (Car Builders' Dictionary, 1879)

Journal Bearings and Lubrication

In the family of vehicles, the railroad car is a simple mechanism. It has few moving parts, unlike the automobile's and the locomotive's hundreds of bearings and wear points, all potentially troublesome and all requiring regular attention to ensure satisfactory operation. A car has only eight main bearings. If the axle boxes are properly serviced, the car should run trouble-free, always excepting the auxiliary equipment such as heating or lighting systems. Yet the simplicity of the car's basic running gear is deceptive. The possession of few moving parts does not necessarily guarantee trouble-free operation. To the contrary, the axle bearings were a notorious source of trouble to both railroad workers and the traveling public, particularly in the early period. Overheated journals caused delays and in some instances frightful disasters. The term "hotbox" became part of the railroad vernacular. The perfection of a reasonably dependable axle bearing has occupied the car builder's attention since the industry's infancy. Enormous progress has been made over the years, but the search continues for a perfect style of bearing.

Through almost the entire history of the American railroad car, bearings have followed one basic design (Figure 7.23). The housing, or box, was outside of the wheels. The axle ends protruded through the wheel and into the box. The smoothly finished ends were called journals. The single half bearing, a crescent-shaped bronze casting, rested on top of the journal. Properly called journal bearings, these pieces were frequently referred to as brasses, even though the metal was a bronze alloy. The lower portion of the box was a cavity used for an oil reser-

voir. It was packed with loose strands of heavy thread, called waste, which acted as a wick to wipe oil on the underside of the journal. It was a simple, self-lubricating bearing that in general proved very satisfactory.

The precise form of the earliest journal bearing is not known, but the oil reservoir, or oil cellar, was used in this country before 1832.[49] In some instances it had a floating cork in place of waste, though wicking was probably also used during the same period. Brass bearings were the standard in Britain at the time; the oil cellar, however, appears to be of native origins. Unfortunately, no detailed drawings can be found for the pioneer period of the American axle box. The general drawings that are available only hint at the nature of the arrangement. It is apparent from the illustrations in Chapter 1 (for example, Figures 1.2, 1.10, and 1.76) that the earliest boxes were very small and that some were undoubtedly more like grease boxes than waste-filled oil reservoirs.

Plain bearings offer too much resistance to the free rolling of the train, according to some pioneer mechanics, who sought to introduce frictionless running gears. When the American railroad era opened, the most promising effort in this direction was a device patented by Ross Winans on October 11, 1828. At the time Winans was an unknown farmer from Vernon, New Jersey. The spectacular performance of his so-called friction wheel bearing made him an instant celebrity in American and British railway circles. Although the device ultimately proved a failure, it did launch its originator on a distinguished and occasionally stormy engineering career. Winans proposed to substitute rolling for rubbing friction and to reduce the surface area at the contact

Figure 7.24   Winans's short-lived "friction wheel," or roller bearing, was patented in 1828. (B & O Annual Report, 1831)

point of the secondary axle. He accomplished it by providing an outside, cup-shaped bearing wheel. The axle bore against the inside rim of the friction wheel, which itself turned on a small-diameter shaft (Figure 7.24). Thus the main bearing was a rolling one. Actually, the axle was floating. Winans said that this was an advantage because there would be less flange wear in traversing curves, since the axles, and hence the wheels, were free to depart slightly from their normally fixed parallelism. He did not discuss what effect it might have on the car's steadiness—nor did he explain how brakes might be applied.

Not many months after the patent was approved, Winans exhibited a large-scale model of his friction wheel in Baltimore. The tests astonished all who saw them: the 125-pound car, carrying a load of 860 pounds, was propelled by a piece of twine running over a pulley with a ½-pound weight at its end.[50] Friction was reduced by 40 percent. A larger car was built, and it performed in an equally astonishing manner. In February 1829 four full-sized cars were at work, hauling away spoils from a deep cut being made for the railroad near Baltimore.[51] A single horse could move 15 tons over the rough temporary track. News of the great technical breakthrough reached other fledgling railroads. The South Carolina, the New Castle and Frenchtown, and the Ponchartrain were all eager to try Winans's scheme. Excitement over the friction wheel peaked when Winans demonstrated a car in England at the famous Rainhill Trials. He was now in the company of such masters as Stephenson and Hackworth. The prestige of American engineering was advanced by the remarkably free-wheeling handcar that Winans propelled back and forth over the tracks of the Liverpool and Manchester Railway.[52] His triumph was short-lived, however. The directors of the B & O had adopted the invention, and in service the friction wheel exhibited a fatal defect: the journal and inner periphery of the friction wheel wore very rapidly.[53] White-metal liners were tried, but to no avail. Its extra weight and high cost were against the Winans bearing as well. By 1831 the B & O was replacing the friction wheels with plain bearings.[54]

The B & O, still willing to experiment in spite of the friction wheel's failure, adopted the invention of an associate of Winans, John Elgar (1780–1854), who used new materials to build a better plain bearing. In a patent specification dated October 1, 1830, he proposed to retain the outside-bearing scheme introduced by Winans, but he envisioned a conventional bearing made from more durable materials. The axle ends would be "steeled," that is, a wrapper of steel would be forge-welded around the journal portion of the axle and would form a bearing shaft 2 inches in diameter by 4½ inches long. He would use chill-hardened cast-iron bearings rather than brass. An oil box with a capacity of ½ gill and a dust seal were other features of Elgar's patent. Many years afterwards Benjamin H. Latrobe recalled that Elgar's bearing was used on the B & O for many years with good results.[55] The B & O conducted a test in 1844 which demonstrated the superior antifriction qualities of soft, white-metal-lined brass journals, but the Elgar bearing was so durable and so cheap that it continued as the standard.[56] Other roads also used chilled-iron journal bearings. Von Gerstner reported seeing them on cars of the Boston and Worcester. The South Carolina tested Elgar's iron boxes and steeled journals in 1838 with such good results that within two years all the eight-wheel cars on the line were fitted with them.[57] The materials were at first purchased from the New Castle Manufacturing Company, but after a few years they were made at the railroad's own shops. The boxes showed very little wear and flourished on cheap tallow lubricants. Elgar's style of bearing performed well so long as very slow speeds and extremely light cars were in fashion. Just when it was abandoned is not recorded, but it was undoubtedly unsuitable for passenger train service by the Civil War period, when operating conditions began to change dramatically.

Meanwhile the conventional journal box was being improved. New features outlined in a patent (No. 1390) granted to John H. Tims of Newark, New Jersey, on October 31, 1839, may even be the source of what soon became the basic American railway box. It had such modern components as an inverted half bearing and an oil cellar with wicking brushing the underside of the axle (Figure 7.25). The inventor admitted that underside lubrication of journals was already known, but he claimed that the use of a sponge (his preference) or other absorbent materials to convey the oil to the journal was an original idea. If this was true, Tims's patent contained a basic innovation that came into universal use.

Other features of his design were not retained, however. The oil was introduced through a petcock. The front of the box could not be opened, so that it was necessary to remove the rear dust seal and pull the bearing out through the rear opening. A more awkward and troublesome maneuver is difficult to imagine. An opening in the front of the box with a hinged or bolt-on lid made the bearing more accessible for inspection or replacement, but extracting the bearing was still difficult. It was necessary to remove the truck, disassemble the box, or take other extreme measures.

John Lightner, master car builder for the Boston and Providence, suggested a simple solution of the problem. He placed a thin wedge or spacer block above the bearing (Figure 7.26). A jack under one corner of the car or the bolster would remove the weight so that the wedge could be slid out, leaving a free overhead space. The bearing could be tipped up and over the end lip or collar of the axle and thus removed after a few minutes' work, with no displacement of the car or the truck. Lightner received a patent (No. 5935) on November 21, 1848. The merit of the scheme was quickly recognized, and by 1855, 150 railroads were using it.[58] Within a very few years the wedge became a standard feature of American railway boxes—and one that remains in use to the present day. Yet apparently Lightner received a rather small reward for his contribution to railway science. In 1850 he sold the rights to the patent to other parties, who decided to demand a license fee of $5 a year for each car fitted with the axle box wedge.[59] The revenues from such a fee would have been enormous, and the users went to court rather than pay it. They claimed that the idea had actually been developed a decade before Lightner's patent by Seth Boyden, an ingenious mechanic in Newark. Boyden, best known for such innovations as malleable iron and patent leather, also found time to manufacture locomotives and cars. Some of the cars he produced for the Morris and Essex Railroad in 1838 were said to have axle boxes with bearing wedges.[60] The scheme was not patented, but one of the original boxes was recovered from an old car in 1860 and introduced as evidence to support the claim.

Perhaps even more important than the general arrangements for car boxes perfected by Tims, Lightner, and Boyden were the improvements being made in bearing metals. All agreed that bronze was the best material for bearings. The precise formula followed was a matter of individual opinion. The Syracuse and Utica preferred a mix of 8 parts copper to 1 part tin, while the Union Pacific held that a more exotic combination of copper, tin, lead, and antimony was necessary for a proper car brass.[61] Other

roads began to favor phosphor bronze, an alloy that first received attention in the late 1850s.[62] Phosphorous proved to be a deoxidizing agent that produced a tough, homogeneous bronze, well suited to the hard service imposed on railway car bearings. After 1870 its commercial potential was recognized and it became something of an industry standard.

Years before the coming of phosphor bronze, railroads recognized the benefits of soft lead- and tin-based liners for locomotive and car brasses. A new journal was smooth and rode snugly in its polished brass bearing, but in time it became scored and rough. Wrought iron was the only material available for axles in the early period, and being fabricated, it was full of flaws and tiny seams which opened up in time, forming an irregular surface. Turning the journal was troublesome and expensive. A soft-metal liner, however, was self-fitting, even on a rough journal surface. It would swage or mold itself in and around the ridges and grooves of the worn journal. The basic idea is generally credited to Isaac Babbitt of Taunton, Massachusetts, who introduced the scheme in 1839. Babbitt advocated a high-tin alloy that proved very effective but was too expensive for such large-scale applications as car brasses. The industry soon substituted a cheaper alloy made of 80 percent lead and 20 percent antimony, although the term "Babbitt metal" persisted. Car builders, while willing to compromise on the ingredients, were not slow to adopt Babbitt's idea. In 1841 the Philadelphia, Wilmington, and Baltimore's annual report credited the soft-metal-lined boxes with reducing friction and saving oil. The Philadelphia and Reading was employing Babbitt bearings, according to a note in the Directors' Minute Books for April 2, 1842. By the mid-1870s, 1/2-inch-thick soft-metal liners were widely adopted. The Master Car Builders Association made them recommended practice in 1893 and standard practice in 1915, and so they remain to this day.

The consistency over the years in the general design and materials of journals did not hold true for box size. This grew steadily decade by decade until about 1920, when passenger car size reached maturity. The earliest car builders believed that they could reduce friction by employing the smallest possible journal, and some shops constructed journals as small as 1¾ inches in diameter by 3¼ inches long.[63] The size was grudgingly increased as larger cars appeared, but the belief in the small bearing remained in vogue during the pre–Civil War years, when 3- by 5-inch bearings were considered adequate. Around 1860 practical engineers began to realize that small bearings ran hot because too much of a load was concentrated on the bearing surface.[64] The pressure was so great that the lubricant was forced out and friction was increased, not reduced. Then large bearings came into favor. In 1869 the Master Car Builders recommended a standard bearing size of 3¾ inches by 7 inches. Some builders thought that it was overly generous when the standard was first introduced, but the rapid growth of American cars rendered it too small within a period of twenty years. In 1889 the 4¼- by 8-inch journal became standard. Within seven years it gave way to the 5- by 9-inch journal. Standard sizes in 1920 are given in Table 7.2.

The M.C.B. action in setting journal standards was a futile

Figure 7.26 John Lightner's 1848 patent added a space plate, K, which permitted easy removal of the bearing, B. This modest invention has remained a basic feature of car journals.

effort to effect a rapid conversion to a single car bearing. In reality the conversion was slow and erratic. Some roads did not like the M.C.B. design. Among them was the North Western, which continued to install a peculiar style of brass that it claimed ran cooler than any other.[65] Some railroads agreed on the sizes specified by the M.C.B. but not on the other critical dimensions, so that brasses could not be freely exchanged. The situation remained chaotic as late as 1900, when the *American Railroad Journal* reported:[66]

One hundred and eight patterns of car brasses make quite a formidable array. It is a silent commentary upon the meaning of the "M.C.B. standard," when such a lot of miscellaneous and grotesque metal shapes must be carried in stock on a single railroad to meet the combined requirements of home and foreign cars. Upon investigating a reason for carrying one of these patterns, we found that a certain prominent road is responsible for putting out brasses and wedges bearing the symbol "M.C.B.," of which the wedges are too narrow to work in standard M.C.B. journal boxes without allowing the brass to tilt over on its side and with side lips so deep that the lugs bear upon the lugs of the brass and prevent the top of the brass from touching the bottom of the body of the wedge. The brass will not work with an M.C.B. wedge and the wedge will not work with an M.C.B. brass. In justice to the railroads the name of this road ought to be made public.

Just when or how completely this problem was resolved has not been uncovered; presumably it was gradually rectified by more rigid enforcement of the interchange rules.

A lubricated bearing requires five times less power to turn than a dry one: the friction is reduced that much by a simple slick of oil or grease. Engineers have continually searched for more effective lubricants and better ways to keep the journal well coated with oil. During the first decades of railroading only natural oils were available. There are few animal and vegetable derivatives that were not tried at some time or other in a car journal box. The Boston and Worcester ran some road tests in October 1838 to determine the best lubricant available locally.[67] The line was using sperm oil at $1 a gallon and wanted to find a cheaper substitute. Tallow, which cost only 10 cents a pound, worked admirably in the summer but froze nearly solid in winter. Common whale oil, priced at 35 cents a gallon, won the contest. Many years later the Michigan Central became a convert to whale oil after testing it against mineral oil.[68] An express train ran eight trips for a total of 1,584 miles. Boxes on one side were lubricated with whale oil and wood shavings, while those on the other side were filled with mineral oil and ordinary waste packing. The whale oil boxes ran cool, whereas there were five hotboxes on the mineral-oil side. Whale oil cost 60 cents a gallon, mineral oil $1.34 a gallon—a clear victory for the fishing fleets of New Bedford.

Palm oil was another widely used car lubricant. The Newcastle and Frenchtown reduced costs by diluting it with tallow.[69] Boxes would run without attention for eight months. The B & O managed nicely on plain tallow, as did many Southern lines. The Housatonic Railroad recommended pork fat, noting that it cost only 30 cents a year per axle box.[70] Animal fat had its hazards, however. A Southern traveler in the 1840s was astonished by the numbers of hogs that gathered at every station when the train pulled in. Then at one stop he noticed that "gaunt and hungry hogs sauntered up to cars, and with their snouts raised the lids of the axle-boxes, and greedily drained the tallow."[71]

Even after the introduction of petroleum, some railroads stayed with natural lubricants. In the late 1860s the Camden and Amboy concluded that "earth oils" were unsuitable for journal bearings and reverted to its old standards of lard and sperm oil.[72] In 1875 the Master Car Builders report continued to exhibit some skepticism about petroleum, although admitting that because of its cheapness, most lines were employing mineral or "West Virginia oil." It was thought safer to mix the "unnatural" petroleum derivatives with something dependable like tallow or lard. Some lines also believed in additives like "polar grease." The Texas and Pacific claimed that this cooling compound improved the oil so that a quart would last for 2,600 miles.[73] Others argued that adding expensive cooling compounds to cheap oil cost more than buying good oil in the first place.[74] The purchasing agent was singled out as the villain here for his false economy in seeking out the cheapest lubricant. His accounts looked good, but the operating department was left to contend with an epidemic of hotboxes. Adding to the confusion was a blind dependence on brand names such as Black Oil, Well Oil, Stock Oil, and Cleveland Cup.

Laboratory testing was the only sure way to secure a reliable product. In 1875 the Pennsylvania established a chemical laboratory for this purpose.[75] New supplies were tested on a monthly basis, and inferior or adulterated oils were rejected. In some instances the lab discovered that 50 percent cottonseed oil had been added, making the lubricant nearly useless. By the late 1880s many major lines employed their own chemists.[76] The freewheeling days of the shady oil dealer were over, at least so far as journal bearing lubricants were concerned.

The process of journal oiling was to go through another important reform: winter- and summer-weight oils were adopted. Because lubricants suitable for one season were not right for the other, railroads began to use thin oil in winter and heavy oil in summer. On the New York Central the changeover began every November 15.[77] The old waste and oil were removed and set aside for the spring, and a tin tag was attached to the journal box lid noting the date and place of the oil exchange. Regular inspec-

Table 7.2 Standard Journal Sizes Adopted in 1920 by the Master Car Builders

Journal Size, in.	Axle Capacity, lbs.	Car Weight, tons
A. 3¾ × 7	12,500	33
B. 4¼ × 8	20,500	56½
C. 5 × 9	27,000	68
D. 5½ × 10	34,000	84½
E. 6 × 11	42,500	105
F. 6½ × 12	51,000	125½

Figure 7.27   A typical car journal with the front lid or cover removed. The space plate, bearing, axle end, and waste are clearly shown.

tion was another feature of the New York Central rules. Through cars were oiled at New York City and Buffalo. Inspectors touch-tested the boxes at intermittent stops but had orders not to open them or add oil unless they were running hot. Local cars travel-ing shorter distances at lower speeds were oiled on a mileage basis—once every 3,000 miles.

The quality and the handling of waste were as important to journal lubrication as the oil itself. Waste is a tangled mass of threads, usually salvaged from short ends or remnants of cops, a by-product of textile mills (Figure 7.27). They were a haphazard mixture of cotton, wool, and hemp. Waste was clean, absorbent, and of course, very cheap—just the thing for wiping up around the machine shop or stuffing into a journal oil box for wicking. Pure-wool carpet yard was considered the best waste, whereas mixed waste, which was often nothing more than floor sweep-ings, was regarded as undesirable.[78] It was important to have long strands, so that the waste would intertwine to form a pad with few free loose ends that could work up between the brass and journal. While wool and cotton were popular, some lines mixed in cured hair or other fillers.[79]

A few lines experimented with the felt or cotton pad as early

as 1875, but they were criticized for extravagance. Some mechan-ics also claimed that pads threw so much oil around the inside of the box that it found even more places than usual to leak out.[80] Enough oil was already lost through the notoriously inefficient rear dust seal. Waste remained the standard packing for many generations, but interest in developing a practical lubricating pad continued. Advertisements for the latest device regularly ap-peared in the trade press, but in general the railroads were not buying. Following World War II the industry at last conducted a serious investigation of the subject, with results both favorable and unfavorable to the pad. Because of better oil distribution, pads proved effective in holding bearing temperatures as much as 40 degrees below those possible with waste.[81] Yet none of the pads tested would stand up in regular service. Synthetic materi-als did much to improve the situation, however, and within a few years long-lasting pads were available. The most revolutionary conversion was made in the freight car fleet, because by that time a large part of the passenger car fleet had roller bearings. In 1957 pad lubricators were required on all new freight cars.[82] By 1961 loose waste was a thing of the past, and during the follow-ing year no cars so equipped would be accepted for interchange service, according to a rule established by the Association of American Railroads.

Improved lubrication techniques all but eliminated bearing failures. In 1950 99.95 percent of railroad cars were running trouble free on this account, yet the hotbox remained "the scourge of railroading."[83] Since the beginning of the steam railroad, operating men had lived with this bogeyman that confounded schedules and demolished equipment. One passenger described the everyday scene in nineteenth-century America:[84]

The advent of a "Hot Box" was heralded by an "aroma" modest at first but growing more insistent as the train sped along and causing travelers to sniff until an outspoken one called across the aisle to a friend—"Hot Box?" and received a nod in reply. Then came rattling of bell cord and quick answer from whistle. The train drew to a halt. Many male passengers swung to the ground and clustered about the fuming journal. Into the throng strode our brakeman with his surgical equipment—a pail of water in one hand—in the other a pail slopping with a dreadful looking mess. His short-hooked poker scalped the journal lid and exposed the heated parts belching smoke. He hooked out the smouldering waste and dashed water inside which forthwith rushed away in sizzling bursts of steam. Then he poulticed the feverish interior with that dreadful looking dressing poked in by rod and hand. Back he slapped the lid and back to their seats scrambled the passengers. The whistle blew. The train resumed its journey while the brakeman busied himself in cleaning his filthy hands.

The great majority of hotboxes ended on this cheerful note—a mere diversion to break the tedium of travel. But some boxes were not so easily cured, as illustrated by the halting passage of the Pacific Express in 1870.[85] A half hour was lost in Harrisburg while two journals on a front-end car were doused and repacked. The train was forced to stop again at Lancaster to quench a blazing hotbox on another car. Three more stops were necessary before reaching the final terminal in Philadelphia. Hours were lost, tempers frayed, other trains sidetracked, and the whole main line put into a furor because of a simple bearing failure. And a hotbox posed an even greater threat than the possibility of alienating passengers or disrupting schedules. First, any breakdown on the road was a setup for a rear-end collision. Second and still more dangerous, a hotbox occurring at high speed could actually burn the journal off, causing a derailment. Steam service had hardly begun before such an accident took place. In November 1833 a train on the Camden and Amboy was running fast to recover lost time. An axle on one of the cars overheated and broke off, causing a wreck that killed two people and injured many others. Among the passengers were former President John Quincy Adams and Cornelius Vanderbilt. By far the gravest disaster attributable to a hotbox occurred 110 years later, near Philadelphia.[86] In 1943 the Congressional Limited was whipping along at nearly 60 miles an hour when a journal on the seventh car in the sixteen-car train burned off. A colossal derailment followed, resulting in 79 deaths and 129 injuries. It was one of the worst rail disasters in American history. An examination of the defective journal showed that its surface reached 1400 degrees before the failure occurred. This disaster cost the Pennsylvania

Railroad millions of dollars. And less sensational hotbox failures also caused a heavy drain on the line's treasury; for example, in the early 1950s the P R R claimed annual losses of 4 million dollars due to hotboxes.[87] The vast majority, of course, occurred on freight cars, which received less care than the passenger fleet and were subject to overloading. Unfortunately, no separate cost figures seem to have been maintained for freight and passenger trains, but the 4-million-dollar total noted above for a single major trunk line gives some idea of the staggering cost to the entire industry.

A long, sad list of causes of the hotbox problem can be assembled. Inferior oil was an obvious villain. Cheap brasses made from an indiscriminate mixture of scrap bronze were blamed by some authorities.[88] Loose bearing liners were another cause—a problem that was ultimately solved by a change in the white-metal alloy and more careful fluxing. Imperfect truck design and badly maintained, out-of-square trucks contributed to hot bearings. Overheated journals also resulted from too small a bearing for the normal load or, in the case of freight cars, serious overloading. Ineffective dust seals at the front and rear of the box permitted grit to enter and mix with the oil. Because the boxes were at track level, dust swirled around the trucks in a furious storm, yet it was not uncommon to see freight cars with missing journal box lids.[89] The rear seal was a wooden board with a hole cut in it so that the axle might pass through into the box. Sometimes it split and fell out, and its absence was not noticed for months because of its inaccessible position.

Even the weather was blamed for hotboxes. According to some mechanics, spring and summer were the worst seasons, because it was then that the ballast was renewed and the tracks were at their dirtiest. Other mechanics dreaded the winter, when the oil thickened and moved sluggishly through the waste.[90] Snow could blow through a partially cracked journal lid and emulsify the oil. In the extreme climates of New England and the Northwestern states, the oil and waste would freeze into a solid rock-like mass, ending all possibilities of lubrication.

Waste grabs were long considered a major cause of hotboxes. Loose threads tended to travel up under the edge of the bearing on the rising side of the journal.[91] They would work their way between the bearing and the journal, overheat because of the intense friction and pressure, and burn, setting the box ablaze. A freshly packed box was often the worst risk, because in new waste there are so many lint ends that can be easily picked up by the journal. Buffering impacts encountered in ordinary switching were often sharp enough to cause the bearing to jump up off the journal. A strand or two of waste might then flop up and be caught between bearing and journal when the bearing reseated itself. Waste grabs were not eliminated until lubricating pads were finally adopted.

There was a belief in the railroad industry that all these causes of the hotbox problem were secondary compared with the human element. According to many managers, the careless workman was the major cause of hotboxes. Carmen, oilers, dopers, car tenders—whatever they were called, they were all the same: a lazy, indifferent, dollar-a-day breed of laborer.[92] Even

Figure 7.28   A Timken roller bearing assembly of about 1950.

with the commercial beginnings of roller bearings and represent a late effort to perpetuate the old-fashioned friction bearing for passenger car service. In 1933 the New York Central's West Albany testing laboratory began work on a smoke-odor capsule that would ignite, calling attention to a hot bearing before a serious rise in temperature occurred.[98] Two small capsules with soft plugs which melted at 320 degrees were placed inside the box. One was a stink bomb, the other a smoke bomb. The device was used in some cars during the next year, and in 1936 the bombs were credited with uncovering sixteen hotboxes. The Central began to install the alarms on all its passenger equipment. By 1950, 3,500 cars were carrying them.[99] Despite their effectiveness, the Southern Pacific was the only other line to show much interest. Later the industry was drawn to an even more promising hotbox detector: in the late 1950s heat-sensitive electronic units could be set up at the side of the track to detect a warm journal on a passing train. By 1969 1,500 units had been installed, and some roads were placing them every 25 miles along the track.[100]

Centuries before the introduction of steam railways, mechanics sought to perfect something superior to the plain bearing. The use of rollers and wheels were a daily reminder of the great friction savings possible in rolling rather than sliding a load. The ancient Romans employed roller bearings in the first century A.D.[101] Fourteen hundred years later, Leonardo sketched roller bearings of strikingly modern appearance. Large weather vanes, such as the one atop Independence Hall, were mounted on roller bearings during the eighteenth century. The Patent Office began to receive applications for many ingenious types of antifriction bearings late in that century. In 1802 the tapered roller bearing was patented. The notion that ball and roller bearings are a recent invention is clearly false, but it is true that they did not come into commercial use until relatively modern times. The railroad industry was slow to adopt them, and was perhaps even more reluctant than other segments of American business to accept the antifriction bearing. It was not until about 1940 that the railroads seemed ready to utilize the free-rolling products of Timken, S.K.F., and Hyatt (Figures 7.28 and 7.29).

Part of this reluctance can be explained by the railway itself: the metal wheel running on a metal rail was in effect an antifriction bearing. The rolling friction of a railway car is very small compared with a highway vehicle. The small gain possible with antifriction bearings on the wheel axles would be offset by their extra cost. Furthermore, the first generations of ball and roller bearings did not prove very dependable for heavy machinery. They may have been adequate for a sewing machine or a bicycle, but they could not handle the crushing tonnage of a railroad car. Railroad men were content with solid bronze and Babbitt metal journals and saw little reason to change.

Among those eager to wipe out this prejudice was Ross Winans, whose friction-wheel invention brought him fame but ultimately proved impractical. For nearly fifty years the railroads apparently ignored any other inventor so foolhardy as to propose a similar experiment. Street railway men were more receptive,

when they tried, few oilers could pack a box properly. In their zeal they would overstuff the box so that it wiped the oil off rather than brushed it on, or they would pack it so tightly in the back of the box that the oil could not penetrate and only the front half of the bearing received lubrication. Managers also charged that oilers did not even look inside the boxes because before about 1870, it was necessary to remove two bolts to gain access to most journals.[93] After this time the so-called Fletcher lid came into use. A small coil spring on one bolt allowed the cover to be snapped open for easy inspection.[94] This simple improvement probably saved many a bearing. In time hinged lids came into favor.

Careful management, combined with good materials, could reduce the incidence of hot journals. Why, it was asked, did the Lake Shore have 1,100 hotboxes a month, the Boston and Albany only 2? Indifference was not a fault restricted to simple car tenders; it was the superintendent's responsibility to see that his men knew their jobs and to discipline them if they did not follow orders. On a casually run line cars might go four and five years without repacking; on a well-run road they were repacked every month.[95] The Delaware and Hudson concentrated on hotbox reduction and showed what regular maintenance, periodic checks, and better employee training could do.[96] In reviewing its passenger car fleet alone, the line reported 215 hotboxes in 1915. By 1920 the number had dropped to 45. Ten years later only 10 hotboxes occurred. And the entire industry was making progress. In 1928 a passenger car that ran 100,000 miles between hotboxes was considered a record setter.[97] By 1934 some cars ran as many as 1,000,000 miles between hotbox failures. The problem was at last all but eliminated by the roller bearing.

Hotbox alarms and detectors were introduced coincidentally

Application of Timken Bearings for 5"x9" size axles. Wheels can be turned without removing Bearing from axle. Dot and dash line shows contour of A. R. A. standard axle. Note how oil level is checked.

Figure 7.29   A Timken roller bearing assembly drawing of about 1950.

and in the 1870s and 1880s a number of antifriction bearings were tested on horsecars.[102] Results were encouraging on these very light but underpowered vehicles and naturally suggested applications in the larger arena of the steam railroad. In the late 1880s several hopeful inventors began to seek permission from various roads to test their wares. In 1886 Charles D. Meneely of Troy, New York, tried some of his tubular roller bearings on the Delaware and Hudson Railroad.[103] The first sets were a failure because of alignment problems. By the early 1890s Meneely had a new design that proved more serviceable, and the D & H applied it to four cars.[104] The inventor claimed such startling advantages as a 90 percent saving in starting power, a 25 percent saving in fuel, an 80 percent saving in lubrication, and an "absolute" elimination of hotboxes. Meanwhile, the Rock Island was trying the Sharpneck frictionless roller bearing. Tests began in May 1888, and again gratifying results were at first reported.[105] After 38,000 miles on the experimental car in suburban service out of Chicago, several more suburban cars were fitted with the bearings. It was said that a single man could push one of these cars on the level. In the East, the Tripp Manufacturing Company of Boston prevailed upon the New York and New England to try a set of its bearings, which had been used with great success on streetcars.[106] Tripp was more moderate in its claims of friction savings over plain bearings, but the company did maintain that its journal could run for two years or more on the original lubrication.

The optimistic statements of the antifriction advocates were soon contradicted by railway mechanical men. One reported that a patented bearing ran trouble-free for five months and then seized up without warning, never to turn again. The inventor had

meanwhile disappeared. M. N. Forney, probably the most respected man in railway mechanical affairs in nineteenth-century America, spoke out against the roller bearings. He noted that the friction and fuel savings claimed were nonsense, because journal friction represented the smallest portion of the train's resistance. This resistance could actually be measured, chiefly by calculating the weight of the train, the gradients, the curves, and the air resistance.[107] Others objected to the size of the journals (Sharpneck's, for example, required a box nearly 12 inches square) and to the fact that they could not be fitted into a standard-sized journal box. Their first cost—$50 each in the late 1880s—was far more than that of a plain bearing.[108] And of course, a basic argument against roller or ball bearings is that they are no more friction-free than plain bearings except at very slow speeds. They were beneficial in starting the train—an important advantage, to be sure; but once the train was rolling their usefulness diminished.

Roller-bearing manufacturers found other markets outside the railroad industry that led to the perfection of their products for commercial use. With the early roller bearing, design and basic form were not the problems; what the makers needed to discover were the correct materials and manufacturing methods. At an early date they found out that the balls or rollers must not be allowed to rub against one another or a "grindstone" effect was created; they had to be separated and held parallel by a cage or ring. Ordinary mild steel was not up to the demands of the design, for very hard, precise surfaces were needed for a durable bearing. Thus bearings required better tooling, like centerless grinders, and more specialized heat-treated steels, together with refined production methods to ensure accuracy. After these elements were combined, antifriction bearings were successfully manufactured on a commercial basis. This appears to have taken place in the 1890s, although more research is necessary to establish the course of development.

The roller bearing was once again being considered for railway work during the early years of the twentieth century. In 1906 the Strang gas-electric car was equipped with antifriction bearings. William McKeen, mechanical chief of the Union Pacific, adopted roller bearings for his gasoline rail cars. In 1911 the Bangor and Aroostook began testing the bearings on a medium-weight coach.[109] A few years later the Boston and Albany tried them on three passenger cars. More applications were contemplated, but the bearing was slightly oversized, requiring new truck pedestals. These experiments only suggested what was to take place in the next two decades.

Its large-scale application in automobiles had helped to establish the roller bearing as an economically feasible device. The antifriction bearing maker was now marketing a thoroughly practical product, and the railroads represented a lush, untapped field for new sales. In 1921 the Swedish-based firm S.K.F. outfitted six passenger cars on the Pennsylvania.[110] The success of this installation led the road to purchase bearings for 166 cars five years later. These were rugged fixtures that did not burn up after a few months of running. Some of the original sets, in fact, traveled 2.6 million miles before being retired late in 1946.

After a few years the Milwaukee Road followed the Pennsylvania in trying roller bearings but surpassed it in enthusiastic acceptance of the antifriction idea. In 1924 the road began testing.[111] Bearing resistance was calculated at 54.4 pounds with plain bearings but only 7.5 pounds with rollers. Lubrication was managed by an annual greasing. In 600,000 miles of running, no failures occurred. An unexpected benefit was discovered during brake tests: roller-bearing-equipped cars showed less tendency to slide during emergency applications. The reason, mechanical men theorized, was that the action of the bearings eliminated the transfer of weight characteristic of plain bearings during fast stops. There was ample reason to convert, and late in 1926 the Milwaukee began to equip 127 cars with Timken bearings. The New Haven and others began to follow the lead of the Pennsylvania and the Milwaukee, so that by 1929 there were 850 roller-bearing-equipped passenger cars operating in the United States. In 1930 nearly 500 passenger cars were fitted with the new style of journal bearing.

During the next few years, however, the pace of conversion fell off sharply. Strict economy was necessary because of the Great Depression. And some mechanical men continued to question the benefits of roller bearings. A bronze bearing complete with wedge and box cost only $14.40; an equivalent roller bearing cost $260 to $400, depending on the manufacturer.[112] Skepticism about the value of roller bearings was illustrated by the reluctance of some lines to use them even for new lightweight cars. The first stainless-steel streamline equipment purchased by the Santa Fe in 1936 and 1937, for example, had conventional bronze friction bearings.

But in a few more years there was a massive shift of opinion on the subject. Between 1935 and 1940, despite the bad times, an average of 300 roller-bearing-equipped cars joined the passenger fleet each year through the purchase of new cars or the rebuilding of existing stock.[113] Another national emergency, World War II, again interrupted the switch from plain to roller bearings. After the war's end, however, almost no new passenger cars were ordered with plain bearings, according to the *Car Builders' Cyclopedia* of 1961. The higher cost of roller bearings, which had once been a convincing argument against them was now demonstrably offset by the wonderful economies realized after several years of service. Cars ran 30,000 miles a month, trouble-free.[114] It was not unusual to receive ten years' service before replacement was necessary. Long, trouble-free operation is probably what really sold roller bearings to practical railroaders. Free rolling was apparently seen as something of a fringe benefit. And while passenger cars were converted to antifriction bearings rapidly in the early postwar years, work was just beginning on a similar substitution for the larger fleet of freight cars.

Axles

Axles were once the weakest of the components forming the railway car's undercarriage. The previous section discussed axle journal end failures as the hotbox problem, but experienced car masters knew that the entire iron shaft was a source of unremitting trouble. Bent and broken axles littering the yards and shops of every major line in the country gave silent testimony to countless accidents, delayed trains, and sometimes major disasters. Yet small axles were favored because they reduced dead weight and cost. During most of the nineteenth century few car axles were made much larger than 4 inches in diameter. These slender, 350-pound shafts were expected to safely support high-speed express trains or heavily burdened freight cars. Most were fabricated from scrap iron for the sake of economy. The soundness of their manufacture was left largely to mechanics whose knowledge of metallurgy was based on tradition and the superficial appearance of things. If the finished product looked good, it was sent out on the road. If its core was rotten, this could only be discovered when it broke. Even though statistics available for the early years are sketchy, the record of axle failures was a dismal chronicle.

In 1834 the South Carolina Railroad stated in its annual report that broken axles were the principal cause of accidents on its line. A few years later the record books of the Boston and Worcester list a series of wrecks due to breaking axles.[115] In one car, three journals broke off. In June 1837 a broken axle on one car caused a wreck that damaged five other cars. In 1843 a Long Island Railroad train was delayed two hours for the same reason.[116] Such mishaps were apparently so common that they were reported only incidentally, and no one organization or publication made a complete record of them. In 1856 the 150-mile-long Philadelphia and Reading Railroad was breaking 300 axles a year, and other roads were reportedly doing no better.[117]

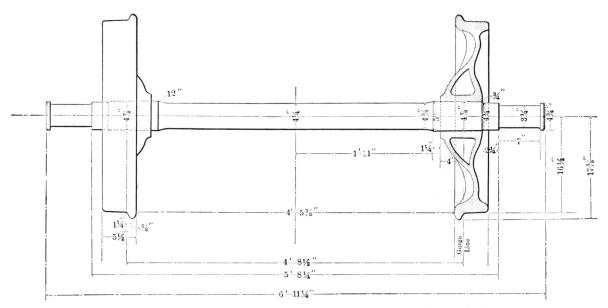

Figure 7.30 A typical axle and wheel set of the 1880–1890 period. Standardization of this assembly was a major accomplishment of the Master Car Builders Association. (William Voss, Railway Car Construction, *New York, 1892)*

Between July 1848 and November 1852 the Reading Railroad had 1,061 accidents because of broken axles.[118] In later years the *Railroad Gazette* served as the industry's semiofficial compiler of accident statistics, but because it depended on facts casually collected from scattered newspapers, its totals can only be viewed as a sampling. Even so, between 1873 and 1882 the *Gazette* reported an average of roughly 35 broken axles a year.[119]

It can be argued that most of these failures did not end in tragedy; yet a few did. In 1850 three passengers were killed on the Western Railroad because of a broken axle.[120] Three years later an accident on the Boston and Maine resulted in the death of President Franklin Pierce's son. A broken axle derailed the train. The shaft measured only 2½ inches in diameter and showed evidence of an old crack. A grimmer wreck at New Hamburg, New York, in 1871 demonstrated the dangers of axle failures even on freight cars.[121] An axle on a tank car broke, derailing it and several other cars on a bridge just as the Pacific Express was passing on the opposing track. The pileup resulted in twenty-two deaths.

Clearly there was an urgent need to improve the strength and dependability of car axles. Increasing their size was an obvious precaution. In its report for 1832 the South Carolina noted that it was enlarging its axle diameter size from 2¼ to 2⅞ inches. In 1839 Von Gerstner reported that the lines running into Boston were using 3- to 3½-inch axles. In the early 1850s several New York railroads noted that they were employing 4-inch-diameter axles.[122] By the turn of the century the Master Car Builders Association was specifying axles for passenger cars with a 5⅜-inch diameter at their centers and 6⅜ inches at the hubs.

Car axles were made smallest in diameter at their centers for several reasons (Figure 7.30). The taper saved iron where the full diameter was least needed. It also allowed the axle to flex or spring slightly, thus helping to dissipate the damaging lateral shocks received when the flanges struck the rail on entering a curve or a switch. The taper was subtle, often no more than ³⁄₁₆ inch. Axles were made thickest at the hub; during the nineteenth century they were a full inch larger in diameter at this point than at the center. This was necessary because axles tended to break at the hub of the wheel. The hub acted as a fulcrum. Since the operation of the wheel results in a flexing of the axle (with the

axle acting as a long lever), a bending movement occurs just inside the rear of the wheel's hub. Here is the point of fracture and the commonest point of breakage.

The direct path to safer axles was better standards of manufacture and stronger materials. These improvements did come about gradually, but early mechanics could not wait for the industrial millennium; they required safer axles in their time. Since complete reform was not immediately attainable, an ingenious compromise was worked out. That axles were liable to break was accepted as a fact of everyday railroading. The worst consequences could be averted by offering the broken axle some fallback position—some auxiliary means of support until the train could be stopped. This idea was developed by Joseph S. Kite, superintendent of the Philadelphia, Germantown, and Norristown Railroad. A pair of beams was fastened to the truck frame inside of the wheels. Oblong iron brackets loosely encircled the axles. In normal service the axles could gyrate freely inside the safety beam brackets, but should they break, the brackets offered a secondary point of bearing. The simple and effective plan was patented by Kite on July 14, 1834. Good reports on the invention began to appear in the trade press. A car with a broken axle made four trips on the Philadelphia and Columbia line without incident.[123] Only then was it discovered that the axle was being held in line by the safety beam stirrups. A similar event was reported on the P W & B, prompting the *American Railroad Journal* to congratulate Kite on the effectiveness of his patent in "these days of accidents and disasters."[124] The safety beam arrangement remained a part of American car design until the end of the wooden truck.

Since time in service was also found to have an effect on safety, axle-wheel sets were retired from passenger service after one or two years, or 50,000 miles of running.[125] They could then be used on freight cars. These wheel sets might continue in freight service for as much as twenty years, during which time the journals might wear down from 3¾ to 2⅝ inches.[126] By the late eighties, however, such long service was not considered economic or safe; a journal showing ⅛ inch of wear was considered unfit for use.[127] Most axles could be expected to run for four or five years and deliver 150,000 miles of travel. A few even doubled that mileage.

Inferior material was the chief cause of unsound axles, yet for generations wrought iron was the best metal available. Really high-quality wrought or puddled iron made a good axle, but too often cheap grades were accepted. There was a great variation in price and quality. In the mid-1850s a car axle might cost as little as $40 or as much as $100.[128] The cheap axle contained a high proportion of cinders and was imperfectly forged. Hot-blast anthracite iron was regarded as decidedly inferior to charcoal iron. Even when good iron was used, the fabrication was not always well performed. Few furnaces could produce blooms large enough to make a car axle. It was necessary to stack several muck bars (a term for raw wrought iron after its first rolling) and reroll it to form a blank large enough for forging. The bloom was too often drawn at two heats when it should have had six. It was then hammered into its final form, ready for machining. Frequently the product was a coarse, loose-grained iron that was subject to crystallization, "cold shuts," breaks, intermixtures of raw iron, and other flaws.

Poor as muck bar axles might appear, many railroads purchased cheaper ones made from scrap. Some authorities condemned scrap axles, while others felt that they were superior to axles made from new iron.[129] Imperfect welding was the major drawback of scrap axles, for bits of steel or cast iron could ruin a weld. The long fibers necessary for a strong shaft were possible only with carefully selected scrap and the use of only a few short pieces in the pile. Much sounder scrap axles could be produced by upsetting the piles and devoting enough time to the furnace and the hammer. The process is described here as it was practiced around 1900:[130]

The scrap is made up into packages or small piles, and placed in a furnace and heated to a welding heat, and then hammered by a steam hammer into a slab. This slab is cut nearly in two, then doubled over and returned to the furnace for a second heat; it is then hammered into slabs about five inches wide by two inches thick. Three of these slabs are again heated to a welding heat and hammered into a single slab, one-half of which is finished into axle shape and the other half (on account of not being hot enough to finish) is again reheated and hammered. After being allowed to cool, the metal is made the required length by a cutting-off machine.

Buyers in search of safe axles in the first decades of railroading relied largely on the reputation of the maker. Good materials and workmanship were his trademark. Car men later turned to sample testing as a more positive method of securing the best-quality product. For every lot of axles purchased, one or two were picked out at random for a physical test. At first the only practical method of assessment was the drop hammer test. Such methods were the subject of a report in the transactions of the British-based Institution of Civil Engineers in March 1850. Several years later a series of similar tests were conducted here in Detroit, where a 150-pound, 12-foot drop hammer was set up.[131] Some axles broke after only 11 blows. The winner of the series was E. Corning & Company of New York, which produced a 4½-inch-diameter axle that withstood 193 blows. Similar tests were conducted by railroads or axle makers, as often as not to the standards established by the individuals involved. The Master Car Builders Association considered the question of a standard testing procedure for nearly thirty years. Finally in 1899 the body agreed on an industry-wide testing formula. The drop hammer remained the single test, and any lot of axles was rejected if the sample ruptured or fractured in any way. But the lot could also be condemned if the sample registered a deflection greater than 8⅛ inches for a 4¼- by 8-inch journal-sized axle. (The permissible deflection varied, of course, for each axle size.) The full test involved five blows, a 1,640-pound hammer, a fall of 16 feet for an iron axle, and a 23-foot fall for a steel axle.[132]

In the twentieth century axle testing and design moved out of the forge shop into the laboratory. More accurate testing machines led to the production of better metals and a more scientific design. Simulated road tests conducted in 1938, for example, inspired new designs that proved 60 to 80 percent more resistant to metal fatigue than conventional patterns.[133] Magnaflux, and in more recent years ultrasonic, inspection has ensured quality control for the quantity production of railway car axles.

The efforts to produce safer car axles through better materials, design, and testing did not satisfy a radical element within and without the industry. This faction contended that the conventional design was all wrong, that the entire concept of wheels solidly mounted on the axle should be discarded. In theory this school had some convincing arguments. Conventional wheel sets are suitable only for absolutely straight railroads. On a curve the inside wheel must drag or slip, because it moves a shorter distance than the outside wheel. This causes unnecessary wear of the wheel and rails and creates an extra drag of 30 percent on the train as it passes through curves.[134] The axle is twisted because of the torsion produced by one wheel attempting to revolve faster than the other. To eliminate these disastrous conditions, various mechanics proposed that one or both wheels be loosely mounted on the axle. Railroad men rejected their recommendation because of the danger of a wobbling wheel, which would be inevitable once the bearing had worn. A more attractive alternative was to split the axle so that each wheel could turn independently of the other. Axles of this type became known as divided, joined, or compound axles.

As early as 1834 James Stimpson patented a divided railway car axle. Little more attention was given to the idea until the 1850s. Josiah Copley of Kittanning, Pennsylvania, devised a hollow divided axle that was tried on the Cleveland and Pittsburgh Railroad. In 1854 Samuel L. Denney patented a divided axle, which he advertised in the *American Railroad Journal*. But like the inventions of his predecessors, Denney's axle did not seem to stir enthusiasm among railroad men, who probably viewed it as a gadget that traded one set of problems for another. A bearing box at the axle's center was just another new trouble spot. Located under the car, surrounded by the truck, it would be a miserable thing to inspect, lubricate, and repair. In 1872 another inventor, G. W. Miltimore, took up the idea. His divided axle

MILTIMORE AXLE.

REFERENCE:

A. *Wheels.*   H. *Pedestal Blocks.*
B. *Inner Axle.*   I. *Bolt through axle and Block.*
C. *Revolving Sleeve.*   K. *Oil Reservoir.*
D. *Wheel Seats on D?.*   L. *Oil Passage, horizontal.*
E. *Oscillating Boxes (Brass)*   M. *D?   D?.   vertical.*
F. *Box Ring.*   N. *Oil Drip.*
G. *Nut screwed to end of Sleeve.*   O. *Hollow Screw Plug.*

Figure 7.31 *Miltimore's divided axle of about 1875 was devised to eliminate axle torque.* (Institution of Civil Engineers, *vol. 53, 1878*)

was tried on the Central Vermont and the Burlington and was quickly pronounced a failure.[135] In four years Miltimore was back with a new design far more elaborate—and costly—than any compound axle yet devised (Figure 7.31). Yet its creator felt certain of success, and indeed it was given a good reception at the U.S. Centennial Exhibit, in company with the other novelties that are an indigenous part of world's fairs. It was even tried on the New York Elevated and the Boston and Albany Railroad. The Master Car Builders, however, continued to show distrust, and eventually the road tests indicated that Miltimore's device became noisy and friction-bound after a few months' service. Railroad history demonstrates that old ideas, no matter how foolish, have a way of being periodically reintroduced: early in 1970 an advertisement appeared in *Railway Age* for a compound axle, with a text that could well have been written by Miltimore himself.

A definitive solution to the car axle problem seemed to have arrived with the introduction of cheap steel in the late 1850s. It was a dense, homogeneous, and elastic metal with a greater tensile strength than wrought iron. With the advent of the Bessemer process, steel became available in large quantities at prices the railroads could afford. Yet they did not make a rapid conversion to the new metal. The resistance to steel axles that lasted for nearly fifty years after their introduction can be explained by more than thick-headed conservatism, for the much advertised superiority of steel over iron axles did not prove itself in actual service. Steelmakers found a friendly reception for their products among track superintendents, a slower acceptance among locomotive builders, and positive rejection from car builders. Steel rails were widely used by the 1870s.[136] Steel made only moderate progress with locomotive men, except for driving wheel tires, which became widely used by 1870.[137] All-steel boilers did not generally appear for another generation. But the persistent holdouts were the car builders, who showed no enthusiasm for steel axles until after 1900.

This is not to say that steel axles did not receive an early trial. Even in 1856 the Buffalo and Erie and the Michigan Southern railroads were testing them, but the experiment seems to have attracted few converts.[138] In 1861 it was reported that steel axles were being used in Europe but not here.[139] Apparently only the Pennsylvania continued to test them on this side of the Atlantic. The road's annual report for 1866 noted that 1,907 steel car axles had been placed in service beginning the previous July. In 1867 the mechanical department was reported to be convinced that steel axles were superior to iron. Steel had performed well in actual service, and all passenger cars would be equipped with it. In time the road hoped to convert freight cars also.

Practical car builders elsewhere, however, emphasized the defects of steel axles. The early Bessemer steels had too much carbon, and the quality was highly variable.[140] Not enough care was taken in their manufacture, so that some were brittle and others were roughly finished or had sharp corners that invited cracks. They broke easily and unpredictably. As late as 1894 a car superintendent delivered a paper before the Northwestern Railroad Club declaring that steel axles were a menace to railway travel.[141]

Other factors contributed to car builders' preference for iron. Steel journals wore out more rapidly than iron. In one test a Bessemer axle gave only 84,000 miles of service, compared with 120,000 for iron.[142] Steel journals were said to be more "liable to heat."[143] Cost was a major consideration as well. Although according to one report, the price of steel between 1865 and 1878 fell dramatically from $140 to $55 a ton, steel car axles were still twice as expensive as iron axles.[144] Mass production of steel lowered its price, but iron held the cost advantage insofar as car axles were concerned. It remained cheaper then steel at least until the late 1880s, and perhaps even later.[145]

Steelmakers persisted, however, and in time a cheaper, more dependable axle was manufactured. Milder and more uniform steel alloys made a great difference. Improved steel had an elastic limit more than double that of iron, and while wrought iron was more ductile, it deteriorated with the constant flexing characteristic of railroad service. More time under the hammer was necessary for a sound axle.[146] Heat treatment normalized the steel-forged axles. But the improvement in their quality was not enough to break down the long-standing prejudice against steel axles. In 1904 builders were still arguing in favor of iron.[147] Ironically, it was the scarcity of cheap, good-quality wrought iron that at last persuaded car builders to accept steel. In 1908 Marshall Kirkman observed that steel axles were now the most common type, not because they were believed superior but only because they were more readily available.[148] Steel won at last, but only by default.

After the reluctant acceptance of the steel axle, there were few important developments except for the growth in axle size, which was treated in the previous section on roller bearings. One event late in the history of car axles was the invention of a hollow or tubular axle. It was considerably lighter than the conventional solid variety, offering not only less total dead weight, but more important, less unsprung dead weight. Hollow axles were tried early—around 1835—but did not receive much serious attention until the 1940s.[149] The idea was revived by H. C. Urschel, whose design was promoted and manufactured by the Pittsburgh Steel Company.[150] The 9¼-inch-diameter main body of the axle was a thin-wall tube. The walls were considerably thicker at the wheel seat and the journals. The hollow construction saved 275 pounds in a standard 5½- by 10-inch journal-sized axle, which amounted to a weight saving per car of more than ½ ton. The manufacturer claimed a 50 percent improvement in fatigue strength. Urschel's axle was no one-time experiment; it was seriously considered by the trade despite its greater cost. Thousands were manufactured for main-line service. Despite its advantages, however, the Urschel axle fell rapidly from favor when its propensity for cracking was discovered.[151] It was eventually outlawed for interchange service.

Wheels

Wheels are fundamental to land vehicles, for to move is to roll. If railway mechanical men seemed preoccupied with wheels, it was because they understood that good wheels were fundamental to the railway's success. The terrible demand imposed on car wheels makes them the most highly stressed single element connected with railway service. They support the car and its cargo; they steer the train along the track; they are subjected to terrific torsional stress when rounding curves, and they must withstand the heating stresses caused by the brake shoes.

The first railroad car wheels were patterned after common wooden-spoked wagon wheels. They had a flanged tire fastened over the wooden felloe or rim. Europeans remained loyal to the tired wheel, although they soon devised stronger centers in place of the wooden spokes. American mechanics, however, discarded the tired wheel for one-piece cast-iron wheels, which became the standard by about 1840. Around 1875 tired wheels again came into favor here for passenger cars, although cast-iron wheels were retained for freight cars until relatively recent years. After the coming of the steel car around 1910, one-piece wrought-steel units replaced the fabricated-tire style of wheels. They were all but universally adopted within a decade and remain the standard railway passenger car wheel today.

Since the beginning of the modern steam railway, wheels have been rigidly mounted on the axles. This practice naturally created difficulties in rounding curves, because the outer wheel was traveling further than the inner one. If the wheels were loose on the axles they could adjust their rate of travel without difficulty, as highway vehicles do, but the solid axle fastening forces both to travel at a uniform rate. To negotiate curves one wheel must slip, and part of the resulting torsion must be absorbed by the axle. To relieve the undesirable twisting action as much as possible, a coned or tapered wheel tread was introduced. Being sloped, the

wheel tread now had many diameters, although the difference was only about ¼ inch from one extreme to the other. Advocates of the cone claimed that it was effective in allowing the wheel set to seek a more equal rate of travel on curves, because the outer wheel rode on its largest diameter while the inner rode on a smaller diameter. The benefits of this technique may sound dubious to the novice, but tapered wheel treads have stayed in fashion to the present day. Such eminent railway men as M. N. Forney have argued against the coned tread, insisting that flat or cylindrical treads are as good or better.[152] Other authorities have challenged the coned wheel by noting that it does nothing to relieve flange wear on curves, because in actual tests the outer flange always bears hard against the rail no matter what degree of taper is used.[153] An even more serious defect of the cone tread is that it promotes "hunting" on straight tracks as the wheels wander back and forth on their unequal diameters. This characteristic does reduce flange wear, but at the expense of a steady ride. So far, however, no one has managed to overthrow the coned wheel tread.

James Wright of Baltimore is said to have invented the coned wheel in 1829. This assumption is based on a patent issued to Wright in September of that year.[154] Actually the idea can be traced to the tramways of Shropshire nearly a century earlier.[155] Horatio Allen reported on coned car wheels in England during his visit of 1828.[156] If its origins are in dispute, so is the exact degree of taper that a coned wheel should have. The B & O favored a steep taper of 1 in 14, while the nearly level New York Central employed a gentle taper of only 1 in 38.[157] In 1906 the Master Car Builders Association adopted the taper of 1 in 20, which has continued as the industry standard to the present time.

The overall size of railroad car wheels has remained remarkably constant over almost a century and a half. With few exceptions, freight cars have ridden on 33-inch-diameter wheels for the entire period. (The exceptions today are the thousands of "low-level" TTX cars for TOFC and auto-rack service, which use 28-inch wheels, and all 100-ton-capacity cars, which require 36-inch wheels. The biggest eight-wheel cars, with 125-ton capacity, use 38-inch wheels.) Many passenger cars have employed 33-inch wheels, although larger wheels were generally used. In the 1830s, 36-inch wheels became popular and, except for a brief period late in the nineteenth century when 42-inch wheels were in vogue, remain the standard of the present day.

Big wheels offer several advantages. They make fewer revolutions than a small-diameter wheel, resulting in less tire wear and cooler running brasses. They ride better and are said to pass through frogs and crossovers more easily.[158] Large wheels deflect or flatten out less on the rail under a given load and hence require less power to draw. Their larger tread and flange surface provide more braking area, and their greater mass absorbs and dissipates heat from the brake shoes more readily.[159] Of course, many arguments were made over the years against large wheels: they weigh and cost more; they tend to bend axles; and when they are made in the largest size of 42 inches, they interfere with air and steam pipes and brake rigging. But the only really con-

Figure 7.32   The cone or taper of a modern car wheel is shown in this drawing. The idea dates from the beginning of American railroading.

A.R.A. STANDARD
TREAD AND FLANGE.

vincing argument is the last one, against the 42-inch size. Around 1900 very large wheels gave way to the compromise size of 36 inches, which is still the standard for American passenger cars.

The earliest wheels measured 4½ inches wide overall at the tread and flange. They seem to have quickly grown another inch and to have remained at the 5½- to 5⅝-inch girth to the present. The profile of the tread and flange has also shown remarkable constancy, at least since the 1880s.[160] Except for a reduction in flange depth by ⅛ inch for better frog clearance, the dimensions of a modern tire are nearly identical with those of its ancestors (Figure 7.32). An exception to this pattern of consistency was the broad-tread wheel that enjoyed a decade or so of popularity following the close of the Civil War. Much of the Midwestern railway system was built to 4-foot, 9-inch and 4-foot, 10-inch gauges and variables in between. The gauges were just enough at variance to prohibit the interchange of cars. A Pittsburgh businessman, William Thaw, reasoned that broad-tread wheels would allow free intercourse between the Northern lines, allowing shipments east and west in one car.[161] The cars might track poorly on these compromise wheels, being too loose on one and too tight on the other, but at the very slow speeds of the time, it should work well enough. Thaw's scheme was first tried in July 1864, and it was not many years before 175,000 cars—all, or mostly all, freight—were wobbling across the country on treads 5 inches wide.[162] By the late 1870s, however, most of the 4-foot, 10-inch lines had regauged. The cars with broad-tread wheels gradually disappeared, much to the relief of the railroads that had had to use them.

The first wheels in American passenger service were either imported from Britain or made after similar designs. It was natural that the pioneer car builders borrowed from existing technology and modeled the first car wheels on those used by highway vehicles. A flanged tire was substituted for the smooth roadway tire, and solider cast-iron hubs took the place of the traditional wooden ones. One surviving first-generation wheel is today in the collection of the Smithsonian Institution. The iron parts of this 36-inch-diameter wheel were uncovered during some excavation at the old shops of the New Jersey Railroad and Transportation Company in Jersey City. The missing wooden parts were restored, and the wheel, which is believed to date from about 1834, was given to the Smithsonian in 1890. It is currently on loan to the B & O Transportation Museum in Baltimore. The cast-iron hub has six U-shaped arms, which are bolted to every other spoke (Figure 7.33). The axle opening is bored out to a diameter of 3 inches; note the keyway. The flange is worn thin and sharp. The tire measures 4½ inches wide. It is probable that a secondary plain tire was fitted to the wheel to facilitate remounting of the flanged tire, but it is not present on the restored wheel.

Wooden spoked wheels worked satisfactorily with the light, horsedrawn passenger cars operated by many pioneer railroads, but once steam operations began, higher speeds and heavier cars rendered them hopelessly inadequate. Finding a dependable substitute was not easy. A great variety of designs and materials

were tried. Isaac Dripps, onetime master mechanic of the Camden and Amboy, described how this was accomplished on his line:[163]

In 1834 commenced to make a wrought iron wheel for the driving wheels for locomotives and wheels for passenger cars. These wheels had round wrought iron spokes riveted to flanges on a cast iron hub, with outer end of spokes riveted to two iron rings forming the periphery of wheel center; these rings are kept separate the proper distance by blocks and studs riveted between them. On these rings are shrunk a wide wrought iron band, on which is shrunk and bolted to it the wrought iron tire as shown by accompanying drawing marked A.

These iron wheels proved in practice but very little better than the wooden wheels as the rivets would jar loose, and found it difficult to keep them tight. So that in 1835 the cast iron center with wrought iron tire was tried and finally adapted for wheels for locomotives. The wooden and wrought iron wheels were still continued in use under passenger cars. About the year 1838 we could not procure any wrought iron tire, and were compelled to try the experiment of using chilled cast iron tire for both locomotives and car wheels. At this time we also used chilled wheels for freight locomotives, but found a great difficulty in getting them properly chilled.

The first chilled tire we tried had counter sunk holes cast in them for receiving the bolts to fasten tire to the body of the wheel, but with the very light chilled tire we then used they broke through the bolt holes. To remedy this and make the tire secure from further damage, should it break, I cast a flange or rim projecting inwards 1½ inch below the metal of tire on outside of tire.

This flange also added material strength to the tire, and was also used to bolt it to the body of wheels, the bolts passing through this flange and wheel with the nuts screwing up on inside of wheels, so that if the chilled tire did break no portion could leave the wheel. I consider this plan the best way of securing any kind of tire to the wheel center. In time we found both the wooden and wrought wheels very troublesome under passenger cars, and next tried a new wrought iron wheel with flat wrought iron spokes, radiating from a cast iron hub with flanges cast on to receive these spokes. With this wheel was used a tubular axle made of ³⁄₁₆ iron, 10 inch diam., secured to the hubs by rivets. On the outside of these arms or spokes was shrunk on and bolted the wrought iron tire. The arms or spokes were also rivetted to the cast flanges on hub. This wheel is shown by accompanying drawing.

This wheel stood well, proved a strong and efficient wheel, but proved too expensive and was soon abandoned.

The next experiment in wheels for passenger coaches was a solid wooden center with eight sectional pieces fitted into a cast iron hub, extending about half the diameter of solid center, with screw bolts passing through a wrought iron ring in front through the wooden pieces and securely bolted to this flange. On the outside of wooden center is a wrought iron band which held the wooden center firmly together so that the tire could not be removed at any time without injuring the center pieces. On some of these wheels were shrunk an

Figure 7.33   A wooden, cast-iron, and wrought-iron 36-inch-diameter wheel used on the New Jersey Railroad and Transportation Company around 1834.

ordinary wrought iron tire, but on others were used cast iron chilled tires, having on the outside of tire a flange projecting inward toward center of wheel with bolts passing through this flanged wooden center, and a wrought iron on back of wooden pieces, thereby fastening the chilled tire very securely to the wooden center.

By this arrangement of fastening there could be no risk or damage should the chilled tire break, as no part of it could possibly get out of place, being held there by 8 ⅞ bolts. After a fair trial this plan of wheel was adapted as the standard wheel for passenger coaches. These wheels were used on an ordinary wrought shaft, the tubular shaft being found too expensive. The back part of hub of these wheels was cast a sleeve extending over shaft 4 inch., so that in case of an axle breaking this sleeve preventing the parts from separating; and in practice has prevented many serious accidents to trains. The ordinary chilled cast iron wheel has been used on the road for freight cars from the commencement of the road, and in practice were found to stand well and give satisfaction, and were finally adapted and used under passenger coaches.

Philadelphia, September 6, 1885   ISAAC DRIPPS

Other railroads sought other solutions. The Petersburg Railroad, after a poor experience with wooden-center and cast-iron wheels, turned to British makers for wrought-iron spoked wheels on the pattern used by the Liverpool and Manchester Railway.[164] They were more expensive than those available locally, but their excellent safety record (in 1835 not one broke during the entire year) made them a wise investment. In the following year the Rensselaer and Saratoga Railroad obtained some creditable domestic wheels which were described in a newspaper account of the day: "The wheels were cast in Troy by Mr. Starbuck, with patent rolled-iron rimmed tire, well annealed and wrought, being put on the cast wheel [center] while red hot, the cooling of the tire and the contraction of the iron, renders it impossible to be ever deviated from its place, the whole wheel is then turned in a steam lathe by machine tools, rendering the circle of the wheel, perfect from its centre, which is a great desideratum."[165] In February 1837 M. W. Baldwin corresponded with one of his customers on the subject of railroad cars.[166] He mentioned that iron-tired wheels with wooden centers—he did not say whether they were spoked or solid—were available in place of the cheaper cast-iron wheels. They would be guaranteed for two years' service. Though not directly stated, the implication was that the tired wheels were superior to

cast-iron wheels. But whatever Baldwin's true opinion, the tired form of wheel was about to disappear from American practice.

One-piece cast-iron wheels were not new, nor were they first made in the United States. British tramways had been using them since the mid-eighteenth century, but when the steam railway was developed, British mechanics felt that cast wheels were not only antique but inadequate for high-speed service. Americans did not harbor the same prejudice, and they worked to perfect a better cast wheel. Cast-iron wheels were admittedly primitive and in some ways inferior to fabricated wrought-iron wheels, but they were very cheap. Here again, capital-poor American lines were forced to devise less expensive supplies if they were to survive and expand. It was also true that American roads were not attempting any fast running in the early years, and a poor grade of wheel might serve their needs adequately.

It was found that cast wheels had several surprising advantages in addition to their low first cost. The very hard chilled tread and flange provided a long-wearing running surface, and one that could withstand the grinding friction of brake shoes. The center or plate of the wheel was not chilled and remained tough, fibrous, and able to absorb normal road shocks without cracking. The hub stayed soft for easy machining. A cast wheel weighed only five-eighths as much as an equivalent tired wheel. A large part of the original cost could be recovered through the high scrap value, for worn-out wheels could easily be broken up with a hammer and remelted. Recovering the metal from a wrought-iron wheel was difficult before the acetylene torch was perfected.

Cast wheels had some disadvantages, of course. They were more likely to fracture than wrought wheels because of the essentially brittle nature of cast iron. Hidden flaws are possible in any casting. Furthermore, the entire wheel was ruined once the chilled tread wore through; it was a "one wear" wheel. And all cast wheels are slightly out of round. Before large grinding lathes were commonly available, there was no practical way to machine the hard running surface of the casting. These defects were serious but not catastrophic, and American railway men came to champion the chilled cast-iron wheel even while they admitted its failings.

What made the cast wheel so attractive was the incredibly durable running surface offered by the chilled tread. Anyone who has worked with a raw iron casting is familiar with its tough outer skin, which always makes the first cuts difficult when the

36 inch 475 lb. Wheel

Used on South Carolina & other Southern Roads

Figure 7.34   *Cast-iron car wheel produced by Ross Winans with a chilled tread and split hub. The design dates from about 1835.* (American Railroad Journal, *June 26, 1847*)

piece is turned or filed. This outer surface is caused by the molten metal's quick cooling upon touching the cold surface of the mold. The phenomenon was well known to all foundrymen. The wheel maker attempted to capitalize on the normally undesirable chilled surface by intensifying it for the purpose of strengthening the tread and flange of the wheel. By making the rim of the mold from cast iron rather than sand, an extremely deep chill could be obtained: anywhere from ½- to 1-inch deep, depending on the quality of the iron and the skill of the foundry. The metal would cool and crystallize almost instantly into a silvery white metal almost as hard as diamonds. Chilled iron was for generations the hardest metal commercially available. It could not even be cut with high-speed steel; only the grinding wheel could penetrate its tough hide. Simply stated, the chill is formed by the fast cooling, which occurs so rapidly that the carbon does not have time to separate but remains combined with the iron. This produces a steel-like metal with a very high carbon content of $3\frac{1}{2}$ percent. In the nonchilled portion of the wheel the carbon separates, taking the form of graphite, which results in ordinary soft grey iron.

The origins of the chilled wheel are unknown. The 1888 *Car Builders' Dictionary* said simply, "The name of the inventor is not preserved." Cast-iron wheels were in general use on British mine tramways by 1780, and by 1812 mention is found of "case hardened" cast-iron wheels.[167] Losh and Stephenson obtained a British patent in 1816 that mentions chilled treads. The idea appeared in America a decade later. Richard P. Morgan mentions this style of wheel in his August 29, 1829, patent covering six-wheel cars. During the following year Horatio Allen reported

on chilled wheels in a letter to John B. Jervis.[168] In later years the idea was claimed on behalf of various American inventors. Henry Mooers of Ithaca, New York, maintained that he cast wheels of this type in 1827 for the Ithaca and Owego Railroad.[169] One might doubt his statement, however, because the line did not open until 1834. And of course, although no facts seem to support it, the honor has been ascribed to Ross Winans, the man who invented everything, according to Zerah Colburn.[170]

On July 29, 1834, Phineas Davis received a patent for casting car wheels. Davis hoped to improve the chill and strengthen the wheel by inserting ½-inch-diameter reinforcing rods inside the wheel castings. The rods were located in the tread. After Davis's death in 1835, Ross Winans took up the manufacture of this form of wheel and continued to produce it until at least 1847. An illustrated advertisement appeared in the *American Railroad Journal* during the same year (Figure 7.34)[171]

Jonathan Bonney claimed to have made chilled wheels for the B & O at the Mt. Savage Iron Works in western Maryland in 1829.[172] A few years later he was in the same line of work in Wilmington, Delaware, with a firm that became a leading producer of chilled wheels in this country, Bush and Lobdell. By the mid-1830s many American foundries were producing chilled car wheels. Among them were locomotive builders such as M. W. Baldwin, Thomas Rogers, and Henry R. Dunham. Dunham was not a great success in the manufacture of locomotives, but he did do a lively trade in the car wheel business and held at least one patent.[173]

Car wheel makers were encouraged to expand the production

Figure 7.35   Cast-iron car wheels of about 1835 used on the Allegheny Portage Railroad.

LOBDELL
1838

EDDY
1845

SINGLE PLATE
1850ᶜ

ATWOOD
1847

WASHBURN
1850   1859

WHITNEY
1847

LOBDELL
1859

MOORE
1858ᶜ

WHITNEY
1859

Figure 7.36   A selection of solid-plate cast-iron car wheels. (Traced by John H. White, Jr.)

of the cast-iron pattern by a growing enthusiasm for the product. In its annual report for 1834 the Baltimore and Ohio spoke of the excellent service provided by chilled wheels: They "have stood the test of experience and have come up to our most sanguine expectations."[174] Another indication of their growing popularity was the increasing number of patents: by 1840 a dozen had been granted for cast-iron wheels. They appear to have become the standard form in America by this time.

Spoked wheels were at first favored (Figure 7.35). Some roads, like the Reading, continued using them into the 1850s, but

elsewhere they appear to have been replaced by solid disc wheels during the previous decade.[175] The chill was deficient opposite the ends of the spokes, and these soft spots wore flat, rendering the wheel useless long before the rest of the tread was worn out.[176] It was also found that the rim broke away between the spokes. Moreover, the flat spokes acted as a fan to stir up the dust along the tracks. Because of the uneven contraction of the casting, the hub could not be cast solid; it had to be cast with three or more splits or gaps.[177] Iron bands were shrunk on to bind the hub together, and the gaps were filled with molten zinc. The problem of the broken hub was not restricted to spoked wheels; it was a difficulty experienced with all early forms of cast-iron wheels.

The disc or plate wheel overcame some of the major difficulties of the spoked wheel. It was stronger, and because of the even distribution of material around the rim area, a deeper, more uniform chill was possible. Commercial manufacture of disc wheels began in the late 1830s. Baldwin is said to have produced some as early as 1836, but no patent was taken out; the evidence comes from a patent case some forty years afterwards.[178] On March 17, 1838, Jonathan Bonney, Charles Bush, and George Lobdell patented a double-plate wheel and began production for the trade. The original design was not satisfactory, but it was probably the first widely used plate wheel in the country (Figure 7.36). The design was subsequently refined and remained popular for many years. It continued to be produced late in the nineteenth century but was used only where extra strength was needed, such as for locomotive leading wheels. The original pattern required split hubs because of the uneven contraction between the fast-cooling chilled rim and the slower-cooling hub and plate. Some of this differential was evened out by adding a wave or S curve to the plate. But the Lobdell wheel, as it came to be called, required a large center core, making it expensive to manufacture. Other designs were soon introduced. In 1847 Asa Whitney secured patents on several car wheel patterns after his so-called corrugated design, which carried the wave concept to its ultimate limit. The plate was dished in a series of sharp radius curves, almost like a drapery of cloth, with the idea of relieving the uneven contraction of the castings. In time Whitney was required to simplify his design (Figure 7.36). Other inventors proposed schemes for single-, double-, and even triple-plate wheels that exhibited both graceful and contorted profiles in every imaginable combination of convex and concave curves. By 1860 there were nearly 100 patents covering cast-iron wheel shapes alone.[179]

The single-plate style was simple and cheap to make, because it required no core. But it was not regarded as a first class wheel and seems to have been little used for passenger car service. It was adopted to some extent for freight cars and streetcars before 1880. Double-plate wheels were considered the best form, combining as they did strength and light weight. Triple-plate wheels were a needless complication. Of the various double-plate wheels, none could approach the Nathan Washburn design for popularity (Figures 7.36 and 7.37). Washburn was a practical

Figure 7.37   *Washburn's double-plate car wheel, patented in 1850, was the most popular cast-iron wheel in America. This scene was photographed in Oakland, California, in 1869.* (Society of California Pioneers)

foundryman whose major business undertakings were in Worcester, Massachusetts. His design, patented on October 8, 1850 (No. 7710), was pleasing in appearance and exceptionally strong. The S-shaped brackets on the rear of the rim strengthened the flange, making it a very serviceable wheel. Wheel makers liked it because of its simple shape and the fact that it was relatively easy to cast. It quickly became the standard style of cast-iron wheel. In 1879 it was declared the strongest and most reliable made; 70 percent of all car wheels in service were said to be of this pattern.[180] Naturally other patentees tried to cash in on Washburn's success. Claiming that he had patented the same wheel in 1848, Anson Atwood wanted a license fee of 25 cents for every Washburn-style wheel made—a royalty that would have amounted to a fortune. But Atwood's bulbous and ill-conceived design bore little resemblance to Washburn's elegant profile (Figure 7.36). Even after the cast-iron wheel disappeared from passenger car service after 1910, it remained the standard form of freight car wheel until 1928, when it was at last displaced by a new design.

Several years before the Washburn wheel was introduced, foundrymen discovered a technique which permitted the casting of solid hubs. The weak and expensive bands and zinc filler blocks were at last eliminated. Controlled cooling was the solution to the dilemma of uneven contraction. Some foundries simply buried the hot wheels in sand, ashes, or fine charcoal, allowing them to cool slowly.[181] Asa Whitney devised a heated, airtight soaking pit, where the hot castings were kept for three to four days. The temperature was gradually lowered during this period. The process, which did not adversely effect the chill, was described late in 1847.[182] A patent was granted to Whitney in the spring of the following year.

Wheel makers also sought to improve the safety and durability of their product by employing the best metal available. Most early ironmasters had no real idea of the physical characteristics of the metals used; instead they depended on the reputation of the pig iron they bought.[183] Charcoal iron was the only acceptable material for wheels, but again the early wheel men did not really understand why it produced a superior wheel. They selected the right material for the wrong reason, concluding that pig iron free from sulphur and phosphorous made the best wheels, when in fact these very elements were effective in producing a good chill. The undesirable ingredient in a mix was silicon, too much of which had a negative effect on the chill.[184] But even without a scientific understanding of metallurgy, the early-nineteenth-century founders produced good wheels if for no other reason than that excellent grades of cast iron were available to them. Pig iron made by the charcoal process had a tensile strength of 23,000 pounds, compared with ordinary pig iron's tensile strength of 18,000 pounds.[185] Some furnaces produced extra-fine charcoal irons with tensile strengths up to 41,000 pounds.[186] It would be difficult to make a weak casting from such strong iron.

Wheel makers naturally sought the most celebrated brands of iron available and emphasized the trouble they took to secure the very best for their customers. According to Paul R. Hodge, an early locomotive designer, wheel makers of the 1830s and 1840s found that a mix of Baltimore and New Jersey ores gave the best results.[187] Bush and Lobdell of Wilmington, Delaware, imported pig iron from North Carolina.[188] The Ramapo Wheel Company of New York boasted of the quality iron dug from its own mines in Berkshire County, Massachusetts, and Litchfield, Connecticut. By 1880 some makers were mixing a variety of irons for the best alloy. Iron ore from Maine, Michigan, Maryland, and Pennsylvania was stacked in separate piles around the foundry yard.[189] Samples were tested to determine strength and chilling properties. The ironmaster decided on the right balance, taking into account as best he could the adverse effects of the scrap wheels introduced into the furnace. Daily samples were pulled from the furnace and tested to ensure a quality iron. According to the *Railroad Gazette*, much depended on the judgment of the men involved. No textbook formula was enough; it required a knowledgeable foreman whose "care, mature experience and vigilant inspection" could alone produce a first-class product.[190]

Wheel makers were forced to learn more about their art as the price of charcoal iron increased after the Civil War. It was more expensive than coke iron, for even in the early 1870s it was selling for $62 a ton, which was $20 more than coke iron.[191] William G. Hamilton, once superintendent of the Jersey City locomotive and car wheel plant and later employed by the Ramapo Wheel Company, found that poorer grades of cast iron could be greatly improved by mixing in 5 percent of scrap steel.[192] Worn-out steel rail could be purchased cheaply for the purpose. The iron was strengthened by 17 percent. Hamilton patented his "steeled" wheel alloy process in 1868 and 1873. It was widely used, although some makers, such as Whitney, found that the same results could be obtained by adding wrought-iron scrap. Some years after Hamilton's process was announced, other engineers suggested the addition of a trace of manganese.[193] In 1911 Robert C. Totten claimed to have solved the car wheel metal problem by adding chrome and nickel to the alloy.[194] These metals counteracted the bad effects of combined carbon that was present in the chilled portion of the scrap from old car wheels. Totten's wheel was said to be 300 percent stronger than the ordinary cast-iron wheel.

As mentioned earlier, the tread and flange were cast against a smooth iron surface and therefore did not require machining. This portion of the wheel came from the mold like a die casting (Figure 7.38). Unfortunately, a large iron casting stays in a semimolten state for so long that it does not remain absolutely true. Few cast-iron wheels were perfect circles. In 1883 it was reported that car wheels were anywhere from $\frac{1}{16}$ to $\frac{5}{16}$ inch out of round.[195] By 1939, however, techniques and standards had advanced so much that a wheel over $\frac{1}{32}$ inch out of true was not acceptable, and one-third of all wheels were only $\frac{1}{64}$ inch out.[196] Faster pouring (10 to 14 seconds) may have helped increase accuracy.

The safety of cast-iron wheels was another major concern of both the maker and the purchaser. European engineers could not understand the American loyalty to the cast-iron wheel. They found it incredible that this brittle material could provide safe service under hundreds of thousands of cars—cars far heavier than any running outside North America. The Institution of Civil Engineers in England expressed wonder at the extensive use of cast-iron wheels in America. One member stated that they were "but poor expedients" when compared with the wheels used in England.[197]

By the late 1890s cast-iron wheels were outlawed in all European countries except Austria.[198] As already explained, the reason for the success of the cast wheel in America was the availability of high-strength cast irons. And they had a remarkably good safety record in this country: in 1883 there were 5,600,000 cast wheels in service on U.S. lines, and only 28,000 were condemned each year.[199] Many of these were discarded because of cracks and thus cannot be classified as total failures. Wheel failures were controlled in the winter by reducing train speeds. The Boston and Providence said that attempts to maintain normal schedules in cold weather were both reckless and expensive.[200] The line's rolling stock had as many breakdowns in the three winter months as it did in the rest of the year, filling the shops with crippled locomotives and cars. A 25 percent reduction in train speeds proved of some benefit, however.

The introduction of power brakes posed a new threat to wheel safety. The friction of the brake shoes against the tread and flange could raise the temperature of this portion of the wheel to 1350 degrees F.[201] The combination of such high temperatures and rapid cooling by snow and frigid winds crystallized the metal and shelled out the wheel treads.[202] Winter continues to play havoc with car wheels even today because of this problem. It is one reason why some roads have adopted disc brakes.

The most intelligent method of improving wheel safety was to determine the soundness of the wheel before it entered service. Hidden flaws, blowholes, and hairline cracks were difficult to detect visually, but it was possible to uncover a flaw by striking the casting with a hammer. A defective piece would not ring true, and in the case of an extreme fissure it might literally fly to pieces. The sledgehammer was the metal crafter's time-honored testing instrument. At Whitney's plant each wheel was subjected to several heavy blows with a hammer before leaving the shop, and 10 percent were found defective by this simple test.[203] As early as the spring of 1843 drop-hammer tests were conducted in Boston.[204] Several makers submitted their products for examination. A 131-pound hammer with a fall of 10 feet was employed.

Figure 7.38 A mold for a cast-iron Washburn wheel. The chilled rim and sand mold for the front tread and flange is suspended by a gimbal on the right side. The bottom or rear mold stands on the floor in front of the men.

Winans's spoked wheel cracked at the rim after one blow, and the entire wheel shattered after eight falls of the hammer. The broken casting revealed an array of blowholes. Several local Boston foundries made only a slightly better showing than their Baltimore rival. Bush and Lobdell, however, put a 496-pound, double-plate wheel on the anvil that survived the hammer admirably. After nine blows the wheel was unaffected; it began to break up only with the tenth.

Late in the nineteenth century, wheel testing became more formal and ingenious. Two wheels from each lot of one hundred were set aside for testing.[205] The sample was expected to withstand fifteen blows from a 140-pound drop hammer. In addition, every wheel in the lot received a heavy blow from a 6-pound hammer. By 1929 a 250-pound ball replaced the old-fashioned hammer.[206] The fall varied from 9 to 13 feet, depending on the size of the wheel being tested. Each piece was expected to withstand a dozen blows.

The second trial was a thermal test to establish the wheel's ability to withstand heating caused by the brake shoes. A ring of molten iron was ladled around the wheel's rim, and the time required for this severe shock to crack the casting was measured. The thermal test was introduced by the Griffin Wheel Company in the early 1890s and was soon adopted by the Master Car Builders Association.[207] At first a wheel that could survive the ordeal for two minutes was considered adequate for normal service, but by the late 1920s the length of the test had been increased to five minutes.[208]

Another check on wheel safety was the record kept on every car wheel in interchange service. The date and the maker or place of manufacture were cast in raised letters on each wheel. If the wheel failed, the nature of the defect was noted and other wheels in the same lot were investigated. In some cases the faulty wheel or its fragments were returned to the foundry for a more careful analysis. The Pennsylvania Railroad began keeping such accounts sometime before 1876.[209] The records specified the part of the cupola charge from which the iron was taken, the depth of the chill, the weight, the size, the wheel's length of service, and its behavior in service. Special entries were made in the event of a failure. The records grew into a prodigious file. In 1883, for example, there were 5.6 million wheels in service, and some twenty-five years later there were 18 million on the books.[210] The clerical time was considered well spent, however. In addition to promoting safety, the records made it possible to enforce the wheel maker's guarantee. For a time the guarantee was on a mileage basis of 40,000 to 60,000 miles, but by the turn of the century the guarantee was limited to a time span of four years.[211]

The yeoman work of wheel inspection was done in the field by a grimy car knocker, who quickly bobbed under the train at every major station stop. His only apparatus was a hand lantern and a long-handled hammer. The uninitiated passenger sat and wondered at the dim light and clanging hammer as the inspector went over every wheel looking for a telltale crack or a muted plate that did not ring like solid metal.

Although the cast-iron wheel was the dominant form of American car wheel during most of the nineteenth century, there was always an undercurrent of interest in tired wheels. The fabricated wheel had several advantages. The entire wheel need not be scrapped just because the tread or flange was worn out; a new tire made it like new. It was salvageable, whereas an old cast-iron wheel could only be broken up for scrap. Supporters' claims that tired wheels were safer and more economical encouraged American railway men to give the tired wheel yet another chance. The engineering trade press was undoubtedly also responsible for periodic revivals of tired wheels. British journals were naturally filled with notes, drawings, and patent descriptions of the latest ideas in this area, and some of the material was reprinted by American magazines. The *Journal of the Franklin Institute*, for example, regularly featured such items even during the 1840s.[212]

Records of actual road tests are also available for tired wheels during the nineteenth century. The Hudson River Railroad made an attempt to use them in 1849.[213] The line was atypical in that it began with an avowed purpose of high-speed passenger service—a necessity if it were to compete with the swift Hudson River steamers. Most American lines of the time were content with operating speeds of 25 mph or less. The new road announced its intention of matching the best English trains with speeds of 50 to 60 mph. In emulation of these trains the line imported a large number of British car wheel sets at the princely sum of $500 each. Even on the relatively straight Hudson River track, however, the flanges were rapidly cut away by the curves. Many years after the road had discarded the British wheels, their remains could be seen at the Thirty-first Street Station in New York City.[214]

Several lines running out of Boston gave the British tired wheels a trial in about 1850.[215] The Boston and Providence devised its own form of cushioned, tired wheel, which had a cast-iron center with 2-inch-thick wooden wedges between it and the wrought-iron tire. It worked well and was put on all the road's passenger cars. In 1857 the B & P's master mechanic, George S. Griggs, patented the same style of wheel for locomotives. In 1853 two lines in New York were also trying wrought-iron tired car wheels.[216] The Utica and Schenectady was disappointed by the experiment, for the wheels cost three times more than the cast variety and their flanges quickly wore thin, causing derailments. Despite the bad experience of the U & S, the Northern Railroad was willing to give the scheme a trial because so many of its cast wheels broke during the winter season.

In England the Mansell wooden-center wheel, introduced in 1848, was receiving great acclaim. The solid wooden center was built up of wedge-shaped pieces of teak, which produced a resilient and quiet-running cushioned wheel. The Camden and Amboy was reported to have used a similar design, with red cedar instead of teak, for some years before (and presumably some years after) 1861.[217] By the late 1860s several American lines were said to be trying Mansell wheels. The Michigan Southern adopted them for the sleeping cars of its Excelsior Line.[218] The Boston and Lowell and the Hudson River railroads reported the experimental use of Mansell's design. Steel tires seemed to solve the problem of rapid flange wear, but the Boston and Lowell abandoned the test because the tires worked loose.[219] Pullman and the Gilbert Car Works also experimented with Mansell wheels during the same period, but no-large scale purchases were made by American railroads despite the exceptional performance of this design in England.

Success came to the tired wheel from an unlikely source and in a most unlikely form. The inventor was a former locomotive engineer, Richard N. Allen (1827–1890), who had drifted from one position and occupation to another.[220] He was persuaded by his brother-in-law to buy into a paper mill at Pittsford, Vermont. The plant produced a common grade of cheap strawboard for which the market was glutted, and Allen soon found himself the sole owner of a profitless business. Rather than close down, he set out to find other uses for strawboard besides its customary role as covers for inexpensive textbooks. Somehow from his railroading experience he thought of adopting the weak paperboard for use in car wheels. How this paradoxical idea came to him is not recorded, but it was greeted with ridicule.[221] A railroad car wheel made of paper? Even after Allen's wheel was in common usage, the term "paper wheel" aroused puzzlement and mirth in most people. Actually the basic idea was nothing more than the substitution of compressed paper for wood at the wheel's center. The paper center was attached to the tire by front and back metal plates securely bolted to the paper disc by twenty-four or more bolts (Figure 7.39). The paper center, even though greatly compressed and nearly as hard as ivory, was spongy enough to cushion the ride and deaden the sound of the wheels grinding over the rails.

Allen began work on his idea in 1869, and after some difficulty he persuaded a local railroad to provide a car for testing.[222] Skeptical of the bizzare experiment, the line was not about to trust a valuable piece of rolling stock to a crank. It provided

ALLEN PAPER CAR WHEEL COMPANY,
General Offices: 239 Broadway, N. Y.

Especially adapted for Sleeping and Drawing-Room Cars, Locomotive and Tender Trucks. Steel Tire with Annular Web—Strongest, Most Durable, and Most Economical Wheel in use. 74 manufactured in 1877; 60,000 manufactured to date; present facilities, 36,000 per year. Works at Hudson, N. Y., at Pullman, and at Morris, Ill.
W. H. FENNER, JR., President.   J. C. BEACH, Treasurer.   C. H. ANTES, Secretary.

Figure 7.39   The Allen paper wheel, introduced in 1869, had a center core built up from strawboard. Steel or iron plates were bolted on the front and rear. It was widely used in this country between 1880 and 1915.

Allen with an old freight car that carried wood between fueling depots. To everyone's astonishment, the car ran 5,000 miles trouble-free. In the spring of 1870 Pullman purchased a set of Allen wheels for one of his sleeping cars.[223] The wheels performed well, and Pullman bought more and eventually adopted them as a standard. Now that he had the patronage of the rising Pullman empire, Allen was no longer viewed as a foolish tinker. The paper wheel became an everyday fact of luxury railroad travel, making its inventor famous. Allen himself, however, appears to have been squeezed out of the firm by the early 1880s.

The growing interest in tired wheels in this country sprang from several sources. In 1910 George L Fowler, onetime associate editor of the *Railroad Gazette*, recalled that in the early 1870s railroad mechanical men developed "a feeling of insecurity" about the continued use of cast-iron wheels because of the increases in train speed and passenger car weight. These men accepted steel tired wheels from a genuine belief in their greater safety. Others, however, adopted them for more cynical reasons. According to Fowler, some men championed various patented forms of steel tired wheels merely because of their advertising value. This aspect of technical history is one that deserves a study of its own. Pullman himself was perhaps as much interested in the paper wheel's promotional value as he was in its mechanical merits. He knew the value of publicity and pursued it eagerly, particularly in the early years of the sleeping-car business. The paper wheel captured attention, and the early notices of Pullman cars always seemed to mention the wonderful paper wheels which silently and securely transported the car and its occupants across the country.

In one of his master strokes of publicity, Pullman became associated with Frank Leslie's 1877 tour of the United States and may have been one of its backers. The publisher returned the favor with generous coverage of Pullman in *Frank Leslie's Illustrated Newspaper*. One engraving shows Pullman, using his walking stick as an instructor's pointer, explaining the wonders of the paper wheel to an attentive bystander.[224] The accompanying article said:

While our party was viewing the exterior of the vehicle, Mr. George Pullman himself strolled up. Pointing to the wheels, he made the somewhat alarming announcement that they were made of paper! In proportion to its weight, he said, good paper, properly prepared, is one of the strongest substances in the world. It offers equal resistance to fracture in all directions. While the toughest woods are sometimes liable to crack and split under severe trial, and ordinary iron becomes brittle from the constant jarring on the smoothest of steel rails, paper possesses a certain amount of elasticity very desirable in a car wheel. Paper wheels, he said, were subjected to an enormous hydraulic pressure and, when surrounded with a flange of steel, were the most perfect wheels yet invented.

Extravagant endorsements of the Allen wheel were also made by A. B. Pullman, doubtless because of the interest that his brother G. M. Pullman held in the firm.[225] After 1881 Allen's main plant was located on the grounds of Pullman's mammoth Chicago works.

The manufacture of these pasteboard wonders began with circular paper sheets that were glued together with ordinary flour paste.[226] In 1882, 117 sheets were used, but by 1893, 200 were required. The discs were compressed by a 650-ton press for three hours. They were then dried and cured in a warm room for six to eight weeks to ensure the evaporation of all moisture. The seasoned discs were turned in a lathe to size. Bolt holes were drilled, the outer ¼-inch-thick iron plates were put in place, and the steel tire was bolted on. The finished paper core for a 42-inch-diameter wheel weighed 185 pounds. A complete wheel of this size weighed 1,115 pounds.

In its infancy the paper wheel was an insignificant thing, associated almost exclusively with Pullman sleepers. In 1877 Allen produced only 74 wheels.[227] Within a year his company claimed that nearly 1,500 were in service, but this represented only a tiny fraction of passenger car wheels.[228] In 1883 the firm said that

ATWOOD HEMP CAR WHEEL CO.,

MANUFACTURERS OF

ATWOOD PATENT HEMP-PACKED STEEL-TIRED WHEELS FOR CARS AND ENGINES.

No Bolts,   **Fewest Possible Parts,**   **No Rivets,**

Tires put on cold and held by interlocking of Hemp Packing, which acts as a cushion, cuts off
Metallic Connection, and prevents Rigidity.

Thoroughly Tested, Safest, Cheapest and Most Noiseless of all Steel-Tired Wheels.

JOS. W. DREXEL, President.   ROBERT F. SHEPARD, Sec'y and Treas.

OFFICE: — 59 LIBERTY STREET, NEW YORK.

C. W. LEAVITT & CO., SELLING AGENTS, 161 BROADWAY.

PAIGE'S PATENT WROUGHT METAL WHEELS.

Office, 211 Superior St., Cleveland, O.

J. E. FRENCH, President.

W. S. DODGE, Secretary and Treasurer.

Adapted for Sleeping and Drawing Room Cars, Locomotive and Tender Trucks. Steel Tires with ½
inch Plates, securely bolted, making to a perfectly Safe, Durable and Noiseless Wheel.

SNOW'S PATENT STEEL-TIRED METAL WHEELS,

WITH INTERCHANGEABLE HUB,

Annular Web to the Tire, Wrought-Iron Plate and Retaining Ring

Durability of Construction, Noiseless in Service

—AND—

*Manufacture especially for all Classes of Equipment
where Safety and Economy are the principal considerations.*

Address: W. W. SNOW, - - - New York.

RAMAPO, - - - New York.

THE ELLIS STEEL-TYRED WHEELS.

MANSELL RETAINING RINGS.

CAST-IRON CENTERS (Double-Plate,
Single-Plate or Spoke).

WROUGHT-IRON CENTERS.

W. R. ELLIS, 18 BROADWAY, NEW YORK.

---

*Figure 7.40 A great variety of tired wheels were produced in the United States follow-
ing the success of the Allen wheel. This collection of advertisements is a sample of the
types offered.*

30,000 were running on 150 railroads.[229] Since there were roughly 200,000 wheels in passenger car service at this date, Allen's wheels were on less than 15 percent of the fleet. Statistics are not available for the other makes of tired wheels available at the time, but it is clear that cast-iron wheels were no longer the exclusive bearers of the traveling public. During the 1880s more major lines began to adopt paper or some other form of tired wheels. The Santa Fe announced in 1880 that it would equip all its passenger cars and locomotive leading trucks with paper wheels.[230] The Milwaukee Road made a similar commitment in 1882, and the Northern Pacific followed suit the next year.[231] In 1886 Allen's firm announced that 60,000 paper wheels were in service.[232] In seven years the number increased to 115,000.[233]

Allen's success encouraged a band of imitators. Some produced cushioned wheels; others simply offered a demountable steel tired wheel. By the mid-1880s the advertising pages of the *Railroad Gazette* and other trade journals were crowded with notices by Snow, Pecham, Ellis, Miltimore, Boies, Munton, Allston, Paige, Thurber, and Coopers (Figure 7.40). Some of these were marginal firms that probably produced only a few wheels, but others were sponsored by major railway equipment producers. The Thurber wheel was marketed by the Brooks Locomotive Works, the Boies wheel by the Dickson Manufacturing Company, the Snow wheel by the Ramapo Wheel Company, and the Paige wheel first by the Wason car works and later by an independent firm in Cleveland. Production figures are not available, but it would seem reasonable to assume that even the combined totals would amount to only a fraction of Allen's production. At least one British manufacturer tried to grab a piece of the American market, offering its product through the agency of W. R. Ellis & Company of Boston and New York.[234] It was called the Brunswick Steel Tired Wheel and featured flat metal spokes bent in the shape of a star. The Boston and Albany had been using domestically manufactured tired wheels of this design since 1870 and claimed that they gave excellent service.[235]

A number of Allen competitors tried to capitalize on the idea of the cushioned wheel. Atwood wrapped hemp ropes around a cast-iron center between the tire and rim. A firm in Indianapolis inserted a layer of rubber.[236] The Boies wheel featured flexible center plates. None of these schemes appears to have gained wide acceptance, and in modern times the whole concept of the cushioned wheel has fallen into disrepute.[237] The so-called sandwich wheels, built up from alternate plates of steel and rubber, became fashionable for transit cars in the mid-1930s, but ultimately the industry concluded that they rode no better or more silently than conventional one-piece wheels. They tend to come apart in service and are suspected of creating a "torque effect" that corrugates the rails.

During the era of the tired wheel, its advocates made much of its superior safety, increased mileage, and greater economy, while detractors claimed precisely the opposite. In 1875 the Boston and Albany expected its tires to last for 370,000 miles, or seven times longer than the average cast wheel.[238] The road said that the tired wheel ran for 134,000 miles initially, 186,000 addi-

tional miles after the first turning, and another 50,000 miles after the second turning. Several years later the Grand Trunk said that it had tired wheels running up to 474,000 miles and that it hoped to get 500,000 miles out of them.[239] British lines realized 18 to 20 years' service from a tire.[240] The improved mileage was attributable to the growing popularity of weldless steel tires by the late 1870s.[241] The largest producers were in England and Germany—Allen, for example, bought heavily from Krupp. Such tires were made in the United States as early as 1867 by the Nashua Iron & Steel Company. Some wheel centers wore out one or two tires and were still good for more service.[242]

Critics of tired wheels acknowledged that these mileage figures were impressive. Some were even ready to admit that cast wheels would not last a year in express-train service, even though they were good for seven years under a freight car. However, the longevity of the tired wheels came at a high price. They actually cost more money than they saved. The price of a tired wheel was $80, while a cast wheel sold for $12. One engineer figured that the potential interest from the excess investment on a set of paper wheels would buy a new set of cast wheels every six months.[243] To this must be added the expense of periodic turning and the remounting of new tires. Even the friends of the steel tire agreed that its high first cost was the major drawback. Yet they maintained that the extra cost was more than justified on grounds of safety. They spoke of the tired wheel as "absolutely safe"—an assertion quickly, and rightly, challenged.[244] On November 1, 1876, a paper wheel under the Pullman sleeping car *Woodbine* broke, derailing the Niagara Express on the North Pennsylvania Railroad.[245] Several passengers were killed. A suit brought against Pullman contended that "paper wheels" were unsuitable for railway service. The suit was disallowed after several years of wrangling, because the plaintiffs could not prove their case. If they had been successful, the paper wheel might have disappeared from the American railroad at an early date. No overall statistics are available, but again isolated reports indicate that tired wheels were not in fact absolutely safe. In 1883 it was reported that three failed in service, although these failures caused no deaths or serious injuries.[246] Ten years later the Wagner sleeping car *Alva* was thrown into the ditch because of a broken tire.[247] The failure of other wheels used under Wagner cars is documented by a series of accident photographs in the possession of Lee Rogers, a railroad collector of Washington, D.C. One of these views is reproduced in Figure 7.41.

The complaints against tired wheels amounted to more than faultfinding, but the charges, real and imagined, in no way discouraged such enthusiasts as Pullman. Among their other virtues, he claimed, tired wheels produced a substantial reduction in truck repairs because of the cushioning effect of the paper centers.[248] Others clearly shared this faith, for in 1897 all the important manufacturers united as the Steel Tired Wheel Company.[249] The monopoly was composed of nine firms, which included Allen's paper wheel company. The company prospered for a decade and then went into a decline. Steel cars were too great a burden for the fabricated wheels. In the process of brak-

Figure 7.41   Steel tired wheels had a good safety record, but they could fail, with devastating results. (Lee Rogers Collection)

ing, the tires became overheated, causing them to expand and loosen.[250] By 1915 they were declared unsafe by a spokesman for the Interstate Commerce Commission.[251] Eight years later the Allen plant at Pullman, Illinois, was abandoned. Production had apparently ceased some time before, and the paper wheel was already spoken of as if it were some almost forgotten antiquity. Apparently no wheels of this type were then on main-line trains, although some were used in and around the Pullman plant as shop trucks. In the mid-1960s paper wheels served the same purpose at the St. Louis Car Company.

The third or modern phase of the railway car wheel began in the first decade of this century, when the rolled- or wrought-steel wheel came into favor. It was a one-piece, single-plate wheel superior to the chilled-iron and fabricated tired style of car wheel. It was stronger and cheaper than the tired pattern, although it could not compete with the cast-iron wheel on the basis of first cost. In 1917 a wrought-steel wheel cost $19, a chilled wheel only $9.[252] It was calculated that $350 million would be needed to reequip all American freight and passenger cars with steel wheels.[253] It was argued that such an extraordinary expenditure was really unnecessary because of the progress being made on the chilled wheel. The wheel makers formed an association in 1908 which opened a laboratory for the sole purpose of improving chilled-wheel design and manufacture. The same group supported a similar program at the University of Illinois.[254] By increasing the wheel's mass by 40 percent, the researchers were able to increase its capacity by 140 percent. The efforts of the Association of Manufacturers of Chilled Car Wheels prolonged the life of the cast-iron wheel for another half century. Its obituary appeared in the February 22, 1901, issue of the *Railroad Gazette*; yet it continued to serve as the standard

freight car wheel for decades. In 1930 it was estimated that 95 percent of all U.S. freight cars, 25 percent of passenger cars, and 66 percent of locomotive tenders were equipped with cast-iron wheels.[255] Faith in the chilled wheel, however, declined permanently following World War II, when larger freight cars and faster schedules proved too much for the old standard. In 1958 the Association of American Railroads Mechanical Division voted to outlaw chilled wheels on new cars.[256] Five years later manufacture of the wheels was abandoned, and in 1968 they were banned from interchange service.

The prolonged acceptance of cast wheels for freight cars was based on what constituted reasonable risk weighed against first cost. They were considered safe enough for this service, and their very low first cost was a definite attraction. But after the coming of the steel car, the chilled wheel was not deemed good enough for passenger service by many engineers. It might carry the weight, but speed was a factor as well, and fast running imposed unsafe stress on the brittle cast-iron disc. The fact that as many as one-quarter of American passenger cars rode on chilled wheels in 1930 (although another source says that only one-fifth were so equipped in 1924) probably means that these cars were either lightweight suburban units or obsolete branch-line cars not in high-speed, main-line service.[257]

Inventors were working on one-piece steel wheels generations before they came into general use. On July 11, 1854, George B. Hartson of New York City received a patent (No. 11,428) for machinery to produce a one-piece rolled wheel. The proposed machinery was not unlike that used in actual production many years later. Another American, Samuel Van Stone, received a patent in 1863 for a different style of machine to roll and forge car wheels. A year earlier Krupp was producing cast-steel car wheels, which were advertised as being available through his

Figure 7.42 Solid rolled-steel wheels were introduced commercially around 1903 and became popular for passenger cars within a few years. Cast-iron wheels continued to be used under freight cars until very recently.

American agent.[258] No reports have been uncovered on the use of these wheels in North America, although by 1881 Krupp claimed to have manufactured 100,000 of them. In 1868 several U.S. lines were reported to be testing cast-steel wheels produced by the British firm of Vickers.[259] It is unknown whether these wheels were literally "cast" or whether they were forged wheels made from cast steel—a term then current to differentiate Bessemer steel from that made in the traditional manner. A year later the Black Diamond Steel Works of Pittsburgh was producing steel wheels for the Pennsylvania on an experimental basis.[260] The first wrought-steel wheel in this country was displayed by a Belgian firm at the U.S. Centennial Exposition of 1876.[261]

A decade later, H. W. Fowler of Chicago patented a process for making a solid, one-piece, rolled-steel wheel.[262] Production began in 1888, and several thousand were turned out for some twenty railroads over the next two years. They were made by heating, rolling, and annealing cast-steel blanks. The blanks, however, proved porous and spongy, resulting in a defective wheel. Most of the wheels were quickly abandoned, but some remained in service for fourteen years and ran as much as 800,000 miles.

There were other experiments during the time of the Fowler wheel, such as a Pittsburgh concern's attempted manufacture of a cast-steel wheel invented by F. W. Webb of England.[263] The steel wheel was cast in a rotating mold, reheated, and rolled. It was said to withstand seventy-three blows during a drop-hammer test, but like Fowler's wheel, it attracted no long-term customers. It would take big money to perfect a practical steel wheel; a rich, patient, and determined entrepreneur was needed who could outlast the period of rejection that all new products must undergo. The man was Charles T. Schoen, a pioneer manufacturer of steel freight cars.

In 1898 Schoen began producing rolled-steel wheels with a process devised by Henrik V. Loss.[264] By 1903 the method had been perfected, and Schoen organized a separate plant for wheel making (Figure 7.42). In 1904 he sold 1,134 of his wheels; in 1905, 22,332; and in 1906, 58,590. Production dropped back to 54,467 in 1907 because of the financial panic, but in 1908 the business was so promising that Carnegie bought Schoen out. The following year the plant produced 269,000 wheels. By this time Midvale, Standard Steel, and the Forged Steel Wheel companies were also in production. In 1912 5 percent of American passenger cars had wrought-steel wheels, a figure that climbed to an estimated 80 percent in twelve years.[265] Today they are still the standard passenger car wheel.

The production of wrought wheels involves the following steps. A thick slab is cut from a 15-inch-diameter ingot. The slab is heated and placed in a 12,000-ton press for rough shaping. It is reheated and undergoes a second forging. The center axle hole is punched out. It is reheated again and placed in a huge rolling machine, where the tread flange and plate are given their final shape. The wheel is put in a press to dish the plate. It is then quenched, reheated, and allowed to cool slowly to relieve internal stress developed by the forging and rolling processes. At last the wheel is machined for final finish.

Beginning in 1930, heat-treated wheels came into favor.[266] The tread and flange were hardened to improve mileage. In this way the body of the wheel could remain as low-carbon steel—a relatively cheap material that was not prone to thermal cracking. Yet the running surfaces were hard and long-wearing. Most roads preferred multiple-wear wheels; even though single-wear wheels were lighter, the multiples were considered more economical over the long run. Machine tolerances were made much more exact (0.003 inches) after about 1940 to ensure a smoother ride and better balance. Ordinary machining in an engine lathe was no longer good enough; wheel sets were now finished by a precision grinding machine.

In 1941 the A.A.R. Mechanical Division began to classify wrought-steel, heat-treated wheels under three broad groups, designated A, B, and C.[267] Class A is the softest wheel of the group, with a relatively low carbon content (up to 0.57 percent).

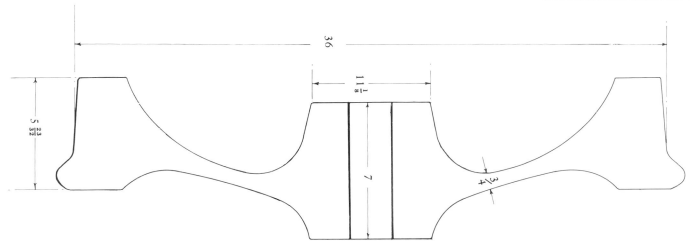

Figure 7.43 Cast-steel wheels are rarely found in passenger service, though they are common on freight cars. This wheel was produced by Griffin around 1970. (Traced by John H. White, Jr.)

These wheels are the fastest-wearing, but they are also the safest, because they are the most resistant to shelling and thermal cracking. They are accordingly favored for service with heavy-tread (brake shoes) braking loads. Class B has a greater carbon content (0.57 to 0.67 percent), wears longer than class A wheels, and is suitable for high-speed service with moderate braking conditions (or disc brakes) and high wheel loads. Class B wheels are preferred for heavyweight cars. Class C has the highest carbon content (0.67 to 0.77 percent) and is the least used because of the tendency to develop thermal cracks, particularly when used with shoe brakes. These are the longest-wearing wheels and could be used successfully with nonthread or disc brakes.

At the time that class A, B, and C wheels were classified, the Mechanical Division attempted to reduce the variety of wrought-steel wheels in production.[268] Prior to 1941 there were about 500 designs in use. This number was cut to 58 in 1942, and by 1967 the number of standard designs was down to 29.

Brakes

In 1882 an English writer observed, "No one can look upon one of our modern locomotives going at the rate of 60 miles an hour without . . . breathing a prayer for the safety of the living freight behind it."[269] Indeed engineers had concentrated on making the train go; swift locomotives and free-rolling cars were the products of their ingenuity. But they had invested less energy and imagination in getting the train to stop. The mechanism for retarding its motion was a miserable collection of light, manually powered rods and levers that was in no way comparable to the elegant machinery of propulsion. Early locomotive superintendents, in fact, wanted nothing to do with brakes. Before about 1875, few American locomotives were equipped with brakes of any kind. The tender had a set of shoes mounted on one truck, but this was mainly for the purpose of holding the engine while the train was at rest in the engine house or at a way station. Brakes, as every good engineman knew, would force the driving boxes and side rod bearings out of line. The reasoning seemed to be that stopping the train was a troublesome job for the car department to handle. In fact, railroad men seemed to feel generally indifferent on the subject. All the major advances in braking technology were made by men outside the field.

The history of railway brakes can be divided into three sections that form a convenient outline of its evolution. These are the hand brake, the continuous brake, and the power brake. Between 1830 and 1875, hand or manually powered brakes were the basic system employed. A lever or hand wheel on the end platforms caused a linkage of chains, rods, and levers to force a block or shoe against the tread of the wheels. The resulting friction retarded the train's motion. The kinetic energy of the train was converted into heat. This technology was directly adopted from highway vehicles. Around 1850 the efficiency of the hand brake was doubled when the brakes on both trucks were interconnected, so that all eight wheels could be braked by turning the wheel at one end of the car.

The second major development, continuous brakes, actually overlaps the eras of both the hand and the power brake. The attempt to devise a continuous brake was an effort to connect the brakes of all cars in the train so that they could be worked in concert. No really satisfactory continuous brake was created, however, until the power brake itself appeared. Various forms of power brakes were introduced at an early date, but none were accepted into everyday practice until about 1870. The schemes offered included steam, electric, vacuum, and compressed-air brakes. In the course of less than a decade the compressed-air system came to prevail, and it remains the standard form to the present day.

When the railroad era opened, brakes were regarded by some managers as an unnecessary appendage. It was claimed that one pioneer line in the South used the following method of pulling up to a station.[270] On approaching, the engineer would signal by raising the safety valves. Slaves then rushed forward to seize hold of the engine and cars and pull back, while the station agent thrust a stick of wood between the spokes of a wheel to complete the stop. The credibility of this story can be questioned, but there is other evidence of the lack of brakes on some infant American railways. In 1840 the Austrian engineer Von Gerstner noted the general absence of brakes on the cars running between Greensville and Roanoke, Virginia, even though the line had some steep grades.[271] Fifteen-car freight trains had brakes on only the first three or four cars. Passenger trains depended entirely on the tender brake. The Boston and Maine followed the British system of using brakeless freight cars with a "brake van" at the end of the train.[272] In the late 1840s the Michigan Central depended on tender brakes, according to the recollections of an

Figure 7.44   A double-acting brake devised by Henry R. Campbell about 1834 Brakemen operated it from roof seats.

old employee.[273] A long lever operated the brakes, and at times it was necessary for the fireman to stand on the lever to exert enough pressure.

Most lines, however, recognized the need for brakes on each car. Illustrations of the earliest American cars uniformly show some form of braking system (for example, Figures 1.2, 1.3, and 1.61). The favored style was the lever brake, copied from road coaches of that day. The B & O, the Camden and Amboy, the New York and Harlem, and other early roads used this form. The brakeman, seated on the roof, exerted his powerful leg muscles to push the long lever. The mechanical advantage was terrific, and for light cars traveling at slow speeds, it was an adequate system. A double-acting arrangement was devised around 1834 by Henry R. Campbell, the engineer for the Philadelphia, Germantown, and Norristown Railway.[274] Cross rods connected both brake shoes on either lever (Figure 7.44). Campbell's plan was copied by the Boston and Providence. But even with a reasonably efficient hand brake, stopping the cars was an uncertain business—as illustrated by an accident that occurred during the opening ceremonies of the New York and Harlem Railroad.[275] The mayor of New York City and other dignitaries were loaded into two horse-drawn cars. To ensure that all went well, the lead car was driven by the railroad's own vice president. A sober and experienced hack driver was engaged to operate the second car. All went well until the vice president made a stop, in part to show the perfection of the car's brake. The hack driver, following closely, pulled back on the reins and called to his team to stop, but in a moment of confusion he forgot the brake. In an ordinary highway coach the horse could draw up and stop the vehicle, but the rail car was heavier and had a nearly frictionless track. It did not stop. The pole of the second car rammed into the rear panel of the lead car with a great crash, and out tumbled the mayor and his friends. No one was hurt, but the railroad had little success afterwards in convincing city officials that street railways were perfectly safe.

The greatest single failing of the manual system was the brakeman himself. He was said to be an indifferent class of worker, reckless, "full of muscle—small brains—inactive" with little or no experience.[276] And where was he when the whistle sounded? Never at his station, but comfortably seated in the cars like a passenger, or loitering around thinking about everything but his duty. In 1853 a writer asked how the railroads could justify hiring such a poor class of men for a job so vital to the public safety. In this "age of lightning and steam," with speed the new god, only the most dependable of men should be manning the brake staff.[277] But the railroads were determined that it should remain a low-paying job. The lines around Boston employed Irishmen exclusively, most likely immigrants fresh from the docks.[278] It was said that native Americans were too enterprising to accept such a poor salary; they aspired to better jobs.

The general ineptness of brakemen was not the sole reason for rough and uncertain stops. Some of the blame is the managers', for they asked too much of the brakeman. He was a janitor as well as a train stopper. If he was frequently away from his sta-

tion, he was often sweeping out the cars, tending to lamps, firing the stove, hooking up the bell cord, calling out the next station, or helping with the luggage. A traveler describes the harried brakeman during a journey of 1865:[279]

Through the aisle scurried the brakeman. Dumping coal into the tall black stove and thumping with the poker—shouting the name of the next station—tearing through the door and slamming it behind him at the whistle's call for "Brakes!", his work was never ended. How interesting it was to watch him set the brakes! His lithe body bending and swaying as he twisted the wheel, one foot pressed against the foot pawl to hold the winding chain—then slinging himself across the gap to the other platform to whirl the wheel there—back again to the first wheel to add a bit more tightening—then like a flash to the other to give equalization.

The passenger train brakeman may have had an impossible job, but it was easy compared with the miserable situation of a freight brakeman. That worker rode atop the cars in all seasons; he had almost no opportunity to get out of the weather. He was expected to jump from one swaying roof to the next, and if it was icy or dark he could easily fall under the wheels or down an embankment. In 1881 it was calculated that at least ten brakemen were killed in this country every day.[280] Many more were maimed. Insurance was not available for such work. And even after the coming of automatic couplers and air brakes, the brakeman's job remained highest on the list of the thirty most dangerous trades, which included the work of electricians, miners, and sea captains.[281]

How well did hand brakes work? Most attempts to assess them quote the biographer of Westinghouse, who contended that the old fashioned "armstrong" method would halt a passenger train traveling at 30 mph in 1,600 feet.[282] Actual tests made during the period, however, indicate that brakemen achieved far shorter stops. An account published in 1855 says that an alert crew stopped a 119-ton train going 30 mph in only 450 feet.[283] A train of equal size going 28 mph was stopped in 376 feet.[284] The popular, 1,600-foot statistic at first seems a gross exaggeration, yet it must be remembered that the smaller figures from the tests are the result of staged stops. The brakemen were at the wheels—tense, alert, and waiting for the whistle to call "Down brakes." During normal operations a brakeman's showing would not have been so good. Suppose that he was not poised over the brake wheel but working inside a car. When the whistle sounded, it might take him 10 to 15 seconds to get out on the platform to man the wheel. At 30 mph the train would already have traveled 325 to 500 feet before the shoes first touched the spinning wheels. That stop would be made in, say, 800 feet. But if the brakeman was half asleep in a seat, or was blocked for a moment by passengers standing in the aisle of the car, it would probably have been 20 or even 30 seconds before he reached his post. Therefore stops of 1,600 feet may not be such an exaggeration.

The number of brakemen per train apparently varied widely, doubtless determined to a large extent by speed and terrain. A report made in 1838 which encompassed most of the major Northeastern lines indicated that it was common to have one

HODGE-BRAKE.

STEVENS-BRAKE. *Plan.*

TANNER-BRAKE. *Plan.*

Figure 7.45   Three popular brake rigs or foundation plans were the Hodge, intro-duced in 1849, the Stevens, introduced in 1851, and the Tanner (actually designed by W. I. Nichols), introduced in 1846–1847. (Car Builders' Dictionary, 1879)

brakeman for every two cars.[285] However, a report issued in 1853 concerning New York railroads showed a wider varia-tion.[286] One line had two brakemen on each train, with a con-ductor who helped when necessary. The number of cars was not indicated. Another line had one man for each car, and another had only one for every four cars. Freight trains were more poorly manned; the Canada Southern employed only two brakemen for each fifty-car train.[287] Of course, the speeds were much less ambitious on freights. The average brakeman could exert 150 pounds' pressure on the brake wheel.[288] The wheel itself multi-plied this to 900 pounds, and the levers, depending on their size and arrangement, could compound it to as much as 4,500 pounds. It was this great pressure that pushed the shoes against the wheels.

The effectiveness of hand brakes was materially improved about the middle of the century, when the double-acting founda-tion plan came into widespread use. It may be difficult to believe, but brakes before then were normally fitted to only one truck. This provided only 50 percent braking ability per car and re-quired the brakeman to run the length of the car to reach the wheel of the next car. The double-acting plan was a relatively simple system of levers and rods that connected the brakes on both trucks to the wheel at either end of the car, permitting all eight (or twelve) wheels to be braked at once. It also allowed the brakeman to easily set the brakes on two cars by stepping between the platforms. The idea was so obvious that it appeared at least as early as the late 1830s, or about the time when eight-

wheel cars were becoming popular.[289] As mentioned previously, the system was already in use on four-wheel cars. Compound or double braking was also an integral part of the earliest continuous braking schemes, such as those tried by Griggs and others. And yet the scheme was resisted on the grounds that such elaborate foundation plans were a needless complication and expense. This argument was reasonable for the first decade or so of American railroading, but by 1850 it was simply an excuse to justify a false economy. Double brakes had become a necessity because of the increase in train weights and speeds.

While many mechanics had considered the problem, the first to devise a long-term solution was Willard I. Nichols, car super-intendent on the Hartford and New Haven Railroad. Nichols, who thought of the double brake more as a laborsaving than a lifesaving device, perfected his foundation plan in 1846 and 1847.[290] He died not long afterwards without patenting the invention. Nichols's brake, which became known in later years as the Tanner brake, was widely used (Figure 7.45).

Meanwhile other inventors were active in the field. Charles B. Turner of Buffalo obtained a patent in November 1848 for a bumper brake that included connecting brakes for both trucks. The bumper feature was a failure, but an associate of Turner's, Henry L. Tanner, saw the potential profit in the scheme for double-acting brakes.[291] Tanner, who described himself as a "general trader," was not a mechanic or a railroad man but a shrewd speculator who knew a good thing when he saw it. He modified Turner's patent and began selling rights to a double-

Figure 7.46 *Turner's foundation plan was patented in 1848. The two upper drawings show the original design; the lower perspective sketch indicates Tanner's modifications in 1849–1850.*

acting brake totally unlike the device described in the patent (Figure 7.46). During his travels in the Midwest he saw the Nichols brake and realized that it was a foundation plan far superior to the reformed Tanner brake. Because it was common property, Tanner decided to take Nichols's plan for his own and reissue the Turner patent to cover the idea. By then both inventors were conveniently dead. For unexplained reasons (perhaps due to resistance on the part of the Patent Office), Tanner changed his tactics on reaching Washington. Instead of revising the Turner specifications, he bought rights to a brake patent that had been pending since 1847. Lafayette F. Thompson and Asahel G. Batchelder, both of Massachusetts, had applied for a patent on a bumper brake described as a wretched contrivance that refused to work. Apparently its novelty was also in question, because the government would not issue a patent. (Several buffer or bumper brake patents were already in existence.) Thompson and Batchelder, one a mechanic, the other a carriage maker, were probably open to offers after five years of waiting. Tanner canceled their drawings, specifications, and model and substituted new ones in Thompson and Batchelder's names to cover the Nichols designs. The Patent Office cooperated nicely, granting Tanner a patent in the name of Thompson and Batchelder on July 6, 1852. Tanner now set out to sell licenses for the use

of his brake, his customary fee being $5 per mile of line and a free pass. Where he met resistance, he entered suit, usually seeking a small claim. A victory against the Erie and the Hudson River railroads in 1853 helped to establish the validity of his claim. It was the same technique that Winans had used so effectively.

In 1855 Tanner sold his rights to Thomas Sayles of Lansingburg, New York. By this time double-acting brakes were common on passenger cars, though it would be some years before they were installed on much freight equipment. Tanner acted as Sayles's agent, and his vigorous pursuit of license fees soon made him anathema to the industry. Roads that did not pay were taken to court. In 1907 Angus Sinclair mentioned that Tanner's name was of "litigious memory," but Tanner and Sayles were interested in extortion, not in being kindly remembered. During the first term of their patents they collected over $53,000 in fees. During the period of the first renewal (1866–1874) they took in another $80,000. At least, these are the available figures; the exact total, which was thought to be far greater, was never revealed. Even after the patents expired and a second renewal was denied, Tanner pressed old claims amounting to 90 million dollars. In 1877 he was involved in some two hundred suits. In 1881 his partner Sayles died. Tanner's attorneys were suing the

Chicago and North Western for 15 million dollars, but by this time the case was going poorly and he settled for 500 dollars. Finally in 1882 the U.S. Supreme Court put an end to the swindle, ruling that no equity could be claimed after a patent had expired.

Tanner was not the lone purveyor of double-acting brake foundation plans. He had competition from two other major patentees, Nehemiah Hodge of North Adams, Massachusetts, and Francis A. Stevens of Chicago. Hodge's gear was patented in 1849 and became the most popular single form of double-acting brake in the United States. By 1874 it was said to have been used by 75 percent of the members of the Master Car Builders Association. On first inspection the rig looks overly complicated, with its six levers and seven rods, but it combined well with the air brake and had good compensating characteristics which equalized brake shoe pressure between both trucks (Figure 7.45). Stevens devised a much simpler form of rigging which was patented in 1851.[292] It employed no floating or center levers but depended only on brake beam mounted levers. It was popular in the West and, like the Hodge brake, remained in favor until modern times.

The braking of all wheels on each car was accomplished by the double-acting brake. This great advance was apparently in wide use by 1855. But railroad men envisioned an even better form of brake that would not depend on the uncoordinated efforts of several brakemen but would control every brake on the train through a single mechanism. Such a brake would be operated by the engineer. It would respond more certainly and quickly in the event of danger. It would produce a more even stop through relatively light pressure on all wheels rather than hard braking of a few. This would cause less wear and tear on the machinery and largely avoid the problem of flattened wheels. Inventors' first ideas on how to power such a brake centered on capturing the momentum of the train itself. In theory it was an appealing plan, because it conserved energy and required no auxiliary power source, such as steam or compressed air. The idea fascinated railroad mechanics throughout the nineteenth century, and countless plans were tested. But like so many alluring concepts, the notion of using the train's own power was to remain an unfulfilled dream.

The momentum brake went under several names: buffer, bumper, and compression brake. Most depended on the motion of the draft gear. Every railway man had observed the gathering in of the slack when the engineer shut off steam preparatory to a stop. Harnessing this enormous energy through a contrivance of levers and rods leading to the brake gear seemed beautifully simple. Stephenson was probably the first engineer to try to create a buffer brake. His plan, tested on the Liverpool and Manchester Railway in 1832, was a failure. Three years later, John K. Smith of Port Clinton, Pennsylvania, patented a buffer brake that was tested on the Little Schuylkill Railroad during the same year.[293] Reportedly it worked well, but no other roads are known to have used it.

In 1839 a novel form of momentum brake appeared on the Boston and Providence Railroad.[294] It remained in service there for at least a decade and was tried on several neighboring lines. Part of its longevity must be credited to the fact that it was the invention of the presiding master mechanic, George S. Griggs. Under the watchful eye of its creator, almost any mechanism can be nursed along indefinitely. Griggs did not employ the motion of the buffers; rather, he utilized the motion of the train to drive a windlass which wound up a rope or chain. A control rope ran the length of the train. A tug on the rope moved a lever on each car that engaged a clutch on a windlass, or drum, which was belted to one of the wheel axles (Figure 7.47). Thus all brakes on the train could be set at one time. Griggs's brake was said to be six times as powerful as the ordinary hand brake. During one test a sixteen-car freight train going downgrade at 15 mph was stopped in its own length, yet only one car on the train was fitted with Griggs's brake.[295] The cost of the apparatus was estimated at $25 per car. This momentum brake was one of the few that enjoyed early and prolonged success, but it was hardly widespread.

In the 1850s many inventors offered plans for buffer brakes. Turner's patent, as well as the patent of Thompson and Batchelder, have already been mentioned. In 1852 a Philadelphian, T. G. McLaughlin, convinced the Camden and Amboy to try his new self-acting buffer brake.[296] Seven years later another advocate of compression gear, W. R. Jackson, made tests on the B & O.[297] These and similar designs all failed; the universal defect was inability to control the brake's action. When the train speed slackened for any reason, the brakes came on whether they were wanted or not. This made running slowly a tricky affair, and switching a near impossibility. It was necessary to disconnect the brake to back up. In addition, most brakes took hold unevenly. The slack running in at the head of the train would jam on the front brakes hard, while the rear cars were rolling free. Joseph Olmsted, a Chicago inventor, attempted to master the control problem by introducing an electromagnetic clutch with a basic mechanism much like Griggs's old plan. Olmsted's momentum brake was tried here and in England in 1869 and 1873, and then like so many others, it disappeared. After this date there appears to have been no more interest in momentum brakes for passenger service. Power brakes had already proved themselves, but some optimistic mechanics seemed to believe that a workable buffer brake was possible for freight cars. In the 1880s when the industry was debating over the desirability and the possible form of power brakes on freights, the ancient compression scheme was revived as an economic alternative to the costly Westinghouse apparatus. These schemes raised many false hopes and resulted in as many patents, but the inherent defects of the compression brake were not overcome by the new generation of inventors.

At the outset of the new wave of compression brake fiascos, simpler forms of continous brakes were being employed. Some of the coal roads in eastern Pennsylvania were having success with manually powered continuous brakes that involved little more than tying together the brake levers on a series of small four-

Figure 7.47 Griggs's momentum brake of 1839 was the first continuous, mechanically powered brake to be used in everyday service. In the lower figure, C is the gear drive, d the clutch, e the gear-drive frame, f the pinon gear, g the pulley, and h the drive belt. (American Railroad Journal, *September 15, 1841*)

wheel cars with a long wooden rod. The train of cars was under the control of a single brakeman.[298] The gravity road at Mauch Chunk was running fourteen car trains in this fashion in 1838. A more sophisticated scheme had been worked out for passenger cars by the B & O three years earlier. A tall brake staff at one end of the car was actuated by an overhead rope that ran forward to the engine. When the rope was pulled, the lever was jerked off a perch, allowing a counterweighted lever to fall and apply the brakes on all eight wheels. The device is pictured in the German drawing of the B & O Washington branch car reproduced in Figure 1.70. The plan was credited to the brother of the B & O's first president, Evan Thomas. A history of the company printed in 1853 claims that a patent was issued, but no record of it can be found.[299]

These primitive continuous brakes must have suggested the possibility of even better train-stopping systems. Many mechanics believed that a chain connecting all the car brakes together would be the solution. English mechanics seemed particularly attracted to the chain brake idea and began trying to perfect it in the 1840s.[300] They were moderately successful in adapting the chain brake to everyday operations, and a few lines, such as the London and North Western, had some in service as late as 1892. The earliest such brake recorded in an American patent was invented by Lucius Stebbins of Hartford and patented in April 1848. The chain was drawn taut by a friction drum powered by the locomotive driving wheels. No record exists on the adoption of Stebbins's design, but an identical system, patented in 1855 by William Loughridge (d. 1890) of Weverton, Maryland, was in rather wide use for about fifteen years (Figure 7.48). Loughridge's apparatus was in effect a form of momentum brake, because the power drum was driven by the motion of the

train. A pulley fastened to a countershaft was pressed against one of the locomotive's rear driving wheels. The inventor claimed that a series of movable pulleys overcame problems of correct chain tension and brake shoe pressure. He called his system the graduating car brake. Within two years it was in use on the Cincinnati, Hamilton, and Dayton; the Ohio and Mississippi; and the Mad River and Lake Erie.[301] In 1859 the Pennsylvania adopted the Loughridge brake—such a triumph that the inventor reproduced the letter of acceptance in an advertisement.[302] A few years later the railroad claimed that Loughridge's lifesaving device made the P R R the safest route possible, and that after three years of use and the transport of over 3 million passengers, not a single fatality had occurred.[303]

But as critics of the chain brake pointed out, while the inventor claimed that brake shoe pressure was uniform on all cars and that no wheels could slide, in actual service the wheels locked on the front car and there was almost no pressure whatever after the third car.[304] Loughridge's pulleys and movable bearings could not compensate for the varying length of the train, a measurement which fluctuated considerably because of the huge amount of slack allowed by link and pin couplers. The friction wheel was cut away by the flange and developed flat spots, making its operation unreliable.[305] Stops tended to be rough. Also, in extreme winter weather the chain would freeze around the clusters of pulleys and become inoperable. Even Loughridge must have recognized the defects of the chain brake, for in the 1860s he began to experiment with steam and air. Commenting on the problems of the chain brake, a British author said that after so many efforts to perfect the system, "it would have become, ere now the universal brake of the world" if that had been possible.[306] In spite of its deficiencies, the Pennsylvania did not

Figure 7.48   William Loughridge's chain brake was patented in 1855. A drum, powered by the rotation of the rear driving wheel, wound up the chain to apply the brakes. (Smithsonian Neg. X2401)

abandon the chain brake until late in 1870. For nearly fifteen years the constant jangling of heavy chains supplied a noisy counterpoint to the grinding and clicking of the wheels on many of the road's better trains.[307]

Most continuous braking systems were designed to function for both service and emergency stops. Thus most were intended to replace the brakeman, or as one Victorian writer observed, to make him "a victim of evolution." But William G. Creamer of New York City was less ambitious; he introduced a spring-powered brake meant only for emergency stops. The brakeman was to make regular or service stops as before, but in a crisis an auxiliary system would be available that could instantly apply all the brakes for a fast halt. Creamer selected a large spring to power his apparatus (Figure 7.49). The spring and its case were geared to the brake wheel shaft. A rope ran the length of the train to the engine. A pull of the rope lifted the catches on the spring cases, thus releasing the springs and setting the brakes. The engineer, conductor, or any trainman (passengers were advised not to touch the apparatus) could set the brakes anywhere on the train by pulling the control rope. Once the rope had been pulled and the Creamer brake had stopped the train, it was necessary to rewind the springs. The brake was not a laborsaving device, but it did solve one of the gravest problems not met by the manually powered brake—the occasional need for an emergency stop.

Creamer patented his device in December 1853; he was issued a second patent in November 1856. In 1855 he organized the U.S. Railroad Car Brake Company to exploit the scheme.[308] An enthusiastic first customer was the Hudson River Railroad. The spring brake added a measure of much needed security for the line's rapid express trains. During one test a train moving at 24 mph was stopped in 277 feet.[309] Creamer's brake was adopted by other lines such as the New York Central, the Lake Shore, and the Central Pacific. Contemporary photographs dating from the late 1860s clearly show the apparatus on the C P's passenger equipment. Creamer brakes were installed on the coaches and Silver Palace cars. Even the Central Pacific's emigrant sleepers were so equipped, as illustrated in Figure 6.27. Precisely how widely used Creamer's brake was cannot be determined, but as early as 1861, 1,000 were reported in operation.[310] In the following year the inventor attempted to sell the spring brake to British railways.[311] He persisted for at least five years, but apparently only the South Eastern Railway purchased it.[312]

With the coming of power brakes, Creamer's brake was obso-

lete. Heavier cars pushed train weights beyond the capacity of spring power. When the Creamer brake was introduced, trains were light enough that the springs worked as well as or better than the hand system, but by 1870 this brake was doing well to stop a train in 1,000 feet.[313] Even so, some lines still believed that the Creamer brake was a desirable extra precaution in the days before the automatic air brake. The Lake Shore, for example, advertised in its 1873 calendar that it was using the Creamer brake. Creamer himself believed in the future of the spring brake, for as late as 1870 he applied for a patent extension. Patent officials denied the request, stating that the estimated $80,000 profit realized by the inventor was adequate compensation.[314] Some railroad officials were already disenchanted with the Creamer brake as a safety device. Brakemen often failed to rewind the drums after the brake had been used, thus rendering it useless for the next emergency.[315] Ironically, what finally discredited Creamer's invention was a terrible wreck—exactly the sort of disaster that it was intended to prevent. A train on the Hudson River line plowed into a derailed freight train, killing twenty-one persons. It was declared a "historic failure," and much criticism was directed at both Creamer and the railroad for depending on so absurd a system as a collection of clock springs to stop an express train.[316]

In the last major epoch of the railway brake, mechanical power replaced human effort. The introduction of the power brake was a direct outgrowth of increasing speed and traffic density. With the rise in traffic came the additional revenue to pay for more elaborate safety apparatus. Many systems seemed suitable to drive the power brake: steam, compressed air, vacuum, electricity, and hydrostatic pressure were all suggested and tried. All were intended to replace the brawny but brainless brakeman with a more powerful and dependable means for retarding travel. All, except for certain forms of the steam apparatus, were meant to be continuous brakes.

Naturally steam was the first power source tried. When any engineering problem was being considered in the nineteenth century, steam was always offered as a sure solution. If it could drive the wheels of commerce, why could it not stop them as well? A good supply of steam was available from the locomotive; it need only be harnessed to a cylinder and the brake shoe. In 1833 Robert Stephenson patented a steam brake for locomotives. It was an intelligently arranged device, but it was a locomotive and

Figure 7.49   Creamer patented his spring brake in 1853. It was an emergency brake that could be activated by the engineer. (Engineering, July 12, 1867)

not a train brake.[317] Fifteen years later steam brakes were tried on the Boston and Providence, but a boiler explosion gave them a bad name, even though they were in no way connected with the accident. In 1855 Henry Miller of New York City offered a steam brake with cylinders under each car and a rubber hose between the platforms to draw steam from the locomotive.[318] His device was tested on the Michigan Central and the New Haven. It stopped a 104-ton train running at 30 mph in 700 feet. Other inventors active in this area included Wendell Wright,

who patented a brake in January 1855 that could work on steam or air. The following year Thomas W. Smith, a partner in the Alexandria, Virginia, firm of Smith and Perkins, introduced a steam brake that was given flattering testimonials by several local and Northeastern lines.[319] Among the lively succession of inventions at this time, apparently none could overcome the defects of steam for train brakes. High-pressure steam lines between and under the cars were dangerous. Low-pressure steam did not have enough power and would condense into water that could freeze inside the line and block passage of the vapor (as it also did in heating lines, mechanics later discovered). Steam brakes on the locomotive were simple, efficient, and widely used around 1880 to 1900, but how to convey steam to the train of cars was a problem never to be solved, at least so far as braking systems were concerned.

Water was another familiar power source in nineteenth-century America. It propelled grist mills, elevators, cotton presses, and washing machines. As early as 1859 Lewis Kirk, onetime master mechanic for the Reading, suggested a hydrostatic railway brake. His patent, granted in May of that year, specified a water cylinder under each car. Pressure was maintained by a buffer-actuated pump, giving each car a self-contained system. In 1874 two Philadelphians, W. H. Henderson and Thomas McBride, independently devised hydrostatic systems, both of which were tested on the Philadelphia and West Chester, a suburban line with many stops.[320] Henderson used a large tank under the engine cab floor with a head of steam to maintain the pressure. He thus claimed to have avoided the expensive and troublesome steam pump necessary for other systems, such as those of his competitors, McBride and Westinghouse. Hoses and cylinders much like the equipment for the air brake carried the water back under the cars. An obvious objection to the water system was that the braking fluid would freeze in cold weather, but the inventors suggested a glycerine mixture. The water brake made little progress, however, and was never used in regular service so far as is known. The French or Le Chatelier water brake, which was adopted fairly widely in the American West late in the century, was not a water brake in the true sense. It was a locomotive compression brake that overcame the usual problems of braking the train by reversing the locomotive and allowing the cylinders to work as an air pump. Le Chatelier closed off the exhaust passages so that cinders were not drawn into the cylinders. He also introduced a spray of water to lubricate the cylinders. However, his was strictly a locomotive and not a car or train brake.

The air brake was to sweep away all rival braking systems, though it did not seem like such a sure thing during the first years of the Westinghouse brake. At the time a competing form of atmospheric brake was being vigorously promoted, with considerable success. It was efficient, reliable, and cheaper than the compressed air brake. It required no pump and could maintain full power even with frequent stops, an advantage not offered by the air brake. It provided a quick release and a fast application. This was the vacuum brake, a product that initially showed great promise but was widely purchased only in Great Britain, where it remains a national standard.

Simplicity was the great virtue of the vacuum brake. A vacuum was created in the train line(a system of pipes and hose running the length of the train) by a steam ejector on the locomotive. The pressure of the atmosphere then pushed a piston or caused a diaphragm to collapse, depending on the design employed. The motion was transmitted through a system of rods and levers to the brake shoes. The apparatus required no compressor or stor-

age tanks. The negative pressures involved in its operation made seals a relatively small matter.

Vacuum brakes also had some serious shortcomings. Before 1883 there was no American vacuum system with an automatic safety feature, such as the air brake had incorporated for a decade. If a hose broke or an ejector failed, the system was useless. The problem was solved in time, but meanwhile Westinghouse enjoyed an enormous advantage during a critical period in the history of power brakes. Once a railroad had adopted the air system, it was not likely to scrap the equipment just because an equally efficient vacuum system was made available. Another and perhaps even more important defect was the vacuum brake's lack of power. The ejector could create an effective pressure of $8\frac{1}{2}$ to $11\frac{1}{2}$ psi.[321] This figure probably reflects the vacuum's maximum efficiency at sea level. It would fall off with each foot of elevation, reducing its effectiveness on mountainous routes. The air brake, with pressures originally of 70 and later of 110 psi, was a much stronger brake, and power obviously became more vital as train speeds and weights increased. For short or light trains, the vacuum brake was very satisfactory, and it consequently flourished briefly during the first years of the power brake in America.

Vacuum brakes were patented as early as 1844 in England. Sixteen years later Nehemiah Hodge, already mentioned in connection with foundation gears, obtained a U.S. patent for a vacuum brake. Hodge created a vacuum with a steam ejector on the locomotive. Accordion-like rubber bags mounted under the cars would pull together or collapse, creating the motion to apply the brakes. It is not known if Hodge's brake was actually used. It was a precedent, and Westinghouse thought enough of it to purchase the patent and have it reissued in 1879.

The vacuum brake idea remained dominant until the success of the air brake prompted a general upswing of interest in the subject of power brakes. John Y. Smith, a partner in the locomotive manufacturing firm of Smith and Porter, lived in Pittsburgh, the city where Westinghouse introduced and perfected the modern air brake (Figure 7.50). Smith, an Englishman, had been active in American railway mechanical affairs since the time of the Civil War.[322] He was familiar with the requirements of the industry and must have recognized the potential market for power brakes that Westinghouse's system had only begun to satisfy. Smith selected the vacuum system and began public tests in 1871.[323] It is impossible to say with certainty whether or not he was aware of Hodge's 1860 patent, but his apparatus copied that scheme part for part. Beginning in July 1872, he obtained several patents which provided details and improvements, but the basic scheme remained Hodge's.

Smith did well in selling his apparatus. By the spring of 1873 it was in use on the Boston and Lowell, the New Haven, and the Central Railroad of New Jersey.[324] Within a year several more lines had found it satisfactory. A test on the Central of New Jersey showed that it was equal to the air brake: a twelve-car train with a 35-ton engine going 35 mph was halted in 555 feet.[325] The *National Car Builder* recommended the Smith

Figure 7.50 J. Y. Smith's vacuum brake, patented in 1872, was later marketed by Westinghouse. (Car Builders' Dictionary, 1879)

Figure 7.51 The Eames vacuum brake, introduced in 1874, had a moderate success in North America. (Car Builders' Dictionary, 1879)

apparatus, saying that while it was not as powerful as the Westinghouse brake, it was adequate for the requirements of most U.S. lines.[326] Its first cost was lower, and it did not require a "mint of money and a portable machine shop to keep it in good order." This statement was an exaggeration, but it contained enough truth to prompt Westinghouse to buy rights to Smith's patents, apparently in 1875 or 1876.[327] Westinghouse produced vacuum brakes on Smith's design for several years under his own name.

Smith continued to market his apparatus in England, with enormous success. In 1876 he was said to be taking over the British market, where his system was more widely used than the Westinghouse brake.[328] A few years later nearly 1,900 British cars were equipped with the Smith brake.[329] However, the fortunes of the Smith brake began to decline almost as rapidly as they had risen. A modern British railway historian has declared it "a pernicious thing" and has blamed Smith's apparatus for a series of accidents, culminating in the 1889 Armagh disaster on the Great Northern of Ireland in which eighty people died.[330]

The Smith brake was on the market for only a year before another major vacuum brake was offered by Frederick W. Eames of Watertown, New York. Eames received his first patent (No. 153,814) on August 4, 1874. His brake incorporated no remarkable features, although he did use a cup-shaped brake cylinder (Figure 7.51). The large cast-iron cup was about 30 inches in diameter. A rubber diaphragm was bolted to the periphery of the cup and collapsed into it when the vacuum was created. Like Smith and Hodge, Eames produced the vacuum with an ejector.

Eames took over an old water-powered stone mill building on Beebee's Island in the Black River at Watertown and began to manufacture his apparatus around 1876.[331] He had already run tests on the nearby Rome, Watertown, and Ogdensburg Railroad. Within three years, twenty-nine U.S. railroads were using Eames's brake. By 1881 sixty-seven different railroads in nine countries had adopted it.[332] The brake was particularly well received in South America; in Brazil, for example, nine out of eleven lines installed it.[333] But the true showcase of the Eames brake was the New York Elevated railways. Some 200 locomotives and 500 cars hurried up and down Manhattan Island; 3,500 trains made 70,000 stops daily, demonstrating the efficiency of the Eames brake before an enormous audience.[334] Here the vacuum brake made a good showing, because the trains were extremely light and the stops frequent. It had plenty of power to stop the diminutive locomotives and cars, and unlike the air brake, the vacuum system had no difficulty in recovering from a series of stops.

Eames knew that British railway operations were ideally suited to the vacuum brake. Many manufacturers, including his American rival, J. Y. Smith, were already demonstrating their wares in England. Seeking a way to dramatize his product abroad, Eames found a flashy express locomotive built in 1880 by the Baldwin Locomotive Works. The high-speed, single-axle engine had been repossessed by its maker and was awaiting a buyer.[335] Eames had the engine refurbished and made ready for road trials. (He

named it in honor of his father, Lovett Eames, a fact that has caused some confusion as to the inventor of this form of vacuum brake.) After a test on the New Haven, the demonstrator was shipped to England. In 1881 it appeared on several British lines and at the Exhibition of Life Saving Appliances. British railroads were under pressure from the Board of Trade, a powerful government agency, to adopt some form of continuous brake.[336] Legislation seemed imminent that would require such action, and Eames hoped to capture at least a portion of the windfall orders for power brake gear. But his expectations were not realized, for only two lines, the Lancashire and Yorkshire and the Rhymney railways, seem to have chosen his gear.[337] The English were perfecting their own vacuum brakes and did not need outside help. After all, they had gifted mechanics such as John Aspinall, who had perfected an automatic vacuum brake in 1879, years before any such device appeared in America.[338]

Eames returned home to develop an automatic vacuum brake. In order to be competitive with the Westinghouse and Aspinall systems, it must incorporate the fail-safe principle; that is, if the train parted or the ejector failed, the brakes must come on automatically. His first effort, a complicated two-pipe system devised in 1881 or 1882, was not dependable.[339] Early in 1883 he tested an improved duplex-automatic brake on the New Haven and acquired new capital for its manufacture.[340] Meanwhile a strike occurred at the plant and Eames, attempting to enter the works, was shot and killed by a striker on April 20, 1883.[341]

His successors in the company realized that the vacuum brake was not likely to survive in America. The export trade was doing well, but domestic sales fell off as more lines adopted the Westinghouse system. As late as 1886 the Eames company claimed that 200 railroads were using its equipment and reported one new customer, the Indianapolis, Decatur, and Springfield Railroad.[342] Two years later, however, the Car Builders' Dictionary stated that vacuum brakes were used only to a very limited extent in this country. The Smith brake was nearly extinct, being found only on the Long Island, and Eames's last stronghold was the New York Elevated.[343]

In 1885 the Eames company's new chief mechanical engineer, A. P. Massey (1842–1898), devised an air brake.[344] Westinghouse naturally sued, but the Watertown firm continued to push into the air brake field, encouraged by several railroads who wanted to block Westinghouse's growing monopoly. In 1890 the Eames company was reorganized as the New York Air Brake Company, and since the earliest Westinghouse patents had now expired, the new firm was free to copy their designs.[345] As air brake technology advanced, and as the need for complete interchange of cars became acute, Westinghouse and New York Air Brake were again in the courts. The litigation was at last resolved in 1912, when Westinghouse and its competitor agreed on a cross-licensing accord which covered all air brake patents.[346] Under this arrangement New York Air Brake's business expanded to a yearly capacity of 50,000 brake sets. Among its customers were such major roads as the New York Central, the Seaboard Air Line, and the Union Pacific railroads. Of the total air brake market, New York Air Brake held—and probably still holds—one-quarter; the great bulk of orders go to Westinghouse.[347]

The air brake is an important bench mark in the history of railroad technology. It came into being simultaneously with the opening of the transcontinental railroad, the introduction of the dining car, and Janney's first coupler patent—events which marked the opening of a new railroad era. There is a tendency to ascribe the invention of the pneumatic brake to George Westinghouse, Jr., alone. He was unquestionably an outstanding engineer and businessman; it may not be an exaggeration to call him a genius. Nevertheless, some admirers have overstated his contributions. In the dedication of *Development of the Locomotive Engine* (1907), Angus Sinclair eulogized Westinghouse as "The preserver of the Travelling Public whose great invention of the Automatic Air Brake has preserved more human lives than any military general or tyrant ever succeeded in destroying." Westinghouse was not a latter-day Prometheus who seized a lifesaving miracle from the heavens to save mankind. The modern railway brake was the work of many men. It evolved through countless successes and failures. Westinghouse benefited directly and indirectly from the experiments of his predecessors, although he claimed to have been unaware of their efforts and to have repeated past mistakes before arriving at the belief that compressed air was the answer.

The air brake was not Westinghouse's idea, nor was it a novelty when he began to work on his system in 1868. Numerous British and American inventors had previously patented and tested such apparatus. In fact, compressed air as a power source was almost as old as the locomotive, for compressed-air engines and vehicles had been proposed as early as 1800.[348] An Englishman, D. Crawford, patented an air brake in 1845; a fellow countryman, S. C. Lister, did likewise in 1848. Twelve years later John McInnes actually tested an air brake on a British line.[349] Meanwhile U.S. mechanics and inventors were trying to develop the same idea. In 1852, for example, Cutler and Rapp of Buffalo proposed a compressed-air brake that included all the basic features of the first Westinghouse system except for the steam pump on the locomotive.[350] Their pump was driven by the motion of the train. Several other designs were patented before Westinghouse even began to study the problem.

His attraction to the subject of railway brakes after he had been in a train wreck, his chance reading of a magazine article about digging railway tunnels with compressed-air machinery—these stories about Westinghouse have been often told. The important facts about Westinghouse are his background and his relationship with railway officials of the day. Westinghouse was barely twenty when he began to think about railway brakes, but he already had a good knowledge of practical mechanics. His father was a manufacturer of farm machinery, and George had worked in the family shop since he was a boy. A serious and ambitious youth, he took to the road in his late teens to promote a car replacer he had patented. During his travels he became acquainted with several Eastern railway officials. They liked and trusted the sober young man from Schenectady, and when he approached them later with an idea for a pneumatic brake, they did not dismiss him as foolish or impertinent. Two local railway superintendents, W. W. Card of the Pan Handle and Robert Pitcairn of the Pennsylvania, figured large in the early success of the Westinghouse brake. Their help was not forgotten; both became officials in the Westinghouse Air Brake Company.[351]

After he had considered chain and steam brakes, the idea for a pneumatic brake matured in Westinghouse's mind by July 1868, when he applied for a patent.[352] A reciprocating steam pump on the locomotive supplied compressed air to a tank located under the cab floor. When a valve was opened in the cab, air was sent back to the cars through a pipe, with flexible hoses between the platforms. The pipe directed the air into cylinders attached to the brake rigging. Although the scheme was hardly original, no one had managed to translate it into a dependable, working system. Westinghouse found backers in Pittsburgh, Robert Pitcairn among them. A Worthington feed water pump was rebuilt into an air compressor, and the other necessary parts were fabricated in a local machine shop. W. W. Card gave Westinghouse permission to fit his apparatus to locomotive Number 23, with four coaches, and make a road trial on a local train, the Steubenville Accommodation.[353] The test run began in September 1868, and the train had barely cleared the Pittsburgh terminal when the brake was thrown into emergency.[354] A team and wagon straddled the track; they were spared only because of the fast stop possible with the air brake.

Although the brake worked, it needed refinements before it was ready for the market. The steam pump valve gear was a troublesome detail that required reworking. While Westinghouse struggled to perfect the mechanism and refine the design for commercial manufacture, his patent (No. 88,929) was issued on April 13 1869. More tests and demonstrations were needed to introduce the brake to the industry. Just a year after the first trial, the brake was shown to members of the Master Mechanics Association. A special train was run from Pittsburgh to Altoona, where the efficiency of the air brake could be exhibited at its best over the mountainous divisions of the Pennsylvania Railroad. In November of the same year, a ten-car train demonstrated the merits of the air brake to the directors of the railroad. It was then sent to Chicago, where its performance inspired enthusiasm. During one trial the train, moving at 30 mph, stopped in 19 seconds, or 380 feet.[355]

Orders began to flow into Pittsburgh from the West, New England, and elsewhere. In September 1870 the Westinghouse brake was declared beyond the experimental stage and was regarded "as a decisive solution" to the problem of brakes on passenger trains.[356] Eight railroads were now using the apparatus on 355 cars. Of these, 200 were on the Pennsylvania. In less than two years 85 railroads and 4,000 cars were equipped with the Westinghouse brake.[357] By 1874 the number of cars so fitted had nearly doubled. At the time of the Centennial, nearly three-quarters of the passenger equipment in this country had air brakes.[358] So rapid a conversion is almost without precedent in an industry famous for its wait-and-see attitude.

Several factors explain Westinghouse's success. These were the industry's need to improve the safety and speed of railroad service and each road's need to outperform its competitors. The need for a dependable power brake was obvious. Traffic density and train speeds had advanced markedly since the Civil War. The public was calling for greater safety measures. In a letter written in 1875 to John Garrett, president of the B & O, Mrs. Ella M. Williams said that she and some friends were returning to Baltimore from the West, and that before reaching Chicago their train stopped short of what might have been a major collision, thanks to the wonderful new brake invented by "Mr. Westinghouse."[359] Her party had planned to take the B & O for the final portion of their journey, but on learning that the Westinghouse apparatus was not employed by that road, they selected an alternate route. Mrs. Williams was from Baltimore and wanted to support her home-town railroad, but considerations of personal safety ruled out such loyalty. She asked Garrett, "Oh, why don't you make your cars safe too!" The message was clear: to keep passengers it was necessary to adopt the latest safety apparatus. When competing lines advertised such safety measures as Miller platforms, Baker heaters, and air brakes, there was nothing to do but follow their lead. Some political pressure was also brought to bear. In 1873 the Michigan legislature made air brakes compulsory on all passenger trains.[360] Other state governments showed similar inclinations, although state laws appear to have had much less influence on the railroads' decision to patronize Westinghouse than the pressure of public opinion itself. The passenger car fleet was rapidly and voluntarily converted to air brakes. It was the far larger and more costly outfitting of freight cars that required Federal legislation.

Service was another major reason for Westinghouse's success. He did not merely devise a workable mechanism; he instituted a systematic process of manufacture, quality control, and service in the field.[361] Repairs were simplified because parts were precisely made. Westinghouse was a meticulous businessman whose careful handling of orders and accounts was a relief to railroad managers weary of dealing with brilliant but disorganized inventors. Westinghouse was a marketer as well as an inventor. He made sure that dependable field representatives were available to answer technical questions, advise on repairs, and see that parts were available. He was careful not to sell apparatus that had not been absolutely proved. Devices that worked well on the test rack might fall to pieces under the rigors of everyday service or in the hands of unskilled workers. He ran continuous experiments to test the durability of his apparatus even after its efficiency seemed irrefutable. The triple valve, for example, was subjected to 460,000 applications before Westinghouse was satisfied that it was ready for production.[362]

Even during the early years of his success, Westinghouse knew that the original straight air system was defective. Conceptually it was pedestrian, a system that almost any mechanic might devise. In practice, however, it suffered from several serious problems. Because air had to be pumped from the locomotive the full length of the train, it took too long to reach the rear cars. So

much time elapsed that the front brakes were pulling hard before the rear ones began to take hold. This was an inherent problem of the air brake. The slow response of the rear brakes was due to the slow rate of travel of compressed air, which will move at 1,100 feet per second in the open but only 900 feet per second inside a pipe. There was no real way to resolve the difficulty, but the situation could be improved by tinkering with the mechanism, or by increasing the pipe size to speed the air flow. On short trains, the slow response was not critical, and because the air brake was used almost exclusively on passenger trains for its first fifteen years, Westinghouse could well have ignored the problem. But he was not satisfied; he wanted the braking to be uniform. This meant regulating the air volume and pressure at both ends of the train, something that could not be achieved with the old straight air system.

There was another and more pressing reason to revise the air brake: it was not really all that safe. If a hose burst, if the train broke in two, if the main air reservoir burst, or if the pump failed, the train would be without brakes. Westinghouse wanted a "fail-safe" system, so that if any of these events occurred, the brakes could still be applied. Moreover, in an ideal system they should come on automatically. At this point Westinghouse broke away from conventional thinking and began to evolve a truly epochal invention: the automatic air brake.

There is possibly even a third reason that Westinghouse felt a new design was necessary. The straight air system was hardly patentable; competitors could easily point to prior inventions. A growing business such as Westinghouse's would attract rivals, and he needed a specialized design to protect his valuable monopoly. Late in 1871 Westinghouse filed for a patent to cover his first work on the automatic system.[363] It involved an independent air apparatus for the cars, with an auxiliary tank and control valve under each car. If the train broke in two, the brakes would automatically set for an emergency stop. The cars might even have an axle-driven pump. The engineer would have no control over this secondary system, and the normal service stops would be made, as before, with the straight air system. Two patents were issued to Westinghouse in 1872. But he was not satisfied with the emergency system; he wanted the safety feature to work with the normal braking operation. And then in 1872–1873 came his inspiration for an air-actuated brake. In operation it was exactly opposite to the old system. The brakes were activated by discharging rather than charging the train line. A brain was needed: a pressure-sensitive valve that could admit air from the train line into the reserve tank, readmit air from the same tank to the brake cylinder, and exhaust the air to release the brake. Each car must have such a valve as well as its own air tank (Figure 7.52). This lively, sophisticated valve was called the triple valve because of its three basic duties. It was a remarkable chunk of cast iron no larger than a grapefruit, and so amazing to mechanics of its day that some of them half expected it to speak.

Thus the triple valve made two contributions. First, it achieved more uniform braking, because full pressure was available in

Figure 7.52   The Westinghouse brake, put on the market in 1868–1869, is shown here as it was improved in 1873, with a reserve tank under each car and the triple valve which automatically applied the brakes in case of an air failure. (Railroad Gazette, April 14, 1876)

each car's reserve tanks. Second, it provided a fail-safe operation, because any major reduction in the pressure of the train line (as would happen if it broke in two) caused the triple valve to pop open to full emergency. Normal service stops were made by small, controlled reductions in the train line pressure: the triple valve was designed to respond in 2½ ratio. That is, if the engineer exhausted 10 pounds from the train line, the triple valves would feed 25 pounds of air into the brake cylinders. Before about 1900, the standard line pressure was 70 psi.

The first triple-valve patent was issued to Westinghouse in February 1873, and an improved version was patented on May 13, 1873, (No. 138,827). All subsequent air brake developments were an outgrowth of the 1873 automatic system. It was, of course, refined over the years to improve its transmission speed and power. The triple valve itself disappeared, to be replaced by a more complex and sensitive control valve. But even the most recent form of air brake follows the basic 1873 plan.

Acceptance of the automatic system, despite its olympian superiority over the old form, was not swift. The Pennsylvania managed to convert all its passenger cars by 1879, but the New York Central did not adopt the air brake in any form until 1880.[364] It was the last major railroad to do so and was criticized for this tardiness. There is a frequently told anecdote about the old Commodore scoffing at the idea of stopping a train with a whiff of air. Westinghouse specifically denied the tale in his 1910 recollections, but the Central's slow acceptance does lend plausibility to the story. At the national level, less than half of all air brake equipment was of the automatic design by April 1880.[365] At that time 1,770 locomotives and 5,218 cars were equipped with the new or automatic system, while 2,677 locomotives and 10,244 cars had the old straight air equipment. In 1881, 12,270 passenger cars had automatic brakes, while 11,389 still had the

old system—a great increase in use of the new brake in only one year.[366] But progress was not fast enough. The dangers of the old system were tragically demonstrated by three wrecks on the B & O in 1887. The first and most serious occurred at Republic, Ohio, in January, when an express running at 65 mph crashed head-on with a freight, killing thirteen people. The straight air brake was said to be partially responsible.[367] A few months later, part of a passenger train broke away because of a defective coupler and rolled downgrade on the most mountainous portion of the line, near Terra Alta, West Virginia. Again the straight air brake was blamed. In November an express trying to make up time ran out of control and derailed in Washington, D.C., an accident also attributed to the old-fashioned brake.[368] (It is possible that the B & O was still using the Loughridge air brake, which did not incorporate the patented automatic features of Westinghouse.)

If certain lines were negligent about buying the latest apparatus available, at least they were using some form of power brake. By the early 1880s it appears to have been universal on American passenger cars. In 1882 Westinghouse claimed that 36,218 cars of all types had his equipment.[369] During the next year he claimed 55,000 cars on a worldwide basis, but said that the majority were running on domestic lines.[370] To accommodate this booming trade, he purchased a cotton mill in Allegheny City (north Pittsburgh) for conversion into a new air brake plant.[371] Two years later the facility had to be enlarged, and by the late 1880s the Allegheny works was too small to handle the increasing orders for freight brakes. Westinghouse decided to build an industrial complex in Turtle Creek Valley, east of Pittsburgh. He had ridden countless miles through this valley while testing the triple valve on a local train called the Walls Accommodation. He located his signal company, the Union Switch and

Figure 7.53   The Quick Action brake was introduced in 1887–1888 to speed the braking on long freight trains. It was also used for passenger equipment.

Signal, in the valley town of Swissvale and erected a new air brake plant in Wilmerding. Money was no problem, for the firm paid a 400 percent dividend in 1886.[372] The new plant opened in 1890 and has remained the home of the business, which currently calls itself WABCO.

Westinghouse had a Pullman-like compulsion to dominate the field; he sought to buy out competitors and develop foreign sales. Many of his rivals were self-eliminating but others could only be overcome by absorption. His purchase of the Smith vacuum brake patents was the first acquisition. In 1898 he bought the Boyden Air Brake Company of Baltimore, and not long afterward the American Brake Company of St. Louis.[373] Only the New York Air Brake Company managed to withstand his patent suits and stock offers. In 1872 Westinghouse began to push for overseas sales, organizing an export branch and establishing a shop in England.[374] Seven years later he opened a Paris branch, and in subsequent years formed companies in Germany and Russia.

Meanwhile Westinghouse was also working on the air brake. The railroads at last seemed ready to make the tremendous capital outlay necessary to apply power brakes to the freight car fleet. This would involve not just thousands of units, but more than a million cars. The Master Car Builders scheduled a test near Burlington, Iowa, in July 1886. Of the brakes tested, only the West-

inghouse air brake and the Eames vacuum brake proved successful. But they were hardly perfect. The air brake's old problem of delayed action proved a near-disaster on the fifty-car test train. It took 22 seconds for the rear brake to take hold in an emergency stop, and the rear of the train piled up against the lead cars with a thunderous crash.[375] Westinghouse was aware of the need for quicker action in uniformly reducing the air throughout the system. The solution was found in venting the air at each car for emergency stops. Westinghouse developed a new triple valve that achieved serial venting: rather than bleeding the air off at one point, the apparatus now discharged it at fifty points. During the second Burlington tests of 1887, the action of the brake was improved from 22 seconds to only 6 seconds. Westinghouse was determined to do even better. He made other changes in the quick-action triple valve and provided a larger train line—¼ inch greater in diameter than the old standard 1-inch pipe. These improvements helped speed the action on a fifty-car train to a miraculous 2½ seconds. The story of the quick-action brake is normally told exclusively in relation to freight cars, but it was also the successor to the original 1873 automatic brake for passenger cars (Figure 7.53). The development of this design was the last creative association that George Westinghouse had with air brake technology. Hereafter his associates, such as W. V. Turner, took charge of design improvements.

Figure 7.54 The LN brake, which appeared in 1908, was one of a series that improved the power and responsiveness of the air brake. (Car Builders' Dictionary, 1919)

The industry's languid application of the quick-action brake to the freight car fleet is well documented in the railroad trade press and subsequent writings on the 1893 safety appliance act. But the applications to passenger cars are less well recorded. Some indication of the pace of conversion may be gained from the example of the Pennsylvania Railroad, then one of the most progressive in the country. Late in 1893 it was reported to have nearly completed the conversion from automatic to quick-action equipment.[376] If it took the most liberally disposed roads five years to make the change, how long must it have taken the most conservative?

The quick-action apparatus had no sooner become a standard than the industry discovered that a more powerful system was needed for high-speed passenger trains. In the early 1890s some very fast trains were introduced, such as the Empire State Express. But the speed of these trains was no advantage unless they could be safely controlled. In 1894 Westinghouse introduced the high-speed brake, which increased train line pressure from 70 to 110 psi.[377] A reducing valve was attached to the brake cylinders. In an emergency stop it permitted very high-pressure air to work the brakes; as the train slowed, it gradually cut back on the pressure so as not to lock the wheels. This device was based on a discovery made in 1878 by George Westinghouse and Douglas Galton during a series of railway brake tests conducted in England. They found that the friction between the brake shoe and wheels is considerably *less* when the wheels are revolving rapidly than when they are revolving slowly.[378] Hence initially, a very high brake shoe pressure is important to stop the train in the shortest possible distance. However, the pressure must be carefully and automatically controlled to avoid rough stops, flattened wheels, and skids. Sliding wheels adversely affect the braking process. The new high-speed brake functioned well and registered stops 450 feet shorter than the quick-action brake.[379]

From this time forward, passenger and freight car brake apparatus went separate ways. The K (1905) and later the AB (1933) brakes became the standards for freight service. During the same period passenger cars went through a half-dozen major changes in brake styles, each representing a refinement upon a refinement. All aimed for greater power and smoother stops.

Moreover, all these systems had to be designed to work together, and while they might be quite different in details, all had to be capable of functioning interchangeably, considering the mix of cars possible. That is, an aging baggage car at the head of the train might have a PM style of brake, while the club car would have the latest HSC gear.

In 1902 an improved version of the high-speed brake was placed on the market.[380] It featured a new style of triple valve and air cylinder; otherwise it was much the same as the 1894 design. In 1908 the LN brake was introduced, again with a new triple valve and cylinder (Figures 7.54 and 7.55).[381] It featured a quick recharge that permitted successive brake applications with no material loss in the braking force. A second air tank aided in this operation.

With the coming of the steel car, trains suddenly outgrew their brakes, for train weight tripled. In 1890 a 280-ton train going 60 mph could be stopped in 1,000 feet.[382] In 1914 a 920-ton train traveling at the same speed took 1,760 feet to stop with the 1890 style of brake. Westinghouse was called upon for a quick solution. Increasing the line pressure much above 110 psi was not considered practical, nor was an increase in cylinder size, which was already 18 inches in diameter.[383] Multiplying the leverage of the foundation gear beyond the 9 to 1 ratio resulted in dragging brake shoes, slow releases, and added maintenance. Clasp brakes were of some benefit but did not really solve the problem. The best interim solution was to mount two brake cylinders per car. Thus in 1910 Westinghouse introduced the PC brake, which could stop a train of steel cars traveling at 60 mph in 1,100 feet (Figure 7.56). This was an improvement of almost 40 percent over the older forms of high-speed equipment.

In 1914 the PC brake was replaced by a more powerful successor, the UC brake, which incorporated three air tanks and a control valve so radically different that it was called a universal rather than a triple valve (Figure 7.57). This brake could stop a heavyweight train traveling 60 mph in 860 feet.[384] The UC was a longtime standard for passenger equipment in this country.[385]

Just before the opening of the lightweight era, a new electrically controlled brake, the HSC, was placed on the market. Electric pneumatic brakes had appeared at the Burlington tests in

Figure 7.55   A better understanding of the brake rig and apparatus can be gained from this rare 1913 underbody photograph. LN apparatus is shown.

1887–1888. For a time Westinghouse had reluctantly believed that electric control was a necessity, but he did everything in his power to avoid the introduction of a secondary system that might fail and jeopardize the basic operation. His company successfully avoided electricity for many years, but at last it became apparent that the relatively slow-acting, purely pneumatic system was not good enough. To guarantee safety, however, the HSC brake was not dependent on the electrical control. Without electricity its action slowed down, but it could still function pneumatically if

the circuit failed.[386] The HSC was first used in 1932. Its superior power and extra-smooth stopping characteristics made it a standard for the lightweight streamline trains then coming into service. Electric control improved stopping capability by 40 percent.[387]

An even better form of electropneumatic brake, the D-22, was introduced in 1935. It broke away from traditional running-gear design by mounting the brake cylinders on the trucks rather than under the body of the car (Figure 7.58). This eliminated the

Figure 7.56   *In response to the need for more powerful brakes on steel cars, the twin-cylinder P.C. brake was developed in 1910. (L. K. Sillcox,* Mastering Momentum, *New York, 1941)*

Figure 7.57   *The UC, placed on the market in 1914, became an industry standard. (Car Builders' Dictionary, 1919)*

foundation gear. Two or four small brake cylinders attached to each truck's side frame were connected directly to minimal brake rigging of the truck. This saved weight and the lost motion of the long levers and rods typical of the old system. The smaller cylinders also used less air.[388]

Because of the decline in railway passenger travel, recent developments have been sparse in this area of railroad technology. One of the latest improvements was a pneumatic-hydraulic brake introduced by the New York Air Brake Company in 1969.[389] It features a high-pressure hydraulic system (800 psi) controlled by, and interchangeable with, conventional air brakes. The apparatus has been applied to over 120 electric suburban cars built for the Illinois Central Railroad in 1971. It is possible that this brake or some equally radical departure will at last displace the present standard D-22.

Brake Shoes

Train stopping involves some basic principles of physics. Because energy cannot be destroyed, the kinetic energy or motion of the train must be converted into some other form of energy. When we think about energy conversion, we tend to imagine the transformation of heat into energy. In braking, however, just the opposite occurs. The heat generated by the brake shoes bearing against the wheels dissipates the train's forward motion. The brake shoes and, to a lesser degree, the wheels are ground away in the process. The shoes, being more expendable, are made of a comparatively soft material such as cast iron. The ideal brake shoe is also an efficient radiator that can dissipate large amounts of heat without rapid self-destruction. Again, cast iron proved well suited for this purpose. Practical as it was, however, cast iron did not produce the perfect brake shoe, and car builders tried a variety of other materials.

Cast iron appears to have been used only rarely during the first two generations of American railroading. Wood was the principal brake shoe material then. A half-dozen pioneer mechanics, including Wilson Eddy, Reuben Wells, and John Kirby, were interviewed on the subject in 1891.[390] None could recall the use of cast-iron shoes much before 1855, though after that time they became an industry-wide standard. In the early period white oak was favored, but beech and hickory were also used. Some lines

Figure 7.58   Lightweight equipment and faster trains called for more powerful pneu-matic brakes, such as the model D22 shown here. The black silhouette is a control system called a decelostat that prevented wheel skids by automatically adjusting the air pressure. (Car Builders' Cyclopedia, 1953)

nailed leather to the brake heads or blocks, as shoes were often called at this early date. The great defect of the wooden shoe was its tendency to wear or char. As train speeds and weights increased, wooden shoes would simply burn off. The smoke was frightening to passengers, and the flaming shoes endangered the entire carriage and its occupants during the wooden age. The burnt-off shoes posed an even greater threat, because they left the train without brakes.

To overcome this problem the Boston and Providence tried stone shoes in 1853, but they proved unsatisfactory, presumably because of excessive tread wear. The Madison and Indianapolis, which had a surplus of strap rail after rebuilding its tracks with T iron, began shoeing its wooden brake heads with the short pieces of the discarded rails. The straps became so hot, however, that the blocks charred, the shoes loosened in time, and the experiment was abandoned. The road then adopted cast-iron shoes. And what happened in Indiana happened elsewhere, with very few exceptions. The B & O installed wrought-iron shoes and continued to use them on passenger cars until 1882, when the line at last made a tardy conversion to cast iron. The Hudson River Railroad alone remained loyal to wooden shoes into the 1870s.[391] As late as 1894, efforts were made to revive the wooden shoe.[392] The Wharton Switch Company placed a cast-iron shoe on the market that had end grain oak inserts. The Wharton shoe was reported to be widely used on street railways, where it accumulated mileages up to 6,000 miles, but it never seems to have been accepted by any steam lines.

Cast shoes were a national standard before 1870. They were a simple article that any foundry could produce, but it was not so easy to make a really good shoe. Indifferent makers turned out shoes with hard chilled surfaces or poorly mixed irons that produced hard spots and cut grooves in the wheel tread.[393] George M. Sargent, a major shoe manufacturer, admitted that after fourteen years in this specialty he was unable to determine the optimum mixture for the perfect brake shoe. He knew that uniformity of the metal was paramount, and that it called for a degree of quality control not found in any but the best-managed

foundries. So long as purchasing agents sought the cheapest supplier, he said, the railroads would be plagued with poor shoes.

The rapid wearing away of the shoes required frequent and troublesome renewals. The "car knocker" had no more miserable job than to crawl under the trucks and fit on a new set of shoes. In the wintertime especially, he could think of the most imaginative excuses for putting it off. Some roads were inclined to stay with wrought-iron shoes because they lasted five or six times longer than cast iron.[394] But wrought iron tended to warp and curve when heated. The metal rolled and bunched up at the ends, and the surface became hardened and slippery. Wrought-iron shoes might outlast cast iron, but they were less effective in stopping the train.

It occurred to Isaac H. Congdon (1833–1899), master mechanic for the Union Pacific, to combine wrought and cast iron in a brake shoe. The body was made of cast iron, with wrought-iron segments cast in the front wearing surface. These composite shoes were said to outwear the common pattern by three to four times.[395] They cost 4 cents a pound, or only 1 cent more than conventional shoes. Congdon patented the idea in March 1876. Manufacture began the same year in Chicago by George M. Sargent, who developed Congdon's invention into a widely used variety of brake shoe.[396] Congdon shoes were still being produced in 1909.[397]

A variation on the Congdon shoe was developed by James Meehan, master mechanic on the Cincinnati, New Orleans, and Texas Pacific. Meehan embedded pieces of hardened tool steel in the face of the shoe to cut away the surplus metal rolled over the wheel tread's outer edge. Such shoes helped to maintain the correct tread profile on locomotive and steel tired passenger car wheels. They partially eliminated the costly procedure of the wheel turning lathe. It was still necessary to turn the wheel sets after extended mileage, but the Meehan shoes were effective in maintaining the profile for long periods of time. Meehan patented the idea in 1884. In later years it was known as the Ross-Meehan shoe.

More important than these improvements was the introduction

Figure 7.59 The Christy brake shoe, patented in 1865, permitted quick replacement by removal of the key that held the shoe to the brake head. (William Voss, Railway Car Construction, *New York, 1892)*

of demountable brake shoes by James Christy of Philadelphia in 1865. Christy offered a quick-change shoe held to a bracket or head on the brake beam by a long, tapered spike-like key (Figure 7.59). It greatly simplified the changing of shoes—the job could be done in one minute. By 1885 fifty to sixty railroads were using the Christy shoe. A year afterwards it was adopted by the Master Car Builders Association as a standard. The Christy patent involved another very desirable feature: the back of the shoe and the head were separated so that air could circulate and cool the shoe.

Ordinarily only one brake shoe was used per wheel. This remains true for freight cars, but during the first decade of the twentieth century the need for greater braking power prompted experiments with double or clasp brakes. A shoe was placed on the opposite sides of the wheel. Their vise-like grip not only doubled the braking surface but also helped to hold the axles and boxes in alignment. The idea was suggested at an early date, but no serious efforts to perfect it were made until 1887, when the Pennsylvania tried it on two cars.[398] The test failed because of a poorly designed brake gear. The clasp brake made little progress until 1909, when Thomas L. Burton of the Reading began to devise a better foundation plan.[399] A car fitted with the double brake stopped in two-thirds of the distance that a conventional car required. The moderated pressure resulted in less shoe wear and added to wheel life by lowering tread pressure. Shoes lasted for 25,000 miles in some cases. Shoes were so cool after emergency stops at 80 miles per hour that they could be touched with the bare hand. By August 1912 clasp brakes were standard on the Reading and the Central of New Jersey. As the heavyweight cars grew in number, so did the acceptance of the clasp brake.

A more radical departure was offered as railroads moved into the lightweight era. The time-honored shoe bearing against the running edge of the wheels was discarded. In its place was an axle-mounted drum brake not unlike that used for automobiles. It was felt that the wheels were doing enough to guide and support high-speed trains; they should not be required to stop the train as well. High-speed braking was destructive to car wheels, because the resulting high tread temperatures resulted in thermal cracks and shelling. These bad effects could be entirely avoided by installing a separate "wheel" or drum for braking. As described in Chapter 1, drum brakes had been tried in the 1830s on the Mohawk and Hudson Railroad (Figure 1.61). It was a scheme forgotten for nearly a century, until in 1936 Edward G. Budd asked one of his engineers, C. L. Eksergian, to develop a modern car brake.[400] Eksergian's first idea was to adapt the automotive drum brake, but the shoes, being inside of the drum, could not dissipate the heat quickly enough. The design was not practical for so large a vehicle as a railroad passenger car. An open disc with shoes pressing on both sides offered the cooling characteristics needed. The disc was actually a thick cast-iron drum with ribs or blades at its center (Figure 7.60). It was a combination drum and blower, whose 7,000 square inches proved effective as a heat exchanger. The drum was bolted to the wheel hub. Small air cylinders, one for each drum, operated tong levers

fitted with asbestos composition shoes. After a series of test-stand experiments, the disc brake was tried on a Burlington coach early in 1938. During the next year the lightweight motor train General Pershing, also on the Burlington, was fitted with the new form of brake. The results were so encouraging that the Santa Fe and the Milwaukee railroads ordered several sets. Budd claimed that the disc's cooling action was so effective that no checking of the drum or dusting of the brake linings could be detected.[401] Stops were made quietly—the usual screeching was eliminated. This was of great benefit to the passengers, for tests showed that they often complained about a train's poor ride when their discomfort was actually caused by noisy brakes.[402] Budd gave the following figures on the brake's efficiency in train stopping:

> 60 mph   Train stopped in less than 1,000 feet.
> 80 mph   Train stopped in less than 1,600 feet.
> 100 mph Train stopped in less than 2,500 feet.

It was claimed that the shoes lasted 100,000 miles, while ordinary metal shoes wore out in 6,000. Wheel life was increased by 70 percent. And the dead weight of the conventional brake rigging was eliminated.

Remarkable as these advantages appear, there was no rush to adopt the disc brake. In 1946, nearly eight years after its commercial introduction, only 200 cars were equipped with it.[403] More interest was evident in the postwar years. By 1949, 500 cars carried the Budd brake, and by 1960 the number climbed to 2,200, a sizable portion of the passenger car fleet.[404]

The resistance to the disc brake can be explained in part by the railroads' normal rejection of new ideas, but it was also due to improvements being made in the conventional brake shoe. In the mid-1950s Westinghouse introduced a high-friction composition shoe composed of rubber, asbestos, and powdered iron. It offered faster stops and more mileage than cast-iron shoes. Several years later the Griffin Wheel Company marketed a similar shoe that outlasted iron shoes by four to five times and required only one-third to one-half the braking force for a stop.[405]

Figure 7.60   Shoe or tread brakes were standard on American lines until 1938, when disc brakes were introduced by the Budd Company. (Pullman Neg. 64008)

Couplers

If railroad cars were permanently fastened together, the simplest form of attachment would be possible. But cars are constantly being coupled and uncoupled in the normal course of railway operations. Trains are made up at the terminals before departure. Cars are then picked up and set out during the run, and the train itself is generally dismembered upon arrival at the last station. Despite the obvious need for an efficient and convenient coupling device, American railways were long content with the most primitive methods for connecting cars. As late as 1870 the link and pin coupler remained common on passenger cars. And this primeval device predominated on freight cars until almost the end of the nineteenth century.

Not much thought was given to the very first car couplers; a few links of chain were considered good enough. Figure 1.2 in Chapter 1 shows such rudimentary couplings. The Mohawk and Hudson cars of 1831 had chain links that were dropped over pins fastened to the car's frame. A hasp-like iron strap was hinged (or possibly sprung) over the pin to keep the chain from losing its purchase. The plan was simple, but the links provided too much slack, resulting in whiplash starts and thunderous stops. The shortcomings of the chain coupler were recorded by a passenger

who described the inaugural trip of the Mohawk and Hudson:[406]

When he [the conductor, Mr. Clark] finished his tour, he mounted upon the tender attached to the engine, and, sitting upon the little buggy-seat, as represented in our sketch, he gave the signal with a tin horn, and the train started on its way. But how shall we describe that start, my readers? It was not that quiet, imperceptible motion which characterizes the first impulsive movements of the passenger engines of the present day. Not so. There came a sudden jerk, that bounded the sitters from their places, to the great detriment of their high-top fashionable beavers, from the close proximity to the roofs of the cars. This first jerk being over, the engine proceeded on its route with considerable velocity for those times, when compared with stagecoaches, until it arrived at a water-station, when it suddenly brought up with jerk No. 2, to the further amusement of some of the excursionists. Mr. Clark retained his elevated seat, thanking his stars for its close proximity to the tall smokepipe of the machine, in allowing the smoke and sparks to pass over his head. At the water-station a short stop was made, and a successful experiment tried, to remedy the unpleasant jerks. A plan was soon hit upon and put into execution. The three links in the couplings of the cars were stretched to their utmost tension, a rail, from a fence in the neighborhood, was placed between each pair of cars and made fast by means of the packing-yarn for the cylinders, a bountiful supply being on hand. . . . This arrangement improved the

Figure 7.61   Link and pin couplers were the basic passenger car union until the 1870s, and they prevailed in freight service into the 1890s. The top four drawings are of a typical coupler; the bottom four show the New Jersey Railroad and Transportation Company's standard. The drawings date from 1859. (A. Bendel, Aufsätze Eisenbahnwesen in Nord-Amerika, Berlin, 1862, plate 11)

order of things, and it was found to answer the purpose, when the signal was again given, and the engine started.

A better scheme was soon evolved that remained a standard on American railways for many generations. A hollow iron tube about 5 inches square was fastened to the underside of the end platform. The end that projected out a few inches from the platform had a face plate. A hole was bored through the top and bottom of the drawhead to receive a pin. The pin held a link inserted through the face plate. It was a loose, rudimentary sort of fastening that a country blacksmith might have devised (Figure 7.61). Pictorial evidence indicates that this form of coupling, commonly called the link and pin, was in use by the mid-1830s. In the drawing of the B & O's Washington branch car of 1835, the coupler is placed well back under the platform, the pin being dropped through the floor (Figure 1.70). A very long link would be required. It should also be noted that the drawheads were

connected to an iron drawbar running the full length of the car. This device protected the body and frame from the draft strains. A more conventional link and pin coupler is shown in the Boston and Providence eight-wheel car drawing of 1838 (Figure 1.75). It was the type commonly used until the coming of the automatic coupler.

An ordinary link was made of 1½-inch-diameter bar stock, slightly flattened. It measured about 13 inches long by 5 inches wide. Links weighed 6 to 8 pounds each. The pin was made of the same-sized wrought-iron bar and also generally measured just over 1 foot in length. These rugged chunks of iron were intended to withstand the heavy draft strains as the train pulled and surged along the tracks. Some mechanics thought that it served its office too well under emergency conditions, and that it might be better if the link were designed to break under an unusually heavy strain such as a derailment. If the link snapped at the juncture of the derailed car, the wreck would be con-

Figure 7.62   An ironbound wooden coupler link used on the Philadelphia, Germantown, and Norristown. Links of this type were introduced in the 1830s. (Smithsonian Neg. 16204)

tained. The rest of the train could remain on the track instead of being pulled off to the uncertain end that awaited the errant car or locomotive. Sometime before 1840 the B & O began using coupling links made from wood, of all improbable materials.[407] They functioned admirably by snapping off at the critical moment, leaving the nonderailed cars standing unharmed on the tracks. One report stated that this safety coupling saved a train of cars when a cow derailed a locomotive near Cumberland. It was regrettable, the *American Railroad Journal* said, that so few lines were using the superior wooden fastening. The Reading and its subsidiary, the Philadelphia, Germantown, and Norristown, also adopted the wooden link.[408] A link from the P G & N is in the Smithsonian Institution's collections (Figure 7.62). It is made from an oak plank measuring 15½ inches long by 4 inches wide by 2 inches thick. Two ¼-inch iron bands are riveted around its edge, but note that they do not meet in the center. The link is weakest at the point of separation. The pinholes are lined with small iron bands to increase their service life. Such links were in use until the early 1870s, when the air brake's slow releases put such a strain on the wooden couplings that they would break when the engine started up.

The link and pin coupler remained popular because of its low first cost and simplicity. But was it actually economical? Certainly not when its record of human misery is considered. In maiming and killing it was without equal; crushed fingers, hands, and ribs were its specialty. Trainmen could recognize one of their fraternity anywhere by his missing finger or mutilated hand. And the link and pin was also uneconomic from an operational point of view. Even without the accident claims, railroad managers began to realize that the coupler was expensive because of the prodigious loss of links and pins. In 1875 an official of the New York Central said, "Millions [of pins] are made every year but no one knows what becomes of them."[409] A relatively small Indiana trunk line found that it lost over 43,000 links and pins annually.[410] The Lake Shore calculated the cost of such losses at over $52,000 a year, or $5.20 a car.[411] This material did not just innocently disappear; it was stolen. The Atlantic and Great Western found that couplers were being sold to local junkmen.[412] In one yard investigators discovered twelve barrels of links and pins that had been labeled as old iron. In a nearby lumbering camp they saw a huge logging chain, worth $1,500, made from coupling links.

Railroads sought ways to overcome this financial drain, but there seemed no efficient way to police the thieves. Cars were exposed in yards and on sidings, and much of the thievery was probably carried out by railroad employees. The Pennsylvania, which retained the old-style coupler for passenger cars long after most roads had switched to Miller couplers, solved the problem by securing the link to the drawhead with a secondary rivet (Figure 7.63).[413] Other roads searched for ways to lower the cost of manufacturing links and pins. The Central Pacific devised a machine that turned out 1,000 links a day, whereas only 200 could be made in a day by the old hand method.[414] The process was further automated by the nation's leading maker of such

materials, the Pittsburgh Forge and Iron Works. In 1880 it had a machine that would cut, bend, and scarf (bevel the ends of) 6 to 7 links a minute.[415] The weld was formed under a steam hammer in ¼ minute. Another forging machine pounded out a matching pin every 1¼ minutes. Six years later a new pin-making machine was producing 35 pins a minute, and an improved link maker, about ready for use, was scheduled to complete 25 links a minute.[416]

A better way of reducing the financial drain was to eliminate the link and pin coupler itself. An automatic coupler would have no loose pieces to be pilfered, and the savings thus realized would ultimately pay for the new device. Furthermore, there was a positive passion among inventors during much of the nineteenth century to create an automatic railroad coupler. In 1897 1 percent of all U.S. patents involved car couplers.[417] Of 28 patents related to railroad devices issued during the week of September 5, 1882, 11 were for car couplers. The total number climbed year by year. In 1873 there were 800 car coupler patents; by 1875, 900; by 1884, 2,300; by 1887, over 4,000; by 1897, 6,500; and by 1930, long years after the Janney coupler's adoption, 11,813.[418] Coupler patentees became a nuisance that railroad men avoided at all cost. Mechanical officers were exhausted by their entreaties, models, visits, and prospectuses. The enthusiastic inventors were said to represent "baffled hopes and misdirected labor" more fully than any other single segment of the population. The scheme of one rustic mechanic was likely to be mistaken for a threshing machine component. Another resembled a torpedo, and the car coupler of Syrus Washington Wright of Eden, Alabama, was construed as an unduly complicated steam engine valve gear. Unfortunately, sensible ideas were often buried in the overwhelming debris of foolish plans. It seems fair to say that the very intensity of interest in the subject delayed rather than facilitated the introduction of automatic couplers.

Suggestions for a self-acting coupler appeared as early as the late 1840s. In 1847 A. G. Heckrotte of Washington, D.C., began to promote a device that a Baltimore paper felt would "no doubt supersede the old fashioned bolts and chains."[419] The drawbar was a box, open at one end, with a spring trundler much like that of a gun lock, and a roller inside. A side lever released the trundler. A solid iron bar in place of the usual link made the connection. It had notches at both ends which snapped in between the jaw formed by the trundle and roller (Figure 7.64). The device was actually tested on the B & O but was not adopted.

David A. Hopkins (1824c.–1889) of Elmira, New York, patented a semiautomatic coupler on August 1, 1854 (No. 11428), That drew some favorable attention (Figure 7.64). It involved a modified link and pin drawhead with a block inside that prevented the pin from falling in place until the block was pushed back by a link from the opposing coupler. The pin was then free to drop. The scheme was simple, and because it closely resembled the standard form of coupler instead of some outlandish contraption, it was rather well received. The Hopkins coupler was used on the Erie for nearly five years.[420] The inventor claimed that it was not more widely adopted because of defec-

Entire length, 18½ inches. Iron, 4 by 1 inch.   No. 1 Face Plate.

Figure 7.63   The link in the coupler here was permanently attached to prevent its removal or loss. The engraving dates from about 1875. (DeGolyer Foundation Library)

tive manufacture rather than poor design.[421] It is also probable that the extra cost discouraged sales. As one railway manager said, "Your device would be more expensive than Paddies are."[422]

Just a year after Hopkins's coupler appeared, J. T. England, a minor official of the B & O, unveiled a semiautomatic coupler that had some limited use. England followed the same general idea of the Hopkins coupler, but instead of a spring-actuated block he employed an iron ball. The ball blocked the pin, holding it up. When the link entered the drawhead, the ball was pushed back into a cavity out of the way and the pin was free to fall in place (Figure 7.64). England received a patent (No. 13869) on December 4, 1855. Endorsements were offered by several railway mechanical men, but the device seems to have been an ingenious curiosity rather than a standard product.[423]

Literally thousands of other schemes were patented; the few that have been described represent some of the better-received early automatic coupler designs. Their lack of success can be ascribed as much to an indifferent market as to their mechanical defects. But now that the railroads were becoming the dominant public carrier, railroad operations and safety records were ever more important and ever more a matter of public concern and criticism. By 1860 railroads were no longer an infant industry but a major force in American transportation.

At this juncture an ebullient figure, Colonel Ezra Miller, came forward with a solution to one of the railroad's worst safety problems. In the mid-1860s Miller devised a combination trussed end platform, draft gear, and coupler that diminished the likelihood of telescopes and eliminated the link and pin coupler. Within a decade it was widely used on American passenger cars.

Miller had already had careers in politics, the New York State Militia, and civil engineering. He was described by one admiring biographer as "a cultivated and experienced" engineer notable for his purity of character and animated conversation.[424] Miller had just the right combination of talents to create and sell a new form of car coupler. While traveling in the Midwest to survey land, he was impressed not so much by the inferiority of the car couplers on the trains he rode as he was by the weakness of their end platforms. Miller rightly felt that car builders had made a major error by dropping the end platforms below the floor level. The main frame and the platform should be in line and strongly fastened together to withstand end buffering. Miller also decided that the ends of the car should be drawn tightly together to diminish the likelihood of one car mounting the end of the car in front. With the loose, slack link and pin arrangement, cars bounced so freely that they could jump up onto an adjacent platform—a sure invitation to the much dreaded telescoping accident (Figure 7.65). The old system of loose links also permitted the cars to sway so much that it added substantially to the general misery of travel, except for passengers who happened to be sailors.

Miller first began to think about the problem during his travels in 1853. The standard biographical accounts of his life say that the idea took nearly a decade to mature. It is recorded that he re-

Heckrotte **1847**   PLAN VIEW

Hopkins **1854**

England **1855**

Figure 7.64   Some early efforts toward developing an automatic coupler. All were tested, and the England device was put into rather wide use on the B & O. (Traced by John H. White, Jr.)

ceived his first patent on March 31, 1863 (No. 38057), and two others in January 1865 (No. 46126) and July 1866 (No. 56594).[425] However, in 1887 it was revealed that he may have been inspired by the work of Joseph Miller, believed to be a cousin, who patented a hook style of coupler in 1854.[426] Four other hook couplers were on file at the Patent Office before Miller received his first patent. Their principle was simple: two drawbars with spear-shaped ends were sprung at their ends so that when they passed by each other, they would snap together and hold fast (Figure 7.66). A lever on the end platform was used to pull them apart for uncoupling.

Miller wisely included a trussed platform and a central buffer in his general design, thus giving the railroad a neat package that solved several problems at once. The scheme was so attractive that it met with almost immediate success. The Erie adopted it in 1866, the year that Miller's third patent was granted.[427] The Burlington was using it on many trains by the following year.[428] In 1869 the Philadelphia, Wilmington, and Baltimore was re-

Patented March 31, 1863, January 31, 1865, and July 24, 1866.

Figure 7.66   Miller's hook-style coupler was widely adopted in this country for passenger cars between 1875 and 1900. For most of that period it was a national standard.

Figure 7.65   Miller combined a strong antitelescoping platform with his passenger car coupler. The weakness of the conventional drop platform is illustrated in the upper two details of the drawing.

ported to be putting it on all new cars.[429] By 1870 it was extensively used in the Middle West and in the Western states, although the more conservative New England lines showed little interest. But railroads in all regions were soon persuaded to install Miller's platform by the reports of its miraculous performance. In May 1870 a day express on the Allegheny Valley Railroad rammed into a slide at 40 mph, but the sturdy Miller platforms prevented telescoping.[430] A more spectacular accident on the Boston and Providence two years later again proved the effectiveness of Miller's lifesaving invention.[431] The train struck a wagon and team, the remains of which fell under the cars and sheared off the trucks. The wheelless cars slid along the track, then went off the right of way and down a small embankment into a field. But the Miller hooks remained fast, holding the train together until it came safely to a stop amidst the weeds. One nervous couple on their way to the 1876 Centennial Exposition were reassured by a fellow traveler:[432]

"Now, madam," continued our interlocutor, "there is no possible danger of the accident you fear, the ordinary oscillation, by which a car is swung from the rail, being very nearly quite removed; and, my friend," addressing me, "there is equally small danger, in the middle cars, of 'telescoping;' the train is made up of cars provided with the trussed platforms of Colonel Miller, which absolutely prevent both the kinds of casualty you have feared."

What possible excuse could be offered for not adopting the Miller platform and coupler? So many lines began using it that by December 1874 Miller could claim that 85 percent of all U.S. and Canadian railroads were carrying his apparatus.[433] Of the 110 railroads that had not yet converted, only the Pennsylvania was a major line; the others were small roads with little passenger traffic. The passenger car fleet had abandoned the link and pin coupler in something like eight years. It was a swift and deliberate transition, free of the painful delays and deliberations that attended the conversion of the freight car fleet. Of course, the freight changeover involved more than a million cars and a vast expenditure compared with the relatively small passenger car fleet. There was also more public pressure to make passenger trains safe. Fatalities and injuries among trainmen were viewed as a secondary evil; the statistics were even separately reported by the I.C.C.

Figure 7.67   *A Miller coupler of about 1885. The single buffer, immediately above the hook, maintained tension between the cars so that the couplers would not separate.* (Voss, Railway Car Construction, *1892*)

Figure 7.68   *This drawing represents an underside view of a Miller coupler and platform of about 1885.* (Voss, Railway Car Construction, *1892*)

Figure 7.69   An end elevation view of a Miller coupler and platform. The large lever on the right is used to pull the drawhead to one side for uncoupling. Note that the end of the drawhead is slotted to receive a conventional link and pin. (Voss, Railway Car Construction, 1892)

But in fact, the often criticized pace of installing automatic couplers on freight cars was not all that slow, for it was accomplished in little more than a decade. Tourists in Russia are told with some pride that the Soviet Union completed its conversion around 1957. And in 1971 the large-scale application of automatic couplers was just getting underway in Western Europe, according to some firsthand reports.

The success of the coupler made Miller a rich man. He was said to be one of the few inventors who personally realized a fortune from his patents.[434] In one of his earliest catalogs he discussed the question of his growing net worth.[435] "Is my charge too high? I am told that the improvements will earn for me a fortune! I often hear this remark. Granted. Should I not have it? If I earn *hundreds* of fortunes for the railroads of America, should I not have *one* for myself and family?" He claimed that his basic license fee was $100 per car, but in practice it proved far less. As it worked out, the actual fee fell between $25 and $53.[436] Some roads got by more cheaply by paying a flat amount. Among the Burlington papers is an agreement with Miller dated June 11, 1879, that grants use of the patents until they expire for a mere $800.[437] Miller also profited from the sale of his gear, though he does not seem to have been directly engaged in its manufacture. In a letter of September 14, 1877, he quotes the following prices: coupling hooks, $24.50; buffers, $11; uncoupling lever, $2.[438]

Miller's design was praised by William Voss, an expert in railroad car construction, who said around 1890 that the coupler was little changed from the original pattern introduced nearly twenty-five years before.[439] The parts were admirably proportioned in both size and shape. Stronger materials had been substituted in some instances, however. The original hooks had been fabricated from $\frac{5}{8}$-inch flat bars, the skeleton filling pieces being of cast iron (Figures 7.67 to 7.69). But now hooks were made from solid wrought-iron forgings or steel castings. Voss went on to say that Miller's apparatus was used on the majority of American passenger cars, but that it faced an uncertain future because of the recent (1888) adoption of the Janney coupler for freight cars. Voss admitted that the Miller hook had some notable failings. It was difficult to uncouple, particularly on curves.[440] The single central buffer was ineffectual in stabilizing the car's rock and roll, because the single faces were located at the axis of the motion. The hooks tended to uncouple unexpectedly, because they were only 6 inches deep, and variations in the level of cars, height of the draw bars, or condition of the track surface could cause them to slip apart. Such occurrences were more common than the public knew. The deeper jaws of the Janney coupler solved this problem. The introduction of the vestibule created another difficulty for the Miller coupler: there simply was no space for the long throw of the uncoupling lever. It was also necessary to slam the cars together in order to couple the hooks.

These failings were one reason for the Miller's gradual decline, but the main cause was the adoption of the Janney coupler for freight cars. The operation of mixed trains was common on branches, but even on the main line it was occasionally necessary

Figure 7.70 In 1873 Janney patented what has become the basic coupler of today. This drawing, published in 1874, shows the device in its first form. (National Car Builder, November 1874)

to connect all varieties of rolling stock. And at many terminals, switching engines handled both freight and passenger cars. A common standard was desirable. Since the Pennsylvania had never used the Miller Hook, its subsidiaries and connecting lines made the conversion to the Janney at an early date. In the spring of 1890 the Burlington changed from Miller to Janney.[441] Ironically, in the following year the I.C.C. noted that the Miller was at its peak: of some 28,000 passenger cars, 21,350 had Miller hooks.[442] Within two years, however, the number was down to 17,830; and by 1899, the last year for which statistics were compiled, it had dropped to 11,600.[443] Even this figure represents a sizable portion of the American passenger car fleet. Some lines obviously were not bothered by the interchange of passenger cars and saw no need for a hasty conversion. It is also true that a combined Janney-Miller coupler which had been developed sometime in the 1880s extended the useful career of the obsolete hook coupler. Just when the last Miller hooks disappeared from regular service does not seem to have been recorded, but it is safe to assume that some were in operation into the early twentieth century.

The device that replaced Miller's coupler was invented by Eli H. Janney of Alexandria, Virginia. By all odds Janney should have remained one of the thousands of anonymous amateur inventors whose couplers rested untried in the files of the Patent Office. He had no mechanical or railroad background. He came from a farming family, served in the Confederate Army, and worked as a dry goods clerk during the hard times following the Civil War. Superficially he was the very stereotype of the "coupler crank." But Janney somehow stumbled upon the plan for a rotating jaw coupler that in time became the American standard. He tinkered and whittled his way toward a major advance in railroad technology. Just why or how he was so motivated is not really known; presumably he was drawn to the coupler problem by the distressing stories of injuries to trainmen in the nearby yards. In April 1868 he received a patent for a drop-hook coupler. It was a poorly conceived plan that justifiably received little attention, although some popular writers have mistakenly assumed that it was the basis of the modern coupler. Five years later, on April 29, 1873, Patent No. 138,405 was issued to Janney for an open-jaw or knuckle coupler that incorporated the elementary design of the present-day style. A clenched fist may have

suggested the idea: when closed, it forms a hook; when opened, its grasp is broken. The coupler had a bifurcated head, representing the hand, and a hinged or revolving jaw, representing the fingers. The jaw was locked closed by a pin to effect the coupling. When the pin was released, the jaw could swing open to release the coupling (Figure 7.70). Because the connection (or jaw) was on edge, it was called the "vertical plane coupler," a phrase contrived by the M.C.B. to characterize the general style of coupler devised by Janney and his imitators.

In 1874 Janney formed a partnership with two local backers to promote his 1873 patent. Late that year five railroads were already testing the device.[444] Of these lines the Pittsburgh, Fort Wayne, and Chicago, a subsidiary of the Pennsylvania, was the most significant for Janney's future. From the initial performance of the coupler on eight coaches in 1874, the road was so impressed that within four years it equipped all 152 passenger cars with Janney's device.[445] The parent company watched with interest. Its allegiance to the archaic link and pin was an embarrassment, but for some reason the Pennsylvania continued to reject the Miller hook. Janney's device was at that time "in the inchoate model stage," but the Pittsburgh, Fort Wayne, and Chicago's good experience convinced the Pennsylvania that the design could be perfected.[446] In 1876 the road applied Janney's coupler to 55 cars and showed an inclination to install it more widely. With such an endorsement, Janney lost his amateur status.

In 1877 Janney broke off with his original partners and made an agreement with William McConway and John J. Torley of Pittsburgh to manufacture his coupler.[447] These men were experienced mechanics who specialized in malleable iron castings. Janney now had the expert help he needed to manufacture and market a practical coupler. The Pennsylvania made good its promise and applied the Janney coupler to its passenger cars sometime before 1883.[448] Progress was slower elsewhere, however, because of the prevailing loyalty to Miller.

Meanwhile Janney worked to improve his product (Figures 7.71 and 7.72). In a third patent of 1879 he altered the shape or contour of the knuckle in order to smooth its engagement and disengagement. Three years later Janney developed a coupler specifically for freight cars. The spring catch that was suitable for passenger cars was considered too weak for freight trains, so he

Figure 7.71   Janney improved his coupler with the help of practical advice from McConway and Torley, malleable-iron makers of Pittsburgh. (Railroad Gazette, January 10, 1879)

Figure 7.72   Janney combined his coupler with a safety platform and buffer arrangement much like that offered by Miller. (Railroad Gazette, January 10, 1879)

used a drop pin to hold the knuckle closed. This arrangement was soon applied to both freight and passenger car couplers, although many spring catches remained in service as late as 1905. By the end of the 1880s the Janney coupler had evolved into a rugged, serviceable piece of apparatus.[449] Less the draft gear, it weighed 210 pounds. It measured 33¼ inches long. The shank or barrel was a hollow casting 5 inches in diameter and ⅞ inch thick. The head and barrel were a malleable iron casting, and the knuckle was forged iron or cast steel. The drop pin was malleable iron, while the knuckle pin was crucible steel 1⅜ inches in diameter. A pair of passenger car couplers sold for $190.

The merits of Janney's design slowly gained recognition during

Figure 7.73 By 1882 Janney was offering couplers with pin catches in place of the original spring latch. The three lower engravings show how the slotted jaw could accommodate the old-fashioned link and pin couplers.

the eighties. By 1888 it was acknowledged the best possible form of freight car coupler: in that year the Master Car Builders adopted it as their standard after years of debate and testing.[450] Janney won the M.C.B. adoption by a large majority, but he was forced to waive his rights to the eighth and ninth claim of his patent 212,703 (Feb. 5, 1873), which covered the knuckle profile. This permitted other manufacturers to produce a vertical plane coupler that would couple with those manufactured by McConway and Torley. The M.C.B. did not ask Janney to abandon patent claims on other details of the coupler, such as the drop pin lock. But none of Janney's patent covered the general plan of the open-jaw coupler, and other makers were thus free to produce devices of a very similar pattern. During the year of the M.C.B.

ratification, 150 American railroads were already using the Janney coupler.[451]

The early years of the coupler were not without difficulty, for the knuckles broke too easily. The Boston and Albany claimed that 296 fractured in just two months, and the line's master car builder, the quarrelsome Fitch Adams, said that this dismal record vindicated his objections to the Janney coupler.[452] A broader survey showed that Adams's contention was incorrect, because the annual national breakage rate for Janney knuckles was only 3 percent, and overall Janney coupler failures ran no more than 1 percent.[453] Much of the problem was attributable to the necessity for cutting a large slot in its face and drilling a hole through the knuckle so that the link and pin connections

Figure 7.74   The adoption of the Janney-style coupler in 1888 did not usher in an age of standardization. Each manufacturer had an individual plan for the jaw and latch pin assembly. While all could couple together, parts were not interchangeable. (Railroad Car Journal, July 1899)

could be made (Figure 7.73). By 1899 the transition period was nearly over, most cars had automatic couplers, and stronger, solid knuckles would soon be universal.[454]

Maintenance and road repairs were another grave problem during the early years of the M.C.B. coupler. In its concern to avoid a monopoly, the association had created an even more undesirable situation by encouraging a proliferation of coupler makers, each with his own pattern. All were of the same general dimensions and all would work together, but each make was unique insofar as individual parts were concerned. No two made the same locking pin or knuckle. In 1891 there were nineteen major coupler makers producing Janney-type couplers.[455] Every division repair shop was obliged to carry parts for Smillie, Gould, Hinson, Trojan, and the other brands. By 1896 there were twenty-

Figure 7.75 *The type D Janney coupler was introduced in 1916. Basically intended for freight service, it remained the industry standard until 1931. (Car Builders' Cyclopedia, 1928)*

four makers, and by 1899 nearly eighty styles of knuckles were in interchange service.[456] A chart of this bewildering variety of parts is shown in Figure 7.74. It has been claimed that the problem was solved in 1916 when the type D coupler was adopted by the Master Car Builders (Figure 7.75).[457] This coupler was designed essentially for freight cars and in time did bring a measure of standardization to that portion of the car fleet. However, the same was not true for passenger cars. The *Car Builders' Cyclopedias* showed a variety of coupler styles made by Symington, Gould, Sharon, and other makers (Figure 7.76).

If complete uniformity in design was beyond the reach of the car masters, they did succeed in strengthening the Janney coupler. Cast steel replaced those parts formerly made of malleable iron. All dimensions were enlarged except for the contour of the knuckle opening. In 1904 the Master Car Builders voted to increase the thickness of the knuckle's hub by 1 inch.[458] Because so much metal was added at this critical point, it was necessary to change the standard profile that had been in effect since 1888. In 1911 the Car Builders decided to restudy coupler design and develop something beyond the piecemeal reforms made during the past twenty years. The heavy steel cars that were then coming into favor required a basic overall improvement in couplers. Finally in July 1916 the M.C.B. voted to adopt the new pattern of coupler, which was designated as the type D.[459] It weighed 400 pounds, or one-third more than its predecessor. Its strength was improved by 100 percent, and its designers claimed that it would extend service life by 300 percent.

The type D incorporated an improvement devised by Eli Janney the year before his death. It was the knuckle pin protector, which strengthened the knuckle, relieved the knuckle pin from side strain, and reduced leverage on the knuckle by moving the

fulcrum point forward. A patent was issued posthumously to Janney in 1912.

The type D was perfect for freight cars, but it was adopted for passenger cars as well.[460] Yet the type D was hardly accepted before planning was initiated for a successor. The new coupler, type E, was a modification of type D. Again the individual parts were slightly enlarged to increase overall strength.[461] Knuckle depth, for example, was increased from 9 to 11 inches. The type E was adopted as the new standard in 1931.

Meanwhile efforts were underway to perfect a distinctive coupler design especially suited for passenger cars. The conventional Janney style of coupler effected a relatively loose connection between the cars. This was satisfactory for freight equipment, but the resulting end play adversely affected the ride. A tighter, more secure coupler was required for passenger cars. This need led to the introduction of the tightlock coupler (Figures 7.77 and 7.78). It had a massive head with a side-wing pocket. A mating, wedge-shaped lug on the opposing coupler fitted into the pocket and held the coupler heads rigid. The actual connection was made, as before, by a Janney-type open-jaw coupler. Charles H. Tomlinson (1869–1938) pioneered this style of coupler.[462] He had some experience in the field through his association with the Ohio Brass Company of Mansfield, Ohio, a producer of electric railway apparatus. Tomlinson designed some of the company's most successful traction car couplers. In 1928 he interested the B & O in testing a tightlock coupler that he hoped to perfect for main-line railroad service.[463] Tomlinson envisioned a fully automatic device that did not simply couple the cars but also connected the steam heating, air brakes, air signal, and electric lines. The industry was not ready for a fully automatic coupler, but it did like the tightlock feature, and over the next few years this

Figure 7.76   A passenger car coupler of 1910. Cast steel had replaced malleable iron by this period.

design was revised for commercial use. Its extra strength and its antitelescoping and antijackknifing characteristics were regarded as adding significantly to the safe operation of fast, lightweight cars. By 1937 the tightlock coupler was an alternate standard.[464] Its popularity grew in the postwar period, and in 1954 it was made an industry-wide standard for passenger cars.

In recent years interest in a fully automatic coupler has revived. Manual handling of the brake hoses, signal hoses, and steam connections is labor-intensive. It is also dangerous. The coupler itself is often a problem when it is not positioned properly, for it tempts foolhardy trainmen to step in toward an approaching car and kick one coupler. Occasionally the pin, when lifted, will not cause the knuckle to swing open fully. Again, a trainman will sometimes step inside the rails and pull it open by hand. These mechanical failings, together with the carelessness of operatives, have led to the death and injury of many men. As recently as 1965, 7 trainmen were killed and 944 injured in coupler-related accidents.[465]

Nearly a quarter century before Tomlinson's efforts to introduce a fully automatic coupler in 1928, Westinghouse attempted to market a combination car, steam, and air coupler.[466] The device was not widely adopted, but this did not discourage others from trying to interest the railroads in similar contrivances. The *Car Builders' Cyclopedias* issued in the 1920s contain illustrations of many such designs. In 1969 the Midland-Ross Corporation reintroduced the scheme and ran road tests on the Union Pacific.[467] More recently the Association of American Railroads began a research effort toward this end in response to a strong expression of interest on the part of railroad managers.[468] However, the prospects for fully automatic couplers have received a setback because of the failure of the control line on the Metroliner cars to stay in contact. Trains occasionally broke in

two while running. The problem was solved by installing conventional jumper lines between the cars. But even if a practical design were perfected, it is not likely that the industry would attempt a rapid conversion. The cost would be prohibitive, at least as far as freight cars are concerned, although the conversion of the now relatively modest passenger car fleet would be more likely.

Draft Gears

Couplers are not rigidly fastened to the cars; the mounting includes a spring or friction mechanism to serve as a shock absorber. This apparatus, attached to the rear of the coupler's shank, is called the draft gear. Its main function is to protect the car structure from coupling impacts. It is expected to absorb, or at least cushion, the forces between colliding cars and to equalize the speed of moving cars when coupled together as a train. Draft gears perform well in impact speeds up to about 5 mph. Conventional gears can handle blows up to 300,000 pounds, but beyond this point (2¾ inches of travel), the springs are totally compressed and the car frame must absorb the collision. If the force is great enough, of course, the frame will give way.[469] The compression of the draft gear is called buff.

Draft gears serve a second major function: the stretching, or slack action, considered necessary for train operations. This was regarded as essential to the starting of heavy trains because the engine was not required to pull all the cars at once. When stopping, the slack would bunch in, so that when the locomotive started up, it was pulling against one car at a time until the slack pulled out. By the time the draft gear on the last cars was elongated, the forward part of the train was rolling at a relatively

Figure 7.77 The tightlock coupler, first tried in 1928, eventually became standard for passenger cars because of its greater strength and antitelescoping characteristics. (New York Central)

fast gait. If too fast a start was made, the couplers on the rear cars were endangered. Inexpert handling resulted in many draft gear failures. It was the major car repair expense; estimates of repair costs arising from draft gear and coupler failures ranged from 38 to 70 percent of total car repairs.[470] The great majority of these problems, however, were freight-car-oriented. Passenger cars were relatively free from the failures because trains were short, and they received more careful handling. Also, they were not run through hump yards, where much of the buff damage took place. One car designer compared passenger equipment to a pampered house pet and freight cars to a stray dog.[471] The analogy is a good one, for passenger cars rarely left the home road, while freight cars wandered about the country on foreign roads, enduring a rough-and-tumble life in interchange service. So long as a freight car could roll, no one seemed to care whether the draft gear functioned or not.

While passenger car draft gears did not undergo the abuse that freight gears received, they were an important adjunct to passenger train operations because they could dramatically affect the ride. A defective draft gear could cause a surging, or rapid

forward jerking, of the car that a nonexpert might well blame on the trucks.[472] In such a case the springs were not necessarily broken; they might simply have grown soft with age, so that the changeover between the full compression of the springs and the friction gear was sharp rather than smooth. As the train's speed fluctuated, which was inevitable because of grade and power variants, the cars developed an annoying jerking motion. The draft gear had to be well maintained to perform optimally.

The desirability of draft gears was recognized during the first decade of American railroading. Charles Davenport, a pioneer car builder of Boston, patented a device on September 9, 1835, to ease shocks and cushion the motion of the train at the drawbars. A full leaf spring was attached to the end of the coupler shank. This style of draft gear is illustrated by the English drawing of an American passenger car of 1845 which is reproduced in Figure 1.78. In 1838 P. Alverson, car builder for the Hartford and New Haven, began testing a spiral spring draft gear.[473] The arrangement was shown in a contemporary engraving, which also included the wooden style of coupling link described in the previous section (Figure 7.79). The inventor said that he pre-

574

Figure 7.78   The type H tightlock coupler was made the industry-wide standard for passenger cars in 1954. (Car Builders' Cyclopedia, 1953)

ferred this form of coupling link and suggested that it be used in combination with the spring draft gear. It is not directly stated that he was the originator of the wooden link, but he may have been. Alverson's gear was said to have performed well after two months' service, and the *American Railroad Journal* expressed the hope that the elastic connection would entirely replace the old fashioned stiff ones. The editorial went on to say: "We hope that after this, we shall not hear of those unpleasant jerks—one of the greatest sources of nuisance in Railroad travelling—and the more reprehensible, because they can so easily be remedied."[474]

Both Davenport and Alverson offered the industry a simple solution to the draft-cushion problem that was largely followed in succeeding generations of railroad cars. Steel springs were about as direct a means of dealing with the matter as could be

found, but other inventors envisioned more complicated methods. As early as 1835 air cylinder draft gears were tried on the Boston and Worcester.[475] The scheme was not successful, however, for one of the air springs failed on the Whitmore car, leaving the remainder of the train behind. A few other roads, notably the B & O, insisted that the solid, inelastic draft gears were best after all. Those of this persuasion championed the continuous drawbar. The scheme was introduced on the B & O in 1835 by Jacob Rupp.[476] A heavy rod or flat bar ran the full length of the car between the couplers, thus entirely relieving the frame of draft and buff. The New Jersey Railroad and Transportation Company tried the scheme around 1840.[477] It put a spring in the center to cushion the draft, but found the scheme heavy and unnecessary. The Lehigh Valley tried a massive continuous drawbar measuring 3 by 7½ inches, but even this huge plate

Figure 7.79 Cushioning devices, called draft gears, fasten couplers to the car frames. The 1838 scheme shown here used coil springs. Note also the wooden link in detail 6. (American Railroad Journal, January 1, 1839)

buckled in service, throwing the couplers so far out of line that they could not be joined. The dead weight was another disadvantage. At last even the B & O conceded that the scheme' was not such a good idea, and by 1875 the road was using continuous drawbars only on freight cars.[478]

As already mentioned, rubber springs were introduced for truck suspension during the mid-1840s. They were also applied to draft gears in place of conventional steel springs (Figure 7.80). They undoubtedly exhibited the same advantages and deficiencies that they showed in truck service. As Figure 7.61 suggests, they appear to have remained popular for draft gears until the 1860s. In 1929 the idea was revived by the Waugh Equipment Company for passenger cars and became so popular that after 1940, rubber springs were applied to freight cars as well.[479]

During the last quarter of the nineteenth century, coil steel springs were the basic form of draft gears. Often a double spring was used. Some were made with a smaller spring inside the main one, but in other instances the springs were placed in tandem on a common shaft.[480] Miller used volute springs on the buffer and coupler hook, while Janney used large coil springs in his

more elaborate buffing system (Figures 7.67 and 7.71). The matter of drawsprings was so important that the M.C.B. established a standard capacity for them as early as 1879. In that year a drawspring had to have a capacity of 13,000 pounds. As car weights increased, the limit was increased: to 18,000 pounds in 1884 and to 22,000 pounds in 1893. Such increases did much to save cars from destructive buffing shocks, but as springs grew in strength, so did the problem of recoil. The kickbacks produced jolts almost as great as the shocks that produced the kickbacks. How to dampen the recoil was a major problem to be overcome.

Westinghouse was attracted to the problem during the course of the Burlington, Iowa, brake trials.[481] He worked out a plan with a triple wedge that compressed an interlocking set of friction plates (Figure 7.81). The action of the wedge and plates dissipated the spring's recoil. The friction draft gear also increased the overall capacity of the device, well beyond what the ordinary coil spring gears could offer. A patent (No. 391,997) was issued on October 30, 1888, just about the time that the device was placed on the market. The first models were rated at 60,000 pounds, although some individual units were said to ab-

Figure 7.80   Rubber springs were used for draft gear cushions during the 1840s. This engraving is from an advertisement appearing in the American Railroad Journal, September 16, 1848.

Figure 7.81   A Westinghouse friction draft gear patented in 1888 offered an alternative to older designs which depended entirely on springs. O-N is the wedge and P-B the friction plates. (Engineering News, March 1, 1890)

sorb as much as 100,000 pounds.[482] The Westinghouse apparatus made a good showing during demonstration tests. Fourteen couplings were made with two cars fitted with friction gear at Altoona early in 1889.[483] The speed was gradually increased after each coupling. By the twelfth coupling, with the speed at 28 mph, some of the blocking gave way behind one of the buffers, and one truck broke loose and stripped away a body bolster. When the speed went up to 30 mph, even heavier damage was sustained by the cars, but the Westinghouse gear was retrieved from the wreck undamaged. Despite such good results, the industry was slow to adopt friction gears. By 1899 Westinghouse had boosted the capacity of his apparatus to 170,000 pounds, but only 3,000 freight cars and few locomotive tenders were using them.[484] The extra cost was one reason for the rail-

roads' resistance. There were also complaints about the gear's stiffness; under ordinary draft and buff movements, it tended to act more like a solid block of iron than a spring.[485] The introduction of steel cars made friction gears a necessity, and despite their failings they became the basic forms of draft, at least for freight cars, around 1920. By 1943 85 percent of the U.S. freight car fleet had friction gears.[486] They were extensively used for passenger cars as well.

In 1927 a radically new form of draft gear was introduced by Otho Duryea.[487] Conventional draft gears, he observed, moved a scant 2 to 3 inches and then went solid. Duryea envisioned not only a longer travel of 7 inches but also an independent draft gear that would, in effect, float within the car's center sill.[488] A large bank of coil springs absorbed the impacts. Duryea's cushion underframe ran the full length of the car inside the center sill and had couplers attached at both ends. It was something of a super, continuous drawbar, with a capacity of 64,000 pounds. Duryea intended the device primarily for freight cars, but it was used for passenger cars as well. Apparently the first passenger use was on the B & O's lightweight trains of 1935, the Abraham Lincoln and the Royal Blue.[489]

As already discussed, draft gears posed a relatively minor problem for passenger cars. In the area of freight cars, however, they amount to an obsession among car designers. The problem has been partially solved in recent years by various forms of hydraulic cushioned gears. The first appeared commercially in 1955, and it is assumed by many that they are a modern improvement. But hydraulic draft gears were used in 1893 on some passenger cars built for the New York Central Railroad.[490] Perhaps if railroad managers were better students of their own history, the long-standing problem of damage claims might have been resolved generations ago.

CHAPTER EIGHT

Self-Propelled Cars and Motor Trains

THE HISTORY OF THE MOTOR CAR revolves about the basic question of cheap transportation. It originated in the search for small, self-propelled units to service marginal runs that did not require a full passenger train. Traffic on many branch and suburban runs would barely fill a single coach, yet it was necessary to operate a coach, a baggage car, and a locomotive. Even such a minimal train required a full crew, and costs easily outran revenues. Conversely, on heavily traveled suburban lines conventional steam trains could not efficiently move the traffic. Because of the poor acceleration inherent in the steam locomotive, the frequent stops slowed the flow of commuters to and from the great cities. Multiple-unit, self-propelled electric cars were better adapted to handle high-volume traffic economically. On both the lightly traveled branch line and the crowded commuter run, the story of the rail car is dominated by economics rather than mechanics. Its major theme is the hard facts of operating costs.

Self-propelled, or motor, cars are at both ends of the passenger traffic spectrum. One tends to think of them only in terms of the branch-line shuttle, but the electric multiple-unit cars of the New Haven and Illinois Central are also self-propelled cars. And they are engaged in moving vast crowds along the heavily patronized commuter routes leading out of New York and Chicago.

The great majority of self-propelled cars are electrics, although there have been some quixotic experiments with steam and internal combustion over the years. In general the motive power methods that have been tried for self-propelled cars form a rough chronology, with steam first, then gasoline or gas-electric, straight electric, diesel and diesel electric, and lately, gas turbine. The chapter will follow this general outline, keeping the focus on standard, main-line developments. Street railways, rapid-transit cars, inspection vehicles, and experimental or home-made branch-line cars will receive only brief mention.

Self-propelled cars were tested on American railroads before the Civil War in an effort to reduce operating costs for lightly patronized or marginal trains. The attempt to introduce such equipment was persistent but largely unsuccessful during the nineteenth century. The railroads faced no serious competition in this period, and the abundance of obsolete or surplus equipment, which had little book value, was suitable for secondary runs. The expenditure of large sums for expensive motor cars did not seem necessary. After the new century opened, however, the situation began to change rapidly. Street railways and interurbans had already cut into suburban and branch-line traffic. The automobile promised to draw away even larger numbers of short-haul passengers.

Railroad men began to worry. They became more receptive to radical cost-cutting methods. If the compact internal combustion engine was so efficient for the highway, why not use it for light rail vehicles as well? Several hundred gasoline rail cars were purchased by American railroads between 1905 and 1915. The cars worked fairly well, but they were sensitive and proved cantankerous unless given special treatment. The mechanism was alien to the average railroad shopman. The operating crews distrusted the motor car's springy, uncertain, "I am about to go off the track" ride. The raucous, evil-smelling engine added little to the charm of the gasoline car. The prejudice against such vehicles was evident in the scornful names given them by railroaders and the traveling public: "doodle bug," "galloping goose," and "wooden shoe" do not describe conveyances of much style or comfort. There was a general retreat from the rail car; after a brave beginning it was left to wither and die.

However, the economic crisis following the end of World War I persuaded railroaders to reexamine the question. The war had demonstrated the reliability of the internal combustion motor. Spark plugs and carburetors had become everyday words and devices. The transport of the future, the airplane and automobile, was gasoline-powered. If the railroads were to hold any portion of the short-haul market, they must do so with economical, small-capacity units. This ruled out conventional steam trains with their large operating crews. One- or two-man motor cars were seen as the solution. Between 1921 and 1930 the railroads purchased approximately 800 such units, but the Depression and World War II curtailed new orders. During the 1950s the diesel rail car became popular, although it was cut short of its full potential by the general decline in rail travel. After 1960 nearly all motor cars were in suburban commuter service.

The general trend of main-line electric cars is somewhat different. Mechanically, they were a direct outgrowth of the electric street railway. The rolling stock was far heavier—on a scale with existing main-line trains. Main-line electric, self-propelled cars were also normally outgrowths of terminal electrifications, such as the New York Central's Grand Central project. A spin-off of this scheme was the installation of multiple-unit cars in 1906 on the Harlem division to service heavy commuter runs in and out of Manhattan. Unlike the free-ranging motor car, the electrics were tied to short lengths of suburban trackage and rarely ran more than 50 miles from the home terminal. The self-propelled electric car reached its apogee in 1969 with inauguration of the high-speed Metroliners between New York and Washington. The multiple-unit vehicle had become part of first-class, long-distance train travel.

Steam Cars

Beginning around 1850, there was an intermittent effort spanning eighty years to introduce steam-powered self-propelled passenger cars on American railroads. It seemed to run in ten-year cycles, with occasional short periods of intense activity. After a series of failures, the idea would fade away. But after a short hiatus it would be revived, presented as a fresh solution to an old problem, and rejustified by claims for lower fuel and labor costs. The memory of past failures was forgotten or ignored; the steam car taught no lesson except the moral that it is unwise to disregard the lessons of history. Over a hundred steam cars were built for service in this country, and of the total only a few could be classified as more than a marginal success.

Steam cars were first tried in England on the Eastern Counties

Figure 8.1   Moore and Parrott's steam car, patented in 1851, was operated on the Boston and Worcester. Similar cars were later used on the Pittsburgh, Fort Wayne, and Chicago Railroad. (U.S. Patent Office)

Figure 8.2   The Baldwin Locomotive Works built the steam car Novelty in 1860 for the Pittsburgh, Fort Wayne, and Chicago. The driving wheel was a solid, plate-style wheel. (Traced by J. H. White, Jr.)

Figure 8.3 William Romans constructed the Economy *in the Columbus, Piqua, and Indianapolis repair shops in 1861. (Railway and Locomotive Historical Society)*

Railway. Late in 1848 W. Bridges Adams built the *Fairfield*, a six-wheel, 31-foot-long combined locomotive and passenger car, at the suggestion of James Samuel, an official on the same line.[1] A year earlier the two men had produced a very small, open steam car, but apparently it was an inspection vehicle not intended for revenue service. In 1849 Adams fabricated an eight-wheel car, the *Enfield*. These units performed acceptably but failed to spark much enthusiasm elsewhere in England, for it was almost a quarter century before another steam car was produced there. The pioneering experiments of Adams and Samuel, however, may have encouraged M. W. Baldwin to build similar machines in the United States. Positive information is not available on the point, but in 1846 and 1847 Baldwin did produce some extremely small power units for the Vermont Central and the Georgia railroads.[2] They were so small, even for this early time, that they could not properly be called locomotives. They were about the right size to power a self-propelled car, although no further inference can be drawn.

The first American steam car for which there is positive information was produced in 1850 or 1851 by Joseph H. Moore and William P. Parrott of Boston. On December 2, 1851, the partners received a patent (No. 8561) in which they stated that they had already built a 12-ton test car (Figure 8.1). The eight-wheel model had a vertical boiler, with cylinders mounted on the outside frame of the front truck. A flexible steam pipe connected the boiler and engine. Its weight was less than one-half that of an equivalent locomotive, tender, and coach. A light locomotive and tender were priced at $6,000 and a coach brought $2,000, while the steam car cost only $4,000 and was equal to the full-sized train. The inventors reported speeds of 40 mph with the test car but admitted that there were some mechanical problems. These difficulties were overcome, for a later report states that Moore and Parrott cars operated successfully on several branches of the Boston and Worcester.[3] They were economical and well suited to the light traffic patterns of the tributary lines.

A few years after the Boston cars appeared, J. H. Moore moved west to become superintendent of the Pittsburgh, Fort Wayne, and Chicago. His interest in steam cars remained strong, and in 1860 he took over a car plant in Massillon, Ohio, to produce an improved model of his 1851 patented design.[4] The new, all-iron car was under construction in June 1860. It would seat 94 and carry 3 tons of baggage in its long, 77-foot body.[5] A

separate apartment for mail was also provided. The cylinders measured 8 by 14 inches, the wheels 42 inches. Its weight was light: computed at 16 tons, with fuel and water adding another 3 tons.[6] It cost $6,000.[7] The designer claimed that a cord of wood would propel the vehicle 125 miles, and that the total operating cost would amount to one-third that of a regular train. In the fall of 1860 the car carried the president of the railroad from Pittsburgh through to Chicago, demonstrating its ability to make long runs when necessary. Two sister cars were produced to assist in servicing the New Brighton accommodation run. Moore's success was short-lived, however, for in 1862 the U.S. War Department requisitioned the cars for use in the South. Photographs exist showing them in service in the Atlanta and Nashville area during the war years.[8]

At the same time that Moore was building his huge iron locomotive-car, M. W. Baldwin was preparing a more modest steamer of his own design (Figure 8.2). The six-wheel wooden car measured just under 32 feet in length and weighed only 7½ tons.[9] The front compartment, housing the vertical boiler, occupied only 30 inches. The 5- by 12-inch cylinders propelled a solid-plate, 42-inch driver. Longitudinal seats accommodated 38 passengers, and long tanks under the seats stored 100 gallons of water. Exhaust steam heated the passenger compartment. In June 1860 the car, named the *Novelty*, made a test run over the Norristown line at 30 mph. A month later it was in service on the Pennsylvania out of Pittsburgh, exhibiting good fuel economy. A bushel of coal was sufficient fuel for 50 miles. The car was delivered to its consignee, the Pittsburgh, Fort Wayne, and Chicago, which intended it for suburban operation.[10] However, no subsequent accounts of its service life can be found. It must have been a failure, since Baldwin built no more steam cars for another fifteen years.

Late in 1861 the Erie's master car builder, C. A. Smith, produced a 60-passenger steam car of his own design.[11] It had a wooden frame, iron gas pipe posts, and a sheet-metal body. The composite body was powered by two 8- by 12-inch cylinders and an upright boiler. The firebox exhausted out of twin smokestacks. It could pull one car with ease, but with two, it had some trouble starting. Although it ran for a time between Paterson and Jersey City, it was not found suitable for suburban service and was sent west to work between Binghamton and Hornellsville, New York. But its jerky, uneasy ride annoyed passengers, so that the ma-

GRICE & LONG,
114 Walnut Street, Philadelphia, and 94 Wall Street, New-York,

Call the attention of all interested in cheap Railway transportation to their STEAM PASSENGER CAR. A speed of 25 miles per hour, and a traction equivalent to a train of 80 passengers, is attained. The consumption of fuel and running expenses are very light; carries fuel and water for 25 miles, and weighs less than half of the Locomotive. For City Railways the heaviest grades may be overcome, and, by the use of their patent VIBRATING TRUCK, the shortest curves can be turned.

GRICE & LONG'S STEAM PASSENGER CAR.

Reference is made to Messrs. J. W. VAN RENSSELAER, Sup't West Jersey R. R. Co., W. H. GATZMER, Camden and Amboy R. R. Co., JAMES MORNELL, Philadelphia and Trenton R. R. Co.

Figure 8.4   Grice and Long of Trenton, New Jersey, built a number of steam cars for both street and steam railways in the 1860s. Larger units, such as the one shown here, were intended for branch and main-line service. (Scientific American, *October 26, 1861*)

Figure 8.5   The Vulcan Iron Works of San Francisco completed the Napa *in 1865 for the Napa Valley Railroad. It is shown as it was relettered in later years for service on the San Joaquin Valley Coal Mining Company.* (Locomotive Engineering, *October 1891*)

chinery was eventually removed and the car was retained as a smoker. Years later it was dropped beside the tracks near Attica, New York, to serve as a section-gang shelter. It was reported at this site in 1898.

William Romans was another railway mechanic smitten by steam car fever in the 1860s. At the time he was master mechanic for the Columbus, Piqua, and Indianapolis Railroad, with headquarters in Columbus, Ohio. The road was obliged to operate a lightly patronized train to Piqua, 73 miles from Columbus. Romans saw the steam car as the best way to cut costs, and he designed and built one in the railroad's own shop during the

summer of 1861.[12] Appropriately named *Economy*, the 16-ton, 70-foot-long car seated 48 passengers (Figure 8.3). A 6-foot-deep rear platform provided extra baggage space. The rear truck was made so that it could be pulled forward to reduce the overall wheelbase for short turntables. Romans built a miniature self-contained 2-2-0 locomotive for the power unit. It had 7- by 18-inch cylinders, 54-inch driving wheels, and 28-inch leading wheels. The *Economy* was designed to operate at 25 mph, the normal running speed of the time.

Sometime after the *Economy* entered service, Romans produced a second car, the *Express*. Anticipating that others would

Figure 8.6   Joseph P. Woodbury's steam truck of about 1865 was at-tached to the car body by a ball-bearing ring plate. The smokestack is not in place in this construction photograph. (Railway and Locomotive Historical Society)

want duplicates of the Columbus-built cars, he patented his de-sign on January 28, 1862 (No. 34,270). Romans must have been disappointed by the reaction of other railroads, for it was not until 1865 that he received an order. The Minnesota Valley Rail-road commissioned him to build a car for a 10-mile suburban run between St. Paul and Shakopee.[13] The *Shakopee* made four trips a day and became a great favorite with engine crews because of its warm, roomy cab; it was an ideal working station during the bitter Minnesota winters. When the railroad's western terminal moved to a more important junction, however, the steam car could not handle the traffic, even with a trailer car. The *Shako-pee* was stored for a time and used only for an occasional inspec-tion or paymaster's run. Then it was leased to the Northern Pa-cific for similar duties. Late in the 1870s it was dismantled, and the body was salvaged for a peanut stand.

Other hopeful inventors tried to promote the steam car. Prom-inent among these were Robert H. Long and Joseph Grice, car builders of Trenton, New Jersey.[14] Long obtained a patent in January 1860 for a six-wheel steam car that featured an offset boiler and gear drive. The design was intended for street railway service; however, difficulties in selling the idea to the transit industry led Grice and Long to turn to main-line steam roads for a market. Approximately ten railroads bought one or more cars. One of the first was the Camden and Amboy. The *Scientific American* of October 26, 1861, described the car as 37 feet, 6 inches long, with seating for 36 passengers (Figure 8.4). A small baggage room separated the engine and passenger compart-ments. The 11-ton steamer consumed 7 pounds of coal per mile. Similar cars were sold to the West Jersey, the Jacksonville and Lake City, the Huntington and Broadtop, and other lines, includ-ing one in Mexico. Around 1863 Long and Grice introduced a larger eight-wheel car. The Portland and Kennebec Railroad bought one of these 12½-tonners for $8,500. That price may well suggest why steam cars were not more widely used.

The far West also contracted the steam car fever of the 1860s. A number were produced for the street railways of San Fran-cisco, but only two seem to have been built for a steam line. In November 1865 the Vulcan Iron Works of San Francisco com-pleted the 21-ton car *Napa* for the Napa Valley Railroad.[15] It was powered by a hefty 2-2-0 locomotive unit with 9- by 18-inch cylinders and 48-inch drivers (Figure 8.5). The locomotive car was capable of pulling as many as eighteen freight cars, and it could move along smartly with eight.[16] The passenger compart-ment seated 26. The water tank held 750 gallons, the coal bunker 800 pounds. In later years the *Napa* had several owners. It finally came into the possession of the Southern Pacific, which stored it outside the Sacramento repair shops. Around 1925 the old vet-eran was junked. The *Napa's* sister car, the *Calistoga*, had a much shorter career. It was completed in 1867 by Vulcan but in 1875 was rebuilt into a conventional locomotive by the Napa Valley Railroad.

Although the excitement created by the steam car during the 1860s led to no widespread adoption, this failure was certainly not due to lack of enthusiasm by the steam car's promoters. This group was filled with zealots, of whom the most active was Jo-seph P. Woodbury of Boston. His belief in the steam car was absolute; he felt that it was essential to the success of both street and steam railroads. He contended that his cars could do the same work as conventional equipment for half the cost, and at a saving of one-third to one-fifth in dead weight.[17] Woodbury's devotion to the perfection of a superior self-propelled rail car is shown in the five patents that he took out between January 1865 and February 1871. Several ideas described in these papers were gear and direct-drive power units, center door entrances, and roller-bearing mounting of the engine-boiler to the car body. The latter plan was perhaps Woodbury's most original contribution. It permitted the driving machinery and a vertical boiler to be fastened to a four-wheel truck and yet stay free to swivel beneath the car's floor (Figure 8.6). A double iron ring turned easily on iron ball bearings.

In the fall of 1865 Woodbury exhibited a demonstration model built to his designs by the Salem Car Company.[18] The 30-foot car, which could seat 40 passengers, was tested on the Eastern Railroad and other New England lines. The steam, or at least a portion of it, was exhausted into the water tanks both to muffle the car's running noise and to preheat the feedwater.[19] Next Woodbury produced a large, 75-passenger car, its machinery presumably made by the Manchester Locomotive Works. In 1866 Woodbury contracted with the New England Steam Car Com-pany to build cars of his patented design. It cannot be deter-mined which, if any, railroads bought them. But occasional reports, one appearing as late as 1876, indicate that Woodbury was still actively pursuing the idea.[20] A photograph taken around 1890, now in the private collection of Thomas Norrell of Silver Spring, Maryland, shows an aged Woodbury car in switch-ing service at the Bellows Falls Brewing Company, Walpole, New Hampshire.

If the 1860s can be described as the most active period of the steam car in America, the next twenty years were the most inac-tive. The 1880s in particular were a decade of almost complete stagnation. In the 1870s, however, there were a few feeble at-tempts to push steam cars. The most interesting was conducted by William Baxter of Ilion, New York. Baxter was active in many areas of mechanical engineering and was associated with the

Figure 8.7   *The New England Railroad's steam car of 1897 was powered by a steam truck fabricated by the Schenectady Locomotive Works. Note the large water tanks under the car body.* (Railway Age, *October 15, 1897*)

Figure 8.8   *The Erie produced a steam car in 1897 by combining a Schenectady steam unit with an old emigrant car body.* (Alco Historic Photo, *National Railway Historical Society*)

Remington Arms Company, which produced his compact portable steam engine. He introduced a new concept in steam car practice, for it used a compound engine—one of the first such applications in American railway service. Remington built the machinery for the test car, which was ready by the fall of 1872.[21] The equipment was mounted on a 30-passenger horsecar with a vertical boiler on the front platform and the engine directly connected to the wheels without benefit of gearing. The 5- and 8- by 12-inch cylinders were arranged so that the engine could operate as either a simple or a compound unit. When extra power was required, a transfer valve was opened to direct high-pressure steam directly into both the high- and low-pressure cylinder. The car eventually entered service on the Worcester and Shrewsbury Railroad, a short suburban line that connected Worcester, Massachusetts, with a lake resort.[22] It remained on the line until almost 1890. But for all the merits of the car, Baxter appears to have had no success in selling his design to the railroad industry.

In the far West, the Central Pacific's Sacramento shops turned out a new steam car for service on the subsidiary Northern Railway between Woodland and Williams, California.[23] The car did not include Baxter's compound-engine idea or any other notable mechanical features, but it was described as a very neat and compact 30-passenger car. It had cylinders that were 8 by 14 inches and 42-inch driving wheels, and it measured 61 feet in length. This car, like so many of its predecessors, appears to have been an experiment that did not spawn any offspring.

By the late 1890s the past failures of the self-propelled cars had been forgotten, and the idea was basically so appealing that it revived among a new generation of railway managers. The high cost of branch-line train operations and the savings possible with self-contained cars again seemed irresistibly attractive. It was a reenactment of the 1860s—a flurry of activity that in the end produced few lasting results. This late-Victorian revival can be credited to C. Peter Clark, general manager of the New England Railroad and, not so incidentally, the son of the road's president.[24] Clark was disturbed about the diversion of traffic by a new trolley line that paralleled one of his branch lines. In a single year 350,000 passengers abandoned the New England for the trolley.[25] Hardly enough patrons remained to justify continuance of service. Clark made a general study of the problem and found, to his consternation, that local trains running in the Boston area averaged only 47 passengers per trip, hardly enough to justify a locomotive, two cars, and a full crew. A combined coach and locomotive could do the same job for a fraction of the cost.

Clark, who appears to have considered the idea original, named his new concept in transport the Composite car. In the spring of 1897 he sought the advice of several locomotive builders, among whom A. J. Pitkin, general manager of the Schenectady Locomotive Works, was the most responsive. Pitkin took a personal interest in the novel project and devised and patented a ball-bearing, circular, iron-frame boiler mounting, apparently not realizing that the same scheme had been invented thirty years before by Woodbury.

In October 1897 the Composite car was ready for service. To hold down costs and speed delivery, the machinery was installed in an old dining car. The boiler and engine truck were placed in the kitchen end of the car; the original six-wheel truck was removed (Figure 8.7). The compartment was insulated and fireproofed with asbestos board and galvanized sheet iron. In the passenger compartment the tables were replaced with side seats, boosting the capacity to 60 passengers. The boiler, cylinders, and driving wheels were built as a unit. The power truck was attached to the car by a circular iron frame which could turn on a nest of one hundred and twenty-five 1½-inch steel balls.[26] The 52-inch-diameter boiler contained three hundred and eighteen 1¼-inch-diameter fire tubes and was built for a working pressure of 200 psi, which was rather high for the time. The cylinders measured 12 by 16 inches; the drivers were 42 inches in diameter. The 57-ton Composite was a full sized, powerful unit that could be operated by a two-man crew. Oil fuel had been considered, but in the end the firebox was designed for coke or hard coal.

The Composite began service on a 20-mile branch, but the grade proved too much for its limited boiler.[27] It was next tried on Cape Cod, whose dead-level profile did not tax the Composite's steaming capability. Here it pulled six freight cars without difficulty. It often hauled a coach, and on one occasion moved eleven freight cars. It registered speeds up to 60 mph, but this was recognized as too fast for its small drivers, and it was limited to 45 in regular service. The fire could go for 12 miles without attention, and the 1,500-gallon water tanks were good for 60 miles of running. Yet for all these glowing reports, the Composite was not duplicated for the line. The negative decision may have been made by the New Haven, which took over the New England Railroad during this period. The car was not the success that Clark envisioned, but it did remain in service longer than the typical main-line steamer. It worked on the Dedham branch of the New Haven until October 1904.[28] After two years in storage, it was sold to a small line in northern New York.

The Erie also faced a decline in traffic on certain branches because of the encroachment of electric street railways, and like the New England Railroad, it sought ways of cutting costs and increasing the frequency of service. In the spring of 1897 it tried a small steam storage car produced by the Kinetic Power Company.[29] The car seated only 20 passengers and was undoubtedly too small for the job. And then news came of Clark's full-sized steamer. The Erie quickly asked the Schenectady Locomotive Works to build a duplicate power truck (Figure 8.8).[30] It was delivered in October 1897, just weeks after the New England

Figure 8.9 The steam power truck furnished by the Schenectady Locomotive Works for the cars shown in Figures 8.7 and 8.8 (Engineering News, February 6, 1902)

unit was ready. An old emigrant coach provided the body (Figure 8.9). The assembly cost the Erie $10,000. It was considered a bargain, however, because operating costs were pared from 27 cents a mile for a locomotive, combine car, and crew, to 16 cents a mile for the Composite and its crew. The road calculated that the steamer would pay for itself in just over three years. The Erie talked of buying another five or six, but in the end it acquired only one more.[31] Purchased in 1898, the car carried the road number 680 and was lettered for the Erie subsidiary, the New Jersey and New York. The anthracite coal burner was fitted with a power unit of the same size and design that Schenectady had produced for the New England Composite and for the Erie 1897 emigrant conversion unit. But the Erie's enthusiasm was short-lived, for even though the car ran fourteen trips a day over the 4-mile branch, breakdowns were frequent and the ride was rough, more like a locomotive's than a coach's.[32] In addition, the boiler was too small to produce the necessary quantity of steam for the schedule. The car was retired and dismantled sometime before 1902.

During the same period the Baldwin Locomotive Works decided to enter the steam car competition. Its main rival was Schenectady, and it could not allow its opponent to enter a new field without challenge. Also, Baldwin's chief, Samuel Vauclain, was confident that he could build a superior steam car. At the time, he was pushing his own compound locomotive patent, and a successful steam car might be a novel way to advertise the system. Compounding was then regarded as a cure-all for the steam locomotive's poor fuel economy. It might prove a double benefit for steam cars, since a smaller boiler would be possible because the steam was used twice. Vauclain also felt that a con-

Figure 8.10   Baldwin's self-stoking vertical boiler was designed in 1897 to eliminate the need for a fireman on steam cars of its manufacture. (Engineering News, February 6, 1902)

Before the *Woodsdale*'s disgrace, Baldwin managed to sign on two more Midwestern customers for larger versions of Vauclain compound steam cars. The Detroit and Lima Northern's Number 100 was more on the scale of a full-sized coach, measuring 58 feet in length.[36] Mechanically it was similar to the *Woodsdale*, except for the lack of a condenser (Figure 8.11). Another variation was the enclosure of the valve gear to protect it from dust. But apparently the Number 100 never even entered regular service on the D & L N; it was sold to the Fairfield Traction Company of Lima, Ohio, which subsequently remodeled it into a combine car.[37] The steam unit was used as a small switcher.

The third Vauclain car was produced for the Pittsburgh, Cincinnati, Chicago, and St. Louis Railway in 1899.[38] It incorporated the compound-engine, condenser, and magazine-boiler features in a 58-ton car. The 30-foot passenger compartment seated 40. The frame was a composite of steel channels and timber. The magazine held 2,500 pounds of coal. According to the Baldwin order books, the car was intended for service on the Cincinnati, Lebanon, and Northern branch, but its only record of service was on the Xenia to Springfield, Ohio, branch. There it ran for a month, only to be withdrawn and sold to the Clearfield Coal Company of Pennsylvania. Poor steaming and problems with the coal magazine were responsible for its retirement.[39]

Thus Baldwin and Schenectady both failed in their effort to introduce a successful steam car. Neither was ready to admit defeat, and both would try again within a few years, but for the time being the self-propelled car was again discredited. Many reasons were given for its failure. The vertical boiler was a major problem because of its limited steaming capacity, at least in sizes practical to enclose within the confines of a car body. Also, deposits settled on the lower tube sheet of such boilers, causing them to overheat and thus loosening the tubes.[40] To maintain steam, small boilers were worked very hard, resulting in rapid deterioration. Too much time was required to water, coal, and fire up a cold boiler. And there was the inherent defect of the small steam plant, where efficiency seems to be directly related to size. The most productive steam units are ocean vessels and power plants, which do not have width, height, or weight limitations. Steam cars are relatively small units that proved high in maintenance expense and low in thermal efficiency.

The huge water tanks under the car were a weight problem in themselves, but the heavier floor frames needed to support them added substantially to the car's dead weight. To propel this extra tonnage required a larger engine and boiler—again more unwanted weight. Condensers partially solved the problem, but in themselves they were heavy enough to make the cars sway because of the rooftop mounting. Axle loadings were heavy, requiring outsize bearings and extra care in maintenance to avoid hotboxes. The cars rode poorly, being part locomotive and part car. It was in general an unhappy combination, with all the poorest qualities of both components. The "Little Wonders," "Economies," and "Lilliputians" failed to outperform conventional steam trains even in operating costs, for when depreciation and maintenance were figured in, their record was unimpressive. It

denser would materially reduce the size of the water tanks, thus cutting dead weight. It also offered the side benefit of muffling the exhaust. Of course, these auxiliaries amounted to something of a trade-off because of the extra weight, cost, and complexity of the subsidiary systems. A peculiar self-firing, vertical boiler was another part of the Baldwin plan (Figure 8.10). Coal was housed in a central magazine, which was loaded from a rooftop hopper. The fuel fed itself into the firebox by gravity and by the car's motion, thus freeing the engineer to concentrate on his operating duties. At least, that was Baldwin's intention.

In June 1898 the first Baldwin car went into service on the Cincinnati, Hamilton, and Dayton's Middletown, Ohio, branch.[33] Named the *Woodsdale*, the 24-ton steamer looked like an oversized streetcar (Figure 8.11). It seated 28 passengers and had a 6-foot-long baggage room. The 48-inch-diameter boiler supplied steam at 180 psi to cylinders measuring 5½ and 9 by 12 inches.[34] During tests, the car ran for 38 miles with no attention to the fire and at speeds of 42.5 mph. The C H & D paid $5,000 for the *Woodsdale* and put the car in service between Hamilton and Middletown, Ohio, on a schedule of six trips a day, with stops spaced out at 2 miles. After a few months, however, the *Woodsdale* was withdrawn from service and placed in the storage line outside of the shops. The road said that many problems had forced this early retirement.[35] The boiler was too small. The water capacity was inadequate. The coal magazine was difficult to load. The fire needed frequent clearing. The water capacity of 300 gallons was insufficient, even with the condenser. In addition to these difficulties, the car rode poorly.

Figure 8.13 The American Locomotive Company completed this oil-burning steam car in 1908. (Alco Historic Photo, National Railway Historical Society)

Figure 8.14 The grotesque power unit of the Rock Island's steam car Number 2552 was hidden inside the front compartment.

Figure 8.15   This formidable steam car was produced by the St. Louis Car Company in 1906 to the designs of W. G. Wagenhals. (Scientific American Supplement, *December 22, 1906*)

Surely the most spectacular steam passenger car of this period was the Wagenhals-Kobusch car built by the St. Louis Car Company in 1906 for the Missouri Pacific.[55] It was a Leviathan among steam cars, measuring over 82 feet long and weighing 59 tons. Its designer, W. G. Wagenhals, apparently believed that an increase in size would solve the steam car's problems. Wagenhals had been general manager of a street railway in Cincinnati before taking the same job with the Ohio River and Columbus Railroad. A man who liked mechanics, he had built a gasoline automobile in 1891 and a small inspection car for the O R & C. With this background he recognized the potential of the self-propelled rail car. In 1905 he ordered a steam car from the St. Louis Car Company. That firm's president, George J. Kobusch, was impressed with the idea and encouraged Wagenhals to develop the ultimate steam car.[56] The result of this partnership was the mighty Number 1000 (Figure 8.15). The wooden body was supported by a steel frame with truss rods for extra stiffness. The forward portion of the body was metal-sheathed to help fireproof the engine room. An oil-fired, fast-steaming, torpedo-boat style of boiler furnished steam at 250 psi. The cylinders were mounted inside the wheels on the front truck, with a gear drive to one pair of drivers. The second set of wheels was driven by connecting rods. The 1000 was scheduled for a cross-country tour before delivery to the Missouri Pacific. The demonstrator won only two orders for the St. Louis Car Company: two smaller cars, presumably duplicates, were produced in 1908 for the St. Joseph and Grand Island and the Crystal City and Uvalde (Texas) railroads.[57] The 40-footers seated 41 and weighed 33 tons. The machinery was similar to the 1000's, except that the side rod drive was eliminated. The service history of these cars has not

been uncovered. They could not have been notably successful, because no more were produced.

After 1910 the gasoline engine was the unrivaled motive power unit for small vehicles. By this time it had been developed into a compact, dependable power source, as demonstrated by the growing numbers of automobiles and trucks in everyday service. The lesson was obvious, and the railroad industry began to accept gasoline rail cars. Yet an undercurrent of resistance was evident among railroad men. They knew and trusted steam and saw little to admire in "doodle bug" gasoline cars. They still hoped that the past failures of the steam car might be reversed by a more refined and scientific design. Much had been accomplished in steam automotive design; perhaps these techniques could be applied to the self-powered rail car. The Stanley brothers of Newton, Massachusetts, were leaders in the steam auto field. They were the only domestic manufacturers to mass-produce steamers over an extended period. They had a genius for light steam power design, and they devised reliable, practical vehicles. The Stanleys were confident that they could redeem the failure-ridden steam car. Late in his life, Francis E. Stanley took personal charge of the project.[58] He was about to retire from the automotive business, and he apparently viewed the rail car as a challenging retirement project.

Francis Stanley took established designs of the Stanley Company and adapted them directly to the rail car. They included a high-pressure (up to 1,000 psi) fire-tube boiler wound with piano wire for extra strength and light weight. The engine, a faithful copy of the standard auto design, was a simple noncompound, two-cylinder unit. Kerosene was used to fire the boiler. A test car, the Number 100, was produced around 1915 for service

Figure 8.16 *Several Unit steam cars were built between 1915 and 1922 after designs prepared by the famous steam automobile manufacturers, the Stanley Brothers. The car here was made for the Canadian National in 1922.* (Railway Review, September 30, 1922)

on the Boston and Maine. In 1916 a second car, the Number 101, was produced for the White River Railroad of Vermont.[59] The 50-foot, 44-passenger car body was built by the Laconia Car Company. The Stanley Company provided the machinery. The squat vertical boiler contained 600 tubes ¾-inch in diameter. The fire was automatically regulated by the rise and fall in boiler pressure. The 15-ton car could operate at speeds of 45 to 60 mph—a rate more than adequate for any branch line. A roof condenser was installed to avoid the need for large water tanks or frequent water stops. The manufacturers claimed that 1 gallon of fuel (kerosene or heating oil) would propel the car 5 miles, but in actual service fuel economy was only about one-quarter of this figure.[60] In 1922 the car was reported to be very satisfactory and dependable. It often pulled a trailer, and on occasion it was harnessed to a freight car. Since the 101 could develop up to 1,000 horsepower, at least for short intervals, it was capable of maintaining a satisfactory schedule on hilly lines.

The project was just getting established when Francis Stanley was killed in an auto accident on his way to a meeting involving the rail car. The Rushmore Company of Boston then acquired control of the business.[61] The new owners adopted the trade name Unit Car. They built cars for the Boston and Maine, the Canadian National, and the Uruguayan government in South America. These cars resembled the 100 and the 101, except that they were heavier. The 1922 Canadian National car, for example, weighed nearly 30 tons.[62] The piano-wire boiler was replaced by a larger watertube unit (Figure 8.16). The engine design remained unchanged. However, the Unit Car failed to revolutionize the industry and seemed to have little effect on the growing popularity of its gasoline-powered competitors. The reasons for its demise are unknown, but no more cars were produced after the early 1920s. The original demonstrator was not scrapped until 1941 or 1942, after being stored for several years.[63]

At the time that the Unit Car was introduced, the International Harvester Company began to study the possibilities of a steam car.[64] The scheme matured slowly, for it was not until 1927 that the first pilot model entered service. The Locomotor, as it was called, was a massive 73-foot combine car seating 63 passengers. It had a forced-circulation watertube boiler furnishing steam at 600 psi. Two 8-cylinder uniflow engines, mounted under the floor, powered the trucks through automotive-type drive shafts. The roof condenser's work was accelerated by power-driven fans visible from the car's end. The Locomotor moved 1½ miles per gallon of fuel oil—a respectable figure considering its size and weight. In 1929 the Milwaukee Road purchased a combined baggage–Railway Post Office Locomotor.[65] The 58-ton unit had no passenger compartment; trailers were used for this purpose. The B & O and the New York Central were reported ready to purchase similar units. However, only the Milwaukee ordered a second car. It was produced in 1931 by the Ryan Car Company of Chicago, which had meanwhile taken over the Locomotor's assets.[66] This car weighed 63 tons and had two 225-horsepower, eight-cylinder uniflow engines (Figure 8.17). The two Locomotors covered Milwaukee runs, mainly in Iowa, before being retired in 1938 and 1939.[67]

The Locomotors were working out their last years when George D. Besler, son of a former Central Railroad of New Jersey president and owner of the Davenport Locomotive Works, began to develop the ultimate steam railway car. Around 1934 Besler was experimenting with an advanced, high-pressure, steam power package. He gained considerable notoriety by producing several vehicles, including an airplane, powered with his engine. The New Haven decided that the young engineer had created more than a popular science novelty and agreed to cooperate in the fabrication of a two-car train (Figure 8.18). Two semilightweight commuter cars were selected for the conversion,

Figure 8.17   The Locomotor Company delivered this powerful eight-cylinder steam car to the Milwaukee Road in 1931. Except for the roof-mounted condenser, it looked like a conventional gas-electric car of the period. (William Baldridge, Cedar Rapids, Iowa)

Figure 8.18   Besler's high-pressure steam train was placed in service on the New Haven in 1936. It represented the last effort to produce an American steam rail car. (Railway Gazette, March 12, 1937)

Figure 8.19 A three-cylinder gasoline engine powered the belt-driven Eureka. *It was tested on the New York Central and Pennsylvania railroads in 1899. (*Railroad Gazette, *May 5, 1899)*

and the work was carried out by the Budd Company.[68] Streamlined front and rear cabs were installed. Body skirts and a blue paint scheme helped create a more modern appearance. The interior was refurnished as well, and new seating was provided for 152 passengers. The power plant featured a forced-circulation boiler with a continuous tube tapering from ¾ inch to 2¼ inches in diameter.[69] Set for 1,500 psi, it was unquestionably the highest pressure ever used for a steam car. The long-wheelbase power truck had a cast-steel frame with two pairs of compound cylinders cast integrally. They measured 6½ and 11 by 9 inches. The drive was directly connected to cranks on the axle ends; there were no gears or transmission. A roof condenser was used. The 1,000-horsepower Besler train cruised comfortably at 70 miles per hour. It could be operated from the rear cab by pneumatic controls.

Besler's car entered service late in 1936, running six trips daily between Bridgeport and Waterbury, Connecticut.[70] The train remained in service for about eight or nine years. It was reasonably satisfactory, but the fact that it was never duplicated indicates that it was no competition for diesel rail cars of the time. It was heavy, weighing nearly 152 tons. Although much of the excess weight can be blamed on the use of old cars and the added burden of air conditioning, the power machinery alone weighed 45,200 pounds—far more than a comparable internal combustion plant. The 14-foot, 6-inch wheelbase lead truck must have caused heavy flange wear. And because everything about the power plant was exotic and custom-made, maintenance was itself a custom process. Perhaps the Besler car was too sophisticated to succeed, or perhaps it was not given a fair trial. Whatever the reasons for its failure, it was the dramatic final chapter in the history of the American steam car.

Internal Combustion Motor Cars

Long before the final steam car entered revenue service, railroad mechanics had begun to explore other power sources that might prove suitable for rail car service. Actually it was the street railways, rather than the railroad industry itself, that showed the most willingness to test exotic types of propulsion units. Perhaps their need was greater, since horsecars were slow and inefficient. The search for a compact mechanical drive led to experiments with steam, soda, compressed air, electric batteries, and gasoline cars. Steam failed miserably, even though it was familiar and highly successful in other applications. Its deficiencies caused mechanics to try out power units then bordering on the visionary, such as the internal combustion engine. Gas and gasoline engines

were being installed on horsecars during the middle and late eighties. In Europe such engines were being successfully applied to road carriages, but America was slow to accept the automobile. The street railway was the gasoline engine's first proving ground on this side of the Atlantic. Early in 1886 the Connelly Gas Motor Company began work on its first car, which was tested in Brooklyn in January of the following year.[71] At least two more Connelly cars were produced and tested during the next two years.

Even more important experiments were started in 1888 by William H. Patton in Pueblo, Colorado.[72] The first car had a mechanical transmission that proved to be its weak point. Patton recognized this defect, and while seeking to overcome it, apparently became the first person to discover the happy marriage of the internal combustion engine and electric transmission. He embellished the idea by interposing batteries between the engine-generator and the traction motors, because he believed that a "flywheel" effect was necessary. This part of the scheme proved extraneous, but the basic concept later came into wide-scale use with the advent of the diesel-electric locomotive. Between 1890 and 1898 Patton built several gas-electric streetcars and small locomotives. The apparatus was impressive enough to gain the backing of George M. Pullman. By the time the first car began sputtering around the yard of the Pullman plant, however, the street railway industry was firmly committed to electric traction. One branch line in Maine purchased a Patton car, but the inventor was unable to interest the larger steam railroads in trying one.

Other mechanics also attempted to introduce gasoline rail cars during this period, but only one is known to have built a unit that a steam railroad considered worthy of testing.[73] In 1899 the Vimotum Hydrocarbon Car Company of Chicago sent an eight-wheel passenger car produced by the Jewett Car Company to Indianapolis for tests on the New York Central and the Pennsylvania railroads. The 41-foot-long unit was more on the order of an interurban car than a full-sized steam road coach. A three-cylinder Wolverine engine generated 45 horsepower to propel the 21-ton car at speeds up to 41 mph. The transmission included a variable-speed pulley with a huge flat belt carrying power to the wheels on one truck. The car does not appear to have measured up to its name, *Eureka*, for the Vimotum Company retreated into an early and probably well-earned obscurity (Figure 8.19).

The twentieth century was destined to be the internal combustion age. Many progressive engineers and businessmen understood why the compact, oil-burning engine would prevail, yet few seemed ready to try it on a rail vehicle. One was Edward H. Harriman, a financier and railroad executive of remarkable en-

Figure 8.20   William R. McKeen, superintendent of motive power for the Union Pacific, began producing gasoline-powered rail cars in 1905. This, his seventh car, was completed in 1906. (American Railroad Journal, May 1906)

Figure 8.21   McKeen's six-cylinder engine powered only the rear wheel of the front truck through a chain drive.

ergy and vision. His forte was corporate and financial planning at the highest level, but he also seemed to find time for mundane management problems, such as how to cut costs on marginal passenger operations. During the first years of the new century, Harriman was impressed by the performance of high-speed torpedo boats being tested by the Navy.[74] It occurred to him that the light, yet powerful engines used in such vessels could be placed in a passenger coach. Or better yet, a railroad could build a special car on the order of the sleek, knife-edge torpedo boats. He turned the idea over to William R. McKeen, Jr., superintendent of motive power for the Harriman-controlled Union Pacific Railroad. McKeen was not an old-guard master mechanic but one of the new, college-trained breed of engineers, still in his

early thirties. This in itself did not guarantee an openness to new ideas, but McKeen also happened to be an independent and free-thinking man who reveled in the unorthodox. Although he had no experience in the field of gasoline engines, he was intrigued by the novelty of the assignment and took to it with enthusiasm.[75]

The resulting McKeen car represented the real beginning of the main-line gasoline rail car. It was the first to be built in any significant number, and it marked the origin of a serious interest in such rolling stock by an American railroad. It was also one of the most distinctive and imaginative rail vehicles ever produced. McKeen literally followed Harriman's idea for placing a torpedo boat on the rails (Figure 8.20). The nautical look of the slender, prow-shaped body was later reinforced by the use of porthole windows. The steel body was novel if not exactly a pioneer. The wind-splitting, aerodynamic design was futuristic, albeit scientifically incorrect. Later experiments showed that the wedge-shaped prow should trail the car, not lead it.

McKeen's power train was as novel as the body plan (Figure 8.21). The engine (200 to 300 horsepower) was carried directly on the front truck. A 5-inch-wide Morse silent chain powered the forward or large-diameter wheel set; only one set of wheels was driven. The six-cylinder engine was separated at the middle into two three-cylinder units. The chain drive sprocket was mounted on the crankshaft at this central point, so that the chain could run to a corresponding sprocket on the forward axle. A friction clutch could disengage the chain and the low gear. The gear was used to start the car, but at a speed of 10 mph the chain was shifted to direct drive.[76] The direction of travel was changed by reversing the engine. The engine was started by compressed air stored in tanks, which were replenished by an air compressor driven by the main engine.

Designs for the first car were ready early in 1904. It was the only single-truck, wooden-bodied car built by McKeen. The Number 1 was produced at the Union Pacific's Omaha shops, as were all the other McKeen cars, totaling approximately 150. The 31-foot experimental model seated 25 and was powered by a 100-horsepower engine.[77] It made a trip from Omaha to Portland, Oregon (about 2,000 miles), and returned without difficulty, or so read the official news releases. Years later an old U P mechanic recalled that the long journey involved many problems, and that Motor Car Number 1 was pushed into more than one terminal.[78] After its return the Number 1 was assigned to a branch line, where it pulled a mail car. It eventually caught fire (thus demonstrating one of the hazards of gasoline vehicles) but was rebuilt with a box cab.

A second, larger car was completed in September 1905.[79] It was mounted on eight wheels, measured 55 feet in length, seated 57, and weighed 28 tons. The steel body was scientifically patterned after a bridge truss, and the diagonal members of the frame necessitated small windows—one drawback to the plan. Since light weight was a paramount consideration, the trucks and wheels were designed to eliminate as much excess metal as possible. The interior was beautifully finished, with antique mahogany paneling and a cream-white ceiling embellished with gold and sepia designs. The seats were covered in leather, and the car was brilliantly lit with acetylene lamps. The exterior was painted a rich maroon with gold stripes and lettering. The trucks and undercarriage were olive green.

McKeen was now ready to tool up for production, and four cars similar to the Number 2 were quickly rolled out of the shop. In 1906 came the Number 7—the first McKeen car to display the distinctive porthole windows and drop-center door. With the next car, McKeen began to manufacture his own gas engines. In 1908 a separate corporation was set up for the production and sale of gasoline motor cars. All work continued to be performed within the Union Pacific shop complex. Orders were solicited and received from other railroads. The curious torpedo cars began to appear on the Erie, the Norfolk Southern, the Alton, and numerous Western lines (Figures 8.22 and 8.23). By November 1909 fifty-six McKeens were reported in service on U.S. railroads (mostly on the Union Pacific), with one in Mexico.[80] Some had 70-foot bodies, and nearly all were powered by 200-horsepower engines.

McKeen presented papers and furnished data to railway clubs in an effort to promote his car.[81] His material was, of course, laudatory and intended to allay fears that such units would always be experimental. He portrayed them as reliable, economical, and safe, and pointed out that they were smoke-free, easy on the track, and instantly ready for service. McKeen documented his claims with specific examples. The cars on the Union Pacific ran mainly on branches in Nebraska; some traveled 390 miles a day, and many hauled trailers. During its first year's service, car Number 19 had had only two engine failures. The cars were as reliable as a steam locomotive, and were available twenty-three days out of the month. One unit on an isolated branch with no repair facilities ran for six months without a single failure. It performed splendidly during a blizzard that froze the steam locomotives at the same terminal. And for economy the cars had an outstanding record, according to McKeen. In regular running they averaged 3 miles to the gallon. Total operating costs, including normal running repairs, were figured at 12 cents a mile. And the two-man crew saved $195 a month in wages.

McKeen also boasted about the wedge-shaped front end which, he explained, had been suggested by earlier experiments in Germany and by a demonstration electric car shown at the St. Louis World's Fair in 1904.[82] He claimed that a car with a flat front was subjected to wind resistance equal to 87.5 horsepower, while resistance to the wedge front amounted to only 21.9 horsepower. Just how McKeen arrived at these figures is not known, but in light of later streamlining experiments, their validity is questionable. In addition, the normal top speed of the cars was 50 miles an hour, and even at this velocity, it is unlikely that the prow-shaped nose would offer any material saving of fuel. The wedge front did have at least one practical, if unintended, advantage: it served as a collapsible buffer in accidents. In one case, two McKeen cars hit head-on with a combined impact speed of 75 to 85 mph.[83] Although both front ends were demolished, the engines were undamaged and no one was seriously injured or killed. The cars were not even derailed. In another incident, a McKeen car hit a steam locomotive at 30 mph with no damage beyond a dented front end.

The reports issued by McKeen were not entirely objective. His critics admitted that his design was original and distinctive, but not everyone found the cars' appearance pleasing or their performance satisfactory. A contemporary observer objected to their "warlike-looking submarine design." Ralph Budd, onetime president of the Burlington and no enemy of motor cars, recalled that the arched-roof McKeens were called potato bugs.[84] Aesthetic objections aside, the design was also attacked for mechanical shortcomings. The engine was balky and could be operated

Figure 8.22   McKeen built approximately 150 cars, the majority of which operated on the Union Pacific and its related roads. The example here was produced for the Baltimore and Ohio.

only by expert workmen. The camshaft and timing parts were prone to rapid wear.[85] The compressed-air starting system was troublesome, and the storage tanks were inadequate. There was no reverse gear. To back up, it was necessary to stop the engine, reverse it, and start again—a cumbersome process. The engine, even when functioning well, was noisy, particularly in climbing a grade. Mounting the engine directly on the front truck was considered a poor idea, because it was subject to every shock and jar that the roadbed had to offer.

These defects did not encourage sales. Except for the closely associated Union Pacific, few repeat orders were placed. Most lines bought one or two cars as an experiment, and after 1917 very few new McKeens were built.[86] Three years later, the com-

pany was liquidated. But McKeen's demise may have been occasioned by more than engineering defects, for it is possible that the market was simply saturated by the 250 cars produced by McKeen and General Electric. The industry may have felt that only so many runs were suitable for motor cars and that no more were required. This speculation is supported by the fact that both manufacturers suspended production simultaneously.

A few McKeen cars remained in service thirty years after the U P's Omaha shops suspended production. Some were rebuilt with squared-off front ends, new front trucks, and gas-electric drives. The Chicago Great Western rebuilt three into a deluxe motor train, the Blue Bird.[87] The lead unit was given a new 300-horsepower engine, and trailing cars were remodeled with buffet,

Figure 8.23   Because of its unorthodox appearance and troublesome power plant, most McKeen customers bought only one or two experimental cars. (Railway and Locomotive Historical Society)

sleeping, and lounge sections. McKeen cars remained most conspicuous on the Union Pacific, which had thirty-two in service in 1929.[88] The majority had run more than a million miles by then. In later years some were rebuilt as gas-electrics, and two were in use as late as 1947. Possibly the last McKeen to operate was the unit belonging to that museum of American railroading, the Virginia and Truckee. The car was in service as late as 1946 and looked very much as it did in 1910, when it was sold to the Nevada short line for $24,000.[89]

General Electric decided to enter the self-propelled rail car market just months after McKeen began design studies for his first car. The project was officially launched late in 1904, but G.E. officials had been considering it for several years.[90] They were motivated by two considerations: the success of gasoline cars in Europe, and the prospect of a new outlet for electrical apparatus. They envisioned a self-energized trolley car that could operate on ordinary steam railroads without the colossal capital cost associated with conventional electrification. Patton had explored an idea that G.E. engineers now hoped to perfect. They felt that a major drawback to the larger gasoline rail cars was the mechanical transmission. The gearshift and clutch might be suitable for light highway vehicles, but it could not withstand the strains of a large gas engine and a heavy rail vehicle. An electric transmission could handle a prime mover of any horsepower. Moreover, it would relieve the engine of torsional strains and shocks caused by crossings, switches, curves or rough track that were ordinarily transmitted by a direct mechanical connection between the engine and the wheels. Because there was no fixed relation between the engine and the car's speed, the engine could be turning over at its optimum revolutions per minute. Hence the engine's maximum power could be utilized when the car was starting up or when it was climbing a grade. Conversely, the engine-generator

set could be loafing while the car sped over a level portion of the line. The electric transmission offered the possibility of an infinite number of speed changes. It was easily reversed. Double end controls were readily installed. If desired, every wheel could be powered by a separate motor. In short, G.E. felt that it had a salable engineering concept. It also produced a stock line of generators, motors, controllers, and their associated electric street railway apparatus which could be adapted for gasoline rail cars.

The first G.E. car was something of a makeshift (Figure 8.24). A 65-foot-long wooden body with arched windows and a steel underframe was furnished by Barney and Smith. A six-cylinder, 140-horsepower engine was delivered from England in mid-1905. Assembly proceeded slowly, for the car was not ready for testing until January 1906. Subsequent trials on the Delaware and Hudson were disappointing. The 68-ton gas-electric was heavy, underpowered, and undependable. It was returned to its maker, where the engineers began on it afresh. Two years would pass before a second G.E. car was ready for a trial.

Meanwhile McKeen was already in production. His lightweight steel cars proved a model for his competitor. General Electric turned to the Wason Car Company for assistance. The Massachusetts car builder came forward with a body design that in effect translated McKeen's thinking into a more conventional form. The sharp prow gave way to a rounded snout, and the porthole windows were replaced by conventional sash. However, the general concept of a light, arched-roof steel body was retained, as was the center door entrance and the rounded rear end. G.E.'s own design team took over the planning of the new engine. The result was a compact V8 weighing less than one-third of the imported English engine. The combination of Wason body and new engine resulted in a lithe, 50-foot motor car that weighed only 31 tons, or less than half as much as the first test car.

Figure 8.24   In 1905–1906 General Electric built its first experimental gasoline-electric car, which was fitted with an English engine. (Jim Shaughnessy Collection)

Figure 8.25   After constructing several experimental units, G.E. introduced a production-model gas-electric car in 1909. The two cars for the Southern shown here were completed during the following year. (Amer can Railroad Journal, February 1910)

Although the second G.E. car was vastly superior to the original wooden demonstrator, it did not fully meet the expectations of the designers, for a third demonstrator was soon underway. It was not ready until early in 1909. General Electric had clearly lost the race with McKeen, who had already produced thirty-five cars and had another twenty underway. But G.E. doggedly pressed on. The Number 3, somewhat larger than the Number 2, began a series of road tests that covered 50,000 miles. The 35-ton demonstrator toured every railroad that would allow it on the property and made an appearance at the Master Car Builders Convention in Atlantic City.[91] Its 125-horsepower plant delivered about 2 miles to the gallon and could propel the 52-passenger car at a steady 60 mph. Its good performance finally proved that General Electric was ready to produce self-propelled cars on a commercial basis. In October 1909 the Southern ordered two 55-foot-long cars that were near-duplicates of the Number 3 (Figure 8.25). Three other railroads placed orders as well, and deliveries began the following year.

Figure 8.26   Interior of a G.E. gas-electric of about 1910. Most of the bodies were built. by the Wason Car Company (General Electric Company)

The initial performance reports were encouraging. Exploiting the testimonials of its customers, G.E. reported a dazzling series of successes in a presentation before the New York Railroad Club.[92] The 107-mile-long Minneapolis-based Dan Patch Line was serviced exclusively by gasoline motor cars. The cars ran the full length of the road in only three hours and twenty-five minutes, including stops. They used the cheapest fuel (naphtha at 6 cents a gallon), even on days when the temperature was 16 degrees below zero. They ran 3,900 miles a month at an average cost of 18 cents a mile. The equivalent cost for a steam train was figured at 33 cents a mile. The gross yearly saving came to $6,970, or roughly 25 percent of the car's original cost. Repair figures were equally impressive. The gas-electric's repairs cost 4¼ cents per mile, a locomotive's 8½ cents. The cars were rugged and could be pushed well beyond their rated capacities. They were used to switch at terminals and could pull as many as twenty freight cars. The Frisco, another large-user of G.E. equipment, pulled two trailers and 415 passengers with one of its gas-electrics.

The industry reacted enthusiastically, during the year of these sensational reports, G.E. delivered 31 cars—the greatest number produced in any twelve-month period (Figures 8.26 to 8.28). Several models were offered, the largest being a 71-footer weighing 49 tons and seating 91 passengers. But product variety and customers' praises were not enough to sustain purchases. Orders fell off after 1912. Deliveries slowed to sixteen in 1913, and then to only two by 1916. No definite explanation can be found for the abrupt decline that both G.E. and McKeen experienced after

1915. The possibility that the market was saturated has already been suggested. In addition, there was prejudice against gasoline motor cars in railroad circles. Although such attitudes are difficult to document, R. S. Zeitler does mention the "disparaging view of the internal combustion engine" held by railroaders at the conclusion of his paper on the self-contained railway motor car appearing in the *Cyclopedia of Engineering*.[93] Under pressure, many old-line master mechanics could be persuaded to buy a motor car for testing. But even if management was ready to give it a fair trial, a hostile operating crew could sabotage the interloper. Only a few embarrassing road failures were needed before its enemies proclaimed that the damfool thing was a failure, just as they had predicted. There was no question that motor cars could succeed only under favorable conditions. Repair and operating crews had to be trained in their care and handling. Shops, tools, and parts must be available. A gas-electric car could not be maintained properly by a blacksmith or a boilermaker.

There were other more direct reasons for the gas-electric's initial failure. G.E. expected that the advantages of electric transmission would sell the motor car, but the system was not without faults. It was heavy and expensive: electric transmission cost $10,000 to $12,000 more than a mechanical drive and weighed 10 to 20 percent more.[94] The first cost of the cars, ranging from $20,000 to $30,000, discouraged larger sales. The same money would buy a locomotive, which represented a profit-making tool, while the gas car was viewed as a marginal cost-cutter. In addition, G.E. management shifted policy. Rather than promoting gas-electric cars, it decided to aim for a larger market by en-

*Figure 8.27   Engine-room compartment of a G.E. gas-electric car of about 1910. The
V-8 engine developed 125 horsepower. (General Electric Company)*

couraging main-line electrification. This would involve the sale of generating, substation, and overhead-wire hardware in addition to motor and control equipment. In 1917 the decline in orders and the war emergencies provided good reasons to close out the gas car division.

While G.E. and McKeen dominated the first generation of motor cars, a number of smaller firms also attempted to introduce them. For example, the Kuhlman Car Company of Cleveland, Ohio, built an eight-wheel gasoline motor car for the New York Central Railroad in 1906.[95] The 43-foot, 9-inch car was powered by a 220-horsepower Chase engine and registered speeds up to 65 mph. Another streetcar builder, Brill, constructed several heavy gas-electric-battery cars between 1906 and 1909 for the Strang Gas Electric Car Company of New York City.[96] They were tested in the Midwest but apparently did not seem worth serious study to any major railroad. Fairbanks, Morse, & Company, a major producer of handcars and other railway supplies, began to make four-wheel streetcar-like gasoline cars around 1906. Of the limited number built, most went to short lines, but one was purchased in 1909 by the Pennsylvania for a branch in Delaware.[97] In the Far West the Hall-Scott Company of Berke-

ley, California, began rail car production in 1909.[98] The firm was already established in the gasoline engine business. Steel bodies, 150-horsepower, six-cylinder engines, and automotive-style drive shafts to the rear truck were typical of the Hall-Scott design (Figure 8.29). Most of the cars were built for Western customers, but at least one unit was sent as far east as Indiana and a few were turned out for Canadian lines.[99] Rail car production apparently continued until 1926, when the firm became associated with the American Car and Foundry Company. After that time it produced engines for Pullman and Brill rail motor cars.[100]

Besides the cars built by small commercial suppliers, all kinds of home-made gasoline rail cars were fabricated by American short lines. Documentation is scarce, but enough pictures have been printed in the railroad hobby publications to indicate that the practice was fairly widespread. Although the variety and crudeness of these contraptions defy any general analysis, all were born from a "root hog or die" necessity. The Rio Grande Southern, a poverty-stricken Colorado narrow-gauge line, constructed a distinctive group of rail busses between 1931 and 1937 that were locally known as the Galloping Geese.[101] Awkward in

Figure 8.28   G.E. produced one of its larger cars in 1914 for the Illinois Central. The 71-foot-long car seated 86 passengers and included a separate compartment for Negroes. (DeGolyer Foundation Library)

appearance and bumpy in ride, these homely vehicles helped keep the line in business during the worst years of the Depression (Figure 8.30). Elsewhere, local mechanics combed the junkyards in search of just the right piece to complete a rail car. The parts available—a truck chassis, a tractor engine, or perhaps a discarded streetcar body—dictated what form the car would take. Most had simple mechanical drives, but at least one rather elaborate gas-electric was assembled in 1919 by a Kentucky short line, the Flemingsburg and Northern. A 90-horsepower gas engine powered a 55-kilowatt generator.[102] The body appears to be made from two four-wheel streetcars.

Technically the improvised short-line motor cars were of no importance. In the general field of transportation they were also fairly inconsequential. They are not represented in I.C.C. statistics, which recorded only railway motor cars that were operated by class-one lines. Yet it is worth noting the existence of this motley fleet of home-made cars, because at one time hundreds of them picked their way over the shaky, weed-grown rail lines of rural America (Figure 8.31).

The motor car was an idea that had been introduced ahead of its time. By 1915 it was considered a failure; almost no new units were being ordered, and most railroads seemed only too pleased to return to traditional methods of operation. Yet just five years

later motor cars were again warmly received. Much had changed during this short period. World War I had profoundly affected national attitudes as well as the national economy. The shift to an automotive society was accelerating, with a corresponding erosion of short-haul rail passenger traffic. At the same time, inflation was boosting wages and the price of materials. Operating costs soared. Some roads figured that branch-line steam train costs per mile were as high as $1.50, while revenues amounted to no more than $1 and sometimes as little as 20 cents per mile.[103] Formerly, branch-line trains could be expected to make a small profit or at least break even. A small loss could be absorbed or charged off to good will, but in the postwar period a 50-mile branch could lose $15,000 a year per train.[104] One Western line calculated its annual losses on short-haul passengers at 7 million dollars. In 1923 the B & O reported that 30 percent of its passenger trains (all were short runs) failed to cover expenses and that only the main-line expresses showed a profit. The railroads were caught in the double squeeze of declining revenues and rising costs. Every cost-cutting method was being investigated. Now even the despised motor car began to look rather inviting.

At first the industry tended to overreact by adopting extremely small units. These cars were actually city busses outfitted with flanged wheels. They were a sensible investment for the most lightly traveled branch lines, but they did not have the capacity

Figure 8.29   The Hall-Scott motor company of Berkeley, California, built gasoline rail cars for the Nevada Copper Belt Railroad in 1910. The 36-ton car seated 32 passengers.

Figure 8.30   The Rio Grande Southern fabricated a gasoline rail car powered by using a 1936 Ford V-8 truck engine. This car and its sisters were locally known as the Galloping Geese.

Figure 8.31   Some poor short lines produced ungainly home-made gasoline rail cars. This one was photographed on the Collins and Glennville Railroad in 1926.

Figure 8.32   J. G. Brill, a major streetcar and motor bus builder, entered the rail car field in 1921. Its first model was an adaptation of a standard 30-passenger bus. (George Krambles Collection)

or stamina for main-line service. There was no question, however, that the rail bus could make a better showing than the steam train. In 1921 Brill began to produce diminutive, 6-ton rail busses that ran 14 miles per gallon and cost no more than 30 cents a mile to operate.[105] The 29-foot-long vehicle seated 31 passengers and cruised at speeds up to 30 mph (Figure 8.32). It was the ultimate compact, which could provide rail transportation at the minimum cost. Comfort and safety were also minimal, however. The featherweight bus bounced and rolled and offered its human cargo little protection in a smashup. Yet setting a production-line highway bus on flanged wheels was so cheap that several firms were soon engaged in their manufacture. Mack, White, and F.W.D. (Four Wheel Drive), already established as truck and bus makers, entered the rail bus market in the early 1920s. Brill, the nation's leading producer of streetcars, became so enthusiastic about rail cars that it organized an automotive car division in 1923 to specialize in their production.[106] These firms were challenged by such unknowns as Edwards, Russell, and Sykes, all of whom perished after building a few units except for the Edwards firm of Sanford, North Carolina. Edwards produced nearly 100 cars, some for such major roads as the Burlington, before it too expired in the late 1940s.[107]

Most rail busses were produced for short lines, but a few big roads found employment for them. The New Haven purchased one of the first Macks to come off the assembly line in 1921. The 32-footer seated 35, and it remained in service until 1939.[108] Edwards's first production car, built in 1922, was sold to the B & O for use on a branch in West Virginia.[109] It was a 34-passenger, 60-horsepower car that pulled a tiny, 9,350-pound, eight-wheel trailer car (Figure 8.33). According to the Scientific American of June 1922, the Edwards car was trimming costs beyond any reasonable expectation. The steam train servicing the Green Springs branch cost 43 cents a mile, the rail car

only 8. This included wages, fuel, maintenance, and depreciation.

Highway-type rail cars continued to be supplied to main-line roads in later years. In 1933 the Cotton Belt purchased a 10-ton Astro-Daimler rail bus.[110] Called Eagle of the Rails, the 38-foot-long car seated 40 and was said to give a smooth, silent ride at speeds up to 80 mph. At the same time the Cotton Belt converted a highway bus for rail service in its own shops. Fifteen years later, six more rail busses appeared in Texas on the Houston North Shore branch of the Missouri Pacific.[111] These Twin Coach units remained in service until about 1960. A more modest example of a late-model rail bus was the 28-passenger unit built by the Superior Coach Company of Lima, Ohio, in 1948 for the Mississippi and Skuna Valley Railroad (Figure 8.34). The stock school-bus body was fitted with a small Railway Post Office compartment.

As already noted, the rail bus never really flourished on the main line, but some units of this type were supplied to the New Haven in the 1950s. In 1951 Mack furnished a 21-ton double-ender mounted on P.C.C. (President's Conference Committee) streetcar-style trucks.[112] The body resembled a standard city bus of the time. The 41-foot, turquoise and cream bus did well in its tests, and another nine units were ordered in 1953 for service on the New Haven's Cape Cod lines. Deliveries were made during the following year, but the railroad management had changed and the new administration put the busses in storage. Few ever turned a wheel in revenue service. Two were sold to the Sperry Company in 1958 for conversion to rail inspection vehicles. A few years later others were sold to Spain, but they were lost at sea when their cargo ship sank during the crossing.[113] The ill-starred Mack rail busses were never really able to demonstrate their worth.

Under certain conditions, the ideal bus would be a combination highway-railway vehicle. It would be able to use the

Figure 8.33   The Edwards chain-drive rail bus of 1922 was powerful enough to haul a trailer. The motor train was retired in 936.

Figure 8.34   Twenty-eight passengers, as well as mail and express packages, were jammed into the 27-foot, 8-inch length of this rail bus completed in 1948 by the Superior Coach Company of Lima, Ohio. (Superior Coach Company)

Figure 8.35 The Evans Auto-railer began operation in the northern Virginia suburbs of Washington, D.C., in 1936. The bus could run on railroad tracks and on the highway with equal ease. (Horace Thorne)

public streets for access to the central city and yet thread its way through the suburbs on a private right-of-way. The scheme was tried as early as 1932 by the Twin Coach Company.[114] Two years later the Public Service Company of New Jersey outfitted three city busses with retractable flange wheels for use between New Brunswick and Trenton. Meanwhile the Evans Products Company of Detroit, a major railroad supply firm, produced a vehicle that it named the Auto-Railer. Its chief selling points were its flexible operations and its smooth, resilient ride, a result of the fact that the pneumatic tires actually carried the bus on the rails. In 1936 sixteen passenger coach Auto-Railers were sold to the Arlington and Fairfax Railway for operations in the Washington, D.C., area (Figure 8.35). The stubby cars, resembling light delivery trucks, were unable to save the Arlington line, nor did they convince the industry that combination rail and road vehicles offer any special advantages. The pneumatic-tire-on-steel-rail scheme did have a wider appeal, and a few such units were produced during the early 1930s by Fairbanks, Morse in cooperation with the Chrysler Corporation and the Budd Company.

Rail busses had only a limited market on the main line because they were simply out of scale for the job. Larger cars with generous seating and baggage space were required. And they had to be substantial enough to withstand the buffing and occasional collisions that were inevitable on a heavily traveled class-one road. Seating for 40 to 50 passengers, together with a 15-foot express baggage space, called for a body somewhat smaller than a full-sized coach but at least equal to a commuter car. An engine of 100 to 200 horsepower was about the minimum requirement if the car was to keep up with the prevailing schedules. The undersized, pigmy-powered rail bus could not meet these demands, and the supply industry soon developed an enlarged style of gasoline car especially suited to the needs of the main line.

In 1923 and 1924 Brill introduced several standard models for branch- and main-line service. The model 55, the smallest unit offered, was a 43-foot, 38-seat car powered by a four-cylinder, 68-horsepower engine (Figure 8.36).[115] The 14½-ton model's normal running speed was 38 mph. The model 65 was essentially the same car as the 55, except that it had a larger six-cylinder engine for extra power (Figure 8.37). The model 75 featured a 55-foot-long body, seating for 59 passengers, a 190-horsepower engine, and a total weight of 25 tons (Figure 8.38).[116] All the cars used a mechanical transmission with six speeds forward and three in reverse. An auxiliary transmission was incorporated in the cast-steel bolster of the forward truck. Only the forward truck was powered. The engine was mounted to the car frame, with a short drive shaft connecting the main transmission to the secondary gear box on the truck. These cars were not made from highway busses; the design was specially developed for railway service. The trucks had inside bearings, a feature uncommon in American rolling stock. The wheels were held on by large nuts so that they could be removed to allow access to the auxiliary gear box for servicing.

At the same time that the gas mechanicals appeared, Brill

offered a 250-horsepower, 60-foot, 37-ton gas-electric car (Figure 8.39). Mechanical transmission was not considered adequate to power plants over 200 horsepower. Some model 250s had elongated 73-foot bodies with seating for 84 passengers.[117] Between 1921 and 1937 Brill built approximately 250 motor cars in all styles, from miniature rail busses to heavy diesel-electrics for domestic railroads.[118] Trailers and foreign orders swelled production by another 70 units. In general the Brill cars performed well, but they rode poorly and tended to "nose," or sway noticeably back and forth, even on a straight track.[119] This swaying was characteristic of most rail cars, because the engine was placed at the extreme front of the body, adding as much as 10 tons to the forward end.

The Edwards Railway Motor Car Company, already mentioned in connection with light rail busses, began to offer a larger car better suited to heavy branch- or main-line operations. In 1922, just as Brill was tooling up for large-scale gas car production, Edwards built the Burlington a 9½-ton car that was 32 feet, 7 inches long.[120] The 39-seater was propelled by a 60-horsepower Kelly-Springfield engine with a normal top speed of 45 mph. During the next four years the Burlington purchased another eight cars from Edwards. All were larger than the first unit, and most conformed to the maker's popular model 20. This design measured 43 feet long, seated 39, and weighed just under 21 tons (Figures 8.40 and 8.41). A 100-horsepower engine and four-speed mechanical transmission made up the power plant. One of the best and most original features of the cars was the truck mounting of the engine. This placed the engine under the floor, freeing the forward compartment for more baggage space. In 1925 Edwards produced an extra-large car, Number 552, for the Burlington. It measured 65 feet, 7 inches in length and had a 100-horsepower Buda engine on each truck.[121] The Colorado and Southern, a subsidiary of the Burlington, considered buying a model 20 for $16,500, but decided instead to purchase a second-hand unit from the C B & Q.[122] The car performed splendidly at first, paying for itself during a single year's operation. But it was a mechanical disappointment, with a particularly troublesome transmission. The C & S also found the car too small and too light for the run from Fort Collins to Greeley, Colorado. It was sold in 1929. Presumably the parent corporation also found the Edwards cars unsatisfactory, for all were sold or retired between 1931 and 1939. These units were better suited to short-line operations, where they gave good service. One of them is still running on the California Western Railroad.

The most astute effort to reintroduce the motor rail car was made by Harold L. Hamilton (1890–1969).[123] His experience as a railway mechanic and later in the engineering and sales offices

Figure 8.36   Brill built this gasoline mechanical-drive Model 55 around 1924. The 15-ton car seated 38. (Historical Society of Pennsylvania)

of the White Automobile Company suggested to him the potential of the railway motor car. Hamilton admired the G.E. gas-electric car and believed that the giant firm had foolishly abandoned a splendid engineering concept. He decided to conduct a survey to discover exactly what features were most wanted by railroad managers.[124] Around 1921 he began gathering data, which revealed a clear pattern of railroad requirements that was eventually translated by Hamilton's small staff into a practical rail car. His chief engineer, Richard Dilworth, was a former G.E. gas-electric mechanic, who would later play a major role in the perfection of the diesel-electric locomotive.

During the course of the survey and design work, Hamilton organized the Electro-Motive Corporation with offices in Cleveland. Again Hamilton moved carefully. Rather than rushing to acquire a plant, machinery, and staff, he avoided large capital outlays and the obligations of an expanded payroll by jobbing the manufacture of the cars out to subcontractors. G.E. supplied the electrical gear and the Winton Engine Company the prime mover. The St. Louis Car Company fabricated the body and trucks and performed the final assembly. Electro-Motive's small sales and engineering staff entered the uncertain rail car market with little more than a post office box and a roll of blue prints. The first car was ready for testing by the firm's first customer, the Chicago Great Western, in the summer of 1924.[125] The 57-foot, 38-ton unit was powered by a six-cylinder, 175-horsepower engine specially designed for railway service by Winton (Figures 8.42 and 8.43). The passenger compartment seated 44; folding seats in the baggage room could accommodate 8 more. The engine room was very compact, measuring only 8 feet long. Traction motors powered only the front-truck wheel sets. The car was no beauty, but its square, rivet-encrusted steel body had a rugged, utilitarian look that was somehow rather appealing. It

performed as efficiently as its no-nonsense appearance promised, and without the noise and vibration characteristic of most rail cars. At 4 miles to the gallon, its fuel consumption was considered satisfactory.

The Northern Pacific also took delivery of an E.M.C. car in 1924. Orders began to flow in, stimulated by reports of good service. The Colorado and Southern, for example, saved $14,000 per year by replacing a steam train between Cheyenne and Wendover, Wyoming, with an economical E.M.C.[126] The car itself was not cheap—$50,022.32, or as much as a moderate-sized steam locomotive—but any piece of capital machinery that can pay for itself in less than four years is a bargain. During its second production year, Electro-Motive sold 36 cars. The business continued to grow during the prosperity of the twenties (105 units were sold in 1928), and by the end of the decade E.M.C., which continued to depend on outside contractors, absolutely dominated the rail car market (Figure 8.44).[127] The Depression ended Electro-Motive's boom, but before rail car production was suspended in 1932, 500 cars had been completed. The firm's potential was so promising that General Motors purchased it and its chief supplier, the Winton Engine Company, in 1930.[128] G.M. channeled its new subsidiary away from rail cars and into diesel locomotives. Electro-Motive's golden touch continued, and the firm came to control the domestic motive power market—an exceedingly rich prize during the years when American railroads converted from steam to diesel on a wholesale basis. From E.M.C.'s humble origins in the gas-electric rail car field, both an industrial giant and a revolution in railroad operations had evolved. If only in this transitional role, the motor car achieved real significance.

The many marginal producers of rail motor cars even included the Pullman Company. As the nation's largest producer of pas-

MODEL 65 GASOLINE CAR

Length of baggage compartment	12 ft. 4 in.
Width over posts	8 ft. 4 in.
Total seating capacity	41
Weight	31,000 lb.

MODEL 75 GASOLINE CAR

Length of baggage compartment	16 ft. 3 in.
Width over posts	9 ft. 6 in.
Total seating capacity	59
Weight	50,000 lb.

GAS-ELECTRIC CAR

Length of baggage compartment	19 ft. 3 in.
Width over posts	9 ft. 6 in.
Total seating capacity	65
Weight	75,000 lb.

Figure 8.37   Three Brill gasoline rail cars from an advertisement of 1925.

senger cars, Pullman had dabbled in motor car construction since the time of the Patton gas-electrics in the late 1880s. During the twenties the firm built gas-electrics in limited numbers for the Santa Fe, the Burlington, and other lines. Many of these were produced for Electro-Motive. Two Pullman rail cars were notable for their size and mechanical arrangement. They were built in 1926 for the Detroit, Toledo, and Ironton.[129] The 72-foot, 6-inch-long units weighed a moderate 66 tons because of the liberal use of aluminum in the body structure. The engine-generator mounting was below the floor. Each car carried two 150-horsepower Hall-Scott engines, one of which turned clockwise and the other counterclockwise in an effort to dampen vibrations. The big cars rolled along at speeds up to 60 mph. Pullman's share of the market was relatively small, but it soon acquired two subsidiaries, Standard Steel and Osgood Bradley, that were more active in this field.

All gas-electrics tended to look alike: in the tradition of the heavyweight era in steel car architecture, they were boxy and plain. This superficial standardization was further reinforced by Electro-Motive's dominance. The surface uniformity was undermined, however, by a trend to increased horsepower and to the introduction of extra-large units and diesel power. By the late 1920s horsepower had advanced to 300 or more, and electric transmission had become common. Rail cars were designed to pull trailers for peak loads, but the original units of 175 horsepower were generally not powerful enough for such assignments. In addition, more than one trailer might be needed on certain runs. The Lehigh Valley, for example, wanted a motor car capable of towing three trailers, and late in 1926 Brill produced four 500-horsepower motor cars to meet this requirement.[130] The 70-foot, 65-ton cars had two engines apiece. Although two engines were common for the larger gas-electrics, a few triple-engine cars

Figure 8.38   Brill's model 75, delivered in 1925, was a gasoline mechanical car. The wooden trailer was an ex-Staten Island Railroad commuter car. (Baltimore and Ohio Railroad)

Figure 8.39   Brill delivered this gas-electric rail car in 1926. (Westinghouse Electric Corporation)

Figure 8.40 Edwards delivered this Model 20 gasoline mechanical car in 1924. It measured 43 feet in length and weighed 17 tons. (Horace Thorne)

were built by Mack. In 1928 Mack sold the Reading Company a 76-foot, 7-inch triple motor car that had three 120-horsepower motor generator sets. Two years later the B & O purchased a 600-horsepower car from Standard Steel (Figure 8.45). It was really more locomotive than car and represents, as do the other high-horsepower rail cars, a move to replace steam trains with motor trains. In 1930 Brill built a 550-horsepower gas-electric car and three cars (one of them an observation coach) for the Gulf, Mobile, and Northern. The Santa Fe, the New York Central, and other roads also acquired powerful rail cars. The trend to greater power was evident in cars produced during 1931: of thirty units, all gas-electrics, that were delivered for domestic service, twenty-three had 400 horsepower, one had 900, and the remaining six had less than 400.[131] Rail cars were being used for faster and more heavily traveled runs, but internal combustion power was coming to the railroads surreptitiously. Rather than being accepted directly in a locomotive, it was introduced for hauling trains in the disguise of the rail car.

Nearly all the rail cars constructed during the twenties were powered by conventional gasoline engines, but builders were also looking for engines capable of utilizing cheaper fuels. It was possible to burn a low-grade petroleum by-product, called distillate, in a gasoline engine with at least moderate success. By the late twenties gasoline was up to 15 cents a gallon, while distillate cost only 4 cents.[132] Taxes and the growing demand created by highway vehicles were responsible for the escalating gasoline prices. But distillate could not be efficiently burned in a low-compression gasoline engine. Combustion was incomplete, smoke abundant, and power low. The railroads were urged to try diesel engines, which were well suited to the consumption of cheap fuel. Diesel engine design had not been perfected for vehicle use, however. The engine remained a heavy, bulky mechanism suited to stationary power plants and little else.

Even so, some attempts were made to harness the diesel to rail cars. The Canadian National was probably the first North American line to try the diesel-powered rail car. It acquired the first unit in 1925 and during succeeding years purchased more of the cars, which were fitted with English engines, while the vast majority of rail lines remained loyal to the conventional gas-electric.[133] In the United States the Rock Island appears to have been the leader in testing diesel-electric power plants, for in 1925 and 1926 the road repowered four McKeen cars in this fashion.[134] In 1927 the Milwaukee Road installed a Foos diesel in an aging General Electric rail car.[135] During the following year four diesel-electrics were ordered by the Pennsylvania and the Reading railroads.[136] The 300-horsepower engines were built by

Figure 8.41   Edwards' best design feature was in mounting the engine and transmission on the power truck. (Horace Thorne)

Westinghouse. The car bodies were produced by Pullman for the Pennsylvania and by Bethlehem Steel for the Reading (Figure 8.46). Ingersoll-Rand also tried to sell the railroads on diesel-electric motor cars by producing a 600-horsepower demonstrator in cooperation with the St. Louis Car Company.[137] The car could pull four trailers and still maintain a schedule, and it achieved the mileage of a 400-horsepower gas-electric with a cheaper fuel. But the railroads were still skeptical, and the gas-electric remained the clear favorite through the early 1930s.[138]

Gasoline motor cars were not confined to one region of the country. They sputtered and smoked through the pines of New England, the wheat fields of the Great Plains, and the mountains and deserts of the Southwest. The New Haven was one of the first major lines to seriously consider equipment of this type. In 1921 the road began to purchase light rail busses, but it soon realized that larger motor cars were required for most trains.[139] By 1931 the New Haven had thirty-six motor cars in operation.[140] By 1927 the neighboring Boston and Maine was covering 130 schedules with motor cars on 60 percent of its system.[141] Twenty-four cars ran a total of 3,000 miles daily, with such success that the line hoped to replace more steam trains with gas-electrics. They were popular with the public, whose only reported complaint concerned the smell of the exhaust. The Penn-

Figure 8.42   Electro-Motive did more than any other firm to revive the gas-electric rail car. Its first rail car, produced in 1924, is shown here. (Electro-Motive Division, General Motors Corporation)

sylvania also found the motor car a necessity and acquired over fifty such units by 1930. The Burlington had the largest fleet (fifty-seven cars), which covered 27 percent of its passenger mileage.[142] The Burlington's motor cars rolled up 3.3 million miles annually. The Southern Pacific carried its Texas-Louisiana passengers in gas-electrics that paid for themselves in twenty-two months.[143]

Some smaller lines abandoned all steam passenger operations in favor of motor cars. The Gulf, Texas, and Western netted $22,555 from gross passenger and mail revenues of $35,558.[144] Its Brill rail car, which cost about $17,000, paid for itself in less than a year. In 1925 the Cincinnati Northern, a New York Central subsidiary, replaced all its steam trains with motor cars and thereby saved 29.3 cents per mile.[145] The year before, the Gulf, Mobile, and Northern had begun a similar program on a far grander scale. The 800-mile railroad was a major trunk line that operated the full range of passenger service, from plush limiteds to seedy locals. At first, motor cars were assigned to the lesser trains, but in 1930 the St. Tammany Limited was reequipped as a four-car motor train. The new train was hardly super-deluxe, but it had nice furnishings and was used for through service. In 1935 the last steam passenger operations on the Gulf, Mobile, and Northern were abandoned in favor of motor trains.[146]

During a single decade the motor car had grown from an unloved infant into an admired adult. Since the early 1920s it had proved a reliable transportation tool, able to run for millions of passenger miles at remarkably low cost. Much of its success is attributable to the perfection of the internal combustion engine. Such power plants could now operate for 6,000 hours, or 150,000 miles, between major overhauls.[147] The fussy, undependable

motors that had so frustrated the efforts of McKeen and the other motor car pioneers were now an antique memory. The growth of the motor car fleet is outlined in Table 8.1.

Table 8.1   Self-Propelled Rail Cars on Class-One Railroads

Year	Electric	Gasoline	Gas-Elect.	Diesel	Not specified	Trailers & misc.	Total
1915	—	—	—		—	—	1,647
1925	1,462	←	(112)°	→	12	293	1,879
1930	2,541	227	581	6	6	2,233	5,594
1935	2,451	277	611	19	9	1,178	4,545
1940	2,290	192	550	28	6	1,041	4,107
1950	2,178	134	252	137	0	944	3,645
1960	—	—	—	—	—	—	2,975
1970	2,444	←	(174)°	→	15	—	1,879

° Breakdown by gasoline, gas-electric, and diesel motive power is not given. The figure covers all three categories.
Source: I.C.C. statistical annual reports for the years noted, except for the 1915 figures, which are from *Railway Mechanical Engineer*, June 1915.

The motor car's triumph was cut short in 1930 by several factors. The Great Depression following the 1929 panic dried up passenger traffic. In 1920 each American traveled an average of 440 miles a year by rail; in 1934 that figure was down to 146 miles.[148] The most noticeable decline was on short-haul trains, where traffic decreased by 84 percent. Thus the motor car's chief market had nearly vanished, since the people who continued to travel were going by highway. Automobile registration continued

Figure 8.43   Electro-Motive's early rail cars weighed 38 tons and were powered by 175-horsepower Winton gasoline engines. (Railway Review, *September 6, 1924*)

to climb; even during the worst years of the Depression, 1 to 3 million new cars crowded onto the highways annually. In addition, intercity buses expanded their operations. By 1930 the Greyhound Corporation operated over 5,000 miles of routes and contemplated forming a national network.[149] Within three years that goal was more than realized, for Greyhound had expanded its route mileage to 40,000. Other independent bus operators were expanding just as aggressively—all to the detriment of the railroads. Some roads decided to join the competition rather than fight it; the Pennsylvania, for example, bought a large block of Greyhound stock, while the Burlington formed its own motor bus division.

The railroads were forced to admit that although rail cars could reduce operating expenses, they could not reverse America's trend toward the highway. Motor rail cars proved to be only an interim solution. As branch-line and secondary trackage was abandoned (over 15,000 miles disappeared during the 1930s), more and more surplus motor cars appeared on the storage tracks.[150] The orders for new cars dried up: in 1929, 135 were ordered; in 1933, 11.[151] Yet just when the motor car seemed

destined for extinction, an unexpected revival took place in the form of the glamorous motor trains for the long-distance traveler. A series of sleek, radical, lightweight self-propelled cars opened a new era in American railroading and made what was once a dowdy necessity into a star. The coarse utility of the traditional gas-electric gave way to the racy, glossy look of the streamliner. In the words of Ogden Nash: "The Iron Horse has run its course and we ride a chromium weasel."

The streamline motor car era began with an experimental unit designed and built by W. B. Stout (1880–1956), an aeronautical engineer. Stout's turtle-like *Rail Plane* was an ultra-lightweight vehicle which incorporated airplane materials and construction techniques. Its welded tubular frame and aluminum skin helped hold the weight of this 60-foot vehicle to only 12½ tons, or 500 pounds per passenger. A conventional steel coach weighed nearly 5,000 pounds per passenger. The *Rail Plane* was ready for tests by September 1932.[152] It registered speeds of 90 miles an hour and ran briefly on the Gulf, Mobile, and Northern Railroad.[153] Pullman had sponsored the project in the hope that the radical design might lead to a commercially viable product. Indirectly

Figure 8.44 This Electro-Motive car of 1926 developed 220 horsepower. The body was produced by the St. Louis Car Company. (Railway and Locomotive Historical Society)

this goal was realized, for Stout's experiment attracted the attention of W. Averell Harriman, the board chairman of the Union Pacific Railroad. Twenty-five years earlier his grandfather had been inspired by torpedo boat experiments to order the studies that led to the McKeen cars; now the younger Harriman encouraged the development of Stout's lightweight design for everyday service. The project was headed by a member of the Union Pacific's engineering staff, Everett E. Adams. Pullman's own designers and W. B. Stout were also involved.

Preliminary studies quickly elminated the single-unit self-propelled car and concentrated the design effort on a self-propelled car with two trailers permanently attached. The Union Pacific wanted a multiunit, articulated motor train for high-speed main-line service. An order was placed with Pullman in May 1933 for a three-car train, referred to as the M10,000.[154] The first, or power, car contained the engineer's cab, a 600-horse-power distillate engine-generator, and baggage and mail compartments. The trailing cars offered coach and buffet seating for 116. The 200-foot-long train was built largely of aluminum alloys and weighed 85 tons (Figure 8.50). It was 3 feet lower than conventional equipment, and its sheathed underside cleared the

rails by just 9½ inches.[155] The framing, riveted together from plate, channels, and square tubes, was reminiscent of the antique schemes proposed by LaMothe and the other pioneers associated with tubular car construction. But Stout's goal of lightweight construction was partially frustrated by the safety requirements. Heavier framing members were needed to match the strength of conventional equipment, because the smaller cross section of the motor trains resulted in less stiffness.[156] In addition, overall weight could not be reduced much more than 25 percent, even with high-tensile-strength metals. Because thin sheets tend to buckle or deflect, actual strength is below the theoretical figures. For these reasons the remarkably low weight per passenger achieved in the *Rail Plane* could not be duplicated in the M10,000.

When delivered in February 1934, the M10,000, or City of Salina, was a sensational brown and yellow caterpillar on rails. It looked like a monstrous airplane fuselage without wings, tail, or wheels. It was heralded as "Tomorrow's Train Today" during a 13,000-mile, twenty-two-state tour. President Roosevelt personally inspected the train during its Washington stop, and it was said that the Federal government shut down so that the humblest

Figure 8.45   Standard Steel Car Company constructed this powerful gasoline rail car in 1928 for the Baltimore and Ohio. It was converted to diesel power in 1935 and was retired in 1950. (Baltimore and Ohio Railroad)

civil servant might retrace F.D.R.'s steps. Finally in January 1935 it entered regular service.[157] It continued to operate until late 1941 when, after nearly 900,000 miles, it was sold for scrap. Its limited capacity had always been a handicap, and now that the engine was worn out, installing a new diesel was not considered economic. In addition, the demand for scrap aluminum was great because of the war emergency. For all these reasons the M10,000 was condemned to early retirement.

Even before the first motor train was completed, Harriman began to order more aluminum streamliners from Pullman. Each one grew in consist and in overall physical size until 1936, when the last of the motor trains, the City of Denver, was delivered (Figures 8.51 and 8.52). The equipment had swollen almost to the full configuration of conventional rolling stock. All were named for principal municipalities on the Union Pacific system: the City of Portland, the City of Los Angeles, the City of San Francisco, and the City of Denver. The City of Denver had two locomotives totaling 2,400 horsepower, ten cars with seating for 182, and a combined weight of 666 tons. This train, which was

equal to full-sized equipment, ran an impressive total of 5.6 million miles before being retired in 1953.[158]

Pullman produced only one more motor train on the general design of the M10,000. The Green Diamond, a five-car articulated unit, was delivered to the Illinois Central in the spring of 1936.[159] The 230-ton train was propelled by a 1,200-horsepower diesel-electric plant. Unlike the Union Pacific automotive trains, the Green Diamond was made from Cor-Ten steel instead of aluminum, which accounted for its greater weight. The bodies were riveted rather than welded fabrications. The first car was devoted entirely to the machinery of locomotion. The second was a combined Railway Post Office–baggage car. The third and fourth cars were coaches, while the trailing observation had a kitchen as well as dining and lounge compartments. The 328-foot, 6-inch-long train seated 146, including the occupants of the diner-observation car. The Green Diamond represented the Illinois Central's effort to capture a larger share of the Chicago–St. Louis traffic, for which it competed with the Alton, the Chicago and Eastern Illinois, and the Wabash. Mileage and speed on all

Figure 8.46   The Reading Company's Number 74 was among the first diesel-electric rail cars in the United States when it entered service in 1928. The 73-foot car weighed nearly 69 tons. (Westinghouse Electric Corporation)

Figure 8.47   Plain interiors were characteristic of gas-electric cars, because they were normally assigned to short secondary runs.

Figure 8.48 *The engineer's cab of gas-electrics was compact and noisy. The engine is visible to the left.*

Figure 8.49   The B & O modernized a twenty year-old Brill gas-electric in 1946 with a fancy paint job and a 250-horsepower diesel engine. The Number 6040 was retired in 1953.

Figure 8.50   The Union Pacific's City of Salina was the first streamlined motor train to enter service in the United States. Pullman completed it in 1934. (Pullman Neg. 37922)

Figure 8.51   Because the City trains were articulated, car bodies spanned a common pair of trucks at each end. (Pullman Neg. 38747)

roads was about equal, but the Illinois Central hoped to lure patrons away from its rivals by capitalizing on the public's infatuation with streamliners. The ploy seems to have worked very well, for the train ran full, and a second section was often needed to accommodate the disappointed passengers who could not crowd aboard the Green Diamond. Reservations had to be made a week in advance, and business was so good that the I C regretted not having bought a ten-car motor train. In the first six months of 1938, the train was netting $110,000, or nearly 40 percent of gross ticket receipts.[160] This was an excellent return, considering its first cost of roughly $425,000. The line's experiment with motor trains was amply repaid before the Green Diamond was retired in 1949 and scrapped one year later.

Pullman's rivals were eager to gain a share of the streamline motor train market. Budd, in fact, had entered the car-building business by this means even before the City of Salina's delivery. But an even more daring attempt to introduce a new dimension in railroad passenger equipment was made by the Goodyear-Zeppelin Corporation of Akron, Ohio. The firm was looking for work, because the Depression and the disaster to the airship *Macon* had left its plant idle.[161] The company decided that it

had the technical experience and the staff to design and build a radical motor train by applying existing aircraft materials and techniques. The bankrupt New Haven showed an interest in the idea, and with a $250,000 loan from the Public Works Administration, the road signed a contract early in 1934 with the airship manufacturer. The general plan followed that of Pullman's M10,000 for a three-car articulated aluminum train (Figures 8.53 and 8.54). It was about equal in size to the M10,000, for the Comet, as the New Haven named it, seated 160 and had a power plant at either end of the train, which contributed appreciably to the total weight of 126 tons.[162] The train was equipped with 400-horsepower, six-cylinder Westinghouse diesels that were used in connection with the usual electric transmission. The body was nearly square in cross section, and the underbelly was fully covered for increased strength and better aerodynamics. The lower exterior body panels were machined in a decorative whorl or fish-scale pattern. Light- and dark-blue bands and a gray roof added to the exterior finish.

The *Comet* was ready for testing in late April 1935. Its high-speed run of 110 mph between New Haven and Boston beat the fastest regularly scheduled train by fifty-five minutes. But the

Figure 8.52 *The undersized Union Pacific motor trains were compared with a conventional coach in this 1936 drawing.* (Railway Mechanical Engineering, September 1936)

Figure 8.53 *The Goodyear-Zeppelin Corporation completed the motor train* Comet *for the New Haven in 1935.* (Railway Gazette, February 1937)

Figure 8.54   The aluminum, diesel-powered Comet *of 1935 was a mechanical success,
but it was too small for main-line needs. (Association of American Railroads Neg. 659)*

Comet had been purchased for the more prosaic operation between Boston and Providence, which did not include such a
record-breaking pace; the average speed was 60 mph. Patronage
was not as good as expected, and additional local stops were
added that further slowed the schedule. In all, the New Haven
seems to have purchased an elegant racehorse for draft-horse
service. Even so the road had no complaints, for the Comet was
available 89 percent of the time and produced a respectable net
revenue. After three years of service and 405,000 miles, a general
overhaul showed that it was in excellent condition.[163] There was
some engine wear and the train had to have new wheels; otherwise few repairs were required. The silvery, lizard-shaped rail
car rolled along satisfactorily until its retirement in 1951; yet it
remained the sole rail car produced by Goodyear-Zeppelin.

If Goodyear-Zeppelin offered the most exotic design, American
Car and Foundry produced the most conservative. In shape,
general arrangement, styling, and materials A.C.F.'s first motor
train, the Rebel, was the opposite of the trains designed by Pullman, Goodyear-Zeppelin, and Budd. A.C.F. was the last to join

in the streamline craze and may have concluded that the time for
extravagant designs was past and that more utilitarian plans
were needed to save railroad passenger traffic. Furthermore, the
purchaser, the Gulf, Mobile, and Northern, had no illusions
about extraordinary service or high-speed operations. Aerodynamic design was hardly needed for a train that would stop
every 10 miles and would average only 35 mph. The Rebel trains,
as they were called, were intended to replace standard steam
trains on existing schedules. It was hoped that they would attract
new passengers because of their fresh appearance and air conditioning. The road also expected, in view of the good record of
gas-electric cars, that the motor trains would reduce operating
expenses. The Rebels might be called the sensible man's streamliners—solid, practical, devoid of foolish experiments.

The Gulf, Mobile, and Northern ordered two complete four-
car Rebel trains for delivery in 1935 (Figures 8.55 and 8.56).
A fifteen-year loan from the Public Works Administration made
the purchase possible. Construction was undertaken at A.C.F.'s
Berwick, Pennsylvania, plant, the former Jackson and Woodin

Figure 8.55 The Gulf, Mobile, and Northern's Rebel trains were built by A.C.F. in 1935. Unlike most of the other early streamlined motor trains, they were not intended for high-speed service.

Car Company. The 73-foot motor car of the Rebels had a baggage and mail compartment.[164] Two coaches measured 75 feet long; each seated 62 and 71 passengers respectively, and each had separate sections and toilets for white and colored passengers. A small buffet in the first coach offered light food service, but no separate seating for diners. The 77-foot-long observation car had six open-section sleeping berths, a single stateroom, and a lounge seating 18. The cars were not only slightly shorter than conventional cars, but also somewhat lower and narrower. The eight-wheel cars weighed about 62 tons each—an unimpressive achievement, considering that most full-sized lightweight cars weighed about 50 tons. The blunt-nosed Rebels had less flare and style than other streamliners, but at least they had smooth sides. The exterior Cor-Ten steel sheets were welded to framing members. The paint scheme was also a plain, straightforward silver, with two broad red and gunmetal stripes.

Actually, the Gulf, Mobile, and Northern had every reason to be happy with the Rebels, for they were well suited to the job at hand. Traffic grew while expenses declined. The double-truck cars proved much more practical than the articulated trains. An extra coach operated between Jackson, Mississippi, and New Orleans, which would have been impossible with an articulated train. The equipment worked so well that the road bought a third train in 1938. The Rebels continued to run into New Orleans until 1954.[165] Then, after the trains had stood for a while in storage, the Alco engines were removed from the power cars and sold to a shipping company, and in 1962 the cars were scrapped.

Even before the time of the Rebel, A.C.F. had been considering the manufacture of a more advanced style of motor rail car. The general planning called for a single-unit, air-conditioned streamlined car with a hydraulic transmission and power plant under the floor. Quantity production was also contemplated, but the number of orders received never permitted it. Late in 1934 two cars were delivered to the Norfolk Southern by A.C.F.'s Brill

Figure 8.56 The Cor-Ten steel Rebel cars were undersized but not remarkably lightweight. (Railway Mechanical Engineer, July 1935)

Figure 8.57   A.C.F.'s motor rail car Number 2025 seated 57 and weighed 26 tons. It was delivered in 1935. (Seaboard Airline Railroad)

Figure 8.58   A.C.F. hoped that its streamlined Motorailers would develop quantity sales, but relatively few were sold. The Miss Lou was delivered in 1940. (Alan R. Lind Collection)

plant.[166] The low, well-shaped bodies were not unlike the streamlined electric suburban cars produced by the same plant for the Norristown suburban line. Two more were acquired by the tidewater line soon afterwards. The good service rendered by these units prompted the Seaboard Airline to purchase three A.C.F. rail cars.[167] These were slightly larger than the Norfolk Southern units: the body was stretched out to 64 feet, the total weight registered 26 tons, and seating was provided for 57 (Figure 8.57). Because the cars were not intended to pull trailers, no center sill was required. Couplers were provided, however, for emergency towing in the event of a breakdown. The bolsters were a box-girder weldment. The use of high-tensile steel alloys and aluminum, together with the light frame, helped trim unwanted pounds from these units. A.C.F. supplied similar cars to the Chicago and Eastern Illinois and the Missouri and Arkansas.

In 1940 A.C.F. introduced a larger, more powerful streamlined diesel rail car: the Motorailer. Again quantity production was envisioned, but again only a limited number were made—fewer than a dozen for domestic use. During the first year of production, two units were delivered to the New York, Susquehanna, and Western, a short line operating through several New Jersey bedroom communities in the environs of New York City.[168] The cars were 76 feet long, weighed 38 tons, and seated 80. They featured controls at either end of the car (eliminating the need for a turnaround at the end of the run) and center door entrances. Later in 1940 two similar Motorailers named Miss Lou and Illini were placed on the Illinois Central (Figure 8.58). Like the Susquehanna cars, they were built of Cor-Ten steel. Because they were intended for intercity rather than commuter service, baggage and buffet facilities were included. The Illinois Central also acquired a two-car Motorailer, the Land O'Corn, for service between Chicago and Waterloo. Seating for 109 passengers was supplemented by 16 places at the lunch counter situated in the trailer car. None of the trio proved entirely successful, and in 1943 and 1947 they were sold to the Susquehanna. They ran on that line until the 1950s, when they were replaced by Budd cars. Not long after the Land O'Corn entered service, another Motorailer was produced for the Missouri Pacific. Except for a large order of 22 cars sold to the South American republic of Colombia in 1941, A.C.F.'s Motorailer was indifferently received. Production was, of course, cut short by World War II, but the company's failure to reintroduce the design afterwards indicates that the product was not in demand. Moreover, A.C.F.'s management had become enchanted with another fleeting jack-o-lantern, the Talgo train, and possibly saw no need to refine the Motorailer for the postwar market.

While Pullman and A.C.F. were producing their first lightweight motor trains, a Philadelphia auto parts manufacturer decided to enter the rail car field. The Budd Company had been associated with the French Michelin firm for many years, and the success of its overseas partner in introducing rubber-tired rail cars encouraged Budd to try a similar course. The Depression had dried up normal business, and any new product line was worth investigating. Michelin produced its first rail car in 1929

and had nine in service within three years.[169] The use of pneumatic tires distinguished these vehicles from all previous rail cars. The ultimate in cushioned wheels was said not only to give a superlative ride but to increase traction by 35 percent. An internal aluminum safety ring allowed the wheel to fall only $\frac{5}{8}$ inch in the event of a puncture. A flange was also applied to each wheel to guide the vehicle in case of a failure, but under normal conditions the flange did not touch the railhead, because the soft rubber tire hugged the rail. Budd hoped to make the rubberized rail car even more appealing by offering a strong, light, rustproof body made of stainless steel. The advantages of this material and of the flawless method of fabrication, Shotwelding, were described in Chapter 2.

The first car, ready in early 1932, was a boxy twelve-wheeler seating 40 passengers. It was powered by an 85-horsepower Junker diesel engine. A smaller car, La Fayette, was next built for testing in France.[170] In November of the same year a 50-foot, 12-ton car was delivered to the Reading Company for use on its New Hope branch.[171] The 47-seater could rarely be coaxed to speeds beyond 45 mph. It was prone to derail, a failing attributed to its bouncy ride. These deficiencies, coupled with its price of $75,000, discouraged any widespread interest in the Budd-Michelin rail car. Only two more were built for use in this country, although Michelin produced thirty rubber-tired cars in 1937 for the Eastern Railway of France. Such vehicles have also been successfully used for subway service in Europe, Canada, and Mexico in recent years.[172] In 1933 the Pennsylvania acquired a two-car gas-electric, which generally followed the Reading car in overall size and design. But the pneumatic wheels worked no better than they had on previous Budd cars, and they were replaced with conventional steel wheel sets. The Number 4688 and its trailer were a balky and never-to-be-repeated experiment on the Pennsylvania, but they did remain in service longer than any of the Budd-Michelin units. The Pennsylvania kept them until 1948, when they were sold to a Virginia short line, the Washington and Old Dominion Railroad. The second owner came to regret the purchase, however, and retired them after less than five years' service.

Late in 1933 the final Budd-Michelin car was delivered to the Texas and Pacific.[173] The two-car, air-conditioned train measured 140 feet in length and weighed 52 tons. The power car had a knife-edge leading prow, two 240-horsepower gasoline engines, and conventional steel wheel trucks. The extremely light 12-ton trailer seated 76 passengers and was carried by sixteen rubber tires. It was Budd's largest rail vehicle to date, and would prove the biggest failure. Before any revenue passengers were carried on its projected Fort Worth–Texarkana route, the Silver Slipper derailed on a test run.[174] The trailer was fitted with new trucks, but the power car proved unreliable, and after only a year the Silver Slipper was scrapped. The costly failure was extremely embarrassing to Budd and forced the firm to abandon its touted rubber-tired rail vehicles.

Yet the failure of the Michelin scheme did not persuade Budd to quit the rail car field, because the firm rightly felt that its

Figure 8.59   The Pioneer Zephyr under construction at Budd's plant in 1934. The stainless-steel motor train remained in service until 1960. (Budd Company)

stainless-steel car bodies were superior to anything available in the field. Its next train, the Zephyr, confirmed Budd's faith in his product. It was such an outstanding success that the previous disappointments were quickly forgotten and Budd became a leader in the passenger car field. The Zephyr was based on an optimistic belief that rail passenger service could be saved by a fresh approach. The name itself means the west wind of spring-time promise. The Zephyr was a three-car motor train ordered in 1933 by the Burlington as a countermeasure to the Union Pacific's growing fleet of streamliners (Figures 8.59 and 8.60). The entire train was only 196 feet long and seated 72 passengers.[175] Food service was limited to a buffet. Yet because of its small size a 600-horsepower engine could propel it, and do so at a fraction of a conventional train's operating cost. Like the other pioneer streamliners, the Zephyr was an effective traffic generator. From the beginning of regular service late in 1934, the seats were well filled. Business was so good that there was a waiting list. In three years the Zephyr ran thousands of miles with very few failures.[176] The train paid for itself in thirty months, and misgivings about its safety and durability were silenced by its performance. The Zephyr struck several animals and was involved in four grade-crossing accidents (in two of

these the vehicles were trucks) without any major damage.[177] It was predicted that the train would derail easily, but the accidents proved this fear unfounded. However, a head-on collision with a steam locomotive in October 1939 did cause heavy damage to the power car.[178] It was collapsed for 28 feet, and the engine was driven back into the mail room, resulting in two deaths and thirty-eight injuries. But the train was rebuilt and ran another 3.2 million miles before its retirement in 1960.

The Zephyr's success encouraged the purchase of more Budd motor trains. In 1935 the Boston and Maine bought the Flying Yankee, a slightly enlarged articulated train. Like the Zephyr, it was an extremely durable first-generation streamliner; it ran 2,735,600 miles before retirement in 1957.[179] To cover more intensively patronized routes, the Burlington decided to duplicate the Zephyr on a larger scale, with heavier, bigger, and wider cars, and more cars per train. The Mark Twain, the Twin Cities, and the Denver Zephyrs followed in quick succession. Although these motor trains reflected great credit on their maker, Budd's prewar motor rail car manufacture was to end in failure. Actually, Budd had dropped motor car production to concentrate on fabricating conventional cars, but in 1941 it received an order from the Denver and Rio Grande Western to produce a pair of

Figure 8.60   The Pioneer Zephyr began service on the Chicago, Burlington, and Quincy Railroad in 1934. (Railway Mechanical Engineer, May 1934)

two-car self-propelled trains.[180] It built the Brigham Young and the Heber C. Kimball, which weighed just under 132 tons each and measured 153 feet in length. Two 192-horsepower diesels powered each train. But Budd had grievously miscalculated, because the cars were underpowered for the stiff grades between Denver and Salt Lake City. The engines were so overworked that breakdowns were frequent and a steam locomotive was on standby service to tow the trains in. After only seven months, the Budd twin units were retired.[181]

Perhaps the humiliating failure of these prewar trains eventually prompted Budd's engineering staff to prove that they could design a dependable rail car. Whatever the reason, the company decided to produce one for the postwar market. The project seemed ill-advised, for no such cars had been built for years, and the industry's rejection of the Motorailer appeared to demonstrate that even a modern rail car was not wanted. Streamlined motor trains proved a short-lived phenomenon as well. American railroads chose instead to buy full-sized lightweight locomotives and cars. Rail cars and motor trains were a vanishing breed. But Budd felt that it could succeed where others had failed, because a dependable lightweight rail car could provide a cost-cutting tool that would attract the railroads. A high-horsepower unit built with economical off-the-shelf components would be more than a streamlined gas-electric. In fact, electric transmission would be replaced by a hydraulic transmission to save weight and first cost. Two 275-horsepower G.M. diesel engines, designed for rugged duty in army tanks, supplied the snap and ginger unknown to the old gas-electrics. Horsepower was figured at 9 per ton, or about double that of the older generation of rail cars.[182] The engines were placed below the floor; a jack shaft transmitted power to one pair of wheels on each truck. The engines were mounted in a cradle that could be quickly rolled out from under the car for inspection or servicing. If one should need major repairs, a skilled crew could replace it in just over an hour with a standby engine.

The 59-ton cars were designed for multiple-unit operation, so that two or more might be coupled together and operated by a single driver. Disc brakes made stops possible in 2,300 feet at 85 mph.[183] Generous 90-passenger seating made the cars ideal for commuter and accommodation trains. And they were attractive, with shiny stainless-steel bodies and bright modern interiors. The entire package was thoughtfully conceived and well executed. The original selling price of $128,750 was not much greater than the charge for a lightweight coach.[184] Operating costs were equally attractive: in 1950 they were calculated at 56 cents per mile. Diesel fuel was cheap at 8 cents a gallon, and the thrifty Budd car realized 2 to 3 miles per gallon. In later years the selling price crept up to $160,000 and operating costs rose to "less

than 72 cents per mile," but all other costs had gone up as well, and the Budd rail car remained a bargain.[185] Maintenance cost 26 cents a mile, or about one-quarter that of an equivalent steam train.

Budd named its new product the RDC for Rail-Diesel-Car. A demonstrator was ready in July 1949, and the first production car went to the Boston and Albany in May of the following year.[186] Within three years 116 were in service on railroads in all regions of the country, as well as in Australia, Cuba, and Saudi Arabia. When the last RDC left the Red Lion plant in 1962, a total of 398 had been produced. Of the four production models offered, the RDC-1, a simple coach, was the most popular (Figure 8.61). It accounted for two-thirds of the cars produced for domestic service. The RDC-2 and RDC-3 had baggage and mail space, making them suitable for long-distance, main-line, or branch-line service where head-end business was a factor. The RDC-4, the least popular model, had no passenger space, for it was devoted entirely to baggage and mail (Figure 8.62). It was 11 feet shorter than the 85-foot standard length of the other models. In addition to these four, an odd class, the RDC-9, was produced for the Boston and Maine in 1956.

The RDC's performance on the Western Pacific was an early and much publicized success.[187] The railroad operated a steam train, mainly for the accommodation of its employees, between Oakland, California, and Salt Lake City. The route ran 924 miles through sparsely settled country that was a traffic manager's nightmare. The train cost as much as $310,000 a quarter to operate, and because most of the passengers were deadheads, the best the line could hope to do was cut its losses. Two RDC-2s purchased in 1950 achieved that goal by trimming operating costs to a manageable $76,900 a year. Thus a net loss of $60,000 became a profit of $207,000. The cars proved economical and dependable, performing without failure through subzero weather and driving snowstorms so fierce that the glow of the headlight was obscured.

Eastern lines made good use of the RDCs as well. With their help the B & O cut operating costs on local trains between Baltimore and Washington by 51 percent. Later the road operated RDCs in over-the-line service. The cars were rarities in that they offered dining facilities which went beyond minimal buffets. Six tables seating 24 passengers were an unexpected luxury for patrons of the normally bare-bones rail motor car.[188] The B & O also demonstrated the Budd car's capacity for speed by staging a fast run between Washington and Chicago that took twelve hours and twenty-nine minutes. This was not typical, however; most RDCs operated on leisurely schedules.

The two major roads serving Boston were the largest operators of RDCs. In 1953 the New Haven had forty of them running 1.25

SK-22833

RDC-1 is strictly for carrying passengers. It seats 89, on walkover seats.

Weight, light	108,000 pounds
Weight, ready to run	112,800 pounds
Normal maximum weight (including 89 passengers)	126,600 pounds

Figure 8.61 Budd's RDC, a diesel mechanical-drive unit, proved one of the most popular and successful rail cars produced in the United States. Nearly 400 were turned out between 1949 and 1962. (Budd Company)

million train miles a year. They replaced sixteen locomotives, sixty-four cars, and four old-model rail cars. But the Boston and Maine had an even larger fleet—the largest in the country. How and why the road came to acquire 109 Budd cars between 1952 and 1958 was explained by a former member of its mechanical department, Ellis Walker.[189] He said that the state public utilities commission had pressured the line for years to retire its aging fleet of wooden commuter cars. The railroad pleaded poverty but finally agreed that if certain trains, all notorious money losers, could be abandoned, the old cars would be taken off. The deal was in the public interest, and after some negotiation apparently an unofficial understanding was reached. The B & M bought secondhand steel cars as well as new Budd rail cars, both of

which finally replaced the last wooden cars serving the American traveler.

The last RDC was produced in 1962 for the Reading Company, yet hundreds of them continue in daily service. Today most are assigned to commuter runs. Many have been scrapped, others have been sold to Canadian roads, and in the case of the Rock Island, several have had their engines removed and been converted into ordinary locomotive-hauled coaches. But even as they drop away one by one, the RDCs are remembered as one of the most successful styles of self-propelled rail car ever constructed.

The other rail car designs of the postwar era present a marked contrast to the RDC. Fanciful concepts were promoted in place

Figure 8.62   This RDC-3 seated 48 and carried 5,000 pounds of mail and 8,250 pounds of baggage. It was delivered in 1950. (Lawrence L. Williams)

of sound engineering. The low-center-of-gravity panacea became fashionable despite an embarrassing series of failures. Recent history was ignored as this superficially appealing scheme was resold to another generation of railroad presidents, whose ignorance of the technical problems led them to waste millions of dollars. The Talgo, Xplorer, Train X, and their various cousins, sisters, and aunts were costly failures that briefly captured public attention because of their radical appearance and their promoters' hard-sell tactics. But they failed to deliver safety, comfort, or reliability. Their modest operating savings were more than offset by their high first costs, large repair bills, and operating difficulties.

The leader of the low-center-of-gravity revival was the American Car and Foundry Company. Once a major builder of passenger equipment, the firm had lost its number-two position in the industry to Budd even before Pearl Harbor. A.C.F. apparently hoped to regain its position by marketing a revolutionary concept in passenger train design that had been advanced by a Spanish engineer. It was a low-slung, ultra-lightweight design that A.C.F. hoped would complete the transition to the automotive train. Some staff members may even have viewed the success of the lightweight car as only an intermediate phase in railroad development. Perhaps they hoped to lead a large-scale conversion to automotive trains before their rivals could perfect their own designs. Whatever the real motives of A.C.F., the firm spent ten frustrating years trying to sell the Talgo concept in America.

As early as 1837 W. B. Adams, an English mechanic, had proposed low-slung, articulated, two-wheel railway carriages. The idea was reintroduced more than a century later by A. A.

Goicoechea, a Spanish engineer who named his design Talgo as an acronym for train, articulated, light, Goicoechea, and Oriel—the last being the name of the family who helped finance the project. The inventor proposed a two-wheel triangular frame or running gear, coupled together so as to lead the trailing units into curves. A prototype was built in Spain for test purposes, but it was realized that a serviceable train must be produced by an established car builder. The job was given to A.C.F., and so began America's misadventure with the Talgo.[190] By April 1949 the first of three duplicate trains was ready (Figure 8.63). Two were sent to Spain, and the third was retained by A.C.F. for testing and display.[191] The 167-foot-long train was made up of an 810-horsepower diesel electric locomotive, a baggage car, and four coaches. Except for their width, the squat coaches were equal in weight and size to a nineteenth-century horsecar. The basic construction was aluminum, which helped to hold the 20-foot-long cars to a weight of 3½ tons. The wheels were of the sandwich-core type made of rubber and steel, much like those used on P.C.C. streetcars. The air-powered hydraulic brakes had automotive-type brake drums. The two trains that were shipped to Spain entered regular service in July 1950 after prolonged testing. They cut 3½ hours from a 396-mile run that traversed mountainous territory between Madrid and the French border. They were found to be somewhat underpowered and difficult to operate in reverse except at extremely low speeds.[192] An American writer reported that the ride was poor, with a pronounced fore-and-aft motion, but that the trains went around curves smoothly and with little roll. The Spanish were pleased with the Talgo and eventually purchased more, though not from A.C.F.

Figure 8.63   The low-slung, articulated Talgo train was a Spanish design first manu-factured in this country by A.C.F. in 1949. (National Archives)

Meanwhile A.C.F. was unable to create much enthusiasm among American customers. In 1953 the company built a second, improved demonstrator in which the individual coach units were lengthened from 20 to 34 feet.[193] Three units were coupled together to form a six-wheel car that could be cut in or out of the train, thus answering the usual objection to articulated trains by American operating men. The running gear was redesigned to facilitate operating in reverse, and the framing was strengthened to meet the 400-ton buff load requirements. Yet the cars were still low-slung and undersized. The new Talgo was indifferently received at first. But a shake-up in top railroad management brought men into power who were attracted to innovative schemes such as the Talgo. Robert R. Young had gained control of the New York Central, and one of his admirers, Patrick B. McGinnis, was running the New Haven. Both faced critical fiscal problems because of passenger train losses. In former years, Young had expressed faith in exotic equipment as a solution to the passenger train dilemma. He envisioned a vast new fleet of cars for the Chesapeake and Ohio and dreamed of taking over the Pullman Company. None of these plans were realized, but he did finance experiments with a Talgo type of design in 1947.[194] The C & O was a coal road and passenger traffic was a minor consideration, but the Central was a major passenger carrier, and its heavy losses made desperate measures appropriate. Young, who was a showman, hoped to reverse the irreversible by introducing a futuristic train.

It seemed that the Talgo would shed its Cinderella role, for orders began to appear in 1955. The New Haven purchased two, the John Quincy Adams and Dan'l Webster. The Adams was delivered by A.C.F. in 1956.[195] It had five three-unit cars seating 480 passengers. The Webster was produced by Pullman and seated only 392. The Boston and Maine ordered a train similar to the Adams but did not name it. The Rock Island purchased another A.C.F. Talgo, also in 1956, and called it the Jet Rocket. The New York Central showed good faith by purchasing a Pullman-built Talgo named the Xplorer. These trains cost anywhere from $788,000 to $1,200,000 each. They were scheduled for main-line, long-distance travel, and according to their publicity, would usher in a new era of railroad travel.[196]

But once the press parties and champagne excursions were over, the super-lightweight bubble burst. The Talgos really offered nothing new. A low center of gravity provided no measurable advantage in the ride or in passenger comfort. Just the reverse—passengers complained about the poor ride and high noise level.[197] The low ceilings, small seats, tiny toilets, and box-lunch dining service were decidedly substandard, and the public was not impressed. Both the Jet Rocket and the B & M Talgo were quickly withdrawn from through-train service and downgraded to commuter runs. The New Haven trains were disaster-prone. The Dan'l Webster caught fire on an inaugural press run, and it and its companion cars regularly broke off their third-rail shoes when entering the Grand Central electrified trackage.[198]

Figure 8.64   The United Aircraft Turbo Train of 1967 was a space-age fizzle. All U.S. units were retired in 1976. (Amtrak)

They were retired after several years, and the Xplorer was laid off after two years.

The Talgos were another costly failure for the railroads, but other industrial giants also took losses on the extra-lightweights. General Motors thought it could build a cheap lightweight by adapting stock components for the purpose. EMD (the Electro Motive Division) built two Aerotrains in 1956. The 1,200-horsepower locomotives were duplicates of the unit fabricated for the Jet Rocket. The cars were standard 40-foot G.M. bus bodies widened by 18 inches.[199] They were carried on four wheels and suspended on motor-bus air springs. The 16-ton cars seated 40 passengers. The trains had ten cars in their consist. One train was sent East for a year's trial on the Pennsylvania. The second train operated on several Western lines. The Union Pacific leased it for a time, but rapidly became disenchanted with its poor performance and high repair costs.[200] After a brief Mexican tour, the Aerotrain was sold in 1958 to the Rock Island for commuter service. At low speeds the train worked fairly well, but passengers complained about the annoying sideways motion that sometimes became violent.[201] The Aerotrain was retired in 1965 and was subsequently given to the National Museum of Transport in St. Louis. There it now stands, a forlorn monument to a well-intended but ill-conceived effort to revive passenger train travel in America. The Talgos and their second cousin, the Aerotrain, were clearly experiments introduced after their time rather than before it. In the words of David P. Morgan, editor of Trains

magazine, they "proved spectacularly unsuited" to the needs of the railroad industry.

The most recent development in rail motor cars is the gas turbine car. Actually its roots are in the Train X, Talgo era, though it has been represented as something very new. The publicity once again claimed that the car had achieved a marriage of aircraft and railroad technology. Supporters spoke of lightweight aluminum construction, low centers of gravity, wind-tunnel tests, and other supposedly revolutionary ideas that had in fact been knocking around for more than forty years. The designer of the Turbo Train was Alan R. Cripe, an engineer who had been involved in the planning of Train X and its successors. He held patents on many components used on these trains and was a firm believer in guided axles and controlled banking systems.[202] In 1963 the prospects for more experiments in this area seemed remote, but Cripe had faith in the concept and delivered a paper on it that was heard by an official of United Aircraft. The U.A. executive was impressed and convinced others in his company to support Cripe's research. The decision proved timely, for the high-speed ground transportation act was passed in 1965 and Federal funding for experiments of this nature was now available. A contract was signed the following year, and a test train was rushed through production and completed in May 1967. The double-ended, welded-aluminum train, reminiscent of the motor train built nearly thirty-five years earlier, measured 203 feet long and weighed 105 tons (Figure 8.64). It seated 144 passengers and stood 35 inches lower than a regular train. Four 400-horsepower turbine engines propelled the train; a fifth turbine ran the generator for heat and lighting. The train was complex, sophisticated, and admittedly experimental, and understandably it proved troublesome. It was not ready for regular service until almost two years after delivery, and even then mechanical problems kept it out of service 54 percent of the time. This was a dismal record compared with an ordinary diesel passenger locomotive, which registered a downtime of 22 percent.[203] The transmission, lubrication pumps, air suspension, and axle guidance system were particularly temperamental.[204] A ride on the train in 1970 was memorable for the tangle of loose wires, gauges, and electrical test gear in the passageways of the power units, the entire maze being monitored by a worried young man who was presumably a United Aircraft engineer. The train rode poorly at slow speeds but smoothed out remarkably as the pace quickened. The seats were full, and according to the crew, the train was very popular with the public.

By the end of 1970 the Turbo train had logged 500,000 miles. Better insulation had reduced the noise level, and an additional car was being added to boost seating.[205] It was kept running, though not without difficulty, and the pitiful breakdown record of the five turbos that had been built for Canada did little to restore faith in United's ability to produce an economical or dependable rail car. When it came time to order more turbine rail cars in 1972, Amtrak selected a French contractor. These units were straightforward double-truck rail cars, unencumbered by exotic suspensions, that had operated successfully in France since 1967. When a second order was given to the French in 1974, United Aircraft decided to quit the rail car field. It seems regrettable that no American manufacturer developed a practical turbo rail car to replace the RDC. Still, it is more than likely that the Talgo–Train X scheme will snare yet another sponsor, for like the monorail, it is an alluring engineering concept whose charms never seem to fade.

Electric Self-Propelled Cars

The growth and concentration of suburban travel to and from major American cities led to the adoption of electric self-propelled cars on several North American railroads. (It will not be possible to discuss all these installations here, nor is it necessary to do so, since they have been well covered in a volume by William D. Middleton.[206] Also, this section will not treat interurban and rapid transit lines even though some, such as the Chicago, South Shore, and South Bend, operated equipment of main-line proportions.) Beginning as early as 1835, Boston became a center for commuter traffic.[207] Within three years four railroads were hauling commuters into the hub; some even offered season tickets. By 1860, 143 daily trains were serving the outlying towns. Twenty years later, 403 trains were necessary as Boston jobholders sought housing in Newton, Lynn, and other nearby villages.[208] The outskirts of New York, Philadelphia, and

Chicago were growing in the same way. In time this traffic swelled to overwhelm the big-city terminals: the locals were crowding out the main-line trains. In the early years of the twentieth century suburban traffic was growing at 20 to 25 percent a year, and while four or five tracks could handle the through trains even in major cities, another twenty-five tracks were needed to accommodate the suburbans.[209] By 1925, 364,000 commuters—more than the entire population of Arizona—moved in and out of New York every day.

Some lines built auxiliary downtown stations to handle the traffic; others continued to expand their main terminal. After a time, however, no more property was available, or else it was not available at a price any railroad could afford. Managements made various efforts to improve the efficiency of suburban trains. The Boston and Albany and the Illinois Central purchased special double-ended locomotives with high tractive power to hustle the trains in and out of the station more quickly. Because they could run equally well in forward or reverse, no time was lost in turnarounds. But such measures were not enough in the more extreme cases of traffic density, and some railroads decided that electrification was ultimately cheaper than endlessly expanding terminal trackage. What the electrics offered was rapid acceleration—three times that of a steam locomotive. The effect on terminal capacity is obvious. But electrification involved more than cars and trolley wire: expensive substations and the mechanics of power distribution added materially to costs, which even in the 1920s could reach nearly a million dollars per mile. Sometimes the cost of electric suburban operations was justified on a road because of its traffic volume, yet its management could not or would not raise the capital to do the job. The lines running out of Boston were cases in point. Elsewhere, electrification was adopted only after municipalities passed smoke abatement laws, for clean air was another major advantage of electric suburban service.

Electric railroad technology was worked out in all details by the street railway industry almost a decade before any main-line electric operations took place. Traction motors, gearing, controllers, and power generation and distribution were refined and proved for dependable everyday use when the Baltimore and Ohio opened its Baltimore tunnel line in 1895. Although some additional problems were involved, main-line electric railroading was essentially trolley engineering on a larger scale. Sensibly, the steam railroads borrowed as many ideas as possible from the city systems. They took the 600-volt DC plan and its appurtenances, and they even employed engineers, such as Frank J. Sprague or George Gibbs, who had made their reputations in the transit field. Sprague offered what was possibly the single most important contribution to electric railway engineering: the multiple-unit system, which enabled one man to operate two or more motor cars. The M.U. system saved more than labor alone. Because the motors were under each car, power was spread throughout the entire length of the train. And because the ratio of power to weight was generally higher than that of a locomotive, acceleration was greater. This was a prime requirement on

Figure 8.65 The cluttered underside of a standard-era, multiple-unit electric rail car is shown here. Most of these pipes are electric conduits. (Pullman Neg. 29720)

lines where traffic was congested; the faster that trains could get in and out, the greater the capacity of the railroad. Sprague's system was first used on the Chicago elevated in 1897. It quickly established itself as a basic transit tool and is used today by every electric railway that runs multiple-unit equipment.

Electric cars have other common characteristics. They are heavier, more expensive, and more complicated than an ordinary coach. Their seating capacity is greater, their luxuries fewer than their locomotive-hauled counterparts. Toilets are small or non-existent. Interior decor is spartan and the seating elementary. They are arranged to carry the maximum loads with the minimum comfort. Because the trips are short, the fares substandard, and the prospects of a profit ludicrous, most patrons bear the miseries of commuting with resignation. Mechanically the cars are a forest of wiring and underbody gear (Figure 8.65). Where to hang all the controls, breakers, resistors, transformers, air compressors, and air brake gear are problems that baffle successive generations of engineers. To keep the body space free for

the maximum number of passengers, everything possible is placed under the floor. The operator's cab is a tiny closet crowded into one side of the vestibule. Electrical gear easily adds 10 tons to the weight of the car, which makes it necessary to use heavier axles to carry the load and withstand the torque of the traction motors. Because of all the extra mechanical-electrical apparatus, M.U. cars cost roughly 25 percent more than regular coaches.

Multiple-unit cars were commonly coupled together semipermanently as married pairs—sometimes as motor-trailer combinations, but also as twin motor cars. It was considered prudent to have controls in the trailer car to facilitate switching operations. Large entrances and sliding doors were also common. The newer cars were patterned ever more on subway vehicles and often had center entrances and no vestibules. Toilets were eliminated wherever local health regulations did not interfere, and separate toilets for the sexes were rare. Good ventilation was a necessity because of the heavy loads carried at rush hours. Air conditioning came last to the suburban commuter, and it arrived in a big

Figure 8.66   The Pennsylvania Railroad designed its class P-54 suburban coach in 1907 for conversion to electric service. Figure 8.72 shows a later example. (Railway Review, June 8, 1907)

way only after government subsidies were offered for new equipment.

The first large-scale use of M.U. electric rail cars in America began in 1905 on the Long Island Railroad.[210] Traffic was booming and plans were underway for a direct connection into the Pennsylvania's Manhattan station, which was then under construction. The program was started none too soon, for between that date and 1919, passenger volume climbed by 353 percent. Eventually about 40 percent of the line was electrified, and these nearby sections carried over 70 percent of the traffic.[211] The first cars were designed by Gibbs and duplicated those planned for the New York subway. In 1905–1906, 134 were built. They measured only 51 feet long and were made decidedly undersized in all dimensions in order to fit the subway loading gauge. Sometime afterwards it was decided to build a separate full-sized tunnel, and the next cars were modeled after the parent road's standard steel suburban coach, the class P-54. Because the electric cars were self-propelled rather than locomotive-hauled, they were classed as MP-54 (Figure 8.66). In 1909 and 1910 A.C.F. turned out 222 MP-54s for the Long Island.[212] They measured 64 feet, 6 inches long, seated 72 passengers, and weighed 54 tons. Two 225-horsepower motors drove each car. Around 1915 the Long Island adopted a Gothic arched roof in place of the squat clerestory. The design was not particularly pleasing and gave the cars a more subway-like appearance (Figure 8.67). Hundreds more were built on this plan; by 1930 the road had 1,000 MP-54s in service.[213] This was the largest M.U. fleet in North America and represented a triumph in standardization.

In 1932 the Long Island acquired an experimental double-deck car, the Number 200, produced in the parent road's Altoona shops. The MP-54s were dependable workhorses, but in view of their limited seating and heavy weight, it seemed as if a better design must be possible. The prototype had a split-level, or seat-over-seat, arrangement; it was not a double-decker in the sense of the modern gallery car. The Number 200 was a trailer measuring 72 feet long and weighing 36 tons.[214] Two more were constructed in 1937, but further production was delayed until 1947 because of the war. Sixty were built in Altoona in 1947–1948 at a cost of $143,000 each.[215] They were 80-foot-long motor cars equipped with air conditioning (Figures 8.68 and 8.69). They seated 132, or nearly double the number of the old MP-54s. But they were not popular with the public; the steps to the upper seats were a hazard, the many nooks and corners made them difficult to clean, and women disliked to occupy the upper seats for fear of encouraging peeping toms.[216] The railroad decided that the cars were too expensive and that single-level cars with

two-and-three seating were a more economical alternative. No more bilevels were ordered, and by 1971 a number had been retired. During the period 1953–1955, 160 new M.U.s were purchased that featured two-and-three seating. These larger-capacity cars permitted the M.U. fleet to be reduced to 718.[217] At the same time, hundreds of old cars were given a thorough mechanical refurbishing, modern interiors, and new seating. Air conditioning was considered too expensive, but powerful fan-ventilating systems were installed.

In 1906 as part of its Grand Central improvement program, the New York Central began the second main-line electrification to make large-scale use of M.U. cars. The third rail was eventually extended 34 miles to Croton and 24 miles to White Plains.[218] Electric locomotives hauled through trains in and out of Grand Central, but the suburban service was to be maintained by 125 motor cars furnished by A.C.F. These short, arched-window cars had a distinctly interurban look (Figure 8.70). They measured 62 feet long over the couplers, seated 64, and weighed 53 tons. The trucks were fabricated by the American Locomotive Company. The side frames were forged from square bars much like a miniature steam locomotive. There were 600-volt DC motors that produced 200 horsepower per car. The cars had both electric and gas lighting. Steam and electric heating were provided as well. The dual systems were necessary in the event that the cars operated outside of the electric zone. Between 1910 and 1929 the Central acquired over 200 additional motor and trailer cars. These units were about 3 feet longer than the original cars and looked more like conventional coaches. Express trains consisted entirely of motor cars, ranging from two to ten per train. Locals ran with a mix of two motor cars to each trailer, with three to twelve cars per train.

In 1926–1927 the motor cars were cut in half for the addition of an 8-foot section, which increased the weight by 15 tons and the seating by 12.[219] The art-glass arched windows in the older cars were plated over. Many of the trailers were motorized, and the electrical systems of all units were reworked so that trains could be operated in multiples of fifteen cars. Except for ten new cars delivered by Standard Steel in 1929, the New York Central M.U. car fleet remained static until 1950, when St. Louis began deliveries of 100 new motor cars.[220] The 85-foot-long, smooth-sided, welded steel cars weighed 75 tons and carried 130 passengers in two-and-three seating (Figure 8.71). These arched-roof cars followed lightweight car design. They were a welcome supplement to the Central's aging M.U. fleet, although technically they were unremarkable. The old-fashioned 600-volt DC system was dependable but obsolete. In 1962 fifty-three new motor cars

Figure 8.67   The Long Island Railroad modified its parent road's class P-54 suburban coach with a distinctive arched roof. The drawing here dates from about 1925. (Master Car Builders Report, 1926)

Figure 8.68   The Altoona shops produced this electric double-decker in 1947. The 80-foot car seated 132 passengers. (Pennsylvania Railroad)

Figure 8.69   The interior of a Long Island Railroad bilevel commuter car. (Pennsylvania Railroad)

Figure 8.70 *The New York Central's short electric cars of 1906 were among the first steel cars used in main-line service in this country.* (Car Builders' Dictionary, 1906)

Figure 8.71 *The St. Louis Car Company produced this electric M.U. car for the New York City suburban service out of Grand Central Station in 1950.* (St. Louis Car Company)

Figure 8.72   A late-model class MP-54 electric M.U. car built at the Altoona shops.
(Altoona Public Library)

made possible the retirement of a large portion of the older cars.[221] Only the Bronx locals were serviced by the prewar heavyweights after this time.

Some historians have contended that the New Haven pioneered electric self-propelled rail cars in the United States. It is true that the road began to operate electric cars on their Nantasket Beach line in 1895.[222] Several other New Haven branches in Connecticut were electrified in the period 1896–1898. The experience gained from these operations was valuable, but it did not lead to any immediate main-line electrification, nor was it technically more advanced than anything that was being done on hundreds of suburban and interurban electric lines elsewhere in the country. Actually, the New Haven had no main-line M.U. operations until 1910, some five years after such service began on the Long Island and the New York Central. Again the scene of operations was the New York City area. The road's four-track main line from Manhattan to Stamford, Connecticut, was the most heavily traveled in the nation. Electric locomotive-hauled trains began running in 1907, but commuter traffic was so dense that the advantages of M.U. trains were obvious. In February 1910 motor cars began operations between Grand Central Station and Port Chester.[223] The first four motor cars were delivered by the Standard Steel Car Company. Their open platforms and clerestory roofs gave them an antique appearance, but electrically they were the most advanced electric cars in America because of their 11,000-volt AC electric system. The circuitry was further complicated by secondary wiring for 600 volts DC, so that the cars could operate on the Grand Central's trackage.

With their four 150-horsepower motors, they were undoubtedly also the most powerful electric cars in the country and could pull two trailers. They measured 70 feet long, seated 76, and weighed just under 87 tons.

More cars were acquired as traffic doubled during the next decade. By 1922 the New Haven was transporting 16.3 million passengers a year over this short section of railroad.[224] Nine years later 129 motor cars and 92 trailers shuttled to and from Grand Central. The New Haven equipment remained undistinguished until 1938, when thirty lightweight cars were received from Pullman's Worcester, Massachusetts, plant. These units were probably the first lightweight M.U.s built for domestic service, yet the trade press was curiously silent about them.

The Pennsylvania Railroad cautiously studied the M.U. systems of the other Eastern lines. Its Penn Station electrification, opened in 1910, remained a locomotive-powered tunnel operation, but during the next year the road began to consider its chaotic Philadelphia suburban service.[225] Local traffic had overwhelmed the Broad Street Station. When it was opened in 1881, the eight-track terminal comfortably handled 160 trains a day. But the mounting number of commuters required two enlargements in 1890 and 1894 that resulted in a sixteen-track station. By 1910, 574 trains a day created a traffic snarl beyond its capacity. Delays were frequent during bad weather, yet further expansion of the facility was considered impossible because of the downtown setting. It was decided to electrify the two most heavily traveled commuter routes, Paoli and Chestnut Hill, and thus speed up train movements. Faster in and out speeds, and

Figure 8.73   The Illinois Central began large-scale electric suburban operations out of Chicago in 1926. The cars were constructed by Pullman. (Illinois Central Photograph, Trains magazine collection)

the elimination of the turnaround switching movements that were necessary with steam locomotives, improved the situation immeasurably. With no additional tracks, the terminal was thought to have an actual capacity of 1,000 trains a day after electrification was completed.[226]

In 1915 the first M.U.s began to run out of Broad Street. Ninety-three class P-54 steel suburban cars were converted, and

indeed had been designed for conversion to M.U. service. The low overall height permitted the mounting of pantographs. Two 225-horsepower motors were fitted on the front truck. The 64-foot-long cars seated 72 passengers. As the electric operation expanded from Philadelphia to Norristown, to Wilmington (and later to Washington), and to Harrisburg, hundreds more MP-54s joined the roster (Figure 8.72). Eventually more than 500 were

Figure 8.74 In all, 280 steel motor and trailer cars were built for the Illinois Central's electric service. The design followed conventional steam railroad practice. (Master Car Builders Report, 1926)

placed in service.[227] Standardization was carefully maintained, but some improvements, such as roller bearings, were applied to the newer units. In 1950 a major program of remodeling was started at the Wilmington shops, which installed new trucks, windows, frames, and interiors. Eight years later the Pennsylvania belatedly began to replace the MP-54s. (The new cars will be discussed later in this chapter.)

Outside the Northeast, the only main-line M.U. operation of any size was the Illinois Central's Chicago suburban line. Historically it was an early and increasingly important route, serving the residential-industrial communities to the south of Chicago. It began in 1856 with four daily trains between Chicago and Hyde Park.[228] By 1880 traffic had grown so much that two separate tracks were built and reserved exclusively for suburban trains. Double-ended locomotives and light cars, the latter modeled after those used by the New York Elevated, were purchased. Sixteen trains a day were run, but the *Railroad Gazette* predicted that "there is plenty of room for growth and the growth is sure."[229] Thirteen years later 62 locomotives and 380 cars carried an astonishing 9 million passengers to the Columbian Exposition grounds in the space of five months. Yet the traffic continued to grow. In 1925, 24 million passengers were carried; between 1900 and 1929 the traffic increased by 350 percent. One-third of the total Chicago suburban service was funneled into the Illinois Central, and the lakefront sky was blackened by the daily passage of more than four hundred trains.[230] The city had forty railroads with 1,700 steam locomotives working within its border,

but the Illinois Central's suburban operation was the most exposed and hence received the most criticism.

In 1891 talks began on the advantages of electric operations. Definite planning was initiated in 1911, but the project languished until the city passed an act in 1919 requiring electric operations by February 1927.[231] The conversion involved a major reconstruction of the entire 30-mile line (actually 250 miles of track) which included grade separations and new stations. The total cost of the project came to 24 million dollars.[232] Electric operations began ahead of schedule: the first M.U.s were running in August 1926. The electrics cut as much as 30 percent from the old schedules and were credited with an acceleration rate six times that of the steam trains. In addition to clean and rapid service, the electrics proved themselves a prudent investment, for the steam trains were losing $337,000 annually while the M.U.s made a profit of $400,000.[233]

In 1921 when the electrification program was getting underway, the Illinois Central purchased its first steel cars.[234] They were intended as prototypes for the electrics and were used later as trailers. A worldwide study was made of suburban rolling stock. This information was used in designing cars that set the basic pattern for 280 motor and trailer units built between 1921 and 1929 (Figures 8.73 to 8.75). They were conventional-appearing, heavyweight steel cars, arranged in married pairs of a motor and trailer car. They measured 72 feet in length and seated 84. The first group of trailers weighed 94,800 pounds, but in the second order a total of 7,600 pounds was saved by substi-

Figure 8.75   *The clean interior of an Illinois Central electric suburban car just before it entered service in 1929. Rattan seats were durable but not very popular with passengers. (Pullman Neg. 33068)*

tuting aluminum for the roof sheets, doors, and interior sheathing. An additional saving of 6,100 pounds was made in subsequent orders by using aluminum for the conduits and junction boxes. The motor cars weighed 141,200 pounds. The bodies and frames were made from copper-bearing steel. The interiors were considered bright and spacious because of the large windows, high-gloss white ceilings, wide aisles (31 inches), and the absence of luggage racks. Wide, 48-inch doors and electric "door engines" speeded loading and unloading, as did the platform-level stations. In rush hours the rated seating of 84 was augmented by an additional 109 standees. The motorman was protected by safety glass in the end windows. The exteriors were painted dark olive green, with black roofs and gold lettering. Four 250-horsepower, 1500-volt DC motors per motor car pushed the trains along at 60 mph. They accelerated at 1.5 mph per second. The traction type of Tomlinson coupler was used. For forty-five years these Pullman heavyweights unfailingly carried millions of Chicago commuters. In 1971 new lightweight cars began to replace the Pullmans, but a number of the old units remain in service today, a testament to their durability and solid construction.

The first generation of American M.U. cars actually lasted fifty years. The emphasis was on compatibility of electrical equipment, which made good economic sense but retarded technical advancement in the area. This philosophy was engendered by the long life characteristic of traction equipment and by the economics of rail commuter operations. Poor earnings do not encourage substantial new investment. Body styling and interior appointments changed slightly with each new order of cars, but the basic electrical systems remained largely unaltered. The new cars purchased in 1949 by the Reading for its Philadelphia suburban lines were equipped to duplicate the original cars delivered in 1931.[235] The Long Island, an even more extreme example, was receiving cars as late as 1963 with performance levels equal to those of its 1908 equipment.[236]

The advantages of the AC-DC electrical sytem had been considered for many years, and when it was finally adopted after 1950 it represented a major technical advance in this conservative field. Alternating current had excellent characteristics for long-distance power transmission, while direct-current motors were superior to AC motors. Because 600-volt DC was an early standard on street railways, this type of traction motor became highly refined and mass-produced. It was adopted for diesel-electric locomotives as well, which established its position even more securely. In addition to being cheaper than AC motors because of quantity manufacture, it was more powerful at low speeds—by as much as 47 percent, one estimate claimed.[237] What was needed was a compact, dependable unit aboard each car to convert AC to DC. Motor generators were far too bulky. In 1913 Westinghouse tested a mercury-vapor tube rectifier on a M.U. car.[238] However, the rectifier interfered with telephone service, and the tube itself had to be replaced every three months. In 1949 a more durable ignitron mercury-arc rectifier was tested on a Pennsylvania M.U. car. This trial was so successful that the New

Haven specified ignitron converters on 100 new electric cars built by Pullman's Worcester plant.[239] Deliveries of the stainless-steel-sheathed cars began in February 1954 (Figures 8.76 to 8.79). The 86-foot-long coaches seated 120 and were powered by four 100-horsepower motors. Because of this larger seating capacity, 150 old M.U.s were retired. Four years later the Pennsylvania purchased six Pioneer III-style coaches with ignitron rectifiers.[240] These were the lightest (45 tons) and fastest (90 mph) multiple-unit cars produced up to that time. The trucks were of an advanced inside-bearing design. The small, 30-inch-diameter wheels, the air springs, and the gear box motor drive were all new features (Figure 8.79).

The electrified Pioneer IIIs were such a success that Budd modified the design slightly for the first of the Silver Liners that have since become the standard electric car in the Philadelphia suburban lines of both the Pennsylvania and the Reading railroads. Rather than carrying mercury-arc rectifiers, the new cars were equipped with rectifiers of silicon diode, which were more compact and did not require firing circuits or cooling systems. The first 55 Silver Liners were delivered in 1963 at a cost of 14 million dollars.[241] They weighed 50¾ tons and seated 124 passengers in cars with toilets and 127 in those without. No interior paint was used; sheet and molded plastic shapes covered the entire surface. By 1974 over 200 Silver Liners were in service and another 144 were under contract.[242] The last group was produced by General Electric's Erie, Pennsylvania, plant. Each car had 8,600 parts and 3½ miles of wiring.

The success of the Silver Liners encouraged the development of a more modern style of car for 600-volt DC commuter lines. In 1966 Budd began designs for Long Island's new Metropolitan class. These 85-foot-long cars had side doors and no vestibules. Their curved-side bodies helped create the image of a giant subway vehicle. They weighed 46 tons and had molded plastic interiors.[243] The exteriors were smoothed (rather than fluted)

Figure 8.76 A major advance in electric car technology was the rectifier, or AC-DC equipment. The Number 4674 and its sisters were the first production models to be so equipped when delivered by Pullman in 1954. (E. B. Luce Corporation Photograph)

Figure 8.77 A New Haven electric rectifier car interior of 1954. (E. B. Luce Corporation Photograph)

stainless sheets accented by blue stripes. It was one of the largest orders on record: 620 cars, at a cost of 130 million dollars. Despite the size of the order, it was not a happy one for the supplier or the purchaser. The cars did not perform well at first, and this proved an embarrassing and costly problem for the Budd Company. In fact, the firm resolved to quit the car-building field, although new orders which were subsequently received caused management to reverse its decision.

The latest innovation in electric self-propelled cars appeared on the Illinois Central in the spring of 1971.[244] Planning that

was begun in 1967 produced designs for double-deck or gallery cars similar in their general arrangement to those running on other nonelectrified Chicago suburban lines. The presence of overhead wires inhibited the use of high cars. However, the Illinois Central justified its choice of galleries, rather than single-level cars with two-and-three seating, on the grounds of passenger comfort. Two-and-two seating was preferable and allowed space for wider aisles. An order for 130 cars was given to the St. Louis Car Company. The cars cost $307,000 each and, like most recent commuter equipment purchases, a large share of the

Figure 8.78   Most M.U. cars were outfitted as coaches, but a few, such as this 1954 New Haven club car, had more luxurious furnishings. (E. B. Luce Corporation Photograph)

funding was public. In this case two-thirds of the money was Federal.

The 85-foot-long units stood 15 feet, 10 inches high, weighed 67 tons, and seated 156 passengers (Figure 8.80). The front portion of the roof was depressed 22 inches to accommodate the pantograph. Center and end entrances were provided. The Cor-Ten steel bodies were painted silver gray with orange and black highlights. Vinyl-covered, foam-cushion seats replaced the old-fashioned rattan of the heavyweight cars. Tinted safety glass eliminated the need for shades. Air conditioning and electric heat ensured passenger comfort, while four 160-horsepower motors guaranteed a rapid acceleration rate of 1.9 mph per second. New York Air Brake furnished the latest electropneumatic-hydraulic brake equipment. The old Tomlinson style of coupling was retained, however. The trucks were similar to those used on the Metroliners.

The new cars, named Highliners, were on all counts a splendid improvement over the old equipment, but not enough were acquired to fully service the line. During peak rush periods it was

necessary to operate some heavyweight trains, and the dangers of this practice were soon tragically demonstrated. On October 30. 1972, one of the new gallery trains overshot the 27th Street Station.[245] The crew then backed into the station—a flagrant violation of the Standard Code. Meanwhile, a heavyweight train following behind had already passed the signal guarding the station and was thus unaware that another train was in the same block. The old train tore through its modern replacement, telescoping the rear car and causing 45 deaths and over 300 injuries. The new cars met the 400-ton buff standard, but the end framing was not sufficient to withstand the impact of a 400-ton, seven-car train at 30 to 40 mph. The difference in end profiles also facilitated telescoping, and once through the end stanchions, the light sheet-metal body of the Highliner offered little resistance to the terrible progress of the heavyweight train.

The Metroliner, America's most spectacular electric rail car, is essentially a multiple-unit car despite its sophisticated engineering, its claims to space-age technology, and its interiors that are surely the ultimate in plastic aircraft decor. It is simply the fast-

Figure 8.79   Budd's Pioneer III–style truck was adopted for electric M.U. cars in 1958. Note the disc brakes and massive cup-shaped mounts for air springs. (Lawrence S. Williams)

est, most powerful, and best-furnished electric self-propelled car yet constructed. The Metroliners were introduced when the North American passenger train seemed to have no future. They did much to restore faith in the railroad's capacity to move passengers efficiently—at least over moderate distances—but technically they proved troublesome and may be the last of their breed.

The Pennsylvania's New York–Washington line was one of the most heavily patronized in the world. Even after the World War II peak, it carried 130,000 passengers daily. But after 1946 business began to fall away as travelers returned to the highway. New highway networks stimulated the decline. And in 1961 Eastern Airlines began the low-cost air shuttle between New York and Washington. Frequent schedules and a flying time of under one hour drew thousands of commercial travelers away from the Pennsylvania. The railroad trimmed its schedule by half an hour and installed some rebuilt cars, but it was a losing battle. Patronage continued to decline, so that by 1958 it was only one-

half of the 1946 level. To reverse this trend would require drastic action, but the railroad felt that it was possible because of the interurban-like nature of the route: Baltimore, Wilmington, Philadelphia, and Trenton, located between the financial and political capitals of America, were on-line traffic generators. Faster schedules and new equipment were an old formula that had worked in the past and might just succeed again. The railroad's interest was reinforced by the public statements of Senator Claiborne Pell, who voiced concern over the growing congestion of the Northeast corridor and suggested that it could best be relieved by upgrading existing rail service rather than building more highways and airports.

In 1958 the railroad ran some high-speed tests with its new Pioneer III electrics which indicated that M.U. equipment would be suitable for long-distance travel.[246] It was viewed as the best choice on both engineering and economic grounds. New electric locomotives would be costly to design and build, and comparable cars would still be needed. By combining the loco-

Figure 8.80   In 1971 the Illinois Central began to replace its standard-era steel cars with new bilevel cars produced by St. Louis. (Illinois Central Gulf Railroad)

Figure 8.81   The Metroliners are the ultimate in electric self-propelled rail cars. Budd began deliveries in 1967. (Herbert H. Harwood, Jr.)

Penn Central **Metroliner built by the Budd Company looking in through the front.**

A—Supplementary springs   C—Reading lights   E—Continuous lighting fixture   G—Main air duct
B—Exhaust air duct   D—Parcel rack   F—Air distributor

Figure 8.82   The stainless-steel Metroliners are distinguished by their bulging sides, small windows, and airplane-like interiors. (Car and Locomotive Cyclopedia, 1970, p. 490)

motive and car, Pennsylvania engineers felt that faster deliveries and lower costs would follow. M.U. equipment offered better acceleration, full power no matter how many cars were in a train, and increased dynamic braking. Track maintenance would be eased by eliminating heavy locomotives. More studies followed, resulting in a detailed report that called for major track and

overhead wire improvements as well as new rolling stock. For 55 million dollars' worth of improvements, it was estimated that 150-mph schedules could be offered. At this speed the railroad could overwhelm its competition.

As an act of good faith, the High Speed Ground Transportation Act of 1965 pledged the Federal government to share in the costs of upgrading rail service in the Northeast corridor. The railroad, however, was expected to pay the major bills. The participants spoke of a renaissance of rail travel and optimistically announced that new service would begin in October 1967.

Fifty stainless-steel cars were ordered from the Budd Company in May 1966 at a cost of $435,000 each.[247] The coaches seated 80, the snack bar coaches 60, and the parlors 34. All measured 85 feet long by 10 feet, 6 inches wide over the curved sides (Figures 8.81 and 8.82). The coaches weighed 84 tons. Four 300-horsepower motors delivered an acceleration of 1.2 mph per second and were capable of propelling the cars at 160 mph. Because of the contemplated fast running, the underframes were covered over to protect electrical gear from flying ballast. The electrical system was complex and fragile. The controllers were minicomputers that automatically took over and precisely regulated the progression of power points, once the operator manually moved the lever to on. Deceleration was handled in the same fashion, and controls were so designed that dynamic braking (electric braking through the motors) would normally stop the train. The air brake was to be used only in emergencies. Although these elaborate systems were intended to ensure smooth operations, they were also prone to malfunction. The trucks were comparatively conventional, 8-foot, 6-inch wheelbase units with coil springs and equalizers produced by General Steel Industries. But even they proved troublesome in regular service.

The projected opening date came and passed with no Metroliners ready for service. Miles of wiring, resistors, and transistors proved fruitful sources of bugs. Most designs were untried and untested. No American firm had ever before attempted to build a high-speed electric rail car. The Metroliners worked satisfactorily at 120 mph, but developed serious problems in higher ranges.[248] Yet the specifications called for 150- to 160-mph operations, and the railroad was not about to accept the cars until they performed smoothly at that speed. It was unrealistic to expect such a complicated design to be worked out on the production line. More time and a prototype would have been the rational approach, but the Metroliner was not a simple matter of engineering. The Federal government had a hand in the project and was demanding an early and, in the opinion of some observers, unrealistic completion date. The axiom that politics and engineering rarely mix was proved in this case.

Finally on January 16, 1969, the first Metroliners were ready for service. By October enough cars were running to cover six round trips a day.[249] The public was pleased, and 700,000 passengers rode the trains during their first year. But the operating department remained unhappy. Breakdowns were frequent; 20 percent of the fleet was normally out of service. Sometimes as many as one-half of the cars were in the shop.[250] Even at the fastest regular operating speeds of 120 mph, the cars rode poorly. The speed was dropped back to 100 mph, which is the rate currently being observed. Track conditions were the primary cause for the orders to go slow. The multimillion-dollar track rebuilding program was abandoned after the Penn Central bankruptcy in June 1970. Limited maintenance did not help matters, but the cars' ride cannot be blamed entirely on the roadbed. The trucks were unequal to the job.[251] The coil springs were too stiff, and it was not possible to install anything much better within the confines of the side frames. Journal box wear compounded the problem. The secondary air springs caused the body to surge and did not cushion the ride.

In 1970 studies were begun to upgrade the electrical system and ride of the Metroliners. Several reworked cars were ready in 1974 that showed tangible improvements in both problem areas. Rewiring and roof-mounted exhaust grids for the dynamic brakes improved dependability. The original Metroliners could be expected to run 18,000 miles a month, while the improved model would cover 25,000 miles.[252] A monitor helped speed diagnoses of electric problems, many of which were simple enough to be corrected at the station platform rather than in the shop. But the major problem with the Metroliners remains how to find out what is wrong. Short of replacing the trucks, there seems no way to improve the ride. The most suitable substitute is a Swiss truck, the SIG type M, used successfully in Switzerland and France since the early 1960s. It features long, soft springs, bell crank axle boxes, and disc brakes. But these improvements are expensive: rewiring and new trucks would come to $500,000 per car, or more than the original purchase price.[253]

In retrospect, the Metroliner can be viewed as the greatest success and the greatest failure achieved by the multiple-unit cars. Metroliners proved a success in rebuilding rail passenger service, and they remain Amtrak's brightest star. Yet their large amount of downtime and high maintenance cost seem to have discredited the M.U. concept for high-speed train service. It is also ironic that so much money and effort was expended on special equipment that does not operate any faster than conventional trains. Amtrak's recent decision to purchase new electric locomotives and nonpowered cars is a demonstration of its disenchantment with the original Metroliners.

APPENDIX A: BIOGRAPHICAL SKETCHES OF CAR DESIGNERS AND BUILDERS

The lives of famous locomotive builders are comparatively well recorded. Anyone who has even sampled railroad literature recognizes the names of M. W. Baldwin, Richard Norris, and William Mason. But except for George Pullman and Webster Wagner, the names of car builders are unfamiliar and puzzling. Who was Benjamin Welch, for example, or James Leighton? The following sketches provide some answers, though this roster does not pretend to be definitive. In many cases the available information is incomplete or conflicting, but the notes offered here are at least a start and may encourage more investigation by other researchers.

In addition to the major figures in the design and production of the American passenger car, these selections cover less prominent men who were simply representative of their craft. Some of them devoted long years of service to their work although they made no spectacular contributions. In my opinion, material progress is not exclusively the creation of heroes and inventors; ordinary men are also necessary to the flow of the world's work.

Documentation is not offered in this section of the book, but the principal sources were the railroad trade journals. Among the standard biographical dictionaries that were consulted, the most helpful was the *Biographical Directory of Railway Officials of America* (first edition published in 1885; later titled *Who's Who in Railroads*). Obituaries in the *Master Car Builders Reports* were informative, and on occasion state and local histories provided data.

EVERETT E. ADAMS (1881–1954) was the designer of early and imaginative streamlined trains such as the City of Salina. He was born in Watertown, Massachusetts, and graduated from the University of California. He began railway service in 1905 on the Southern Pacific. Between 1913 and 1933 he worked for the Union Pacific, after which he joined the Pullman Car and Manufacturing Company. In addition to streamlined motor trains, he is credited with designing the duplex sleeping car.

FITCH D. ADAMS (1822–1904), one of the great personalities associated with American car building, was active in the trade for nearly fifty years and seemed to have had an opinion on every facet of it. At meetings of the Master Car Builders Association he could be expected to comment upon all papers given and generally seemed to delight in taking an opposite view from that of the speaker. He was a president of the organization and missed only one of its annual meetings.

Adams was born in Canterbury, Connecticut, and worked as a carpenter until 1848, when he entered the car-building shop at Norwich. In 1853 he joined the Buffalo Car Company as a subcontractor. He and his employer were wiped out by the 1857 panic. After a year in the lumber business, he reentered railway work at the Buffalo and Erie's car shop. He was soon made

master car builder and stayed with the road until 1867. Then he became superintendent of the Ohio Falls Car Company, which failed in the spring of 1870. Adams joined the Boston and Albany and remained head of its car department at Allston, Massachusetts, until his retirement in 1896.

He built some of the lightest wooden cars ever constructed and was a firm believer in the arch truss. He was an equally outspoken advocate of the steel tired wheel, and seemed in general a man who was open to new ideas.

RICHARD N. ALLEN (1827–1890), inventor of the famous Allen paper car wheel, was born near Springfield, Massachusetts. His early years were spent as a locomotive engineer on lines in the South and Midwest. After the Civil War he began to try various occupations, such as inventing and well drilling in the Pennsylvania oil regions. Then his brother-in-law induced him to invest in and manage a strawboard mill in Pittsford, Vermont. The business was a poor one, and Allen began to seek new uses for his products. Somehow the unlikely idea of a paperboard-center steel tired wheel occurred to him. The first was produced in 1869. The invention was ridiculed at first, but Pullman endorsed Allen's design and the paper wheel became a success. Eventually two plants, in Hudson, New York, and Pullman, Illinois, were needed to fill the demand for Allen wheels.

WILLIAM C. ALLISON (1817–1891), who was born into a Quaker family of modest means in Chester County, Pennsylvania, built a great manufacturing business and fortune in Philadelphia. After apprenticing himself to a wheelwright, Allison opened his own wagon shop when he was only nineteen. He began building railroad cars around 1840 and developed his company into one of the largest car plants in the nation. In 1851 a blacksmith, John Murphy, who had supplied ironwork to Allison was taken in as a partner. The plant eventually became known as the Junction Car Works. Passenger car production was suspended in 1866, but the firm continued to build freight cars for many years.

WILLIAM P. APPLEYARD (1857–1905) was born in upstate New York and, unlike most early car builders, received a university degree. He graduated from Notre Dame's architecture school and worked in that trade for several years. In 1888 he joined Pullman as an inspector and became superintendent of repairs. Five years later he was made foreman of the New Haven's Old Colony division car department. In 1895 he was promoted to master car builder for the entire system. At about the same time he began work on his curious copper-sheathed passenger cars. In 1904 he returned to Chicago and the Pullman Company. When he was only forty-nine years old, he was run down and killed by a train at the Illinois Central Station.

WILLIAM C. BAKER (1828–1901), the originator of the renowned hot-water car heating system bearing his name, was born in Dexter, Maine. He went to New York City around 1850 and was soon attracted to the study of central heating. One of his earliest successful designs was for a hot-water heating system, patented in 1866, which was almost universally adopted for all first-class railway cars in North America. During his life Baker obtained over forty patents, most of which covered car heaters. He was also active in developing steam heating apparatus. He was accidentally killed by a train while visiting his daughter in the New York suburbs.

CHARLES E. BARBA (1877–19?), an important figure in steel passenger car design, was born in Freemansburg, Pennsylvania. In 1902 he joined the Pennsylvania Railroad engineering staff and worked with William Kiesel on designs for the P-70 class of passenger cars. He left Altoona in 1915 and later worked for the Midvale Steel Company and the Watertown Arsenal. Between 1922 and 1925 he was superintendent of the Osgood Bradley Car Works. During the following five years he served as mechanical engineer with the Boston and Maine Railroad. Barba prepared two valuable papers on passenger car construction. The first appeared serially in the 1907 *American Railroad Journal*; the second was published by the American Society of Mechanical Engineers in 1932.

JOHN H. BARKER (1844–1910) made the firm of Haskell and Barker into a major car builder. Born in Michigan City, he was the son of one of the founding partners of the firm and became president in 1878 after his father's death. Under his leadership the plant expanded until it was the largest single manufacturing establishment in the state of Indiana. It eventually became part of Pullman's car-building empire.

ELIAM E. BARNEY (1807–1880), one of the founders of the Dayton Car Works (better known by its later title of Barney and Smith), was born in Sackets Harbor, New York. He started his career as a teacher of classical literature and managed several schools in the East and Midwest. He settled in Dayton, Ohio, in 1843 to head the local academy, but his health failed after a few years and he resigned. Oddly, he decided to enter the sawmilling business as a cure. During these years he met Ebenezer Thresher, another intellectual turned industrialist. The two liked one another and determined to form a partnership in 1849. The result was the car works that was renamed Barney and Smith in 1864. The firm succeeded beyond the greatest hopes of its founders. For many years it was the second-largest car builder in North America and at one time employed 2,000 men. Barney's sons took an active part in the plant's operations. In 1892, however, the family turned the management over to outside interests, and the business went into a decline. Production ended in the early 1920s, and the firm was liquidated in 1923.

JACOB N. BARR (1848–1904), renowned for his contributions to cast-iron wheel technology, was also active in many other areas of rolling-stock design. He was born in Lancaster, Pennsylvania, and graduated from Lehigh University. He worked in the mechanical departments of several railroads, including the B & O and the Erie, but spent much of his career with the Milwaukee Road. Barr held sixteen patents, including a plan for a contracting cast-iron wheel mold and a design for a passenger car vestibule.

CHARLES BILLMEYER (?–1876) was a railroad car manufacturer of York, Pennsylvania, and a major partner in the Billmeyer & Small Company, proprietors of the York Car Works. (See David E. Small for more details.)

THOMAS A. BISSELL (1835–1902), one of the masters of wooden passenger car construction, was born near Erie, Pennsylvania. He was educated for the ministry but decided to go west and work on the railroad. He started as a bridge builder on the Burlington and soon graduated to the Aurora repair shops as a draftsman. In 1872 he joined Pullman and helped reorganize the car plant in Detroit. The following year he succeeded C. F. Allen as Pullman's mechanical superintendent and continued in that post until 1881, when he became superintendent of Barney and Smith. Five years later he took charge of the Wagner Sleeping Car Company's main shops in Buffalo. Declining health forced his retirement in 1895.

Bissell, a gifted car designer, did much to perfect wooden car framing. He had several patents covering vestibules, platforms, and other features. He was active in the Master Car Builders Association and served as a vice president. As the *Railroad Gazette* said at the time of his death, "His influence on the art of passenger car building has been of the first importance."

WILLIAM F. BIXBY (1857–1931) is credited with forming the American Car and Foundry Company. He was born in Adrian, Michigan, and began his railway career in 1873 as a baggageman. Twelve years later he entered the car-building trade as a lumber agent for the Missouri Car and Foundry Company of St. Louis. Within two years he was promoted to vice president and later took the presidency. As head of the company he became merger-minded. First he engineered a consolidation with the Michigan Peninsular Car Company of Detroit; then in 1899 he directed the combination of eighteen car builders into A.C.F. He became the first president of the new industrial giant, but was soon made chairman of the board and retired altogether in 1905.

During his retirement years Bixby earned two college degrees, helped to organize the American Red Cross, acted as a receiver for the Wabash Railroad, and took part in many educational and philanthropic enterprises.

JOHN E. BRADLEY (1860–1938), a grandson of Osgood Bradley, brought the Bradley Car Company successfully into the modern steel age. He left college in his second year and in 1882 joined the family business. After his uncle's death in 1896 he was made general manager, and when his father died five years later he became president and owner of the company. Recognizing that the firm must have a modern plant if it were to survive, he reorganized it as a joint stock company. The new shop was completed in 1910. Twenty years later, Pullman took over the business and Bradley retired.

OSGOOD BRADLEY (1800–1884), founder of the great car works in Worcester, was born in Andover, Massachusetts. His English ancestors came to the area during the time of the first Puritan settlements. After he completed high school, he was apprenticed to a carriage maker in Salem. In 1822 he moved to Worcester and opened his own shop. Like other carriage builders, he was drawn to railroad car construction, and by 1837 he was specializing in that line.

Despite the fact that his business survived to be handed on to his children, Bradley was in financial difficulties all his life. He failed several times and was considered a poor credit risk, if not a poor businessman, by Dun and Bradstreet. The company's fiscal affairs appear to have been managed more carefully after his passing.

ALBERT BRIDGES (1811–1881) was a partner in the car-building firm of Davenport and Bridges between about 1837 and 1851. For the next quarter century he operated a large New York City railway supply house, which specialized in all varieties of railroad-related hardware. (More on his car-building career is given in the sketch on Charles Davenport.)

JOHN G. BRILL (1817–1888) founded the Philadelphia-based J. G. Brill Company, which became the largest streetcar builder in North America. Brill was born near Cassel, Germany, and came to the United States in 1840 after learning the cabinetmaker's trade in Bremen. He found employment at Murphy and Allison's Philadelphia car works, where he rose to the position of foreman of the streetcar shop. In 1868 he left Murphy and Allison and started his own car plant. Within a few years he was building railroad and city cars.

The firm continued to grow and after 1900 began to acquire other car plants, including the Wason Car Company of Springfield, Massachusetts. After the coming of steel construction, Brill appears to have abandoned the manufacture of equipment for main-line service, except for the self-propelled rail cars described in Chapter 8. It continued as an important streetcar fabricator until 1941. Motor bus production continued until 1954, when the firm suspended the manufacture of transportation equipment in any form.

EDWARD G. BUDD (1870–1946) did much to advance the lightweight passenger car and stainless steel fabrication. He was born in Smyrna, Delaware, received a high school education, and worked in various metal-manufacturing plants in the Philadelphia area. In 1912 he opened a small shop to make automobile parts and soon began turning out complete auto bodies for such major firms as the Dodge brothers. Later he went into the automotive wheel business. By the time Budd decided to enter railway car manufacturing in the early 1930s, he was a major industrialist. Within a few years he was accepting substantial orders for railway cars, much to the indignation of the established builders. In 1945 a huge plant built in Red Lion, Pennsylvania, by the government for war production was taken over by Budd's company for passenger car manufacture. Budd died the following year, but his firm produced such famous postwar trains as the California Zephyr. Car orders fell off dramatically after 1960, and in 1972 Budd announced its intention to close the Red Lion plant. However, new orders have since reversed that decision.

CHARLES DAVENPORT (1812–1903), an important pioneer passenger car manufacturer, had a plant in Cambridgeport, near Boston. He was born in Newton, Massachusetts, and became apprenticed to a carriage maker. In 1832 he opened his own shop under the firm name of Kimball and Davenport to produce coaches and carriages. He entered the car-building trade about two years later. Within a few years his company was probably the largest single car manufacturer in the United States. In 1837 or 1839 Davenport's boyhood friend, Albert Bridges, became a partner. Some ten years later Lewis Kirk of Reading, Pennsylvania, joined the firm to supervise the construction of locomotives, a new sideline. This venture was not successful, however, and the business was reorganized with Charles Davenport as the sole partner. The firm recovered, but many other car plants were now active. In 1855 or 1856 Davenport decided to close the plant and retire with a comfortable fortune. He claimed to have produced 4 million dollars' worth of cars for fifty different railroads.

WILLIAM A. DAVENPORT (1831–1888) operated a large car plant in Erie, Pennsylvania. He was born in Schuyler County, New York, and began life modestly as a clerk. In 1866 he opened a car wheel foundry and two years later established the Erie Car Works. The plant was a major freight car builder but did not, so far as is known, make any outstanding contributions in the area of passenger car design. Davenport took an active part in the affairs of the Master Car Builders Association, an interest which was curiously rare among his rivals in the car building field.

OTHO C. DURYEA (1880–1941) was the inventor of the cushion draft gear underframe. He was born in Wyoming, Illinois, the younger brother of J. Frank and Charles E. Duryea, the American automotive pioneers. He began work on the cushion underframe around 1925. The device was used on passenger equipment as early as 1935, but its great importance came in more recent years, for jumbo freight cars could not have been built without it.

FREDERICK W. EAMES (1843–1883) created the vacuum brake bearing his name. He was born in Kalamazoo, Michigan, the son of an inventive mechanic-farmer, Lovett Eames. Because Frederick named an experimental locomotive, which he used to demonstrate the vacuum brake, after his father, some careless writers have ascribed the brake to the father rather than the son. Eames received his first patent in 1874 and began to market the apparatus about two years later. (Other details are given in Chapter 7).

Eames was shot and killed while attempting to enter his Watertown, New York, factory during a strike. The firm survived and was reorganized in 1890 as the New York Air Brake Company.

ORSAMUS EATON (1792–1872) was a partner in the prominent car-building firm of Eaton and Gilbert, proprietors of the Troy Car Works, which was for many years a major passenger car producer. Eaton opened a carriage shop in 1820, and Uri Gilbert became a partner ten years later. The railroad cars that they added to the business a few years afterwards soon became their main product. Eaton retired in 1862, but the business continued under the able management of Gilbert.

JULIUS E. FRENCH (1836c.–1910), a prominent railway supply manufacturer, was born in Perry, Ohio. In the early 1870s he helped establish the Winslow Car Roofing Company in Cleveland. In 1882 he organized another firm in the same city to produce a steel tired wheel designed by W. H. Paige, and fifteen years later he helped consolidate the major manufacturers of tired wheels into the Steel Tired Wheel Company. The firm was reorganized in 1902 as the Railway Steel Spring Company.

LEANDER GARY (1827–1886) was a founder of the Master Car Builders Association and the longtime head of the New York Central's car department. He was born in York County, Maine. He followed his father into the carpentry trade, and when he was about twenty-four entered the service of the American Car Company in Seymour, Connecticut. He worked briefly for the Naugatuck Railroad and in 1855 was made master car builder for the New York and Harlem Railroad. In 1873 he was appointed the head of all car shops on the New York Central, a position calling for unusual administrative ability. This quality Gary demonstrated as president of the M.C.B. between 1874 and 1885. Although he was an inventive mechanic, he was not given to patenting his work and it is difficult to document his innovations. Because of illness, Gary retired early in 1885 and died in November of the following year.

J. TAYLOR GAUSE (1823–1898) was born in Chester County, Pennsylvania. He went to work at Harlan and Hollingsworth in 1843 as an errand boy and filled many minor positions there before beginning to receive advancement. Finally in 1858 he could afford to buy an interest in the firm. In 1867 he was made a vice president, and he became president with the death of Samuel Harlan in 1883. He was a careful manager, and the business maintained its leading position under his direction. He retired in

1896 but was recalled to head the plant two years later, just months before his death.

GEORGE GIBBS (1861–1940) is best remembered in railway car circles as a pioneer in the introduction of steel passenger cars. His designs for the first New York subways did much to convince steam railways that all-metal cars were a sound investment. Although he considered electrical engineering his major field, he was understandably proud of his pivotal role in passenger car design.

Gibbs was born in Chicago into a prosperous mercantile family. He graduated from the Stevens Institute of Technology and worked for Thomas A. Edison for several years. Then he took a position with the Milwaukee Road, where he did important work in train-board electric lighting and steam heating. In 1897 he joined the Baldwin Locomotive Works, which recognized the potential of heavy electric traction. In 1912 he established the firm of Gibbs and Hill, Consulting Engineers, and continued his work in the area of railway electrification.

URI GILBERT (1809–1888) was the major figure in the important Troy, New York, car-building firm of Eaton and Gilbert, later known as the Gilbert Car Works. A carriage maker by trade, Gilbert also proved to be a gifted manager, and after 1862 he became the main partner in the firm. The plant produced a great variety of cars, both freight and passenger. It had a large export business, and it took special pride in the elegant equipment it produced for the Wagner Sleeping Car Company. After Gilbert's death the firm went into a decline. The panic of 1893 was the final blow, and two years later the great plant closed down.

EDWARD E. GOLD (1847–1931), a key figure in the introduction of steam car heating, was born in Waverly, Illinois. He began manufacturing central heating apparataus in 1867, and in 1882 he invented a system of steam car heating that was widely used in North America. Gold took out 100 patents during his lifetime. (Whether he was related to Egbert H. Gold, inventor of the vapor system of car heating, has not been determined.)

JAMES GOOLD (1790–1879), who built some of the first passenger cars produced in this country, was born into a farming family in Granby, Connecticut. He worked briefly as a bookbinder but found carriage making more to his liking. In 1813 he opened his own shop in Albany, New York, where he produced the first cars for the Mohawk and Hudson Railroad in 1831. Though carriages were his principal product, he continued to build railroad cars until about 1870. He also constructed streetcars between about 1865 and the time of his death. The business was continued for many years by his descendants but was finally closed in 1913.

CHARLES A. GOULD (1849–1926) was a manufacturer of couplers and electric train lighting apparatus. He was born in

Batavia, New York, and worked for several years as a merchant. Apparently Gould was not a mechanic, but he had the knack of recognizing marketable items for the railway supply trade and developed an immense business in this area. He bought into a steam forge business in 1884 and began to turn out car couplers. Several years later he acquired rights to the platform and vestibule patented by Thomas A. Bissell of the Wagner Car Company. Next he opened a steel plant in the Midwest, and he built a giant factory for the production of car parts, locating it adjacent to the New York Central shops at Depew, New York (near Buffalo). When electric car lighting came in, Gould began to manufacture batteries. His properties were taken over by the Symington Company of Rochester, New York.

THOMAS A. GRIFFIN (1852–1914) created the huge car wheel manufacturing combine that remains the largest single producer in that line today. His father, Thomas F. Griffin, emigrated from Ireland as a boy and started a foundry in Rochester, New York, where he began making car wheels in 1847. In 1870 the father moved the business to Detroit. Thomas A. Griffin broke away from his father, opened an independent wheel foundry in Chicago, and in time bought out his father's company. He took over or established new wheel factories in St. Paul, Kansas City, Denver, Tacoma, Boston, and Los Angeles. Griffin was a pioneer in profit sharing and helped organize the Association of Manufacturers of Chilled Car Wheels in 1908.

JOHN M. HANSEN (1873–1929), a gifted mechanical engineer and businessman, was born in Sandy Lick, Pennsylvania, and attended the Western University of Pennsylvania. He took a job in the engineering department of Schoen Pressed Steel Company, later the Pressed Steel Car Company, where he worked on the designs of early steel freight cars. In 1902 with the backing of James B. (Diamond Jim) Brady and the Mellon family, he organized the Standard Steel Car Company, with a plant at Butler, Pennsylvania. In 1906 a second shop was opened in Hammond, Indiana. Within a few years Hansen acquired control of the Middletown (Pennsylvania) Car Company, the Baltimore Car & Foundry Company, the Osgood Bradley Car Company, and the Keith Car Company. He also opened plants in France and Brazil. Hansen retired in 1923, and in the year of his death his car-building empire was taken over by Pullman.

SAMUEL HARLAN, JR. (1808–1883) was a partner in the car-building firm of Harlan and Hollingsworth of Wilmington, Delaware. He was born in Glen Cove, Long Island, and spent his early manhood in Wilmington learning the trade of cabinet-making. In 1836 he became superintendent of Betts and Pusey, owners of a new railroad car shop—one of the first companies to organize with this specific product in mind. A year later Harlan became a junior partner, and by 1849 he was the main partner in a business that had flourished since its founding. It expanded to include a shipyard, and a new partner, Elijah Hollingsworth, gave the firm its mellifluous corporate name that was to become

a standard-bearer in the field. By 1860 Harlan and Hollingsworth employed over 600 men and had become one of the biggest car builders in America.

After Harlan's death the firm continued to be well managed, but it seems to have concentrated more on its ship-building business. In 1904 Bethlehem Steel acquired Harlan and Hollingsworth. Car production was suspended in 1939, but ship building continued for many years at Bethlehem's Wilmington, or "Harlan," Works.

WILLIAM T. HILDRUP (1822–1909), longtime superintendent of the Harrisburg Car Works of Pennsylvania, was born in Middletown, Connecticut, and began his career as a carpenter. Around 1842 he entered the employ of Osgood Bradley in Worcester, Massachusetts, and learned the car-building trade. Ten years later he transferred to the Elmira, New York, car plant, but during the following year he helped establish a shop in Harrisburg with the financial backing of eight other partners. Although the plant originally produced all styles of cars, around 1865 it began to specialize in freights. The works prospered for several generations, then failed in 1890 and does not appear to have reopened.

ELIJAH HOLLINGSWORTH (1806–1866) was born of a Quaker family near Wilmington, Delaware. He learned the machinist's trade in his native city, and when he was only in his middle twenties he was made foreman of Baldwin's new locomotive shop in Philadelphia. Returning to Wilmington in 1841, he became a partner in the car-building firm of Harlan and Hollingsworth.

RICHARD IMLAY (1784–1867), one of the first names associated with American passenger car construction, was born in Hartford, Connecticut. His father was in the importing business. How Imlay entered the carriage- and coach-building trade is uncertain, as are most facts about his life. Around 1828 he opened a shop in Baltimore and two years later began producing cars for the B & O. He appeared to have relocated in Philadelphia during the following year. In 1835 he built an eight-wheel car, the Victory, which included several remarkable features, including a buffet and a water closet. Two or three years later he built for the Cumberland Valley Railroad what is probably the first sleeping car ever produced. Despite these innovations, Imlay's business failed during the depression following the panic of 1837. His shop apparently closed in 1840, and he seems to have had no subsequent connection with railway cars except for a patent suit involving his alleged invention of the truck center plate.

JOB H. JACKSON (1833–1901) was a founding partner of Jackson and Sharp, operators of the Delaware Car Works in Wilmington. This plant, in combination with the Harlan and Hollingsworth factory, made Wilmington the car-building center of America before Pullman opened his great Chicago plant in 1881.

Jackson was born into a family of modest means in Chester County, Pennsylvania. He quit school at an early age and worked as a tinsmith and mechanic. In 1863 he and Jacob Sharp opened a small car shop, in which Jackson managed the office while Sharp, an experienced car builder, supervised the shop. In 1870 Sharp retired, and a joint stock company was formed. Jackson, a shrewd businessman, acquired most of the shares within a few years and developed the Delaware Car Works into a formidable enterprise. He added a shipyard in 1875 and became involved in the management of the Woodruff Sleeping Car Company. By the late 1880s the Jackson and Sharp works was producing 400 passenger cars a year. It also built streetcars and produced sash work.

The works was taken over by the American Car and Foundry Company in 1901. Within a few years A.C.F. concentrated its export work at the Wilmington plant and produced hundreds of wooden narrow-gauge cars. The plant closed around 1945.

ELI J. JANNEY (1831–1912), who invented the present-day car coupler, was born into a farming family in northern Virginia near Washington, D.C. He served in the Confederate Army and returned home so poor after the war that he gave up his farm and went into the dry-goods business. Apparently he became interested in coupler design when he heard about injuries suffered by a local trainman. Although Janney knew little of mechanics, he began whittling models and thinking about the problem. In 1868 he obtained a patent for a design that led nowhere, but five years later he received a second patent that clearly foreshadowed the hinged, knuckle type of coupler. In time the Janney coupler (sometimes erroneously called the "Jenny" coupler in the popular literature) was perfected and accepted by the industry.

JOSEPH JONES (1818–1879), recognizing the need for an association of car builders, called together a group of car masters in 1864 to discuss problems of mutual interest. He suggested that a more formal organization could be beneficial, and from this beginning the Master Car Builders Association emerged in 1867.

Born in South Wales, Jones came to this country when he was twelve years old. He was apprenticed to a carpenter and went to work for the Utica and Schenectady Railroad in 1846. He became head of the New York Central's main car shops at West Albany in 1865. Two years later he was given the honor of affixing the first signature to the constitution and by-laws of the M.C.B. He retired from railroad work in 1875 but remained active in the carpentry trade until a week before his death.

WILLIAM F. KIESEL, JR. (1866–1954), an important figure in the design of steel passenger cars, was born in Scranton, Pennsylvania. In 1887 he graduated from Lehigh University as a mechanical engineer, and in 1888 he began a forty-eight-year career with the Pennsylvania Railroad's mechanical department. In 1899 he was made chief draftsman and within ten years was promoted to mechanical engineer.

He was active in locomotive and freight car design, but it was his work in developing the P-70 steel coach that made him a notable contributor to the development of the American passenger car. Kiesel held 135 patents for improvements in railroad rolling stock. He retired in 1936.

JOHN KIRBY (1823–1915) served as master builder on the Lake Shore and Michigan Southern for twenty-two years. He was born in Oxfordshire, England, and was trained to be a coach maker. He emigrated to the United States in 1848 and worked in the car shop of the Syracuse and Utica Railroad for six years. Then he went west to take a job in the Michigan Southern's shops at Adrian, Michigan, where he became foreman after two years and head of the car department in 1870. In addition to repair work these shops constructed many new cars, including some elaborate palace sleepers.

Kirby was an early member of the Master Car Builders Association. He served one term as its president just before his retirement in 1892 and was elected treasurer for several terms between 1900 and 1909.

JAMES T. LEIGHTON (?–1892) was a second-echelon car builder whose wandering career typifies that of many railroad men. No record of his early life can be discovered, except for the fact that he was born in Portland, Maine. In 1873–1874 he took out several patents for improvements in sleeping cars. He and his brother founded the New Haven Car Company, where they built cars of all types. They also operated their sleeping cars for Wagner and for at least one Southern line until about 1879. Leighton ran his own line of sleeping cars in the New Haven area until about 1885, when he went to work for Jackson and Sharp. In later years he was employed by the Baker and Sewall car heating companies.

JOHN S. LENTZ (1847–1935) was employed by the Lehigh Valley Railroad's engineering department for a record seventy years. He was born in Mauch Chunk, Pennsylvania, and began work on the Lehigh Valley in 1865. He was promoted to the position of master car builder in 1881 and held the job under various titles until 1928. After this time he acted as a consultant to the car department. Lentz was also active in the Master Car Builders Association and served as its president for the 1894–1895 term.

JOHN LIGHTNER (1811–1896) was employed in the car shops of the Boston and Providence Railroad for fifty-five years. He was born in Baltimore and learned the trade of carriage making. After working briefly for the B & O, he was hired in 1833 by the Boston and Providence as a car repairman. In 1835 he was promoted to master car builder at the road's Roxbury shops, a position he held until retiring in 1888. Lightner made many mechanical contributions, the most important of which was a journal bearing box patented in 1848.

GEORGE G. LOBDELL (1817–1894) was a principal partner in the prominent car wheel firm of Bush & Lobdell. He was born in New York City and was apprenticed in 1832 to his uncle, Jonathan Bonney, a pioneer car wheel maker. When Bonney died in 1838, Lobdell succeeded him as a partner in the Wilmington, Delaware, firm. He became a leading manufacturer and a recognized authority in his field.

EDWIN LOCKWOOD (1810–1881), master car builder for the Camden and Amboy Railroad, was born in Orange County, New York, and began his working life as a wheelwright. In 1832 he was hired by the Stevens family, who owned the C & A, to work on the Hoboken ferry. A year later they transferred him to the C & A shops at Bordentown, New Jersey. In 1837 he was made foreman of the car department, a position he held until his retirement in 1875. The cars of the Camden and Amboy differed from those of most other early U.S. lines because of the exposed truss framing beneath their bodies and the use of six-wheel trucks many years before they were common elsewhere.

WILLIAM McCONWAY (1842–1925) was a leading figure in the introduction and manufacture of the Janney coupler. He was born in Ireland but came to the United States as a boy. He entered the malleable iron casting business and by 1866 was a junior partner in a Pittsburgh iron firm. Four years later he and John J. Torley established their own company, which still produces materials for the railroad industry. McConway became involved with Janney during the formative years of the automatic coupler, and in 1884 he incorporated the Janney Car-Coupling Company. All the shares were owned by McConway and his associates; ironically, Janney was not a member of the firm. Without the technical expertise of McConway and Torley, however, the Janney coupler might not have become a practical reality.

WILLIAM R. McKEEN, Jr. (1869–1946) did outstanding work in developing early gasoline-propelled rail passenger cars. He was born in Terre Haute, Indiana, the son of a prosperous banker. He received degrees from Rose Polytechnic Institute and the University of Berlin. Entering railway service in 1891, he worked for several Midwestern lines until 1898, when he joined the Union Pacific. In 1908 he was made superintendent of machinery. Three years earlier he had built a gasoline rail car which proved so promising that it was put into commercial production. Over the next twelve years about 150 of his motor cars were manufactured. McKeen was unquestionably a mechanical genius; one account claims that he was granted 2,000 patents.

EZRA MILLER (1812–1885) invented a car coupler and platform that was popular on American passenger cars in the 1870s and 1880s. He was born in Pleasant Valley, New Jersey, and studied and practiced civil engineering. He joined the New York State Militia and rose to become a colonel—a title he enjoyed using for the rest of his life. In 1848 he moved to Wisconsin to survey Western land. A few years later he began to study the car coupler problem, an interest presumably aroused during the extensive rail travel required by his work. His invention slowly matured, and he obtained patents for it in 1863 and 1865. The device proved successful and was widely used before the introduction of the Janney coupler. Miller's outgoing, politically minded personality was an asset in achieving the rapid and widespread acceptance of his device.

KARL F. NYSTROM (1881–1961) made important advances in the design of lightweight passenger cars. He was born in Sweden and worked his way through school, earning a degree in mechanical engineering in 1904. The following year he came to the United States. He worked as a draftsman and engineer for Pressed Steel, Pullman, A.C.F., and several railroads before joining the Milwaukee Road in 1922. By this time he was concentrating on car design. His work in developing lightweight, all-welded cars is described in Chapter 2. He retired in 1949, a recognized leader in his specialty.

WILLIAM H. PAIGE (1842–1885), for many years superintendent of the Wason Car Company, held several patents for sleeping-car improvements that were used by Wagner. Around 1881 he resigned his post at Wason to devote his time to a firm organized to produce steel tired wheels of his design.

GEORGE M. PULLMAN (1831–1897), the only name connected with railroad cars that is known to the general public, is best remembered for something he did not invent—the sleeping car. As the success story of a self-made man, his life can hardly be rivaled; it was filled with achievement and reward. He did much to upgrade and civilize railway travel. He created a vast and intricate business empire that built, serviced, and operated a huge fleet of sleeping cars. He was a man of incredible vision, energy, and judgment. It is curious that his fame rests on his modest mechanical skills, while his great gifts as an organizer and capitalist receive scant attention.

Pullman was the third of ten children born in Brocton, New York, to a house builder and mover of small means. George left school when he was fourteen and worked as a carpenter and cabinetmaker. He started his own house-moving business when he was 22 and did well in it. A few years later he transferred the business to Chicago, where he entered the sleeping-car trade and began the brilliant career that is described in Chapter 3.

E. J. W. RAGSDALE (1885–1946) was chief engineer of the Budd Company's railway car division. Born in San Francisco, he traveled widely in the Far East and led an adventurous life before joining Budd shortly after World War I. He developed the Shotwelding process vital to the manufacture of rivetless stainless-steel cars. Ragsdale was an advocate of disc brakes and did much to encourage a departure from traditional passenger car design.

WILLIAM E. RUTTER (1812–1882) operated a car plant in Elmira, New York, for a brief period and worked as a car builder for several railroads. He was born in Providence, Rhode Island, and learned the carpentry trade in Baltimore. In the early 1830s he took a job with the Boston and Providence and in later years became master mechanic on the Stonington Line. In 1851 he moved to Elmira and opened a car repair shop to service the newly constructed New York and Erie Railway. A year later he began to build cars. In 1858 he sold out to the Erie but stayed on as the manager. In 1870 the works was leased to the Erie and Atlantic Sleeping Car Company, an Erie-sponsored corporation that was quickly absorbed by Pullman. The Elmira facility was used by Pullman for repairs until 1886.

Rutter's son, James H. (1836–1885), became a favorite of the Vanderbilts and served as president of the New York Central.

CHARLES T. SCHOEN (1844–1917), who was prominent in the development of steel freight cars, also made a notable contribution to passenger equipment by introducing wrought-steel wheels. Schoen was born in Concord, Delaware, and worked with his father in the cooper's trade. He went to Philadelphia at the end of the Civil War to learn the metal-fabricating business. In 1892 he established a plant in Pittsburgh that made pressed-steel car parts, and soon he began to manufacture freight cars that revolutionized freight car design. The business was reorganized as the Pressed Steel Company in 1898 and became an important producer of steel passenger equipment in later years. Schoen was forced to resign from the car plant in 1901. He opened a wheel plant at Butler, Pennsylvania, but sold out to U.S. Steel in 1907.

CHARLES A. SELEY (1856–1939) drafted the basic specifications for the railway post office car. Their aim was to set minimum-strength standards that would safeguard railway mail clerks, but the specifications were eventually adopted by the railroad industry for all passenger cars.

Seley was born in Wapella, Illinois, and entered railway service as a draftsman in 1876. By 1899 he was a mechanical engineer with the Norfolk and Western. He was elected president of the Western Railroad Club in 1907. In later years he went into the supply trade.

HENRY H. SESSIONS (1847–1915) is popularly (but incorrectly; see Chapter 5) remembered as the inventor of the vestibule. He was born in Madrid, New York, the son of Milton Sessions, who was master car builder for the Central Vermont Railroad and later superintendent of the Taunton Car Works. The son served an apprenticeship under his father and then struck out on his own, working for several railroads in New York, Iowa, and Texas. In 1885 he was made superintendent of the Pullman plant in Chicago and served in this post until 1892. For several years he worked as a consultant to Pullman, then began to devote full time to the affairs of the Standard Coupler Company.

Although he was not the originator of the vestibule, Sessions

made important contributions to wooden passenger car design during its last decades. His ingenious metal reinforcing scheme did much to strengthen this form of construction.

JACOB F. SHARP (1815c.–1888) was a partner in the car-building firm of Jackson and Sharp. He was born in Hunterdon County, New Jersey, and worked as a carpenter and bridge builder before taking a job in 1840 at the Harlan and Hollingsworth car plant in Wilmington, Delaware. In 1863 he and Job Jackson founded the Delaware Car Works. However, Sharp left the firm in 1870, several years before it emerged as a major car builder.

DAVID E. SMALL (1824–1883) was the leading partner in the York, Pennsylvania, car-building firm of Billmeyer and Small. His father was in the construction and lumber business. It was not a major transition to expand the sash mill into a car shop, and in 1852 Charles Billmeyer was taken in to help organize the new firm, originally called the Etna Car Works but later better known as the York Car Works. The plant grew rapidly into a major car builder, a position it held until about 1885. For many years it specialized in narrow-gauge cars, and members of both founding families remained active in its affairs. Car production was suspended around 1910.

CALVIN A. SMITH (1822–1903) may have been the inventor of the side-panel truss for wooden passenger cars. Born in Newfield, Maine, Smith first worked as a carpenter. Around 1840 he found employment at a car shop in Lawrence, Massachusetts. In 1851 he began to work for the Erie and was soon made foreman of the main car shop at Piermont, New York. He is said to have built one of the first parlor cars, the *Metropolis*, sometime before leaving the Erie in 1878. Between that date and his retirement in 1902, he served as superintendent of the Union Tank Car Line.

PRESERVED SMITH (1820–1887) was a partner in the Dayton Car Works, also called Barney and Smith, between 1864 and 1877. Like the other major partners in this important firm, Smith was an educated man rather than a practical mechanic. A native of Warwick, Massachusetts, he was active in railroad management before joining the car building of Barney and Smith.

WILLIAM W. SNOW (1828–1910), a prominent car wheel maker, was a founder of the famous Ramapo Car Wheel Works. He was born in Heath, Massachusetts. Snow considered becoming a minister, instead went into the book-binding trade, then in 1848 joined his brother as an iron founder. Soon he was deeply involved in the mysteries of car wheel making. In 1859 he became general manager of the Union Car Wheel Works in Jersey City, which was a major wheel manufacturer. Six years later he opened his own plant at Ramapo, New York. The business was a success, and it grew into one of the largest independent car wheel makers in North America. Since Snow was dedicated to the manufacture of a first-class product, he turned to making track materials when competitors drove down the price of wheels so that it was no longer possible to offer top-quality merchandise. Snow was a founder of the Steel Tired Wheel Company and the American Brake Shoe and Foundry Company.

JOHN STEPHENSON (1809–1893), who is best remembered as a major producer of streetcars, also manufactured steam railroad cars earlier in his career. He was born in northern Ireland and was brought to this country two years later. His parents had him educated for a mercantile position, but he persuaded them to apprentice him to a carriagemaker when he was nineteen years old. Two years later he opened his own shop, specializing in the construction of omnibuses, and this naturally led to the manufacture of railroad cars. Around 1837 he introduced a peculiar body style that used a lattice frame.

The long financial depression following the panic of 1837 forced Stephenson to close his shop in 1842. He blamed his ruin on the worthless notes issued by his railroad customers, and vowed that he would never again build a railroad car. Technically he remained true to his word, but he did eventually reenter the trade by producing streetcars.

WALTER V. TURNER (1866–1919), an air brake expert who made many contributions in this field, was born in England. He worked first in the woolen textile trade, then emigrated to the United States in 1888 to take a job with a New Mexico ranch. Turner went into ranching for himself but failed, and he found employment as a car repairman with the Santa Fe Railway. From this unlikely beginning he rose to the top of his profession as a leading authority on pneumatics. He became fascinated with the air brake, quickly mastered its theory and mechanics, and began to patent improvements. His ideas were so impressive that in 1903 George Westinghouse hired him and within seven years made him a chief engineer. After about 1890 Westinghouse himself was no longer active in air brake design; it was associates like Turner who devised improvements in this area. For example, Turner developed the K triple valve, a longtime standard for freight cars. He also perfected the LN passenger car brake and the universal valve that replaced the triple valve. In all, Turner acquired some 400 patents.

WILLIAM VOSS (18?–19?) was the only practical car designer to write a book on the subject. He was born in Germany and graduated from a technical school. Later he came to the United States and worked as master car builder on the Burlington, Cedar Rapids, and Northern Railroad. In 1889 he became an assistant engineer with the Fox Pressed Steel Company. Next he served as assistant superintendent of the Barney and Smith Car Works, and in 1898 was appointed superintendent of Jackson and Sharp, where he worked until at least 1908. Little more information seems to be available on his life and career, even though his book, *Railway Car Construction* (1892), was the only one published in this field for many years.

WEBSTER WAGNER (1817–1882), sleeping-car pioneer and founder of the second-largest sleeping-car company in America, was born into a farming family in Palatine Bridge, New York. He learned the wagon-making trade from his elder brother, with whom he formed a partnership. Their business failed, however, and Wagner became a station agent in his home town. In 1860 he quit that job to devote more time to some sleeping cars that he had begun to manage on the New York Central Railroad. He became a protégé of Cornelius Vanderbilt, soon took over the management of the sleeping cars on the huge Vanderbilt system, and within a few years became a wealthy man.

In 1870 Wagner was elected to the New York State Assembly, and later he gained a seat in the State Senate. He looked the part, since he was "a white-haired, white whiskered but fresh and vigorous man," according to one description. Ironically, he was killed aboard one of his own cars in a terrible rear-end collision at Spuyten Duyvil, New York.

NATHAN WASHBURN (1818–1903) designed the standard pattern of the double-plate cast-iron car wheel. He was born on a farm in Stafford, Connecticut, and worked as a carpenter and iron founder in his early years. He organized his own foundry in 1840 and operated it in several locations. By 1860 Washburn was a well-established manufacturer with facilities in Worcester, Massachusetts, Troy and Schenectady, New York, and Toronto, Canada. These works produced pulleys, cotton machinery, railroad rails, and gun barrels, in addition to car wheels. After 1865 he decided to concentrate on car wheels. Later he acquired property with iron ore on it and developed a steel tired wheel.

THOMAS W. WASON (1811–1870) was born in Hancock, New Hampshire, the son of a carpenter and farmer who was too poor to give his children many years of schooling. Wason found work in Massachusetts as a carpenter, but in 1845 he resolved to start his own business in Springfield. His brother Charles joined him, and together they began to build freight cars with their own hands. The enterprise grew slowly. In 1848 the brothers leased a portion of the defunct Springfield Car and Engine Company. Charles left the firm in 1851 to open his own car plant in Cleveland, and Wason took in several new partners. His business expanded until it had become a major car builder by the time of Wason's death. The Wason Car Company continued as a major producer for many years but appears to have largely dropped out of steam railroad car building by 1900. Brill purchased the firm in 1907, and production was concentrated on street railway vehicles until the plant closed in 1932.

BENJAMIN WELCH (1827–1913), a pioneer West Coast car builder, carried the traditions of the East to the workshops of California. He was born in Peaks Island, Maine, and entered railway service in 1846. He worked briefly for the Portland Company, a maker of locomotives and cars, before setting out for the California gold fields. Welch failed as a miner and was glad to find a position in the small car shop of the Sacramento Valley

Railroad. In 1863 he took the job of car builder for the Central Pacific, whose Sacramento shops became the largest manufacturing establishment in the Far West. Scores of passenger cars, including some elegant private cars, and hundreds of freight cars were built under Welch's direction. He retired in 1902.

GEORGE WESTINGHOUSE, JR. (1846–1914) did not invent the air brake so much as introduce and develop a workable version of it. He was born in Central Bridge, New York, of German and Dutch-English parents. His father operated an agricultural-tool factory in Schenectady, and George received a practical schooling in the shop. He was a remarkably large, strong, and intelligent young man with an unmistakable gift for mechanics. When he was only nineteen years old he received the first of nearly 400 patents that he would take out during his lifetime. The early success of the air brake financed Westinghouse's explorations in electricity, natural gas, railway signaling, and steam turbines. He made significant contributions in all these areas.

Westinghouse was not only an imaginative and original engineer but also a skillful businessman. His great manufacturing empire failed during the panic of 1907, and a few years later he lost control of two major divisions. However, Westinghouse remained active in the air brake and signal divisions until the time of his death.

ASA WHITNEY (1791–1874), a major manufacturer of car wheels, was born in Townsend, Massachusetts, the son of a blacksmith. He followed his father's trade and worked in various small shops and factories in the Northeastern states. Around 1830 he entered the employ of the Mohawk and Hudson Railroad to make machinery and cars. Within three years he became superintendent of the line, and his subsequent progress was swift. He was appointed to the office of canal commissioner for the state of New York. In 1842 he became a partner with the Philadelphia locomotive manufacturer, M. W. Baldwin. Four years later he left Baldwin to open a car wheel plant, which prospered and became for a time the largest in North America. He served briefly as president of the Philadelphia and Reading Railroad but was forced to retire in 1861 because of failing health.

WILLIAM H. WOODIN (1868–1934) was for many years an official of the American Car and Foundry Company. His grandfather had been a founding partner of the old car-building firm of Jackson & Woodin of Berwick, Pennsylvania. After this plant merged with A.C.F. in 1899, Woodin worked for the parent corporation and in 1916 became its president. He also headed the American Locomotive Company, and he remained on the boards of both concerns for many years. Woodin, a prominent figure in business and financial circles, served under President Franklin Roosevelt as Secretary of the Treasury.

JONAH WOODRUFF (1809–1876), the elder brother of the sleeping-car pioneer Theodore T. Woodruff, also played a leading role in sleeping-car development. He was born in Watertown, New York, left school at the age of ten, and spent his early years on the family farm. He took up portrait painting with some success, and appears to have made a living at it. In 1838 he served as a colonel in the Canadian Rebellion's "patriot army" and was often called by this title afterwards. During the late 1850s he joined his brother's sleeping-car business and eventually showed a greater talent for the management of the firm than its founder. He helped organize the Central Transportation Company, which for a time seemed destined to dominate the sleeping-car trade in North America. After Pullman leased the Central Transportation Company in 1870, Woodruff decided to start over again and organized his own line. He had already been involved in sleeping-car design, and during his career he obtained several patents in the field. He is best remembered for devising the Silver Palace and Rotunda sleeping car designs.

Woodruff's new company, while it was not a threat to Pullman's empire, was expanding when his health gave way in the mid-1870s. Early in 1876 he went to Bermuda hoping to recover, but he died there a few months later.

THEODORE T. WOODRUFF (1811–1892) is, after Pullman, the popular choice as inventor of the sleeping car. Born in Watertown, New York, of New England parentage, he stayed on the family farm until he was sixteen years of age. Things mechanical were his interest, and he was pleased to leave the land to his parents' care and become a wagonmaker's apprentice. Around 1830 he took work in a local foundry. Expanding his knowledge of the mechanical trades, he demonstrated an inventive bent by designing a mowing machine. When the railroad was built as far as Watertown, Woodruff found a new outlet for his skills. Eventually he gained a practical knowledge of car building and constructed his first sleeping-car model. He went west in 1856 to take a job as master car builder for a railroad in Illinois. His real interest was in promoting his sleeping-car design, however, and within a year he left his new position. He secured patents, built a test car, and organized his company, as described in Chapter 3.

After retiring from the sleeping-car business in 1864, Woodruff lived the life of a wealthy gentleman for a time and built a gorgeous mansion in Mansfield, Ohio. But the old life was more to his taste, and in 1870 he returned to Philadelphia to take over a foundry in Norristown. The panic of 1873 ruined the business, however, and Woodruff was forced into bankruptcy. He bore his failure heavily and spent his last years in a flurry of patents and business schemes, vainly attempting to remake his fortune. A white-haired old man of eighty-five, he was on a business trip to promote a ship's propeller when a train struck and killed him in New Jersey.

APPENDIX B: STATISTICS

The most frustrating problem encountered during the research for this book was the lack of reliable statistics on the American railway passenger car. A scarcity of figures covering the early cars was to be expected, but discrepancies in the statistics available for the modern period make it very difficult to trace the history of special types of cars, such as the parlor. Certainly a large body of statistical data is available for the years after 1880; sources include the Census reports, Interstate Commerce Commission reports, *Poor's Manual of Railroads*, the railroad trade press, and the American Railway Car Institute. Yet they rarely agree, and many of the discrepancies are not small (see Table B.2). For example, there is a difference of 85,000 between the 1880 Census report of 455,000 freight cars in the U.S. fleet and the 1880 *Poor's* report of 540,000. Presumably *Poor's* included all service, work, and privately owned freight cars, while the Census counted only revenue cars owned by operating railroads. There is no way to be certain, however. Passenger car totals are likewise confused by various sources, inclusion or exclusion of Pullman, baggage, express, and mail cars. Also, some sources incorporated Canadian cars in U.S. totals; *Poor's* often did so. In general, government statistics (the Census and the I.C.C.) showed smaller totals than private reporters such as *Poor's* or the *Railroad Gazette*. This might be explained by the government agencies' placid acceptance of "information received," whereas the private publications, having more contacts within the industry, apparently compiled more complete reports. By the early years of the twentieth century, more general agreement prevailed. The discrepancies among sources were still noticeable, particularly in regard to figures on new cars produced. For example, in 1946 the A.A.R. reported delivery of 481 new passenger cars, while the figure given by the American Railway Car Institute was 1,329.

No official statistics are available for the early years of the railroad industry. Table B.1's totals of 900 passenger cars for 1840, 3,000 for 1850, 10,000 for 1860, and 13,000 for 1870 are simply estimates projected from known railroad mileages. During this period several crude reckonings were attempted, but they rarely agreed. The 1855 New York State Railroad Commission Report estimated that there were 10,120 passenger cars and 41,786 freight cars. *The American Railway Review* of September 18, 1859, claimed that there were 5,900 passenger cars and 114,500 freight cars. On April 18, 1861, the same journal revised the figures to 12,272 passenger and 114,750 freight cars, based on a rule of thumb of about one-third of a passenger car and four freight cars per mile of track. Perhaps this rule was reasonably accurate for the early period, but it does not agree with later mileages and known car figures. Compared with 1880 figures, for example, the formula overestimates passenger cars by 13,000 and underestimates freight cars by nearly 200,000! Albert Fishlow attempted to resolve the problem in his economic history of American railroads before 1860.[*] His totals are based on the passenger revenue figures presented in the published reports of the Massachusetts and New York railroad commissions for 1849. He concludes that there was a national total of 500 passenger cars in 1839, 1,400 in 1849, and 4,900 in 1859. But these totals seem too low, particularly when compared with the mileage formula given above. And both methods may produce data wide of the actual totals.

Perhaps some more accurate measure of our pioneer fleet can be gained from using a representative trunk line as an example. The Baltimore and Ohio's annual reports gave far more data on its rolling stock, including breakdowns by types of cars, than other roads did. Following are some statistics from selected years:

B & O Report, 1836, with about 120 miles of line (p. 74).
1,062 freight cars (82 of which are on 8 wheels)
46 coaches (27 of which are on 8 wheels)

B & O Report, 1847, with about 220 miles of line (pp. 46–47).
983 freight cars
54 passenger cars (23 coaches, 9 ladies' cars, 3 smokers, 7 baggage, 1 sleeper, 3 mail, 2 emigrant, 6 express 6-wheelers)

B & O Report, 1851, with about 250 miles of line (table H).
1,443 freight cars
78 passenger cars (34 coaches, 11 ladies', 7 smokers, 12 baggage, 7 mail, 7 excursion)

B & O Report, 1865 (p. 56).
3,398 freight cars
164 passenger cars (130 coaches, 24 smokers, baggage, & mail, 8 office, 1 Post Office, 1 pay car)

B & O Report, 1876 (p. 58).
12,072 freight cars
353 passenger cars (237 coaches, 58 baggage, 13 Post Office, 6 palace [parlor?], 6 sleepers, 10 office, 9 emigrant, 2 pay cars, 12 combines)

The tables given in this appendix are only general indicators; their figures should not be taken as absolute. Because of the conflicting nature of the evidence, no effort was made to determine exact numbers. Instead, the highest figure was generally accepted and rounded off to the nearest tenth or hundredth. Besides the fact that the numbers are approximations, the tables may contain statistical defects, for the author does not pretend to be an expert in this area. As they stand, the tables represent the most satisfactory arrangement of figures that could be assembled by a layman threading his way through "lies, damn lies and statistics."

[*] *American Railroads and the Transformation of the Ante-bellum Economy*, Cambridge, Mass.: Harvard University Press, 1965, pp. 419–421.

Table B.1   American Passenger Cars by Type, 1840–1970

YEAR	COACH	PARLOR	DINER	BAGGAGE	MAIL°	SLEEPING	COM-BINE	OFFICE	SELF-PRO-PELLED	OTHER	TOTAL
1840											900e
1850											3,000e
1860											10,000e
1870	9,000e			3,500e		400†e				100	13,000e
1880	12,500	130‡	40	4,000		1,300†				430	18,400
1890	19,000	370‡	160‡	6,700	3,000e	2,100†	350			720	32,400
1900	20,000	500‡	350	6,000	3,500e	3,500†	4,400	600		250	38,200
1910	25,800	1,350†	950	8,220	1,500e	4,500†	5,700	780		3,000	52,300
1920	29,300	1,570†	1,350	12,100	1,240	7,100†	5,680	950		4,610	63,100
1930	25,100	1,760†	1,760	12,900	950	8,500†	5,270	880	3,660	2,320	63,900
1940	15,200	1,370†	1,530	12,000	1,500	6,400†	2,500	570	4,100	30	45,200
1950	14,200	730†	1,800	13,600	1,750	6,100†	2,000	560	2,650	110	43,500
1960	10,250	300	1,170	11,400	830	2,600†	750	400	2,900		30,600
1970	4,030	210	570	3,710	320	800	80	210	2,630		12,560

° Mail car figures are confused by designation of R.P.O. and combined mail-baggage cars as mail cars.
† Includes Pullman
‡ Less Pullman
e Estimate

Source: Data from I.C.C. statistical reports, Association of American Railroads yearbooks and statistics on class 1 railroads, *Poor's Manual of Railroads,* and other sources mentioned in the opening statements of this appendix.

Table B.2   American Railroad Car Statistics from Various Sources, 1880–1960

YEAR	U.S. CENSUS	POOR'S	AM. RLY. CAR INST.	A.A.R.	I.C.C.
1880	16,805 P	17,575 P			
	455,450 F	539,255 F			
1890	26,820 P	30,211 P			
	918,491 F	1,061,970 F			
1900	34,713° P	34,995° P			34,713° P
	1,365,531 F	1,350,258 F			1,416,125 F
1910	47,179° P	51,158 P	47,179° P		47,095° P
	2,148,478 F	2,297,620 F	2,135,121 F		243,236 F
1920	56,102° P		56,102° P	53,501 P	56,102° P
	2,388,424 F		2,417,999 F	2,322,122 F	2,388,424 F
1930	53,584° P		53,584° P	52,130 P	53,584 P
	2,322,267 F		2,352,046 F	2,276,867 F	2,322,267 F
1940	38,308° P		38,308° P	37,817 P	38,308° P
	1,684,171 F		1,706,387 F	1,653,663 F	1,706,387 F
1950	43,372 P		37,359° P	37,146 P	37,359° P
	1,745,778 F		1,769,983 F	1,721,269 F	1,769,983 F
1960	25,746 P		25,746° P	25,655° P	26,947° P
	1,690,000 F		1,709,549 F	1,676,629 F	1,690,396 F

P—Passenger cars
F—Freight cars
° Less Pullman

Source notes: U.S. Census figures are from *Historical Statistics of the United States: Colonial Times to 1957,* 1960 (see also supplement to 1962). Figures offered disagree from those in original published reports of the Census. (In 1890, for example, the original reports claim 27,653 passenger cars.)

Poor's Manual of Railroads figures for 1880 are from 1892 edition; figures for 1870, 1900, and 1910 are from 1911 edition.

Table B.3   Passenger Car Prices, 1831–1960

TYPE	YEAR	R.R. OR BUILDER	SOURCE	PRICE OR COST*
4-wheel coach	1831	B & O	Annual report	$ 700
4-wheel coach	1840c.	All	Von Gerstner†	800–900
8-wheel coach	1840c.	All	Von Gerstner†	2,000
8-wheel coach	1843	All	Baldwin papers, Hist. Soc. of Penna	1,700
8-wheel coach	1845	Eastern	A.R.R.J., Oct. 9, 1845	2,200
Coach	1848	Louisa	R.&L.H.S. Bulletin 65 (1945)	1,350
Coach	1850	P R R	J. E. Thomson letter book, P R R Library	1,850
Sleeper	1860	Wagner	Am. Rly. Rev., May 31, 1860	3,500
Sleeper	1863	Pullman	West. R R. Gaz., July 18, 1863	6,500
Sleeper	1866	Central Transportation	R.R. Record, Jan. 18, 1866	8,000
Parlor	1867	Lake Shore	R.R. Car Jour., Nov. 1894	17,500
Private	1875	D. O. Mills	N.C.B., Jan. 1876	15,000
Coach	1875	All	N.C.B., May 1875	4,800
Baggage	1879	All	N.C.B., Sept. 1879	2,200
Coach	1880	P R R	Rly. Age, Nov. 11, 1880	5,500
Sleeper	1882	Pullman	Rly. Rev., Aug. 12, 1882	14,000
Coach	1882	Pullman	Rly. Rev., Aug. 12, 1882	7,500
Diner	1886	All	R.R.G., Jan. 1886	12,000
Coach	1887	Southern Pacific	N.C.B., Sept. 1887	4,500
Coach	1892	All	R.R.G., Aug. 5, 1892	5,000
Drawing Rm.	1892	All	R.R.G., Aug. 5, 1892	12,000
Coach	1906	All	R.R.G., Dec. 28, 1906	8,000
Coach	1907	All	R.R.G., Dec. 27, 1907	8,500
Coach	1915	All	Rly. M.E., June 1915	12,800
Mail	1915	All	Rly. M.E., June 1915	11,000
Sleeper	1915	All	Rly. M.E., June 1915	22,000
Business	1915	All	Rly. M.E., June 1915	15,000
Motor	1915	All	Rly. M.E., June 1915	20,000
Baggage	1915	All	Rly. M.E., June 1915	8,500
Coach	1920	A C L	Rly. Age, Jan. 7, 1921	34,036
Baggage	1920	Ill. Cent.	Rly. Age, Jan. 7, 1921	26,650
Diner	1920	Ill. Cent.	Rly. Age, Jan. 7, 1921	51,005
Postal	1920	C & N W	Rly. Age, Jan. 7, 1921	27,500
Coach	1925	N Y C	Rly. Age, Jan. 2, 1925	27,625
Diner	1925	F E C	Rly. Age, Jan. 2, 1925	48,500
Baggage	1925	M P	Rly. Age, Jan. 2, 1925	18,168
Business	1925	Reading	Rly. Age, Jan. 2, 1925	84,104
Coach	1930	All	Military Eng., Sept.–Oct., 1930	27,000–33,000
Diner	1930	All	Military Eng., Sept.–Oct., 1930	43,000–51,000
Baggage	1930	All	Military Eng., Sept.–Oct., 1930	18,000–20,000
Rebel motor train	1936	G M & O	Rly. Age, Aug. 22, 1936	184,175
Rail bus	1936	—	Rly. Age, Aug. 22, 1936	24,611
Coach	1936	Milwaukee	Rly. Age, Aug. 22, 1936	24,649
Coach	1940	All	Am. Rly. Car Inst., 1940	53,905
Diner	1940	All	Am. Rly. Car Inst., 1940	68,851
Baggage	1940	All	Am. Rly. Car Inst., 1940	19,945
Sleeper	1940	All	Am. Rly. Car Inst., 1940	98,642
Combine	1946	—	Rly. M.E., May 1946	84,541
Diner	1946	—	Rly. M.E., May 1946	98,076
Tavern-Lounge	1946	—	Rly. M.E., May 1946	96,936
Coach	1947	P R R	Wash. News, April 17, 1947	90,000
Coach	1950	All	Am. Rly. Car Inst., 1950	112,413
Diner	1950	All	Am. Rly. Car Inst., 1950	138,275
Baggage	1950	All	Am. Rly. Car Inst., 1950	43,632
Sleeper	1950	All	Am. Rly. Car Inst., 1950	143,240
Dome	1955	Great North.	Trains, Dec. 1959	243,738
Bilevel commuter	1954	—	R.R. Mag., March 1954	140,000
Coach	1958	All	I.C.C. Report, 1958	235,934
Sleeper	1958	All	I.C.C. Report, 1958	307,497
Diner	1958	Great North.	Trains, Dec. 1959	242,569
Pioneer III Coach	1956	Budd Co.	Trains, Nov. 1956	90,000
Coach	1960	All	Am. Rly. Car Inst., 1960	154,137
Diner	1960	All	Am. Rly. Car Inst., 1960	184,668
Baggage	1960	All	Am. Rly. Car Inst., 1960	36,355

* Most of these figures are prices, but in cases where the cars were built by the railroads, the amount given represents the cost.

† F. A. Ritter Von Gerstner, *Die innern communication der vereinigten Staaten von Nord-America*, Vienna, 1842–1843.

Table B.4   Cars Owned by the Pullman Company and Its Competitors

YEAR	SLEEPING	DINING	PRIVATE RENTAL	PARLOR	MISC.	TOTAL PULLMAN	WAGNER, WOODRUFF, ETC.	RAILROAD-OPERATED SLEEPERS
1870						300[e]	100[e]	
1880						700[e]	300[e]	277
1890						2,135	600[e]	516
1900	3,213	24	24	466	152	3,879		393
1910	4,094	25	34	629	65	4,847		490
1920	6,559	0	27	1,019	121	7,726		558
1930	8,263	18	22[e]	1,186	312	9,801		223
1940	6,320	1	17	479	93	6,910		110
1950	5,993	0	0	227	6	6,226		148
1960						2,650		523
1970						0		677

[e] Estimate

Source: 1936 I.C.C. Report on the Pullman Company, published annual reports of the Pullman Company, and data in *Poor's Manual of Railroads*. Figures for 1900 are from a Pullman publication of Feb. 1, 1901, entitled *A Descriptive List*.

Table B.5   Passenger Car Production, 1871–1960

YEAR	NO. BUILT	NO. ORDERED	YEAR	NO. BUILT	NO. ORDERED
1871[a]	1,507		1920	751	1,613[a]
1872–1879	(No data)		1921	1,162	308
1880[b]	395		1922	908	2,425
1881	407		1923	1,888	2,191
1882	458		1924	2,332	2,704
1883	587		1925	2,312	2,140
1884	374		1926	2,798	1,987
1885	259		1927	2,007	1,542
1886	456		1928	1,549	2,287
1887	663		1929	2,419	2,315
1888[c]	2,471		1930	1,520	679
1889[d]	(No data)		1931	243	11
1890[e]	1,654		1932	39	39
1891	1,640		1933	9	16
1892	2,195		1934	275	421
1893	1,986		1935	202	126
1894	516		1936	189	448
1895	430		1937	621	564
1896	474		1938	420	268
1897[f]	494		1939	273	319
1898[g]	669	1,046	1940	250	370
1899[h]	1,201	1,590	1941	363	546
1900[i]	1,515		1942	393	4
1901[j]	1,949	2,249	1943	675	1,720
1902	1,948[k]	2,510	1944	995	740
1903	2,007[k]	2,153	1945	928	2,993
1904	2,144[k]	2,055	1946	1,329	1,238
1905	2,551[k]	2,947	1947	858	316
1906	3,167[k]	2,716	1948	951	506
1907	5.457[k]	1,475	1949	939	109
1908	1,645[k]	1,065	1950	1,078	102
1909	2,698[k]	3,891	1951	183	261
1910	4,136[k]	3,073	1952	195	520
1911	3,362	2,466	1953	329	164
1912	2,509	2,888	1954	351	610
1913	2,654	3,535	1955	412	1,078
1914	3,589	1,494	1956	391	269
1915	1,513	1,878	1957	232	78
1916	1,344	2,006	1958	143	63
1917	1,684	1,118	1959	70	193
1918	750	6	1960	251	264
1919	126	290			

[a] 1871 figures from *Railroad Gazette*, Oct. 4, 1873.

[b] 1880–1887 from *R.R.G.*, Jan. 6, 1888 (Est. of major builders and R.R. shops).

[c] 1888 from *R.R.G.*, Jan. 4, 1889.

[d] For 1889 *R.R.G.*, Jan. 10, 1890, reported on passenger car production but gave no figures.

[e] 1890–1896 from *R.R.G.*, Jan. 1, 1897 (9 major commercial builders).

[f] 1897 from *R.R.G.*, Dec. 31, 1897.

[g] 1898 from *R.R. Car Journal*, January 1899, and *R.R.G.*, Dec. 30, 1898.

[h] 1899 from *R.R. Car Journal*, January 1900.

[i] 1900 from *R.R.G.*, Dec. 28, 1900 (exclusive of R.R. shops).

[j] 1901–1960 from *Railroad Car Facts*, 1944 and 1961 eds. (Data include express refrigerator and milk cars.)

[k] Figures include foreign output. (From *Railway Age*.)

APPENDIX C: AMERICAN RAILROAD CAR CHRONOLOGY

1830   A typical four-wheel passenger car, seats 15, weighs 2 tons, and measures 15 feet long.

1831   Mail is carried on a car of the South Carolina Railroad.
The *Columbus*, the first eight-wheel or double-truck passenger car, begins service.

1833   The B & O is using reversible seat backs.
The need for separate baggage cars is recognized.

1834   J. S. Kite patents the safety beam as a safeguard against broken axles.

1835   The passenger car *Victory*, made by Richard Imlay, features a toilet, a buffet, and a clerestory roof.

1837–1838   Richard Imlay builds the first sleeping car, the *Chambersburg*, for the Cumberland Valley Railroad.

1838c.   The double-truck passenger car is now standard in America.

1838   Bonney, Bush, and Lobdell patent the double-plate, cast-iron wheel.

1839   John Tims patents a journal bearing box.
The Western Railroad (of Massachusetts) provides a special car for its director.
A sixteen-wheel passenger car is built by Davenport and Bridges.

1840   The typical coach is 35 feet long and 8½ feet wide; it weighs 8 tons and seats 50.

1841   The swing bolster truck is patented by Davenport and Bridges.

1845c.   Twelve-wheel cars appear on the Camden and Amboy and are adopted by several other lines within two years.

1845   Equalizers are used for passenger car trucks.
The Eastern Railroad is operating parlor cars.

1848   John Lightner patents a journal bearing box.

1849   At least eight railroads have operated one or more sleeping cars.

1850c.   Double-acting brakes are introduced.
Side-panel truss permits long bodies and lightweight construction.
Some lines begin using iron body bolsters.

1850   Nathan Washburn patents a double-plate car wheel that will remain popular until 1928.
Rubber springs have become popular for passenger car trucks.

1851   The Hudson River Railroad tests gas lighting for cars.

1852   Charles Waterbury patents a platform canopy that foreshadows the vestibule.

1853   The Hudson River is operating stateroom cars.

1856   The Illinois Central begins running stateroom cars.

1857   T. T. Woodruff builds his first sleeping car on patents received the year before.
The Spear hot-air car stove is patented.

1858   Webster Wagner enters the sleeping-car business.

1859   George M. Pullman enters the sleeping-car business.
An all-iron passenger car is built after a design patented in 1854 by B. J. La Mothe.
Webster Wagner reintroduces the clerestory roof.

1860   Primitive lunch counter cars appear on a few lines.

1862   The Central Transportation Company, the first large-scale sleeping-car firm, is organized.
A Railway Post Office car fitted for en route sorting appears on the Hannibal and St. Joseph Railroad.

1862–1863   The Pittsburgh, Fort Wayne, and Chicago acquires about twenty-four all-iron passenger cars.

1864–1865   Pullman's famous sleeper, the *Pioneer*, enters service.

1865   Dark exterior colors replace the yellow-cream tints formerly favored for passenger cars.
Kerosene lamps are growing in popularity for car lighting.
The Baker hot-water car heater is introduced. It will be in wide use within a decade.

1866   The New York Central Sleeping Car Company is formed by Webster Wagner.

1867   The Pullman Palace Car Company is incorporated.
Wagner begins parlor car service.
Limited food service is offered aboard Pullman's hotel cars.

1868   The first full-sized dining car, the *Delmonico*, enters service.

1869   Allen's paper car wheel is introduced.
The Master Car Builders adopt a standard bearing.
Westinghouse receives his first patent for an air brake (patents had been issued to other inventors for air brakes in earlier years).

1870   Pullman leases the Central Transportation Company and begins to gain control of the nation's sleeping-car business.

1870	Soft white-metal bearings become popular for wheel journals.
1873	G. F. Chalender patents an iron-plate, side-panel truss for wooden cars. Eli Janney's second patent embodies the basic plan of the modern car coupler. The Westinghouse automatic air brake is introduced.
1875c.	Parlor cars are now relatively common on many U.S. railroads. The vacuum brake and the air brake compete for the power brake market.
1875	The Miller coupler is widely used on passenger cars.
1879	The Central Pacific introduces emigrant sleeping cars. An air-activated water system for cars is patented, though it will rarely be used until about 1890.
1880	Serious attempts begin to develop a practical steam car heating system.
1881	Pintsch gas lighting for cars is introduced into the U.S. Pullman opens a giant car works south of Chicago.
1882	The Pennsylvania Railroad tests electric car lighting. Webster Wagner is killed aboard one of his palace cars. The Pennsylvania begins lounge car service.
1883	The Mann Boudoir Car Company is incorporated to reintroduce the stateroom sleeping car to the American public. Buffet cars appear on the West Shore railway.
1885c.	Dining cars are becoming common on most U.S. lines.
1885	A patent is issued for the Forney car seat—an early attempt to fit the seat to its occupant.
1887	The first Sessions vestibule enters service.
1888	The quick-action air brake is introduced.
1889	The Robbins steel passenger car is completed after a design patented in 1868.
1890	Iron and steel reinforcing is devised to strengthen wooden car construction.
1893	The New York Central introduces a class of wooden cars remarkable for their 72-foot-long bodies. Pullman patents the wide vestibule.
1894	Railway Mail Service has grown enormously since 1869. Over 3,000 cars are carrying the mail. The high-speed air brake is introduced.
1897	George M. Pullman dies.
1899	The American Car & Foundry Company (A.C.F.) is formed from a merger of seventeen smaller firms. The Pullman Company takes over the Wagner Sleeping Car Company.
1900	Steam heating is now standard on first-class cars. Steel beam end platforms are becoming common for the better class of passenger car.
1904	The first steel baggage car is delivered to the Erie. Vapor (low-pressure steam) car heating is developed. Cast-steel trucks are tested on the Big Four railroad.
1905	The Garland car ventilator is introduced. The first steel mail car enters service.
1907	The first production-model steel passenger car is built. The first steel Pullman, the *Jamestown*, is completed.
1910	Mass production of steel sleeping cars begins with Pullman's *Carnegie*. Steel axles are now standard; the iron axle has reluctantly been abandoned. Large-scale production of wrought-steel wheels begins. Electric car lighting is finally accepted as practical.
1911	The Santa Fe tries an ice cooling system on its deluxe train. The Post Office establishes minimum strength standards for its cars. The type D Janney-style coupler is adopted—the first coupler with interchangeable parts.
1913	The last wooden cars are ordered by a U.S. railroad.
1920	In the entire passenger car fleet, 60 percent of the cars are still wooden.
1921	The Pennsylvania begins using roller bearings for passenger cars.
1922	Pullman and Haskell & Barker merge their car plants.
1923	The first large order of aluminum passenger cars is completed for the Illinois Central Railroad by Pullman.
1927	Pullman tests an electromechanical air conditioner on one of its sleepers.
1929	Pullman takes over the Standard Steel Car Company.
1930	Heat-treated (hardened) car wheels are introduced.
1931	The B & O inaugurates the first fully air-conditioned train. The type E Janney-style coupler is adopted.
1934	*City of Salina*, the first streamline motor train, enters service on the Union Pacific.

1934　An all-welded passenger car is produced by the Milwaukee Road shops.

1935c.　Lightweight arched-roof passenger cars are now accepted as the new standard design.
Triple bolster trucks are introduced by Pullman.

1936　There are 5,800 air-conditioned cars in service.

1937　The Southern Pacific begins to purchase articulated, standard-sized passenger cars.
The first Roomette sleeping car enters service.

1938　The tightlock style of coupler is adopted.

1938–1939　Budd introduces disc brakes, though they will not be widely used until about 1955.

1939–1940　The Southern Pacific purchases triple-unit dining-car sets.

1940c.　Floating bolsters for car trucks are introduced.

1942　The private car *Ferdinand Magellan* is remodeled for the use of President Franklin D. Roosevelt.

1945　The Sleepy Hollow car seat is developed.
The first Vista Dome car begins service on the Burlington.
Roller bearings are now standard for new passenger cars.

1945–1946　Railroads place huge orders for new cars—3,000 in all.

1947　Pullman sells its sleeping-car division to an association of U.S. railroads.

1949　The first all-electric dining car appears on the Illinois Central.

1950　Only about 15 percent of the passenger fleet consists of modern lightweight cars.
Gallery, or double-deck, cars are adopted for suburban trains in the Chicago area.

1956　Budd produces a cheap production-line coach, the Pioneer III, in a vain effort to revive new car orders.
The first Slumber Car—an economy sleeper—enters service.

1963　The Railway Mail Service, which had already been much reduced, is cut heavily.

1965　An order of new coaches is delivered to the Kansas City Southern. No new conventional passenger cars will be ordered for nearly a decade.

1966　Metroliners are ordered from the Budd Company.

1968　The Pullman Company, which has been owned since 1947 by an association of railroads, suspends sleeping-car operations altogether.

1969　Metroliners begin regular service.

1971　Nearly all passenger cars are acquired by Amtrak, a government-sponsored corporation, organized to operate long-distance passenger service in the United States.

1973　Amtrak orders 57 new cars from Budd—the first conventional passenger equipment purchased in nearly ten years.

1974–1975　Amtrak adds another 435 units to its 1973 order.

1975–1976　Amtrak orders 284 bilevel cars from Pullman.

1977　The last railway post office car operates between New York and Washington.

NOTES

CHAPTER ONE

1. Edward Hungerford, *Daniel Willard Rides the Line*, New York: Putnam, 1938, p. 240.

2. The *Pioneer* has been pictured in various accounts as a simple clapboard box-body with three windows. This rendering arose from an artist's erroneous reconstruction, which was based on a sketch in Smile's biography of George Stephenson that supposedly showed the first passenger car on the Stockton and Darlington Railroad. And that sketch itself considered purely imaginary by British authorities.

3. August Mencken, *The Railroad Passenger Car*, Baltimore: Johns Hopkins Press, 1957, hereafter referred to as Mencken. A notice from the *Baltimore Gazette* of Aug. 14, 1830, is given on p. 10. A nearly identical account from the *Baltimore American* is given in vol. I of Edward Hungerford's *The Story of the Baltimore and Ohio Railroad*, New York: Putnam, 1928.

4. Minutes of Newcastle and Frenchtown Railroad, Nov. 12, 1831. The original book was in the possession of the late C. L. Winey; its present location is unknown to this writer.

5. *American Railroad Journal*, March 3, 1832, p. 149. Hereafter cited as *A.R.R.J.*

6. Ibid., April 28, 1832, p. 288, and May 12, 1832, p. 307. Lithographic views showing Imlay-style cars on the P G & N and the Philadelphia and Columbia Railroad are given in Figs. 238 and 241 of Seymour Dunbar's *A History of Travel in America*, vol. III, Indianapolis: Bobbs-Merrill, 1946.

7. *Scribner's Magazine*, August 1888, p. 197.

8. In his two-volume study of American railroads (*Die innern communication der vereinigten Staaten von Nord-America*, Vienna, 1842–1843), F. A. Ritter Von Gerstner consistently described three-section compartment cars whenever he discussed four-wheel passenger cars in enough detail to indicate the body style.

9. *Poulson's American Advertizer*, Nov. 11, 1831.

10. E. Harper Charlton, *The Street Railways of New Orleans*, vol. 13, no. 1, Los Angeles: Interurbans, 1955, p. 250.

11. *A.R.R.J.*, Nov. 17, 1832, p. 737.

12. This drawing was probably first used in the 1882 annual report of the Lake Shore and Michigan Southern Railroad. The report contained a short history of its predecessor line, the Erie and Kalamazoo.

13. Alvin F. Harlow, *Road of the Century*, New York: Creative Age Press, 1947, p. 17.

14. Frank W. Stevens, *The Beginnings of the New York Central Railroad*, New York: Putnam, 1926, p. 305.

15. Harlow, *Road of the Century*, p. 56.

16. *National Car Builder*, December 1870, p. 5.

17. *Railway Patent Claim Reporter*, 1862, p. 262.

18. Dunbar, *A History of Travel in America*, vol. 3, p. 1006, gives an account of travel on the Boston and Providence Railroad in 1835.

19. *A.R.R.J.*, Sept. 28, 1850, p. 609.

20. L. T. C. Rolt, *The Railway Revolution: George and Robert Stephenson*, London: Longmans, 1960, p. 46.

21. C. F. D. Marshall, *Centenary History of the Liverpool and Manchester Railway*, London: Locomotive Publishing Company, 1930, p. 59.

22. Alvin Harlow, *Steelways of New England*, New York: Creative Age Press, 1946, pp. 56–57, and plate opposite p. 274.

23. William Whiting, *Arguments of William Whiting, Esq., in the case of Ross Winans vs. Orsamus Eaton et al.*, Boston, 1853, p. 129. Hereafter cited as *Whiting, Winans vs. Eaton.*

24. *A.R.R.J.*, Dec. 8, 1855, p. 773. Imlay made his proposal to the directors of the Newcastle and Frenchtown.

25. *Ross Winans vs. The Eastern Railroad Company*, U.S. Circuit Court, Massachusetts District, October 1853 term (Boston 1854). This 1299-page transcript of the eight-wheel-car patent suit is the basic source for the following discussion of the early B & O eight-wheel cars. Hereafter referred to as *Winans vs. Eastern Railroad.*

26. B. & O annual report, October 1834, p. 32, lists the *Winchester, Dromedary*, and *Comet* as having been constructed during that year. But both Gatch and Winans said that they were built a year or two earlier.

27. Von Gerstner, vol. 1, p. 36.

28. B & O annual report, October 1834, p. 11.

29. John T. Scharf, *The History of Philadelphia*, vol. 3, Philadelphia, 1884, pp. 2259–2260.

30. In *Winans vs. Eastern Railroad*, p. 740, the model is credited to Proctor (1833). In *Whiting, Winans vs. Eaton*, p. 82, it is credited to Fultz (1829).

31. Minute Books, Philadelphia, Germantown, and Norristown Railroad, October 14, 1835, and March 29, 1837. These books are in the possession of the Reading Company, Philadelphia.

32. *American Society of Civil Engineers Transactions*, 1878, p. 206. Here W. M. Roberts reminisces about early engineering.

33. Von Gerstner, vol. 1, p. 129.

34. *Locomotive Engineering*, February 1895, p. 92. Walter De Sanno, son of the P & C master mechanic, recalled the use of drop-center cars by Hall & Company, one of the many private operators on the line in the 1830s.

35. M. W. Baldwin Letters, Historical Society of Pennsylvania, Philadelphia.

36. *Winans vs. Eastern R.R.*, pp. 1242–1243.

37. *Railroad Gazette*, April 4, 1874, p. 124. Hereafter cited as *R.R.G.*

38. *A.R.R.J.*, Nov. 11, 1837, p. 629.

39. South Carolina annual reports, 1839–1842. S. M. Derrick, *The Centennial History of the South Carolina Railroad*, Columbia, S.C.: The State Company, 1930, provided information on the hogshead cars.

40. *A.R.R.J.*, April 1, 1840, pp. 209–210.

41. Von Gerstner, vol. 1, p. 265.

42. The information on the Diamond cars is from Stephenson's own testimony in the printed record of *Pullman vs. Wagner*, 1881.

43. *Rochester Daily Democrat*, Aug. 28, 1839. Richard F. Palmer kindly provided this reference.

44. Dunbar, *A History of Travel in America*, vol. 3, p. 858.

45. The two-page 1839 prospectus is reproduced fully in H. T. Gause, *The Semi Centennial Memoir of the Harlan and Hollingsworth Company*, Wilmington, Del., 1886.

46. *A.R.R.J.*, Nov. 15, 1838, pp. 297–299.

47. *Winans vs. Eastern R.R.*, p. 897.

48. Letter from Ross Winans to Hartford and New Haven Railroad, Jan. 30, 1836. Maryland Historical Society, Baltimore.

49. *Winans vs. Eastern R.R.*, p. 1001.

50. Ibid., pp. 897 and 1005.

51. Ibid., p. 996.

52. *A.R.R.J.*, Sept. 28, 1850, p. 609.

53. The information for the eight-wheel car case was taken from the following sources: *Winans vs. Eastern R.R.*, 1854; *Whiting, Winans vs. Eaton*, 1853; William Whiting, *Twenty Years' War against the Railroads*, Boston, 1860; and mss. transcripts of Winans vs. the New York and Harlem R.R. (U.S. District Court, Southern New York, 1855), and Winans vs. the New York and Erie R.R. (U.S. Supreme Court, 1858): both transcripts are in the U.S. National Archives, Washington, D.C.

54. D. B. Eaton, *Ross Winans against the New York and Erie Railroad Co.; General Statement of Facts*, New York, 1856, pp. 6–7.

55. *Railway Review*, June 9, 1883, p. 318.

56. *R.R.G.*, May 19, 1893, p. 369.

57. *A.R.R.J.*, Oct. 9, 1845, pp. 647–648.

58. *Railroad Car Journal*, August 1894, p. 167. Hereafter cited as *R.R.C.J.*

59. Ibid., January 1894, p. 2.

60. Railway and Locomotive Historical Society, *Bulletin 65*, 1945, pp. 61–63.

61. *National Car Builder*, February and June 1883, pp. 16 and 65.

62. *Locomotive Engineering*, June 1904, p. 248.

63. *American Railway Times*, Jan. 19, 1862, p. 25.

64. *A.R.R.J.*, May 14, 1853, p. 307.

65. *Railroad Advocate*, Aug. 29, 1857.

66. A. Bendel, *Aufsatze Eisenbahnwesen in Nord-Amerika*, Berlin, 1862, p. 36. Hereafter cited as Bendel.

67. Railway and Locomotive Historical Society, *Bulletin 6*, 1923, p. 81. Recollections of J. H. French, a former officer of the Old Colony; reprinted from 1900 New England Railroad Club reports.

68. A history of the Reading was published serially in an employees' magazine, *The Pilot*, between 1909 and 1914. The articles, prepared by J. V. Hare, have been reprinted in book form in recent years (published by John H. Stock, Philadelphia, 1970c.).

69. Hamilton Ellis, *Railway Carriages in the British Isles from 1830 to 1914*, London: George Allen & Unwin, 1965, p. 25.

70. *American Railway Review*, May 31, 1860, p. 328.

71. Ibid., Aug. 9, 1860, p. 71.

72. *American Railway Times*, July 20, 1867, p. 230. Hereafter cited as *A.R.T.*

73. Ibid., January 1861, p. 25.

74. *A.R.R.J.*, May 7, 1864, p. 459, and Jan. 7, 1865, p. 28.

75. *R.R.G.*, April 1887, p. 54.

76. Ibid., Sept. 4, 1875, p. 367, shows a diagram for an X-brace clerestory roof. A similar sketch is given in the 1879 *Car Builders' Dictionary*.

77. The literature of the narrow-gauge railway is large and growing. The best general survey is Howard Fleming's pioneer study, *Narrow Gauge Railways in America*, New York, 1875 (reprinted in 1949). A chapter is devoted to cars. A few of the more recent studies, largely pictorial, that have been published by enthusiasts are cited in the following footnotes.

78. *Cincinnati Daily Gazette*, March 16, 1876, in speaking of the College Hill Railroad.

79. *R.R.G.*, July 20, 1871, p. 201.

80. Fleming, p. 52.

81. *N.C.B.*, November 1875, p. 165.

82. A history of several Carter narrow-gauge cars, together with working drawings, is given in Bruce Mac Gregor's *South Pacific Coast*, Berkeley, Calif.: Howell North, 1968.

83. Two book-length studies of 2-foot-gauge railroads have been written: H. T. Crittenden, *Maine Scenic Route*, Parsons, W. Va.: McClain Printing Company, 1966; and Linwood W. Moody, *The Maine Two-Footers*, Berkeley, Calif.: Howell North, 1959. Crittenden's work is the most valuable to those seeking information on cars because of the drawings included in the appendix.

84. *R.R.G.*, Aug. 4, 1882, pp. 472–473, provided a detailed list of the weights of all materials for a B & A coach.

85. *R.R.C.J.*, December 1898, p. 371.

86. *R.R.G.*, Aug. 3, 1872, p. 338.

87. Ibid., Aug. 17, 1872, p. 361.

88. *Bulletin of the International Railway Congress*, 1904, pp. 1067–1070.

89. *Franklin Institute Journal*, January 1879, p. 62.

90. Ezra Miller suggested this reason for dropped platforms in a prospectus of 1873.

91. *Master Car Builders Reports*, 1884, p. 125.

92. *N.C.B.*, July 1872, p. 8.

93. *R.R.C.J.*, May 1893, p. 195.

94. *A.R.R.J.*, August 1901, pp. 242–243, and *M.C.B.*, 1902, pp. 137–140.

95. *M.C.B.*, 1887, p. 224.

96. *Rly. Age*, July 28, 1881, p. 434.

97. "The Forest and Railroad," *New England Railroad Club*, November 1897, pp. 30–58.

98. *Railway Mechanical Engineer*, February 1919, pp. 85–88.

99. *R.R.G.*, Aug. 5, 1881, p. 431.

100. *R.R.C.J.*, January 1899, p. 1.

101. *Engineering News*, Dec. 14, 1893, p. 471.

102. The Chalender truss is described in the following sources: *R.R.G.*, June 22, 1877, p. 282; *M.C.B.*, 1885, pp. 74–75; *R.R.C.J.*, October 1894, p. 207.

103. *R.R.G.*, Feb. 14, 1890, p. 107.

104. *R.R.C.J.*, May 1892, p. 131.

105. Patent No. 562,343, June 16, 1896.

106. *Rly. Rev.*, Dec. 6, 1890, p. 733.

107. *M.C.B.*, 1895, pp. 258–265, and 1896, pp. 277–281.

108. *Rly. Age*, Feb. 5, 1897, p. 112.

109. *R.R.C.J.*, May 1899, p. 118. Drawings for the Bissell platform are reproduced in the 1906 *Car Builders' Dictionary*, figs. 1931–1940.

110. Statistics for car orders are taken from *Railway Age Gazette* statistical reports for the year noted. Figures on cars in service are from the I.C.C. annual reports.

111. *Rly. Rev.*, May 6, 1893, p. 274.

112. Stevens, *Beginnings of the New York Central*, p. 305.

113. Von Gerstner, vol. 1, p. 88.

114. *Railway and Locomotive Engineering*, May 1902, p. 225.

115. Von Gerstner, vol. 2, p. 87.

116. Ibid. Von Gerstner also said that an iron band was "fastened" near the flange to prevent shattering should the wheels crack. This sounds like Davis's patented wheel, where a wrought-iron rod was cast integrally with the thread.

117. *Winans vs. Eastern Railroad*, p. 1005. Griggs made a similar statement in *Whiting, Winans vs. Eaton*.

118. Von Gerstner, vol. 1, p. 36.

119. Karl Von Ghega, *Die Baltimore-Ohio Eisenbahn*, Vienna, 1844.

120. From a letter of Oct. 9, 1843, by Eaton and Gilbert to M. W. Baldwin. The Baldwin papers, Historical Society of Pennsylvania, in Philadelphia.

121. The perspective exterior view is from *A.R.R.J.*, Aug. 8, 1846, p. 511. The same engraving appeared in the Aug. 25, 1845, issue of *Scientific American*. The mechanical drawing is from *A.R.R.J.*, Aug. 7, 1845, p. 368.

122. *A.R.R.J.*, June 15, 1842, pp. 380–382.

123. *Railroad Advocate*, Nov. 15, 1856, p. 2.

124. These drawings are from Douglas Galton, *Report to the Lords of the Committee of the Privy Council for Trade and Foreign Plantations, on the Railways of the United States*, London: Eyre and Spottiswoode, 1857–1858.

125. *Railroad Advocate*, June 2, 1855, p. 4.

126. Ibid., July 19, 1856, p. 1.

127. *N.C.B.*, February 1873, p. 35.

128. The description and drawing are from Gustavus Weissenborn's *American Locomotive Engineering and Railway Mechanism*, New York, 1871. The first section of this serialized work was published in 1871. The car description (pp. 189–194) was printed in about 1874.

129. Roy B. White, former president of the B & O, was head of the C R R N J when the old cars were sent to the fair. Around 1954 he confirmed their origins to L. W. Sagle, curator of the B & O Transportation Museum, who was suspicious of their pedigree as original B & O equipment. The cars were formally deeded to the B & O in January 1928.

130. *N.C.B.*, March 1884, p. 28.

131. *N.C.B.*, July 1884, pp. 86–87.

132. *A.R.R.J.*, April 1899, p. 114.

133. *N.C.B.*, August 1881, p. 90; *R.R.G.*, Aug. 4, 1882, p. 472; *N.C.B.*, April 1884, p. 40.

134. *A.R.R.J.*, October 1891, pp. 441–442; a St. Charles standard coach is described in *N.C.B.*, February 1891, p. 22.

135. *R.R.G.*, May 19, 1893, p. 370.

136. The drawings and data are from *Engineering News*, Dec. 14, 1893, pp. 470–471.

137. The drawings and description are taken from *Railroad Gazette*, Aug. 16, 1894, p. 193, and *Railway Master Mechanics Magazine*, March 1894, p. 46.

138. The facts presented here were furnished by the late Charles E. Fisher, longtime president of the Railway and Locomotive Historical Society, and by an article on the wooden passenger cars of the New Haven in *Bulletin 103*, October 1960, of the society.

139. More information on the passenger cars of this company is given in John A. Rehor, *The Nickel Plate Story*, Milwaukee: Kalmbach, 1965.

CHAPTER TWO

1. Hamilton Ellis, *Railway Carriages in the British Isles from 1830 to 1914*, London: Allen & Unwin, 1965, pp. 35 and 37.

2. Ibid. Also *American Railroad Journal*, Oct. 20, 1855, p. 665.

3. *Scientific American*, March 19, 1846. The drawing accompanying this article is reproduced in August Mencken, *The Railroad Passenger Car*, Baltimore: Johns Hopkins Press, 1957, p. 36. An early handbill describing Lewis's car is in the collections of the New Jersey Historical Society, Newark.

4. *Scientific American*, Aug. 23, 1851, p. 388.

5. Information about La Mothe is drawn primarily from Walter A. Lucas, "The First Iron Passenger Cars," *Bulletin*, Passaic County Historical Society, vol. 5, no. 5, October 1961, and from the following leaflets:

La Mothe's Patent Life-Preserving Car, New York, 1855, Landauer Collection, New York Historical Society.

How to Make Railroads Pay . . . , New York, 1856, Cincinnati Historical Society.

La Mothe's Patent Iron Railroad Cars, New York, 1859, Roebling Collection, Rensselaer Polytechnic Institute, Troy, N.Y.

The last prospectus includes reprints from *American Railroad Journal*, *Scientific American*, and various city newspapers. Other sources are noted individually.

6. *A.R.R.J.*, March 31, 1855, p. 201, and Oct. 22, 1853, p. 679.

7. See note 5.

8. *A.R.R.J.*, March 31, 1855, p. 201.

9. *Railroad Advocate*, Sept. 20, 1856, p. 2.

10. *American Railway Times*, April 11, 1868, pp. 118–119.

11. *American Railway Review*, Sept. 8, 1859.

12. *A.R.R.J.*, March 3, 1860, p. 190.

13. *National Car Builder*, August 1870, p. 3.

14. *Railroad Gazette*, Aug. 8, 1879, pp. 464–465.

15. Correspondence between D. C. McCallum and the La Mothe Iron Car Company is among the U.S. Military Railroad papers in the National Archives, Washington, D.C. Refer to record group 92.

16. *A.R.R.J.*, July 4 and 11, 1863, pp. 625 and 645; *Western Railroad Gazette*, Aug. 8, 1863, p. 2.

17. *Western Railroad Gazette*, Oct. 27, 1860, p. 2.

18. Annual report, Pittsburgh, Fort Wayne, and Chicago Railroad, 1862.

19. *Amer. Rly. Rev.*, Nov. 15, 1860, p. 293.

20. Annual report, P F W & C, 1863.

21. *Locomotive Engineering*, Dec. 1898, p. 568.

22. George B. Abdill's *Civil War Railroads*, Seattle: Superior, 1961, is a convenient source for these views.

23. *Amer. Rly. Rev.*, March 28, 1861, p. 182.

24. The Roebling papers are housed at Rensselaer Polytechnic Institute, Troy, N.Y.

25. *Railroad Record*, May 18, 1865, p. 161; *A.R.T.*, June 3, 1865, p. 174.

26. *N.C.B.*, August 1892, p. 127.

27. *R.R.G.*, April 20, 1877, p. 181.

28. *Engineering*, April 22, 1870, p. 275.

29. *N.C.B.*, August 1870, p. 3, and June 1874, p. 88.

30. *N.C.B.*, July 1884, p. 90, and September 1884, p. 115.

31. *N.C.B.*, July 1888, p. 104, and *R.R.G.*, Jan. 16, 1891, p. 16.

32. *Engineering News*, Aug. 25, 1888, p. 139; see also *National Cyclopedia of American Biography*, vol. 3, 1893, pp. 68–69.

33. Ibid., vol. 13, p. 508, and trade papers listed in the following notes. Perky organized the 1882 Denver Mineral Exposition and the St. Joseph Cereal Exposition, and he established the National Food Company. His erratic career continued with the founding of a girls' school.

34. Anderson's career is summarized in *A.R.R.J.*, June 1889, p. 293, and in many biographical dictionaries.

35. *Engineering News*, March 26, 1887, p. 203, and *N.C.B.*, January 1888, p. 15.

36. *N.C.B.*, August 1889, p. 113.

37. *Engineering News*, Aug. 25, 1888, p. 139.

38. C. L. Rutt (Ed.), *History of Buchanan County and St. Joseph*, (Chicago, 1904), pp. 232–233 and 250.

39. *Railway Review*, Jan. 17, 1891, p. 44; also *The Zephyr* of February 1939, a magazine published by the Burlington Railroad, which reported on documents found in St. Joseph pursuant to the demolition of the old Federal Building. The location of these papers is now unknown.

40. This description of the final years of the Robbins car is taken from correspondence in 1948 between J. P. Roberts of the National Museum of Transport (St. Louis) and Leo Blondin's widow. Mrs. Blondin sold the museum a German silver lamp from the car. The *Washington Post* of Jan. 23, 1916, also contains useful information.

41. *Railway Age*, July 12, 1890, p. 493.

42. C. B. Hutchins, Dec. 16, 1890, No. 442,894; G. M. Bird, Jan. 24, 1893, No. 490,469.

43. *N.C.B.*, June 1887, p. 80.

44. Prospectus of The American Fire Proof Steel Car Company, 1890c., in the Bureau of Railway Economics Library, Association of American Railroads, Washington, D.C. Other information on the Green and Murison car is from *R.R.G.*, Jan. 4, 1889, pp. 7–8 and 12; *Rly. Rev.*, Feb. 1, 1890, p. 60.

45. *R.R.G.*, Feb. 28, 1890, p. 148.

46. *Rly. Age*, Sept. 13, 1901, p. 232.

47. *A.R.R.J.*, May 1905, p. 152.

48. Abundant published material is available on Gibbs and the early steel cars. The railroad trade press followed these events closely. A summary review is offered in the *American Society of Mechanical Engineers Transactions*, vol. 35, 1913, pp. 93–96. A detailed personal recollection was given in a May 23, 1929, news release by the Pennsylvania Railroad. See also Gibbs's obituary in the *American Society of Civil Engineers Transactions*, vol. 105, 1940, pp. 1840–1844.

49. *R.R.G.*, Aug. 21, 1903, p. 598.

50. Ibid., Sept. 30, 1904, pp. 383–386, offers a detailed description and drawings of the Gibbs subway cars.

51. Ibid., April 7, 1905, pp. 318–319 and 329.

52. *R.R.G.*, June 26, 1903, p. 452.

53. The trade press was generous to the Illinois Central's side-door cars;

see particularly *R.R.G.*, Sept. 4, 1903, p. 630. See also *Master Car Builders Reports*, 1904, p. 207.

54. The three Erie steel cars are described in *M.C.B.*, 1908, pp. 307–308 and 322.

55. *M.C.B.*, 1904, pp. 195–216.

56. *R.R.G.*, Nov. 1, 1907, pp. 530–531.

57. *R.R.G.*, June 22, 1906, p. 686, and *M.C.B.*, 1908, pp. 309–310.

58. *International Railway Congress Bulletin*, vols. 26 and 27, 1912–1913, a five-part series: "The Construction of Iron Passenger Cars on the Railways in the United States of America" by F. Gutbrod. Hereafter referred to as Gutbrod Report.

59. *Western Railroader*, March 1966, pp. 3–9; *A.R.R.J.*, January 1907, pp. 6–8; Gutbrod Report, part 2, pp. 1264–1276.

60. *Railway Age Gazette*, June 12, 1908, pp. 83–84.

61. *A.R.R.J.*, June 1907, p. 232.

62. Information supplied by John S. Fair from Pennsylvania RR mechanical records.

63. *R.R. Age Gazette*, Feb. 18, 1910, p. 367.

64. "The Passing of the Wooden Car from This Railroad," a pamphlet issued by the P R R in June 1928.

65. *A.R.R.J.*, December 1910, p. 489.

66. *Safety on the Railroads: Hearings before a Subcommittee of the Committee on Interstate and Foreign Commerce of the House of Representatives Sixty-third Congress on Bills Relative to Safety on Railroads*, Washington: Government Printing Office, 1914. Hereafter referred to as *Safety on Railroads*.

67. *Railway Mechanical Engineer* (formerly *A.R.R.J.*), February 1922, p. 68; hereafter referred to as *Rly. M.E.*

68. *Safety on Railroads*, pp. 363–366; Gutbrod Report, part 4, pp. 110–114.

69. *Rly. Age Gaz.*, Dec. 15, 1911, pp. 1207–1210.

70. *A.R.R.J.*, November 1909, pp. 447–451; Gutbrod Report, part 4, pp. 114–124 (misidentified as C B & Q).

71. *A.R.R.J.*, June 1911, pp. 208–211.

72. *Rly. M.E.*, September 1928, pp. 506–508.

73. Ibid., January 1928, pp. 23–26.

74. *Rly. Rev.*, Oct. 3, 1903, p. 721.

75. *Rly. Age Gaz.*, 1912, p. 638. Most of the general discussion in this section is drawn from the report *Safety on Railroads* (1914) previously cited.

76. Gutbrod Report, part 3, pp. 36 and 57.

77. *A.R.R.J.*, Jan. 1911, pp. 1–5.

78. A good survey of Stillwell's career is given in *Steam Locomotive and Railroad Tradition*, May 1963, pp. 38–45.

79. *R.R.G.*, June 14, 1907, pp. 831–835.

80. *Rly. M.E.*, July 1917, pp. 387–389.

81. *Rly. Age Gaz.*, Jan. 19, 1912, pp. 84–85. See also R. B. Shaw's *Down Brakes*, London: Macmillan, 1961, for other examples of early steel car wrecks.

82. *Rly. M.E.*, March 1917, p. 187.

83. *A.R.R.J.*, April 1905, p. 112.

84. *Rly. Age Gaz.*, June 16, 1911, p. 1483, and June 15, 1912, p. 1430.

85. *Car Builders' Cyclopedia*, 1928, p. 611.

86. *A.S.M.E. Transactions*, vol. 35, pp. 74–92.

87. Ibid., July 1921, p. 444. See also descriptions of individual cars in various trade journals, 1907–1920.

88. *A.S.M.E. Transactions*, October 1915, pp. 520–521.

89. *Central Railroad Club*, March 1909, pp. 52–67.

90. *Rly. Age Gaz.*, February 1913, p. 58.

91. *International Railway Congress Association Bulletin*, vol. 11, 1929, p. 663. This volume contains an extensive survey of steel passenger cars through the world. Hereafter cited as *Int. Rly. Congress*, vol. 11.

92. *New England Railroad Club*, April 8, 1924, pp. 70–81.

93. Gutbrod Report, part 3, p. 41.

94. Ibid., part 5, p. 325.

95. *Car Builders' Cyclopedia*, 1928, p. 599.

96. *Rly. M.E.*, March 1923, pp. 161–165, contains a summary of a paper read before the Canadian Railway Club in December 1922.

97. Ibid., July 1927, p. 470.

98. *Rly. Age*, Sept. 17, 1921, pp. 518–519.

99. *Int. Rly. Congress*, vol. 11, pp. 602–610 and 664; *Rly. M.E.*, March 1923, p. 164.

100. *Rly. M.E.*, March 1923, pp. 161–165.

101. *New England Railroad Club*, April 8, 1924, pp. 70–81.

102. Gutbrod Report, part 5, p. 331.

103. *Rly. Age*, Feb. 8, 1930, p. 377; Aug. 22, 1931, p. 283; July 22, 1933, p. 147. *Light Metals*, May 1945, p. 223, offers a worldwide survey of aluminum rolling stock; it is clear from the article that the United States was the leader in the early years.

104. Ibid., October 1927, p. 665.

105. *Trains*, August 1951, p. 41.

106. *Rly. Age*, April 22, 1944, p. 762; Nov. 22, 1941, p. 854; Oct. 14, 1939, p. 551. See also *Rly. M.E.*, June 1938, p. 213.

107. *Rly. Age*, Jan. 12, 1946, p. 148.

108. *Rly. M.E.*, October 1934, p. 358.

109. The early and late history of railroad streamlining is treated in Robert C. Reed's *The Streamline Era*, San Marino, Calif.: Golden West Books, 1975. For more detail on Calthrop, see Harold L. Van Doren's *Industrial Design, A Practical Guide*, New York: McGraw-Hill, 1940. For more detail on Adams, see *Dictionary of American Biography*, vol. 1, 1928, p. 58.

110. *Rly. M.E.*, January 1935, p. 23; *Rly. Age*, Nov. 20, 1937, p. 729.

111. *Rly. Age*, Nov. 4, 1939, p. 693.

112. A comparison is made here between the cars of the G M & N Rebel motor train and those of the full-sized Royal Blue. Both were built in 1935 by A.C.F. and both were of Cor-Ten steel.

113. *Mechanical Engineering*, January 1938, p. 29.

114. *Rly. M.E.*, August 1944, p. 356; Dec. 7, 1946, p. 961. *Rly. Age*, Dec. 28, 1946. See also *Int. Rly. Congress*, vol. 24, February 1947, for a worldwide survey of lightweight car construction.

115. Information on late-model heavyweights is from *Rly. M.E.*, March 1938, p. 40, and May 1942, p. 212; *Car Builders' Cyclopedia, 1943*, p. 613; *Rly. Age*, Jan. 10, 1947, p. 47. W. D. Edson of Amtrak reports that as of October 1971, the C & A parlors were still in service on the successor G M & O. The coaches were out of service awaiting a buyer.

116. These figures, like nearly all statistics offered on cars, are guesstimates based on data in *Railway Age*, publications of the American Railway Car Institute, and census reports. None of these sources give clear figures for lightweight cars. See also *Rly. M.E.*, November 1946, p. 614.

117. Articles and advertisements on the merits of aluminum cars appeared in railroad trade journals. See *Rly. Age*, Feb. 8, 1930, Aug. 22, 1931, and July 22, 1933. The author recalls a luncheon in recent years sponsored by a major aluminum maker where the virtues of aluminum freight cars were extravagantly praised.

118. *Rly. Age*, Nov. 20, 1937, p. 729; *Western Society of Engineers Proceedings*, April 1939, p. 69.

119. *Rly. M.E.*, August 1944, p. 358.

120. *Rly. Age*, Dec. 14, 1946, p. 996.

121. The failings of aluminum cars were recorded in a conversation in September 1971 with William D. Edson, chief of equipment for Amtrak and former chief mechanical officer of the New York Central Railroad.

122. *Rly. Age*, Dec. 19, 1955, p. 28.

123. *Rly. M.E.*, December 1950, p. 717.

124. *Rly. M.E.*, June 1933, p. 185; *Light Metals*, May 1945, p. 241.

125. See note 118.

126. *Rly. Age*, May 1935, p. 171; Coverdale and Colpitts "Report on Streamline Trains," 1938 ed., p. 59.

127. The Royal Blue cars were sent to the Alton in 1937 for use on the Ann Rutledge. Remodeled heavyweight cars were used in their place on the Royal Blue.

128. *Rly. M.E.*, March 1940, p. 87.

129. *Rly. Age*, June 14, 1971, p. 14, and Nov. 8, 1971, p. 29.

130. Most of the data on the stainless steel car is from the writings of E. G. Budd and his son:

A Lecture . . . Harvard Graduate School of Business, March 6, 1934. No publisher.

Genesis of An Art by E. G. Budd, Newcomen Society, American Branch, Princeton University Press, 1938.

Edward G. Budd, 1870–1946, E. G. Budd, Jr., Newcomen Society, Princeton University Press, 1950.

"Lightweight Champion," *Railroad Magazine*, October 1949, pp. 12–39.

The best technical article on Budd's methods is given in *Heat Treating and Forging*, May 1936, pp. 236–239.

131. *Rly. M.E.*, March 1936, p. 89.

132. Ibid., July 1937, p. 303; *Mech. Eng.*, January 1938, p. 29.

133. Ragsdale gave a good account of his ideas on passenger car construction in a paper before the Society of Automotive Engineers just eleven days before his death. It was printed in *Rly. M.E.*, March 1946, p. 111.

134. *Rly. M.E.*, April 1939, p. 127.

135. Budd-Pullman rivalry was reported to the author by Arthur Dubin during a conversation in May 1971. Dubin referred to interoffice correspondence among Pullman officials and to his own earlier conversations with employees of both firms.

136. *Rly. Age*, March 13, 1937, p. 418.

137. Ibid., April 1944, p. 773.

138. R. K. Wright, *Southern Pacific Daylight Train 98–99*, Thousand Oaks, Calif.: Wright Enterprises, 1970, p. 510.

139. Conversation with William Howes, director of passenger services, B & O–C & O, May 1970.

140. *Railway Locomotives and Cars*, July 1954, p. 39.

141. *Rly. Age*, Jan. 12, 1946, p. 150.

142. *Int. Rly. Congress*, Report on Lightening Rolling Stock, 14th session, Lucerne, 1947.

143. *Trains*, November 1956, pp. 43–47; *Railway Locomotives and Cars*, August 1956, p. 49.

144. *Trains*, September 1958, p. 10.

145. References to Cor-Ten steel are taken largely from advertisements appearing in *Rly. Age*, *Car Builders' Dictionary*, 1943, pp. 357–360, and *Report of the Mechanical Advisory Committee to the Federal Coordinator*, Washington, D.C., 1935. It should be noted that other steelmakers produced similar products.

146. *Western Society of Engineers Proceedings*, April 1939, p. 72.

147. *Rly. Age*, April 1944, p. 774, in a survey of passenger cars built in the period 1934–1944, states that of 2,319 cars, 1,626 were of Cor-Ten construction. A U.S. Steel advertisement in the 1943 *Car Builders' Dictionary* says that 2,300 passenger cars had been built with Cor-Ten. A similar advertisement in the 1955 *Rly. Age* gives a total of 6,388.

148. *Rly. M.E.*, February 1935, p. 43.

149. Ibid., October 1940, p. 382; December 1942, p. 515. Also *Car Builders' Cyclopedia*, 1943, p. 596.

150. *Steel*, Aug. 22, 1938, p. 51.

151. *Rly. M.E.*, November 1941, p. 476.

152. Ibid., April 1933, p. 118.

153. *Rly. Age*, December 1936, p. 825.

154. *Rly. M.E.*, March 1942, p. 93, contains a detailed sketch of Nystrom.

155. *The Hiawatha Story* by Jim Scribbins, Milwaukee: Kalmbach, 1970, offers a fine pictorial record of the Milwaukee Road's lightweight cars. It contains basic historical data but almost no engineering material. Technical data on Number 4400 are given in *Rly. M.E.*, October 1934, p. 361.

156. *Rly. M.E.*, June 1935, p. 266.

157. *Welding Engineer*, September 1936, p. 30; *Rly. M.E.*, November 1936, p. 467; March 1939, p. 95.

158. *Rly. M.E.*, April 1943, p. 157.

159. *Rly. Age*, May 20, 1939, p. 865; October 14, 1939, p. 586.

160. *Rly. M.E.*, December 1934, p. 444; see also note 157.

161. Ibid., August 1936, p. 331.

162. *Rly. Age*, Feb. 1, 1941, p. 246.

163. *Rly. M.E.*, May 1937, p. 241.

164. *Rly. Age*, Feb. 22, 1947, p. 398.

165. Ibid., Sept. 12, 1936, p. 367.

166. *Trains*, June 1963, p. 3.

167. Ibid., June 1965, p. 9.

168. *Rly. M.E.*, October 1948, p. 584.

169. *Report on Light Weight Trains*, Statement 3639, May 1936, Interstate Commerce Commission. A copy of the original mimeographed report is on file in the I.C.C. Library, Washington, D.C.

170. Robert Shaw, *Down Brakes*, p. 327, and *Rly. Age*, Dec. 17, 1939, pp. 878–80.

171. *Rly. M.E.*, March 1940, p. 107.

172. *Rly. M.E.*, April 1948, p. 173.

173. The A.A.R. specifications are published yearly in the *Manual of Standards and Recommended Practices*. A full set may be seen at A.A.R. headquarters in Washington, D.C.

174. This allegation was made to the author by a former railway mechanical official who had no motive for slandering Budd but who wishes to remain anonymous for personal reasons.

175. *Rly. Age*, Jan. 12, 1946, p. 151.

176. *Rly. M.E.*, March 1946, p. 111, and April 1947, p. 212; *Railroad Magazine*, October 1949, p. 26.

177. *Rly. Age*, March 21, 1955, p. 50.

178. Gutbrod Report, part 2, p. 1250.

179. *Rly. M.E.*, February 1919, p. 81.

180. *Car Builders' Dictionary*, 1919, figs. 2877–2878.

181. *Rly. Age*, Jan. 12, 1946, p. 148.

182. *Rly. M.E.*, September 1950, p. 489.

183. Ibid.

184. Ibid., July 1948, p. 59.

185. *Rly. Age*, Jan. 12, 1946, p. 148.

186. *Rly. M.E.*, April 1943, p. 157.

187. *Report of the Mechanical Advisory Committee to the Federal Co-ordinator*, 1935, p. 599.

188. *New York Railroad Club*, Nov. 21, 1924, p. 7441.

189. *Rly. Rev.*, Sept. 18, 1897, p. 533. *Rly. Gaz.* (British), March 4, 1949, pp. 227–228 claims that a bilevel car was used in India as early as 1862.

190. Information on the Long Island double deckers is from *Rly. Age*, August 1932, p. 221; *The Locomotive Magazine*, Nov. 15, 1933, p. 338, and *Railroad Magazine*, October 1947, p. 80.

191. *Railroad Magazine*, March 1952, p. 58.

192. *Rly. M.E.*, December 1950, p. 713, and Richard C. Overton's *Burlington Route*, New York: Knopf, 1965, pp. 526 and 563.

193. *Rly. Age*, April 5, 1955, p. 10.

194. *Rly. Age*, April 13, 1954, p. 24; April 5, 1955, p. 9.

195. *Trains*, October 1956, p. 21.

196. Ibid., January 1959, p. 16.

197. Ibid., October 1956, p. 25; January 1969, p. 54. R. J. Wayner (Ed.), *Pennsy Car Plans*, New York: Wayner Publications, 1969.

198. Richard C. Overton, *Burlington Route*, p. 200.

199. *Sci. Am.*, May 2, 1891, p. 274.

200. *Rly. Rev.*, Aug. 2, 1902, p. 581.

201. *Rly. M.E.*, September 1945, p. 378.

202. *Rly. Age*, Jan. 12, 1946, p. 14.

203. *Rly. M.E.*, August 1947, p. 395.

204. *Rly. Age*, Oct. 11, 1947, p. 604.

205. *Rly. M.E.*, May 1949, p. 241.

206. *Trains*, August 1956, p. 9.

207. Jim Scribbins, *Hiawatha Story*, p. 114, and *Rly. Locos. & Cars*, February 1953, p. 49.

208. *Trains*, May 1958, p. 19.

CHAPTER THREE

1. Arguments of W. W. Hubbell in William Whiting, *Arguments of William Whiting, Esq., in the case of Ross Winans vs. Orsamus Eaton et al.*, Boston, 1853, p. 3.

2. Seymour Dunbar, *A History of Travel in America*, Indianapolis, 1915, p. 1006.

3. F. A. Ritter Von Gerstner, *Die innern communication der vereinigten Staaten von Nord-America*, Vienna, 1842–1843.

4. Karl Ghega, *Die Baltimore-Ohio Eisenbahn*, Vienna, 1844.

5. Dionysius Lardner, *Railway Economy*, New York and London, 1850, p. 346.

6. *Leslie's Illustrated Newspaper*, Aug. 25, 1877.

7. *Railway Age* (Oct. 20, 1893, p. 773) says that the car *Chambersburg* entered service in the winter of 1836–1837. The most frequently cited date is the winter of 1837–1838, which was given in W. B. Wilson's *History of the Pennsylvania Railroad*, vol. 1, Philadelphia, 1899, p. 395. (The Pennsylvania was the Cumberland Valley's successor.) The son of the Cumberland Valley's pioneer master car builder, Jacob Shaffer, said that the car entered service in the spring of 1838 according to *Locomotive Engineering*, April 1900, p. 150.

8. *George M. Pullman and Pullman's Palace Car Company vs. the New York Central Sleeping Car Company and Webster Wagner. In Equity. Answer as Amended.* U.S. Circuit Court, Northern District of Illinois, 1881. Printed transcript of testimony, 920 pages. Hereafter cited as *Pullman vs. Wagner*.

9. *Pullman vs. Wagner and Pullman's Palace Car Company vs. Jonah Woodruff and J. Hervey Jones; Equity Docket No. 28, November term, 1871*. U.S. Circuit Court, Western District of Pennsylvania. Hereafter cited as *Pullman vs. Woodruff*. Data were copied from ms. records housed in U.S. Federal Building at Pittsburgh in September 1969.

10. *Pullman vs. Wagner*, pp. 395–396.

11. Ibid., p. 410.

12. Ibid., pp. 399 and 408.

13. Ibid., pp. 387–388.

14. Ibid., p. 394.

15. *American Railroad Journal*, Nov. 15, 1838, p. 328, copied from the Oct. 31, 1838, *Baltimore American*.

16. Ibid.

17. *Pullman vs. Woodruff*, 1871.

18. *A.R.R.J.*, March 25, 1854, p. 183.

19. P W & B annual report, 1859.

20. Von Gerstner, vol. 1, p. 129.

21. Ibid., vol. 1, p. 238.

22. *Pullman vs. Wagner*, pp. 229–280.

23. Von Gerstner, vol. 2, p. 305.

24. *Locomotive Engineering*, November 1893, p. 503, and January 1898, p. 14.

25. *Pullman vs. Wagner*, pp. 710–743.

26. Charles S. Sweet, "History of the Sleeping Car," Chicago, 1923, a book-length study that was never published. Sweet's uncle was said to be G. M. Pullman's personal secretary. Through this association Sweet apparently gained access to early records of the Pullman Company and its predecessors. The mss. was in the possession of his daughter, Jane Sweet of Winnetka, Ill., in 1972. Hereafter cited as Sweet's History.

27. *Pullman vs. Wagner*, pp. 320–360.

28. William P. Smith, *The Great Railway Celebration of 1857*, New York, 1858, p. 163.

29. As previously noted, the P W & B had ladies' cars as early as 1838. *A.R.R.J.*, June 15, 1842, described them as equipped with "luxurious sofas . . . washstand and other conveniences."

30. P W & B annual reports 1859, 1860, and 1863.

31. C. H. Corliss, *Main Line of Mid America*, New York, Creative Age Press, 1950, pp. 76–77, and *A.R.R.J.*, Aug. 16, 1856, p. 520. The *Detroit Advertiser* description which follows is from Corliss.

32. *50th Anniversary of Illinois Central R.R.*, 1901, no page numbers. (Souvenir booklet published by the railroad)

33. R. C. Overton, *The Burlington Route*, New York: Knopf, 1965, p. 52.

34. A leaflet dating from about 1859 describing Case's Patent Sleeping Car is in the Harvard Business School (Baker Library) business history mss. collection.

35. Ibid.

36. *Western Railroad Gazette*, April 24, 1869.

37. *A.R.R.J.*, Jan. 1, 1859, p. 12.

38. B & O annual report, 1859, p. 151.

39. A. Bendel, *Aufsätze Eisenbahnwesen in Nord-Amerika*, Berlin, 1862, p. 38, plate E.

40. *A.R.R.J.*, Nov. 28, 1863, p. 1132.

41. I. D. Guyer, *History of Chicago*, Chicago, 1862, rear advertising section.

42. Compiled by Arthur D. Dubin from *Appleton's 1866 R.R. Guide*, in his personal collection.

43. Earlier, some sleeping-car–related patents had been issued on folding furniture for shipboard cabins (such as H. King's of 1841). McGraw's 1838 folding sofa bed, used in the Erie sleeping cars of 1842, was mentioned earlier in the chapter.

44. The U.S. Patent Office has assigned sleeping-car patents to class 105, subclasses 314–326. A total of approximately 500 patents are in this group.

45. *Pullman vs. Wagner*, 1881, pp. 580–656 and 684–707.

46. Siegfried Gideon, *Mechanizaion Takes Command*, New York, W. W. Norton, 1948, 1969, p. 461.

47. *Dictionary of American Biography*, vol. 20, p. 497, provides a good summary of Woodruff's life. Other details are offered by the *National Cyclopedia of American Biography*, vol. 14, pp. 203–204.

48. Joseph F. Wall, *Andrew Carnegie*, Fair Lawn, N.J.: Oxford University Press 1970, p. 141. Hereafter cited as Wall.

49. *National Car Builder*, October 1880, p. 166.

50. *Rly. Age*, Sept. 6, 1889, pp. 586–587.

51. W. S. Kennedy, *Wonders and Curiosities of the Railway*, Chicago, 1884, p. 194; *Appleton's Cyclopedia of American Biography*, vol. 1, p. 605.

52. *Manufacturing and Manufacturers of Pennsylvania*, 1875, p. 185.

53. John A. Haddock, *History of Jefferson County New York*, Albany, 1895, pp. 33–35, and *Rly. Age*, April 5, 1889, pp. 221–222.

54. *Railroad Car Journal*, October 1894, p. 208.

55. *Pullman vs. Wagner*, 1881, pp. 580–660.

56. Wall, pp. 139–140.

57. Railway and Locomotive Historical Society, *Bulletin 59*, October 1942, pp. 34–36.

58. *Rly. Age*, April 5, 1889, pp. 221–222.

59. Wall, p. 199.
60. Sweet's History, p. 128.
61. *Engineer* (Philadelphia), Oct. 4, 1860, p. 63. *Dictionary of American Biography*, vol. 10, pp. 463–464, offer a sketch of Knight's career that includes several factual errors.
62. *A.R.R.J.*, Oct. 10, 1863, p. 969.
63. *Chicago Tribune*, June 5, 1864.
64. Sweet's History, p. 130.
65. *Railroad Record*, Jan. 18, 1866, p. 580.
66. *American Heritage*, December 1970, p. 40.
67. *Scientific American*, Aug. 10, 1867, p. 89.
68. G. M. Best, *Iron Horses to Promontory*, San Marino, Calif.: Golden West Books, 1969, pp. 61–69.
69. *N.C.B.*, May 1875, p. 68.
70. August Mencken, *The Railroad Passenger Car*, Baltimore: Johns Hopkins Press, 1957, p. 161.
71. *Manufacturers of Pennsylvania*, 1875, p. 489.
72. Wall, p. 209.
73. Sweet's History, pp. 130 and 284.
74. Wall, p. 211.
75. Ibid., p. 18.
76. *Rly. Age*, April 27, 1888, p. 269.
77. *Financial and Commercial Chronicle*, June 4 and 11, 1898, pp. 1090 and 1141.
78. Acquisition Reports. During the 1940 Federal suit against the Pullman Company, a series of reports was prepared by the firm's legal department giving financial summaries of the companies taken over by Pullman. Copies were made available to me by Arthur D. Dubin of Chicago. Hereafter referred to as Pullman Acquisition Reports.
79. *Railroad Gazette*, Dec. 3, 1871, p. 382.
80. Ibid., May 3, 1878, p. 223.
81. Sweet's History, pp. 238–242; Dun and Bradstreet records (Harvard Business School), Philadelphia and Pittsburgh volumes; also Pullman Acquisition Reports.
82. Records of this suit are on file in the U.S. District Court of Western Pennsylvania.
83. *Railway World*, Nov. 3, 1877, p. 1046.
84. *N.C.B.*, November 1878, p. 167, and Sweet's History, p. 259.
85. *N.C.B.*, November and December 1879, pp. 170 and 180.
86. C.B.&Q. Railroad papers, Newberry Library, Chicago, boxes 8 and 33.
87. *Railway Review*, Sept. 30, 1882.
88. Ibid., June 21, 1882, and April 8, 1882.
89. *Rly. Rev.*, Dec. 29, 1883, p. 776.
90. Financial data on Woodruff is from Dun and Bradstreet and Pullman Acquisition Reports.
91. Andy Logan, *The Man Who Robbed the Robber Barons*, New York: W. W. Norton, 1965.
92. *Rly. Rev.*, May 9, 1885, p. 227.
93. *Engineering*, April 18, 1873, p. 262, and July 11, 1873, p. 33.
94. Ibid., June 11, 1875, p. 493.
95. *Rly. Rev.*, Aug. 25, 1883, p. 504.
96. *N.C.B.*, January 1884, p. 4; *R.R.G.*, Dec. 14, 1883, p. 819.
97. *Rly. Rev.*, May 9, 1885, p. 227.
98. Sweet's History, p. 252, quoted from Mann Boudoir Director's Minute Books.
99. *Rly. Rev.*, Dec. 18, 1886, p. 675.
100. *R.R.G.*, Jan. 3, 1890, p. 10.
101. *N.C.B.*, January 1884, p. 5.
102. L. M. Beebe, *Mr. Pullman's Elegant Palace Car*, Garden City, N.Y.: Doubleday, 1961, p. 254.
103. Pullman Acquisition Reports.
104. *Rly. Rev.*, Oct. 27, 1888, pp. 621 and 633.
105. *R.R.G.*, Nov. 2, 1888, p. 726.
106. Ibid., March 1, 1889, p. 151.
107. *R.R.G.*, March 30, 1883, p. 202.
108. *New York Elevated Railway Journal*, January 26, 1884.
109. *Rly. Age*, July 16, 1885.
110. *Rly. Rev.*, March 6, 1886, p. 117.
111. The April 1893 *R.R. Car Jour.* (p. 155) mentions Monarch, but the article does not clearly state whether it had suspended operations. Advertisements for the firm appeared in the *Official Guide* through October 1894.
112. Railway and Locomotive Historical Society *Bulletin 59*, p. 32.

113. *N.C.B.*, September 1881, p. 104.
114. *Leslie's Illustrated Weekly*, April 30, 1859.
115. Mencken, pp. 135–141, quotes Walter Thornbury's travel account of 1860.
116. *A.R.R.J.*, Jan. 21, 1882, p. 33.
117. Ibid., Sept. 27, 1879, p. 1067, and *American Railway Times*, April 11, 1868, p. 119.
118. Several biographical sketches are available on Wagner. The most accurate account is offered in *R.R.G.*, Jan. 20, 1882, p. 48.
119. Sweet's History, p. 155i.
120. *R.R. Car Jour.*, January 1892, pp. 53–57, offers a short history of Wagner.
121. *West. R.R. Gaz.*, May 26, 1860, p. 2.
122. Alvin F. Harlow, *Road of the Century: The Story of the New York Central*, New York: Creative Age Press, 1947, quotes the *Schenectady Evening Star* for Oct. 21, 1865.
123. *R.R. Car Jour.*, January 1892, pp. 53–57, and Pullman Acquisition Reports.
124. *N.C.B.*, August 1894, p. 118, and Mencken, p. 77.
125. Sweet's History, pp. 155D–155G; *Reports of the New York State Legislature*, Special Commission on Railroads, 1879, vol. 9, pp. 182–185.
126. Pullman Acquisition Reports, p. 10.
127. Sweet's History, p. 155H.
128. *A.R.R.J.*, July 21, 1882, p. 501.
129. *Rly. World*, Sept. 4, 1875, p. 583, and Oct. 30, 1875, p. 704. Similar accounts are given in Harlow, *Road of the Century*, and Edward Hungerford's *Men and Iron: The History of the New York Central*, New York: Crowell, 1938.
130. *Pullman vs. Wagner*, 1881, p. 56-A, and Sweet's History, p. 155H.
131. Sweet's History, p. 155H.
132. *A.R.R.J.*, July 21, 1882, p. 501.
133. *N.C.B.*, May 1875, p. 68.
134. Alvin F. Harlow, *Steelways of New England*, New York: Creative Age Press, 1946, p. 256.
135. *N.C.B.*, March 1881, p. 32.
136. Ibid., September 1873, p. 220.
137. Ibid., October 1892, p. 158.
138. Ibid., January 1876, p. 6.
139. *R.R.G.*, April 20, 1877, p. 181.
140. Railway and Locomotive Historical Society, *Bulletin 59*, pp. 28–29.
141. *N.C.B.*, October 1880, p. 156.
142. *A.R.R.J.*, April 15, 1882, p. 238.
143. *R.R.G.*, June 8, 1883, p. 359.
144. *N.C.B.*, September 1885, p. 113.
145. New York Central Sleeping Car Company Minute Books for 1886–1899 are housed in the Newberry Library, Chicago. Hereafter N.Y. Central Minute Books.
146. *Rly. Age*, Sept. 13, 1889, p. 599.
147. N.Y. Central Minute Books, Dec. 22, 1886.
148. *N.C.B.*, April 1879, p. 57.
149. *Rly. Rev.*, April 10, 1886, p. 175.
150. Pullman order books—from a transcript prepared by Evelyn Danielson of Pullman–Standard Mfg. Co.
151. *Rly. Rev.*, Jan. 28, 1888, p. 44, and *R.R.G.*, Aug. 12, 1887, p. 526.
152. *Rly. Rev.*, Jan. 28, 1888, p. 44, and *R.R.G.*, March 3, 1888, p. 123.
153. *Rly. Rev.*, Nov. 19, 1887, p. 672, and Dec. 17, 1887, p. 720.
154. *N.C.B.*, April 1887, p. 43.
155. *N.C.B.*, March 1886, p. 29.
156. *R.R.G.*, Oct. 31, 1890, p. 750.
157. Railway and Locomotive Historical Society, *Bulletin 105*, p. 65.
158. *R.R. Car Jour.*, January 1892, p. 57, and April 1893, p. 155.
159. *U.S.A. vs. Pullman*, U.S. Justice Department, 1941, *Complaint*.
160. *R.R. Car Jour.*, November 1896, p. 278, and N.Y. Central Minute Books.
161. *Rly. Rev.*, March 19, 1892, p. 206.
162. *Rly. Rev.*, Nov. 13, 1897, p. 650, and *R.R. Car Jour.*, December 1897, p. 368.
163. *New York Railroad Commission Report*, 1885, p. 1548.
164. N.Y. Central Minute Books, April 10, 1888.
165. *Rly. Rev.*, Aug. 29, 1891, p. 573.
166. *Commercial & Financial Chronicle*, Aug. 12, 1899, p. 332.
167. Ibid.

168. Ibid., Oct. 28, 1899, p. 909, and Pullman Acquisition Reports, p. 27.

169. *Pearson's Magazine*, July 1913, p. 19.

170. *R.R.G.*, Nov. 3, 1899, p. 754.

171. F. W. Pierce, *Field Genealogy*, Chicago, 1901, p. 674; and Sweet's History, p. 157.

172. *The World*, Dec. 23, 1892.

173. Railway and Locomotive Historical Society, *Bulletin 59*, Oct. 1942, p. 30.

174. Joseph Husband, *The Story of the Pullman Car*, Chicago, 1917, pp. 28–32; hereafter Husband. This book gives the recollections of two early employees: Leonard Siebert claimed that the cars were remodeled in 1858, while J. L. Barnes claimed that they were ready for service in September 1859. The date cited by G. M. Pullman in several newspaper interviews was 1859.

175. Husband, pp. 28 and 31.

176. Pullman Collection in Newberry Library: file folder marked History of No. 9.

177. *R.R.G.*, April 6, 1877, p. 153.

178. Ibid., and Sweet's History, p. 186.

179. Stanley Buder, *Pullman: An Experiment in Industrial Order and Community Planning, 1880–1930*. New York: Oxford, 1967. Hereafter cited as Buder.

180. *Colorado Magazine*, 1940, p. 113, and *Pullman News*, April 1941, p. 118.

181. Dun and Bradstreet Reports, mss. in Baker Library, Harvard School of Business Administration, Cambridge, Mass. The first entry is dated July 6, 1863.

182. Husband, p. 33.

183. The story of the *Pioneer's* oversized dimensions can be traced to Pullman's lifetime. *Rly. Rev.*, Oct. 20, 1893, p. 774.

184. Sweet's History, p. 196.

185. Buder, p. 13.

186. *Pullman News*, August 1927, p. 114.

187. D.&B. Reports, July 1866.

188. *Courier-Journal* (Louisville, Ky.), Aug. 22, 1877.

189. Sweet's History, p. 209.

190. *U.S. Justice Department Civil Action 994, Stipulation of Facts*, U.S. District Court of Eastern Pennsylvania, 1940, pp. 82–85.

191. *R.R.G.*, April 22, 1871, p. 39.

192. *Rly. Age*, May 19, 1881, p. 277.

193. Sweet's History, p. 211.

194. Wall, p. 208.

195. Best, *Iron Horses to Promontory*, pp. 69 and 141.

196. Pullman Acquisition Reports, 1940.

197. George Behrend, *Pullman in Europe*, London: Ian Allen, 1962, p. 21.

198. Sweet's History, p. 291.

199. Behrend, pp. 27–30.

200. *A.R.T.*, June 29, 1867, p. 206.

201. *N.C.B.*, March 1874, p. 42.

202. *Transportation Routes to the Seaboard*, U.S. Senate Report, 1874, vol. 1, pp. 150–152.

203. *Report of the Investigating Committee . . . on the Pennsylvania Railroad*, Philadelphia, 1874, pp. 125–128. See also George H. Burgess and M. C. Kennedy, *Centennial History of the Pennsylvania Railroad*, Philadelphia, 1949, pp. 327–329.

204. *N.C.B.*, December 1870, p. 1.

205. *N.C.B.*, January 1878, p. 2.

206. *R.R.G.*, July 7, 1876, p. 302.

207. *Rly. World*, Jan. 2, 1875, p. 10.

208. *R.R. Car Jour.*, January 1892, p. 59.

209. Ibid., Feb. 11, 1882, p. 89.

210. D.&B. Reports, August and October 1875.

211. *R.R.G.*, Oct. 5, 1877, p. 446.

212. *A.R.R.J.*, Nov. 20, 1875, p. 1476.

213. *N.C.B.*, October 1878, p. 154.

214. *A.R.R.J.*, Oct. 4, 1879, p. 1118.

215. *Rly. Age*, Sept. 15, 1881, p. 521.

216. *R.R.G.*, Oct. 23, 1891, p. 756.

217. *N.C.B.*, January 1884, p. 1.

218. C B & Q Papers, Newberry Library, box 8C 5331.

219. Pullman Papers, Newberry Library, contracts file, 1913.

220. Sweet's History, p. 303.

221. Ibid., and *R.R.G.*, Oct. 27, 1893, p. 792.

222. Figures are from Pullman annual reports.

223. Pullman Acquisition Reports, 1940c., report on Association contracts.

224. *Trains*, November 1969, p. 26.

225. *Rly. World*, Oct. 2, 1880, p. 943.

226. *Rly. World*, Jan. 8, 1881, p. 39.

227. *Rly. Rev.*, Feb. 11, 1882, p. 89.

228. *Rly. Age*, June 22, 1888, p. 396.

229. Ibid., Oct. 5, 1882, p. 548.

230. Ibid., Dec. 11, 1891, p. 966.

231. *Rly. Rev.*, Sept. 8, 1883, p. 532.

232. Ibid., March 16, 1889, p. 150.

233. *R.R.G.*, Oct. 21, 1891, p. 796.

234. *Fortune*, January 1938, p. 45.

235. Pullman annual reports, and I.C.C. *Sleeping Car Statistics, 1890–1935*, statement 3631, August 1936; hereafter *I.C.C. Statistics, 1936*.

236. *Pearson's Magazine*, July 1913, pp. 12–23.

237. Justice Department, *Stipulation of Facts*, 1940, p. 93.

238. *Rly. Age Gaz.*, April 15, 1910, p. 987, and Nov. 18, 1910, p. 976.

239. *Boston Globe*, May 1, 1913.

240. J. C. Welliver, Pullman Progress. Proof sheets dated 1929, are held by the Chicago Historical Society.

241. *Moody's Industrials*, 1920, p. 1432.

242. *I.C.C. Valuation Docket 1079, Abstract of the Evidence*, June 25, 1928.

243. *Military Engineer*, September–October 1930, p. 426.

244. *Rly. Age*, April 25, 1925, p. 1053.

245. Ibid., June 11, 1927, p. 1914.

246. Ibid., May 28, 1927, p. 1583.

247. Ibid., Aug. 29, 1936, p. 309.

248. *Fortune*, January 1938, p. 94.

249. *New York Evening Post*, Feb. 9, 1929.

250. *Fortune*, January 1938, p. 94, and *New York Times*, Dec. 28, 1929.

251. *Fortune*, January 1938, p. 42.

252. *I.C.C. Statistics, 1936*, pp. 16–18.

253. *Business Week*, April 16, 1930, p. 14.

254. *Barron's* Nov. 28, 1932, p. 6.

255. Civil Action no. 944 in *U.S. District Court, Eastern District of Pennsylvania, U.S.A. vs. Pullman Company, Answer*, Oct. 15, 1940.

256. *Rly. Age*, Nov. 22, 1941, p. 881.

257. *Pullman News*, October 1945, p. 50; *Trains*, October 1945, p. 4.

258. *Federal Supplement*, vol. 64, 1946, p. 108.

259. *Pullman News*, July 1947, p. 3.

260. *Rly. Age*, Nov. 19, 1949, p. 896, and *Pullman News*, January 1950, p. 29.

261. *Trains*, March 1958, p. 10, and *Daily News* (Chicago), April 16, 1958.

262. *Rly. Age*, Dec. 2, 1968, p. 40.

263. *Pullman News*, October 1929, p. 191.

264. *R.R.G.*, Nov. 6, 1908, p. 1317.

265. *N.C.B.*, October 1873, p. 251.

266. *Rly. Rev.*, April 16, 1887, p. 226.

267. *R.R.G.*, Dec. 7, 1888, p. 805.

268. *Rly. Rev.*, Nov. 9, 1889, p. 645.

269. *R.R. Car Jour.*, December 1892, p. 46.

270. *Pullman Progress*, pp. 44–45.

271. *R.R.G.*, April 19, 1907, p. 636.

272. Ibid., Feb. 4, 1910, p. 287.

273. Ibid., April 29, 1910, p. 1084.

274. *Safety on the Railroads: Hearings before a Subcommittee of the Committee on Interstate and Foreign Commerce of the House of Representatives Sixty-third Congress*, Washington, D.C., 1914, part v, p. 354.

275. *I.C.C. Statistics, 1936*, p. 21.

276. R. J. Wayner, *Pullman Panorama*, New York, vol. 1, 1967, p. ii.

277. William Kratville, *Passenger Car Catalog: Pullman Operated Equipment, 1912–1949*, Omaha: Kratville Publications, 1968, p. 30.

278. *Rly. Age*, April 16, 1932, p. 642.

279. *Pearson's Magazine*, July 1913.

280. *Scientific American*, February 1923, p. 87.

281. *Harper's Weekly*, Feb. 7, 1914, p. 22.

282. *Rly. Age*, Aug. 29, 1935, p. 310.

283. Ibid., April 2, 1927, p. 1071.

284. Ibid., April 16, 1932, p. 640.
285. *Pullman News*, April 1942, p. 133.
286. *Railway Mechanical Engineer*, June 1933, p. 186.
287. *Fortune*, February 1938, p. 78.
288. *Rly. M.E.*, June 1940, p. 241, and *Trains*, August 1960, p. 59.
289. *Rly. Age*, Nov. 17, 1945, p. 797, and *Railroad Magazine*, October 1949, p. 15.
290. *Pullman News*, October 1956, p. 17.

CHAPTER FOUR

1. *Railroad Gazette*, April 28, 1882, p. 254.
2. J. G. Medley, *An Autumn Tour in the United States and Canada*, London, 1873, pp. 70–71.
3. *American Railroad Journal*, Oct. 9, 1845, p. 647.
4. *Scientific American*, Aug. 27, 1853, p. 396.
5. A. Bendel, *Aufsätze Eisenbahnwesen in Nord-Amerika*, Berlin, 1862, plate E.
6. *A.R.R.J.*, Sept. 17, 1864, p. 897.
7. Ibid., Sept. 11, 1869, p. 1043.
8. *Old Time New England*, April 1932, p. 197; *Locomotive Engineering*, April 1893, p. 173.
9. *American Railway Times*, June 9, 1866, p. 182.
10. *National Car Builder*, November 1878, p. 169.
11. *A.R.T.*, March 1, 1868, p. 76.
12. *Loco. Eng.*, April 1893, p. 173.
13. *New York Graphic*, Jan. 17, 1876, p. 603.
14. Alvin F. Harlow, *Road of the Century*, New York: Creative Age Press, 1947, p. 287.
15. *Railroad Car Journal*, November 1894, p. 243.
16. *N.C.B.*, January 1871, p. 3, and May 1875, p. 68.
17. *R.R.G.*, April 22, 1871, p. 41.
18. Ibid., June 26, 1875, p. 269.
19. *Engineering*, April 4, 1875, p. 274.
20. Hamilton Ellis, *Railway Carriages in the British Isles from 1830 to 1914*, London: Allen & Unwin, 1965, p. 94.
21. *N.C.B.*, September 1876, p. 130.
22. *R.R.G.*, April 28, 1882, p. 254, and *A.R.R.J.*, Sept. 9, 1882, p. 664.
23. *R.R.G.*, July 7, 1882, p. 415.
24. *N.C.B.*, December 1885, p. 159.
25. *R.R.G.*, May 16 and Aug. 29, 1884, pp. 380 and 630.
26. Ibid., Aug. 21 and Oct. 21, 1887, pp. 530 and 678.
27. *R.R. Car Jour.*, February 1892, pp. 65 and 74.
28. *American Journal of Railway Appliances*, June 1884, p. 202.
29. Bendel, p. 36 and plate 13.
30. *R.R.G.*, Aug. 20, 1886, p. 577.
31. *N.C.B.*, March 1882, p. 31.
32. Ibid., June 1882, p. 66.
33. *R.R.G.*, May 12, 1882, p. 285.
34. Both the Pennsylvania and the New York Central railroads began operating luxury trains in 1887. The first was the Pennsylvania Special, better known by its later name of Broadway Limited. In the same year came New England's first name train: the Mt. Desert Limited. In 1888 the Golden State Special was introduced; in 1889 the Montezuma Special. The history of these trains is given in A. D. Dubin's *Some Classic Trains* (Milwaukee: Kalmbach, 1964) and L. M. Beebe's two-volume work, *Trains We Rode*, Berkeley: Howell-North, 1965 and 1966.
35. *Rly. Rev.*, April 16, 1887, p. 226.
36. *Railway Mechanical Engineer*, April 1926, p. 221.
37. *Popular Mechanics*, March 1933.
38. *Chicago Daily Tribune*, June 15, 1950.
39. Dubin, pp. 170–171.
40. *Rly. M.E.*, April 1926, p. 221.
41. F. Gutbrod, "The Construction of Iron Passenger Cars on the Railways in the United States of America," *International Railway Congress Bulletin*, vols. 26–27, 1912–1913, part 4, p. 121. Hereafter referred to as Gutbrod Report. Also see Dubin, p. 266.
42. W. W. Kratville, *Passenger Car Catalog: Pullman Operated Equipment, 1912–1949*, Omaha: Kratville Publications, 1968, pp. 82–87.
43. Jim Scribbins, *The Hiawatha Story*, Milwaukee: Kalmbach, 1970, p. 109.
44. *Trains*, September 1956, p. 8.
45. The peak year for dining cars was 1930, when 1,742 were reported in service by the I.C.C.

46. *Popular Mechanics*, October 1925.
47. Ibid.
48. W. C. Bryant, *Picturesque America*, New York: Appleton, 1872.
49. Edward Hungerford, *The Modern Railroad*, Chicago, 1918, p. 304.
50. *R.R.G.*, Jan. 25, 1884, p. 72, and *N.C.B.*, November 1883, p. 125.
51. *Harper's Weekly*, Aug. 25, 1888, p. 643.
52. L. M. Beebe, *Hear the Train Blow*, New York: Grosset and Dunlap, 1952, p. 377, reproduces a Burlington menu of the 1875–1880 period.
53. *Popular Mechanics*, October 1925.
54. *B & O Magazine*, September 1942, p. 64.
55. W. P. Smith, *History and Description of the B. & O. R.R.*, Baltimore, 1853, p. 147.
56. *Rly. Rev.*, March 4, 1893, p. 129.
57. *Western Railroad Gazette*, Sept. 8, 1860.
58. *Engineer*, Philadelphia, Dec. 8, 1860, p. 130.
59. Ibid., Aug. 23, 1860, p. 15.
60. *R.R.G.*, Dec. 19, 1884, p. 893, and E. P. Mitchell, *Memories of an Editor*, New York, 1924.
61. *Lincoln Day by Day*, vol. III, Washington, D.C.: Lincoln Sesquecentennial Commission, 1960, p. 17.
62. J. W. Starr, *Lincoln and the Railroads*, New York, 1927, p. 253.
63. *Models of Hospitals Cars*, a pamphlet published by the U.S. Army Medical Museum in Washington, D.C., 1885.
64. Stanley Buder, *Pullman: An Experiment in Industrial Order*, New York: Oxford, 1967, p. 20.
65. Charles Nordhoff, *California: For Health, Pleasure and Residence*, New York, 1872, p. 29.
66. *Engineering*, July 17, 1869, pp. 46 and 50.
67. *A.R.T.*, April 11, 1868, p. 119, and *Rly. Rev.*, May 7, 1868, p. 1.
68. *N.C.B.*, July 1879, p. 102.
69. Ibid., November 1872, p. 13, and Poor's *Manual of Railroads*, New York, 1871–1872.
70. *N.C.B.*, March 1882, p. 31.
71. *R.R.G.*, Feb. 24, 1882, p. 115.
72. *N.C.B.*, May 1877, p. 68.
73. Ibid., July 1881, p. 84.
74. *Rly. Rev.*, Dec. 10, 1892, p. 777.
75. *Rly. Age*, Jan. 13, 1881, p. 17.
76. *R.R.G.*, May 12, 1882, p. 285.
77. Edward Hungerford, *Men and Iron*, New York: Crowell, 1938, pp. 247 and 305.
78. *Rly. Rev.*, July 19, 1884, p. 378.
79. Ibid., May 7, 1881, p. 253.
80. G. M. Best, *Iron Horses to Promontory*, San Marino, Calif.: Golden West Books, 1969, p. 141.
81. *R.R. Car Jour.*, August 1891, p. 120.
82. Cost figures are from the following sources, in the order given: *R.R.G.*, Aug. 19, 1887, p. 536; ibid., May 19, 1905, p. 540; and American Railway Car Institute *Reports*, 1940 through 1960.
83. *Popular Mechanics*, October 1925.
84. Hungerford, *Modern Railroad*, p. 308.
85. *R.R.G.*, Aug. 5, 1887, p. 517.
86. *Trains*, March 1943, p. 6.
87. Ibid., December 1959, p. 40.
88. *Rly. Rev.*, June 9, 1883, p. 321.
89. *N.C.B.*, August 1886, p. 107.
90. *Rly. Rev.*, Sept. 9, 1899, p. 504.
91. *R.R.G.*, Jan. 15, 1886, p. 35.
92. Ibid., Sept. 22, 1893, p. 700.
93. *R.R.G.*, March 4, 1910, p. 447.
94. *Rly. M.E.*, February 1914, p. 77.
95. Ibid., June 1917, p. 346.
96. Ibid., March 1927, p. 163.
97. W. W. Kratville and H. E. Rank, *Union Pacific Equipment*, Omaha: Kratville, 1969, p. 37.
98. *Mutual Magazine*, August 1940.
99. *Rly. M.E.*, May 1950, p. 265, and *Trains*, September 1951, p. 37.
100. R. K. Wright, *Southern Pacific Daylight Train 98–99*, Thousand Oaks, Calif.: Wright Enterprises, 1970, vol. 1, p. 325.
101. *Trains*, October 1956, p. 24.
102. Ibid., May 1958, p. 19.
103. *N.C.B.*, October 1883, p. 110.
104. *Rly. Rev.*, Nov. 4, 1882, p. 626.

105. *Rly. M.E.*, March 1933, p. 109.
106. *Pullman News*, October 1950, p. 16, and April 1953, p. 2.
107. *R.R.G.*, March 2, 1906, p. 207.
108. *R.R.G.*, Dec. 20, 1895, p. 834, and May 3, 1907, p. 623.
109. Ibid., July 29, 1904, p. 190.
110. *Rly. Age*, Nov. 17, 1945, p. 799.
111. W. W. Kratville, *Steam, Steel & Limiteds*, Omaha: Kratville, 1962, p. 48.
112. Pacific Railroad Club *Proceedings*, August 1920, p. 9.
113. *Trains*, February 1953, p. 18.
114. Wright, *Southern Pacific Daylight*, p. 325.
115. *Mutual Magazine*, August 1940.
116. *Railroad Magazine*, September 1946, p. 88, and *Rly. M.E.*, April 1949, p. 187.
117. *Rly. M. E.*, July 1949, p. 376.
118. Ibid., May 1928, p. 273.
119. *Rly. Age*, April 12, 1930, p. 854.
120. John F. Stover, *Life and Decline of the American Railroad*, New York: Oxford, 1970, pp. 211–212.
121. I.C.C., *Transportation Statistics in the U.S.*, 1961, p. 52.
122. *Popular Mechanics*, October 1925, and J. H. Parmelee, *The Modern Railway*, New York, 1940, p. 116.
123. *Trains*, September 1951, p. 36.
124. John F. Stover, *American Railroads*, Chicago: University of Chicago Press, 1961, p. 242.
125. Wright, *Southern Pacific Daylight*, p. 442.
126. James W. Holden, "Mansions on Wheels," *Railroad Magazine*, November 1939, pp. 6–27; hereafter cited as Holden.
127. *Rly. Age*, Feb. 3, 1881, p. 59.
128. *N.C.B.*, April 1891, p. 56.
129. Holden, p. 26.
130. R. E. Carlson, *The Liverpool and Manchester Railway Project*, New York: Augustus Kelley, 1969, p. 232. See also Ellis, *Railway Carriages in the British Isles*, pp. 19 and 44.
131. S. M. Whipple, *Record of Thompson and Bachelder Center Lever or Tanner Brake*, New York, 1867, p. 20.
132. Harlow, *Road of the Century*, p. 129.
133. *Pullman's Palace Car Company vs. Jonah Woodruff and J. Hervey Jones*, U.S. Circuit Court, Western District of Pennsylvania, 1871, pp. 5 and 9. *George M. Pullman and Pullman's Palace Car Company vs. the New York Central Sleeping Car Company and Webster Wagner*, U.S. District Court, Northern District of Illinois, 1881, p. 786.
134. *N.C.B.*, November 1893, p. 185.
135. William P. Smith, *The Great Railway Celebration of 1857*, New York, 1858, pp. 217–218, and *Railroad Advocate*, Aug. 8, 1857, p. 35.
136. I. S. Trotter, *First Impressions of the New World on Two Travellers from the Old in the Autumn of 1858*, London, 1859, pp. 141–144; and Bendel, p. 38 and plate E, fig. 7.
137. *Am. Rly. Rev.*, Feb. 21, 1861, p. 99; March 27, 1862, p. 398; Nov. 22, 1863, p. 375.
138. *Columbian Weekly Register*, Nov. 24, 1866.
139. W. H. Stennett, *Yesterday and Today: a History of the C.&N.W.*, Chicago, 1910, pp. 71–72.
140. *N.C.B.*, August 1872, p. 10.
141. Best, *Iron Horses*, pp. 110–111.
142. Ibid., p. 48.
143. *N.C.B.*, February 1873, p. 35.
144. *Trains*, November 1949, p. 64.
145. *N.C.B.*, October 1881, p. 117, and Holden, p. 15.
146. Ibid., May 1871, p. 5.
147. Charles S. Sweet, "History of the Sleeping Car," Chicago, 1923, p. 156c. (Unpublished mss.)
148. A Dayton, Ohio, newspaper dated Aug. 2, 1878, speaks of a new car produced for Barnum by Barney and Smith. It is not clear whether this was Barnum's first PV.
149. *N.C.B.*, January 1876, p. 5.
150. Lucius Beebe, *Mansions on Rails*, Berkeley, Calif.: Howell-North Books, 1959, p. 195; hereafter Beebe, *Mansions*.
151. Arthur Dubin of Chicago owns a copy of Pullman's Detroit shop order book. *Pullman News*, April 1936, quotes a Rochester newspaper of Nov. 26, 1877.
152. *R.R.G.*, June 22, 1877, p. 282.
153. *Pullman News*, January 1932, p. 277.

154. *R.R.G.*, Sept. 30, 1887, p. 641, and *Pullman News*, April 1936, p. 117.
155. *Rly. Rev.*, June 10, 1882, p. 336.
156. *R.R.G.*, April 6, 1888, p. 227.
157. Ibid., Sept. 13, 1889, p. 599.
158. *Rly. Rev.*, Sept. 5, 1885, p. 431.
159. *R.R.G.*, Aug. 19, 1887, p. 547.
160. *N.C.B.*, October 1876, p. 150.
161. *Rly. Rev.*, June 10, 1882, p. 336. Beebe claims that the remains of the *Convoy* are in a museum at Jefferson, Texas.
162. *Rly. Rev.*, June 2, 1887, p. 393.
163. Beebe, *Mansions*, pp. 102 and 146–151.
164. *R.R.G.*, Sept. 11, 1885, p. 580.
165. *R.R. Car Jour.*, April 1897, p. 87.
166. Ibid., August 1898, p. 234.
167. *Pullman News*, Jan. 1941, p. 91.
168. Beebe, *Mansions*, p. 67.
169. *Christian Science Monitor*, Oct. 11, 1957, p. 15.
170. Beebe, *Mansions*, p. 204.
171. James Dredge, *The Pennsylvania R.R.*, London, 1879, p. 153 and plate 53.
172. E. Lavoinne and E. Pontzen, *Les Chemins de Fer en Amerique*, Paris, 1882.
173. Beebe, *Mansions*, p. 24.
174. *Scientific American*, October 1931, p. 258.
175. *Rly. Rev.*, May 11, 1907, p. 393.
176. *A.R.R.J.*, January 1910, p. 36.
177. Data on the *Oriental* was furnished by the Adirondack Historical Association, Blue Mountain Lake, N.Y.
178. *Fortune*, July 1930, p. 65.
179. Holden, p. 23.
180. Beebe, *Hear the Train Blow*, p. 391.
181. W. D. Randall, *From Zephyr to Amtrak*, Park Forest, Ill.: Randall, 1972, p. 198.
182. *Cincinnati Enquirer*, Sept. 16, 1962.
183. *Popular Science*, May 1955, p. 153.
184. Beebe, *Mansions*, p. 271.
185. Alvin F. Harlow, *Steelways of New England*, New York: Creative Age Press, 1946, p. 108.
186. *Van Nostrand's Engineering Magazine*, June 1870, p. 657.
187. This 1872 Pullman car rental pamphlet is in the collections of the Newberry Library, Chicago, Ill.
188. *Rly. Rev.*, June 10, 1882, p. 336.
189. *Poor's Directory of Railway Officials* 1887, page x.
190. L. M. Beebe, *Mr. Pullman's Elegant Palace Car*, Garden City, N.Y.: Doubleday, 1961, p. 217, is mistaken on the exact dates involving Worcester. It also incorrectly states that the first car was built new in 1875.
191. Data transcribed on an original print of the *City of Worcester* in the author's possession. See also *Rly. Age*, May 22, 1884, p. 329.
192. *Biographical Dictionary of Railway Officials of America*, New York and Chicago, 1885, p. 159.
193. *N.C.B.*, August 1879, p. 114.
194. *R.R.G.*, July 27, 1883, p. 501.
195. Beebe, *Mr. Pullman's Elegant Palace Car*, p. 218.
196. *Rly. Rev.*, Jan. 10, 1903, p. 20.
197. Ibid., June 19, 1886, p. 319.
198. *A.R.R.J.*, June 1893, p. 285.
199. *Scientific American*, October 1931, p. 258.
200. Holden, p. 26, and a 1939 promotional booklet on Pullman's private rental cars, reprinted in 1972 by Robert Wayner of New York City.
201. *Trains*, November 1949, p. 64.
202. *Railroad Magazine*, November 1945, p. 13.
203. *Lincoln Day by Day*, vol. III, pp. 114, 122, 220.
204. Victor Searcher, *Farewell to Lincoln*, New York: Abingdon Press, 1965, p. 87.
205. W. H. H. Price wrote an account of the Lincoln car for *Locomotive Engineering*, September 1893, p. 415. A similar account by S. D. King appeared in J. W. Starr's *Lincoln and the Railroads*, p. 276.
206. *Lincoln Lore*, May 1957, no. 1431, published by the Lincoln National Life Insurance Company, Fort Wayne, Indiana.
207. *National Republican*, April 25, 1865.
208. *R.R.G.*, Aug. 27, 1886, p. 603.

209. All accounts cited above agree on the U P purchase.

210. E. L. Sabin, *Building the Pacific Railway*, Philadelphia, 1919, p. 282.

211. *R.R. Car Jour.*, April 1897, p. 89.

212. *Denver Tribune Republican*, quoted in *R.R.G.*, Aug. 27, 1886, p. 603.

213. *N.C.B.*, May 1893, p. 80.

214. Advertising materials published by Snow are in the collections of the Chicago Historical Society.

215. *Rly. Age*, Sept. 22, 1905, p. 347.

216. *Minneapolis Journal*, March 19, 1911.

217. *R.R. Car Jour.*, 1897: February, p. 30; March, p. 57; April, p. 95; May, p. 120; July, p. 183.

218. Information leaflet issued around 1969 by the Gold Coast Railroad, Fort Lauderdale, Florida.

219. *R.R.Mag.*, November 1945, p. 24.

220. *New York Times*, April 2, 1961.

CHAPTER FIVE

1. *Ross Winans vs the Eastern Railroad Company*, U.S. Circuit Court, Massachusetts District, October 1853 term, p. 473.

2. *Illustrated London News*, April 10, 1852, p. 285.

3. *Railroad Advocate*, Sept. 20, 1856, p. 2.

4. E. T. Freedley, *Leading Pursuits and Leading Men*, Philadelphia, 1856, p. 320.

5. *George M. Pullman and Pullman's Palace Car Company vs. the New York Central Sleeping Car Company and Webster Wagner*, U.S. District Court, Northern District of Illinois, 1881, p. 80.

6. *American Railway Review*, April 26, 1860, p. 241.

7. *New England Railroad Club*, April 1924, p. 87.

8. *Railroad Car Journal*, November 1896, p. 279; see also *Railroad Gazette*, Dec. 18, 1885, p. 802, and 1888 *Car Builders' Dictionary*, fig. 1139.

9. 1888 *Car Builders' Dictionary*, figs. 1123–1130.

10. *R.R.G.*, June 4, 1886, p. 377, and Dec. 18, 1885, p. 801; *American Railroad Journal*, March 1887, p. 107.

11. 1928 *Car Builders' Dictionary*, p. 645.

12. *Railway Mechanical Engineer*, February 1930, p. 87.

13. Ibid., April 1936, p. 163.

14. Ibid., July 1947, p. 340.

15. Ibid., August 1945, p. 330.

16. *Report of the Mechanical Advisory Committee to the Federal Co-ordinator*, Washington, D.C., 1938, p. 645.

17. *Rly. M.E.*, September 1938, p. 338.

18. *Car Builders' Dictionaries* of all dates offer information on types of seat coverings.

19. Edward Hungerford, *Men and Iron*, New York: Crowell, 1938, p. 143.

20. *R.R. Adv.*, July 19, 1856, p. 3.

21. *R.R.G.*, July 7, 1876, p. 302.

22. Data on physical qualities of mohair were taken from *Car Builders' Dictionary*, 1922, p. 487, and 1928, p. 653.

23. *Pullman News*, April 1935, p. 105.

24. *National Car Builder*, March 1890, p. 46, and *R.R.C.J.*, March 1893, p. 133.

25. *N.C.B.*, August 1870, p. 3.

26. Alvin F. Harlow, *Steelways of New England*, New York: Creative Age Press, 1946, p. 380.

27. *A.R.R.J.*, Jan. 30, 1836, p. 49. McWilliams's patent was issued Nov. 29, 1832.

28. David Stevenson, *Sketch of the Civil Engineering of North America*, London, 1838; F. A. Ritter Von Gerstner, *Die innern communication der vereinigten Staaten von Nord-America*, Vienna, 1842–1843.

29. Von Gerstner, vol. 1, p. 122.

30. August Mencken, *The Railroad Passenger Car*, Baltimore: Johns Hopkins, p. 102, quoting a British travel account of 1839.

31. *R.R. Adv.*, Feb. 7, 1857, p. 4.

32. Ibid., June 2, 1855, p. 4.

33. *Am. Rly. Rev.*, March 27, 1862, p. 398, and *National Cyclopedia of American Biography*, vol. 2, p. 416.

34. From Spear's advertisement, dating about 1875, in Reading Company papers, Historical Society of Pennsylvania, Philadelphia.

35. *Master Car Builders Report*, 1908, p. 340.

36. *Railway Age*, Nov. 27, 1884, p. 731.

37. *A.R.R.J.*, Dec. 12, 1846, p. 793.

38. *R.R.G.*, July 5, 1873, p. 272.

39. *N.C.B.*, January 1878, p. 5.

40. *Engineering News*, April 5, 1890, p. 328.

41. Data on Baker's life and early years are drawn from obituaries appearing in 1901 engineering journals and a sketch published by *Locomotive Engineering*, January 1890, p. 45.

42. *Western Railroad Gazette*, March 21, 1868, p. 1.

43. E. Lavoinne and E. Pontzen, *Les Chemins de Fer en Amerique*, Paris, 1882, vol. 2, p. 26.

44. This description of the Baker system was compiled from *Appleton's Cyclopedia of Applied Mechanics*, 1886 vol. 2, p. 116; *Proceedings of the Institute of Civil Engineers*, 1878, vol. 53, p. 49; and *Car Heating*, an International Library of Technology manual on car appliances, 1912, part I, pp. 1–19.

45. *Pullman News*, April 1954, p. 8.

46. *R.R. Car Jour.*, January 1891, p. 46; *R.R.G.*, March 18, 1892, p. 201; *Railway Review*, Jan. 18, 1890, p. 31.

47. Lavoinne and Pontzen, vol. 2, p. 28.

48. *Rly. Rev.*, July 15, 1882, p. 410.

49. Ibid., April 2, 1892, p. 202.

50. *Railway Master Mechanic*, January 1892, p. xi.

51. *Rly. Rev.*, Dec. 13, 1890, p. 753.

52. *R.R. Car Jour.*, March 1891, p. 72.

53. *Rly. Rev.*, Jan. 15, 1887, p. 31.

54. *A.R.R.J.*, March 11, 1882, p. 163.

55. Ibid., Feb. 8, 1868, p. 145.

56. *R.R.G.*, March 18, 1887, p. 178.

57. *N.C.B.*, November 1875, p. 168.

58. *R.R.G.*, April 1887, p. 54.

59. *Rly. Rev.*, Dec. 13, 1890, p. 753.

60. Eugene S. Ferguson (Ed.), *Early Engineering Reminiscences (1815–1840) of George Escol Sellers*, Washington, D.C.: Government Printing Office, 1965, p. 12, and Hamilton Ellis, *Railway Carriages in the British Isles from 1830 to 1914*, London: Allen & Unwin, 1965, p. 45.

61. *Master Mechanics Magazine*, January 1888, p. 193.

62. *R.R.G.*, March 4, 1887, p. 149.

63. *American Railway Times*, Feb. 11, 1865, p. 46.

64. *Journal of the Franklin Institute*, 1877, vol. 103, p. 196.

65. *American Artisan*, September 1875, p. 229.

66. *Railway World*, Dec. 22, 1877, p. 1214.

67. *Locomotive Engineering*, January 1890, p. 45.

68. *Master Mechanics Mag.*, November 1888, p. 167.

69. *Rly. Age*, May 11, 1882, p. 257.

70. *R.R.G.*, March 8, 1889, p. 162.

71. *R.R. Car Jour.*, May–June 1891, p. 95.

72. *R.R.G.*, Jan. 21, 1887, p. 39.

73. *Engineering Magazine*, vol. 9, 1895, p. 716.

74. *R.R.G.*, Jan. 20, 1888, p. 45.

75. *N.C.B.*, August 1889, p. 115.

76. *The World* (New York City newspaper), Oct. 28, 1888.

77. Testimonial letter by F. K. Hain, printed in 1899 Gold catalog.

78. *Rly. Rev.*, Jan. 15, 1887, p. 38.

79. 1906 *Car Builders' Dictionary*, rear advertising section, p. 33. The figure 40,000 undoubtedly includes street railway cars.

80. L. H. Haney, *A Congressional History of Railways*, vol. 2, Madison, Wis., 1910, pp. 313–318.

81. *A.R.R.J.*, July 3, 1869, p. 737.

82. *N.C.B.*, April 1892, p. 53.

83. *R.R.G.*, Aug. 5, 1887, p. 517.

84. *N.C.B.*, May 1894, p. 75.

85. *R.R.G.*, March 11, 1887, p. 157.

86. Ibid., Dec. 2, 1887, p. 783.

87. Ibid., April 17, 1891, p. 263.

88. *N.C.B.*, November 1893, p. 171.

89. *R.R. Car Jour.*, January 1895, p. 13.

90. *Rly. Age*, Dec. 10, 1921, p. 1163.

91. *Western Railroad Club*, November 1892, p. 4.

92. Ibid., May 1895, p. 382.

93. Ibid.

94. *R.R.G.*, Oct. 12, 1894, p. 707.

95. *Rly. Rev.*, Dec. 5, 1891, p. 790.

96. *M.C.B.*, 1912, p. 365.
97. *Scientific American Supplement*, April 9, 1887, p. 9389.
98. "The Actual Efficiency of a Modern Locomotive," *Record of Recent Construction, No. 60*, pamphlet published by the Baldwin Locomotive Works, Philadelphia, 1907, p. 6.
99. *R.R.G.*, June 12, 1890, p. 410.
100. Ibid., May 19, 1905, p. 566.
101. Marshall M. Kirkman, *The Science of Railways*, vol. 4, Chicago: World Railway Publishing Company, 1904, p. 159.
102. *Rly. M.E.*, April 1934, p. 112.
103. For the Gold vapor system, see *R.R.G.*, June 24, 1904, p. 81, and June 16, 1909, p. 1262; *Rly. Rev.*, Dec. 19, 1908, p. 1026.
104. *Moody's Industrials*, 1920, p. 1936.
105. *Rly. M.E.*, March 1951, p. 52, and *Heating, Ventilating and Air Conditioning Guide: 1956*, 34th ed., American Society of Heating and Air Conditioning Engineers, p. 1089.
106. *Rly. M.E.*, December 1948, p. 712.
107. *Car Builders' Cyclopedia*, 1928, p. 732.
108. *Rly. M.E.*, March 1951, p. 52.
109. *Federal Coordinator's Report*, 1935, p. 640.
110. *Car Builders' Cyclopedia*, 1953, p. 723.
111. *New York Railroad Club*, Jan. 20, 1949, pp. 61–74.
112. Conversation with W. D. Edson, September 1973.
113. *New England Railroad Club*, Oct. 10, 1894, pp. 20 and 24.
114. *A.R.R.J.*, July 8, 1854, p. 428, and June 6, 1857, p. 361; *R.R. Car Jour.*, November 1894, p. 243.
115. *R.R.G.*, Sept. 20, 1895, p. 622.
116. *Railroad Record*, Oct. 5, 1854, p. 538.
117. Fredrick Moné, *An Outline of Mechanical Engineering*, New York(?), 1851, no page numbers; see description of the locomotive *Croton*.
118. *A.R.R.J.*, March 20, 1852, p. 188, and *Rly. Age*, Sept. 6, 1889, p. 587.
119. Harlow, *Steelways of New England*, p. 391.
120. *R.R.G.*, March 29, 1898, p. 309.
121. *R.R. Adv.*, June 13, 1856, p. 2.
122. *R.R.G.*, Oct. 1, 1870, p. 1, and *R.R. Car Jour.*, November 1894, p. 243.
123. *R.R. Car Jour.*, November 1894, p. 243.
124. Lavoinne and Pontzen, p. 24.
125. Lucius M. Beebe, *Trains We Rode*, vol. 2, Berkeley, Calif.: Howell-North, 1966, p. 798
126. Mencken, p. 47.
127. *R.R. Car Jour.*, October 1894, p. 208.
128. *R.R. Adv.*, July 19, 1856, p. 1.
129. *Dictionary of Canadian Biography*, 1972, vol. X, p. 636.
130. *A.R.R.J.*, Sept. 9, 1854, p. 565.
131. Ibid., Feb. 7, 1857, p. 92.
132. Ibid., Aug. 20, 1859, p. 534.
133. Ibid., June 20, 1863, p. 576.
134. *Scientific American*, July 19, 1862, p. 40.
135. *A.R.R.J.*, Feb. 8, 1868, p. 145.
136. *R.R.G.*, Jan. 4, 1873, p. 6.
137. Ibid., June 22, 1877, p. 282.
138. *R.R.G.*, Dec. 14, 1883, p. 819, and 1888 *Car Builders' Dictionary*, p. 191.
139. *R.R.G.*, Aug. 1, 1884, pp. 562 and 568.
140. *A.R.R.J.*, Dec. 15, 1840, p. 354.
141. *R.R.G.*, Jan. 4, 1873, p. 8.
142. *M.C.B.*, 1875, p. 11.
143. *R.R.G.*, Jan. 21, 1887, p. 40.
144. *New England Railroad Club*, Oct. 10, 1894, p. 6.
145. *M.C.B.*, 1875, p. 14.
146. *Rly. Age Gaz.*, Aug. 25, 1911, p. 362.
147. *M.C.B.*, 1908, p. 362.
148. Ibid., p. 340.
149. *New England Railroad Club*, Oct. 10, 1894, p. 24.
150. Ibid. and 1888 *Car Builders' Dictionary*, fig. 1566.
151. *Western Railroad Club*, January 1890, p. 92.
152. *R.R.G.*, March 24, 1876, p. 131.
153. Dudley's ventilator is described in the following sources: *Jour. Franklin Inst.*, July 1897, p. 1; *A.R.R.J.*, June 1901, p. 77; *R.R.G.*, Sept. 30, 1904, p. 392; *M.C.B.*, 1908, p. 341.
154. *R.R. Car Jour.*, May 1894, pp. 80 and 85.

155. Data on Crowder is from *R.R. Age Gaz.*, Aug. 25, 1911, p. 362; *Harper's Weekly*, Feb. 7, 1914, p. 21; J. C. Welliver, Pullman Progress (proof sheets, dated 1929, held by the Chicago Historical Society), pp. 55–57.
156. *Rly. Rev.*, Feb. 22, 1908, p. 147.
157. *R.R. Age Gaz.*, June 20, 1908, p. 244.
158. *Rly. M.E.*, March 1924, p. 158.
159. *Scientific American*, May 21, 1881, p. 323.
160. *R.R.G.*, Dec. 21, 1900, p. 841.
161. *N.C.B.*, September 1891, p. 131.
162. *Loco. Eng.*, August 1898, p. 397.
163. *R.R.G.*, June 20, 1908, p. 245.
164. *Rly. M.E.*, March 1924, p. 158.
165. *N.C.B.*, March 1873, p. 59.
166. *R.R.G.*, April 5, 1889, p. 232.
167. *Rly. M.E.*, November 1930, p. 638.
168. *R.R. Age Gaz.*, Jan. 5, 1912, p. 23.
169. R. S. Henry, *This Fascinating Railroad Business*, Indianapolis, 1942, p. 258.
170. Margaret Ingels, *Willis H. Carrier: The Father of Air Conditioning*, New York: Doubleday, 1952, p. 78. Hereafter referred to as Ingels, *Carrier*.
171. William W. Kratville, *Passenger Car Catalog: Pullman Operated Equipment, 1912–1949*, Omaha: Kratville Publications, 1968, p. 149.
172. *American Society of Mechanical Engineers Transactions*, 1937, p. 733.
173. Ingels, *Carrier*, pp. 78–81, and *Rly. M.E.*, September 1930, p. 508.
174. *Rly. M.E.*, February 1931, p. 108, and November 1931, p. 556.
175. Ibid., May 1932, p. 214.
176. Ibid., September 1930, p. 515, and December 1930, p. 696.
177. Ibid., January 1932, p. 47.
178. Ibid., August 1933, p. 306.
179. Ibid., June 1934, p. 211.
180. Ibid., October 1931, p. 521.
181. Kratville, *Passenger Car Catalog*, p. 149.
182. *M.C.B.*, 1936, p. 254.
183. *Car Builders' Cyclopedia*, 1937, p. 550, and *Fortune*, January 1938, p. 96.
184. *Rly. M.E.*, February 1931, p. 63.
185. *Railroad Magazine*, August 1946, p. 55.
186. *A.S.M.E. Transactions*, December 1937, p. 733.
187. *Rly. M.E.*, October 1931, p. 521.
188. *A.S.M.E. Transactions*, December 1937, p. 737.
189. *Rly. M.E.*, March 1934, p. 76.
190. Ingels, *Carrier*, p. 80.
191. *Rly. M.E.*, October 1931, p. 515, and March 1937, p. 124.
192. *A.S.M.E. Transactions*, December 1937, p. 733.
193. *Trains*, August 1951, p. 13.
194. Statement by W. D. Edson, former car superintendent of Amtrak, January 1971.
195. *Rly. M.E.*, May 1932, p. 173.
196. Ibid., February 1946, p. 90, and March 1947, p. 111.
197. Notes on early lighting were drawn from *Early Lighting: A Pictorial Guide*, Rushlight Club, 1972; L. S. Russell, *A Heritage of Light*, Toronto: University of Toronto, 1968; H. F. Williamson and A. R. Daum, *American Petroleum Industry*, Evanston, Ill.: Northwestern University Press, 1958. The last two works are scholarly studies which have been used elsewhere in this section without specific credit.
198. Mencken, p. 107.
199. Mencken, p. 173.
200. *M.C.B.*, 1875, p. 9.
201. *N.C.B.*, February 1880, p. 27.
202. *R.R.G.*, May 19, 1905, p. 563.
203. *Western Railroad Club*, February 1891, p. 161.
204. *R.R. Car Jour.*, November 1891, p. 15.
205. *Journal of the Franklin Institute*, 1831, vol. 2, p. 76.
206. *N.C.B.*, May 1880, p. 77.
207. Ibid., February 1878, p. 21.
208. *Engineering News*, April 26, 1890, p. 396.
209. *A.R.T.*, Feb. 15, 1868, p. 55.
210. Ibid., March 7, 1868, p. 64.
211. *N.C.B.*, May 1880, p. 77, and *Rly. Age*, Aug. 11, 1881, p. 5.
212. *A.R.R.J.*, April 5, 1834, p. 197.
213. *R.R. Adv.*, Sept. 6, 1856, p. 3.

214. Ibid., Aug. 16, 1856, p. 4, and *A.R.R.J.*, June 11, 1859, p. 377.
215. A. Bendel, *Aufsätze Eisenbahnwesen in Nord-Amerika*, Berlin, 1862, p. 37, and *A.R.R.J.*, October 1899, p. 329.
216. *N.C.B.*, December 1877, p. 180.
217. *Am. Rly. Rev.*, Jan. 17, 1861, p. 21.
218. *R.R.G.*, March 2, 1883, p. 137.
219. *N.C.B.*, June 1871, p. 1.
220. James Dredge, *The Pennsylvania RR.*, London, 1879, p. 151.
221. Lavoinne and Pontzen, vol. 2, p. 19.
222. *N.C.B.*, March 1879, p 38, and *Engineer*, Aug. 29, 1884, p. 164.
223. A. M. Wellington, W. B. D. Penniman, and C. W. Baker, *The Comparative Merits of Various Systems of Car Lighting*, New York: Engineering News Publishing Company, 1892, p. 3.
224. *Rly. Rev.*, Sept. 17, 1887, p. 543.
225. *R.R.G.*, Feb. 29, 1884, p. 164; *Report of International Railway Congress*, September 1900, p. 104.
226. *Rly. Age*, July 24, 1884, p. 5 advertisement.
227. *Rly. Rev.*, Jan. 17, 1882, p. 14, and July 8, 1882, p. 395.
228. *R.R.G.*, April 6, 1883, p. 209.
229. *National Cyclopedia of American Biography*, 1904, vol. 12, p. 189.
230. *R.R.G.*, Jan. 30, 1891, p. 84.
231. *Loco. Eng.*, January 1893, p. 15, and February 1893, p. 52.
232. *Rly. Rev.*, May 7, 1892, p. 293.
233. *A.R.R.J.*, October 1899, p. 329.
234. *Rly. Rev.*, Feb. 10, 1900, p. 70.
235. *R.R.G.*, Oct. 30, 1908, p. 1269.
236. *R.R.G.*, March 26, 1886, p. 210.
237. *N.C.B.*, July 1889, p. 98, and *Western Railroad Club*, February 1891, p. 165.
238. *R.R.G.*, June 7, 1889, p. 371.
239. *Rly. Rev.*, Nov. 5, 1892, p. 699.
240. Ibid., Aug. 25, 1900, p. 476.
241. Ibid.
242. *R.R.G.*, March 29, 1901, p. 219.
243. *R.R.G.*, April 28, 1905, p. 134.
244. *R.R. Age Gaz.*, July 9, 1909, p. 59.
245. *Western Railroad Club*, Feb. 17, 1891, p. 180.
246. *N.C.B.*, February 1879, p. 18.
247. George Behrend, *Pullman in Europe*, London: Ian Allen, 1962, p. 87.
248. *Rly. Age*, Dec. 29, 1881, p. 731.
249. *R.R.G.*, May 31, 1889, p. 355, and Sept. 22, 1882, p. 587.
250. C. W. T. Stuart, *Car Lighting by Electricity*, New York, 1923, p. 2.
251. *R.R.G.*, March 1, 1889, p. 143, and Nov. 29, 1889, p. 786.
252. Ibid., Nov. 1, 1889, p. 718.
253. *Loco. Eng.*, September 1899, p. 394, and *Western Railroad Club*, February 1891, p. 167.
254. *R.R.G.*, May 31, 1889, p. 356.
255. *Rly. Rev.*, Feb. 6, 1886, p. 62.
256. *R.R.G.*, June 13, 1890, p. 410.
257. *R.R. Age Gaz.*, July 9, 1909, p. 56.
258. *R.R.G.*, Feb. 12, 1904, p. 116.
259. Ibid., April 3, 1896, p. 237.
260. Ibid., May 20, 1898, p. 351.
261. Ibid., March 17, 1882, p. 168, and *A.A.R.J.*, September 1884, p. 184.
262. *R.R. Car Jour.*, July 1893, p. 260.
263. *R.R.G.*, June 19, 1896, p. 439.
264. Ibid., April 30, 1897, p. 301, and *A.R.R.J.*, August 1899, p. 275.
265. *Rly. Rev.*, Aug. 25, 1900, p. 471.
266. *R.R.G.*, March 4, 1904, p. 164.
267. *Rly. Rev.*, June 8, 1907, p. 503.
268. *A.R.R.J.*, December 1912, p. 614.
269. Stuart, *Car Lighting by Electricity*, pp. 5 and 55.
270. *Railway Electrical Practice, Part 2*, a handbook published by the American Technical Society, 1942, p. 151.
271. *Rly. Rev.*, Oct. 12, 1907, p. 888.
272. *Rly. M.E.*, August 1944, p. 385, and August 1945, p. 358.
273. Ibid., September 1941, p. 338.
274. *R.R. Age Gaz.*, Jan. 5, 1912, p. 15.
275. *Rly. Age Gaz. M.E.*, October 1915, p. 522.
276. *Rly. Age*, July 29, 1939, p. 175.
277. Illinois Central Railroad papers, Newberry Library, Chicago.
278. Von Gerstner, vol. 1, p. 129.

279. *A.R.R.J.*, June 15, 1842, p. 380.
280. *Harper's Weekly*, Feb. 7, 1914, p. 21.
281. *R.R.G.*, Aug. 17, 1877, p. 375.
282. *N.C.B.*, October 1882, p. 110.
283. *R.R. Car Jour.*, July 1893, p. 256.
284. *International Textbook Library of Technology*, vol. 27, 1912, p. 61.
285. *N.C.B.*, September 1884, p. 121.
286. *Harper's Weekly*, Feb. 7, 1914, p. 20.
287. Railway and Locomotive Historical Society, *Bulletin 65*, October 1944, p. 62.
288. *R.R. Record*, Aug. 11, 1853, p. 377.
289. *Loco. Eng.*, June 1898, p. 287.
290. *R.R. Adv.*, Sept. 20, 1856, p. 2.
291. *A.R.R.J.*, July 1909, p. 299.
292. *Railroad Man's Magazine*, January 1917.
293. *Lippincott's Magazine*, May 1873, p. 521.
294. Bendel, p. 38.
295. *A.R.R.J.*, Sept. 19, 1835, p. 577.
296. *R.R. Car Jour.*, June 1896, p. 109.
297. *History of the Delaware and Hudson Company*, Albany, 1925, pp. 658–659.
298. This statement was reprinted in *Railway and Locomotive Engineering*, May 1910.
299. *New York Tribune*, Sept. 24, 1893.
300. *R.R. Car Jour.*, December 1896, p. 303.
301. R. Haynes, *Troy and Rensselaer County*, New York, 1925, p. 357.
302. *R.R. Car Jour.*, November 1897, p. 349.
303. *New England Railroad Club*, Oct. 11, 1898, p. 10.
304. William Chambers, *Things as They Are in America*, London, 1853; *R.R.G.*, Nov. 11, 1881, p. 631.
305. *R.R. Car Jour.*, November 1892, p. 30.
306. *A.R.R.J.*, Aug. 17, 1850, p. 527.
307. *A.R.T.*, March 16, 1867, p. 86.
308. *R.R. Adv.*, Sept. 20, 1856, p. 1.
309. *Manufacturer and Builder*, September 1870, p. 271; *R.R.G.*, July 5, 1873, p. 272; *N.C.B.*, June 1873, p. 136, and December 1877, p. 178.
310. *N.C.B.*, March 1881, p. 30.
311. *Pullman News*, December 1927, p. 287, and July 1939, p. 6.
312. *N.C.B.*, November 1878, p. 163.
313. *Lippincott's Magazine*, May 1873, p. 521.
314. *Rly. Rev.*, Sept. 3, 1887, p. 516.
315. R. J. Wayner (Ed.), *The Pullman Scrapbook*, New York: Wayner Publications, 1971, p. 34.
316. *N.C.B.*, January 1876, p. 6.
317. *Rly. Rev.*, March 3, 1888, p. 115.
318. *N.C.B.*, November 1892, p. 165.
319. *R.R.G.*, Aug. 5, 1881, p. 431; *Rly. Rev.*, Nov. 8, 1884, p. 576; *A.R.R.J.*, December 1884, p. 292.
320. *R.R.G.*, Jan. 13, 1882, p. 17.
321. *N.C.B.*, June 1884, p. 73.
322. *Rly. World*, May 15, 1880.
323. *R.R.G.*, Aug. 19, 1881, p. 451; William Voss, *Railway Car Construction*, New York: Van Arsdale, 1892, p. 110.
324. *Rly. Rev.*, Aug. 13, 1881, p. 453.
325. Ibid., Nov. 19, 1887, p. 672.
326. Stanley Buder, *Pullman: An Experiment in Industrial Order*, New York: Oxford, 1967, p. 73.
327. *N.C.B.*, June 1884, p. 71.
328. *R.R.G.*, June 8, 1894, p. 397; Cincinnati Historical Society *Bulletin*, Spring 1973, p. 58.
329. *R.R. Car Jour.*, November 1898, p. 339.
330. Russell Lynes, *The Tastemakers*, New York: Grosset and Dunlap, 1949, pp. 95–96.
331. Ibid., p. 96; *R.R. Car Jour.*, November 1900, p. 317.
332. *R.R.G.*, October 22, 1897, p. 754.
333. Ibid., June 30, 1905, p. 773.
334. *Rly. M.E.*, February 1925, p. 97.
335. Ibid., February 1924, p. 91.
336. Ibid., June 1933, p. 191.
337. R. C. Reed, *The Streamliners*, San Marino, Calif.: Golden West Books, 1974.
338. *R.R. Car Jour.*, October 1896, p. 253.

339. *N.C.B.*, January 1881, p. 9.
340. Ibid.; *R.R.G.*, April 5, 1878, p. 178.
341. *Car Builders' Dictionary*, 1888, p. 122.
342. *R.R. Car Jour.*, January 1900, p. 20.
343. Ibid., November 1891, p. 12.
344. *N.C.B.*, September 1891, p. 142.
345. *R.R. Car Jour.*, October 1898, p. 304.
346. *R.R.G.*, Nov. 11, 1881, p. 631.
347. *N.C.B.*, March 1881, p. 30.
348. *Rly. Rev.*, Feb. 6, 1897, p. 79; *A.R.R.J.*, November 1898, p. 387.
349. *R.R.G.*, Sept. 12, 1884, p. 672.
350. *R.R. Car Jour.*, March 1897, p. 75.
351. Ibid., Aug. 21, 1903, p. 598, and March 4, 1905, p. 318.
352. Railway and Locomotive Historical Society, *Bulletin 103*, October 1960, p. 59.
353. *New England Railroad Club*, Oct. 11, 1898, p. 46.
354. *Rly. M.E.*, December 1925, p. 757.
355. Ibid., October 1914, p. 527.
356. Ibid., June 1931, p. 292.
357. Ibid., October 1935, p. 435.
358. On early vestibule devices, refer to *R.R. Car Jour.*, November 1891, p. 18; *R.R.G.*, March 2, 1894, p. 152; *Glasgow Practical Mechanics Magazine*, 1847, vol. 2, p. 204.
359. *R.R. Record*, June 16, 1853, p. 247, and Sept. 22, 1853, p. 471; *A.R.R.J.*, July 1, 1854, p. 406.
360. A rare photograph of a train fitted with Atwood and Waterbury's vestibule is reproduced in L. M. Beebe and Charles Clegg, *Hear the Train Blow*, New York: Grosset and Dunlap, 1952, p. 44. This photograph agrees with the line drawing shown in Figure 5.86.
361. *R.R.G.*, Aug. 26, 1887, p. 557.
362. Ibid., May 25, 1888, p. 341.
363. *N.C.B.*, June 1887, p. 80; 1879 *Car Builders' Dictionary*, p. 194.
364. *Rly. Age*, Sept. 28, 1882, p. 540.
365. *Rly. Rev.*, April 16, 1887, p. 226.
366. *R.R.G.*, Dec. 9, 1887, p. 796.
367. Ibid., Aug. 24, 1888, p. 557.
368. *R.R. Car Jour.*, June 1899, p. 151.
369. *Rly. Rev.*, Oct. 18, 1890, p. 627.
370. *R.R.G.*, Nov. 28, 1889, p. 791, and *N.C.B.*, September 1888, p. 142.
371. *R.R.G.*, March 2, 1894, p. 153.
372. Ibid., Oct. 17, 1890, p. 723.
373. Ibid., March 9, 1894, p. 173.
374. *Rly. Master Mech.*, July 1893, p. 111.
375. *R.R. Mag.*, October 1949, p. 35.
376. *Rly. Age*, Jan. 19, 1946, p. 151.

CHAPTER SIX

1. A. M. Earle, *Stage Coach and Tavern Days*, New York: Macmillan, 1900, p. 236.
2. Memorandum on the history of baggage service dated February 8, 1929, by W. M. Skinner of the American Association of General Baggage Agents.
3. Ibid.
4. *Locomotive Engineering*, February 1895, p. 92.
5. F. A. Ritter Von Gerstner, *Die innern communication der vereinigten Staaten von Nord-America*, Vienna, 1842–1843, vol. 2, p. 96.
6. August Mencken, *The Railroad Passenger Car*, Baltimore: Johns Hopkins Press, 1957, p. 125.
7. A. Bendel, *Aufsatze Eisenbahnwesen in Nord-Amerika*, Berlin, 1862, pp. 38–39.
8. Karl Ghega, *Die Baltimore-Ohio Eisenbahn*, Vienna, 1844, p. 195.
9. E. H. Mott, *Between the Ocean and the Lake: The Story of the Erie*, New York, 1901, p. 397.
10. Alvin Harlow, *Steelways of New England*, New York: Creative Age Press, 1946. The Old Colony stock certificate is reproduced on the endpapers of this volume.
11. *New York Railroad Club*, November 1922, p. 6817.
12. *National Car Builder*, January 1881, p. 5.
13. *Railway Age*, July 6, 1882, p. 372.
14. *Railway Review*, March 2, 1902, p. 146.
15. *Railroad Car Journal*, November 1899, p. 335.

16. *Rly. Rev.*, July 18, 1909, p. 573.
17. *Master Car Builders Report*, 1908, pp. 307 and 322.
18. F. Gutbrod, "The Construction of Iron Passenger Cars on the Railways in the United States of America," *International Railway Congress Bulletin*, vols. 26 and 27, 1912–1913, part 3, p. 64. Also *Railroad Gazette*, June 18, 1908, p. 133.
19. J. W. Shine and F. M. Ellington, *Passenger Train Equipment of the Atchison, Topeka & Santa Fe*, vol. 1, Colfax, Iowa: Railcar Press, 1973, p. 80.
20. Bendel, p. 38.
21. Von Gerstner, vol. 1, pp. 79, 238, 247, and 250.
22. Lucius M. Beebe and Charles Clegg, *The Central Pacific and the Southern Pacific Railroads*, Berkeley, Calif.: Howell-North, 1963, pp. 26 and 29.
23. *Loco Eng.*, July 1898, p. 353.
24. Von Gerstner, vol. 1, p. 121.
25. Illinois Central Papers, Newberry Library, Chicago. Letter from the Buffalo Car Company, May 10, 1855.
26. Michigan Central annual report, 1855, p. 37.
27. *Railroad Magazine*, September 1952, p. 66.
28. *N.C.B.*, April 1879, p. 55.
29. *Rly. Rev.*, Sept. 18, 1883, p. 532.
30. *Rly. Age*, March 1, 1889, p. 144, and *R.R.G.*, Feb. 15, 1889, p. 115.
31. *Rly. Rev.*, May 3, 1884, p. 232.
32. A copy of "Santa Fe Route Tourist Sleeper" of 1900 is in the collections of the Library of Congress.
33. *Railway Mechanical Engineer*, October 1933, p. 380.
34. *Rly. Age*, Nov. 15, 1947, p. 234.
35. *Trains*, June 1947, p. 7.
36. General comments made here and elsewhere about the Railway Mail Service are drawn largely from existing histories on the subject, especially B. A. Long and W. J. Dennis's *Mail by Rail*, New York: Simmons-Boardman, 1951 (hereafter *Mail by Rail*), and Carl Scheele's *A Short History of the Mail Service*, Washington, D.C.: Smithsonian Institution, 1970. Scheele's work (hereafter Scheele) is particularly valuable for its extensive bibliography.
37. *Mail by Rail*, p. 96.
38. *Rly. Age*, July 9, 1938, p. 36.
39. Hamilton Ellis, *Railway Carriages in the British Isles from 1830 to 1914*, London: Allen & Unwin, 1965, p. 42.
40. *Quarterly Review*, vol. 84, 1849, p. 44.
41. Von Gerstner, vol. 1, p. 95.
42. *Rly. Age*, Aug. 6, 1938, p. 223, and C. H. Corliss, *Main Line of Mid America*, New York: Creative Age Press, 1950, p. 76.
43. U.S. Post Office Department, *History of the Railway Mail Service*, Washington, D.C.: G.P.O., 1885, p. 40. Hereafter *History R.M.S.*
44. *American Railroad Journal*, July 28, 1855, p. 470.
45. *R.R. Car Jour.*, November 1894, p. 243.
46. Biographical file on W. A. Davis in the Smithsonian's Division of Postal History.
47. *Beginnings of the True Railway Mail Service and the Work of George B. Armstrong in Founding It*, Chicago, 1906, pp. 74–84.
48. *Railway World*, May 28, 1881, p. 510.
49. *A.R.R.J.*, Sept. 10, 1864, p. 894.
50. *Rly. World*, May 28, 1881, p. 510.
51. *Scribner's Monthly*, June 1873.
52. *Chicago Magazine*, January 1912, p. 350.
53. J. E. White, *A Life Span and Reminiscences of the R.M.S.*, Philadelphia, 1910, p. 12. Hereafter White, *Reminiscences*.
54. *Scribner's Monthly*, June 1873.
55. *R.R. Car Jour.*, August 1892, p. 200.
56. *N.C.B.*, October 1870, p. 5.
57. *Scribner's Monthly*, June 1873.
58. *Rly. World*, Dec. 1, 1877, p. 1142.
59. *R.R.G.*, Sept. 18, 1875, p. 387.
60. *The American Railway*, 1892, p. 318.
61. *N.C.B.*, October 1875, p. 154.
62. *R.R.G.*, Sept. 18, 1875, p. 387.
63. *The American Railway*, 1892, p. 319.
64. White, *Reminiscences*, p. 15.
65. *Scribner's Monthly*, June 1873.
66. *History R.M.S.*, p. 74.
67. *N.C.B.*, August 1882, p. 89.
68. *Rly. Age*, April 30, 1885, p. 283.

69. Post Master General's annual report, 1880–1881.

70. *N.C.B.*, November 1894, p. 173.

71. *Mail by Rail*, p. 136. *Railway Age Mechanical Engineer*, June 1915, p. 299, gives a total of 5,866 mail cars, only 1,566 of which were full R.P.O.s as of December 31, 1914.

72. *N.C.B.*, October 1883, p. 109.

73. E. Lavoinne and E. Pontzen, *Les Chemins de Fer en Amerique*, Paris, 1882, atlas, plate VI.

74. *N.C.B.*, October 1880, p. 164.

75. *Rly. Rev.*, Dec. 19, 1891, p. 816.

76. White, *Reminiscences*, pp. 168–170.

77. *Engineer*, Aug. 31, 1883, p. 163, and *R.R.G.*, Jan. 29, 1886, p. 70.

78. M. McKenna, *History of Fond Du Lac County*, Chicago, 1912, vol. 1, p. 324.

79. *N.C.B.*, May 1885, p. 64.

80. *R.R. Car Jour.*, July 1895, p. 158.

81. *Railway Mechanical Engineer*, April 1913, p. 212.

82. *History R.M.S.*, p. 201.

83. *Mail by Rail*, pp. 141–142.

84. 69th Congress House Hearings (No. 1904), Jan. 31, 1927, p. 3.

85. *N.C.B.*, September 1894, p. 133.

86. White, *Reminiscences*, p. 171, and *Engineering News*, Feb. 20, 1892, p. 173.

87. White, *Reminiscences*, p. 173.

88. *Rly. Age*, Feb. 3, 1911, p. 190, and Feb. 10, 1911, p. 291.

89. *Southern & South Western Railroad Club*, May 1916, p. 11, and *Rly. Age*, Nov. 24, 1911, p. 1049.

90. 69th Congress House Hearings (No. 1904), Jan. 31, 1927, p. 2.

91. *R.R.G.*, June 22, 1906, p. 687, and *Rly. Rev.*, June 15, 1907, p. 522.

92. White, *Reminiscences*, p. 178.

93. *Rly. M.E.*, June 1915, p. 299.

94. Ibid., December 1922, p. 737.

95. Shine and Ellington, *Passenger Train Equipment of the Atchison, Topeka & Sante Fe*, vol. 1, p. 8.

96. House Hearings on Steel Railway Post Office Cars, 69th Congress, First Session, H.R. 4475, April 30, 1926.

97. House Hearings on Steel Railway Post Office Cars, 70th Congress, First Session, H.R. 9678, May 16, 1928.

98. *Car Builders' Cyclopedia 1928*, pp. 515–528.

99. *Railroad Car Facts*, American Railway Car Institute, New York, 1961, p. 73.

100. *Car Builders' Cyclopedia 1953*, p. 472, and *Rly. M.E.*, March 1915, p. 123.

101. *Mail by Rail*, p. 340.

102. Scheele, p. 169.

103. Ibid., pp. 176–177.

104. *Trains*, February 1971, p. 31.

105. *Trains*, October 1960, p. 12.

106. *Year Book of Railroad Facts*, 1972, p. 12, and 1974, p. 10.

107. Shine and Ellington, *Passenger Train Equipment*, vol. 1, p. 10.

CHAPTER SEVEN

1. *Railway Mechanical Engineer*, April 1930, p. 218.

2. *Railroad Advocate*, June 2, 1855, p. 2.

3. *Ross Winans vs. The Eastern Railroad Company*, U.S. Circuit Court, Massachusetts District, October 1853 term (Boston 1854), p. 913.

4. *Railroad Magazine*, June 1942, p. 103.

5. *New York Railroad Club*, April 19, 1900, p. 22.

6. Ibid.

7. *National Car Builder*, January 1881, p. 8, and *Rly. M.E.*, January 1916, p. 27.

8. William Voss, *Railway Car Construction*, New York, 1892, p. 112.

9. *Rly. M.E.*, November 1915, p. 569.

10. *Rly. M.E.*, March 1945, p. 110.

11. *R.R. Mag.*, June 1942, p. 98.

12. *Institution of Civil Engineers Proceedings*, vol. 28, 1869, p. 380.

13. *American Railway Review*, Jan. 19, 1860, p. 33.

14. *Pullman News*, October 1929, p. 191.

15. *N.C.B.*, September 1876, p. 134.

16. Ibid., June 1879, p. 84.

17. *The Engineer*, Philadelphia, Oct. 11, 1860, p. 68.

18. *N.C.B.*, January 1891, p. 4.

19. *Scientific American*, Sept. 11, 1845, p. 10.

20. *New York Farmer & Mechanic*, May 29, 1845.

21. *N.C.B.*, March 1884, p. 32.

22. *Rly. Rev.*, Sept. 7, 1889, p. 524, and *Railroad Gazette*, June 6, 1890, p. 394.

23. *R.R.G.*, May 15, 1896, p. 334.

24. Ibid., Sept. 16, 1898, p. 661.

25. Ibid., June 17, 1904, p. 16.

26. *A.R.R.J.*, March 1906, p. 96.

27. *Rly. M.E.*, November 1915, p. 569.

28. *Car Builders' Cyclopedia*, 1928, fig. 2181.

29. *Rly. M.E.*, March 1945, p. 108.

30. Ibid., March 1946, p. 122.

31. *Railway Age*, July 30, 1949, p. 207.

32. *Car Builders' Cyclopedia*, 1943, p. 1117.

33. Data from W. D. Edson, former car superintendent for Amtrak.

34. *A.R.R.J.*, Oct. 25, 1834, p. 658.

35. William Whiting, *Arguments of William Whiting, Esq., in the case of Ross Winans vs. Orsamus Eaton et al.*, Boston, 1853, p. 126.

36. *Rogers Locomotive and Machine Company*, 1876 catalog, p. 26.

37. *Report of the Entire Testimony in the Case of W. C. Fuller vs. F. M. Ray*, pamphlet published by F. M. Ray, New York, 1850.

38. *A.R.R.J.*, Jan. 5, 1850, p. 11.

39. Ibid., Feb. 28, 1852, p. 136.

40. D. A. Wells (Ed.), *Annual of Scientific Discovery*, Boston, 1857, p. 54.

41. *Car Builders' Dictionary*, 1879, fig. 118.

42. *Journal of the Franklin Institute*, vol. 6, 1843, p. 168.

43. *A.R.R.J.*, Oct. 30, 1847, p. 689.

44. *American Railway Times*, Jan. 19, 1861, p. 24.

45. *Rly. Age*, Dec. 1, 1969, rear cover advertisement.

46. John H. White, Jr., *James Millholland and Early Railroad Engineering*, Contributions from the Museum of History and Technology, paper 69, U.S. National Museum Bulletin 252, Washington, D.C., 1967, p. 7.

47. *Railroad Car Journal*, August 1894, p. 167.

48. *R.R.G.*, June 30, 1876, p. 290.

49. Nicholas Wood, *A Practical Treatise on Rail-Roads*, London, 1832 edition, p. 544, mentions the use of oil cellar boxes on the B & O.

50. *Winans vs. Eastern Railroad*, p. 553, and *Niles Register*, Jan. 8, 1829, p. 299.

51. *Niles Register*, Feb. 21, 1829, p. 432.

52. Ibid., Aug. 29, 1829, p. 2.

53. *R.R.G.*, April 7, 1876, p. 149.

54. B & O annual report, 1831, p. 108.

55. *R.R.G.*, April 7, 1876, p. 149.

56. *A.R.R.J.*, Dec. 8, 1855, p. 777.

57. South Carolina Railroad annual report, 1840, p. 6, and 1843, p. 8.

58. *R.R. Adv.*, June 2, 1855, p. 4.

59. John Cochrane (Ed.), *Railway Patent Claim Reporter*, vol. 1, No. 2, July 1860, p. 37.

60. Ibid., p. 47.

61. *Master Car Builders Reports*, 1875, p. 78.

62. *A.R.R.J.*, May 1899, p. 147.

63. Ibid., December 8, 1855, p. 777.

64. *M.C.B.*, 1883, p. 85.

65. *N.C.B.*, September 1885, p. 113.

66. *A.R.R.J.*, June 1900, p. 171.

67. F. A. Ritter Von Gerstner, *Die innern communication der vereinigten Staaten von Nord-America*, vol. 1, Vienna, 1842–1843, p. 298.

68. *Am. Rly. Rev.*, July 14, 1859, p. 2.

69. *A.R.R.J.*, September 1843, p. 254.

70. Ibid., Nov. 8, 1856, p. 716.

71. Ibid., Sept. 1884, p. 162.

72. *N.C.B.*, December 1874, p. 182.

73. *M.C.B.*, 1875, p. 160.

74. *N.C.B.*, August 1891, p. 122.

75. James Dredge, *The Pennsylvania R.R.*, London, 1879, p. 90.

76. *Rly. Rev.*, June 16, 1888, p. 351.

77. *New York Railroad Club*, Jan. 18, 1900, p. 31.

78. Ibid., p. 12.

79. *M.C.B.*, 1875, p. 76.

80. Ibid., 1875, pp. 76 and 159; 1888, p. 37.

81. *Rly. M.E.*, May 1953, p. 56.
82. *Car Builders' Cyclopedia*, 1961, p. 851.
83. *Rly. M.E.*, September 1950, p. 494.
84. Railway and Locomotive Historical Society, *Bulletin 12*, 1926, p. 48. (Hereafter R.&L.H.S.)
85. *N.C.B.*, September 1870, p. 2.
86. *Rly. M.E.*, November 1943, p. 567.
87. *R.R. Mag.*, October 1953, p. 122.
88. *N.C.B.*, August 1891, p. 122.
89. *New York Railroad Club*, Jan. 18, 1900, p. 5.
90. *Rly. M.E.*, November 1934, p. 409, and May 1946, p. 260.
91. *Rly. M.E.*, March 1943, p. 120.
92. *R.R.G.*, Aug. 11, 1905, p. 121.
93. *M.C.B.*, 1889, pp. 45–52.
94. *N.C.B.*, July 1890, p. 97. Fletcher was not the inventor but rather the advocate of this form of journal box lid.
95. *R.R.G.*, Aug. 11, 1905, p. 121.
96. *Rly. M.E.*, September 1936, p. 405.
97. Ibid., November 1934, p. 407.
98. *Rly. Age*, March 9, 1946, p. 504.
99. *R.R. Mag.*, November 1950, p. 65.
100. *Rly. Age*, Sept. 15, 1969, p. 83.
101. H. T. Morton, *Anti-Friction Bearings* (published by the author), 1954; also *Cassier's Magazine*, May 1897, p. 60.
102. *Cassier's Magazine*, August 1899, p. 392.
103. *R.R.G.*, Nov. 22, 1889, p. 765.
104. *N.C.B.*, June 1892, p. 99.
105. *R.R.G.*, Aug. 30, 1889, p. 575.
106. Ibid., Oct. 5, 1888, p. 653.
107. *N.C.B.*, June 1894, p. 98.
108. *Master Mechanics Magazine*, August 1888, p. 120.
109. *Rly. M. E.*, January 1914, p. 19.
110. Ibid., December 1946, p. 688.
111. Ibid., November 1926, p. 667.
112. *Report of the Mechanical Advisory Committee to the Federal Coordinator*, Washington, D.C., 1935, p. 655.
113. *Railroad Car Facts*, 1961, p. 74.
114. *New York Railroad Club*, Oct. 20, 1949, p. 208.
115. R.&L.H.S. *Bulletin 23*, November 1930, p. 9.
116. *A.R.R.J.*, September 1843, p. 267.
117. *R.R. Adv.*, Sept. 20, 1856, p. 2.
118. *A.R.R.J.*, Feb. 18, 1854, p. 105.
119. *R.R.G.*, Feb. 13, 1883, p. 107.
120. *A.R.R.J.*, Sept. 28, 1850, p. 609.
121. Edward Hungerford, *Men and Iron: The History of the New York Central*, New York: Crowell, 1938, p. 244.
122. *Report on the Causes of Railroad Accidents and the Means of Preventing Their Recurrence*, New York Legislature, January 1853.
123. *A.R.R.J.*, Oct. 15, 1839, p. 256.
124. Ibid., Nov. 15, 1839, p. 289.
125. Charles Paine, *Elements of Railroading*, New York, 1895, p. 113.
126. *M.C.B.*, 1875, p. 134.
127. *R.R.G.*, Sept. 20, 1889, p. 616.
128. *R.R. Adv.*, Sept. 20, 1856, p. 2.
129. *A.R.R.J.*, July 16, 1853, p. 460; *Rly. Rev.*, April 9, 1887, p. 205; *R.R.G.*, April 2, 1886, p. 227; Paine, *Elements of Railroading*, p. 113.
130. Marshall Kirkman, *The Science of Railways*, vol. 4, Chicago: World Railway Publishing Company, 1904, p. 222.
131. *A.R.R.J.*, Aug. 15, 1857, p. 522.
132. Marshall Kirkman, *The Science of Railways. Cars: Their Construction, Handling and Supervision* (supplementary volume on cars), New York and Chicago, 1908, p. 97.
133. Robert S. Henry, *This Fascinating Railroad Business*, Indianapolis: Bobbs-Merrill, 1942, p. 268.
134. *N.C.B.*, February 1879, p. 25.
135. Ibid., November 1872, p. 9; November 1876, p. 165; February 1879, p. 25.
136. Peter Temin, *Iron and Steel in Nineteenth-Century America*, Cambridge, Mass.: M.I.T. Press, 1964, pp. 218–222.
137. John H. White, Jr., *American Locomotives 1830–1880*, Baltimore: Johns Hopkins, 1968, pp. 178–179.
138. *R.R. Adv.*, Sept. 20, 1856, p. 2.

139. *A.R.T.*, Jan. 19, 1861, p. 24.
140. *R.R.G.*, Jan. 20, 1893, p. 47 and Paine, *Elements of Railroading*, p. 113.
141. *Rly. Rev.*, Feb. 11, 1894, p. 106.
142. *R.R.G.*, Dec. 13, 1878, p. 599.
143. *N.C.B.*, August 1875, p. 121.
144. *R.R.G.*, Dec. 13, 1878, p. 599, and *N.C.B.*, July 1878, p. 104.
145. *N.C.B.*, March 1889, p. 43.
146. *Western Railroad Club*, May 1904, p. 270.
147. Ibid., p. 276.
148. Marshall Kirkman, *The Science of Railways. Cars: Their Construction, Handling and Supervision*, New York and Chicago, 1908, p. 94.
149. *R.R. Adv.*, Sept. 20, 1856, p. 2.
150. *Proceedings, A.A.R. Mechanical Division*, 1941, p. 356; and *Car Builders' Cyclopedia*, 1949–1951, p. 1052.
151. Data from W. D. Edson, former car superintendent for Amtrak.
152. *M.C.B.*, 1885, pp. 16–27.
153. *Rly. M.E.*, August 1939, p. 307, and J. E. Muhlfeld, *The Railroad Problem and Its Solution*, New York: Devin-Adair, 1941, p. 193.
154. *A.R.R.J.*, June 15, 1833, p. 369.
155. M. J. T. Lewis, *Early Wooden Railways*, London: Routledge & Kegan Paul, 1970, p. 195.
156. R.&L.H.S. *Bulletin 61*, May 1943, p. 49.
157. *Baltimore Engineer*, August 1928, p. 7.
158. Kirkman, *Science of Railways*, vol. 4, 1904, p. 217.
159. Muhlfeld, *Railroad Problem*, p. 192.
160. *Car Builders' Dictionary*, 1888, fig. 2150.
161. George H. Burgess and M. C. Kennedy, *Centennial History of the Pennsylvania Railroad*, Philadelphia, 1949, p. 324.
162. *N.C.B.*, September 1877, p. 131, and November 1877, p. 169.
163. Memorandum written by Isaac Dripps in 1885. Now in papers of J. E. Watkins, Smithsonian Archives.
164 Petersburg Railroad annual report, 1835, p. 14.
165. *Budget* (Troy, N.Y., newspaper), Sept. 15, 1835.
166. Baldwin mss. collection, Historical Society of Pennsylvania, Philadelphia, letter dated Feb. 17, 1837, to Joseph Johnson.
167. Lewis, *Early Wooden Railways*, p. 195; C. F. Dendy Marshall, *A History of the Railway Locomotive Engine*, London: Locomotive Publishing Company, 1953, p. 126.
168. R.&L.H.S. *Bulletin 65*, October 1944, p. 49.
169. *N.C.B.*, September 1891, p. 140.
170. R.&L.H.S. *Bulletin 32*, October 1933, p. 57.
171. *A.R.R.J.*, June 19, 1847, p. 392.
172. *Baltimore Engineer*, August 1929, p. 5; *Business History Review*, Summer 1968, p. 171.
173. White, *American Locomotives*, pp. 280–286.
174. B & O annual report, 1834, p. 27.
175. *A.R.R.J.*, July 9, 1853, p. 434.
176. *Institution of Civil Engineers*, vol. 28, 1869, p. 383.
177. *U.S. Centennial Reports and Awards*, vol. VI, 1880, p. 41; *A.R.R.J.*, June 1896, p. 108, and September 1896, p. 222.
178. *Anson Atwood vs. Portland Company*, U.S. Circuit Court, District of Maine, Sept. 23 and 24, 1879. Printed copy of pleadings, proofs, briefs, and opinions, in Library of Congress. Hereafter *Atwood vs. Portland*.
179. Zerah Colburn, *Recent Practice in the Locomotive Engine*, London, 1860, p. 53.
180. *Atwood vs. Portland*, p. 42.
181. *Rly. Rev.*, Feb. 2, 1889, p. 64.
182. *A.R.R.J.*, Nov. 6, 1847, p. 705.
183. *Cassier's Mag.*, March 1910, p. 440.
184. *Railway and Locomotive Engineer*, December 1924, p. 367.
185. *R.R.G.*, April 20, 1872, p. 175.
186. Ramapo Wheel & Foundry Company, Ramapo, N.Y., catalog 1887c., p. 14.
187. *Institution of Civil Engineers*, 1869, p. 402.
188. Ibid., 1878, p. 37.
189. *Rly. Rev.*, Feb. 11, 1882, p. 89.
190. *R.R.G.*, April 20, 1872, p. 175.
191. Ibid.
192. Ibid.; *U.S. Centennial Reports and Awards*, vol. VI, 1880, p. 49; *Scientific American Supplement*, July 29, 1876.
193. *R.R.G.*, April 29, 1898, p. 304.

194. Ibid., Feb. 10, 1911, p. 268.
195. *M.C.B.*, 1883, p. 32.
196. *Southern and South Western Railroad Club*, November 1939, p. 37.
197. *Institution of Civil Engineers*, 1878, p. 69.
198. *R.R.G.*, May 22, 1896, p. 364.
199. *Engineer*, Dec. 28, 1883, p. 499.
200. *Report of the Committee Investigating the Affairs of the Boston and Providence R.R.*, Boston, 1856, p. 26.
201. *Rly. M.E.*, July 1947, p. 347.
202. Ibid., August 1933, p. 296.
203. *R.R.G.*, April 20, 1872, p. 175.
204. Bush and Lobdell Catalog, 1851, p. 6. Copy in A.A.R. Library as of 1968.
205. *American Society of Civil Engineers Transactions*, vol. 20, 1899, p. 615.
206. *Baltimore Engineer*, August 1929, p. 9.
207. *R.R.G.*, April 29, 1898, p. 304.
208. Baldwin Locomotive Works, *Record of Recent Construction*, no. 52, 1905, p. 13; *Baltimore Engineer*, August 1929, p. 9.
209. Dredge, *The Pennsylvania R.R.*, p. 83. Data gathered in 1875 and 1876.
210. *Engineer*, Dec. 28, 1883, p. 499; *Cassier's Mag.*, March 1910, p. 440.
211. *R.R.G.*, July 5, 1889, p. 439, and June 19, 1908, p. 211.
212. *Jour. Franklin Inst.*, September 1843, p. 212; October 1845, p. 273; February 1848, p. 119, and September 1848, p. 210.
213. Alvin F. Harlow, *Road of the Century*, New York: Creative Age Press, 1947, p. 146; quotes from a newspaper account dated May 26, 1849.
214. *A.R.T.*, Jan. 19, 1861, p. 24.
215. Zerah Colburn, *The Locomotive Engine*, Philadelphia, 1854, p. 72.
216. *Report on the Causes of Railroad Accidents and the Means of Preventing Their Recurrence*, New York Legislature, January 1853, pp. 51 and 59.
217. *A.R.T.*, Jan. 19, 1861, p. 24.
218. *Western Railroad Gazette*, April 24, 1869, p. 1.
219. *A.R.T.*, April 10, 1869, p. 118.
220. *Rly. Rev.*, Oct. 18, 1890, p. 629.
221. *R.R.G.*, June 3, 1887, p. 373.
222. *N.C.B.*, March 1872, p. 4.
223. Ibid., February 1871, p. 4.
224. *Leslie's Illustrated Newspaper*, Aug. 25, 1877.
225. *Rly. Age*, Sept. 1, 1881, p. 494.
226. *A.R.R.J.*, April 22, 1882, p. 255, and *R.R. Car Jour.*, June 1893, p. 232.
227. Advertising leaflet, 1883c., published by the Allen Paper Wheel Company.
228. *R.R. Car Jour.*, June 1893, p. 232.
229. *R.R.G.*, Jan. 26, 1883, p. 53.
230. *N.C.B.*, July 1880, p. 124.
231. Ibid., July 1882, p. 73, and *Rly. Rev.*, Feb. 10, 1883, p. 83.
232. *R.R.G.*, June 11, 1886, p. xviii.
233. *R.R. Car Jour.*, June 1893, p. 232.
234. *R.R.G.*, June 11, 1886, p. xviii.
235. *M.C.B.*, 1875, p. 140; *N.C.B.*, August 1877, p. 117.
236. *R.R.G.*, Sept. 4, 1891, p. 613.
237. Conversation with George Krambles, general manager of the Chicago Transit Authority, May 1972.
238. *M.C.B.*, 1875, p. 140.
239. Ibid., 1883, p. 58.
240. *Engineer*, Dec. 28, 1883, p. 499.
241. *R.R.G.*, Dec. 13, 1878, p. 599.
242. *N.C.B.*, June 1888, p. 85.
243. *Rly. Age*, Sept. 1, 1881, p. 494.
244. Ibid.
245. *N.C.B.*, November 1879, p. 166.
246. *R.R.G.*, Jan. 26, 1883, p. 53.
247. *N.C.B.*, January 1893, p. 1.
248. *Rly. Age*, Aug. 18, 1881, pp. 473 and 476.
249. *Locomotive Engineering*, April 1897, p. 320.
250. *Pullman News*, April 1923, p. 364.
251. *Engineering News*, July 1, 1915, p. 11.
252. *Rly. & Loco. Eng.*, March 1917, p. 79.
253. *R.R.G.*, May 24, 1901, p. 352.

254. *Southern & Southwestern Railway Club*, March 1917, p. 4.
255. L. S. Marks, *Mechanical Engineer's Handbook*, New York, 1930, p. 1483.
256. *R.R. Mag.*, March 1969, p. 18.
257. *Sci. Am.*, January 1924, p. 35.
258. *A.R.R.J.*, Dec. 27, 1862, p. 1027.
259. *A.R.T.*, Jan. 18, 1868, p. 23.
260. Ibid., July 31, 1869, p. 246.
261. *A.S.M.E. Trans.*, 1881, p. 300.
262. *Rly. Rev.*, June 9, 1888, p. 330; *R.R.G.*, Nov. 21, 1902, p. 889.
263. *R.R.G.*, Feb. 17, 1888, p. 110.
264. *Year Book of the American Iron & Steel Institute*, 1913, p. 367.
265. *R.R.G.*, March 1, 1912, p. 369.
266. *Report of the Mechanical Advisory Committee to the Federal Coordinator*, 1935, p. 650.
267. *Wheel and Axle Manual: Standards, Recommended Practices, and General Information*, 9th ed., Chicago: Association of American Railroads, 1968, pp. 142–145.
268. Ibid.
269. Michael Reynolds, *Continuous Railway Brakes*, London, 1882, p. 1.
270. *Rly. Age*, Dec. 21, 1882, p. 720.
271. Von Gerstner, vol. 2, p. 255.
272. F. C. B. Bradlee, *The Boston and Maine Railroad*, Salem, Mass., 1921, p. 39.
273. *N.C.B.*, April 1890, p. 58.
274. S. M. Whipple, *Appendix. Documents and Depositions . . . in behalf of Respondents before the Senate and House Committees on Patents, in Answer to the Memorial A. G. Batchelder, et. al.* A pamphlet published in 1874 by the Eastern Rail Road Association, Springfield, Mass. Hereafter Whipple.
275. Seymour Dunbar, *A History of Travel in America*, vol. III, Indianapolis: Bobbs-Merrill, 1946, p. 987.
276. *Railroad Record*, Oct. 27, 1853, p. 551.
277. Ibid.
278. Von Gerstner, vol. 1, p. 261.
279. *R.&L.H.S. Bulletin 12*, 1926, p. 48.
280. *N.C.B.*, July 1881, p. 84.
281. George Bradshaw, *Prevention of Railroad Accidents*, New York, 1912, p. 18.
282. H. G. Prout, *A Life of George Westinghouse*, New York, 1922, p. 33.
283. *Report of the American Institute*, New York, 1855, p. 548.
284. *R.R. Adv.*, June 7, 1856, p. 3.
285. Jonathan Knight, *Report upon the Locomotive Engines*, Baltimore, 1838, pp. 4 and 10.
286. *Report on Causes of Railroad Accidents*, N.Y. Legislature, 1853, pp. 43, 67, 88.
287. *M.C.B.*, 1875, p. 24.
288. Voss, *Railway Car Construction*, p. 77.
289. Whipple, p. 44.
290. Ibid., p. 24.
291. Ibid., pp. 1–140; *Rly. Rev.*, Jan. 27, 1883, p. 45.
292. *Rly. Age*, April 7, 1881, p. 188.
293. *Jour. Franklin Inst.*, October 1835, p. 228.
294. *A.R.R.J.*, Sept. 15, 1841, p. 168, and Sept. 28, 1848, p. 691.
295. S. A. Howland, *Steamboat Disasters and Railroad Accidents*, Worcester, Mass., 1840, p. 262.
296. *Sci. Am.*, Dec. 9, 1854, p. 100.
297. *Am. Rly. Rev.*, Oct. 20, 1859, p. 3.
298. M. Chevalier, *Histoire et Description Des Voies De Communication*, vol. 2, Paris, 1841, p. 515; *Hazard's Register of Pennsylvania*, Aug. 2, 1838.
299. *A History and Description of the Baltimore and Ohio Railroad*, no author, Baltimore, 1853, p. 48.
300. Hamilton Ellis, *Railway Carriages in the British Isles from 1830 to 1914*, London: Allen & Unwin, 1965, p. 194.
301. *R.R. Adv.*, Feb. 28, 1857, p. 4.
302. *Am. Rly. Rev.*, July 27, 1859, p. 13.
303. *West. R.R. Gaz.*, Jan. 19, 1861.
304. *Report American Institute*, 1855, p. 552.
305. Paul Synnestvedt, *Evolution of the Air Brake*, New York, 1895, 15.
306. Reynolds, *Continuous Railway Brakes*, p. 41.
307. *R.R.G.*, Dec. 31, 1870, p. 314.

308. *A.R.R.J.*, July 14, 1855, p. 449.
309. *R.R. Adv.*, June 7, 1856, p. 4.
310. *Am. Rly. Rev.*, May 23, 1861, p. 312.
311. Ibid., June 1862, p. 181.
312. *Engineering*, July 12, 1867, p. 32.
313. *R.R.G.*, March 4, 1871, p. 532.
314. *Decisions of the Commissioners of Patents*, Washington, 1870, p. 133.
315. *Master Mechanics Report*, 1872, p. 114.
316. *Master Mechanics Magazine*, May 1898, p. 70.
317. Synnestvedt, *Evolution of the Air Brake*, p. 10.
318. *A.R.R.J.*, March 24, 1855, p. 190.
319. Pamphlet on steam brakes by T. W. Smith, 1856, in Association of American Railroads Library, Washington, D.C.
320. *R.R.G.*, Aug. 14, 1875, p. 338; *N.C.B.*, December 1874, p. 181.
321. G. Weissenborn, *American Locomotive Engineering*, N.Y. 1871, p. 174.
322. *N.C.B.*, August 1890, p. 121.
323. Ibid., October 1871, p. 5.
324. Ibid., March 1873, p. 60.
325. Weissenborn, p. 174.
326. *N.C.B.*, February 1877, p. 24.
327. *R.R.G.*, April 14, 1876, p. 160.
328. *N.C.B.*, October 1876, p. 153.
329. Ibid., February 1879, p. 17.
330. Ellis, *Railway Carriages in The British Isles*, p. 197.
331. *A.R.R.J.*, August 1892, p. 355; *Rly. Rev.*, April 2, 1881, p. 179.
332. *N.C.B.*, June 1881, p. 66.
333. *Rly. Rev.*, July 4, 1885, p. 323.
334. *The New York Air Brake Company Fiftieth Anniversary, 1890–1940*, 30-page pamphlet published by the N.Y.A.B. Co. Hereafter *N.Y.A.B. History*.
335. J. H. White, Jr., *American Single Locomotives*, Smithsonian Studies in History and Technology No. 25, Washington, D.C.: Smithsonian Institution, 1973, p. 7.
336. *Rly. Rev.*, Oct. 9, 1880, p. 528.
337. Ibid., Sept. 30, 1882, p. 561; *Railway Magazine*, London, 1907, p. 468.
338. E. A. Forward, *Handbook of the Science Museum: Land Transport*, vol. 3, part I, London, 1931, p. 89. Hereafter *Science Museum Handbook*.
339. Reynolds, *Continuous Railway Brakes*, p. 92.
340. *N.Y.A.B. History*, p. 3.
341. *R.R.G.*, April 27, 1883, p. 271.
342. *Rly. Age*, Nov. 4, 1886, p. 611; *Rly. Rev.*, March 10, 1886, p. 179.
343. *Car Builders' Dictionary*, 1888, p. 190.
344. *R.R.G.*, July 8, 1898, p. 501.
345. *A.R.R.J.*, Aug. 1892, p. 355.
346. *Moody's Industrials*, 1920, p. 1218.
347. *Fortune*, March 1937, p. 115.
348. John Bourne, *Ensamples of Steam, Air and Gas Engines*, London, 1878, p. 171.
349. *Science Museum Handbook*, p. 89.
350. *Sci. Am.*, Dec. 25, 1852, p. 116.
351. *R.R.G.*, April 10, 1903, p. 272.
352. Prout, *Life of Westinghouse*, p. 24.
353. *Loco. Eng.*, February 1900, p. 75.
354. A.S.M.E. Presidential Address Dec. 6, 1910. George Westinghouse's recollection of origins of the air brake. Hereafter Westinghouse, *Recollection*.
355. *Loco. Eng.*, February 1900, p. 75.
356. *Von Nostrand's Engineering Magazine*, September 1870, p. 327.
357. *Engineering*, May 24, 1872, p. 344.
358. Dredge, *The Pennsylvania R.R.*, p. 170.
359. Garrett papers, Maryland Historical Society, Baltimore.
360. C. Clark, "The Railroad Safety Movement in the United States" (doctoral dissertation, University of Illinois, 1966), p. 74.
361. Clark, "Railroad Safety Movement," 1966, p. 74.
362. *Jour. Frank. Inst.*, April 1874, p. 245.
363. Prout, *Life of Westinghouse*, p. 34.
364. *R.R.G.*, July 23, 1880, p. 403.
365. *N.C.B.*, April 1880, p. 66.
366. *Rly. Age*, Jan. 28, 1882, p. 55.
367. Ibid., April 1887, p. 51.

368. Ibid., Nov. 1887, p. 148.
369. Ibid., March 1883, p. 30.
370. *Rly. Rev.*, Sept. 22, 1883, p. 563.
371. Ibid., Jan. 1, 1881, p. 10.
372. *Fortune*, March 1937, p. 118.
373. *R.R.G.*, Nov. 11, 1898, p. 816.
374. M. Wilkins, *Emergence of Multinational Enterprise*, Cambridge, Mass.: Harvard, 1970, p. 59.
375. Prout, *Life of Westinghouse*, pp. 53–62.
376. *R.R.G.*, Dec. 29, 1893, p. 950.
377. W. V. Turner, *The Air Brake as Related to Progress in Locomotion*, Pittsburgh, 1910, pp. 19–22.
378. Ibid.
379. *R.R. Car Jour.*, August 1897, p. 240.
380. *Rly. & Loco. Eng.*, August 1924, p. 250.
381. *Federal Coordinators Report*, 1935, p. 668.
382. *Sci. Am.*, June 20, 1914.
383. Turner, *The Air Brake*, p. 38.
384. *Sci. Am.*, June 20, 1914.
385. *Car Builders' Cyclopedia*, 1943, p. 951.
386. S. K. Farrington, Jr., *Railroads of Today*, New York: Coward-McCann, 1949, p. 36.
387. *Rly. M.E.*, February 1949, p. 66.
388. L. K. Sillcox, *Mastering Momentum*, New York, 1941, p. 69.
389. *Rly. Age*, Nov. 17, 1969, p. 7.
390. *Loco. Eng.*, October 1891, p. 191.
391. *M.C.B.*, 1874, p. 93.
392. *R.R. Car Jour.*, June 1894, p. 132.
393. *Western Railroad Club*, October 1889, p. 33.
394. *M.C.B.*, 1875, p. 29.
395. Ibid., p. 16.
396. *Rly. M.E.*, Feb. 1913, p. 109.
397. *Car Builders' Dictionary*, 1909, p. 523.
398. *R.R.G.*, Dec. 9, 1887, p. 793.
399. *Rly. Age*, June 12, 1912, Aug. 16, 1912, pp. 1267 and 279. See also R.&L.H.S. *Bulletin* 97, October 1957, p. 65.
400. *N.Y.R.R. Club*, 1949, p. 247.
401. *Rly. M.E.*, December 1947, p. 719.
402. Ibid., March 1946, p. 121.
403. Ibid.
404. *N.Y.R.R. Club*, 1949, p. 247; *Rly. Age*, March 14, 1960.
405. *Car and Locomotive Cyclopedia*, 1966, p. 759.
406. Frank W. Stevens, *Beginnings of the New York Central Railroad*, New York: Putnam's, 1926, p. 42.
407. *A.R.R.J.*, September 1843, p. 285.
408. *Rly. Age*, June 19, 1884, p. 390.
409. *M.C.B.*, 1875, p. 115.
410. *N.C.B.*, June 1876, p. 85.
411. *R.R.G.*, Nov. 21, 1884, p. 829.
412. Ibid.
413. Wilson, Walker, & Company, Pittsburgh. Catalog 1875c., in DeGolyer Library, Dallas, Texas.
414. *N.C.B.*, March 1880, p. 39.
415. *N.C.B.*, December 1880, p. 182.
416. *Rly Rev.*, Dec. 25, 1886, p. 700.
417. Ibid., April 10, 1897.
418. *R.R.G.*, Oct. 4, 1873, p. 403; Sept. 18, 1885, p. 598; March 12, 1897, p. 190; and *R.R. Mag.*, January 1930, p. 233.
419. *R.R. Mag.*, May 1947, quotes the Baltimore *Clipper* for July 31, 1847. See also *A.R.R.J.*, March 18, 1848, p. 177.
420. *Am. Rly. Rev.*, July 27, 1859, p. 15.
421. *N.C.B.*, December 1875, p. 186.
422. Ibid., Nov. 1871, p. 3.
423. *R.R. Adv.*, Dec. 27, 1856, p. 1; *Great Railway Celebration*, Baltimore, 1857, rear advertising section, p. 24.
424. *Industrial America, or Manufacturers and Inventors of the United States*, New York, 1876, pp. 405 and 412.
425. *Dictionary of American Biography*, vol. 12, 1933, p. 625.
426. *A.R.R.J.*, September 1887, p. 401.
427. *R.R. Car Jour.*, November 1894, p. 243.
428. *R.R.G.*, Oct. 2, 1885, p. 630.
429. *A.R.R.J.*, Sept. 11, 1869, p. 1043.

430. Ibid., June 4, 1870, p. 644.
431. Ibid., Nov. 13, 1875, p. 1441.
432. *Industrial America*, 1876, p. 401.
433. *A.R.R.J.*, Dec. 26, 1874, p. 1633.
434. *Rly. Age*, July 16, 1885, p. 456.
435. *Miller's Trussed Platforms*, 1868 published catalog in A.A.R. Library, Washington, D.C.
436. *A.R.R.J.*, June 1887, p. 250.
437. Miller contract is in the collection of the Newberry Library, Chicago.
438. Letter to John W. Garrett, president of the B & O, Sept. 14, 1877.
439. Voss, *Railway Car Construction*, p. 93.
440. *R.R.G.*, April 11, 1890, p. 251.
441. Ibid.
442. I.C.C. Annual Report, 1893, p. 28.
443. Ibid., 1893, p. 28, and 1899, p. 35.
444. *N.C.B.*, November 1874, p. 165.
445. Clark, "Railroad Safety Movement," p. 129.
446. *R.R.G.*, Oct. 2, 1885, p. 630.
447. E. Wilson (Ed.), *History of Pittsburgh*, Chicago, 1898, p. 1048; *The Evolution of Car Couplings*, pamphlet published in 1905 by McConway & Torley, Pittsburgh.
448. *Engineering*, July 13, 1883, p. 33.
449. *R.R.G.*, Oct. 26, 1888, p. 710.
450. *M.C.B.*, 1888, pp. 100–101.
451. *R.R.G.*, Oct. 26, 1888, p. 710.
452. *Eng. News*, May 8, 1890, p. 422.
453. Ibid.
454. *R.R.G.*, March 10, 1899, p. 166.
455. Ibid., Nov. 13, 1891, p. 794.
456. Ibid., Oct. 16, 1896, p. 719 and *R.R. Car Jour.*, July 1899, p. 208.
457. Henry, *Fascinating Railroad Business*, p. 231.
458. *Baldwin Mag.*, July 1928, p. 30.
459. *Rly. M.E.*, Aug. 1919, p. 473.
460. Ibid., August 1920, p. 521.
461. Ibid., May 1932, p. 207.
462. Ibid., September 1938, p. 359.
463. Ibid., March 1931, p. 126.
464. *Car Builders' Cyclopedia*, 1953, p. 738.
465. *Trains*, October 1968, p. 16.
466. *R.R.G.*, Sept. 16, 1904, p. 345.
467. *Trains*, January 1969, p. 31.
468. *Trains*, April 1975, p. 6.
469. Henry, *Fascinating Railroad Business*, p. 233.
470. *R.R.G.*, Nov. 17, 1899, p. 794; *Rly. M.E.*, October 1913, p. 520.
471. *Rly. M.E.*, September 1914, p. 453.
472. Ibid., March 1930, p. 125.
473. *A.R.R.J.*, Jan. 1, 1839, p. 4.
474. Ibid.
475. *R.&L.H.S. Bulletin 23*, November 1930, p. 11.
476. Zerah Colburn, *Locomotive Engineering and the Mechanism of Railways*, London, 1872, p. 97.
477. *M.C.B.*, 1875, p. 108.
478. Ibid., p. 107.
479. *Rly. M.E.*, September 1952, p. 86.
480. *Car Builders' Dictionary*, 1879, figs. 252–258; 1888, figs. 409–643; 1895, figs. 1936–2354.
481. Prout, *Life of Westinghouse*, p. 77.
482. *Eng. News*, March 1, 1890, p. 207.
483. *R.R.G.*, Jan. 25, 1889, p. 61.
484. Ibid., Oct. 13, 1899, p. 705.
485. *Rly. M.E.*, July 1914, p. 341.
486. Ibid., December 1943, p. 591.
487. Ibid., August 1930, p. 445.
488. *Railroad Mag.*, February 1948, p. 68.
489. *Rly. M.E.*, May 1935, p. 171.
490. *R.R.G.*, June 2, 1893, p. 406.

CHAPTER EIGHT

1. R. W. Kidner, *Development of the Railcar*, South Godstone, England: Oakwood Press, 1958, ed. p, 3.
2. J. H. White, Jr., *American Single Locomotives and the Pioneer*, Washington, D.C.: Smithsonian Institution, 1973, p. 10.

3. *American Artisan*, April 19, 1865, p. 394.
4. *Western Railroad Gazette*, June 9, 1860, p. 1.
5. *American Railway Review*, Nov. 15, 1860, p. 293.
6. *Scientific American*, Nov. 24, 1860, p. 340.
7. *American Railway Journal*, Nov. 24, 1860, p. 1039.
8. George B. Abdill, *Civil War Railroads*, Seattle: Superior, 1961, pp. 139 and 162.
9. *A.R.R.J.*, May 26, 1860, p. 442.
10. *Am. Rly. Rev.*, March 8, 1860, p. 144.
11. Ibid., Dec. 5, 1861, p. 168; *American Artisan*, April 19, 1865, p. 394; *Locomotive Engineering*, January 1898, p. 16.
12. *Am. Rly. Rev.*, Aug. 8, 1861, p. 26, and Aug. 15, 1861, p. 42.
13. Railway and Locomotive Historical Society, *Bulletin 95*, October 1956, p. 61; *Loco. Eng.*, March, 1911, p. 101.
14. John H. White, Jr., "Grice and Long: Steam-Car Builders," in Jack Salzman (Ed.), *Prospects: An Annual of American Cultural Studies*, vol. 2, New York: Burt Franklin Company, pp. 24–39.
15. *Western Railroader*, June 1970, p. 16.
16. *Loco. Eng.*, October 1891, p. 205.
17. *American Railway Times*, May 5, 1866, p. III.
18. *A.R.R.J.*, Sept. 23, 1865, p. 917.
19. *A.R.T.*, Aug. 11, 1866, p. 254.
20. *Scientific American Supplement*, June 17, 1876, p. 399.
21. *National Car Builder*, October 1872, p. 1.
22. Ibid., January 1890, p. 14.
23. *Railroad Gazette*, June 13, 1879, p. 323.
24. *New England Railroad Club*, Dec. 14, 1897, pp. 6–54.
25. *A.R.R.J.*, November 1897, p. 367.
26. *Loco. Eng.*, November 1897, p. 817.
27. *Engineering News*, Feb. 6, 1902, p. 102.
28. *Railroad Magazine*, June 1945, p. 129.
29. *R.R.G.*, April 23, 1897, p. 295.
30. *Railway Review*, June 11, 1898, p. 315.
31. *R.R.G.*, April 8, 1898, p. 255.
32. *Eng. News*, Feb. 6, 1902, p. 102.
33. *R.R.G.*, June 24, 1898, p. 463.
34. *Sci. Am.*, June 24, 1899, p. 407.
35. *Eng. News*, Feb. 6, 1902, p. 104.
36. *Rly. Rev.*, June 11, 1898, p. 315.
37. *Eng. News*, Feb. 6, 1902, p. 104.
38. *Rly. Rev.*, April 8, 1899, p. 196.
39. *Eng. News*, Feb. 6, 1902, p. 104.
40. *R.R.G.*, April 13, 1900, p. 238.
41. Ibid., Dec. 30, 1904, p. 688.
42. *New York Railroad Club*, April 19, 1907, p. 688.
43. *R.R.G.*, March 29, 1907, p. 451.
44. *International Railway Congress Bulletin*, 1909, p. 1322.
45. *Rly. Rev.*, July 11, 1908, p. 564.
46. *Car Journal*, September 1909, p. 17.
47. *R.R.G.*, Sept. 17, 1909, p. 511.
48. *R.R. Mag.*, April 1966, p. 16.
49. *New York Railroad Club*, April 19, 1907, p. 685.
50. *Rly. Rev.*, March 24, 1906, p. 213, and Oct. 5, 1907, p. 872.
51. Alvin Staufer and Fred Westing, *Erie Power*, Medina, Ohio: Alvin Staufer, 1970, p. 360.
52. E. D. Worley, *Iron Horses of the Santa Fe Trail*, Dallas: Southwestern Railroad Historical Society, 1965, p. 385.
53. *New York Railroad Club*, May 1922, p. 6743; *Car Review*, October 1909, p. 13.
54. *Car Review*, September 1909, p. 18.
55. *Rly. Rev.*, Sept. 29, 1906, p. 748.
56. *R.R.G.*, June 16, 1905, p. 672.
57. *Car Review*, September 1909, p. 17.
58. R. B. Jackson, *The Steam Cars of the Stanley Twins*, New York: Henry Z. Walck, Inc. 1969, p. 54.
59. *Electric Railway Journal*, Sept. 16, 1916, p. 506.
60. *New York Railroad Club*, May 1922, p. 6745.
61. *Mechanical Engineering*, February 1937, p. 120.
62. *Railway Age*, Oct. 14, 1922, p. 713.
63. *R.R. Mag.*, March 1942, p. 77.
64. *Railway Mechanical Engineer*, December 1927, p. 768.
65. *Brotherhood of Locomotive Firemen and Engineers Magazine*, October 1929.

66. *Rly. M.E.*, May 1931, p. 268.

67. *Trains*, June 1948, p. 52.

68. New Haven Railroad Technical Information Association, Standing Data Sheet 6–9210, 1974.

69. *Rly. M.E.*, November 1936, p. 485.

70. *Railway Gazette* (England) March 12, 1937, p. 495.

71. *Street Railway Journal*, May 1889, p. 112.

72. *Railroad History No. 129*, Autumn 1973, pp. 86–93.

73. *R.R.G.*, May 5, 1899, p. 313.

74. Julius S. Morton (Ed.), *Illustrated History of Nebraska*, vol. 3, Lincoln, Neb., 1913, pp. 473–477.

75. *Trains*, July 1960, p. 30.

76. *New York Railroad Club*, April 1907, p. 669.

77. *A.R.R.J.*, August 1905, p. 294; *Rly. Rev.*, Nov. 23, 1907, p. 1008.

78. *R.R. Mag.*, January 1951, p. 17.

79. *A.R.R.J.*, November 1905, p. 420; *Rly. Rev.*, Oct. 7, 1905, p. 720.

80. *Railway Age Gazette*, Nov. 12, 1909, p. 946.

81. *Master Mechanics Report*, 1909, p. 87; *International Railway Congress Bulletin*, 1909, p. 1358; *New York Railroad Club*, April 1907, p. 669.

82. *Rly. Rev.*, Nov. 23, 1907, p. 1008.

83. *Rly. Age*, Dec. 22, 1911, p. 1267.

84. *Atlantic Monthly*, July 1937, p. 542.

85. *Rly. Age*, April 19, 1924, p. 977.

86. *Trains*, July 1960, p. 39.

87. A. D. Dubin, *More Classic Trains*, Milwaukee: Kalmbach Publishing Company, 1974, p. 350.

88. *American Society of Mechanical Engineers Transactions*, December 1929, p. 29.

89. *R.R. Mag.*, January 1951, p. 23.

90. *Trains*, November 1973, pp. 39–49. All data on G.E. cars are from this source unless otherwise noted.

91. *Rly. Age Gaz.*, June 23, 1909, p. 1451.

92. *N.Y.R.R. Club*, January 1912, pp. 2653–2682.

93. *Cyclopedia of Engineering*, vol. 3, Chicago: American Technical Society, 1926, pp. 385–510. The text appears to have been written for the 1915 edition of this work.

94. *Rly. Age*, Sept. 27, 1924, p. 536.

95. *Rly. Age*, April 6, 1906, p. 625.

96. *Int. Rly. Cong. Bulletin*, 1909, p. 1335.

97. *A.R.R.J.*, November 1909, p. 460.

98. *Rly. Age*, June 19, 1914, p. 1525.

99. Electric Railway Historical Society, *The St. Joseph Valley Line*, Bulletin 16, 1955, p. 24.

100. *Rly. Age*, June 22, 1929, p. 64.

101. *Colorado Rail Annual*, no. 9 (1971), Golden, Colo: Colorado Railroad Museum, pp. 3–49.

102. *Rly. Rev.*, Nov. 14, 1925, p. 769.

103. *Rly. Rev.*, May 27, 1922, p. 741.

104. *Engineers and Engineering*, Philadelphia Engineer's Club, May 1923, p. 133.

105. *Car Builders' Cyclopedia*, 1922, p. 355.

106. *Car Building in Philadelphia*, 1926, p. 11. (Pamphlet published by the J. G. Brill Company.)

107. Edwards Railway Motor Car Company Catalog, 1928. Reprinted in 1970 by Old Line Publishing Company, Milwaukee—contains a short history of Edwards.

108. *Loco 1: The Diesel*, Ramsey, N.J.: Railroad Model Craftsman, 1966, p. 13.

109. *Rly. Age*, July 22, 1922, p. 170.

110. Ibid., Aug. 5, 1933, p. 219.

111. *Motor Coach Age*, December 1969, p. 21.

112. R. L. Kulp (Ed.), *History of Mack Rail Motor Cars and Locomotives*, Allentown, Pa.: Lehigh Valley Chapter, Nat. Rly. Hist. Soc., 1959, pp. 9–11.

113. *Extra 2200 South*, February–March 1969, p. 16.

114. *Motor Coach Age*, April 1968, p. 5.

115. *Car Builders' Cyclopedia*, 1925, p. 357.

116. *Rly. Age*, Dec. 27, 1924, p. 1155.

117. *Railway and Locomotive Engineer*, April 1926, p. 93.

118. Brill production data was furnished by Harold E. Cox of Philadelphia through Louis A. Marre of Dayton, Ohio.

119. *Rly. & Loco. Eng.*, April 1926, p. 93.

120. *Rly. Age*, Aug. 19, 1922, p. 352.

121. *Car Builders' Cyclopedia*, 1928, fig. 1039.

122. E. H. Wagner, Jr., *The Colorado Road*, Denver: Intermountain Chapter, Nat. Rly. Hist. Soc., 1970, p. 260.

123. *Rly. Rev.*, Sept. 6, 1921, p. 343; F. M. Reck, *On Time: The History of Electro Motive Division*, La Grange, Ill.: EMD, 1948.

124. *Rly. Age*, Aug. 30, 1924, p. 371.

125. Ibid., and *Rly. Rev.*, Sept. 6, 1924, p. 343.

126. Wagner, *The Colorado Road*, p. 261.

127. *Our GM Scrapbook*, Milwaukee: Kalmbach Publishing Company, 1971, p. 14.

128. *Rly. Age*, Nov. 5, 1949, unnumbered adv. page.

129. *Rly. & Loco. Eng.*, December 1926, p. 337.

130. *Rly. Age*, Jan. 29, 1927, p. 353.

131. Ibid., Jan. 1932, p. 68.

132. *A.S.M.E. Transactions*, December 1929, p. 30.

133. *Rly. Age*, Oct. 17, 1925, p. 695.

134. *Extra 2200 South*, March–April 1971, p. 21.

135. *Rly. Age*, July 7, 1928, p. 17.

136. *Rly. Age*, Jan. 5, 1929, p. 110.

137. *Rly. Age*, May 26, 1934, p. 773; the car was built in 1932.

138. *Rly. Age*, Jan. 2, 1932, p. 68.

139. *Rly. Rev.*, May 27, 1922, p. 741.

140. *Rly. Age*, Nov. 7, 1931, p. 700.

141. Ibid., April 16, 1926, p. 1195.

142. *Wall Street Journal*, April 24, 1934.

143. *Rly. Age*, Oct. 7, 1933, p. 502.

144. Ibid., Sept. 13, 1924, p. 447.

145. Ibid., Jan. 26, 1929, p. 247.

146. Dubin, *More Classic Trains*, p. 200.

147. *A.S.M.E. Transactions*, December 1929, p. 30.

148. *Atlantic Monthly*, July 1935, p. 535.

149. Art S. Genet, "Profile of Greyhound," Newcomen Society in North America, 1958, pp. 17–18.

150. J.F. Stover, *Life and Decline of the American Railroad*, New York: Oxford, 1970, p. 156.

151. *Rly Age*, Jan. 27, 1934, p. 166.

152. *Pullman News*, October 1933, p. 35.

153. J. H. Lemby, *The Gulf Mobile and Ohio*, Homewood, Ill.: Richard D. Irwin, 1953, p. 137.

154. *Rly. Age*, Dec. 19, 1955, p. 29.

155. H. E. Ranks and W. W. Kratville, *The Union Pacific Streamliners*, Omaha: Kratville Publications, 1974, p. 40. Hereafter *U P Streamliners*.

156. J. E. Muhlfeld, *The Railroad Problem*, New York, 1941, p. 186.

157. *U P Streamliners*, p. 45.

158. Ibid., p. 123.

159. *Rly. M.E.*, May 1936, p. 185.

160. Coverdale & Colpitts, *Report on Streamline, Light-Weight, High-Speed Passenger Trains*, New York, 1938, p. 52.

161. *New York Times*, April 30, 1935.

162. *Rly. M.E.*, May 1935, p. 185.

163. Ibid., February 1939, p. 43.

164. Ibid., July 1935, p. 306.

165. *Trains*, April 1962, p. 14.

166. *Rly. M.E.*, April 1935, p. 136.

167. *Rly. Age*, Jan. 18, 1936, p. 140.

168. *Rly. M.E.*, August 1940, p. 303.

169. Ibid., April 1932, p. 137.

170. *Society of Automotive Engineers Journal*, Feb. 1933, p. 54.

171. *Trains*, April 1973, p. 25.

172. *Modern Transport*, Aug. 7, 1937.

173. *Baldwin Magazine*, January 1934, p. 36.

174. *New York Times*, Jan. 17, 1934.

175. *Rly. Age*, April 14, 1934, p. 533.

176. *Mech. Eng.*, June 1937, p. 402.

177. *Atlantic Monthly*, July 1937, p. 540.

178. *Rly. M.E.*, March 1940, p. 107.

179. *R.R. Mag.*, October 1972, p. 61.

180. *Rly. M.E.*, December 1941, p. 503.

181. Dubin, *More Classic Trains*, p. 291.

182. *Investors Reader*, Oct. 6, 1952, p. 9.

183. *Trains*, March 1953, p. 20.

184. Ibid., May 1950, p. 11.

185. Ibid., March 1953, p. 17, and December 1968, p. 38.

186. Ibid., March 1953, p. 20.
187. *Business Week*, July 7, 1951.
188. *Trains*, January 1957, p. 42.
189. Walker made these statements in a conversation with the author in November 1973.
190. J. R. Day, *More Unusual Railways*, New York: Macmillian, 1960, p. 168.
191. *Rly. M.E.*, July 1949, p. 369.
192. *Trains*, September 1958, p. 22.
193. *Wheels*, May–June 1955, p. 19.
194. *Trains*, July 1956, p. 6.
195. R. J. Wayner, *Car Names, Numbers and Consists*, New York, 1972, p. 246.
196. R. C. Reed, *The Streamlined Era*, San Marino, Calif.: Golden West Books, 1975, p. 255.
197. *Trains*, July 1956, p. 6.
198. Ibid., August 1958, p. 6.
199. Ibid., March 1956, p. 10.
200. *U P Streamliners*, p. 391
201. *Trains*, March 1956, p. 10.
202. Ibid., March 1970, p. 32.
203. *Star News*, April 5, 1974.
204. *7th Report High Speed Ground Transport*, Washington, D.C.: Department of Transportation, 1973, p. 35.
205. *Trains*, June 1971, p. 29.
206. W. D. Middleton, *When the Steam Railroads Electrified*, Milwaukee: Kalmbach Books, 1974. Hereafter cited as Middleton.
207. *Business Hitstory Review*, Summer 1962, p. 153.
208. *R.R.G.*, Sept. 10, 1880, p. 487.
209. Edward Hungerford, *The Modern Railroad*, Chicago, 1911, p. 315.
210. *Railroad Herald*, November 1922, p. 31.
211. *Trains*, December 1957, p. 15.
212. Middleton, p. 266.
213. Ibid., p. 268.
214. *Rly. Age*, August 1932, p. 221.
215. *R.R. Mag.*, December 1971, p. 45.
216. *Trains*, December 1957, p. 26.
217. Ibid., p. 27.
218. *G.E. Review*, November 1914, p. 1025.

219. *Master Car Builders Report*, 1928, p. 283.
220. *Rly. M.E.*, October 1950, p. 580.
221. *Trains*, August 1965, p. 48.
222. Middleton, p. 73.
223. *Electric Railway Journal*, March 26, 1910, p. 518.
224. Middleton, p. 79.
225. *Elect. Jour.*, February 1916.
226. *Railroad Herald*, Nov. 1922, p. 34.
227. *Middleton*, p. 314.
228. J. F. Stover, *History of the Illinois Central*, New York: Macmillian, 1975, p. 295.
229. *R.R.G.*, May 21, 1880, p. 276.
230. *Rly. & Loco. Eng.*, August 1926, p. 239.
231. Ibid., January 1928, p. 25.
232. Middleton, p. 292.
233. *Rly & Loco. Eng.*, December 1927, adv. p. 8.
234. *G.E. Review*, April 1927, p. 212.
235. *Rly. M.E.*, February 1949, p. 89.
236. *Headlights*, December 1968, p. 4. (Published by the Electric Railroaders Association, New York City.)
237. *N.Y.R.R. Club*, March 16, 1950, p. 154.
238. Middleton, p. 334.
239. *Rly. M.E.*, June 1954, p. 78.
240. *Rly. Age*, June 30, 1958, p. 18.
241. Ibid., May 6, 1963, p. 34.
242. Ibid., March 11, 1974, p. 34.
243. Ibid., Dec. 2, 1968, p. 18.
244. *Rly. Age*, April 12, 1971, p. 22.
245. *New York Times*, Oct. 31, 1972.
246. *Trains*, November 1967, p. 21.
247. Ibid., p. 27.
248. *Rly. Age*, June 10, 1968, p. 12.
249. *Trains*, July 1969, p. 27.
250. From a conversation with W. D. Edson, former Amtrak car superintendent.
251. *Rly. Age*, June 11, 1973, p. 20.
252. *Loco. Eng.*, April 11, 1973, p. 3.
253. National Railway Historical Society *Bulletin*, vol. 40, no. 5, 1975, p. 54.

BIBLIOGRAPHY

Manuscript source material was nearly nonexistent for this book. A few original papers covering the eight-wheel car case and the Pullman Company are cited in Chapters 1 and 3.

The trade journals offered the most concrete and detailed information, particularly for the period 1880 through 1915. The book could not have been written without them.

The following books, reports, and pamphlets are not a complete list of all works consulted. General reference books, individual railroad corporate histories, biographies, and local histories have not been specified, although most of them are cited in the footnotes. None of the publications in this bibliography are scholarly works, at least in the formal sense. Few offer more than the most elementary notation of the sources consulted. Nearly all are the product of enthusiasts whose expertise, style, and reliability vary considerably, yet they preserve and collect data that are indispensable to the production of a general survey such as this book. My debt to the Dubins, Wayners, and Kratvilles who pursue railroad car history for pleasure is immense.

BOOKS

Arnold, Ian. *Locomotive, Trolley and Rail Car Builders*. Railway History Quarterly, vol. II, no. 1, January 1965.
A superficial summary that repeats Charlton's data on the subject of car builders in a more condensed form. Of modest value.

Association of Manufacturers of Chilled Car Wheels. *Chilled Tread Wheels*. Chicago, 1928.
An illustrated technical booklet which explains the properties and manufacture of cast-iron wheels. Produced by a trade association, it is obviously prejudiced, but the data provided are valuable.

Beebe, L. M. *The Overland Limited*. Berkeley, Calif.: Howell-North Books, 1963.
An elegant pictorial testimonial to a long-lived Union Pacific–Chicago North Western train. This Chicago–San Francisco train's equipment is well illustrated from the late 1880s to the modern era.

———. *Mr. Pullman's Elegant Palace Car*. Garden City, N.Y.: Doubleday & Company, 1961.
A long, informal history of Pullman and first-class rail travel in America. The author's grandiloquent prose is equal to the subject, but readers are cautioned not to depend uncritically on all the factual data offered. Pictures are abundant and well reproduced.

———. *Mansions on Rails*. Berkeley, Calif.: Howell-North Books, 1959.
Accurately subtitled "the folklore of the private railway car." A delightful book for which the Pullman glass-plate negatives were made available for the first time to any modern author by the custodians of that valuable collection.

Behrend, George. *Pullman in Europe*. London: Ian Allen, 1962.
A long, rather disorganized and rambling account. The focus is on the British Pullman operation, and British Pullman cars are listed and illustrated.

Charlton, E. Harper. *Railway Car Builders of the United States and Canada*. Los Angeles: Interurbans Special 24 (Ira L. Swett), 1957.
A hobbyist-oriented survey. Emphasis is on street railway car builders. Valuable for quick reference, although dates do not always agree with contemporary sources. The author depended on information supplied by correspondence. Early builders are neglected.

Cockle, G. R. (ed.). *Car and Locomotive Cyclopedia*. New York: Simmons-Boardman Publishing Corp., 1974.
The third edition of this work, which is a continuation of the *Car Builders' Dictionary* (see M. N. Forney).

Combes, C. L. (ed.). *Car and Locomotive Cyclopedia*. New York: Simmons-Boardman Publishing Corp., 1966, 1970.
The first and second editions of this work which combined the locomotive and car cyclopedias using the same format established in 1879 by M. N. Forney.

———. *Car Builders' Cyclopedia*. New York: Simmons-Boardman Publishing Corp., 1957, 1961.
The twentieth and the twenty-first editions of this work.

Commault, Roger. *Georges Nagelmackers*. Uzôs, France: La Capitelle, 1970c.
A small, admiring biographical sketch of the founder of the Wagon-Lits company, written by an employee of the firm. Illustrated. The text is in French.

Dorin, P. C. *Coach Trains and Travel*. Seattle: Superior Publishing Company, 1975.
The author falls short of his plan to trace the development of coach travel for the past 175 years. His comments are limited almost entirely to the twentieth century, with heavy emphasis on the post-1930 period. Pictures and text largely repeat what Dubin and Kratvile have already published.

———. *The Domeliners*. Seattle: Superior Publishing Company, 1973.
A pictorial history of dome cars in America that offers a useful, nontechnical survey.

Dubin, A. D. *More Classic Trains*. Milwaukee: Kalmbach Publishing Company, 1974.
A sequel to *Some Classic Trains* that deals with lesser-known trains.

———. *Some Classic Trains*. Milwaukee: Kalmbach Publishing Company, 1964.
A pictorial tribute to the great name trains of America. The Broadway Limited and other luxury trains are shown in excellent photographs and described in trade literature of the period. The focus is on developments between 1890 and 1950. The text is useful for dating equipment and establishing the chronology of name-train service. The author's long interest in the subject and his extensive collection of memorabilia are well displayed in this large volume.

Ellis, Hamilton. *Railway Carriages in the British Isles from 1830 to 1914*. London: George Allen & Unwin, 1965.
A survey history written for the enthusiast. Scholarly readers will criticize the lack of documentation and depth of coverage, but the book remains a valuable and useful introduction to the subject. Well illustrated.

Forney, M. N. (ed.). *The Car Builders' Dictionary*. New York: Railroad Gazette, 1879.
The first volume of this basic series, which was revised and republished every three to four years. It begins with definitions of standard terms and provides engravings of general car types with assembly drawings and details of components. This general format was followed in subsequent editions. The 1879 edition has been reprinted three times since 1949.

Fowler, G. L. *The Car Wheel*. Pittsburgh: Schoen Steel Wheel Company, 1907.
A subsidized work of 160 pages intended to promote the sponsor's product. The author was a respected mechanical engineer and author. The text is technical.

Hitt, Rodney (ed.). *The Car Builders' Dictionary*. New York: Railroad Gazette, 1906.

The fifth edition of this work. Reprinted in 1971 by Newton Gregg of Kentfield, California.

————. *The Car Builders' Dictionary.* New York: Railroad Gazette, 1903.

The fourth edition of this work.

Husband, Joseph. *The Story of The Pullman Car.* Chicago: A. C. McClurg, 1917.

A popular and not remarkably objective or accurate account. It is useful as a starting point, and for many years it was one of the few books available on the subject.

Kidner, R. W. *The Development of the Railcar.* South Godstone, England: Oakwood Press, 1958.

A modest but valuable British publication which compresses considerable data on the subject into a small space.

Kirkman, Marshall M. *The Science of Railways.* Chicago: World Railway Publishing Company, 1904 ed.

A multivolume series written by a railway official and a prolific writer on railroad management and engineering. An unnumbered volume contains a broad but inaccurate history of cars. Volume IV has sections on heating, lighting, wheels, and axles.

Kratville, W. W. *Passenger Car Catalog: Pullman Operated Equipment, 1912–1949.* Omaha: Kratville Publications, 1968.

A descriptive catalog of heavyweight Pullman cars organized by type of car. Names, lot, and plan numbers are given. Photographs and floor plans are included. Part II offers a brief illustrated overview of mechanical equipment that includes a good summary of air-conditioning systems used by Pullman.

————. *Steam, Steel & Limiteds.* Omaha: Kratville Publications, 1962.

A comprehensive pictorial survey of passenger train travel in America during the heavyweight era. A celebration of the wonders of train travel, intended for the enthusiast. The variety and scope of service offered is clearly represented but not analyzed in this 400-page volume.

Kratville, W. W., and H. E. Ranks. *Union Pacific Equipment.* Omaha: Kratville Publications, 1969.

Another textless scrapbook of photographs and diagram drawings that contains a treasury of uncoordinated raw data.

Kulp, R. L. (ed.). *History of Mack Rail Motor Cars and Locomotives.* Allentown, Pa.: Lehigh Valley Chapter, National Railway Historical Society, 1959.

A concise but complete record, well illustrated. An example of railroad-buff literature at its best.

Lister, F. E. (ed.). *The Car Builders' Dictionary.* New York: Railway Age Gazette, 1909.

The sixth edition of this work.

Lucas, W. A. *100 Years of Railroad Cars.* New York: Simmons-Boardman Publishing Corp., 1958.

A popular book of plans and photographs aimed at the model builder. Useful for quick reference. The author-editor was an expert on the subject and served for many years as associate editor of the *Car Builders' Cyclopedia.*

McConway & Torley Company, *The Evolution of Car Couplings,* Pittsburgh, 1905.

An illustrated pamphlet issued by a leading manufacturer of Janney couplers. The work is actually limited to the Janney style of coupler.

McKee, L. A., and A. L. Lewis. *Railroad Post Office History.* Pleasantville, N.Y.: Mobile Post Office Society, 1972.

A hardbound scrapbook of lists, photographs, and cancellation stamp imprints. Many pictures are not captioned, and few are dated. The minimal text consists of reprints of popular articles on the subject. The book is of modest value to the historian.

Master Car Builders Association Report of the Proceedings, 1867–1918.

The earliest reports (1867–1872) were actually not published until 1885. After 1873 the reports reproduced papers and related discussions that offer fascinating insights into the art of railroad car design. However, very little space is devoted to passenger car construction. After 1918 these reports were published as the Association of American Railroads Mechanical Section in combination with the former Master Mechanics Association.

Mencken, August. *The Railroad Passenger Car.* Baltimore: The Johns Hopkins Press, 1957.

This pioneering history of the American passenger car was for many years the only work available on the subject. The modest text concentrates on nineteenth-century developments. No formal documentation is offered, and readers are cautioned not to accept the information uncritically. Reproductions of contemporary travel accounts, which occupy one-half of the pages, form the most valuable portion of the book.

Middleton, W. D. *When the Steam Railroads Electrified.* Milwaukee: Kalmbach Publishing Company, 1974.

An informed and complete survey of main-line electric operations in America, which includes valuable information on multiple-unit operations.

Peck, C. B. (ed.). *Car Builders' Cyclopedia.* New York City, Simmons-Boardman Publishing Company, 1949 and 1953.

The eighteenth and nineteenth editions of this work.

Railway Age Company, *The Modern Railway Car,* Chicago, 1883.

An illustrated booklet describing the private car *Railway Age.* It has an introduction on the history of the passenger car.

Randall, W. D. *From Zephyr to Amtrak.* Park Forest, Ill., 1972.

A listing of lightweight cars and streamlined trains which offers basic data on builders, consists, and service. Illustrated mainly with builders' photographs. Minimal text. See Wayner's *Car Names, Numbers and Consists,* which covers the same subject.

Randall, W. D., and A. R. Lind. *Monarchs of Mid-America.* Park Forest, Ill.: P.O. Box 191, 1973.

A collection of diagram drawings, photographs, and schedules covering Illinois Central Railroad passenger equipment—no text beyond the captions. The book concentrates on twentieth-century developments.

Ranks, H. E., and W. W. Kratville. *The Union Pacific Streamliners.* Omaha: Kratville Publications, 1974.

A remarkably complete study that covers design, construction, and service in both pictures and text. One of the most valuable works yet published on American railroad equipment.

Reed, R. C. *The Streamlined Era.* San Marino, Calif.: Golden West Books, 1975.

A thoughtful analysis of the streamline phenomenon, its origins and applications. The book treats automobiles, ships, and furnishings but the emphasis is on rail vehicles. A good balance of text and illustrations.

Ross Winans vs. The Eastern Railroad Company in Equity. Boston: Moore & Crosby, 1854.

A massive printed transcript that reproduces the testimony and dispositions of the major and minor pioneer master car builders.

This volume is an indispensable source for the early history of the American passenger car.

Rush, R. W. *British Steam Railcars*. South Godstone, England: Oakwood Press, 1971.

A comprehensive, detailed, but undocumented work. Illustrated by modern line drawings.

Scribbins, Jim. *The Hiawatha Story*. Milwaukee: Kalmbach Publishing Company, 1970.

An in-depth study of the Milwaukee Road's premium passenger trains in the streamline era. An informed, readable text, reinforced by excellent pictorial coverage.

Shine, J. W., and F. M. Ellington. *Passenger Train Equipment of the Atchison Topeka & Santa Fe. Volume 1: Head End Cars*. La Mirada, Calif.: Railcar Press II, 1973.

A folio-sized record book containing drawings, photographs, and a brief service history from 1880 to the present time.

Sillcox, L. K. *Mastering Momentum*. New York: Simmons-Boardman Publishing Corp., 1941.

A sophisticated engineering treatise written by an official of the New York Air Brake Company that also treats wheels, axles and truck theory, materials, and design.

Stöckl, Fritz. *Comfort on Rails: Sleepers, Diners, and Saloon-cars on the European Railways*. Basel, Switzerland: Claude Jeanmaire, 1970.

Much on the order of Stöckl's *Rollende Hotels*.

———. *Rollende Hotels*. Part 1. Vienna: Rudolf Bohmann, 1967.

An illustrated history of the International Sleeping Car Company—a solid, well-illustrated popular work. Text in German.

Stuart, C. W. T. *Car Lighting by Electricity*. New York: Simmons-Boardman Publishing Company, 1923.

A textbook written for the practical mechanic. The principal commercial systems are explained and illustrated. A brief historical chapter exhibits more careful preparation than is usual for works of this nature.

Synnestvedt, Paul. *Evolution of the Air Brake*. New York: Locomotive Engineering, 1895.

Treats the development of individual components rather than the general system itself. A small, valuable work, illustrated with line drawings.

Turner, W. V., and S. W. Dudley. *Development in Air Brakes for Railroads, with a Brief Review of Past and Present Operating Conditions*. Pittsburgh: Westinghouse Air Brake Company, 1909.

A learned, clear account written by top experts in the field. The fine narrative is illustrated by numerous tables, drawings, and halftones.

Wait, J. C. (ed.). *The Car Builders' Dictionary*. New York: Railroad Gazette, 1895.

The third edition of this basic work, in which the page size was again enlarged.

Wayner, R. J. *Car Names, Numbers and Consists*. New York: Box 871, Ansonia Station, 1972.

A listing of streamlined and lightweight passenger cars built and operated in the United States. Dates of service and dispositions are generally included in notes. Illustrated by floor plans and diagram drawings. (See Randall's *From Zephyr to Amtrak*.)

———. *New York Central Cars*. New York: Box 871, Ansonia Station, 1972.

A pictorial scrapbook of photographs and captions, mainly twentieth-century views. Wayner has produced similar booklets on the Santa Fe and the Southern Pacific.

———. *Pullman and Private Car Pictorial*. New York: Box 871, Ansonia Station, 1972.

A collection of exterior photographs and captions covering the period 1890–1960.

———. *Rio Grande Car Plans*. New York: Box 871, Ansonia Station, 1969.

Reproduces diagram drawings with caption identifications. Covers the wooden and steel eras. The same author has produced a similar work on the Pennsylvania Railroad.

———. *Pullman Panorama*. Vol. I. New York: Box 871, Ansonia Station, 1967.

A listing of sleeping and lounge cars built at the Calumet shops (Chicago) between 1910 and 1931. Plan and lot numbers, construction dates, car names, and dispositions are given in the book. No illustrations are offered, nor have the subsequent volumes that are described in the introductory text been published so far.

Wellington, A. M. (ed.). *The Car Builders' Dictionary*. New York: Railroad Gazette, 1884. (1888).

A revised and enlarged edition of the first dictionary. The page size was much enlarged. The material was assembled in 1884 but not actually published until four years later. Reprinted in 1971 by Newton Gregg, Kentfield, California.

Wellington, A. M., W. B. D. Penniman, and C. W. Baker. *The Comparative Merits of Various Systems of Car Lighting*. New York: Engineering News Publishing Company, 1892.

A detailed study of car lighting as it was late in the nineteenth century. Theory and practice are carefully studied. Good line drawings are included.

Whittlesey, G. P. *Abridgements of the United States Patents Pertaining to Continuous Car Heating*. New York: Railroad Gazette, 1888.

A specialized booklet that is essentially a listing and summary of patents.

Wright, R. K. *Southern Pacific Daylight Train 98–99*. Vol. 1. Thousand Oaks, Calif.: Wright Enterprises, 1970.

An incredible graphic record of an important West Coast train. Within 650 pages are drawings and photographs that provide construction and in-service views. Both locomotives and cars are included. The book reproduces a complete specification for Pullman-built cars dating from 1936.

Wright, Roy V. (ed.). *The Car Builders' Dictionary*. New York City: Simmons-Boardman Publishing Company.

The following editions were compiled by the same editor: 1912 (seventh), 1916 (eighth), 1919 (ninth), 1922 (tenth); note title change to *Car Builders' Cyclopedia*; 1925 (eleventh), 1928 (twelfth), 1931 (thirteenth), 1937 (fourteenth), 1940 (fifteenth), 1943 (sixteenth), 1946 (seventeenth).

Voss, William. *Railway Car Construction*. New York: R. M. Van Arsdale, 1892.

The finest work ever published on American railway car construction. The text explains concepts, materials, and proportions. It is not a history and was meant to be a handbook for the practical car builder. Both freight and passenger cars are included. Extremely fine engineering drawings and engravings illustrate the work. This book goes beyond *The Car Builders' Dictionary*, which only illustrates what was done; Voss explains why it was done.

Zimmermann, K. R. *The Story of the California Zephyr*. Starrucca, Pa.: Starrucca Valley Publications, 1972.

An intelligent appreciation of this splendid train that includes a long text and abundant pictures.

PERIODICALS

American Journal of Railway Appliances. Vol. 1, 1883–Vol. 21, 1901. Space was devoted to rolling stock declines during 1890s; then the journal evolved into a commentary on machine tools and related matters. Title changed to *Railway Machinery* in 1901. Not a major source.

American Railroad Journal. Vol. 1, 1832, to 1975. The earliest years contain only scattered notes on cars, but more information was included after about 1890. A rich source from that time forward. In 1913 the title changed to *Railway Mechanical Engineer.* In 1953 it changed again to *Railway Locomotive and Cars.* This journal continued to be published under the original volume sequence until January 1975.

National Car Builder. Vol. 1, 1870-Vol. 26, 1895. The earliest years are incomplete; there were few illustrations until after 1880. The name was changed 1886 to *National Car and Locomotive Builder.* It is the single most valuable source for information on the subject.

Railroad Car Journal. Vol. 1, 1890–Vol. 12, 1902. Started as *The Journal of Railroad Car Heating and Ventilating.* Changed to *Railroad Car Journal* in October 1891. Name again changed to *Railroad Digest* in 1901, and the periodical expired during the following year. Well illustrated.

Railroad Gazette. Vol. 1, 1870–Vol. 44, 1908. Always concerned with engineering matters, this paper devoted frequent, if not abundant, space to railroad car items. Occasional drawings were published on the subject. The paper merged with *Railway Age* in 1908. M. N. Forney was the editor for many years.

Railway Age. Vol. 1, 1876 to present. A journal originally devoted to corporate matter; little space was given to rolling stock until about 1890. Then the subject was well covered until about 1955, when it was deemphasized.

INDEX

The Johns Hopkins University Press

This book was composed in Linotype Caledonia text type and Ultra Bodoni and Avant Garde display type by the Maryland Linotype Composition Co., Inc., from a design by Patrick Turner. It was printed on Warren 80-lb. Lustro Offset Enamel Dull by The Maple Press Company and bound by The Optic Bindery, Inc., in Joanna Centennial buckram cloth.